NANOSCALE LIQUID INTERFACES

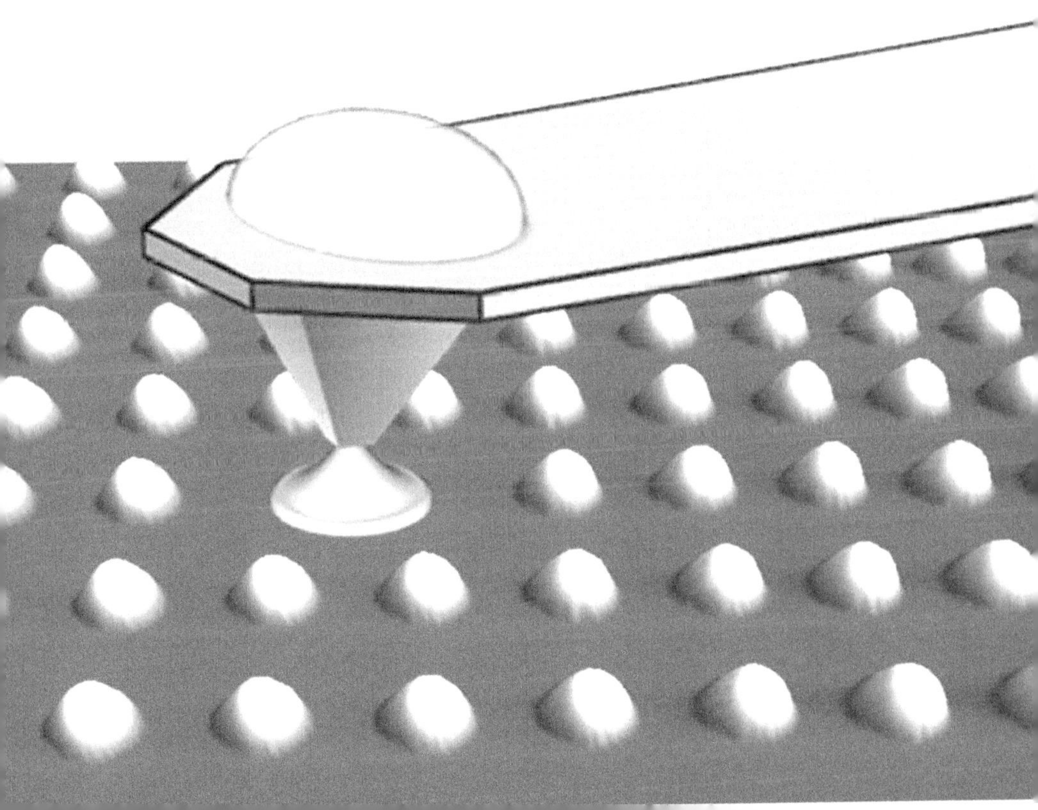

NANOSCALE LIQUID INTERFACES

Wetting, Patterning, and Force Microscopy at the Molecular Scale

edited by Thierry Ondarçuhu | Jean-Pierre Aimé

PAN STANFORD PUBLISHING

Published by

Pan Stanford Publishing Pte. Ltd.
Penthouse Level, Suntec Tower 3
8 Temasek Boulevard
Singapore 038988

Email: editorial@panstanford.com
Web: www.panstanford.com

British Library Cataloguing-in-Publication Data
A catalogue record for this book is available from the British Library.

Nanoscale Liquid Interfaces: Wetting, Patterning, and Force Microscopy at the Molecular Scale

ISBN 978-981-4316-45-3 (Hardcover)
ISBN 978-981-4364-48-5 (eBook)

Printed in the USA

Contents

Preface

Liquids are omnipresent in our everyday life, playing a key role in many domains. Water is, for example, the unique and unavoidable environment necessary for biological activities. In this context, many chemical reactions that sustain metabolic and anabolic activity are performed in confined environments. For instance, protein assemblies define ion channels with nanometre size, which regulate, with a high selectivity, the flow of ions across the membrane of cells. Therefore, many challenging objectives to understand and mimic biological reactions are ahead and require the development of techniques to confine liquids and study their behavior down to the nanoscale. In ambient conditions, capillary effects also become important at micro- and nanometer scale because of capillary condensation that leads to droplets nucleation when a gas phase is tightly confined between two surfaces. Upon downsizing, the capillary adhesion resulting from these liquid bridges scales as r (typical extension of the meniscus) and therefore becomes predominant over volume (r^3 dependence) or surface (r^2 dependence) forces. This size dependence accounts for the reinforcement of sand castles' stability but also for the detrimental effects of stiction in microelectromechanical systems (MEMS) or the imaging artifacts in ambient atomic force microscopy (AFM)

The description of liquids interfaces, in particular in the presence of solid surfaces, is an old question at the confluent between physics, chemistry, and engineering. It was examined by pioneers like Young and Laplace, whose formalism remains the basis of the description of capillarity. The dynamics of wetting has been intensively studied in the second part of the twentieth century, when many experiments and theories were developed. These studies emphasized the importance of a precise description of boundary conditions. For instance, controversial discussions appeared on the necessity to go beyond the classical assumption of a no-slip condition at a solid–liquid interface. In the case of liquid spreading, this condition, coupled with classical hydrodynamics, leads to a stress singularity and a nonphysical stress divergence at the contact line. This problem was circumvented by the introduction of cut-off lengths that are still poorly defined and provoke debate. The development of nanosciences concepts and tools has opened a new field focused on micro- and nanofluidics. New experiments are performed between mesoscopic and molecular scales, the interpretation of which emphasizes the need of a microscopic description of liquid interfaces.

Contrary to solids, which benefit from a large panel of characterization techniques, liquids are more difficult to probe experimentally at the nanometer scale. Owing to their intrinsic dynamic nature and their softness, it is very challenging to investigate them with sufficient spatial and time resolution, while minimizing the influence of the probe. Surface forces apparatus (SFA) represented a real breakthrough, in the late 1960s, as it allowed the direct determination of liquid structure down to molecular scale. Atomic force microscopes later on brought the lateral resolution leading to a broad range of applications. The development of nanotechnology also provided an impressive downscaling of methods, enabling the manipulation of liquids down to nanometer scale, as demonstrated by the fast-growing nanofluidics field. Theoretically, a comprehensive description of liquids spanning from the molecular point of view to macroscopic fluid dynamics remains an open question.

The aim of this book is to provide a state-of-the-art review of the recent advances in the study and manipulation of liquids at submicron or nanometer scale. The chapters gathered in this book give a timely description of what is new in wetting, capillarity, and hydrodynamics and what could be the questions when nanometer scale matters. The book is organized in three parts.

In the first part the main fundamental aspects of the description of liquids interfaces at nanometer scale are discussed. In chapters 1 and 2, two essential elements of the liquid–solid interfaces are described, namely, the liquid structuration and the hydrodynamic slippage that occur at a solid wall. In the case of liquid droplets, and more generally of wetting phenomena, another boundary condition appears at the contact line where three phases—solid, liquid, and vapor—meet. In chapter 3 the basic concepts of wetting are reviewed beyond the usual description of contact line, where Young equation applies, emphasizing the role of long-range forces and wetting films and their influence when scaling down to the nanoscale. The next two chapters are dedicated to thin liquid films, which play a crucial role in many industrial processes, such as coatings or fabrication of microelectronic devices. The study of the stability of liquid films highlights the drastic role of long-range forces in their rupture mechanism (chapter 4). Thin films are also inherent to complete wetting, for example, in the form of a precursor film with molecular or mesoscopic thickness spreading ahead from the macroscopic droplet. The specific behaviors of nematic liquid crystal films due to an additional long-range orientation order are examined in chapter 5. Nanodroplets and nanobubbles obtained when decreasing the lateral size of the liquid structure are then discussed. The statics and dynamics of liquid nanodroplets presented in chapter 6 show a transition from the classical spherical shape to a regime specific to nanometer dimensions where long-range forces dominates. Surface nanobubbles also show intriguing properties, such as their stability, which are less understood than microbubbles, the dynamics of which is perfectly controlled and understood, as reviewed in chapter 7. The peculiar properties of surface nanobubbles can also be probed using acoustic shear waves demonstrating the interest of this method to investigate the mechanical properties of soft interfaces (chapter 8). Another nanoscale form of liquid structure is the nanomeciscus, which appears spontaneously between surfaces due to capillary condensation. The mechanical properties of these menisci are reviewed in chapter 9, mainly focusing on the framework of AFM studies. Nanomenisci play a crucial role in the wetting of nanotubes and nanochannels, an ultimate objective of nanofluidics. The statics and dynamics of liquid flows are presented in chapters 10 and 11, both in the case of synthetic carbon, inorganic nanotubes, or organic nanotubes and in nanochannels fabricated by nanolithography techniques. Many questions arise from these experiments among which the breakdown of the continuum model is at the first place.

The second part of the book is dedicated to the manipulation of liquids to transport, position or modify materials at the nanometer scale. The deposition of molecules or nanoobjects on functional nanodevices is a challenging issue of nanosciences. Since the objects of interest are often in solution, the manipulation of liquids is of crucial importance in the framework of nanopatterning. Conceptually, the simplest way to deposit molecules is to decrease the size of a droplet down to dimensions where it only contains a few molecules. The methods developed with that aim are reviewed in chapter 12, with special emphasis on an AFM-based nanodispensing technique. Liquid nanomenisci spontaneously forming between an atomic force microscope tip and a surface can also be used as a nanoelectrochemical cell, to realize oxidation nanolithography, as demonstrated in chapter 13, where the influence of an electric field on the stability of liquid nanobridges is detailed. In addition to these serial methods, it is shown that the control of wetting processes provides unique methods to pattern surfaces, at a large scale, in a parallel fashion. For example, dewetting leads to self-organized patterns with well-defined shapes and arrangements (chapter 14). Self-assembly resulting from the combination of convective flow and capillary forces can also provide a method to organize colloids on a surface in one-dimensional (1D), two-dimensional (2D), or three-dimensional (3D) arrangements (chapter 15). Finally, the capillary force exerted by a contact line can be used to manipulate high-aspect-ratio nanoobjects such as DNA or nanotubes on a surface. This combing technique is detailed in chapter 16.

The third part of the book is dedicated to AFM experiments investigating biological samples in liquid. The creation of near-field machines, scanning tunneling microscope–atomic force microscope, is most often perceived as the discovery that opened the new area of nanosciences and nanotechnologies. Since the beginning, force microscopy was found as a key instrumentation to study biological samples. In particular, force microscopy and its derivatives provide the opportunity to proceed to a multiscale investigation of structures and functions of biological systems. Therefore, over the last two decades much effort has been dedicated to use force microscopy in liquid environments. These numerous attempts have led to a significant progress in imaging delicate materials such as molecules, filaments, proteins, and membranes. Among those challenges, the sample preparation techniques are crucial, as well as the achievement of a good force balance between the tip and materials to record images with nanometer-scale resolution, as shown in chapter 17. The use of dynamical AFM is currently the most appropriate way to ensure a gentle contact and to minimize the amount of lateral force when scanning a soft object. However, the use of an oscillating cantilever in a liquid environment faces several difficulties, among which is the need to develop new cantilever-driving schemes to ensure a proper oscillating behavior of the cantilever and to keep a good sensitivity down to a few picoNewtons, as described in chapter 18. Biological functions are dynamic in essence; thus we need to increase the scan rate, as is done with high-speed AFM, to access conformational change of active molecular motors or enzymes. Here again locks have to be overcome, especially in liquid, as, for instance, to minimize hydrodynamic forces, still increasing the resonant frequency of the MEMS. Over the last few years, significant improvements have been made, leading to impressive results, as given in chapter 19.

The papers compiled in this book show the impressive achievements performed in the recent years in the study of liquid behavior at nanometer scale, the understanding of which has important implications in a wide range of fields such as biology, coatings, microfabrication technology, corrosion, cavitation, and earth sciences. Moreover, the precise description of liquid interfaces down to nanometer scale may help to answer open questions such as the dissipation processes that rule the spreading of liquids. It also questions the validity of continuum models when scaling down to molecular scale, a topic that should be a source of many advances in the coming years.

Thierry Ondarçuhu
Jean-Pierre Aimé

Chapter 1

Liquid Structure at the Solid–Liquid Interface

Suzanne P. Jarvis,[a,b,*] Gillian Kaggwa,[b] and Khizar Sheikh[b]
[a]*School of Physics, University College Dublin, Dublin 4, Ireland*
[b]*Conway Institute of Biomolecular and Biomedical Research, University College Dublin, Dublin 4, Ireland*
*suzi.jarvis@ucd.ie

In this opening chapter we provide an initial overview of the phenomenon of liquid structuring that can occur in liquids within nanoscale proximity of a solid surface. In this, and subsequent chapters, we explore the nature of nanoscale liquid interfaces, methods for their investigation, and their importance in biological, chemical, and technological contexts.

1.1 Introduction

Since the advent of sensitive techniques for the measurement of surface forces in liquids in the 1970s, it has been apparent that a liquid medium in close proximity to a solid surface is not a structureless continuum as found in bulk, but rather, the discrete molecular nature of the solid and the liquid at the interface leads

Nanoscale Liquid Interfaces: Wetting, Patterning, and Force Microscopy at the Molecular Scale
Edited by Thierry Ondarçuhu and Jean-Pierre Aimé

ISBN 978-981-4316-45-3 (Hardcover), 978-981-4364-48-5 (eBook)
www.panstanford.com

to specific structural forces that can strongly influence interactions in the near-surface region.

The nanoscale physical and chemical properties of the interface both play a role in the nature of the short-range interactions with the surrounding liquid medium. In the case of natural biological surfaces, for example, both physical and chemical strategies have evolved for controlling the interaction with aqueous environments. While many solid–liquid interfaces have been studied, aqueous solutions remain the most important due to the omnipresence of water in most biological and indeed physical interactions at solid–liquid interfaces.

For a biological surface the molecular arrangements of water and solvated ions, their kinetics, and the lateral homogeneity of their structure can have a strong influence on the energy barrier presented by the biological interface, which in turn influences the orientation of approaching molecules, bonding, and self-assembly at the interface. Thus understanding the structure of water at biological interfaces is a critical component in understanding the structure of biological molecules themselves, as well as their recognition of other biomolecules and drugs.

The importance of aqueous solid–liquid interfaces is not limited to biology but is also of major importance in earth science, colloid science, and several areas of surface chemistry. Other solid–liquid interfaces of practical importance, for example, include electrified metal–aqueous interfaces, as these are important for electrochemical reactivity across a number of fields involving corrosion, electrodeposition, and electrochemical energy storage. The extent to which lateral inhomogeneity on a nanoscale is important at these inorganic interfaces is increasing inline with the development of nanoscale textured surfaces and the synthesis of nanocomposite materials.

1.2 Liquids

Liquids can be broadly classified into two groups, polar and nonpolar. Polar liquids are those whose molecules have a permanent electric dipole moment. The permanent dipole moment in polar molecules stems from the difference in the partial charges of the atoms within the molecule. Polar molecules are therefore asymmetrical in terms of electron distribution, in that one region of the molecule carries a partial positive charge, whilst another region carries a partial negative charge. Water is a good example of a polar molecule, as there is an excess of electrons on the oxygen atoms and a lack of electrons on the hydrogen atoms. Polar molecules will attract each other through dipole–dipole attractions between the partial negative charge of one polar molecule and the partial positive charge on another polar molecule.

A nonpolar molecule, on the other hand, is one whose charge distribution is symmetrical when averaged over time. Nonpolar molecules acquire an induced dipole moment in an electric field as the field causes a distortion of the molecule's electronic distributions and nuclear position. This induced dipole is temporary and disappears as soon as the perturbing field is removed. These temporary dipole moments fluctuate rapidly in magnitude and direction, giving rise to intermolecular forces of attraction called London (or dispersion) forces. All molecules, charged or not, polar or not, interact by London forces. To a first approximation, the London force between two molecules is inversely proportional to the seventh power of the distance of separation; it is therefore short range, decreasing rapidly as one molecule moves away from the other.

1.3 Basis of Liquid Ordering at Interfaces

The molecular ordering of liquids at a surface is largely mediated by interactions between the liquid molecules and the surface groups. Depending on the polarity of the interacting surfaces, the interactions can be mediated through dipole–dipole interactions, dipole-induced dipole interactions, or London (dispersion) interactions. Liquid ordering at the solid–liquid interface can also be induced through the constraining effects of two approaching surfaces. Israelachvili[1] presents a detailed discussion on the physical properties of the confined intervening liquid that will be influenced by the liquid structure or ordering at each solid–liquid interface.

For example, the shape, rigidity, and symmetry of the molecules influence the structuring of liquids at the surface. Asymmetric molecules, such as those with side chains or branching, do not order into discrete layers when confined between two surfaces as the molecules remain entangled.[2,3]

1.3.1 *Ion–Dipole Interactions*

The ordering of liquids can be promoted by attractive interactions between an ion and a polar molecule. For example, electrostatic interactions between a positively charged ion (cation) and a water molecule will occur readily in an aqueous environment. This ion–dipole interaction involves the orientation of the partially localized negative charge of the water molecule toward the positively charged cation, forming a solvation shell (hydration shell in the case of water). At close separations, two solvated particles will be repelled from one another due to the bound water molecules that form the hydration layer around the cation. This hydration force is particularly important for ions or molecules in solution as it minimizes their direct contact due to the van der Waals attraction, controlling the stability of the solution.

1.3.2 *Hydrogen Bonding*

Hydrogen bonding is a special case of dipole interaction in which an attractive force exists between a hydrogen atom, attached to an electronegative atom in one molecule and an electronegative atom in a different molecule. Usually the electronegative atom is oxygen, nitrogen, or fluorine, but it can be another atom provided it has readily available electrons. The hydrogen atom then has a partial positive charge. Thus hydrogen bonding is one of the principal mechanisms of hydration of anions in aqueous solution (the bonding of water molecules to the solute species) and also contributes to the hydration of organic compounds containing oxygen or nitrogen atoms. It is the presence of a dipole or hydrophilic group that can induce strong ion-dipole effects that alter the structure of neighboring water molecules.

1.3.3 *Solvation Forces*

The net force between two solids immersed in a liquid has contributions from van der Waals and electrostatic interactions. When the distance between the two approaching surfaces in a liquid is smaller than a few molecular diameters, continuum theories of van der Waals and electric double-layer forces fail to describe the interaction at this length scale. This is a consequence of continuum theories breaking down and the discrete molecular nature of the intervening medium, which leads to interactions that are not accounted for by Derjaguin and Landay[4] and Verwey and Overbeek[5] (DLVO) theory. These non-DLVO forces are often termed structural forces, which are dependent on the specific properties of the intervening medium and also the chemical and physical properties of the surface. These structural forces can be monotonically attractive, monotonically repulsive, or oscillatory in nature and can act with a greater force gradient than the anticipated DLVO forces.

Structural forces that display an oscillatory profile are often termed solvation forces. Solvation forces arise when the liquid molecules behave differently to those in the bulk liquid, ordering into quasi-discrete layers called solvation shells.[1] The origin of the oscillatory force is the consequence of a periodic rise in the force necessary to remove each subsequent layer on approaching the surface. Figure 1.1a shows schematically the molecular layering of a simple liquid at an isolated solid–liquid interface. Here the roughness of the surface is considered to be less than the average diameter, σ, of the molecules. The attractive interactions between the liquid molecules and the solid surface, and the geometric constraining effect of the "hard wall" on the liquid molecules cause the molecules to order into quasi-discrete layers. This is observed as an oscillation in the density profile of the liquid. The oscillations decay within a few molecular diameters from this interface as the layered structure is lost with the liquid density approaching its bulk value (ρ_{bulk}). The layering effect of the liquid molecules is enhanced if two atomically smooth surfaces are brought into close proximity and the layered zones overlap, as shown in Fig. 1.1b, where the liquid is confined within a thin film of thickness, D. Here the constraining

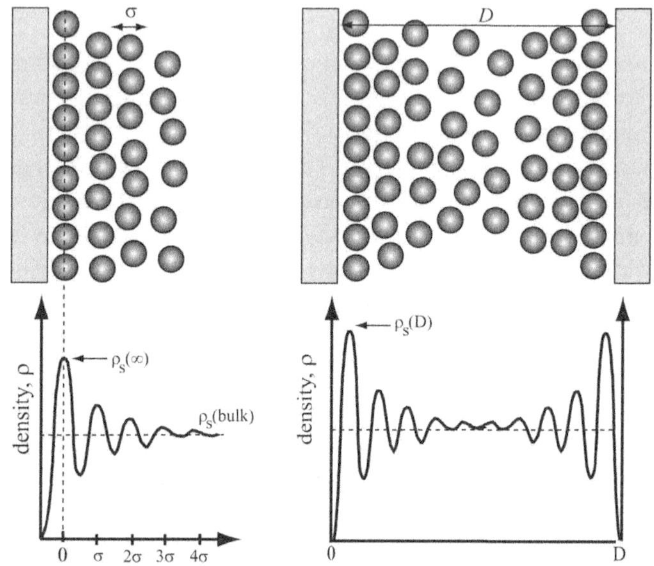

Figure 1.1 Molecular layers of liquid molecules at a smooth solid surface. (a) The molecules close to the interface form layers due to the geometrical boundary of the solid wall. This leads to an oscillation in the density profile near the interface, which decays within a few molecular diameters, σ, toward the bulk of the disordered liquid, ρ_{bulk}. (b) The layered structure is enhanced in a thin film of the liquid confined between two such interfaces separated by the distance, D. (Figure adapted from Israelachvili.[1])

effect induced by two approaching surfaces dominates even in the absence of attractive molecule–surface interactions, leading to stronger oscillations in the density profile across the thin film of layered liquid.

The density of the layer adjacent to the surface, ρ_s, is increased for two approaching surfaces, compared to a situation where only one single interface is present, as shown by the peak value for an isolated surface, $\rho_s(\infty)$ and a confined liquid, $\rho_s(\mathrm{D})$, in Fig. 1.1. Here the liquid molecules reorder themselves to accommodate for the closing gap between the approaching surfaces. The variation in the ordering with a separation of D gives rise to changes in density, as shown in Fig. 1.2. The density of liquid molecules in contact with the surface, $\rho_s(D)$, goes through a maxima and minima. The maxima corresponds the case where the separation reaches a

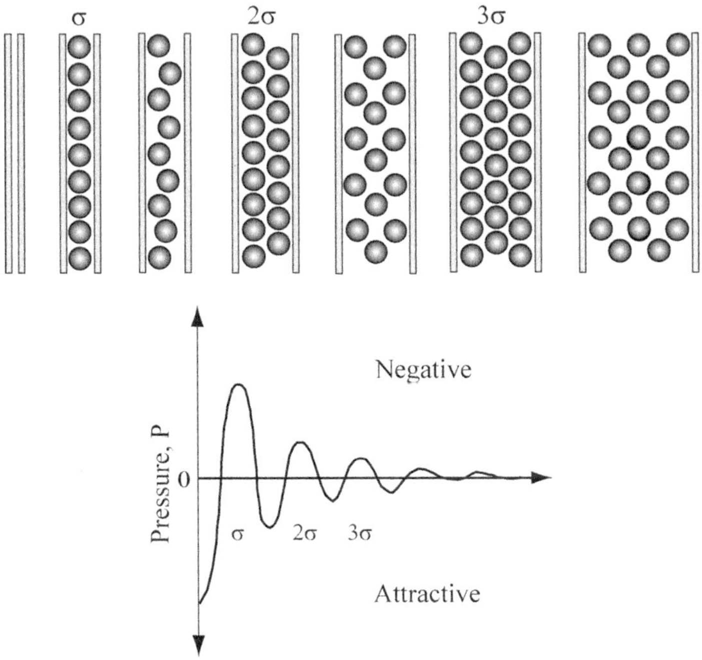

Figure 1.2 The origin of the oscillatory force. The pressure between two surfaces depends on the density of the liquid molecules in the confined film. The pressure reaches a maximum for (ordered) films with a discrete number of molecular layers, leading to repulsive peaks in the pressure/force profile. For intermediate distances the liquid is in a disordered and low-density state, causing a negative pressure and attractive force between the surfaces. (Figure adapted from Israelachvili.[1])

multiple of the molecular diameter, causing the liquid molecules to order, whilst the minima corresponds to a case where the liquid molecular are disordered. In the case of confinement, it has been shown theoretically that this change in the density gives rise to a pressure $P(D)$ between the surfaces[6] that can be described as follows, provided that no interaction exists between the wall and the liquid molecules:

$$P(D) = kT[\rho_s(D) - \rho_s(\infty)] \tag{1.1}$$

where, k is the Boltzman constant and T is the absolute temperature of the system. The result is an oscillating pressure or force that decays in the same fashion as the density profile shown in Fig. 1.2, where the periodicity is close to that of a molecular diameter and becomes zero for distances greater than a few molecular diameters.

1.4 Surface Structure

According to Eq. 1.1, the density or ordering of the boundary layer at the solid wall governs the oscillatory force between two approaching surfaces. Therefore any factors that influence this ordering will in turn influence the force profile. These factors include not only the properties of the intervening medium but also the chemical and physical properties of the surface itself, as discussed next.

1.4.1 *Surface Chemistry*

Surface chemistry plays a role in controlling the interaction between the liquid molecules and the surface and therefore must be taken into account when considering the liquid structure at the solid–liquid interface. For strong attractive interactions between the liquid molecules and the surface, as is the case for wetting, there is a denser packing of molecules at the interface, which is expected to increase the repulsive pressure between approaching surfaces. As a result the oscillations in the force profile are shifted toward higher forces with an increasing effect for small separations. This type of interaction often arises in aqueous solution where hydrogen bond correlation effects give rise to an additional monotonically decaying repulsive force, in addition to any oscillatory forces and DLVO forces.[7] The strength of this repulsive hydration force is dependent on the energy needed to disrupt the water near the surface and dehydrate these surface groups. Between two smooth hydrophilic surfaces the measured hydration force, which is exponentially repulsive, can be altered by changing the pH or ionic composition of the solution.[8]

Water molecules will interact strongly with surfaces containing hydrophilic surface groups, such as hydroxyl, carboxyl, phosphate, and other hydrogen-bonding molecules. Simulation studies investigating the water structure within a few nanometres of a hydrophilic silica surface show that the first layer of ordered water molecules will interact with the surface silanol groups via hydrogen bonding.[9]

When the interaction between the liquid molecules and the surface is weaker than the liquid–liquid interactions, as in the case of nonwetting, then the oscillatory forces profile tends to be shifted downward toward negative (or attractive) forces. The solvation force is still exponential with a decay length similar to the repulsive interaction for two hydrophilic surfaces.[10] These forces extend over a long range, at least as far as the van der Waals interaction, whilst being much stronger, and exist due to the long-range influence of the surface on the structure of the liquid molecule. A recent study investigated the role of surface chemistry on the confinement of water molecules between a surface and a nanosized spherical silicon probe. The surfaces investigated were mica, glass, and highly ordered pyrolytic graphite (HOPG). The authors observed strong oscillatory forces due to confinement on both mica and glass surfaces and less pronounced oscillations when the underlying surface was HOPG.[11]

1.4.2 *Surface Roughness*

The underlying surface roughness plays a role in controlling the solvation force, as it alters the density of liquid molecules at the surface. For two smooth surfaces, the liquid molecules will order into layers, but there will be no lateral ordering within the layers.[12] If the surfaces have a periodic lattice structure, the periodicity will also induce ordering parallel to the surfaces and the oscillatory force will be dependent on the structure of the surface lattices. However, if the two surfaces are randomly rough the oscillatory force will be smoothed out or may disappear altogether and be replaced by a purely monotonic solvation force. This has been demonstrated through surface force measurements involving mica surfaces coated with long-chain molecules in the presence of octamethylcyclotetrasiloxane (OMCTS). In order to elucidate the effects of roughness, experiments were carried out with mica surfaces coated with cetyltrimethylammonium bromide (CTAB) or dioctadecyldimethylammonium bromide (DOAB). The study demonstrated how surface roughness affects the oscillatory force,

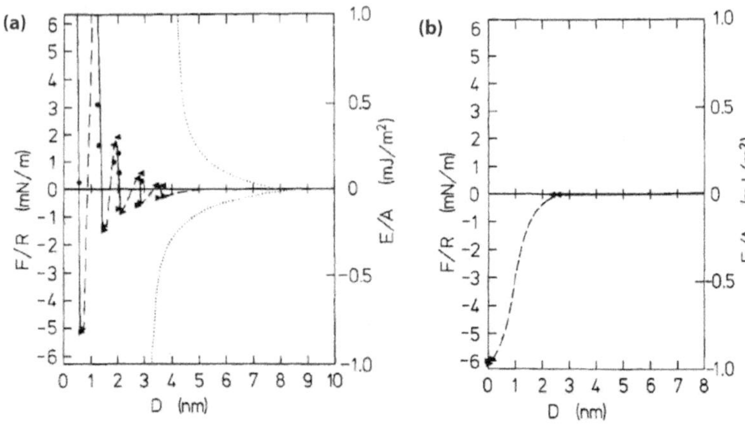

Figure 1.3 Force as a function of separation between mica surfaces coated with (a) DOAB in OMCTS, displaying a reduced oscillatory solvation force as compared to uncoated mica (the dotted lines give the force envelope for uncoated mica), and with (b) CTAB in OMCTS, showing a jump in contact with no oscillatory behavior (reproduced with permission from Ref. 13, copyright 1992, American Physical Society).

leading to a reduced range of oscillations (compared to bare mica surfaces) for DOAB or a complete smearing out in the case of CTAB (as shown in Fig. 1.3). In the later case the interaction resembled a van der Waals–type attraction, whilst for both surfaces the contact adhesion was consistent with Lifshitz theory.[13] In addition, explicit theoretically calculations of the solvation force between rough surfaces showed that for both randomly rough surfaces and surfaces with a discrete distribution of surface topology, the amplitude of the oscillatory force will generally decrease with increasing roughness.[14–16]

1.4.3 *Diffuse Surfaces*

For diffuse surfaces where the surfaces are thermally mobile, the smearing out of oscillations in the force profile due to the liquid structure may occur. This is the case as the thermal fluctuations make the surface dynamically "rough" at a given instant, although they may be smooth over a time average. Diffuse surfaces include polymers, micelles, bilayers, and emulsions. The short-range

(<1 nm) interaction force acting between diffuse surfaces, such as lipid bilayers in an aqueous environment, shows an exponential dependence as a function of distance, with a decay length of 0.1–0.3 nm.[17] Due to the similarity of the decay length to the molecular size of water, the short-range force has been attributed to the force required to remove the ordered water molecules. It has also been claimed that such hydration layers cannot exist due to the disruptive effect of thermally induced molecular-scale protrusions of lipid headgroups and that the repulsive force arising from the overlap of such protrusions ("steric protrusion force") dominates the short-range force. However, recently Fukuma *et al.*[18] showed the existence of intrinsic (i.e., surface-induced) hydration layers adjacent to a supported liquid bilayer in the gel phase when measured over a nanoscale interaction area through direct imaging and force measurements using a low-noise atomic force microscope (AFM) operated in the frequency modulation (FM) mode with an ultrasharp silicon probe (see also chapter 18).

1.4.4 *Constricted Surface Geometries*

The behavior of molecules confined in restricted geometries such as pores has been studied extensively, both experimentally and theoretically. A great deal of work has concerned water confined in nanoporous silicas with pores of a simple geometry[19] and those of complex geometries, such as disordered porous networks. Factors influencing the properties of water in these restricted geometries include surface wettability and the size and geometry of the confining area. For hydrophilic pore walls, the hydroxyl groups on the pore walls will hydrogen-bond to the water and control the distortion of the hydrogen bond network and its surface dynamics. A recent molecular simulation study on the structure of water confined at ambient temperature in hydroxylated silica nanopores with widths $H = 10$ and 20 Å showed that the confined water exhibited different properties than that of the bulk water, showing strong layering of water molecules in the vicinity of the hydrophilic silica.[20] This was observed as density oscillations up to 8 Å from the surface, in the calculated density profiles of the confined water.

1.5 Experimental Studies of the Liquid Structure at Solid–Liquid Interfaces

The experimental tools available for the study of the liquid structure at solid–liquid interfaces have been greatly expanded in recent decades. Langmuir studied the adsorption of gases on glass, mica, and platinum as early as 1918 using simple calorimetric techniques. Subsequently, X-ray and neutron-scattering techniques were developed, which rely on elastic scattering to provide information on the size, shape, and orientation of both bulk liquids and liquids adsorbed to solid surfaces. These techniques showed that hydration layers of thickness 1–3 nm adsorbed to silica surfaces, although it was difficult to decouple these measurements from the roughness of the underlying supports.[21,22]

Not until the development of the surface forces apparatus (SFA) in the late 1960s could experiments be undertaken to study the interaction forces resulting form the liquid structure at these interfaces. The development of the scanning probe microscope in the 1980s finally allowed the study of the liquid structure at solid–liquid interfaces with nanometer-scale lateral resolution.

1.5.1 *Surface Force Apparatus*

The first direct measurements of intermolecular forces between surfaces in liquid at the nanoscale were taken using a surfaces forces apparatus by Tabor and Winterton in 1968 and later by Israelachvili *et al.* in the 1970s.[23,24]

In this technique, depicted in Fig. 1.4, forces are measured between two partially silvered atomically smooth mica sheets mounted on silica cylinders, which are positioned with their long axes at 90 degrees to each other. The vertical separation between the cylinders is controlled mechanically for coarse control and for fine control using calibrated piezoelectric transducers. One of the cylinders is fixed, and the other is mounted on a cantilever spring of a calibrated spring constant, capable of measuring forces as low as 10 nN. The separation of the partially silvered layers is determined to a resolution of 0.1 nm using optical interferometry. The difference between the piezoelectric translation and surface separation allow

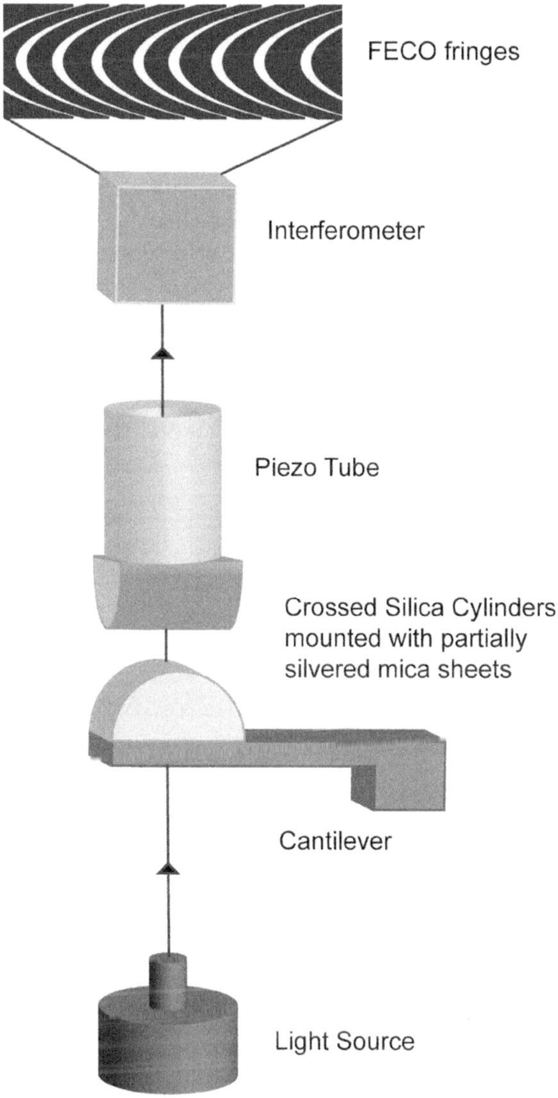

Figure 1.4 Schematic illustration of the surface forces apparatus.

determination of the cantilever deflection, which is converted to force using Hooke's law. The cylindrical geometry used allows the force F measured to be related the interfacial free energy per unit area E between two flat surfaces using Derjaguin approximation:[1]

$$F/R = 2\pi E \tag{1.2}$$

SFA measurements between mica surfaces in an inert silicone liquid, OMCTS, revealed oscillatory solvation forces (section 1.3.3) as the surfaces approached each other, below about 10 nm surface separation. The periodicity of these oscillations corresponded to a molecular diameter of 0.9 nm for the quasi-spherical molecules.[25] These measurements demonstrated that at least five layers of OMCTS molecules were adsorbed to each mica surface (see Fig. 1.5).

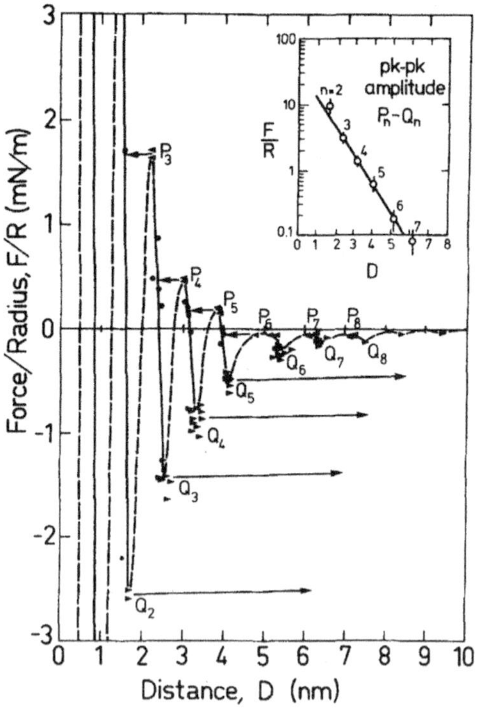

Figure 1.5 Force measured between two crossed cylinders of mica immersed in OMCTS (reproduced with permission from Ref. 25, copyright 1981, American Institute of Physics).

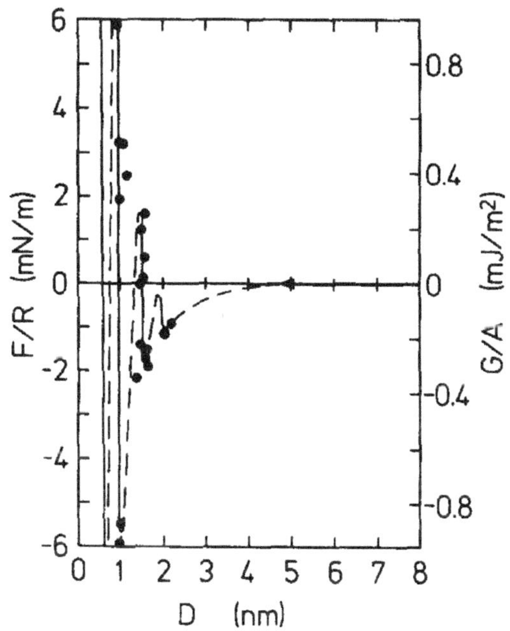

Figure 1.6 Forces measured between two crossed cylinders of mica immersed in n-hexane (reproduced with permission from Ref. 2, copyright 1987, American Institute of Physics).

Similar oscillatory force profiles were also seen in SFA measurements between mica surfaces in liquid organic alkanes[2] (Fig. 1.6). The periodicity of the force minima was 0.4–0.5 nm and did not change with an increase in the length of the alkyl chains, indicating that the alkyl chains had a tendency to orient parallel to the mica surface.

Subsequent SFA force measurements were taken between mica surfaces coated in a variety of surfactant monolayers in symmetric and asymmetric organic liquids. These measurements showed oscillatory force profiles between molecularly smooth polar and nonpolar surfaces in symmetric liquids only, indicating that the oscillatory forces in these cases were due to geometric packing of the molecules on the approach of the surfaces;[3] a summary of these results is shown in Fig. 1.7.

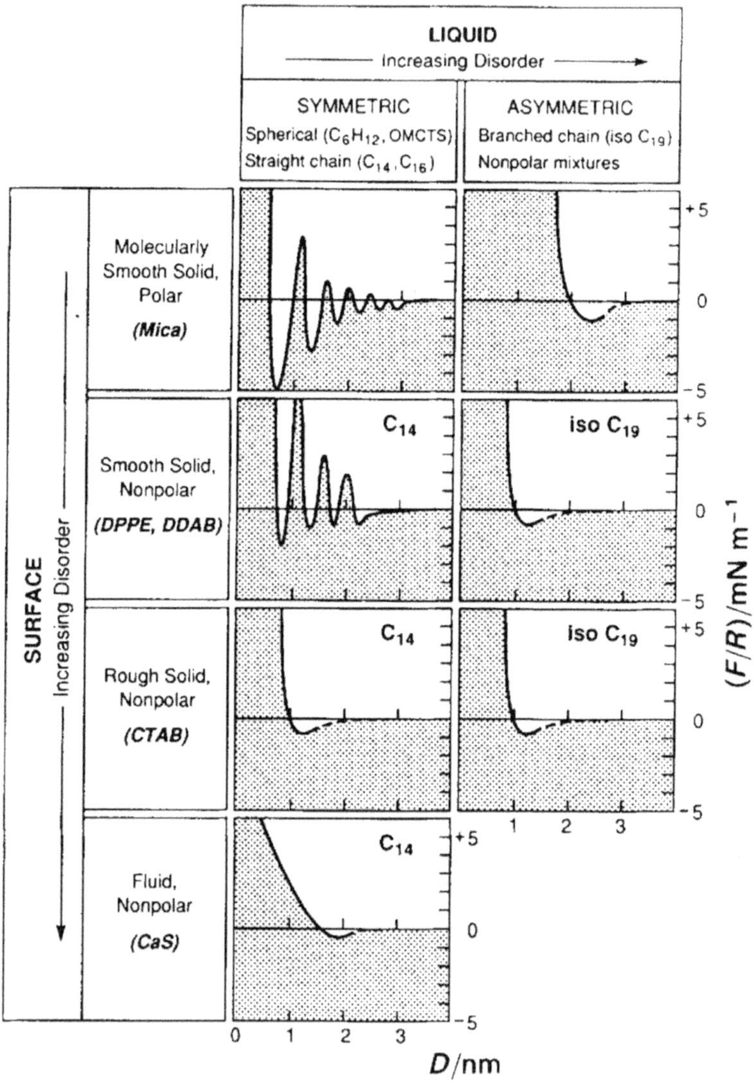

Figure 1.7 Comparison of force measurements taken between different surfactant surfaces in organic liquids (reproduced with permission from Ref. 3, copyright 1990, Royal Society of Chemistry).

SFA force measurements between silica and between mica surfaces in aqueous solutions show a repulsive hydration force below 4 nm. With mica it was found that the magnitude of this repulsive force increased with ionic concentration and varied with the species of cation involved, suggesting that the origin of the repulsive interaction is related to cation binding and dehydration during surface approach (Fig. 1.8).[7] Oscillatory force profiles were also observed below a separation of 1.5 nm, the periodicity of the oscillations being 0.25 nm, the diameter of a water molecule. These measurements showed that at least four layers of water molecules are coordinated to the hydrophilic mica surface. Associated with these hydration layers are cations, the mica surface having a small negative charge. The structure of the water-ion network could not be determined due to the macroscopic nature of the area of SFA surface interaction; however, it is likely that hydrogen bonding (section 1.3.2) plays a key role.

SFA measurements of forces between gel-phase lipid bilayers in monovalent ionic solutions did not show detectable double-layer forces, suggesting that the level of monovalent ion binding to the surfaces was low. Double-layer repulsive forces were, however, seen for divalent ions.[26] Short-range steric hydration forces were observed with monovalent ions. The range of the steric hydration force with monovalent ions did show variation, depending on the cationic species involved (Fig. 1.9), suggesting a complex interaction between the cations and the hydrated lipid headgroups. Due to the macroscopic nature of the SFA measurement technique, no inference could be made as to the lateral organization of the water-ion network adjacent to the bilayer headgroups. This was to change with the subsequent development of the AFM (see also chapter 18).

1.5.2 *Atomic Force Microscope Force Measurements*

The AFM was invented in 1986 by Binnig, Quate, and Gerber.[27] In this technique a sharp tip at the end of a cantilever is used to probe and raster-scan the sample of interest (Fig. 1.10). Probes are typically composed of silicon or silicon nitride with typical radii of curvature from 2 to 50 nanometers. Colloidal microspheres can

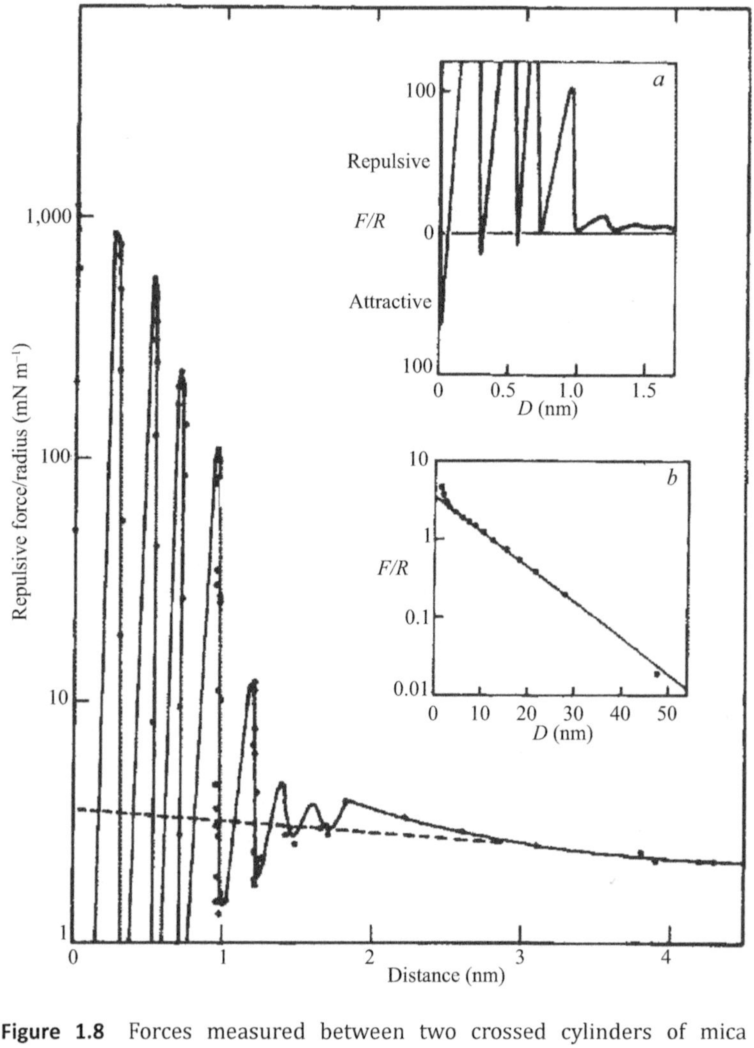

Figure 1.8 Forces measured between two crossed cylinders of mica immersed in a 1 mM KCl aqueous solution (reproduced with permission from Ref. 7, copyright 1983, Macmillan Publishers).

also be attached to the AFM cantilevers to generate probes of well-defined geometry.

The deflection of the cantilever is typically measured using an optical beam deflection system, consisting of a laser, a cantilever, and

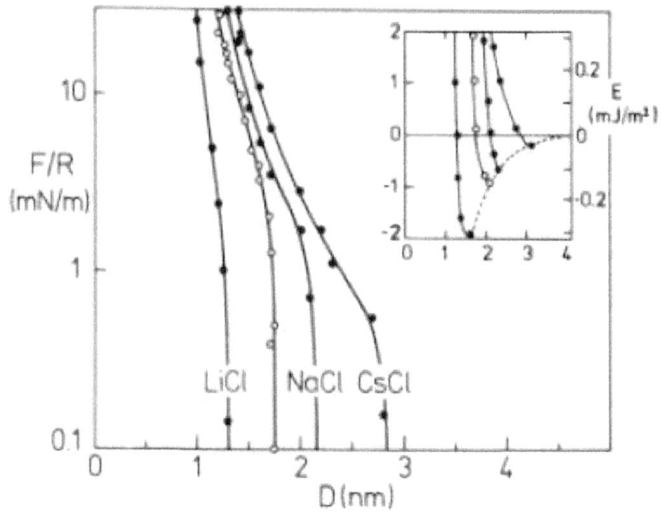

Figure 1.9 Short-range forces measured between two DPPC bilayers on crossed mica cylinders immersed in concentrated aqueous solutions (reproduced with permission from Ref. 26, copyright 1985, American Chemical Society). *Abbreviation*: DPPC, dipalmitoylphosphatidylcholine.

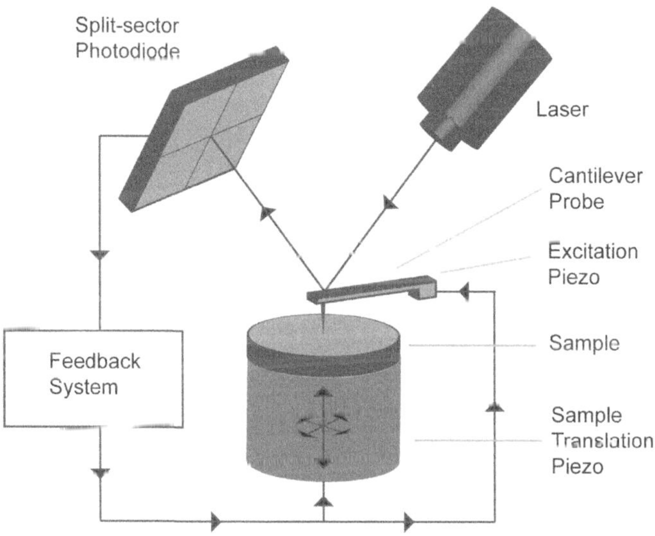

Figure 1.10 Schematic illustration of the atomic force microscope.

a position-sensitive photodiode, although interferometer, capacitive, and piezoresistive deflection-sensing systems can also be used. The sample is translated in the *x*-, *y*-, and *z*-axes using calibrated piezoelectric transducers.

The AFM can be operated in static and dynamic modes. In the static mode, the cantilever deflection (and hence force) is monitored, whilst the surface is translated into contact and then retracted from the probe. A feedback system can be used to vary the sample height to maintain a constant cantilever deflection during scanning. In the dynamic mode of operation, the cantilever is oscillated near its resonant frequency using another piezoelectric transducer near the cantilever. Magnetic and optical excitation systems can also be used. Changes in the tip–sample interaction force during approach and retraction of the tip and sample result in changes in both the amplitude and the frequency of the cantilever oscillation. The feedback system can be set to adjust the tip–sample separation to maintain a constant amplitude during scanning (amplitude modulation mode) or a constant frequency shift during scanning (frequency modulation mode).

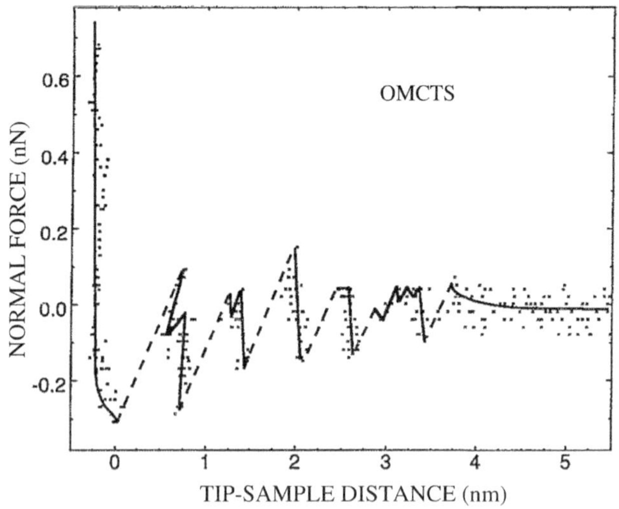

Figure 1.11 The force measured between a silicon nitride AFM tip and an HOPG surface immersed in OMCTS (reproduced with permission from Ref. 28, copyright 1992, American Institute of Physics).

It should be noted that whereas the SFA typically measures interaction forces between areas on the order of millimeters squared, AFM probes are capable of measuring interaction forces using nanometer-scale probes, that is, with higher x–y resolution. Conversely, the geometry of SFA surfaces is better defined than for AFM probes.

The first measurement of oscillatory solvation forces were undertaken by O'Shea *et al.* and were taken is static mode between silicon nitride tips and HOPG surfaces in OMCTS and in dodecanol.[28] In OMCTS at least six force peaks were seen, the magnitude of the peaks increasing as the surface was approached, the periodicity of the peaks corresponding to the diameters of the adsorbed OMCTS molecules, forming layers adjacent to the hydrophobic HOPG substrate (Fig. 1.11). For dodecanol, four transitions were seen upon the tip approach, the periodicity of the transitions indicating that the dodecanol molecules were aligned parallel to the surface (Fig. 1.12).

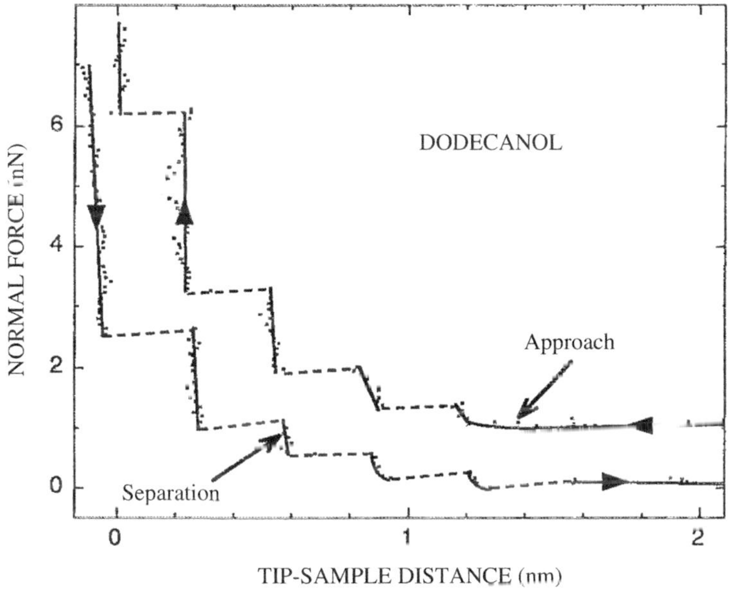

Figure 1.12 The force measured between a silicon nitride AFM tip and an HOPG surface immersed in dodecanol. The approach curve is offset by +1 nN for clarity (reproduced with permission from Ref. 28, copyright 1992, American Institute of Physics).

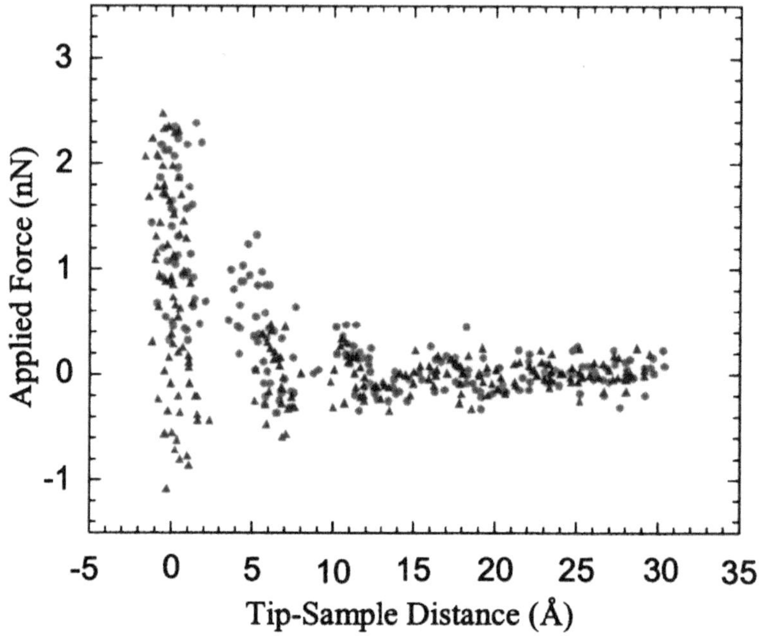

Figure 1.13 The measured approach (dots) and retract (triangles) forces between a silicon AFM tip and an HOPG surface immersed in squalene (reproduced with permission from Ref. 29, copyright 1992, American Physical Society).

A variation on the amplitude modulation technique was also used to measure oscillatory surface forces with increased sensitivity in a branched hydrocarbon, squalene.[29] In this technique, the sample was oscillated with low amplitude (c. 1 Å). and both the cantilever static and dynamic (induced oscillation from a change in the tip–sample force gradient) deflection was measured. Figure 1.13 shows the static force measurement and Fig. 1.14 the dynamic measurement, shown as an interaction stiffness. The sensitivity of the dynamic force measurement is such that the interdigitation of the branched methyl groups can be detected (indicated by arrows in Fig. 1.14).

In 2000 Jarvis *et al.* first applied FM-AFM in liquid and found that the extra sensitivity of the technique enabled them to make the first continuous and stable measurement of structured water.[30] Having

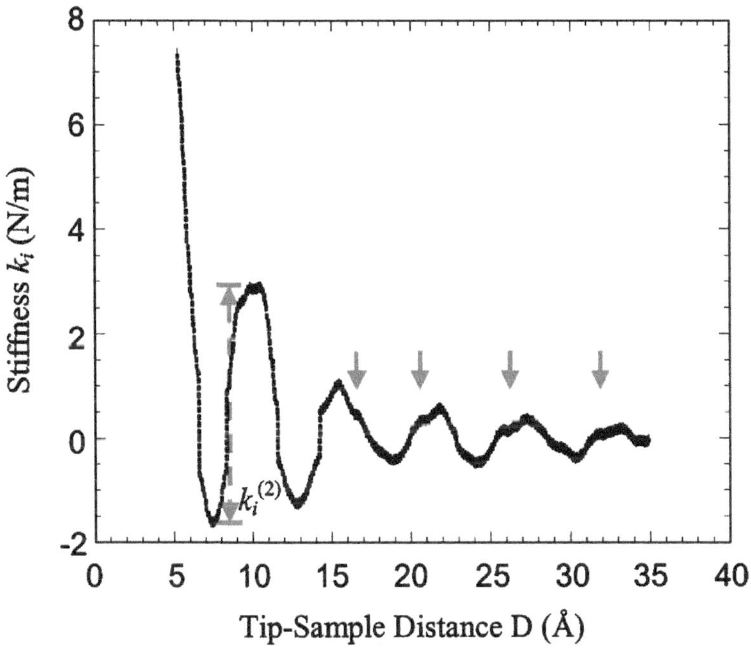

Figure 1.14 The measured interaction stiffness between a silicon AFM tip and an HOPG surface (reproduced with permission from Ref. 29, copyright 1992, American Physical Society).

sufficient sensitivity to measure the localized structure for this small polar module opened up the application of AFM to a much broader range of aqueous–solid interfaces of vast importance in biology, chemistry, and materials science. To implement FM-AFM, they used magnetic activation of the cantilever[31] and combined this with a carbon nanotube probe[32] to reduce the long-range background interaction between the bulk of the AFM tip and a hydrophilic self-assembled monolayer (SAM). With this arrangement they imaged an apparent striped phase of the SAM at high resolution and found evidence of structured water layers in the near contact region (see Fig. 1.15). By utilizing FM-AFM they could correlate highly sensitive force spectroscopy measurements of the water structure extending out from the surface, with a nanoscale lateral characterization of the surface. These initial results indicated the power of high-sensitivity

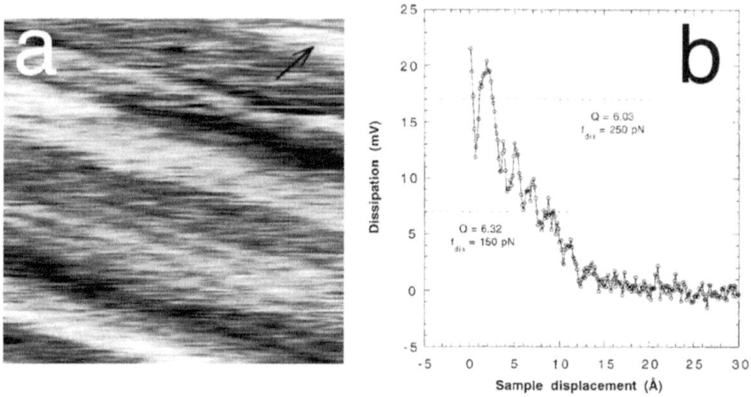

Figure 1.15 High-resolution lateral surface characterization and the associated water structure for a hydrophilic SAM in pure water. (a) 12 × 12 nm^2 area of the SAM showing faint periodicity and striped regions. The region indicated in the top-right corner corresponds to the location of the water structure measurement shown in (b). (b) Dissipation as a function of sample displacement away from the tip (zero on the x-axis is arbitrary). Oscillations in the dissipation signal are separated by approximately 2.2 Å and are associated with the displacement of individual water layers between the tip and the sample (reused with permission from Jarvis *et al.*,[27] copyright 2000, American Chemical Society).

FM-AFM in probing the nature of aqueous environments as a function of the precise lateral position on any surface—a technique that has subsequently been applied to a number of heterogeneous surfaces and biological materials in particular.

1.6 Conclusions

Scattering techniques have been widely applied to the study of the solid–liquid interface at high resolution. However, such measurements obtain data in frequency space and are thus inherently averaged values, which provide no information on the effect of surface inhomogeneities or defects. Nevertheless such techniques have been extensively applied to investigate the binding of water molecules in the interfacial region. While this can provide information about the nature, orientation, and density of surface groups, the structuring of a liquid adjacent to a single interface remains primarily of academic interest as it is the interaction of a secondary object approaching the interface that is of paramount importance in much of biology, earth, and materials science. SFA has enabled the detailed study of the liquid structure at nanoscale separations of two interfaces but has been reliant on atomically smooth mica surfaces in order to resolve structural uniformity, giving rise to net oscillatory forces at nanoscale separations of surfaces of macroscopic dimensions. The successful application of high-resolution AFM to the characterization of nanoscale liquid interfaces has moved research away from studying highly confined liquids trapped in the nanoscale separation between two macroscopic approaching surfaces into the realm of measuring highly localized variations of the liquid structure adjacent to surfaces of nanoscale heterogeneity. The importance of this development, particularly in the case of complex biosystems where interfacial water and solvated ions are considered to play an integral role in biological function, will be examined in greater detail in chapter 18.

References

1. J. N. Israelachvili, *Intermolecular and Surface Forces*. 2nd ed. London: Academic Press. 260–282 (1991).
2. H. K. Christenson, D. W. R. Gruen, R. G. Horn, and J. N. Israelachvili, Structuring in liquid alkanes between solid surfaces: force measurements and mean-field theory, *J. Chem. Phys.*, **87**, 1834–1842 (1987).
3. M. L. Gee and J. N. Israelachvili, Interactions of surfactant monolayers across hydrocarbon liquids, *J. Chem. Soc.: Faraday Trans.*, **86**(24), 4049–4058 (1990).
4. B. V. Derjaguin and L. D. Landau, *Acta Physicochim. (USSR)*, **14**, 633–662 (1941).
5. E. J. W Verwey and J. T. G. Overbeek, *The Theory of Stability of Lyophobic Colloids*. Amsterdam: Elsevier (1948).
6. J. P. Hansen and I. R. McDonald, eds., *Theory on Simple Liquids*. 3rd ed. London: Academic Press (1996).
7. J. N. Israelachvili and R. M. Pashley, Molecular layering of water at surfaces and origin of repulsive hydration forces, *Nature*, **306**, 249–250 (1983).

8. R. M. Pashley and J. N. Israelachvili, DLVO and hydration forces between mica surfaces in magnesium(2+), calcium(2+), strontium(2+), and barium(2+) chloride solutions, *J. Colloid Interface Sci.*, **97**, 2, 446–455 (1984).
9. D. Argyris, N. R. Tummala, A. Striolo, and D. R. Cole, Molecular structure and dynamics in thin water films at the silica and graphite surfaces, *J. Phys. Chem. C*, **112**(35), 13587–13599 (2008).
10. R. M. Pashley, P. M. McGuiggan, B. W. Ninham, and D. F. Evans, Attractive forces between uncharged hydrophobic surfaces: direct measurements in aqueous solution, *Science*, **229**, 1088–1089 (1985).
11. T. Li, J. Gao, R. Szoszkiewicz, U. Landman, and E. Riedo, Structured and viscous water in subnanometer gaps, *Phys. Rev. B: Condens. Matter*, **75**(11), 115415-115421 (2007).
12. J. N. Israelachvili, *Intermolecular and Surface Forces: With Applications to Colloid and Biological Systems*. 2nd ed. London: Academic Press, 260–282 (1985).
13. H. K. Christenson, Interactions between hydrocarbon surfaces in a nonpolar liquid: effect of surface properties on solvation forces, *J. Phys. Chem.*, **90**(1), 4–6 (1986).
14. M. Heuberger and M. Zach, Nanofluidics: structural forces, density anomalies, and the pivotal role of nanoparticles, *Langmuir*, **19**, 1943–1947 (2003).
15. C. Ghatak and K. G. Ayappa, Solvation force, structure and thermodynamics of fluids confined in geometrically rough pores, *J. Chem. Phys.*, **120**(20), 9703–9714 (2004).
16. L. J. D. Frink and F. van-Swol, Solvation forces between rough surfaces, *J. Chem. Phys.*, **108**(13), 5588–5598 (1998).
17. T. J. McIntosh and S. A. Simon, Hydration force and bilayer deformation: a reevaluation, *Biochemistry*, **25**(14), 4058–4066 (1986).
18. T. Fukuma, M. J. Higgins, and S. P. Jarvis, Direct imaging of individual intrinsic hydration layers on lipid bilayers at Ångstrom resolution, *Biophys. J.*, **92**(10), 3603–3609 (2007).
19. S. A. Somers and H. T. Davis, Microscopic dynamics of fluids confined between smooth and atomically structured solid surfaces, *J. Chem. Phys.*, **96**, 5389–5408 (1992).
20. P. A. Bonnaud, B. Coasne, and R. J.-M. Pellenq, Molecular simulation of water confined in nanoporous silica, *J. Phys.: Condens. Matter*, **22**, 284110–284125 (2010).

21. M. A. V. Axelos, D. Tchoubar, and J. Y. Bottero, Small-angle X-ray scattering investigation of the silica/water interface: evolution of the structure with pH, *Langmuir*, **5**, 1186–1190 (1989).

22. B. W. Koenig, S. Krueger, W. J. Orts, C. F. Majkrzak, N. F. Berk, J. V. Silverton, and K. Gawrisch, Neutron reflectivity and atomic force microscopy studies of a lipid bilayer in water adsorbed to the surface of a silicon single crystal, *Langmuir*, **12**, 1343–1350 (1996).

23. D. Tabor and R. H. S. Winterton, The direct measurement of normal and retarded van der Waals forces, *Proc. R. Soc. Lond. A*, **312**, 435–450 (1969).

24. J. N. Israelachvili and G. E. Adams, Measurement of forces between two mica surfaces in aqueous electrolyte solutions in the range 0–100 nm, *J. Chem. Soc., Faraday Trans. 1*, **4**, 975–1001 (1978).

25. R. G. Horn and J. N. Israelachvili, Direct measurement of structural forces between two surfaces in a nonpolar liquid, *J. Chem. Phys.*, **75**(3), 1400–1411 (1981).

26. J. Marra and J. N. Israelachvili, Direct measurements of forces between phosphatidylcholine and phosphatidylethanolamine bilayers in aqueous electrolyte solutions, *Biochemistry*, **24**, 4608–4618 (1985).

27. G. Binnig, C. F. Quate, and C. H. Gerber, Atomic force microscope, *Phys. Rev. Lett.*, **56**(9), 930–933 (1986).

28. S. J. O'Shea, M. E. Welland, and T. Rayment, Solvation forces near a graphite surface measured with an atomic force microscope, *Appl. Phys. Lett.*, **60**, 2356–2358 (1992).

29. R. Lim and S. J. O'Shea, Solvation forces in branched molecular liquids, *Phys. Rev. Lett.*, **88**, 246101,1–246101,4 (2002).

30. S. P. Jarvis, T. Uchihashi, T. Ishida, H. Tokumoto and Y. Nakayama, Local solvation shell measurement in water using a carbon nanotube probe, *J. Phys. Chem. B*, **26**, 6091–6094 (2000).

31. S. P. Jarvis, A. Oral, T. P. Weihs and J. B. Pethica, A novel force microscope and point contact probe, *Rev. Sci. Instrum.*, **64**, 3515–3520 (1993).

32. H. Dai, J. H. Hafner, A. G. Rinzler, D. T. Colbert and R. E. Smalley, Nanotubes as nanoprobes in scanning probe microscopy, *Nature*, **381**, 147–150 (1996).

Chapter 2

Wetting, Roughness, and Hydrodynamic Slip

Olga I. Vinogradova[a,b,c] **and Aleksey V. Belyaev**[a,b]
[a]*A.N. Frumkin Institute of Physical Chemistry and Electrochemistry, Russian Academy of Sciences, 31 Leninsky Prospect, 119991 Moscow, Russia*
[b]*Physics Department, M.V. Lomonosov Moscow State University, 119991 Moscow, Russia*
[c]*DWI an der RWTH Aachen, Forckenbeckstr. 50, 52056 Aachen, Germany*
oivinograd@yahoo.com

The hydrodynamic slippage at a solid–liquid interface is currently at the center of our understanding of fluid mechanics. For hundreds of years this science has relied upon no-slip boundary conditions at the solid–liquid interface that has been applied successfully to model many macroscopic experiments, and the state of this interface has played a minor role in determining the flow. However, the problem is not that simple and has been revisited recently. Due to the change in the properties of the interface, such as wettability and roughness, this classical boundary condition could be violated, leading to a hydrodynamic slip. In this chapter, we review recent advances in the understanding and expectations for the hydrodynamic boundary conditions in different situations, by focusing mostly on key papers from the past decade. We highlight mostly the impact of hydrophobicity, roughness, and especially their

Nanoscale Liquid Interfaces: Wetting, Patterning, and Force Microscopy at the Molecular Scale
Edited by Thierry Ondarçuhu and Jean-Pierre Aimé

ISBN 978-981-4316-45-3 (Hardcover), 978-981-4364-48-5 (eBook)
www.panstanford.com

combination on the flow properties. In particular, we show that hydrophobic slippage can be dramatically affected by the presence of roughness, by inducing novel hydrodynamic phenomena, such as a giant interfacial slip, superfluidity, mixing, and low hydrodynamic drag. Promising directions for further research are also discussed.

2.1 Introduction

Fluid mechanics is one of the oldest and useful of the "exact" sciences. For hundreds of years it has relied upon the no-slip boundary condition at a solid–liquid interface, which was applied successfully to model many macroscopic experiments.[1] However, the problem is not that simple and has been revisited during recent years. One reason for such a strong interest in "old" problem is purely fundamental. The no-slip boundary condition is an assumption that cannot be derived from first principles even for a molecularly smooth hydrophilic (the contact angle [fixed by the chemical nature of a solid] lies between 0° and 90°) surface. Therefore, the success of the no-slip postulate may not always reflect its accuracy but in fact rather the insensitivity of the experiment. Another reason for current interest in flow boundary conditions lies in the potential applications in many areas of engineering and applied sciences, which deal with small-size systems, including micro- and nanofluidics,[2] flow in porous media, friction and lubrication, and biological fluids. The driving and mixing of liquids when the channel size decreases represent a very difficult problem.[3] There is therefore a big hope to cause changes in hydrodynamic behavior by an impact of interfacial phenomena on the flow. For example, even ideal solids, which are both flat and chemically homogeneous, can have a contact angle, which exceeds 90° (the hydrophobic case). This can modify hydrodynamic boundary conditions, as it has been shown yet in early work.[4] Besides that, solids are not ideal yet rough. This can further change, quite dramatically, boundary conditions. It is of course interesting and useful to show how the defects or pores of the solids modify them. But today, the question has slightly shifted. Thanks to techniques coming from microelectronics, we are able to elaborate substrates whose surfaces are patterned (often at the

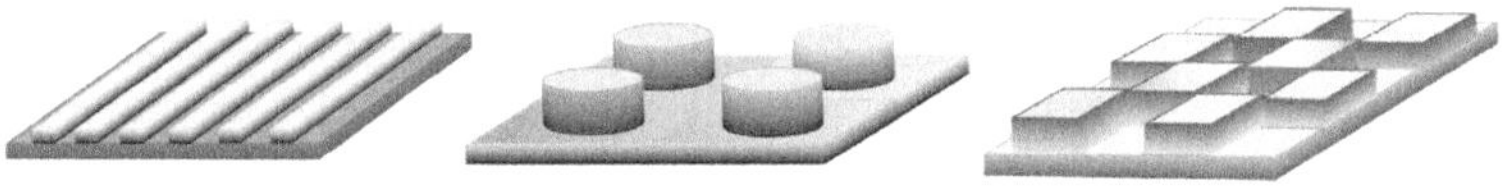

Figure 2.1 Examples of anisotropic (grooves) and isotropic (pillars, chessboard) textures.

micro- and nanometer scales) in a very well-controlled way (see Fig. 2.1), which provides properties (e.g., optical or electrical) that the solid did not have when flat or slightly disordered. A texture affects the wettability and boundary conditions on a substrate and can induce unique properties that the material could not have without these micro- and nanostructures. In particular, in the case of superhydrophobic solids, which are generated by a combination of surface chemistry and patterns, roughness can dramatically lower the ability of drops to stick, by leading to the remarkable mobility of liquids. At the macroscopic scale this renders them "self-cleaning" and causes droplets to roll (rather than slide) under gravity and rebound (rather than spread) upon impact instead of spreading.[5] At the smaller scale, reduced wall friction and a superlubricating potential are almost likely associated with the breakdown of the no-slip hypothesis. This suggests that superhydrophobic surfaces should strongly affect the transport of fluids. This area of research is relatively new but rapidly developing, and it has already attracted scientists from physics, chemistry, mathematics, and engineering.

In this chapter we review recent advances in the understanding and expectations for the fluid–solid boundary conditions in different situations, where hydrophobicity and roughness impact the flow properties, by focusing mostly on key papers from the past decade. Throughout the chapter, we emphasize open questions. We start with a very brief history of the no-slip boundary conditions. Then, after introducing the terminology and models, and describing new developments and instruments, which give the possibility of investigating fluid behavior at the micro- and nanoscales, in the following section we present results obtained for smooth surfaces, by highlighting the role of wettability. Then follow the results for rough hydrophilic and, especially, hydrophobic surfaces. In the latter case we show how roughness can enhance the hydrodynamic slip and thus the efficiency of interfacial transport phenomena. We close with suggestions to guide the next round of experimental and theoretical studies in this expanding area of research.

2.2 Origins and History

The nature of boundary conditions in hydrodynamics was widely debated in the nineteenth century, and many of the great names in fluid dynamics have expressed an opinion on the subject, as discussed in a recent review.[6] The linear slip boundary condition was introduced by Navier,[7] and this remains a standard characteristic of the slip used today. Helmholtz and von Piotrowski were probably the first to report some evidence of slippage at the solid–liquid interface.[8] We refer the reader to a comprehensive review[9] for detailed accounts of early experimental work. The significance of hydrophobicity for the slip phenomenon does not seem to be recognized at this early stage.

The matter was revisited at the beginning of the twentieth century, as reviewed in.[4] Several research groups reported faster flow after they had treated the capillary tube with organic compounds, and interpreted results in favor of slippage of water. There have been, however, some conclusions about complications from surface tension effects in these early experiments. Although not recognized by the authors, their work tried to make a connection between the amount of slip and the hydrophobization of the solid. That is why despite the fact that the early results still did not provide a strong support to the hypothesis of a hydrophobic slippage, we consider these papers to be an important contribution to the subject. The first reliable results for water were probably obtained in the second part of the twentieth century. The measurements by Schnell[10] and Churaev *et al.*[11] in thin hydrophobic capillaries unambiguously suggested a concept of hydrophobic slippage, which allowed to formulate the first theoretical models.[4,12,13] However, despite the increasing body of evidence in favor of hydrophobic slippage, experimental methods and results have been rather limited, and the amplitude of slip and its dependence on various parameters (such as the contact angle and shear rate) still remained an open question.

In the twenty-first century, or during the last decade, the field of hydrodynamic boundary conditions has rapidly advanced, motivated by potential applications in microfluidics, as well as fundamental scientific issues in colloid physics and fluid mechanics. Both hydrophobic and rough interfaces have received enormous attention. Note that some of the first data completely escaped from the theoretical framework, with qualitative (shear rate threshold to slippage; shear rate–dependent slip length) and quantitative (micrometric vs. nanometric slip length) discrepancies,[6] which have caused hot debates. Fortunately, an avalanche development of the field, including experimental methods, allowed one, relatively fast, clarify the situation and highlight reasons for existing controversies. Thus, it is possible to conclude that boundary conditions at the smooth hydrophobic surfaces are currently quite well understood. We also suggest that a rough interface has recently been relatively well understood, although there are still some open questions. Over the past few years hot and promising directions of research seem to be the strategy of a combination of the hydrophobic slip with surface roughness and/or interfacially driven flow, such as electro-osmosis (EO). These new systems of interest could potentially lead to a generation of extremely fast flows in microfluidic devices and stimulated new fundamental studies.[14]

2.3 Terminology and Models

We will refer to as a slip any situation where the value of the tangential component of velocity appears to be different from that of the solid surface. The simplest possible relation assumes that the tangential force per unit area exerted on the solid surface is proportional to the slip velocity. Combining this with the constitutive equation for the bulk Newtonian fluid, one gets the so-called (scalar) Navier boundary condition:[7]

$$u_s = b\frac{\partial u}{\partial z}, \tag{2.1}$$

where u_s is the (tangential) slip velocity at the wall, $\partial u/\partial z$ the local shear rate, and b the slip length. This slip length represents a distance inside the solid to which the velocity has to be extrapolated

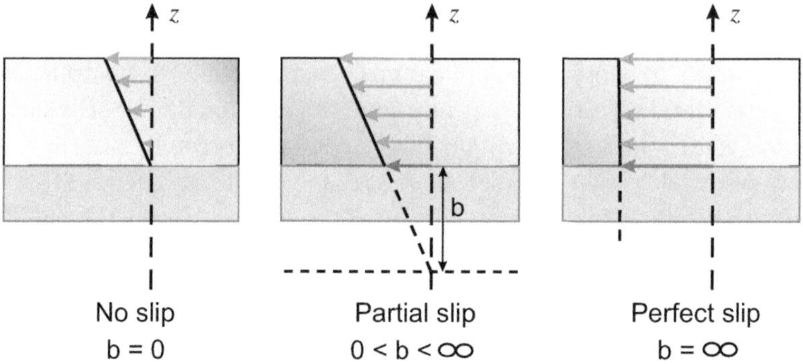

Figure 2.2 Interpretation of the Navier slip length b.

to reach zero. The standard no-slip boundary condition corresponds to $b = 0$, and the shear-free boundary condition corresponds to $b \to \infty$[12] (see Fig. 2.2). In the most common situation b is finite (a partial slip) and associated with the positive slip velocity. It can, however, be negative, although in this case it would not have a long-range effect on the flow.[15]

Obviously the control of slip lengths is of major importance for flow at the interface and in confined geometry. It would be useful to distinguish between three different situations for a boundary slip since the dynamics of fluids at the interface introduce various length scales.

Molecular (or intrinsic) slip allows liquid molecules to slip directly over a solid surface (Fig. 2.3a). Such a situation is not of main concern here since a molecular slip cannot lead to a large b[4,6,16] and its calculations requires a molecular consideration of the interface region.

An important point to note is that although Eq. 2.1 is the most commonly used boundary condition for a hydrodynamic slip, it is not widely appreciated that[7] also postulated the more general relation, $\Delta u = M\tau$ where τ is the local shear stress (normal traction) and M is the constant interfacial mobility (velocity per surface stress). For a Newtonian fluid, $\tau = \eta \partial u/\partial z$, this reduces to (2.1) with $b = M\eta$, where η is the viscosity. Molecular dynamics (MD) simulations have shown that the equation with constant M is more robust than (2.1) with constant b, since the fluctuating slip velocity correlates better with the shear stress (normal forces) than with velocity gradients very close to the surface.[16]

A possible starting point for molecular modeling is a scalar Einstein relation, $M = bkT/A$, relating the M to the "interfacial diffusivity" per unit area A, by analogy with the theory of Brownian motion. This can be recast as a type of a Green-Kubo formula and allows one to prove that the slip length does not depend on the bulk viscosity, η (since it depends on ηD).[16] Rough estimates of the molecular slip suggest $b \propto k_B T \eta D/\epsilon^2$, where ϵ reflects a typical energy of interaction between liquid and solid.[17,18] By relating ϵ and a contact angle, $\cos\theta$, one can then propose a scaling relation $b \propto (1+\cos\theta)^{-2}$, which is consistent with MD results.[17,19] In other words, the static contact angle is the crucial parameter controlling water slippage over smooth hydrophobic surfaces. We remark and stress, however, that for realistic contact angles only b below 10 nm has been predicted. Therefore, it is impossible to benefit of such a slip in larger-scale applications.

The intrinsic boundary condition may be rather different from what is probed in a flow experiment at a larger length scale. It has been proposed[12] to describe the interfacial region as a lubricating "gas film" of thickness e of viscosity η_g different from its bulk value η. Straightforward calculations give the *apparent* slip (Fig. 2.3b):

$$b = e\left(\frac{\eta}{\eta_g} - 1\right) \simeq e\frac{\eta}{\eta_g} \tag{2.2}$$

This represents the so-called "gas cushion model" of hydrophobic slippage,[12] which got a clear microscopic foundation in terms of a prewetting transition.[20] Being a schematic representation of a depletion close to a wall,[21] this model provides a useful insight into the sensitivity of the interfacial transport to the structure of the interface. Similarly, electrokinetic flow displays the apparent slip. Note that recent molecular slip studies[17,19] also suggested a kind of the "vapor cushion model," where $b \propto e^4$, but apparently in a one-component system the value of e is too small to describe most of experimental data, suggesting that a two-component system[12,20] is required. Whether or not the addition of a hydrophobic solute will lead to nonlinear dependence of b on e remains an open question and has to be investigated.

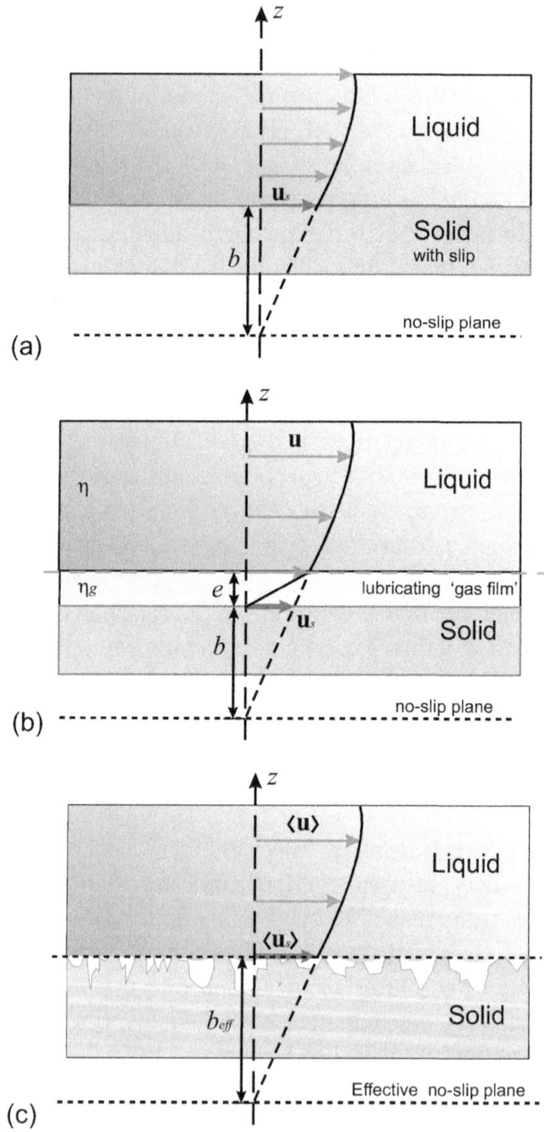

Figure 2.3 Schematic representation of the definition of intrinsic (a), apparent (b), and effective (c) slip lengths (reprinted with permission from Ref. 14, copyright 2011, IOP Publishing).

A modification of this scenario would be a nanobubble-coated surface, that is, a heterogeneous two-phase depletion layer.[22–27] However, as it has been discussed yet in,[4] nanobubbles will reduce the viscosity of the surface layer only if surface tension is small enough (which means that the bubbles deform). If the shape of the nanobubbles will remain a spherical lens, η_g should exceed the bulk viscosity. There remains still an open question connected with the exact expression for the effective viscosity of the layer of surface nanobubbles.

Another situation is that of the *effective* slip, b_{eff}, which refers to a situation where slippage at a complex heterogeneous surface is evaluated by averaging of flow over the length scale of the experimental configuration (e.g., a channel).[3,28–30] In other words, rather than trying to solve equations of motion at the scale of the individual corrugation or pattern, it is appropriate to consider the "macroscale" fluid motion (on the scale larger than the pattern characteristic length or the thickness of the channel) by using effective boundary conditions that can be applied at the imaginary smooth surface. Such an effective condition mimics the actual one along the true heterogeneous surface. It fully characterizes the flow at the real surface and can be used to solve complex hydrodynamic problems without tedious calculations. Such an approach is supported by a statistical diffusion arguments (being treated as an example of commonly used Onsager-Casimir relations for nonequilibrium linear response),[28] theory of heterogeneous porous materials,[29] and has been justified for the case of Stokes flow over a broad class of surfaces.[30] For anisotropic textures b_{eff} depends on the flow direction and is generally a tensor.[28] The effective slip also depends on the interplay between typical length scales of the system, as we will see later. Well-known examples of such a heterogeneous system include composite superhydrophobic (Cassie) surfaces, where a gas layer stabilized with a rough wall texture (Fig. 2.3c). For these surfaces effective slip lengths are often very large compared with the value on flat solids, similarly to what has been observed for wetting, where the contact angle can be dramatically enhanced when the surface is rough and heterogeneous.[31]

2.4 Experimental Methods

The experimental challenge generated a considerable progress in experimental tools for investigating flow boundary conditions using the most recent developments in optics and scanning probe techniques. A large variability still exists in the results of slip experiments, so it is important first to consider the different experimental methods used to measure a slip. Two broad classes of experimental approaches have been used so far, indirect and direct (local) methods.

High-speed force measurements can be performed with the surface forces apparatus (SFA)[32–34] or atomic force microscope (AFM).[35] In particular, in the drainage method[32,35] the end of the spring away from the attached sphere is driven toward the (fixed) plane with a constant driving speed (as shown in Fig. 2.4). The sphere itself, however, does not move at a constant speed, so the spring is deflected as a result of both the surface force (which should be measured separately) and the hydrodynamic forces. The solution of the (differential) equation of motion allows to deduce a

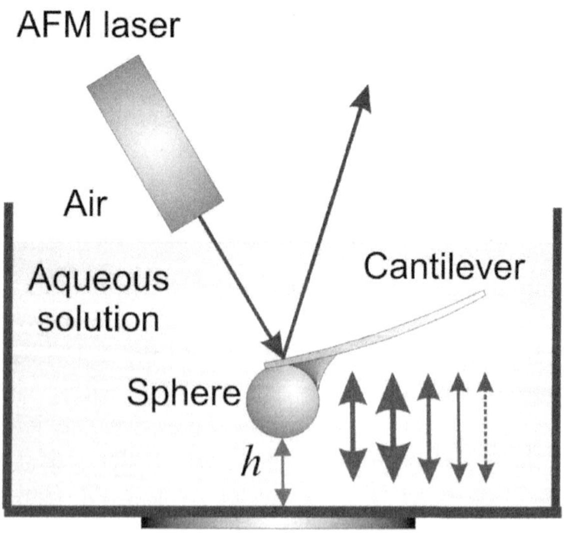

Figure 2.4 Schematic of the dynamic AFM force experiment (adapted from.[38])

drag force, with the subsequent comparison with a theory of a film drainage,[12,36] where the drag force is:

$$F = -\frac{6\pi\eta R^2}{h}\frac{dh}{dt}f^* = F_{\mathrm{T}}f^* \tag{2.3}$$

where $-dh/dt$ is the relative velocity of the surfaces, F_{T} is the Taylor solution for no-slip surfaces (which, however, never appeared in any of G. I. Taylor's publications, as discussed in,[34]) and f^* represents a correction function for slippage, which depends on b/h. For example, for a very convenient experimental configuration of the interaction of a hydrophilic surface with a hydrophobic one, which allows us to avoid the formation of a gas bridge,[37] the correction for a slip takes the form:[15]

$$f^* = \frac{F}{F_{\mathrm{T}}} = \frac{1}{4}\left(1 + \frac{3h}{2b}\left[\left(1 + \frac{h}{4b}\right)\ln\left(1 + \frac{4b}{h}\right) - 1\right]\right) \tag{2.4}$$

Note that when $h \ll 4b$, $f^* \rightarrow 1/4$, so the force is still inversely proportional to the gap but becomes four times smaller than predicted for hydrophilic surfaces. For two similar hydrophobic surfaces, $f^* \rightarrow (h/3b)\ln(6b/h)$ when $h \ll 6b$, so a hydrodynamic drag is only logarithmically dependent on the separation.[12]

Note that AFM force balance incorporates both a (concentrated) force on the sphere and the drag on the cantilever, as shown in Fig 2.1. The drag on a cantilever is neither small nor negligible,[38,39] and its ignorance might cause wrong experimental conclusions.

This approach, being extremely accurate at the nanoscale, does not provide visualization of the flow profile, so these measurements are often identified as *indirect*.

Direct approaches to flow profiling, or velocimetry, take advantage of various optics to monitor tracer particles. These methods include total internal reflection–fluorescence recovery after photobleaching (TIR-FRAP),[40] particle image velocimetry (μ-PIV),[41,42] total internal reflection velocimetry (TIRV),[43] evanescent wave microparticle image velocimetry (EW μ-PIV),[44] and multilayer nanoparticle image velocimetry (nPIV).[45,46] Their accuracy is normally much lower than that of force methods due to relatively low optical resolution, system noise due to polydispersity of tracers, and difficulties in decoupling flow from diffusion (the tracer

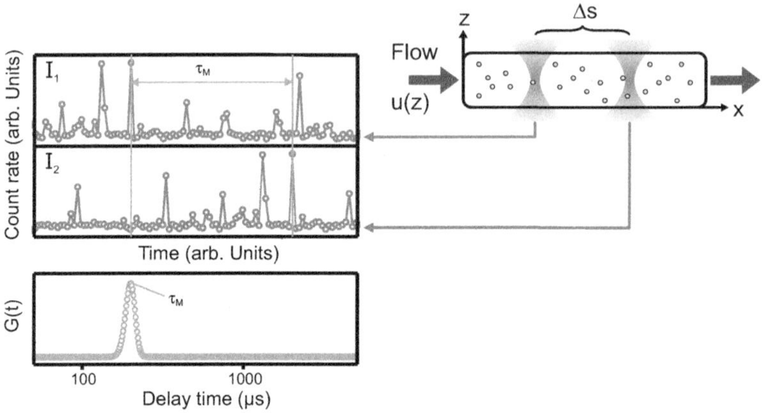

Figure 2.5 Schematics of the DF-FCS method. Two laser foci are placed along the x axis, separated by a distance of a few μm. They independently record the time-resolved fluorescence intensities $I_1(t)$ and $I_2(t)$. The forward cross-correlation of these two signals yields $G(t)$. Two foci are scanned simultaneously along the z axis to probe the velocity profile $u(z)$ (adapted from.[47,48])

distribution in the flow field is affected by Taylor dispersion.[47]) As a consequence, it has been always expected that a slippage of the order of a few tens of nanometers cannot be detected by a velocimetry technique. However, recently direct high-precision measurements at the nanoscale have been performed with a new optical technique, based on double-focus spatial fluorescence cross-correlation (DF-FCS)[47,48] (as is schematically shown in Fig. 2.5). As the fluorescence tracers are flowing along the channel, they are crossing consecutively the two foci, producing two time-resolved fluorescence intensities $I_1(t)$ and $I_2(t)$ recorded independently. The time cross-correlation function can be calculated and typically exhibits a local maximum. The position of this maximum τ_M is characteristic of the local velocity of the tracers. Another example of high-resolution, promising applications of FCS consists in determination of the average transverse diffusion coefficient to probe slippage.[49] Since FCS methods allow consideration of $N \sim 10^6$ particles, this gives a satisfactory signal-to-noise ratio $\sqrt{N}$ of the order 10^3, providing extremely good resolution as compared with other direct velocimetry methods. Coupling with TIRF,[50] which allows the measurements of the distance of tracers from the wall through the exponential decay of an evanescent wave, should further improve the accuracy of the approach.

In the vicinity of the wall the tracers are submitted to a Taylor dispersion, that is, their diffusion combined with shear enhances the migration speed in the flow direction. This phenomenon seriously complicates all velocimetry methods since to extract the velocity of fluid this effect should be modeled precisely. As shown by analysis of DF-FCS data,[47] large observed values of the apparent slip at the hydrophilic wall are normally fully attributed to a Taylor dispersion of nanotracers (see Fig. 2.6). The data obtained with other velocimetry technique still awaits clarification. We suggest, however, that some very large values of a hydrophobic slip[40,41] might reflect a Taylor dispersion too.

Many experiments have been performed on the subject, with somewhat contradicting results. Experimental work focused mostly on a bare (smooth) surface; more recent investigations have turned toward rough and structured surfaces, in particular superhydrophobic surfaces.[51] We refer the reader to comprehensive review articles[6,52] for a detailed account of early experimental work. In our chapter we mention only what we believe is the most relevant recent contribution to the subject of flow past "simple" smooth hydrophobic and rough hydrophilic surfaces, which clarified the existing controversies in the field. We focus, however, more on the implication of micro- and nanostructuring on fluidic transport, which is still in its infancy and remains to be explored.

2.5 Smooth Surfaces: Slippage vs. Wetting

From the theoretical[16,18] and simulation[17,53] points of view, slippage should not appear on a hydrophilic surface, except probably at a very high shear rate.[54] A slip length of the order of a hundred nanometers or smaller is, however, expected for a hydrophobic surface.[12,16,20,55]

On the experimental side, no consensus was achieved until recently. While some experimental data was consistent with the theoretical expectations both for hydrophilic and for hydrophobic

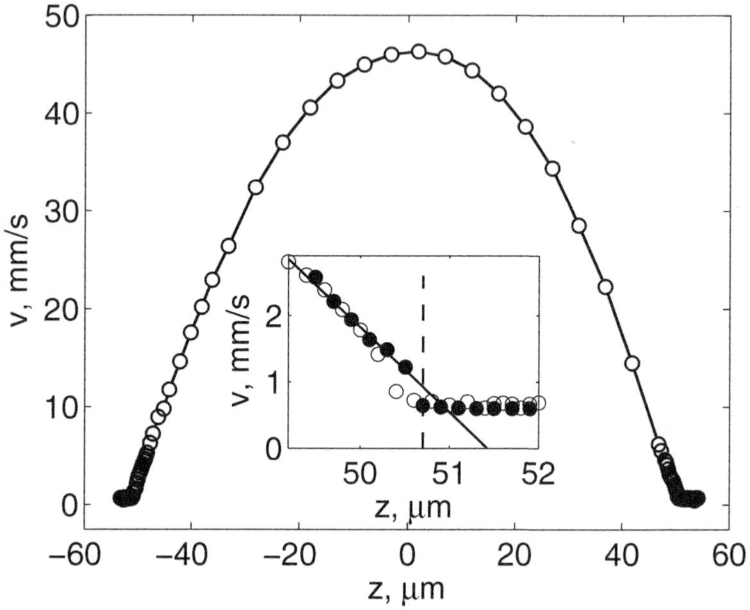

Figure 2.6 Typical velocity profile $v(z)$ (open circles) measured in a ~100 μm channel with 10^{-4} mol/L NaCl solution. Inset shows the same profile in the vicinity of the wall. Error bars were determined from the deviation among repeated measurements. When error bars are absent, uncertainties are smaller than the circles. Dots are simulation results. The dash-dotted line fits data (the apparent slip length, $b_{app} \sim 740$ nm), and the dashed line shows the wall location ($z = 50.69$ μm) (adapted from.[47])

surfaces,[33,35,56] some other reports completely escaped from this picture with both quantitative (slippage over hydrophilic surface, shear rate- dependent slippage, rate threshold for slip, etc.) and quantitative (slip length of several μms) discrepancies (for a review see.[6]) More recent experiments, performed with various new experimental methods, finally concluded that water does not slip on smooth hydrophilic surfaces and develops a slip only on a hydrophobic surface.[38,47,49,57–59] One can therefore conclude that a concept of hydrophobic slippage is now widely accepted.

An important issue is the amplitude of a hydrophobic slip. The observed slip length reached the range 20–100 nm, which is above predictions of the models of a molecular slip.[13,19] This suggests the apparent slip, such as the "gas cushion model," Eq. 2.2. Water glides on air, owing to the large viscosity ratio between water and air (typically a factor of 50). Experimental values of *b* suggest that the thickness of this "layer" is below 2 nm. Another scenario of the apparent slip, such as a nanobubble-coated surface,[22–25] has to be explored in more details.

An important conclusion is that it is impossible to benefit of such a nanometric slip at separations O(μm) and larger, that is, in microfluidic applications. This is why in the discussion of superhydrophobic slippage below we often ignore a slip past hydrophobic solids. However, a hydrophobic slippage is likely of major importance in nanochannels (highly confined hydrophobic pores, biochannels, etc.), where ordinary Poiseuille flow is fully suppressed.

2.6 Rough Surfaces

Only a very few solids are molecularly smooth. Most of them are naturally rough, often at a micro- and a nanoscale, due to their structure, methods of preparation, and various coatings. These surfaces are very often in the Wenzel (impaled) state, where the solid–liquid interface has the same area as the solid surface (Fig. 2.7a). However, even for rough hydrophilic Wenzel surfaces the situation was not very clear, and opposite experimental conclusions have been made: one is that roughness generates an extremely large slip,[60] and another is that it decreases the degree of slippage.[61–63] A conclusion in favor of a slip past rough surfaces is often made based on the fact that the hydrodynamic drag force becomes smaller when one of the surfaces is rough (see Fig. 2.8), which is qualitatively consistent with the model of a slip—Eqs. 2.3 and 2.4. Recent experimental data suggested that the description of flow near rough surfaces has to be corrected. Such a flow is equivalent to expected near a smooth hydrophilic (i.e., no-slip) surface that is located between the top and

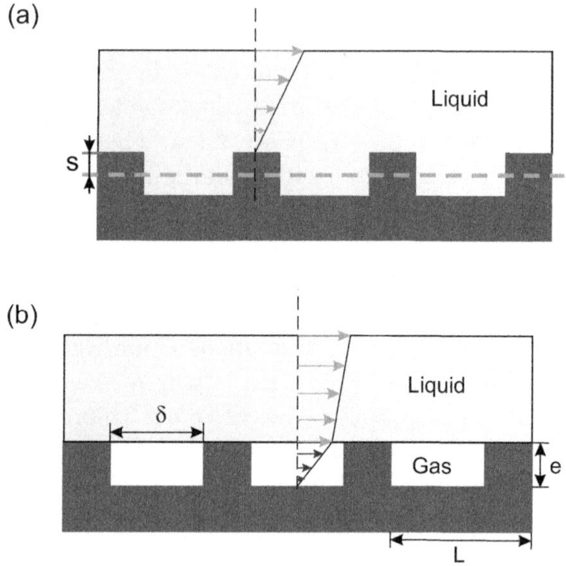

Figure 2.7 Schematic representation of the (a) Wenzel and (b) Cassie pictures with the local flow profiles at the gas and solid areas (reprinted with permission from Ref. 14, copyright 2011, by IOP Publishing).

bottom of asperities.[38] In other words the correction should be done not for a slip but for a separation from a rough surface. It has also been shown that opposing conclusions made by[60–62] simply reflect a different way of a definition of zero separation in the AFM (at the top of asperities) and SFA (at their bottom).

The theoretical description of such a flow represents a very difficult problem. Previous theoretical investigations addressed only a case of a far-field flow, where spatial variations are small compared to the typical length scale of the problem. Most of the articles considered the issue of periodic roughness,[64,65] although some recent work exploited nonperiodic, spatially homogeneous random roughness.[66] It has been concluded that these situations can be modeled by the application of the Navier condition to an equivalent smooth plane deposited on the top of the roughness.[64,67] However, there have been also theoretical (far-field) arguments that this should be rather a smooth equivalent no-slip wall placed

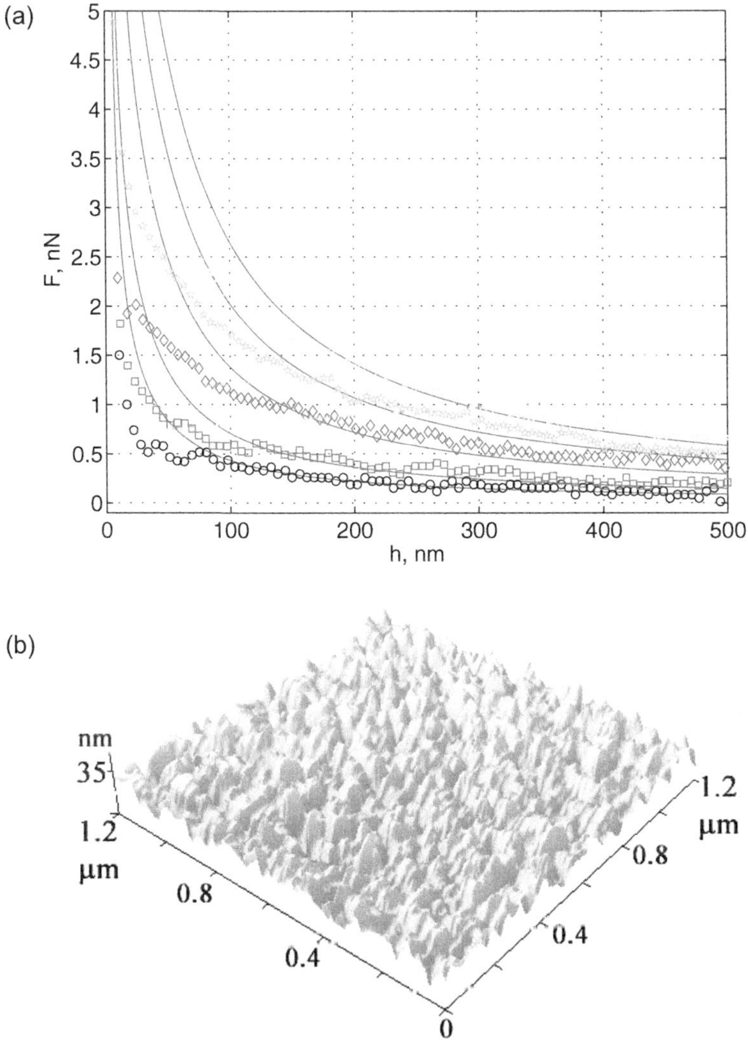

Figure 2.8 (a) Hydrodynamic force acting on a rough sphere ($R = 20$ μm) approaching a smooth plane. From bottom to top is data (symbols) obtained at the driving speed −6, −10, −20, −30, and −40 μm/s. Solid curves show the calculation results obtained for the same speed, by assuming no-slip boundary conditions. (b) An AFM image of an apex of a gold-coated sphere (adapted from.[38])

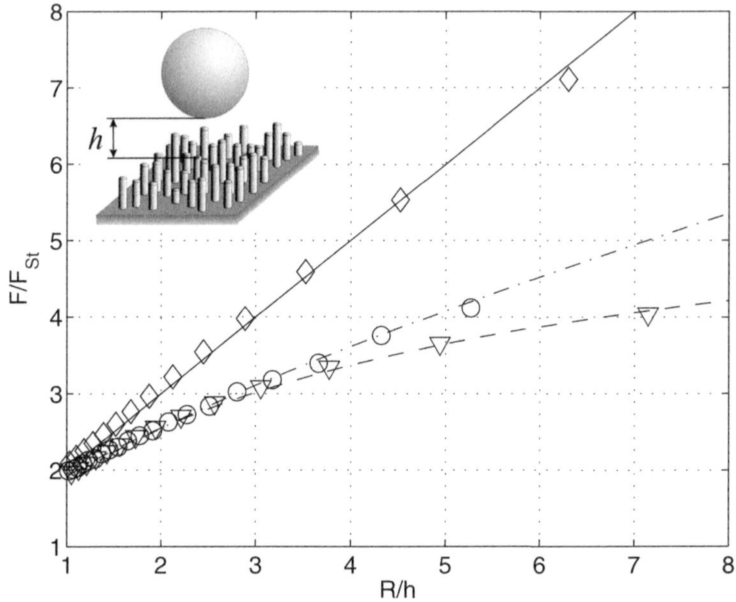

Figure 2.9 Hydrodynamic force acting on a hydrophilic sphere of radius R approaching a smooth hydrophilic (diamonds), smooth hydrophobic (circles), and randomly rough hydrophilic (triangles) wall with $\phi_2 = 4\%$ (adapted from.[53]) Here $F_{\text{St}} = 6\pi\eta RU$ is the Stokes drag. The separation h is defined on top of the surface roughness, as shown in the inset. Simulation results (symbols) compared with theoretical curves: $F/F_{\text{St}} = 1 + 9R/(8h)$ (solid), $F/F_{\text{St}} = 1 + 9Rf^*/(8h)$ with $f^* = f(b/h)$, calculated with Eq. 2.4 (dash-dotted), and $F/F_{\text{St}} = 1 + 9R/(8[h + s])$ (dashed). Values of b and s were determined by fitting the simulation data.

at some distance below the top but still above the bottom of the roughness.[68]

This issue was recently resolved in the lattice Boltzmann (LB) simulation study,[53] where the hydrodynamic interaction between a smooth sphere of radius R and a randomly rough plane was studied (as shown in Fig. 2.9). Besides its significance as a geometry of SFA/AFM dynamic force experiments, this allowed one to explore both far- and near-field flows in a single "experiment." The "measured" hydrodynamic force was smaller than predicted for two smooth surfaces (with the separation defined at the top of asperities) if the standard no-slip boundary conditions are used in the calculation. Moreover, at small separations the force was even weaker and shows different asymptotics than expected if one invokes slippage at the smooth fluid–solid interfaces. This can only be explained by the model of a no-slip wall located at an intermediate position (controlled by the density of roughness elements) between the top and bottom of asperities (illustrated by the dashed line in Fig. 2.7a). Calculations based on this model provided an excellent description of the simulation data (Fig. 2.9).

2.7 Superhydrophobicity and Effective Hydrodynamic Slippage

On hydrophobic solids, the situation is different from that on hydrophilic solids. If the solid is rough enough, we do not expect that the liquid will conform to the solid surface, as assumed in the Wenzel or impaled state. Rather air pockets should form below the liquid, provided that the energetic cost associated with all the corresponding liquid–vapor interfaces is smaller than the energy gained not to follow the solid.[5] This is the so-called Cassie or fakir state. Hydrophobic Cassie materials generate large contact angles and small hysteresis, ideal conditions for making water drops very mobile. It is natural to expect a large effective slip in a Cassie situation. Indeed, taking into account that the variation of the texture height, e, is in the typical interval $0.1 - 10$ μm, according to Eq. 2.2 we get $b = 5 - 500$ μm at the gas area. The composite nature of the texture requires regions of a very low slip (or no slip) in direct contact with the liquid, so the effective slip length of the surface, b_{eff}, is smaller than b. Still, one can expect that a rational design of such a texture could lead to large values of b_{eff}. Next we make these arguments more quantitative.

We examine an idealized superhydrophobic surface in the Cassie state sketched in Fig. 2.7b, where a liquid slab lies on top of the surface roughness. The liquid–gas interface is assumed to be flat with no meniscus curvature, so the modeled superhydrophobic surface appears as a perfectly smooth with a pattern of boundary conditions. In the simplified description the latter are taken as no slip ($b_1 = 0$) over solid–liquid areas and as a partial slip ($b_2 = b$) over gas–liquid regions (as we have shown above, b_1 is of the orders of tens of nanometers, so one could neglect it since b_2 is of the order of tens of μm). We denote as δ the typical length scale of gas–liquid areas. The fraction of solid–liquid areas will be denoted as $\phi_1 = (L - \delta)/L$ and of gas–liquid areas $\phi_2 = 1 - \phi_1 = \delta/L$. Overall, the description of a superhydrophobic surface we use here is similar to those considered in.[29,69-75] In this idealization, some assumptions may have a possible influence on the friction properties and, therefore, a hydrodynamic force. First, by assuming a flat interface, we have neglected an additional mechanism for a dissipation connected with the meniscus curvature.[76-78] Second, we ignore a possible transition toward the impaled (Wenzel) state that can be provoked by additional pressure in the liquid phase.[79,80] Third, we do not take into account the circulation of air in the gas phase[74,75,81] and some possible other effects such as a depinning of the contact line.[82]

Finally, for the sake of brevity we focus next only on the canonical microfluidic geometry, where the fluid is confined between flat plates, and only on the asymmetric case, where one (upper) surface is smooth hydrophilic and another (lower) represents a superhydrophobic wall in the Cassie state. Such a configuration is relevant for various setups, where the alignment of opposite textures is inconvenient or difficult. We also restrict the discussion by a pressure-driven flow governed by the Stokes equations:

$$\eta \nabla^2 \mathbf{u} = \nabla p, \quad \nabla \cdot \mathbf{u} = 0, \tag{2.5}$$

where $\mathbf{u}$ is the velocity vector and p is the pressure. Extensions of our analysis to study other configuration geometries and types of flow would be straightforward.

2.7.1 *Anisotropic Surfaces*

Many natural and synthetic textures are isotropic. However, it can be interesting to design directional structures, such as arrays of parallel grooves or microwrinkles, that consequently generate an anisotropic effective slip in the Cassie regime. The hydrodynamic slippage is quite different along and perpendicular to the grooves.

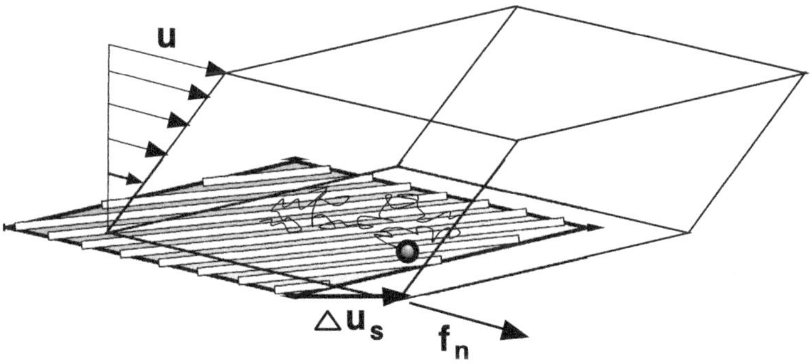

Figure 2.10 Sketch of a tensorial hydrodynamic slip. The normal traction $\mathbf{f}_n$ exerted by the fluid on an anisotropic surface produces an effective slip velocity $\Delta\mathbf{u} = \mathbf{M}\,\mathbf{f}_n$ in a different direction. At the molecular level, the interfacial mobility tensor **M** is related to the trajectories of diffusing particles in the interfacial region, such as the one shown (reprinted with permission from Ref. 28, copyright 2008, Cambridge University Press).

Axial motion is preferred, and such designs are appropriate when liquid must be guided. There are examples of such patterns in nature, such as the wings of butterflies or water striders.

A generalization of the Navier mobility condition is:[28]

$$\Delta\mathbf{u} = \mathbf{M} \cdot (\hat{\boldsymbol{n}} \cdot \boldsymbol{\sigma}) \tag{2.6}$$

where $\mathbf{f}_n = \hat{\boldsymbol{n}} \cdot \boldsymbol{\sigma}$ is the fluid force (normal traction) on the interface, $\boldsymbol{\sigma}$ is the local stress tensor, and **M** is an interfacial mobility tensor. As shown in Fig. 2.10, the effective slip vector is generally misaligned with the force vector for an anisotropic surface, which has been analyzed in a number of studies.[83,84] For anisotropic surfaces, the mobility must be a second-rank tensor $\mathbf{M} = \{M_{ij}\}$. In the case of a Newtonian fluid, $\mathbf{b}_{eff} = \mathbf{M}\eta$, and we get a tensorial version of (2.1), as discussed in:[3,28]

$$\langle u_i|_A\rangle = \sum_{j,k} b_{ij}^{eff} n_k \left\langle \frac{\partial u_j}{\partial x_k}\bigg|_A \right\rangle, \tag{2.7}$$

where $\langle \mathbf{u}|_A\rangle$ is the effective slip velocity, averaged over the surface pattern, and **n** is a unit vector normal to the surface A. The second-rank effective slip tensor $\mathbf{b}_{eff} \equiv \{b_{ij}^{eff}\}$ characterizes the surface anisotropy and is represented by a symmetric, positive, definite 2×2

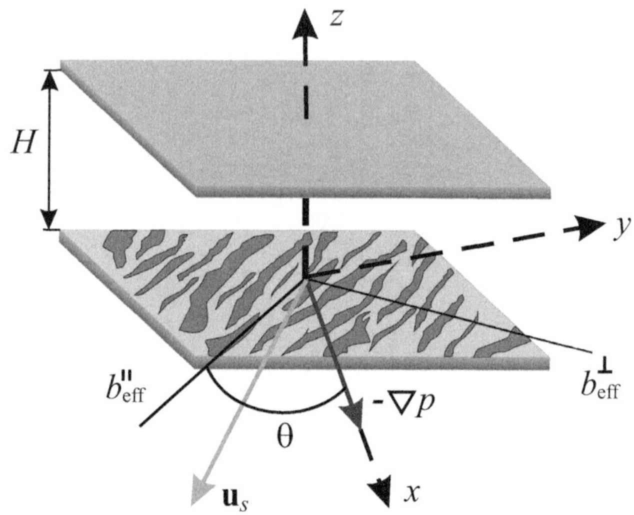

Figure 2.11 Sketch of a flat channel of a thickness H wall with notation for directions along the plates. One wall represents an anisotropic superhydrophobic texture (adapted from.[14])

matrix diagonalized by a rotation:

$$\mathbf{b}_{\text{eff}} = \mathbf{S}_\theta \begin{pmatrix} b_{\text{eff}}^{\parallel} & 0 \\ 0 & b_{\text{eff}}^{\perp} \end{pmatrix} \mathbf{S}_{-\theta}, \qquad \mathbf{S}_\theta = \begin{pmatrix} \cos\theta & \sin\theta \\ -\sin\theta & \cos\theta \end{pmatrix}. \tag{2.8}$$

As proven in[28] for all anisotropic surfaces the eigenvalues $b_{\text{eff}}^{\parallel}$ and $b_{\text{eff}}^{\perp}$ of the slip-length tensor correspond to the fastest (greatest forward slip) and slowest (least forward slip) directions, which are always orthogonal (see Fig. 2.11).

To illustrate the calculation of the slip-length tensor, next we consider the geometry where the liquid is confined between two plates separated by a distance H and one of them represents a superhydrophobic striped wall (Fig. 2.12a).

The effective slip lengths in eigendirections (which are in this case obviously parallel and orthogonal to stripes) has been recently calculated[14] by using the dual series technique suggested in.[70,78] Figure 2.13a shows the typical calculation results (the numerical example corresponds to $b/L = 20$ and $\phi_2 = 0.75$) and demonstrates that the effective slip lengths increase with H and

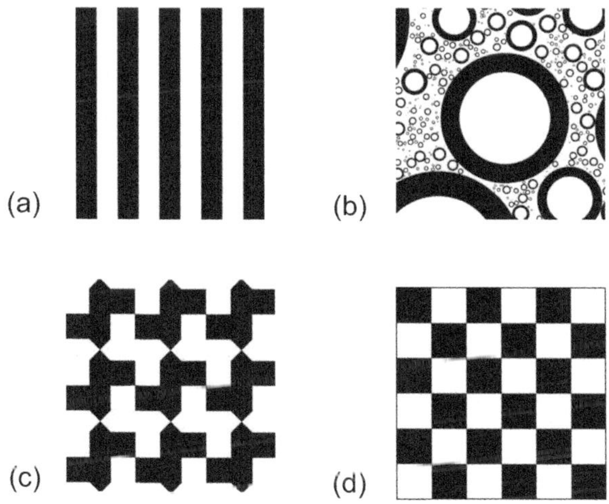

Figure 2.12 Special textures arising in theory. (a) Stripes, which attain the Wiener bounds of the maximal and minimal effective slip, if oriented parallel or perpendicular to the pressure gradient, respectively; (b) the HS fractal pattern of nested circles, which attains the maximal/minimal slip among all isotropic textures (patched should fill up the whole space, but their number is limited here for clarity); and (c) the Schulgasser and (d) chessboard textures, whose effective slip follows from the phase-interchange theorem (adapted from.[85]) *Abbreviation*: HS, Hashin-Shtrikman.

saturate for a thick gap. This points to the fact that an effective boundary condition *is not* a characteristic of liquid–solid interface solely but depends on the flow configuration and interplay between typical length scales, L, H, and b, of the problem. Next we discuss asymptotic limits (of small and large gaps) of this semianalytical solution.[14]

2.7.1.1 Thin channel

In the limit of $H \ll L$ we get:

$$b_{\text{eff}}^{\parallel} \simeq \frac{bH\phi_2}{H + b\phi_1}, \quad b_{\text{eff}}^{\perp} \simeq \frac{bH\phi_2}{H + 4b\phi_1}. \tag{2.9}$$

These expressions are independent of L but depend on H and suggest to distinguish between two separate cases.

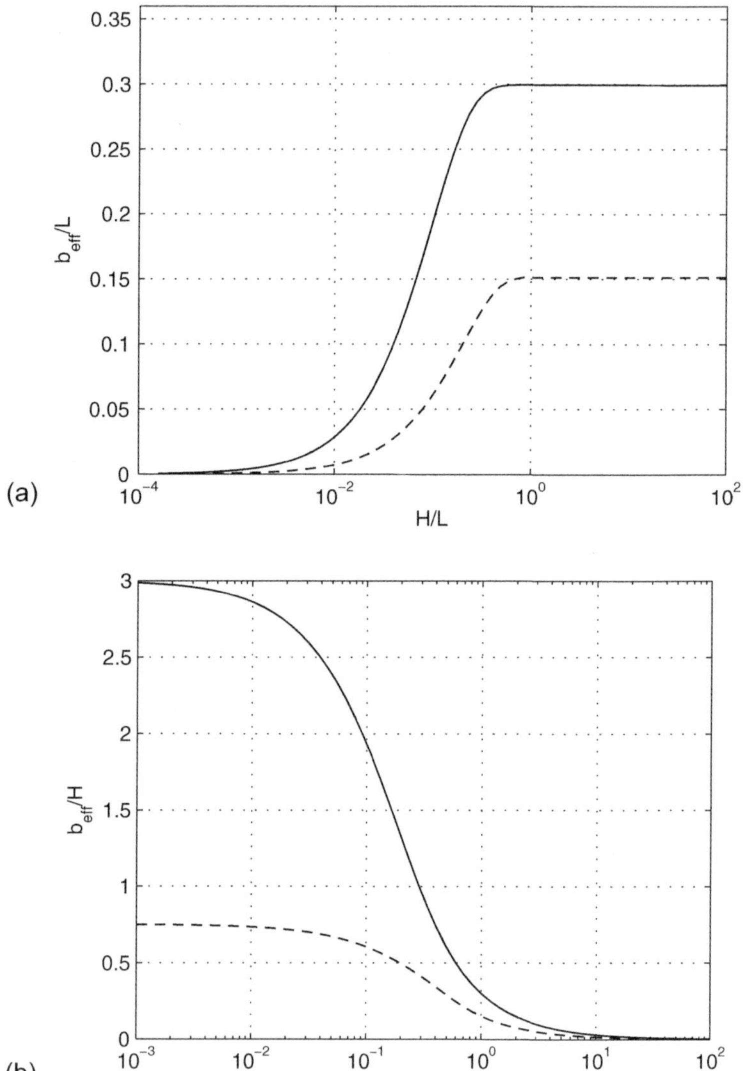

Figure 2.13 Eigenvalues $b_{\text{eff}}^{\parallel}$ (solid curve) and $b_{\text{eff}}^{\perp}$ (dashed curve) of the slip-length tensor for stick-slip stripes of period L with local slip length at the liquid–gas interface $b/L = 20$ and slipping area fraction $\phi_2 = 0.75$ as a function of the thickness of the channel, H (reprinted with permission from Ref. 14, copyright 2011, IOP Publishing).

If $b \ll H$ we obtain:

$$b_{\text{eff}}^{\perp} \simeq b_{\text{eff}}^{\parallel} \simeq b\phi_2, \tag{2.10}$$

so that despite the surface anisotropy we predict a simple surface-averaged effective slip. Although this limit is less important for pressure-driven microfluidics, it may have relevance for amplifying transport phenomena.[86]

When $H \ll b$ we derive:

$$b_{\text{eff}}^{\parallel} \simeq H\frac{\phi_1}{\phi_2}, \quad b_{\text{eff}}^{\perp} \simeq \frac{1}{4}b_{\text{eff}}^{\parallel}. \tag{2.11}$$

The above formula implies the effective slip length is generally four times as large for parallel versus perpendicular pressure-driven flow. Both asymptotic results, Eqs. 2.10 and 2.11, are surprising, taking into account the fact that for anisotropic Stokes flow in a thick channel, a factor of two is often expected as reminiscent results for striped pipes,[87] sinusoidal grooves,[30] and the classical result that a rod sediments twice as fast in creeping flow if aligned vertically rather than horizontally.[88] A very important conclusion from our analysis is that this standard scenario can significantly differ in a thin superhydrophobic channel by giving a whole spectrum of possibilities, from isotropic to highly anisotropic flow, depending on the ratio b/H.

Note that in case of a thin channel the flow can be described by an expression of Darcy's law, which relates the depth-averaged fluid velocity to an average pressure gradient along the plates through the effective permeability of the channel. The permeability, $\mathbf{K}_{\text{eff}}$, is in turn expressed through the effective slip length $\mathbf{b}_{\text{eff}}$, and permeability and slip-length tensors are coaxial. Such an approach allows one to use the theory of transport in heterogeneous media,[89] which provides *exact* results for an effective permeability over length scales much larger than the heterogeneity. This theory allows one to derive rigorous bounds on an effective slip length for arbitrary textures, given only the area fraction and local (any) slip lengths of the low-slip (b_1) and high-slip (b_2) regions.[29,90] These bounds constrain the attainable effective slip and provide theoretical guidance for texture optimization, since they are attained only by certain special textures in theory. In some regimes, the bounds are close enough to obviate the need for tedious calculations of flows over particular textures. In particular, by using the general result of the theory of bounds,[29] one can easily derive Eq. 2.9 and its limits, as well as prove that for a thin channel, longitudinal (transverse) stripes satisfy upper (lower) Wiener bounds, that is, provide the largest (smallest) possible slip that can be achieved by any texture. We remark and stress that according to results,[29] stripes should not be necessarily periodic.

Typical dependence of Wiener bounds for b_{eff}/H on ϕ_2 (at fixed b/H) and on b/H (at fixed ϕ_2) is shown in Fig. 2.14, which well illustrates that the key parameters determining the effective slip in the thin channel is the area fraction of solid, ϕ_1, in contact with the liquid. If this is very small (or $\phi_2 \to 1$), for all textures the effective slip tends to a maximum value, $b_{\text{eff}} \to b$. We can also conclude that maximizing b also plays an important role in achieving a large effective slip.

2.7.1.2 Thick channel

In the opposite case of infinitely large thickness ($H \gg L$), the effective slip lengths are:[70]

$$b_{\text{eff}}^{\parallel} \simeq \frac{L}{\pi} \frac{\ln\left[\sec\left(\frac{\pi\phi_2}{2}\right)\right]}{1+\frac{L}{\pi b}\ln\left[\sec\left(\frac{\pi\phi_2}{2}\right)+\tan\left(\frac{\pi\phi_2}{2}\right)\right]}, \tag{2.12}$$

$$b_{\text{eff}}^{\perp} \simeq \frac{L}{2\pi} \frac{\ln\left[\sec\left(\frac{\pi\phi_2}{2}\right)\right]}{1+\frac{L}{2\pi b}\ln\left[\sec\left(\frac{\pi\phi_2}{2}\right)+\tan\left(\frac{\pi\phi_2}{2}\right)\right]}. \tag{2.13}$$

The above results apply for a single surface and are independent of H. However, these expressions for effective slip lengths depend strongly on a texture period L. When $b \ll L$ we again derive the area-averaged slip length, Eq. 2.10. When $b \gg L$, expressions (2.12) and (2.13) take the form:

$$b_{\text{eff}}^{\perp} \simeq \frac{L}{2\pi}\ln\left[\sec\left(\frac{\pi\phi_2}{2}\right)\right], \quad b_{\text{eff}}^{\parallel} \simeq 2b_{\text{eff}}^{\perp}, \tag{2.14}$$

that coincides with the result obtained by Lauga and Stone[87] for the ideal slip ($b \to \infty$) case. We stress that the commonly expected

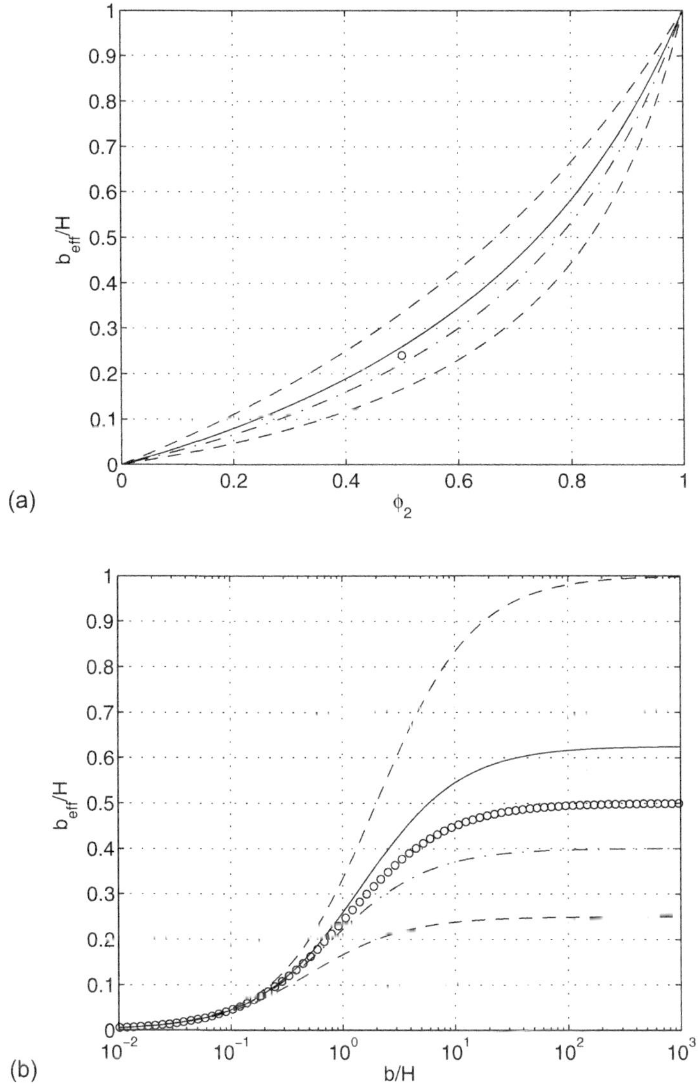

Figure 2.14 Effective slip length, b_{eff}/H, vs. ϕ_2 (for $b/H = 1$) (a) and b/H (for $\phi_2 = 0.5$) (b) in a thin gap limit, $H \ll L$. SH surfaces are anisotropic stick-slip stripes attaining Wiener bounds (dashed curves), isotropic textures attaining HS bounds (upper: solid; lower: dash-dotted curves) and satisfying the phase-interchange theorem (circles) (reprinted with permission from Ref. 14, copyright 2011, IOP Publishing).

factor of two for the ratio of $b_{\text{eff}}^{\parallel}$ and $b_{\text{eff}}^{\perp}$ is predicted only for very large b/L. In all other situations the anisotropy of the flow is smaller and even disappears at moderate b/L.

2.7.2 *Isotropic Surfaces*

As stressed before most solids are isotropic, that is, without a preferred direction. Unfortunately, from the hydrodynamic point of view, this situation is more complicated than considered before. Next we discuss only some aspects of the hydrodynamic behavior in thin and thick channel situations. For a thin channel, arguments are based on the already mentioned theory of transport in heterogeneous media[89] and derived bounds on an effective slip length (effective slip length and permeability tensors are now becoming simply proportional to the unit tensor $\mathbf{I}$) for arbitrary isotropic textures.[29,90] For a thick channel, the only available arguments are based on scaling theory and numerical calculations,[69] which, however, provide us with some guidance.

2.7.2.1 Thin channel

If the only knowledge about the two-phase isotropic texture is ϕ_1, ϕ_2, then the HS bounds apply for the effective permeability, by giving the corresponding upper and lower HS bounds for the effective slip length.[29,90] These bounds can be attained by the special HS fractal pattern sketched in Fig. 2.12b. For one bound, space is filled by disks of all sizes, each containing a circular core of one component and a thick ring of the other (with proportions set by the concentration), and switching the components gives the other bound. Fractal geometry is not necessary, however, since periodic honeycomb-like structures can also attain the bounds. The general solution[29,90] allows to deduce consequential analytical results for an asymmetric case considered in this paper. The upper (HS) bound can be then presented as

$$b_{\text{eff}} = \frac{bH\phi_2(2H + 5b)}{H(2H + 5b) + b\phi_1(5H + b)}, \qquad (2.15)$$

and the lower (HS) bound reads

$$b_{\text{eff}} = \frac{2bH\phi_2}{2H + 5b\phi_1}. \tag{2.16}$$

At small b/H we get Eq. 2.10, and at large b/H these give for upper and lower bounds

$$b_{\text{eff}} = \frac{5H\phi_2}{8\phi_1}, \quad \text{and} \quad b_{\text{eff}} = \frac{2H\phi_2}{5\phi_1}, \tag{2.17}$$

correspondingly.

Finally, phase interchange results[29] can be applied for some specific patterns (Fig. 2.12c,d). The phase interchange theorem states that the effective permeability $\mathbf{K}_{\text{eff}}(b_1, b_2)$ of the medium, when rotated by $\pi/2$, is related to the effective permeability of the medium, obtained by interchanging phases 1 and 2, namely, $\mathbf{K}_{\text{eff}}(b_2, b_1)$:

$$[\mathbf{R} \cdot \mathbf{K}_{\text{eff}}(b_1, b_2) \cdot \mathbf{R}^t] \cdot \mathbf{K}_{\text{eff}}(b_2, b_1) = K_1 K_2 \mathbf{I}$$

where $b_{1,2}$ are the local slip lengths for each medium, $\mathbf{R}$ is the rotation tensor, and $\mathbf{R}^t$ is its transpose. In the particular case of a medium that is invariant by a $\pi/2$ rotation followed by a phase interchange, the classical result follows:

$$K_{\text{eff}} = \sqrt{K_1 K_2}$$

Obviously, $\phi_1 = \phi_2 = 0.5$ for such a medium so that:

$$b_{\text{eff}} = \frac{3H}{4 - \sqrt{1 + \dfrac{3b}{H + b}}} - H, \tag{2.18}$$

At $b/H \ll 1$ we again derive Eq. 2.10, indicating that at this limit all textures show a kind of universal behavior and the effective slip coincides with the average. This suggests that the effective slip is controlled by the smallest scale of the problem,[16,70] so at this limit b_{eff} is no longer dependent on H, being proportional to b only. If $b/H \gg 1$ we simply get:

$$b_{\text{eff}} = \frac{H}{2} \tag{2.19}$$

again suggesting a kind of universality, that is, similarly to anisotropic stripes (cf. Eq. 2.11); in this limit b_{eff}/H for all isotropic textures almost likely scales as $\propto \phi_2/\phi_1$.

Table 2.1 The effective slip length b_{eff} for different textures (shown in Fig. 2.12) in a thin gap limit ($H \ll L$)

Texture	b_{eff}
Wiener upper bound (longitudinal stripes)	$\dfrac{bH\phi_2}{H + b\phi_1}$
Wiener lower bound (transverse stripes)	$\dfrac{bH\phi_2}{H + 4b\phi_1}$
HS upper bound (HS fractal, honeycomb-like texture)	$\dfrac{bH\phi_2(2H + 5b)}{H(2H + 5b) + b\phi_1(5H + b)}$
HS lower bound (HS fractal, honeycomb-like texture)	$\dfrac{2bH\phi_2}{2H + 5b\phi_1}$
Phase interchange patterns (Schulgasser texture, family of chessboards)	$\dfrac{3H}{4 - \sqrt{1 + 3b/(H + b)}} - H$

The results for these special textures are included in Fig. 2.14, which shows that HS bounds are relatively close and confined between Wiener ones. For completeness, we give in Table 2.1 a summary of the main expressions for an effective slip in a thin channel.

2.7.2.2 Thick channel

For this situation the exact solution was not found so far. Nevertheless, some simple scaling expressions have been proposed for a geometry of pillars[16,69] by predicting $b_{\text{eff}} \propto L/(\pi\sqrt{\phi_1})$ (cf. scaling results for stripes $b_{\text{eff}} \propto L/\ln(1/\phi_1)$). This simple result would deserve some analytical justification, which has not been performed up to now, despite some recent approximate analysis.[91] Recent semianalytical and numerical results confirm this scaling dependence,[92] which, in particular, suggests that in a thick channel the array of pillars will give a larger effective slip than longitudinal stripes for a sufficiently small solid fraction. This is opposite to a prediction for a thin channel, where an array of longitudinal stripes provides the largest possible slip that can be achieved by any texture, whether isotropic or anisotropic.

2.8 Other Special Properties of Superhydrophobic Surfaces

As we see before, hydrophobic Cassie materials generate a large and anisotropic effective slippage compared to simple, smooth channels, an ideal situation for various potential applications. A straightforward implication of a superhydrophobic slip would be the great reduction of the viscous drag of thin microchannels (enhanced *forward* flow), and some useful examples can be found in.[28] Next we illustrate the potential of superhydrophobic surfaces and possibilities of the effective slip approach by discussing a couple of other applications. Namely, we show that optimized superhydrophobic textures may be successfully used in passive microfluidic mixing and for a reduction of a hydrodynamic drag force.

2.8.1 *Transverse Flow*

The effective hydrodynamic slip[3,28,30] of anisotropic textured surfaces is generally tensorial, which is due to secondary flows *transverse* to the direction of the applied pressure gradient. In the case of grooved no-slip surfaces (Wenzel state), such a flow has been analyzed for small height variations[93] and thick channels,[94] and herringbone patterns have been designed to achieve passive chaotic mixing during pressure-driven flow through a microchannel.[84,95]

It has recently been demonstrated that similar effects may be generated by a superhydrophobic Cassie surface. The transverse flow due to surface anisotropy is generated only in the vicinity of the wall and disappears far from it, which has already been observed in experiments.[96] In other words, the effective velocity profile is "twisted" close to the superhydrophobic wall (see Fig. 2.15a).

To evaluate the transverse flow the velocity profile has been integrated across the channel to obtain the flow vector:

$$\mathbf{Q} = \int_0^H \langle \mathbf{u}(z) \rangle \, dz \tag{2.20}$$

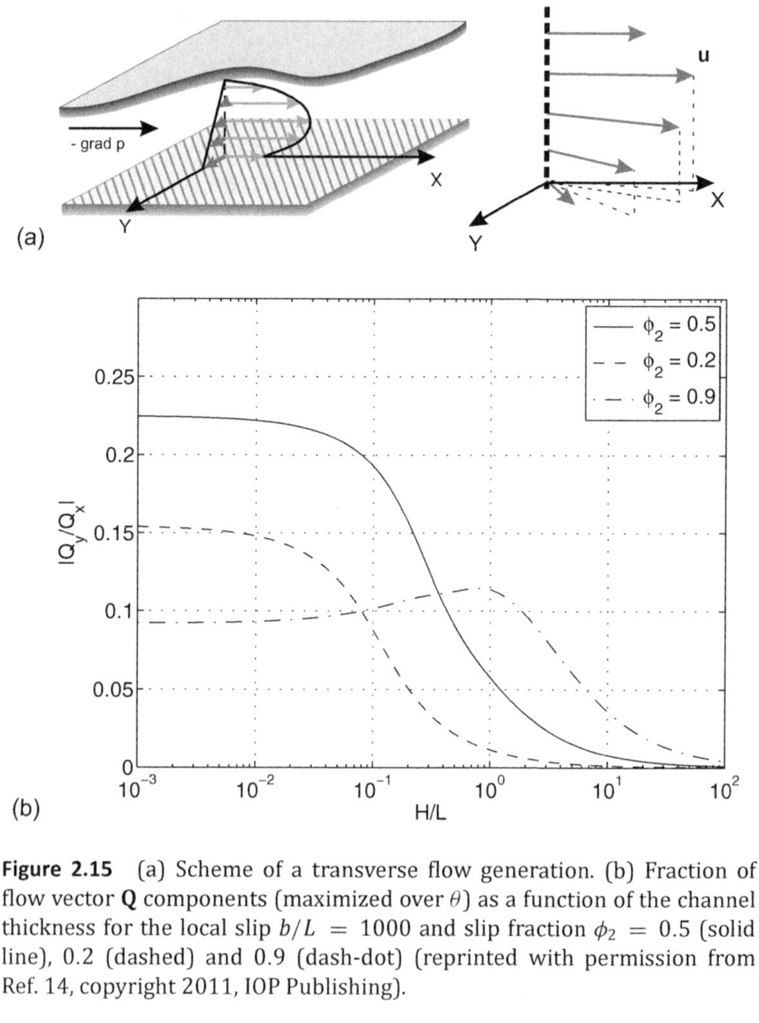

Figure 2.15 (a) Scheme of a transverse flow generation. (b) Fraction of flow vector **Q** components (maximized over θ) as a function of the channel thickness for the local slip $b/L = 1000$ and slip fraction $\phi_2 = 0.5$ (solid line), 0.2 (dashed) and 0.9 (dash-dot) (reprinted with permission from Ref. 14, copyright 2011, IOP Publishing).

with the components:

$$Q_x = \frac{\sigma}{\eta}\frac{H^3}{12}\left[1 + 3\frac{\left(H b_{\text{eff}}^{\parallel}\cos^2\theta + H b_{\text{eff}}^{\perp}\sin^2\theta + b_{\text{eff}}^{\parallel} b_{\text{eff}}^{\perp}\right)}{(H + b_{\text{eff}}^{\parallel})(H + b_{\text{eff}}^{\perp})}\right], \tag{2.21}$$

$$Q_y = \frac{\sigma}{\eta}\frac{H^4}{4}\frac{(b_{\text{eff}}^{\parallel} - b_{\text{eff}}^{\perp})\sin\theta\cos\theta}{(H + b_{\text{eff}}^{\parallel})(H + b_{\text{eff}}^{\perp})}. \tag{2.22}$$

To optimize the texture, channel thickness, and the angle θ between the directions of stripes and the pressure gradient, $|Q_y/Q_x|$ is maximum, providing the best transverse flow.

The maximization in respect to θ indicates that the optimal angle is[14]

$$\theta_{\max} = \pm \arctan \left[\frac{(1 + 4b^{\parallel}_{\text{eff}}/H)(1 + b^{\perp}_{\text{eff}}/H)}{(1 + b^{\parallel}_{\text{eff}}/H)(1 + 4b^{\perp}_{\text{eff}}/H)} \right]^{1/2}. \quad (2.23)$$

and the value of the maximum reads

$$\left| \frac{Q_y}{Q_x} \right| = \frac{1}{2} \left(\tan \theta_{\max} - \frac{1}{\tan \theta_{\max}} \right). \quad (2.24)$$

We conclude, therefore, that since H is fixed, the maximal $|Q_y/Q_x|$ corresponds to the largest physically possible b, that is, the perfect slip at the gas sectors.

To optimize the fraction of the slipping area, ϕ_2, we should now exploit results for effective slip lengths $b^{\parallel,\perp}_{\text{eff}}$ obtained above. Figure 2.15b shows the computed value of $|Q_y/Q_x|$ *vs.* H/L for several ϕ_2. The calculations are made using the value of θ defined by Eq. 2.23. The data suggests that the effect of ϕ_2 on a transverse flow depends on the thickness of the channel. For a thick gap the increase in the gas fraction, ϕ_2, augments a transverse flow. This result has a simple explanation. For an infinite channel $b^{\parallel,\perp}_{\text{eff}}/H \ll 1$ (see Fig. (2.14b)), which gives

$$\left| \frac{Q_y}{Q_x} \right|_{H \to \infty} \simeq \frac{3}{2} \frac{\Delta b_{\text{eff}}}{H}, \quad (2.25)$$

that is, in a thick channel the amplitude of a transverse flow is controlled by the difference between eigenvalues of the effective slip tensor, $\Delta b_{\text{eff}} = b^{\parallel}_{\text{eff}} - b^{\perp}_{\text{eff}}$, which increases with ϕ_2, as follows from the above analysis. We stress, however, that since $|Q_y/Q_x| \propto H^{-1}$, the mixing in a thick superhydrophobic channel would be not very efficient. A more appropriate situation corresponds to a thin channel, as well illustrated in Fig. 2.15b. We see that the largest transverse flow can be generated at intermediate values of ϕ_2. The limit of thin channel has recently been studied in detail by using a general theory of mathematical bounds,[89] and the optimum value of

$\phi_2 = 0.5$ corresponding to a numerical example in Fig. 2.15b has been rigorously derived.[97]

An important conclusion from the above analysis is that the surface textures that optimize transverse flow can significantly differ from those optimizing the effective (forward) slip. It is well known, and we additionally demonstrated before, that the effective slip of a superhydrophobic surface is maximized by reducing the solid–liquid area fraction ϕ_1. In contrast, we have shown that transverse flow in superhydrophobic channels is maximized by stripes with a rather large solid fraction, $\phi_1 = 0.5$, where the effective slip is relatively small.

2.8.2 *Hydrodynamic Interactions*

As a consequence of the superlubrication potential, a hydrophobic texture could significantly modify squeeze film drainage between surfaces. It is of obvious practical interest to minimize the hydrodynamic resistance, F, to the approach of surfaces.

For a Reynolds problem, where a disk of radius R is driven toward (in our case) a superhydrophobic plane with a velocity U (see Fig. 2.16), we should solve Eqs. 2.5 by applying the effective tensorial boundary condition, Eq. 2.7, at the superhydrophobic wall. This allows to derive a general expression for hydrodynamic force satisfying the condition $p = p_0$ at the edge of the disk:[85]

$$F = \frac{3}{2}\frac{\pi \eta U R^4}{H^3} f^*_{\text{eff}} = F_{\text{R}} f^*_{\text{eff}}, \tag{2.26}$$

where F_{R} represents the classical solution of creeping flow equations of the Reynolds lubrication theory,[98] and the correction for an effective slip is:

$$f^*_{\text{eff}} = \frac{F}{F_{\text{R}}} = 2\left[\frac{H + 4b^{\parallel}_{\text{eff}}(H)}{H + b^{\parallel}_{\text{eff}}(H)} + \frac{H + 4b^{\perp}_{\text{eff}}(H)}{H + b^{\perp}_{\text{eff}}(H)}\right]^{-1}. \tag{2.27}$$

Thus the effective correction for a superhydrophobic slip is the harmonic mean of corrections expressed through effective slip lengths in two principal directions:

$$f^*_{\text{eff}} = 2\left(\frac{1}{f^{*,\parallel}_{\text{eff}}} + \frac{1}{f^{*,\perp}_{\text{eff}}}\right)^{-1} \tag{2.28}$$

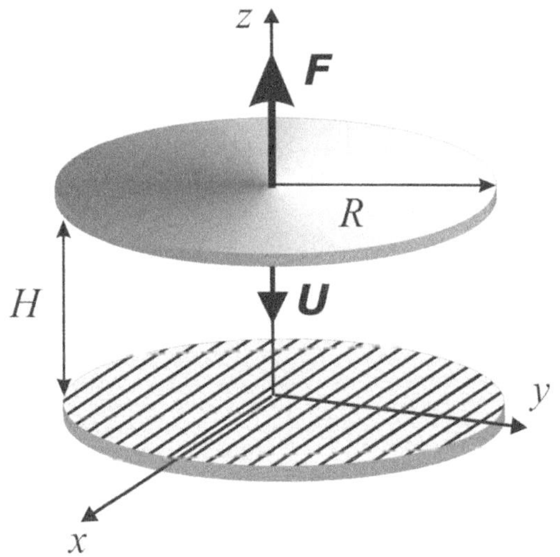

Figure 2.16 Sketch of a hydrophilic disk approaching a superhydrophobic striped disk.

In the case of isotropic textures, all directions are equivalent with $b_{\text{eff}}^{\parallel} = b_{\text{eff}}^{\perp} = b_{\text{eff}}$, so we get:

$$f_{\text{eff}}^{*} = \frac{F}{F_{\text{R}}} = \frac{H + b_{\text{eff}}(H)}{H + 4b_{\text{eff}}(H)} \tag{2.29}$$

Obviously, the case $b_{\text{eff}}^{\parallel} = b_{\text{eff}}^{\perp} = 0$ corresponds to $f_{\text{eff}}^{*} = 1$ and gives the Reynolds formula.

The expression for f_{eff}^{*} is very general and relates it to the effective slip length of the superhydrophobic wall and the gap. In order to quantify the reduction of a drag force due to the presence of a superhydrophobic wall, this expression can be used for *all* anisotropic and isotropic textures, where analytical or numerical expressions for $b_{\text{eff}}^{\parallel,\perp}$ have been obtained.

An important consequence of Eq. 2.27 is that to reduce a drag force, we need to maximize the ratio b_{eff}/H but not the absolute values of the effective slip itself. This is illustrated in Fig. 2.17a, where values presented in Fig. 2.13b were used to compute the correction for the effective slip, f_{eff}^{*}, as a function of the gap. At small H/L our calculations reproduce the asymptotic values predicted by Eqs. 2.11. They, however, vanish at large distances, where $b_{\text{eff}}^{\parallel,\perp}/H$

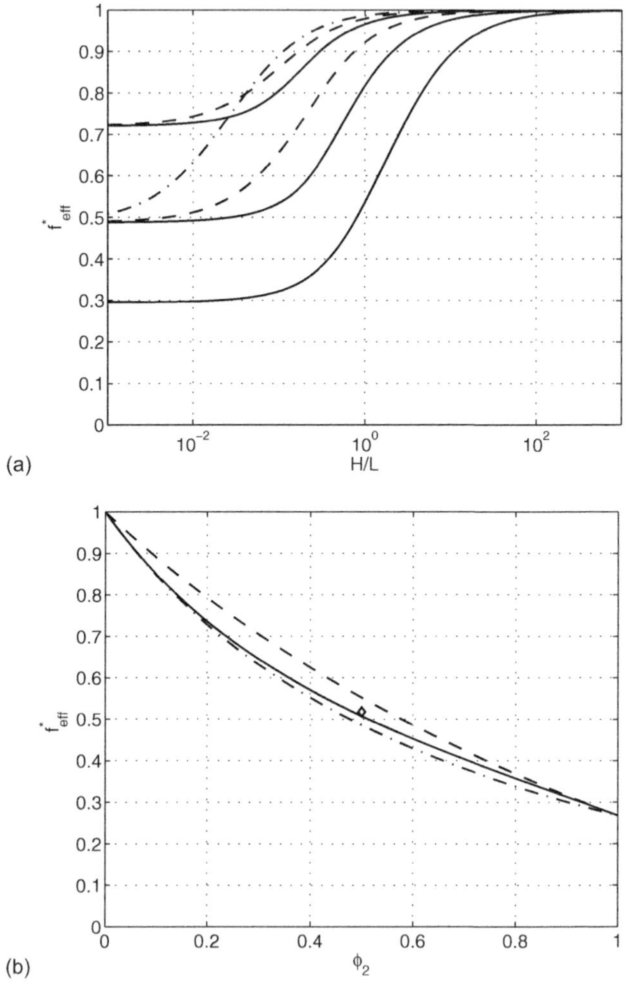

Figure 2.17 (a) The correction factor $f^*_{\rm eff} = F/F_{\rm R}$ for hydrodynamic resistance force exerted on a disk interacting with superhydrophobic stripes vs. a dimensionless gap width H/L. Solid curves correspond to a local slip length $b/L = 10$ (from top to bottom $\phi_2 = 0.2$, 0.5, and 0.9), dashed curves to $b/L = 0.1$ (from top to bottom $\phi_2 = 0.2$ and 0.5), dash-dotted curve to $b/L = 0.01$ and $\phi_2 = 0.5$. (b) The plot of $f^*_{\rm eff}$ vs. ϕ_2 for a thin gap ($H \ll L$) and several superhydrophobic patterns: anisotropic stripes (solid line), isotropic textures attaining HS bounds (dashed and dash-dotted lines), and isotropic Schulgasser structure (diamond), all with a local slip $b/H = 10$ (reprinted with permission from Ref. 14, copyright 2011, IOP Publishing).

Table 2.2 Asymptotic expansions for the force correction factor f_{eff}^{*} in the case of a striped surface

Limiting case	f_{eff}^{*}
$H \gg \max\{L, b\}$	$1 - \dfrac{3(b_{\mathrm{eff}}^{\parallel} + b_{\mathrm{eff}}^{\perp})}{2H}$
$L \ll H \ll b$	$\dfrac{1}{4} + \dfrac{9}{32}\dfrac{\pi H}{L \ln(\sec(\pi\phi_2/2))}$
$b \ll H \ll L$	$1 - \dfrac{3b\phi_2}{H}$
$H \ll \min\{L, b\}$	$\dfrac{2(4 - 3\phi_2)}{8 + 9\phi_2 - 9\phi_2^2}$

Table 2.3 Correction factor f_{eff}^{*} for some specific isotropic patterns in a thin gap limit

Pattern	$b \ll H \ll L$	$H \ll \min\{L, b\}$
HS upper bound	$1 - \dfrac{3b\phi_2}{H}$	$\dfrac{5 - 3\phi_2}{5 + 3\phi_2}$
HS lower bound	$1 - \dfrac{3b\phi_2}{H}$	$\dfrac{8 - 3\phi_2}{4(2 + 3\phi_2)}$
Phase interchange textures	$1 - \dfrac{3b}{2H}$	$\dfrac{1}{2}$

are getting negligibly small. The useful analytical expressions for f_{eff}^{*} corresponding to a configuration of stripes are presented in Table 2.2. Similar estimates for the most important situation of a thin gap can similarly be done for some isotropic textures, and we include these results in Table 2.3.

The results presented in Tables 2.2 and 2.3 show that the key parameter determining reduction of drag is the area fraction of gas, ϕ_2, in contact with the liquid. This is illustrated in Fig. 2.17b, where (using a relatively large b/H), HS bounds for f_{eff}^{*} are plotted versus ϕ_2. If this is very small (or $\phi_1 \to 1$) for all textures, the correction for the slip tends to its absolute maximum, $f_{\mathrm{eff}}^{*} \to 1$. In the most interesting limit, $\phi_2 \to 1$, we can achieve the minimum possible value of correction for the effective slip, $f_{\mathrm{eff}}^{*} \to 1/4$ provided b/H is large enough. We also stress that the results for stripes are confined between HS bounds for f_{eff}^*. In other words, isotropic textures might be the best candidates for a reduction of a drag force.

Another important point to note would be that there exists no universal relationship between f_{eff}^* and the effective Cassie angle, $\cos\theta_C^*$, of the superhydrophobic surface. Indeed, according to the Cassie equation, $\cos\theta_C^* = (1-\phi_2)\cos\theta - \phi_2$,[5] where $\cos\theta$ is the thermodynamic contact angle at the smooth surface. It is well seen already from Fig. 2.17b that the dependence of f_{eff}^* on ϕ_2 is much more complex. In particular, at fixed H it depends on texture. In general, this also reflects the fact that there is no universal relationship between b_{eff} and ϕ_2, despite some recent attempts to establish it.[99]

2.9 Interfacial Transport Phenomena

2.9.1 *Hydrophobic Surfaces*

Besides this drag reduction potential in pressure-driven flows, it has recently been predicted that hydrophobic slippage could be best exploited in surface-driven transport.

Electro-osmosis (EO), that is, flow generation by an electric field, originally developed in colloid science, is currently experiencing a renaissance in microfluidics. It may be considerably amplified by surface slippage, even for nanometric slip length. The reason for this amplification is that the electric double layer (EDL), characterized by the Debye screening length $\lambda_D = \kappa^{-1}$, defines an additional length scale of the problem comparable to b. According to classical formula:[100,101]

$$u = -\frac{q_0 E_0}{\eta\kappa}(1 + b\kappa) = -\frac{\epsilon\zeta E_0}{\eta}(1 + b\kappa) \tag{2.30}$$

where u is the electro-osmotic velocity (outside the double layer), q_0 is the surface charge density, $\zeta = q_0/\kappa\epsilon$ is the zeta-potential across the diffuse (flowing) part of the double layer, ϵ is the permittivity of the solution, and E_0 is the tangential electric field. Therefore, the flow can potentially be enhanced for a thin compared to b EDL, that is, when $\kappa b \gg 1$ (see Fig. 2.18), which has been proven experimentally.[58,101]

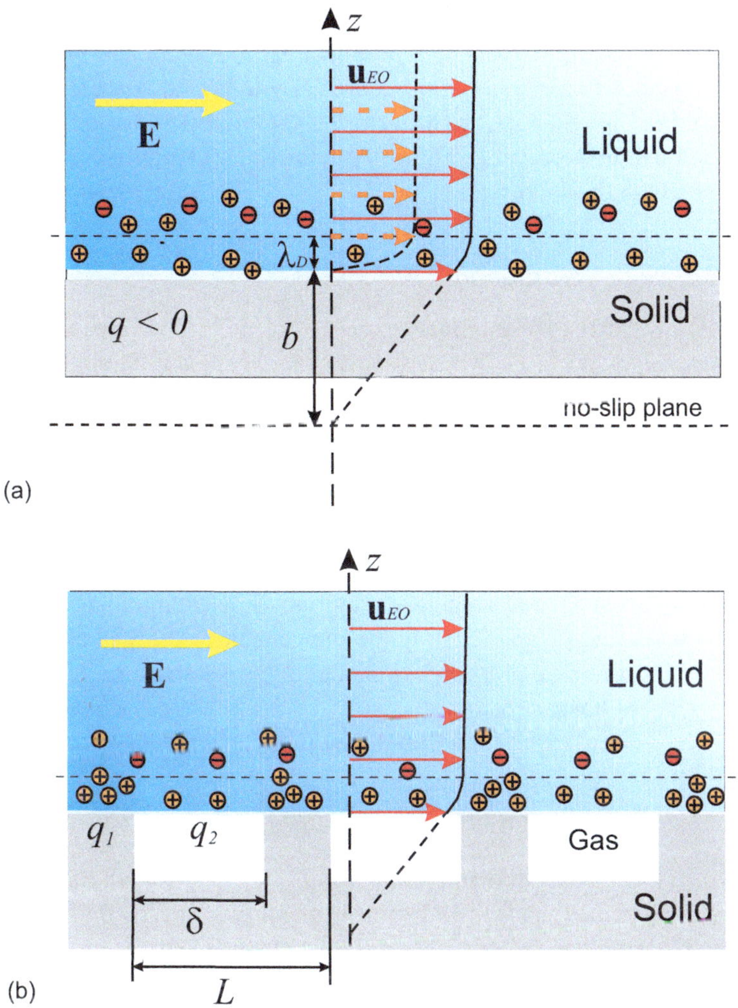

Figure 2.18 Sketch of the influence of hydrophobic (a) and superhydrophobic (b) slippage on electro-osmotic flow.

This expression can be generalized to other interfacial transport phenomena, such as diffusio-osmosis and thermo-osmosis, as it has been well discussed in.[16,86] These correspond to the induction of a flow by the gradient of a solute concentration for the former and by a gradient of temperature for the latter.[102] The flow velocity for these two important phenomena is proportional to the applied gradient of concentration or temperature. Similarly to EO, for both, slippage amplifies the velocity with the factor $(1+b/\lambda)$, where λ is a thickness of a thin interface layer (of the order of the range of interaction of the solute with the solid surface).

2.9.2 *Superhydrophobic Surfaces*

It is now attractive to consider electro-osmotic flow over superhydrophobic surfaces, whose texture can significantly amplify the hydrodynamic slip. Equation 2.30 suggests that a massive amplification of EO flow can be potentially achieved over superhydrophobic surfaces. However, the controlled generation of such flows is by no means obvious since both the slip length and the charge distribution are inhomogeneous and anisotropic. Despite its fundamental and practical significance, electro-osmotic flow over a superhydrophobic surface has so far received little attention. Only recently has such a flow been investigated past an inhomogeneously charged slipping surface in the case of a thick channel ($H \gg L$) and a perfect slip ($b \to \infty$) at the gas sectors.[103,104]

The general result for a thin EDL ($\kappa L \gg 1$) has been formulated in:[103]

$$\mathbf{u} = -\frac{\mathbf{E}_0}{\eta\kappa} \cdot (q_1 \mathbf{I} + q_2 \mathbf{b}_{\text{eff}} \kappa) \tag{2.31}$$

by using the Lorentz reciprocal theorem for Stokes flow. Here $\mathbf{I}$ is the unity tensor, and q_1 and q_2 are the surface charge density at the solid and gas regions, correspondingly. This expression shows that surface anisotropy generally leads to a tensorial electro-osmotic mobility. To highlight the effect of anisotropy, we focus now on a patterned superhydrophobic surface consisting of periodic stripes, where the surface charge density varies only in one direction—likewise the slip lengths. By combining Eq. 2.31 with Eqs. 2.14, we get general results and important limiting situations for stripes summarized in Table 2.4 and discussed next.

In the case of an uncharged gas interface ($q_1 = q_0$, $q_2 = 0$), we predict the simple Smoluchowski formula.[103] In other words, there is no EO flow enhancement, and the flow is isotropic despite

Table 2.4 Electroosmotic slip past superhydrophobic stripes in a thin EDL limit

	$u_{\text{eff}}^{\parallel}$	$u_{\text{eff}}^{\perp}$
General case	$-\frac{E_0}{\eta\kappa}\left(q_1+q_2\frac{\kappa L}{\pi}\ln\left[\sec\left(\frac{\pi\phi_2}{2}\right)\right]\right)$	$-\frac{E_0}{\eta\kappa}\left(q_1+q_2\frac{\kappa L}{2\pi}\ln\left[\sec\left(\frac{\pi\phi_2}{2}\right)\right]\right)$
$q_1=q_0,\ q_2=0$	$-\frac{E_0q_0}{\eta\kappa}$	$-\frac{E_0q_0}{\eta\kappa}$
$q_1=q_2=q_0$	$-\frac{E_0q_0}{\eta\kappa}\left(1+\frac{\kappa L}{\pi}\ln\left[\sec\left(\frac{\pi\phi_2}{2}\right)\right]\right)$	$-\frac{E_0q_0}{\eta\kappa}\left(1+\frac{\kappa L}{2\pi}\ln\left[\sec\left(\frac{\pi\phi_2}{2}\right)\right]\right)$
$q_1=-q_2=q_0$	$-\frac{E_0q_0}{\eta\kappa}\left(1-\frac{\kappa L}{\pi}\ln\left[\sec\left(\frac{\pi\phi_2}{2}\right)\right]\right)$	$-\frac{E_0q_0}{\eta\kappa}\left(1-\frac{\kappa L}{2\pi}\ln\left[\sec\left(\frac{\pi\phi_2}{2}\right)\right]\right)$

anisotropy of a surface (see Fig. 2.19). This surprising result has been confirmed by MD simulations[105] and later analysis.[104] Note that for an uncharged gas interface with a thick EDL ($\kappa L \ll 1$), the results are qualitatively different, and for an uncharged gas interface the EO flow remains tensorial.[104] However, since κL is small and since the electro-osmotic velocity is proportional to the fraction of a charged area, ϕ_1, electro-osmotic flow becomes suppressed (compared to that predicted by the Smoluchowski equation) despite a large, effective slip.[14]

In the case of a charged gas interface (only) a considerable enhancement of electro-osmotic flow is possible. For a uniformly charged ($q_1 = q_2 = q_0$) anisotropic superhydrophobic surface, the expression for electro-osmotic flow can be transformed to:

$$\mathbf{u} = -\frac{\mathbf{E}_0 q_0}{\eta\kappa}\cdot(\mathbf{I}+\mathbf{b}_{\text{eff}}\kappa)\,, \tag{2.32}$$

which might be seen as a tensorial analog of Eq. 2.30. Figure 2.19 includes theoretical results calculated with Eq. 2.32 for a geometry of stripes and is intended to demonstrate that the flow is truly anisotropic and can exhibit a large enhancement from the effective hydrodynamic slip, possibly by an order of magnitude. We stress that such an enhancement is possible even at a relatively low gas fraction, that is, when $\mathbf{b}_{\text{eff}}$ is relatively small (but the amplification ratio, $\mathbf{b}_{\text{eff}}/\lambda_D$, might be huge). Obviously, in the case of an isotropic superhydrophobic surface, Eq. 2.32 transforms to:

$$u = -\frac{E_0 q_0}{\eta\kappa}(1+b_{\text{eff}}\kappa)\,, \tag{2.33}$$

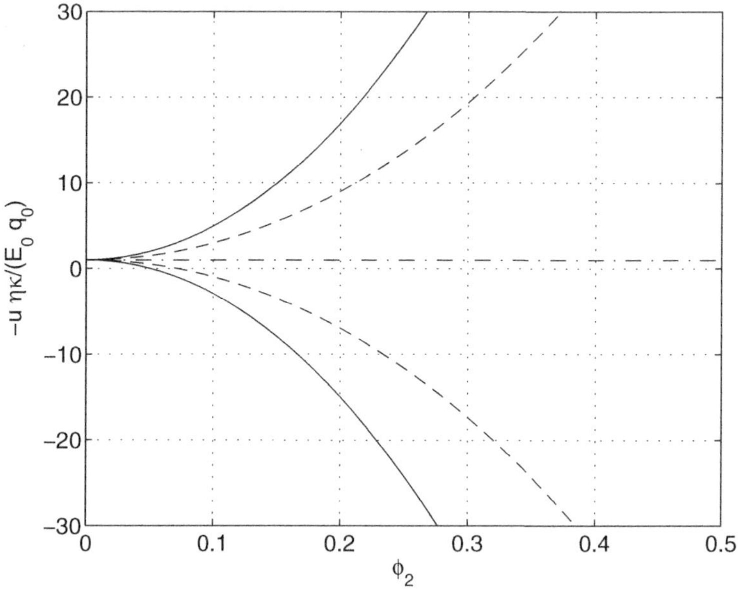

Figure 2.19 The electro-osmotic slip velocities (thin EDL) for eigendirections of superhydrophobic stripes. The dash-dotted line corresponds to the case of uncharged gas sectors $q_1 = q_0$, $q_2 = 0$: transverse and longitudinal EO flows coincide with each other. The upper solid curve corresponds to longitudinal and the upper dashed curve to transverse EO flow for a surface of a constant charge $q_1 = q_2 = q_0$. The lower solid curve corresponds to longitudinal and the upper dashed curve to transverse EO flow for a surface with $q_1 = -q_2 = -q_0$.

and the amplification of EO flow at the isotropic superhydrophobic surface (such as observed in a recent experiment[106]) might serve as very strong evidence in favor of a charge at the liquid–gas interface.

An interesting scenario is expected for oppositely charged solid and gas sectors. In this case Eq. 2.31 transforms to:

$$\mathbf{u} = -\frac{\mathbf{E}_0 q_0}{\eta\kappa} \cdot (\mathbf{I} - \mathbf{b}_{\mathrm{eff}}\kappa)\,. \tag{2.34}$$

The calculation results for this situation are also included in Fig. 2.19 and suggest a very rich fluid behavior. We see, in particular, that an inhomogeneous surface charge can induce EO flow along and opposite the field, depending on the fraction of the slipping area.

Already a very small fraction of the gas sectors would be enough to reverse the effective EO flow. Another striking result is that electroneutral surface ($\langle q \rangle = \phi_1 q_1 + \phi_2 q_2 = 0$) can generate an extremely large EO slip. With our numerical example this corresponds to $\phi_2 = 0.5$.

Of course, the area of research connected with interface transport phenomena is still at its infancy. Thus, the role of surface conductance just started to be probed.[107] Besides that, many assumptions exploited above should obviously be relaxed. For example, in the future, we suggest as a fruitful direction to consider electro-osmotic flow in a thin gap and to assume a partial slip[115] at the gas sectors. It will be very important to investigate transverse electrokinetic phenomena, which could be greatly amplified by using striped superhydrophobic surfaces. Note that if the charge is varied along the direction of the electric field, the fluid close to a superhydrophobic wall is pulled periodically in opposite directions. As a result, the recirculation rolls should develop on a scale proportional to the texture size, L. This should provide an additional opportunity for mixing, similar to that described in[108] but hopefully much faster.

2.10 Conclusion and Future Directions

With recent progress in micro- and nanofluidics, new interest has arisen in determining forms of hydrodynamic boundary conditions.[3,28,30] In particular, advances in lithography to pattern substrates have raised several questions in the modeling of liquid motions over these surfaces and led to the concept of the effective tensorial slip. These effective conditions capture complicated effects of surface anisotropy and can be used to quantify flow over complex textures without the tedium of enforcing real inhomogeneous boundary conditions.

This chapter has discussed the issue of boundary conditions at smooth hydrophobic and rough hydrophilic surfaces and has then given emphasis to effective boundary conditions for flow past hydrophobic solid surfaces with special textures that can exhibit greatly enhanced ("super") properties, compared to analogous flat or slightly disordered surfaces. Research on these superhydrophobic materials during past decades has mostly focused on their extreme nonwettability.[5,31] However, now the field has moved beyond wetting toward transport phenomena.[14,109] An effective slip becomes the main parameter to quantify the effective transport properties near superhydrophobic surfaces, which is in contrast to a traditional approach based on the use of an effective contact angle.

We have discussed formulas describing effective boundary conditions for pressure-driven flow past superhydrophobic textures of special interest (such as stripes, fractal patterns of nested circles, and chessboards). The predicted large effective slip of superhydrophobic surfaces compared to simple, smooth channels can greatly lower the viscous drag of thin microchannels.

The power of the tensor formalism and the concept of effective slippage have then been demonstrated by exact solutions for two other potential applications: optimization of transverse flow and analytical results for hydrodynamic resistance to the approach of two surfaces. These examples demonstrate that properly designed superhydrophobic surfaces could generate a very strong transverse flow and significantly reduce the so-called "viscous adhesion." Finally, we have discussed how superhydrophobic surfaces could amplify electrokinetic pumping in microfluidic devices.

As we have shown, a combination of wetting and roughness provides many new and very special hydrodynamic properties of surfaces, which could be explored with the formalism discussed here. This should allow the local slip tensors to be determined by global measurements, such as the permeability of a textured channel as a function of the surface orientations or the hydrodynamic force exerted on the body approaching superhydrophobic plates. They may also guide the design of superhydrophobic surfaces for microfluidic lab-on-a-chip and other applications.

Despite the impressive advances in hydrodynamic slippage phenomena, many challenges remain, both theoretical and experimental, fundamental and practical. In particular, there is quantitative discrepancy between theory and experiment (measured effective slips are persistently lower than theoretical predictions). This discrepancy reflects gaps in our fundamental understanding of flow past rough and hydrophobic surfaces. Next we briefly discuss promising future directions for research, both experimental and theoretical.

Experiments in thick channels[110,111] have established that hydrodynamic flows are generally slower than one would expect from theory.[14] Current analytical models of the superhydrophobic effective slip are based on the idealized model of a heterogeneous surface with patches of boundary conditions and mostly neglect a number of dissipation mechanisms in the gas phase and at the interface. The effects associated with different aspects of the gas flow and meniscus curvature must be included in the models. Regardless of recent semianalytical and numerical analyses,[78,81] the goal should remain to find simple analytical formulas, with as few adjustable parameters as possible, to fit experimental data.

Drainage experiments conducted using the AFM and SFA, which have the ability to probe fluid films of nano- and molecular thicknesses, may be able to yield detailed information on the superhydrophobic slip. A lot of data is already available.[112–114] The recent theory of hydrodynamic interaction between disks[85] shed some light on what could happen qualitatively, but cannot be used for a quantitative analysis of the hydrodynamic data obtained with the AFM and SFA. The same remark concerns the use of a theory of a film drainage between smooth hydrophobic surfaces,[12] which is not fully applicable to quantify a superhydrophobic slip. We believe that a challenge for a theory would be to extend a theoretical modeling of experimentally relevant sphere versus plane geometry.[116] This will open many possibilities for new experiments and could revolutionize the field.

There are many opportunities to design new experiments and to develop improved theoretical models for electro-osmotic flow past rough, hydrophobic, and superhydrophobic surfaces.[103,104] For example, a systematic study of the effect of surface texture, amplitudes of local slip lengths, and a role of channel thickness would be interesting. It would likewise be interesting to work with conducting but hydrophobic or superhydrophobic surfaces. The study of superhydrophobic Cassie surfaces will naturally pose fundamental questions about a mechanism of EO at the charged gas interface, where both adsorbed ions and their screening clouds are mobile.[104]

Very promising directions are certainly diffusio-osmosis and thermo-osmosis, where solvent flows are induced by gradients of solute concentration or temperature. Being combined with hydrophobic[86] or superhydrophobic slippage, these effects could lead to a giant amplification of flows in microchannels, even if the liquid–gas interface is uncharged, as shown in a recent simulation study.[105] The combination of these two strategies, that is, diffusio-/thermo-osmosis and superhydrophobicity, has to be studied theoretically and experimentally, and we expect a significant expansion into this very interesting area of research.

References

1. G. K. Batchelor, *An Introduction to Fluid Dinamics*. Cambridge University Press (2000).
2. T. M. Squires and S. R. Quake, Microfluidics: fluid physics at the nanoliter scale, *Rev. Mod. Phys.*, **77**, 977–1026 (2005).
3. H. A. Stone, A. D. Stroock, and A. Ajdari, Engineering flows in small devices, *Annu. Rev. Fluid Mech.*, **36**, 381–411 (2004).
4. O. I. Vinogradova, Slippage of water over hydrophobic surfaces, *Int. J. Miner. Proc.*, **56**, 31–60 (1999).
5. D. Quere, Non-sticking drops, *Rep. Prog. Phys.*, **68**, 2495–2532 (2005).
6. E. Lauga, M. P. Brenner, and H. A. Stone, *Handbook of Experimental Fluid Dynamics*. Ch. 19. New York: Springer. 1219–1240 (2007).
7. C. L. M. H. Navier, Mémoire sur les lois du mouvement des fluides, *Mémoire de l'Académie Royale des Sciences de lÍnstitut de France*, **6**, 389–440 (1823).
8. H. Helmholtz and G. Piotrowski, Ueber reibung tropfbarer fluessigkeiten, *Sitzungsberichte der Kaiserlich Akademie der Wissenschaften*, **40**, 607–658 (1860).
9. E. C. Binhgam, *Fluidity and Plasticity*. New York: McGraw-Hill (1922).
10. E. Schnell, Slippage of water over nonwettable surfaces, *J. Appl. Phys.*, **27**, 1149–1152 (1956).
11. N. V. Churaev, V. D. Sobolev, and A. N. Somov, Slippage of liquids over lyophobic solid surfaces, *J. Colloid Interface Sci.*, **97**, 574–581 (1984).
12. O. I. Vinogradova, Drainage of a thin liquid film confined between hydrophobic surfaces, *Langmuir*, **11**, 2213–2220 (1995).

13. J. L. Barrat and L. Bocquet, Large slip effect at a nonwetting fluid-solid interface, *Phys. Rev. Lett.*, **82** (23), 4671–4674 (1999).
14. O. I. Vinogradova and A. V. Belyaev, Wetting, roughness and flow boundary conditions, *J. Phys.: Cond. Matter*, **23**, 184104 (2011).
15. O. I. Vinogradova, Coagulation of hydrophobic and hydrophilic solids under dynamic conditions, *J. Colloid Interface Sci.*, **169**, 306–319 (1995).
16. L. Bocquet and J. L. Barrat, Flow boundary conditions from nano- to micro- scales, *Soft Matter*, **3**, 685–693 (2007).
17. C. Sendner, D. Horinek, L. Bocquet, and R. Netz, Interfacial water at hydrophobic and hydrophilic surfaces: Slip, viscosity, and diffusion, *Langmuir*, **25**, 10768–10781 (2009).
18. L. Bocquet and E. Charlaix, Nanofluidics, from bulk to interfaces, *Chem. Soc. Rev.*, **39**, 1073–1095 (2010).
19. D. Huang, C. Sendner, D. Horinek, R. Netz, and L. Bocquet, Water slippage versus contact angle: a quasiuniversal relationship, *Phys. Rev. Lett.*, **101**, 226101 (2008).
20. D. Andrienko, B. Dünweg, and O. I. Vinogradova, Boundary slip as a result of a prewetting transition, *J. Chem. Phys.*, **119**, 13106 (2003).
21. S. M. Dammler and D. Lohse, Gas enrichment at liquid-wall interfaces, *Phys. Rev. Lett.*, **96**, 206101 (2006).
22. O. I. Vinogradova, N. F. Bunkin, N. V. Churaev, O. A. Kiseleva, A. V. Lobeyev, and B. W. Ninham, Submicrocavity structure of water between hydrophobic and hydrophilic walls as revealed by optical cavitation, *J. Colloid Interface Sci.*, **173**, 443–447 (1995).
23. G. E. Yakubov, H. J. Butt, and O. I. Vinogradova, Interaction forces between hydrophobic surfaces. Attractive jump as an indication of formation of "stable" submicrocavities, *J. Phys. Chem. B*, **104** (15), 3407–3410 (2000).
24. B. Borkent, S. Dammler, H. Schonherr, G. Vansco, and D. Lohse, Superstability of surface nanobubbles, *Phys. Rev. Lett.*, **98**, 204502 (2007).
25. N. Ishida, T. Inoue, M. Miyahara, and K. Higashitani, Nano bubbles on a hydrophobic surface in water observed by tapping-mode atomic force microscopy, *Langmuir*, **16**, 6377–6380 (2000).
26. X. H. Zhang, N. Maeda, and V. S. J. Craig, Physical properties of nanobubbles on hydrophobic surfaces in water and aqueous solutions, *Langmuir*, **22**, 5025–5035 (2006).

27. S. Das, J. H. Snoeijer, and D. Lohse, Effect of impurities in description of surface nanobubbles, *Phys. Rev. E*, **82**, 056310 (2010).
28. M. Z. Bazant and O. I. Vinogradova, Tensorial hydrodynamic slip, *J. Fluid Mech.*, **613**, 125–134 (2008).
29. F. Feuillebois, M. Z. Bazant, and O. I. Vinogradova, Effective slip over superhydrophobic surfaces in thin channels, *Phys. Rev. Lett.*, **102**, 026001 (2009).
30. K. Kamrin, M. Bazant, and H. A. Stone, Effective slip boundary conditions for arbitrary periodic surfaces: the surface mobility tensor, *J. Fluid Mech.*, **658**, 409–437 (2010).
31. D. Quere, Wetting and roughness, *Annu. Rev. Mater. Res.*, **38**, 71–99 (2008).
32. D. Chan and R. Horn, The drainage of thin liquid films between solid surfaces, *J. Chem. Phys.*, **83**, 5311–5324 (1985).
33. C. Cottin-Bizonne, B. Cross, A. Steinberger, and E. Charlaix, Boundary slip on smooth hydrophobic surfaces: intrinsic effects and possible artifacts, *Phys. Rev. Lett.*, **94**, 056102 (2005).
34. R. G. Horn, O. I. Vinogradova, M. E. Mackay, and N. Phan-Thien, Hydrodynamic slippage inferred from thin film drainage measurements in a solution of nonadsorbing polymer, *J. Chem. Phys.*, **112** (14), 6424–6433 (2000).
35. O. I. Vinogradova and G. E. Yakubov, Dynamic effects on force measurements. 2. Lubrication and the atomic force microscope, *Langmuir*, **19**, 1227–1234 (2003).
36. O. I. Vinogradova, Hydrodynamic interaction of curved bodies allowing slip on their surfaces, *Langmuir*, **12** (24), 5963–5968 (1996).
37. D. Andrienko, P. Patricio, and O. I. Vinogradova, Capillary bridging and long-range attractive forces in a mean-field approach, *J. Chem Phys.*, **121**, 4414–4423 (2004).
38. O. I. Vinogradova and G. E. Yakubov, Surface roughness and hydrodynamic boundary conditions, *Phys. Rev. E*, **73**, 045302(R) (2006).
39. O. I. Vinogradova, H. J. Butt, G. E. Yakubov, and F. Feuillebois, Dynamic effects on force measurements. 1. Viscous drag on the atomic force microscope cantilever, *Rev. Sci. Instrum.*, **5**, 2330–2339 (2001).
40. R. Pit, H. Hervet, and L. Leger, Direct experimental evidence of slip in hexadecane: solid interfaces, *Phys. Rev. Lett.*, **85**, 980–983 (2000).
41. D. C. Tretheway and C. D. Meinhart, Apparent fluid slip at hydrophobic microchannel walls, *Phys. Fluids*, **14** (3), L9–L12 (2002).

42. P. Joseph and P. Tabeling, Direct measurement of the apparent slip length, *Phys. Rev. E*, **71**, 035303(R) (2005).

43. P. Huang, J. Guasto, and K. Breuer, Direct measurement of slip velocities using three-dimensional total internal reflection velocimetry, *J. Fluid Mech.*, **566**, 447–464 (2006).

44. C. Zettner and M. Yoda, Particle velocity field measurements in a near-wall flow using evanescent wave illumination, *Exp. Fluids*, **34**, 115–121 (2003).

45. H. Li, R. Sadr, and M. Yoda, Multilayer nano-particle image velocimetry, *Exp. Fluids*, **41**, 185–194 (2006).

46. B. Woolford, J. Prince, D. Maynes, and B. W. Webb, Particle image velocimetry characterization of turbulent channel flow with rib patterned superhydrophobic walls, *Phys. Fluids*, **21**, 085106 (2009).

47. O. I. Vinogradova, K. Koynov, A. Best, and F. Feuillebois, Direct measurements of hydrophobic slipage using double-focus fluorescence cross-correlation, *Phys. Rev. Lett.*, **102**, 118302 (2009).

48. D. Lumma, A. Best, A. Gansen, F. Feuillebois, J. O. Rädler, and O. I. Vinogradova, Flow profile near a wall measured by double-focus fluorescence cross-correlation, *Phys. Rev. E*, **67**, 056313 (2003).

49. L. Joly, C. Ybert, and L. Bocquet, Probing the nanohydrodynamics at liquid-solid interfaces using thermal motion, *Phys. Rev. Lett.*, **96**, 046101 (2006).

50. S. Yordanov, A. Best, H. J. Butt, and K. Koynov, Direct studies of liquid flows near solid surfaces by total internal reflection fluorescence crosscorrelation spectroscopy, *Opt. Express*, **17**, 21150–21158 (2009).

51. J. P. Rothstein, Slip on superhydrophobic surfaces, *Annu. Rev. Fluid Mech.*, **42**, 89–109 (2010).

52. C. Neto, D. Evans, E. Bonaccurso, H. J. Butt, and V. J. Craig, Boundary slip in newtonian liquids: a review of experimental studies, *Rep. Prog. Phys.*, **68**, 2859–2897 (2005).

53. C. Kunert, J. Harting, and O. I. Vinogradova, Random roughness hydrodynamic boundary conditions, *Phys. Rev. Lett*, **105**, 016001 (2010).

54. P. A. Thompson and S. M. Troian, A general boundary condition for liquid flow at solid surfaces, *Nature*, **389**, 360–362 (1997).

55. J. Harting, C. Kunert, and H. Herrmann, Lattice Boltzmann simulations of apparent slip in hydrophobic microchannels, *Europhys. Lett.*, **75**, 328–334 (2006).

56. H. Li and M. Yoda, An experimental study of slip considering the effects of non-uniform colloidal tracer distributions, *J. Fluid Mech.*, **662**, 269–287 (2010).
57. C. D. F. Honig and W. A. Ducker, No-slip hydrodynamic boundary condition for hydrophilic particles, *Phys. Rev. Lett.*, **98**, 028305 (2007).
58. C. I. Bouzigues, P. Tabeling, and L. Bocquet, Nanofluidics in the debye layer at hydrophilic and hydrophobic surfaces, *Phys. Rev. Lett.*, **101**, 114503 (2008).
59. A. Maali, C. Hurth, T. Cohen-Bouhacina, G. Gouturier, and J. P. Aime, Improved acoustic excitation of atomic force microscope cantilevers in liquids, *Appl. Phys. Lett.*, **88**, 163504 (2006).
60. E. Bonaccurso, H.-J. Butt, and V. S. J. Craig, Surface roughness and hydrodynamic boundary slip of a newtonian fluid in a completely wetted system, *Phys. Rev. Lett.*, **90**, 144501 (2003).
61. S. Granick, Y. Zhu, and H. Lee, Slippery question about complex fluids flowing past solids, *Nat. Mater.*, **2**, 221–227 (2003).
62. Y. X. Zhu and S. Granick, Limits of the hydrodynamic no-slip boundary condition, *Phys. Rev. Lett.*, **88**, 106102 (2002).
63. B. Woolford, D. Maynes, and B. W. Webb, Liquid flow through microchannels with grooved walls under wetting and superhydrophobic conditions, *Microfluid. Nanofluid.*, **7**, 121–135 (2009).
64. N. Lecoq, R. Anthore, B. Cichocki, P. Szymczak, and F. Feuillebois, Drag force on a sphere moving towards a corrugated wall, *J. Fluid Mech.*, **513**, 247–264 (2004).
65. W. Jager and A. Mikelic, Couette flows over a rough boundary and drag reduction, *Commun. Math. Phys.*, **232**, 429–455 (2003).
66. A. Basson and D. Gerard-Varet, Wall laws for fluid flows at a boundary with random roughness, *Comm. Pure Appl. Math.*, **61**, 941–987 (2008).
67. C. Kunert and J. Harting, Roughness induced boundary slip in microchannel flows, *Phys. Rev. Lett*, **99**, 176001 (2007).
68. D. W. Bechert and M. Bartenwerfer, The viscous flow on surfaces with longitudinal ribs, *J. Fluid Mech.*, **206**, 105–129 (1989).
69. C. Ybert, C. Barentin, C. Cottin-Bizonne, P. Joseph, and L. Bocquet, Achieving large slip with superhydrophobic surfaces: scaling laws for generic geometries, *Phys. Fluids*, **19**, 123601 (2007).
70. A. V. Belyaev and O. I. Vinogradova, Effective slip in pressure-driven flow past super-hydrophobic stripes, *J. Fluid Mech.*, **652**, 489–499 (2010).

71. C. Cottin-Bizonne, C. Barentin, E. Charlaix, L. Bocquet, and J. L. Barrat, Dynamics of simple liquids at heterogeneous surfaces: molecular-dynamic simulations and hydrodynamic description, *Eur. Phys. J. E*, **15**, 427–438 (2004).

72. C. Cottin-Bizonne, J. L. Barrat, L. Bocquet, and E. Charlaix, Low-friction flows of liquid at nanopatterned interfaces, *Nat. Mater.*, **2**, 237–240 (2003).

73. N. V. Priezjev, A. A. Darhuber, and S. M. Troian, Slip behavior in liquid films on surfaces of patterned wettability, *Phys. Rev. E*, **71**, 041608 (2005).

74. D. Maynes, K. Jeffs, B. Woolford, and B. W. Webb, Laminar flow in a microchannel with hydrophobic surface patterned microribs oriented parallel to the flow direction, *Phys. Fluids*, **19**, 093603 (2007).

75. J. Davies, D. Maynes, B. W. Webb, and B. Woolford, Laminar flow in a microchannel with superhydrophobic walls exhibiting transverse ribs, *Phys. Fluids*, **18**, 087110 (2006).

76. J. Hyväluoma and J. Harting, Slip flow over structured surfaces with entrapped microbubbles, *Phys. Rev. Lett.*, **100**, 246001 (2008).

77. A. M. J. Davis and E. Lauga, Geometric transition in friction for flow over a bubble mattress, *Phys. Fluids*, **21**, 011701 (2009).

78. M. Sbragaglia and A. Prosperetti, A note on the effective slip properties for microchannel flows with ultrahydrophobic surfaces, *Phys. Fluids*, **19**, 043603 (2007).

79. C. Pirat, M. Sbragaglia, A. M. Peters, B. M. Borkent, R. G. H. Lammertink, M. Wessling, and D. Lohse, Multiple time scale dynamics in the breakdown of superhydrophobicity, *Europhys. Lett.*, **81**, 66002 (2008).

80. M. Reyssat, J. M. Yeomans, and D. Quéré, Impalement of fakir drops, *Europhys. Lett.*, **81**, 26006 (2008).

81. C. O. Ng, H. C. W. Chu, and C. Y. Wang, On the effects of liquid-gas interfacial shear on slip flow through a parallel-plate channel with superhydrophobic grooved walls, *Phys. Fluids*, **22**, 102002 (2010).

82. P. Gao and J. J. Feng, Enhanced slip on a patterned substrate due to depinning of contact line, *Phys. Fluids*, **21**, 102102 (2009).

83. A. Ajdari, Transverse electrokinetic and microfluidic effects in micropatterned channels: lubrication analysis for slab geometries, *Phys. Rev. E*, **65** (1), 016301 (2002).

84. A. D. Stroock, S. K. Dertinger, G. M. Whitesides, and A. Ajdari, Patterning flows using grooved surfaces, *Anal. Chem.*, **74**, 5306–5312 (2002).

85. A. V. Belyaev and O. I. Vinogradova, Hydrodynamic interaction with super-hydrophobic surfaces, *Soft Matter*, **6**, 4563–4570 (2010).
86. A. Ajdari and L. Bocquet, Giant amplification of interfacially driven transport by hydrodynamic slip: diffusio-osmosis and beyond, *Phys. Rev. Lett.*, **96**, 186102 (2006).
87. E. Lauga and H. A. Stone, Effective slip in pressure-driven stokes flow, *J. Fluid Mech.*, **489**, 55–77 (2003).
88. G. K. Batchelor, Slender-body theory for particles of arbitrary cross-section in stokes flow, *J. Fluid Mech.*, **44**, 419–440 (1970).
89. S. Torquato, *Random Heterogeneous Materials: Microstructure and Macroscopic Properties*. Springer (2002).
90. F. Feuillebois, M. Z. Bazant, and O. I. Vinogradova, Erratum: effective slip over superhydrophobic surfaces in thin channels, *Phys. Rev. Lett.*, **104**, 159902 (2010).
91. A. M. J. Davis and E. Lauga, Hydrodynamic friction of fakir-like superhydrophobic surfaces, *J. Fluid Mech.*, **661**, 402–411 (2010).
92. C. O. Ng and C. Y. Wang, Apparent slip arising from stokes shear flow over a bidimensional patterned surface, *Microfluid. Nanofluid.*, **8**, 361–371 (2010).
93. A. D. Stroock, S. K. W. Dertinger, A. Ajdari, I. Mezić, H. A. Stone, and G. M. Whitesides, Chaotic mixer for microchannels, *Science*, **295**, 647–651 (2002).
94. C. Y. Wang, Flow over a surface with parallel grooves, *Phys. Fluids*, **15**, 1114–1121 (2003).
95. A. D. Stroock and G. J. McGraw, Investigation of the staggered herringbone mixer with a simple analytical model, *Philos. Trans. R. Soc. A*, **04TA1803**, 1–16 (2004).
96. J. Ou, J. M. Moss, and J. P. Rothstein, Enhanced mixing in laminar flows using ultrahydrophobic surfaces, *Phys. Rev. E*, **76**, 016304 (2007).
97. F. Feuillebois, M. Z. Bazant, and O. I. Vinogradova, Transverse flow in thin superhydrophobic channels, *Phys. Rev. E*, **82**, 055301(R) (2010).
98. O. Reynolds, On the theory of lubrication and its application to Mr. Beauchamp Tower's experiments, including an experimental determination of the viscosity of olive oil, *Philos. Trans. R. Soc. Lond.*, **177**, 157–234 (1886).
99. R. S. Voronov, D. V. Papavassiliou, and L. L. Lee, Review of fluid slip over superhydrophobic surfaces and its dependence on the contact angle, *Ind. Eng. Chem. Res.*, **47**, 2455–2477 (2008).

100. L. Joly, C. Ybert, E. Trizac, and L. Bocquet, Hydrodynamics within the electric double layer on slipping surfaces, *Phys. Rev. Lett.*, **93**, 257805 (2004).

101. V. M. Muller, I. P. Sergeeva, V. D. Sobolev, and N. V. Churaev, Boundary effects in the theory of electrokinatic phenomena, *Colloid J. USSR*, **48**, 606–614 (1986).

102. J. L. Anderson, Colloid transport by interfacial forces, *Annu. Rev. Fluid Mech.*, **21**, 61–99 (1989).

103. T. M. Squires, Electrokinetic flows over inhomogeneously slipping surfaces, *Phys. Fluids*, **20**, 092105 (2008).

104. S. S. Bahga, O. I. Vinogradova, and M. Z. Bazant, Anisotropic electro-osmotic flow over super-hydrophobic surfaces, *J. Fluid Mech.*, **644**, 245–255 (2010).

105. D. M. Huang, C. Cottin-Bizzone, C. Ybert, and L. Bocquet, Massive amplification of surface-induced transport at superhydrophobic surfaces, *Phys. Rev. Lett.*, **101**, 064503 (2008).

106. M. C. Audry, A. Piednoir, P. Joseph, and E. Charlaix, Amplification of electro-osmotic flows by wall slippage: direct measurements on ots-surfaces, *Faraday Discuss.*, **146**, 113–124 (2010).

107. H. Zhao, Electro-osmotic flow over a charged superhydrophobic surface, *Phys. Rev. E*, **81**, 066314 (2010).

108. A. Ajdari, Electroosmosis on inhomogeneously charged surfaces, *Phys. Rev. Lett.*, **75**, 755–759 (1995).

109. L. Bocquet and E. Lauga, A smooth future? *Nat. Mater.*, **10**, 334–337 (2011).

110. P. Tsai, A. M. Peters, C. Pirat, M. Wessling, R. G. H. Lammerting, and D. Lohse, Quantifying effective slip length over micropatterned hydrophobic surfaces, *Phys. Fluids*, **21**, 112002 (2009).

111. J. Ou and J. P. Rothstein, Direct velocity measurements of the flow past drag-reducing ultrahydrophobic surfaces, *Phys. Fluids*, **17**, 103606 (2005).

112. Y. Wang and B. Bhushan, Boundary slip and nanobubble study in micro/nanofluidics using atomic force microscopy, *Soft Matter*, **6**, 29–66 (2010).

113. Y. Wang, B. Bhushan, and A. Maali, Atomic force microscopy measurement of boundary slip on hydrophilic, hydrophobic, and superhydrophobic surfaces, *J. Vac. Sci. Technol. A*, **27**, 754–760 (2009).

114. A. Steinberger, C. Cottin-Bizonne, P. Kleimann, and E. Charlaix, High friction on a bubble mattress, *Nat. Mater.*, **6**, 665–668 (2007).

115. A. V. Belyaev and O. I. Vinogradova, Electro-osmosis on anisotropic super-hydrophobic surfaces, *Phys. Rev. Lett.*, **107**, 098301 (2011).

116. E. S. Asmolov, A. V. Belyaev, and O. I. Vinogradova, Drag force on a sphere moving towards an anisotropic super-hydrophobic plane, *Phys. Rev. E*, **84**, 026330 (2011).

Chapter 3

Wetting Phenomena on the Nanometer Scale

S. Dietrich,[a] M. Rauscher,[b] and M. Napiórkowski[c]

[a] *Max-Planck-Institut für Metallforschung, Heisenbergstr. 3, 70569 Stuttgart, Germany*
[b] *Institut für Theoretische und Angewandte Physik, Universität Stuttgart, Pfaffenwaldring 57, 70569 Stuttgart, Germany*
[c] *Instytut Fizyki Teoretycznej, Wydział Fizyki, Uniwersytet Warszawski, 00-681 Warszawa, Hoża 69, Poland*
dietrich@mf.mpg.de; rauscher@mf.mpg.de; Marek.Napiorkowski@fuw.edu.pl

Equilibrium wetting phenomena constitute prominent examples of surface phase transitions. They depend sensitively on microscopic details that influence macroscopic properties. The corresponding basic concepts of wetting for volatile or nonvolatile fluids are reviewed with a particular emphasis on the morphologies of wetting films forming on chemically and topographically nanostructured substrates and transitions between them. We also review the dynamics of thin films and droplets on the nanoscale, focusing on the relevance of the long range of intermolecular interactions, thermal fluctuations, and hydrodynamic slip at the liquid–substrate interface. We discuss the interplay of these aspects for the dewetting of films as well as for the peculiar dynamics of nanodroplets near chemical and topographical steps.

Nanoscale Liquid Interfaces: Wetting, Patterning, and Force Microscopy at the Molecular Scale
Edited by Thierry Ondarçuhu and Jean-Pierre Aimé

ISBN 978-981-4316-45-3 (Hardcover), 978-981-4364-48-5 (eBook)
www.panstanford.com

3.1 Introduction

In the most general sense the notion of wetting refers to interfacial phenomena occurring as a result of three phases of condensed matter coming into spatial contact with each other. On a homogeneous and flat solid surface acting as an inert spectator phase, these phenomena involve, for example, the formation of thin films or of droplets. The two phases in contact with the substrate can be solid, liquid, or gaseous. Here, however, we focus on fluid phases in contact with homogeneous or structured solid surfaces.

Macroscopically, the wetting properties of homogeneous surfaces are encoded in the equilibrium contact angle θ_{eq}: a small droplet[a] has the shape of a spherical cap, and the contact angle between the droplet surface and the substrate is independent of the droplet size (see Fig. 3.1). Young's law for this contact angle is given by the surface tensions $\sigma_{\alpha\gamma}$ between the fluid α and the substrate γ, $\sigma_{\alpha\beta}$ between the two fluids α and β, and $\sigma_{\beta\gamma}$ between the fluid β and the substrate:[2]

$$\cos\theta_{eq} = \frac{\sigma_{\alpha\gamma} - \sigma_{\beta\gamma}}{\sigma_{\alpha\beta}}. \tag{3.1}$$

Young's law is obtained by minimizing the total free energy of the system; however, more commonly it is interpreted as a statement about the mechanical equilibrium of forces of magnitudes $\sigma_{\alpha\gamma}$, $\sigma_{\beta\gamma}$, and $\sigma_{\alpha\beta}$ acting along the direction of the corresponding interfaces on the three-phase contact line. The phases α and β can be vapor and liquid in the case of a liquid droplet, two liquid phases, or a liquid and a gas in the case of a gas bubble. The surface tensions depend on the combination of materials, the temperature, and the pressure.

On the nanoscale one observes more subtle phenomena. A solid surface in contact with vapor breaks the translational invariance in the direction normal to the surface and can therefore lead to the formation of a wetting film. The thickness of such wetting

[a]This shape of the droplet requires it to be small enough such that surface tension dominates over gravity. This means that the Bond number $\mathrm{Bo} = |\varrho_\beta - \varrho_\alpha|\, g\, R^2/\sigma_{\alpha\beta}$ with the mass density difference $\varrho_\beta - \varrho_\alpha$, the gravitational acceleration g, and the droplet radius R, and the surface tension $\sigma_{\alpha\beta}$ has to be much smaller than one. This typically holds for millimeter-sized or smaller droplets. In any case, θ_{eq} is independent of g.[1]

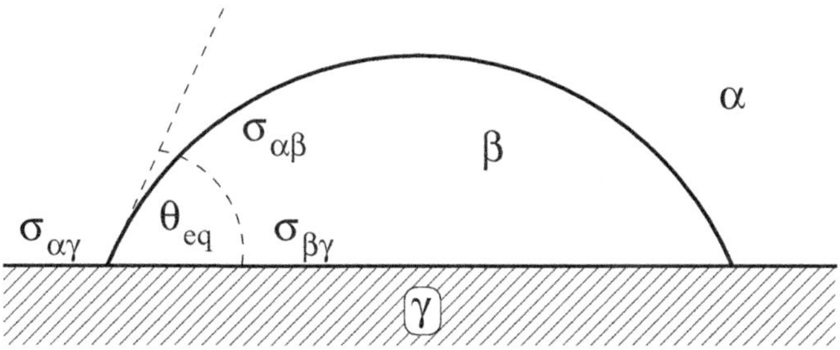

Figure 3.1 A droplet of fluid β (e.g., a liquid) with the shape of a spherical cap on a solid substrate γ surrounded by a fluid phase α (e.g., vapor). The droplet surface meets the substrate at an angle θ_{eq}, which is given by Young's law in Eq. 3.1.

films depends on thermodynamic variables such as temperature and pressure, and thus it is a thermodynamic quantity. One can observe discontinuous or continuous changes of the wetting film thickness upon varying the system parameters. These changes of the thickness reflect bona fide surface-phase transitions.[3]

Volatile fluids cannot form stable droplets: they either evaporate or grow until either the vapor reservoir is exhausted or their growth is limited by gravity or the finite size of the system. However, there are many fluids that exhibit a very low vapor pressure and a small evaporation rate such that on the time scale of a typical laboratory experiment, they behave as if they were nonvolatile. In this sense, such fluids—often polymeric liquids or liquid metals—form stable sessile droplets.

On the nanoscale the shape of such droplets can deviate significantly from the macroscopic shape of a spherical cap due to the finite range of the intermolecular interactions involved. As described later, one can actually relate the macroscopic equilibrium contact angle, the character of wetting transitions, and the shape of nanodroplets to the intermolecular interactions by using, for example, classical density functional theory (DFT). But the long range of intermolecular interactions also affects the motion of droplets. In particular it can lead to lateral interactions between droplets and structures, which are absent on the macroscopic scale.

But not only are the equilibrium shapes of and the forces on nanodroplets different from the ones for large droplets, the character of the dynamics also differs. On the macroscopic scale one typically assumes that fluids stick to solid surfaces, that is, the fluid velocity at the solid surface is equal to the velocity of the solid at this point. This leads to the so-called no-slip boundary condition.[4] However, the structure of fluids is modified in the vicinity of interfaces, which can lead to a significant slip. This is the case, in particular, for systems with very smooth surfaces and large equilibrium contact angles. This slip is characterized by a length that is on the order of up to a few nanometers for simple fluids, but it can reach microns for complex fluids such as polymer melts. Since this slip length is related to dissipation at the liquid–solid interface, it can be calculated via Green-Kubo relations.[5] The hypothesis that slip is a nanoscale phenomenon is supported by the observation that even a surface that is rough on intermediate scales but that exhibits a perfect slip (i.e., vanishing friction between the fluid and the substrate) on the microscopic scale gives rise to an effective no-slip boundary condition on large scales[6] (see also chapter 2).

Finally, thermal fluctuations become relevant for nanosized systems. The field of research that deals with fluid systems in which there is movement and control over liquids in or around objects with at least one dimension of the order of 10–50 nm or less,[a] is called nanofluidics.[7] This includes the study of liquid motion on the nanoscale in and around nanosized[b] objects.[8]

The notion of nanofluidics also refers to the nanoscale version of the technologically relevant lab-on-a-chip concept, which is sometimes referred to as microfluidics. This concept copies the extremely successful miniaturization concept of microelectronics by integrating complete miniature laboratories into small chemical devices.[9,10] In some cases, the techniques used to create nanofluidic devices have been developed originally for producing microelectronic devices. In turn, nanofluidics has some relevance to microelectronics. For example, the thickness of photoresist films used in lithographic processes can be in the submicron range.

[a] Note that a nanoliter droplet has a diameter of the order of 100 μm, and therefore it does not qualify as a nanofluidic system in this sense. A nanodroplet has a volume of the order of zepto liters (zL, i.e., 10^{-21} L).

[b] "Nanofluidics" is not to be confused with the so-called "nanofluids," that is, suspensions of nanosized particles. While understanding the dynamics of nanofluids involves nanofluidics, nanofluidics is a much broader field.

Building a microfluidic device is mainly an engineering problem. Besides the fact that gravity is negligible and flows on the micron scale are in general laminar (i.e., there is no turbulence that would promote mixing), liquids in microfluidic devices behave in the same way as liquids in macroscopic chemical apparatuses. The corresponding flow is described by the Navier-Stokes equations, and wetting properties of surfaces can be characterized completely by the equilibrium contact angle. But a further miniaturization necessarily leads to nanoscale devices, the design of which cannot be linearly extrapolated from the micron scale because the flow in them can no longer be described by macroscopic hydrodynamic equations.

Nanoscale wetting phenomena become particularly important for the miniaturization of open microfluidic systems in which liquid is not transported in closed channels but on lyophilic lanes on otherwise lyophobic surfaces. Flow on these so-called chemical channels is less prone to clogging, the samples can be accessed more easily, and cleaning the device is easy. However, evaporation has to be controlled.

The objective of this article is to review wetting phenomena on the nanoscale in the sense described before. In sections 3.2 and 3.3, we discuss equilibrium and nonequilibrium aspects of nanoscale wetting, respectively. We conclude with an outlook in section 3.4.

3.2 Equilibrium

3.2.1 *Wetting of Homogeneous Substrates*

3.2.1.1 Flat films

Wetting phenomena are paradigms of surface phase transitions.[3,11–15] They are sensitive to both bulk and surface characteristics of the system. The standard scenario of wetting entails a layer of a phase β forming at the interface between the coexisting phases γ and α. Suitable variations of the thermodynamic state of the system can bring about substantial growth of the width of such a wetting layer, which separates the initially adjacent phases γ and α. For example, in the case of a one-component liquid wetting an inert solid substrate, the role of the phase γ is taken by the inert substrate s, the phase α corresponds to vapor v, and the wetting phase β corresponds to the liquid phase l. In the case of a binary liquid mixture, the phase β can be the liquid phase rich in one of the two components of the mixture. It forms between the liquid phase rich in the other component (phase α) and the vapor (phase γ). In each of the above cases the appearance of the wetting phase β leads to the replacement of the single α–γ interface by the two interfaces α–β and β–γ. The mean distance between these interfaces, that is, the average width of the wetting layer, increases upon suitable changes of the thermodynamic state of the system. Theoretically, wetting transitions correspond to an unbounded increase of the thickness of the wetting layer. In experiments, this increase is limited by the actual experimental setup, and layer thicknesses of up to hundreds of nanometers are observed. The increase of the wetting layer is investigated by choosing a particular thermodynamic path in the bulk phase diagram. The wetting layer thickness is macroscopically large at state points located at the bulk α–β coexistence line at temperatures $T_W \leq T < T_C$, where T_W is the surface-specific wetting transition temperature and T_C is the bulk critical temperature (see the schematic μ–T phase diagram in Fig. 3.2, where μ is the chemical potential of a one-component liquid). For studying wetting transitions, usually either one raises the temperature along the bulk $\alpha-\beta$ coexistence line until one crosses the wetting transition temperature T_W at which the thickness of the wetting layer becomes infinite (path *a* in Fig. 3.2) or one isothermally approaches the bulk coexistence line $\mu = \mu_0(T)$ from the α side for $T_C > T > T_W$ (path *b* in Fig. 3.2).

In the first scenario one distinguishes two cases called first-order and critical wetting, respectively. At a first-order wetting transition the mean value $\bar{\ell}$ of the wetting layer jumps to infinity upon crossing T_W from below. At critical wetting the thickness $\bar{\ell}$ increases continuously upon approaching T_W.[3,16,17] This increase follows a power law $\bar{\ell} \sim (T_W - T)^{-\beta_s}$ with the characteristic surface critical exponent β_s. The value of the exponent β_s depends on the character of the interparticle interactions governing the system. To a certain extent, but less than in the case of bona fide critical phenomena, it displays universal properties. First-order wetting is accompanied by a prewetting transition.[3,18–21] This transition occurs

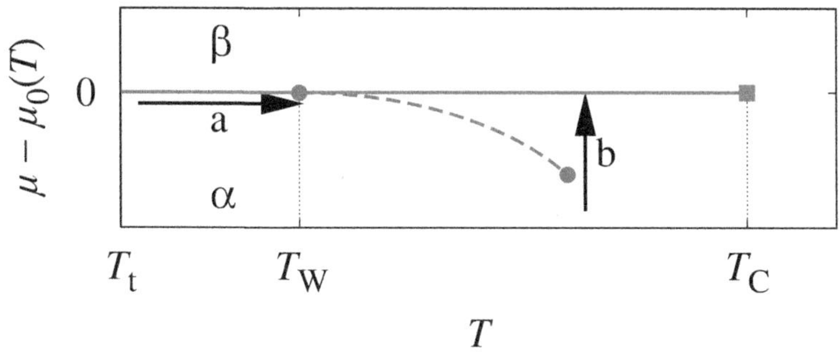

Figure 3.2 Schematic phase diagram of a simple fluid in terms of the chemical potential μ and temperature T. The gas (α) and the liquid (β) phases coexist at $\mu = \mu_0(T)$ up to the critical point T_C. T_W denotes the wetting transition temperature. Path *a* corresponds either to first-order or critical wetting, whereas path *b* corresponds to complete wetting. In the case of a first-order wetting transition, at T_W the so-called prewetting line $\mu = \mu_{pw}(T)$ (dashed line) branches out tangentially from the bulk coexistence curve and ends at the prewetting critical point (T_{pwc}, $\mu_{pwc} = \mu_{pw}(T_{pwc})$) denoted by the full dot. T_t is the triple-point temperature at which the liquid freezes.

upon crossing the so-called prewetting line $\mu = \mu_{pw}(T)$, which branches out tangentially from the bulk coexistence line at $T = T_W$ into the region of thermodynamic stability of phase α, that is, into the region $\mu < \mu_0(T)$ (see Fig. 3.2). It terminates at a prewetting critical point (T_{pwc}, $\mu_{pwc} = \mu_{pw}(T_{pwc})$).[22] The prewetting transition corresponds to a finite jump of the wetting layer thickness; it is often called a thin-thick transition.

In the second scenario, called complete wetting, the continuous increase of the wetting layer thickness occurs as a function of the thermodynamic distance from the bulk coexistence line expressed in terms of the undersaturation $\mu_0 - \mu$. In the case of complete wetting the increase of the wetting layer thickness also follows a power law with an exponent β_s^{co}: $\bar{\ell} \sim (\mu_0(T) - \mu)^{-\beta_s^{co}}$.

Wetting phenomena can be analyzed theoretically on a microscopic, mesoscopic, or macroscopic level, depending on how many details one is interested in. At the mesoscale one explicitly takes into account aspects of the systems that are latent on the macroscopic level. On the macroscopic level the information about interparticle interactions is encoded and subsumed in the coefficients $\sigma_{\alpha\beta}$ and $\eta_{\alpha\beta\gamma}$ of surface and line tensions, respectively. These coefficients can be calculated by microscopic theories as functions of the thermodynamic state of the system and of the parameters characterizing interparticle interactions $V(r)$. The thermodynamic state of the system is specified—in the case of a grand canonical ensemble—by the temperature T and the chemical potential μ. In the case of Lennard-Jones potentials $V_{\mathrm{L-J}}(r) = 4\epsilon[(r_0/r)^{12} - (r_0/r)^6]$, the potential parameters are the potential well depth ϵ and the particle radius r_0. A rough estimate of the size of the system under consideration below which line tension effects are nonnegligible is given by the ratio $\eta_{\alpha\beta\gamma}/\sigma_{\alpha\beta}$. For typical systems one has $\eta_{\alpha\beta\gamma} \sim \frac{\epsilon}{r_0} \sim 10^{-11}$ J/m and $\sigma_{\alpha\beta} \sim \frac{\epsilon}{r_0^2} \sim 10^{-2}$ J/m^2 (away from T_{W} and T_{C}, respectively), so that the aforementioned relevant system size is of the order of nanometers. Thus for nanofluidics line tension effects have to be taken into account.[23-30]

Within macroscopic approaches the surface and line tension coefficients are often considered to be independent parameters that can be varied freely in order to scan the possible states of systems under scrutiny. However, for a given system, that is, for fixed values of the interparticle potential parameters, these coefficients cannot be varied independently. They are certain functions of the thermodynamic state of the system. In the case of a one-component fluid they depend on the temperature only. Although microscopic analyses leading to these functions (in particular $\eta_{\alpha\beta\gamma}$) have been fully implemented only in rather few studies, one should be aware of the aforementioned constraints on the thermodynamic behavior of surface and line tension coefficients. Another important length scale is given by the capillary length $l_c = \left[\sigma_{\alpha\beta}/(\rho_\beta - \rho_\alpha)gm\right]^{1/2}$, where ρ_β and ρ_α denote the number densities of the phases β and α, respectively, g is the gravitational acceleration, and m is the mass of the fluid particles in the case of a one-component fluid. Typically the capillary length is much larger than the length scale below which line tension effects have to be taken into account.

On the mesoscopic level the theoretical description considers the capillary wavelike interface configurations as the relevant degrees of freedom and takes into account their effective interactions. The corresponding effective Hamiltonian determines the free-energy cost related to deviations of the γ–β and β–α interfaces from their planar configurations and takes into account the effective interaction between these interfaces. In the so-called Monge description their instantaneous shapes are given by $z = f_{\gamma\beta}(x, y)$ and $z = \ell + f_{\beta\alpha}(x, y)$, where ℓ corresponds to the distance between parallel reference planes taken as the lateral averages. The effective Hamiltonian $\mathcal{H}([f_{\gamma\beta}], [f_{\beta\alpha}], \ell)$ measures this cost in free energy relative to the reference state consisting of two planar and noninteracting interfaces, that is, for $\ell \to \infty$. It is a functional of the prescribed interface shapes $f_{\gamma\beta}(x, y)$ and $f_{\beta\alpha}(x, y)$ and a function of ℓ, and in addition it depends parametrically on the thermodynamic state of the system. In general it is a nonlocal functional of the interface configurations, and its structure depends, inter alia, on correlations in the system. Within a crude but nonetheless very useful approximation, one neglects these correlations and replaces the actual fluid density distribution by a function that is piecewise constant in each of the adjacent phases and varies discontinuously upon crossing any of these interfaces. (This is known as the so-called sharp-kink approximation.) Within this approach, for small spatial variations of the interface configurations, the effective Hamiltonian can be approximated by

$$\mathcal{H}([f_{\gamma\beta}], [f_{\beta\alpha}], \ell) = \int dx \int dy \Bigg[\frac{1}{2}\sigma_{\gamma\beta}(\nabla f_{\gamma\beta})^2 + \frac{1}{2}\sigma_{\beta\alpha}(\nabla f_{\beta\alpha})^2 + \omega_{\gamma\beta,\beta\alpha}(\ell + f_{\beta\alpha} - f_{\gamma\beta}) + \lambda_{\gamma\beta,\beta\alpha}(\ell + f_{\beta\alpha} - f_{\gamma\beta}) \times \nabla f_{\gamma\beta} \cdot \nabla f_{\beta\alpha} + \dots \Bigg]. \tag{3.2}$$

In the above equation the effective interface potential $\omega_{\gamma\beta,\beta\alpha}(\ell)$ represents the effective interaction between the two corresponding planar interfaces separated by a distance ℓ. The term $\lambda_{\gamma\beta,\beta\alpha}\nabla f_{\gamma\beta} \cdot \nabla f_{\beta\alpha}$ describes that contribution to the effective interaction between the two interfaces that stems from their corrugation. The ℓ dependence of the functions $\omega_{\gamma\beta,\beta\alpha}(\ell)$ and $\lambda_{\gamma\beta,\beta\alpha}(\ell)$ reflects the decay of the microscopic pair potentials governing the system. For long-ranged dispersion forces, which for large distances r decay $\sim r^{-6}$, both these functions decay $\sim \ell^{-2}$ for large ℓ. The derivation of the effective Hamiltonian from a more basic theory, for example, from Landau-Ginzburg-Wilson theory, is still a matter of intense discussion in the literature.[31–35] The key developments in this discussion were sparked by Monte Carlo simulations of Ising systems[36–39] aimed at a microscopic verification of the renormalization group predictions[40–44] based on the effective Hamiltonian for the nonuniversal wetting behavior of three-dimensional systems with short-ranged forces.

Applying the machinery of statistical mechanics to the effective Hamiltonian in Eq. 3.2 yields the free energy of the system and the mean values of the relevant observables. In particular, the knowledge of the mean positions of the interfaces as functions of the thermodynamic state allows one to analyze the wetting transitions. This can be carried out either in the case of volatile liquids, for which the amount of the wetting phase β is not constrained, or in the case of nonvolatile liquids under the constraint of their fixed amount. If correlations in the system are negligible one can invoke mean field theory, according to which the mean values $< f_{\kappa\delta} >$ of the interface positions are taken to be the configurations $\bar{f}_{\kappa\delta}$ minimizing the effective Hamiltonian. The corresponding mean field free energy $\mathcal{F}$ is equal to the effective Hamiltonian evaluated for these minimizing configurations, that is, $\mathcal{F} = \mathcal{H}([\bar{f}_{\gamma,\beta}], [\bar{f}_{\beta,\alpha}], \bar{\ell})$.

In the case of a planar γ–β interface corresponding, for example, to a planar solid substrate (s) wetted by a liquid (l), one has $\nabla f_{\rm sl} = 0$. Accordingly the effective interaction between the substrate–liquid interface at $z = 0$ and the liquid–vapor interface at $z = f(x, y)$ is given by the effective interface potential $\omega(f) = \omega_{\rm sl,lv}(f)$. The corresponding effective Hamiltonian reduces to a rather simple functional form:

$$\mathcal{H}[f] = \int dx \int dy \left[\frac{1}{2}\sigma_{\rm lv}(\nabla f)^2 + \omega(f)\right]. \qquad (3.3)$$

The effective interface potential $\omega(\ell)$ depends on the thermodynamic state of the system, and its dependence on ℓ is determined by the underlying microscopic structure of the system. At liquid–vapor coexistence, for small ℓ it describes the entropic repulsion between the interfaces, while for large ℓ it vanishes, that is, $\omega(\ell \to \infty) = 0$.

The structure of the effective interface potential depends on the type of the intermolecular forces involved. Two generic types of the asymptotic behavior of $\omega(\ell \to \infty)$ have to be distinguished. For long-ranged intermolecular forces decaying algebraically, one has

$$\omega(\ell) = \delta\mu\,\ell + a\ell^{-p} + b\ell^{-q}, \tag{3.4}$$

where $0 < p < q$ depend on the specific type of the forces. For example, for $d = 3$ and (nonretarded) van der Waals forces, one has $p = 2$ and $q = 3$. For short-ranged intermolecular forces (i.e., exponential decay or strictly finite range), one has

$$\omega(\ell) = \delta\mu\,\ell + a\exp(-\ell/\xi) + b\exp(-2\ell/\xi), \tag{3.5}$$

where ξ denotes the bulk correlation length in the wetting phase β. In both cases $\delta\mu = (\rho_l - \rho_v)(\mu_0(T) - \mu) \geqslant 0$ measures the thermodynamic distance from the bulk coexistence line. The coefficients $a(T)$ and $b(T)$ are Hamaker constants,[31,32,45] which depend on temperature and also on the specific material properties of the system. In particular, for continuous wetting transitions $a(T) \sim T - T_W$ changes sign at T_W, while $b(T = T_W) > 0$.

Interface fluctuations are characterized by the parallel correlation length $\xi_\parallel$, which determines the spatial decay of the height–height correlation function of the interface for large lateral distances $|\mathbf{R} - \mathbf{R}'|$, $\mathbf{R} = (x, y)$:

$$< [\ell(\mathbf{R}) - < \ell(\mathbf{R}) >]\,[\ell(\mathbf{R}') - < \ell(\mathbf{R}') >] > \sim \exp(-|\mathbf{R} - \mathbf{R}'|/\xi_\parallel). \tag{3.6}$$

The parallel correlation length $\xi_\parallel$ depends on the thermodynamic state of the system and diverges upon approaching the critical wetting point at which the interface unbinds from the substrate. This singular behavior leads to a power law describing $\xi_\parallel$, and it is specific for the particular wetting scenario. For continuous wetting transitions at bulk liquid–vapor coexistence, one has $\xi_\parallel \sim (T_W - T)^{-\nu_\parallel}$ characterized by the critical exponent $\nu_\parallel$. In the case of complete wetting the corresponding power law $\xi_\parallel \sim (\delta\mu)^{-\nu_\parallel^{co}}$ is characterized by the critical exponent $\nu_\parallel^{co}$. The values of the exponents $\nu_\parallel$ and $\nu_\parallel^{co}$ depend on the type of decay of the intermolecular forces acting in the system.[3,46]

Close to wetting transitions the width

$$\xi_\perp = \left[< [\ell(\mathbf{R}) - < \ell(\mathbf{R}) >]^2 >\right]^{1/2} \tag{3.7}$$

of the interfacial region (often also denoted as the interfacial roughness) follows a power law $\xi_\perp \sim (T_W - T)^{-\nu_\perp}$ for critical wetting and $\xi_\perp \sim (\delta\mu)^{-\nu_\perp^{co}}$ for complete wetting. The roughness $\xi_\perp$ and the parallel correlation length $\xi_\parallel$ are related via the so-called wandering exponent $\zeta_\pi(d)$ so that asymptotically $\xi_\perp \sim \xi_\parallel^{\zeta_\pi(d)}$, which implies relations between the exponents $\nu_\perp$ and $\nu_\parallel$[47,48] as well as $\nu_\perp^{co}$ and $\nu_\parallel^{co}$. The lower index π indicates that the quantity in question refers to a planar and homogeneous substrate geometry. In $d = 3$ and in the generic presence of van der Waals forces (see Eq. 3.4 for $p = 2$ and $q = 3$), one has $\nu_\parallel = 5/2$, $\nu_\parallel^{co} = 2/3$, $\nu_\perp = 0$, and $\nu_\perp^{co} = 0$ (see later).

Within mean field theory the equilibrium interface configuration $\bar{f}(x, y)$ minimizes $\mathcal{H}[f]$, leading to

$$\sigma_{lv}\left[\frac{\partial^2 \bar{f}}{\partial x^2} + \frac{\partial^2 \bar{f}}{\partial y^2}\right] = \omega'(\bar{f}), \tag{3.8}$$

where $\omega'(\ell) = \frac{d\omega(\ell)}{d\ell}$. The derivative of the effective interface potential on the right-hand side (RHS) of Eq. 3.8 is directly related to the concept of the disjoining pressure $\Pi(\ell) = -\omega'(\ell)$ introduced by Derjaguin *et al.* in the thirties of the twentieth century[49-51] in order to obtain equations for equilibrium interface profiles on the basis of the Young-Laplace equation.[25]

The mean field predictions for the thickness of the wetting layer, which are valid for van der Waals forces in $d = 3$,[3] are encoded in the structure of the effective interface potential $\omega(\ell)$, which depends on the thermodynamic state of the system. One can distinguish the following three generic types of behavior of the effective interface potential corresponding to different types of wetting phenomena:

- first-order wetting (see Fig. 3.3). The effective interface potential $\omega(\ell)$ displays two minima, one at a finite thickness $\bar{\ell}(T)$ and one at infinity. They are separated by a local maximum and $\omega(\ell \to \infty) = 0^+$. For $T < T_W$ the minimum at $\bar{\ell}(T)$ is the global one, that is, $\omega(\bar{\ell}(T)) < \omega(\infty) = 0$. At $T = T_W$ the two minima are equally deep so that $\omega(\bar{\ell}(T_W)) = 0$, and for $T > T_W$ their roles are interchanged, that is, $\omega(\ell = \ell_{lo}(T)) > 0$, where $\ell_{lo}(T)$ is the position of the local minimum. For first-order wetting and within mean field theory, one can determine the ranges of metastability of each of the surface phases and

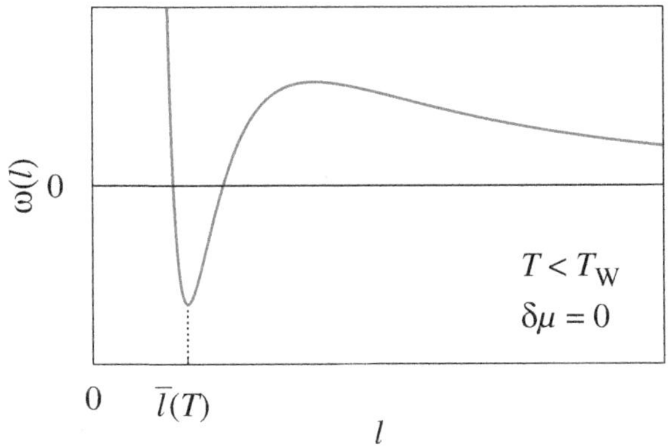

Figure 3.3 Effective interface potential $\omega(\ell)$ corresponding to first-order wetting at bulk coexistence. For $T < T_W$, $\ell = \bar{\ell}(T)$ is the global minimum. For $T > T_W$ the global minimum is at $\bar{\ell}(T) = \infty$. At $T = T_W$ the layer thickness $\bar{\ell}(T)$ varies discontinuously via a macroscopicly large jump.

the corresponding spinodals at which the maximum of the effective interface potential vanishes.

- critical wetting (see Fig. 3.4). For $T < T_W$ the effective interface potential $\omega(\ell)$ has its global minimum at $\bar{\ell}(T) < \infty$ so that $\omega(\bar{\ell}(T)) < 0$. For $T \to T_W$ one observes an unlimited increase of the layer thickness, that is, $\bar{\ell}(T \to T_W) \to \infty$ and correspondingly $\omega(\bar{\ell}(T) \to \infty) = 0^-$ vanishes continuously for $T \nearrow T_W$. As a function of temperature $\bar{\ell}(T)$ increases according to the power law $\bar{\ell}(T \to T_W) \sim (T_W - T)^{-\beta_s}$. For $T > T_W$ the global minimum is located at infinity. In the case of long-ranged forces (Eq. 3.4), one has $\beta_s = 1/(q - p)$, and for short-ranged forces (Eq. 3.5), $\beta_s = 0$, that is, $\bar{\ell}(T \to T_W) \sim -\ln((T_W - T)/T_W)$.
- complete wetting (see Fig. 3.5). In this case the bulk of the fluid is in the vapor phase and thus off the bulk liquid-vapor coexistence, that is, $\delta\mu > 0$, and the coexistence line is approached along an isotherm at $T > T_W$. Distinct from first-order and critical wetting, for large ℓ here the effective interface potential increases linearly, that is, $\omega(\ell \to \infty)) \sim (\mu_0(T) - \mu)\ell$. The position of the global minimum $\bar{\ell}(T, \mu)$ varies with $\delta\mu$ at fixed T ($T_W < T < T_C$) such that

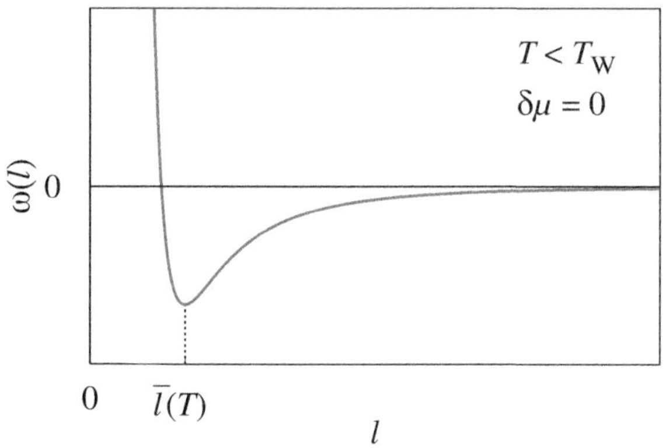

Figure 3.4 Effective interface potential $\omega(\ell)$ corresponding to critical wetting at bulk coexistence. For $T < T_W$ the equilibrium film thickness $\bar{\ell}(T)$ is finite. It diverges continuously for $T \to T_W$ so that $\bar{\ell}(T) = \infty$ is the equilibrium film thickness for $T \geq T_W$.

$\bar{\ell}(\delta\mu \to 0, T > T_W) \sim (\delta\mu)^{-\beta_s^{co}}$. For van der Waals forces $\omega(\ell \to \infty) \sim \ell^{-2}$ so that $\beta_s^{co} = 1/3$.

Within mean field theory the critical exponents $\nu_\parallel$ and $\nu_\parallel^{co}$ pertinent to the lateral correlation length can be inferred from the expression $\xi_\parallel^{-2} = \omega''(\bar{\ell})/(k_B T)$, where $\omega''(\ell)$ denotes the second derivative of $\omega(\ell)$. In the case of critical wetting one obtains $\nu_\parallel = 1$ for short-ranged forces and $\nu_\parallel = (q+2)/(2q-2p)$ for long-ranged forces. In the case of complete wetting $\nu_\parallel^{co} = 1/2$ for short-ranged forces and $\nu_\parallel^{co} = (p+2)/(2p+2))$ for long-ranged forces.

Below the so-called upper critical dimension $d \leqslant d_u$, the above mean field picture of wetting is modified if fluctuations are taken into account. For critical and complete wetting with short-ranged forces, one has $d_u = 3$, while for long-ranged forces $d_u = 3 - 4/(2+p)$ for complete wetting[52] and $d_u = 3 - 4/(q+2)$ for critical wetting.[53] The type of modification depends on the dimensionality of the system and the range of the forces. For long-ranged forces and $d < 3$ the interface fluctuations and the entropic effects resulting from collisions of the fluctuating interface with the substrate give rise to an additional contribution to the effective interaction between the interface and the substrate that decays $\sim \ell^{-\kappa}$, where $\kappa = 2(d-1)/(3-d)$ and thus can compete with the

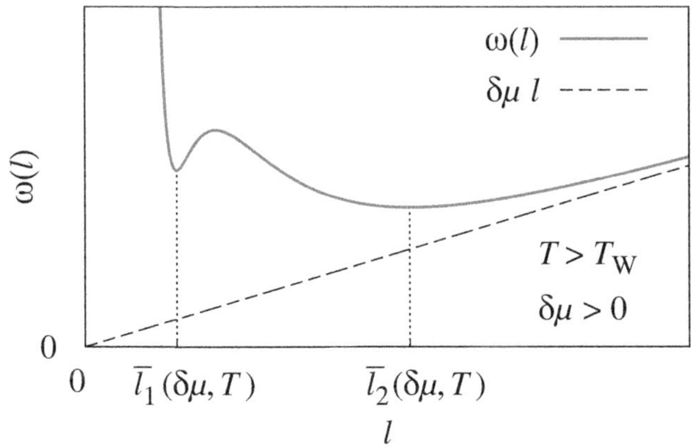

Figure 3.5 Effective interface potential $\omega(\ell)$ corresponding to complete wetting for a system exhibiting first-order wetting at two-phase coexistence $\delta\mu = 0$. For $T > T_W$ and sufficiently small $\delta\mu$, the global minimum is at $\ell = \bar{\ell}_2(\delta\mu, T)$. For $\delta\mu \to 0$ one has $\bar{\ell}_2(\delta\mu, T) \sim \delta\mu^{-\beta_s^{co}}$ with $\beta_s^{co} = 1/3$ for dispersion forces; correspondingly $\omega(\bar{\ell}_2(\delta\mu \to 0, T)) \to 0$. For $\mu < \mu_{pw}(T)$ and $T_W < T < T_{pwc}$ (see Fig. 3.2), the global minimum is located at $\bar{\ell}_1(\delta\mu, T)$. At the thin-thick transition between them one has $\omega(\bar{\ell}_1(\delta\mu_{pw}, T)) = \omega(\bar{\ell}_2(\delta\mu_{pw}, T))$, and the global minimum changes discontinuously between $\bar{\ell}_1(\delta\mu_{pw}, T)$ and $\bar{\ell}_2(\delta\mu_{pw}, T)$.

terms already present in the effective interface potential (Eq. 3.4). This competition leads—besides the above mean field regime valid for $q < \kappa$—to two additional fluctuation regimes, the weak-fluctuation regime for $p < \kappa < q$ and the strong-fluctuation regime for $\kappa < p$. The actual values of the critical exponents depend on the type of regime. In the weak-fluctuation regime they are nonuniversal in the sense that they depend, besides on κ, on the exponents p and q.[54,55] In the strong-fluctuation regime the fluctuations additionally shift the wetting transition temperature as compared with the mean field prediction (valid for this quantity also in the weak-fluctuation regime) and lead to universal critical exponents.

In the case of three-dimensional systems with short-ranged forces, the role of fluctuations has been investigated via renormalization group theory.[40–42,44] It predicts nonuniversal values of the critical exponents that depend on the dimensionless parameter $\mathcal{W} = k_B T_W/(4\pi\xi^2\sigma_{lv})$. However, these predictions were not confirmed by Monte Carlo simulations of the semi-infinite Ising model.[36–39,43] This discrepancy generated intensive discussions, which, inter alia, deal with the structure of the effective Hamiltonian[33,35,56–58] and point at the significance of its nonlocal character.

3.2.1.2 Droplets

The wetting transitions discussed in the previous section relate to volatile fluids in thermodynamic equilibrium with their vapor phase. On homogeneous surfaces droplets of volatile fluids are unstable. If the chemical potential of the vapor (i.e., the vapor pressure) is kept constant, the droplets either evaporate or grow until they reach the system size or gravity becomes relevant. But for nonvolatile fluids, that is, fluids with a negligible vapor pressures and evaporation rates, the volume of fluid in the film or droplet can be considered to be constant on the time scale of interest. As a consequence, nonvolatile fluids can explore a much richer configuration space for mechanical equilibrium; for example, they can form stable droplets on homogeneous substrates. However, to a large extent the underlying physical concepts are the same for both volatile and nonvolatile fluids.

The corresponding Euler-Lagrange equation, Eq. 3.8, has been used very successfully to quantitatively describe the nanoscale shapes of droplets[23] and rivulets.[59] Within this approach $\delta\mu$ in Eqs. 3.4 and 3.5 plays the role of a Lagrange multiplier, which allows one to fix the fluid volume.

Macroscopically, the wetting properties of homogeneous surfaces are encoded in the equilibrium contact angle θ_{eq}. It can be related to the effective interface potential by solving Eq. 3.8 corresponding to the effective interface Hamiltonian in Eq. 3.3 for $\delta\mu = 0$ with the following boundary conditions for a macroscopically large drop: $f(x) \to \bar{\ell}$ and $\partial f/\partial x \to 0$ for $x \to -\infty$ and $f(x) \to x \tan\theta_{\mathrm{eq}}$ for $x \to \infty$ (the latter condition is tied to $\delta\mu = 0$). As a result one obtains

$$\cos\theta_{\mathrm{eq}} = 1 + \frac{\omega(\bar{l})}{\sigma_{\mathrm{lv}}}, \tag{3.9}$$

which is consistent with Young's law (3.1) with $\sigma_{sv} - \sigma_{sl} = \sigma_{lv} + \omega(\bar{\ell})$ (see Refs. [3] and [11]).

By solving Eq. 3.8 one can obtain the shape of nanodroplets on homogeneous substrates, as shown in Fig. 3.6. Here, the effective interface potential in Eq. 3.4 with an additional term $c\,\ell^{-8}$ with $c > 0$ was used so that $0 < \bar{\ell} < \infty$. Apparently the shape deviates significantly from the macroscopic spherical cap shape, in particular near the three-phase contact line where the influence of the effective interface potential is strongest. In line with that, the contact angle of the nanodroplet deviates from the limiting macroscopic value θ_{eq}. But we note that there is no unique definition of the contact angle on the nanoscale. Two possible definitions are that either one takes the angle of a spherical cap fitted to the droplet apex with the substrate surface (see Fig. 3.6) or one takes the slope of the droplet surface at the point of its inflection. For the first definition the resulting contact angles depend on the precise definition of the position of the substrate surface, that is, whether, for example, one takes the

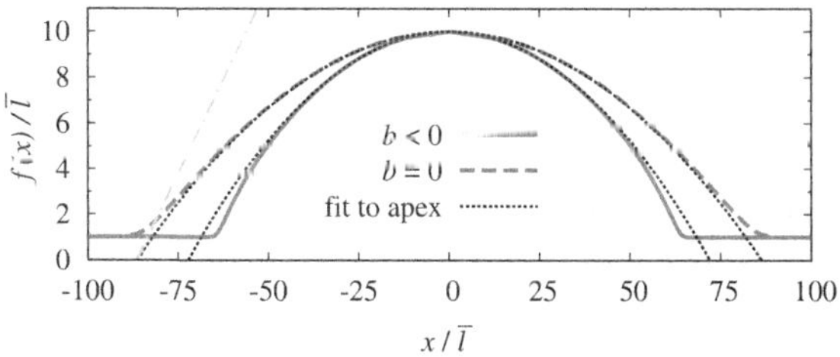

Figure 3.6 Profiles of two nanosized rivulet-like droplets on a solid substrate for an effective interface potential, as given in Eq. 3.4, with $p = 2$, $q = 3$, $a = \frac{23}{30}\,\sigma_{lv}\,\bar{\ell}^2$, and $b = -\sigma_{lv}\,\bar{\ell}^3$ (full red line) and for $a = -\frac{1}{15}\,\sigma_{lv}\,\bar{\ell}^2$ and $b = 0$ (dashed blue line), as obtained numerically from Eq. 3.8 for an apex height of $10\,\bar{\ell}$, corresponding to cross-sectional areas of $960\,\bar{\ell}^2$ and $1150\,\bar{\ell}^2$, respectively. Phenomenological repulsive contributions $\frac{11}{60}\,\sigma_{lv}\,\bar{\ell}^8/\ell^8$ and $\frac{1}{60}\,\sigma_{lv}\,\bar{\ell}^8/\ell^8$, respectively, have been added, which render a global minimum at $\ell = \bar{\ell}$ (with $\bar{\ell} < 1$ nm) of depth $-\sigma_{lv}/20$ in both cases. Also shown are parts of circles that are fitted to the apex of the droplets (dotted black lines). The contact angles of the circles with the substrate surface at $z = 0$ are 15.77° and 13.20°, while the slopes at the points of inflection lead to 18.23° and 9.92°, respectively. Thus both methods yield significantly different values, and they deviate from the macroscopic contact angle of 18.19° (indicated by the dash-dotted green line).

position of the center of mass of the last fluid atoms or of the center of mass of the first solid atoms or a position halfway in between. Thus in order to be well defined, experimental data for contact angles of small droplets must be accompanied by a clear statement about the procedures for its measurement. As can be seen in Fig. 3.6 the results of various procedures can deviate significantly. The shape of the droplets depends on details of the interface potential and not only on the values of $\bar{\ell}$ and $\omega(\bar{\ell})$, that is, the thickness of the wetting layer and the macroscopic equilibrium contact angle. As an example in Fig. 3.6 we show two droplets with $a, c > 0$ and $b < 0$ and with $a < 0$, $b = 0$, and $c > 0$, respectively. For $a > 0$ the droplet is narrower than the spherical cap fit, and for $a < 0$ it is wider.

The difference between the actual contact angle and the macroscopic contact angle is due to the line contributions to the free energy mentioned in section 3.2.1, in particular the line tension τ_{lvs}, which influence the contact angle. Phenomenologically, the force balance (Young-Laplace) equation (3.1) for a mesoscopic droplet with a circular base of radius R (i.e., the radius of the circular wetted area) can be written as

$$\cos(\theta_{\mathrm{eq}}) = \frac{\sigma_{\mathrm{sv}} - \sigma_{\mathrm{ls}}}{\sigma_{\mathrm{lv}}} - \frac{\bar{\tau}_{\mathrm{lvs}}}{\sigma_{\mathrm{lv}}\, R}. \tag{3.10}$$

A detailed analysis, however, shows that this equation suffers from an internal inconsistency, which is related to the freedom in positioning the interfaces. For example, a shift in the radius of the spherical liquid–vapor interface leads to changes in the contact angles, which are of the same order as those stemming from the presence of the line tension. This inconsistency is resolved by taking into account additional terms related to the Tolman length[60] and to the stiffness of the contact line.[29,61] Together with the line tension coefficient τ_{lvs}, they contribute to $\bar{\tau}_{\mathrm{lvs}}$ on the RHS of Eq. 3.10. Thus the correct form of Eq. 3.10 depends on subtleties such as the Tolman length and the definitions for the positions of the various dividing interfaces between the three phases. Thus, in order to be able to infer line tensions from experimental data via Eq. 3.10, one has to provide this additional information. The line tension coefficient τ_{lvs} (which corresponds to a straight contact line) can be calculated from the interface potential using the de Feijter and Vrij expression[62,63]

$$\tau_{\rm lvs} = \sqrt{2\,\sigma_{\rm lv}} \int\limits_{\bar{\ell}}^{\infty} \left[\sqrt{\omega(\ell) - \omega(\bar{\ell})} - \sqrt{-\omega(\bar{\ell})} \right] d\ell, \qquad (3.11)$$

which is based on the effective interface Hamiltonian in local approximation Eq. 3.3. For an expression taking nonlocal effects into account as well as additional contributions that stem from the asymptotes of the surface profile and are not captured by Eq. 3.3, see Refs. [64] and [65]. Thus the line tension corresponds to a macroscopically large drop, which implies $\delta\mu = 0$. This expression has been used to experimentally determine (neglecting, however, the aforementioned caveats) the line tension of a droplet by measuring the droplet shape with an atomic force microscope (AFM).[23,24,27] For a precise description of this issue please see chapter 6. The particular case of liquid crystal films is also discussed in chapter 5. For a quantitative assessment of $\tau_{\rm lvs}$ and of the interface shape near the three-phase contact line, see Ref. [65]. In contrast to interface tensions, the line tension coefficient can be negative. In the latter case one might be tempted to expect a buckling of the three-phase contact line.[66–68] But the corresponding upper threshold wavelength is comparable with the width of the three-phase contact region, and a careful analysis of the effective interface model in Eq. 3.3 shows that within a truly microscopic theory, this instability does not occur.[69]

However, the deviations of the actual contact angle of the rivulets shown in Fig. 3.6 from the expected equilibrium contact angle cannot be attributed to the line tension $\tau_{\rm lvs}$ as given by Eq. 3.11. Unlike droplets, rivulets have a straight three-phase contact line, and formally according to Eq. 3.10 the contact angle should be $\theta_{\rm eq}$. However, the two three-phase contact lines of a rivulet do interact if they are close to each other. This interaction gives rise to a line contribution to the free energy, too, and leads to a modified contact angle. This has been studied quantitatively in the case of volatile fluids in contact with a structured substrate, as will be discussed later.[70,71]

3.2.2 Structured Substrates

Structured substrates are ubiquituous in nature, and understanding their wetting properties is important from the point of view of numerous applications. It also poses interesting challenges for basic research. There are two generic types of substrate structures, chemical and topographical. In the first case one considers planar substrates that are chemically inhomogeneous. Typically, they consist of regions that differ in chemical composition and thus have different wetting properties. The simplest example is provided by an unbounded substrate consisting of two chemically different semi-infinite parts, each being planar and homogeneous. These two parts form a straight chemical step. Another simple example consists of a single stripe-like inhomogeneity embedded in an otherwise chemically homogeneous planar substrate. A more complicated structure consists of a system of parallel stripes. Both the width and the chemical composition can vary from stripe to stripe, the simplest case corresponding to a periodic array with all stripes having the same width and alternating chemical compositions. Other examples of chemical inhomogeneity are provided by a planar, checkered substrate formed by two sets of perpendicular stripes, by disc-like chemical inhomogeneities placed on the planar surface regularly or randomly,[72–74] or by a ring-like chemical inhomogeneity.[75]

The range of possible topographic substrate structures is very broad.[76,77] Spheres and cylinders are among the first nonplanar substrates that have been studied.[78–81] Although they do not admit macroscopically thick wetting layers due to an increasingly high free-energy cost of unbinding spherical or cylindrical liquid–vapor interfaces from the substrate, one can observe curvature-induced thin-thick transitions. Shapes and morphological phase transitions of droplets on cylindrical fibers offer an interesting interplay between curvature and wetting phenomena.[80,82,83] Another simple example of a nonplanar substrate with interesting wetting properties is provided by a wedge formed by two semi-infinite planar substrates meeting at a pivot with an opening angle 2φ. Substrates with parallel grooves have been analyzed both experimentally and theoretically. The groove cross section can be triangular (resembling a wedge of finite depth), can be rectangular, or can have other, more complicated shapes, such as microhoodoos, which render omniphobic surfaces.[84,85] Another form of topographically structured substrates is provided by a regular array of pillars, each of which has a specified height and cross section. In a still different setting the pillars are replaced by elastic "hairs," which may deform due to the presence of a liquid layer formed at such a substrate, leading to a coupling of elastic and capillary forces. Substrates can display both geometrical and topographical structures, for example, a chemically inhomogeneous wedge.[86] Nanoscale structures with various patterns can be prepared experimentally, and their wetting and adsorption properties have been investigated both experimentally and theoretically.[14,87–90] Already simple macroscopic arguments point at the significant role of periodic substrate shapes in determining morphologies of fluid interfaces and transitions between them.[91–95] These conclusions can be refined by resorting to descriptions based on effective interface Hamiltonians.

Wetting and adsorption on chemically structured substrates can be analyzed macroscopically in terms of Cassie's and Wenzel's rules.[96–101] They introduce position-dependent contact angles $\theta^{\mathrm{C}}(\mathbf{R})$ and $\theta^{\mathrm{W}}(\mathbf{R})$, respectively. In the case of a chemically inhomogeneous substrate consisting of two types of regions, denoted as (1) and (2), Cassie's rule has the form

$$\cos\theta^{\mathrm{C}}(\mathbf{R}) = f^{(1)}(\mathbf{R})\cos\theta^{(1)} + f^{(2)}(\mathbf{R})\cos\theta^{(2)}, \qquad (3.12)$$

where the angles $\theta^{(i)}$, $i = 1, 2$, are given by Young's law

$$\cos\theta^{(i)} = \left(\sigma_{\mathrm{iv}} - \sigma_{\mathrm{il}}\right)/\sigma_{\mathrm{lv}}, \qquad (3.13)$$

and $f^{(i)}(\mathbf{R})$ denotes the ratio of the area of type i and of the total area in the region around point $\mathbf{R}$ located near the three-phase contact line of the liquid–vapor interface and the substrate i forming the contact angle $\theta^{(i)}$; $f^{(1)}(\mathbf{R}) + f^{(2)}(\mathbf{R}) = 1$. A topographically structured substrate is described by Wenzel's rule

$$\cos\theta^{\mathrm{W}}(\mathbf{R}) = s(\mathbf{R})\cos\theta, \qquad (3.14)$$

where $s(\mathbf{R})$ is the ratio of the true area of the substrate and of its planar projection corresponding to the geometry of the lateral neighborhood of the contact line located at $\mathbf{R}$; the associated contact angle θ is given by Young's law (Eq. 3.13). The results obtained within this macroscopic approach provide valuable reference data for applying more detailed nanoscopic approaches. The Cassie and Wenzel approaches have been applied frequently to nonvolatile liquids of fixed volume deposited on structured substrates.[68,72–74,95,102] Here, we are mostly concerned with volatile fluids connected to a reservoir with their adsorbed volumes depending on the thermodynamic state.

Naturally, the analysis and understanding of wetting phenomena on structured substrates have evolved from simple to more complicated systems. This increasing level of complexity has two aspects. Together with more complicated substrate structures more complex fluids, for example, molecular liquids and polymers, have been considered. This approach departs in a systematic way from the paradigm of a simple fluid wetting a planar, chemically homogeneous substrate. From a theoretical point view the concept of the effective interface Hamiltonian has turned out to be useful for analyzing wetting of nanostructured substrates as it allows one to encompass the emerging interface morphologies in a more detailed way than within the macroscopic approach. However, there are cases of practical importance, for which its application is limited by substantial bendings of the interface imposed by the substrate structure. On the one hand, these nonplanar configurations contain essential informations on wetting and adsorption occurring in such systems. On the other hand, the relevant interface configurations can substantially differ from those corresponding to planar and chemically homogeneous substrates for which the effective Hamiltonians are constructed. This situation presents a challenge for creating new theoretical tools with broader ranges of applicability.

3.2.2.1 Chemically structured substrates

Planar but chemically inhomogeneous substrates are encountered in various practical application, in particular those related to open nanofluidics.[7,8,14,103] One tries to understand the wetting and adsorption properties of such systems in terms of the wetting properties of separate, chemically homogeneous regions. A simple chemical step[65,104–107] consists of two semi-infinite planes, each composed of a different material, say, (1) for $x < 0$ and (2) for

$x > 0$. Each part of the substrate can be characterized by its own wetting transition temperature $T_W^{(1)}$ and $T_W^{(2)}$, respectively. The morphological properties of adsorbed wetting films can be analyzed conveniently either by DFT[108–110] or by the effective Hamiltonian approach in its local, mean field form. In the latter case the effective interface potential $\omega(\ell(x), x)$ depends explicitly on x, and for the chemical step, it is modeled by the sum of the contributions from each half-plane:

$$\omega(\ell(x), x) = \Theta(-x)\,\omega^{(1)}(\ell(x)) + \Theta(x)\omega^{(2)}(\ell(x)). \qquad (3.15)$$

The effective interface potential $\omega^{(i)}(\ell(x))$ corresponds to a planar and homogeneous substrate of type i, and $\Theta(x)$ is the Heaviside function. Accordingly, the effective Hamaker constant $a^{(i)}(T)$ in Eqs. 3.4 and 3.5 depends on the corresponding wetting transition temperature $T_W^{(i)}$ and in the case of critical wetting $a^{(i)}(T) \sim T - T_W^{(i)}$. In a refined description $\omega(\ell, x)$ does not vary step-like but smoothly across $x = 0$, which reflects the smooth variation of the total substrate potential in spite of a step-like variation of its chemical composition.[65,104,105,111]

The chemical inhomogeneity runs along the line $x = 0$, $z = 0$, where the two half-planes meet. The thickness $\bar{\ell}(x)$ of the wetting film varies along the x-axis. It approaches its asymptotic values $\bar{\ell}_\pi^{(1)}(T, \delta\mu)$ and $\bar{\ell}_\pi^{(2)}(T, \delta\mu)$ for $x \to \mp\infty$, respectively. The rate of variation upon crossing the borderline $x = 0$ depends on the thermodynamic state of the system. It is particularly pronounced for thermodynamic states corresponding to wetting, either continuous or complete, of one of the substrates. On each side of the step one can identify a transition region of characteristic extent $\Delta^{(i)}$, $i = 1, 2$, the size of which is proportional to the corresponding lateral correlation length $\xi_\parallel^{(i)}$ of the height–height correlation function (see Eq. 3.6). Therefore the width $\Delta^{(i)}$ diverges according to a power law that depends on whether a complete or a critical wetting transition is approached. Close to these phase transitions the correlation functions of the interface height exhibit scaling properties with singular scaling functions such that at the wetting transitions the film profiles diverge algebraically as a function of x. For example, for temperatures $T_W^{(2)} < T < T_W^{(1)}$, $\delta\mu = 0$ and the effective interface potential parameters $p = 2$, $q = 3$ (see Eq. 3.4), one finds $\bar{\ell}(x \to \infty) \sim x^{1/2}$.[65,104] The variation of the wetting film thickness entails a line contribution $\Omega_l(T, \delta\mu)$ to the total free energy $\Omega(T, \delta\mu)$. For $\delta\mu = 0$ and $T \nearrow T_W^{(2)}$, that is, for a situation in which the right part of the substrate on its own becomes wet, while the left part remains nonwet, the variation of $\bar{\ell}(x)$ becomes more pronounced (see above). This is reflected in the cusp-like singularity of $\Omega_l(T, \delta\mu = 0)$ as a function of T at $T = T_W^{(2)}$, and its logarithmic divergence for $\delta\mu \to 0$ at $T_W^{(2)} < T < T_W^{(1)}$: $\Omega_l(T, \delta\mu) \sim \ln(\delta\mu)$.[104] One of the findings concerning the interface morphology across a chemical step is that its analysis based on local interface Hamiltonians provides results that are in both qualitative and satisfactory quantitative agreement with those based on a nonlocal interface Hamiltonian.[64,65]

Another simple example of a chemically inhomogeneous substrate is provided by a flat substrate with an embedded chemical heterogeneity forming a straight stripe of width L.[70,71,106] This system is translationally invariant in the direction along the stripe, and its structural properties vary in the x direction perpendicular to the stripe. In particular, the thickness of the wetting film $\bar{\ell}(x)$ varies along the x-axis with distinct saturation values across the stripe, composed of material 2, and outside it, that is, above the substrate composed of material 1. Within mean field theory the equilibrium interface configuration $\bar{\ell}(x)$ minimizes the grand canonical free-energy functional $\Omega[\ell(x)]$. The lateral inhomogeneity of the thickness of the liquid-like layer is tied to the line contribution $\Omega_l(T, \delta\mu; L)$ to the free energy of the system. There are two types of shapes of the liquid–vapor interface $\bar{\ell}(x)$. One configuration corresponds to the interface being bound to the stripe and the other to the interface being repelled from it. Depending on the thermodynamic state of the system and on the width L of the stripe, there is either a first-order or a continuous transition between these configurations.[106] The morphological states of the system form a three-dimensional phase diagram spanned by the variables $T, \delta\mu$, and $1/L$. Therein the surface of first-order morphological transitions terminates at a line of tricritical transitions such that beyond and for appropriately chosen thermodynamic states, the transition of the system from the bound to the repelled phase can proceed continuously. However, thermal fluctuations alter the properties of this effectively one-dimensional system. They smear out the sharp first-order morphological phase transitions and round off the continuous ones.

The parameters of the systems with embedded stripes can be chosen such that the corresponding fully homogeneous substrates of type i undergo critical wetting at temperatures $T_W^{(i)}$, with $T_W^{(2)} < T_W^{(1)}$. For such systems, at $\delta\mu = 0$ and in the limit of large L, one can extract that contribution $\delta\Omega_l(T, L)$ to the line free energy $\Omega_l(T, 0; L)$, which represents the effective interaction between the two straight interface inhomogeneities forming at the two borderlines between the stripe and the remaining parts of the substrate. The interaction $\delta\Omega_l(T, L)$ is attractive and vanishes for a large L, with the type of decay depending on the thermodynamic state of the system. For short-ranged forces and temperatures $T < T_W^{(2)}$, this interaction decays exponentially, that is, $\delta\Omega_l(T, L) \sim \exp(-L/\xi_\parallel^{(2)})$, where $\xi_\parallel^{(2)}$ is the parallel correlation length corresponding to the interface fluctuations above the unbounded substrate (2). In the case of long-ranged forces with $p = 2$, the corresponding decay is algebraic with $\delta\Omega_l(T, L) \sim 1/L^2$.[105] However, for $T > T_W^{(2)}$ this interaction changes its character from exponential to power law decay. In the temperature range $T_W^{(2)} < T < T_W^{(1)}$, that is, for temperatures at which the type 2 substrate forming the stripe is wetted, within mean field theory one finds a *power law* decay $\delta\Omega_l(T, L) \sim L^{-1}$ with a universal amplitude (i.e., independent from the specific surface properties).[70] This change of the decay from exponential to power law is also confirmed by two-dimensional transfer matrix calculations, which take into account the interface fluctuations.[71] These fluctuations change the exponent of the power law from the above mean field result $\Omega_l(T, L) \sim L^{-1}$ to $\Omega_l(T, L) \sim \ln L$.

The above change of the type of effective interaction between linear inhomogeneities upon crossing the critical wetting transition temperature $T_W^{(2)}$ can be considered as a facet of the critical Casimir effect.[112] Here the role of the critical phase is played by the interface fluctuating above that substrate, which is composed of material of type 2; for $T > T_W^{(2)}$ it is depinned from the substrate and the relevant parallel correlation length is infinite. Accordingly this critical state gives rise to an algebraically varying free-energy contribution of the system confined by the stripe width L. The above effect could be experimentally observed in a system in which the stripe consists of a compressible Langmuir-Blodgett film.[105] For such a system the change of the type of the effective interaction between linear interface inhomogeneities would result in a change of the law according to which their mutual equilibrium distance depends on temperature.

Besides various other types of chemically structured substrates, one can also consider special choices of the thermodynamic states of the system for which wetting is studied. In particular, in the case of critical adsorption[113] on chemically structured substrates, one obtains universal scaling functions, which describe the behavior of the modulated excess adsorption[114] near the bulk critical point T_c.

In contrast to the case of volatile fluids, for which the film thickness on straight chemical channels is invariant along the channel axes, one can observe morphological phase transitions for nonvolatile fluids residing on such structures. For macroscopic systems, the corresponding relevant control parameters are the equilibrium contact angles on the stripe as well as on the surrounding substrate (which should be larger than θ_{eq} on the stripe in order to be able to confine the liquids to the channel) and the reduced liquid volume V/L^3 where V is the droplet volume and L the stripe width. For small volumes and small but nonzero contact angles on the channel, the fluid spreads into a cigar-shaped rivulet along the channel. Upon increasing the fluid volume for equilibrium contact angles on the channel below $\theta_{eq}^{(c)} \approx 39.2°$, one observes a first-order morphological phase transition to a state in which the fluid contracts into a bulge-like droplet (see Fig. 3.7). The values $\theta_{eq} = \theta_{eq}^{(c)}$ and $V/L^3 \approx 2.85$ form the end point of a line of morphological phase transitions. On substrates with a finite contact angle this bulge spills onto the surrounding substrate.[102,115] Up to now this kind of phenomenon has not yet been investigated on nanoscale channels.

On a completely wetting chemical channel ($\theta_{eq} = 0$), one can observe axially homogeneous rivulets only if the contact angle θ of the rivulet (which is pinned at the channel edge by the chemical step if θ is smaller than the equilibrium contact angle on the surrounding substrate) is smaller than a certain critical angle. Macroscopically

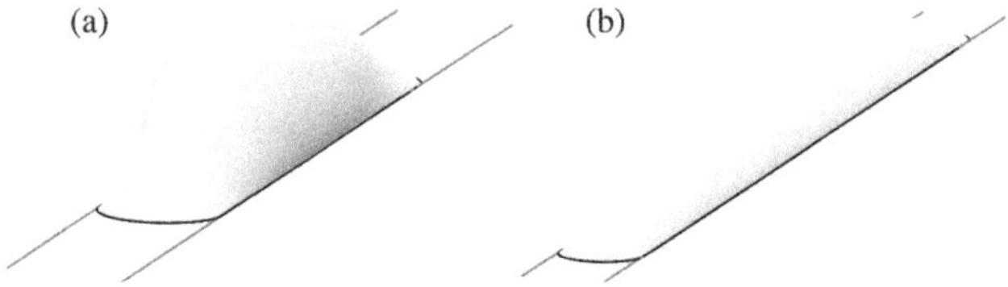

Figure 3.7 Stable bulge-like (a) and metastable ridge-like (b) drop morphology on a chemical channel with sharp chemical steps. The equilibrium contact angle on the channel and the reduced fluid volume are $\theta_{eq} = 38^\circ$ and $V/W^3 = 4.0$, respectively. The channel edges, at which the contact line is pinned, are indicated by red lines, and the free three-phase contact line on the channel is indicated by the black line. The figure is provided from Ref. [116] (Fig. 4.3) by courtesy of the author.

this angle equals 90°, see Ref. [117]. If θ is larger the rivulet is linearly unstable and it breaks up into a string of droplets. This is a Plateau-Rayleigh type, surface tension–driven instability. If $\theta < 90^\circ$ the pressure in the fluid increases for increasing filling level, but it decreases for larger θ. In this case a part of the rivulet that happens to be slightly thinner inflates the rest. Indeed one can show that homogeneously filled channels are stable with respect to pearling if the pressure increases with the filling level. This holds even if the effect of long-ranged intermolecular forces is taken into account in terms of the effective interface potential[118] and also if thermal fluctuations as well as the finite size of the molecules are considered in a molecular dynamics (MD) simulation.[119,120]

Although a homogeneously filled, straight chemical channel of macroscopic length can be linearly stable, it does not necessarily represent the state with the lowest free energy: collecting all the liquid into a single macroscopically large drop always reduces the surface area of the liquid–vapor interface and, if the equilibrium contact angle on the channel is nonzero, also reduces the free energy. In this case, the diameter of the drop is large as compared with the channel width.

Morphological shape transitions of droplets have been observed, both experimentally as well as in simulations, on straight channels[92,121] and on rings.[75,122–124] In the latter case, the ratio of the ring diameter and the stripe width provides an additional control parameter and one can observe two transitions. For small volumes the rivulet on the ring is cylindrically symmetric upon increasing the volume. This configuration undergoes a surface tension–driven, symmetry–breaking Plateau-Rayleigh type instability toward a single bulge. For very large volumes the fluid assumes again a symmetric configuration with a big drop spanning the whole ring, including the nonwetting disc in its center. Up to now these phenomena have not been studied on the nanoscale. In particular the influence of the long-ranged part of the dispersion forces and of thermal fluctuations might lead to qualitatively different results.

3.2.2.2 Topographically structured substrates

A wedge with opening angle 2φ, $0 \leqslant \varphi \leqslant \pi/2$, is a simple paradigmatic example of a substrate, the wetting properties of which are modified by its geometry[125–131] as compared with the planar case $\varphi = \pi/2$. In addition to the standard thermodynamic variables T and $\delta\mu$, the wedge opening angle 2φ plays the role of a geometric control parameter. Here the analog of the wetting transition is called the filling transition, and it corresponds to the liquid layer thickness $\bar{\ell}_{\rm w}(T, \delta\mu, \varphi)$ in the center of the wedge, becoming macroscopically large; in this limit, packing effects in the corner[132] become irrelevant. The filling transition occurs at a φ-dependent characteristic temperature $T_{\rm F}(\varphi) \leqslant T_{\rm W} = T_{\rm F}(\pi/2)$. The filling transition temperature $T_{\rm F}(\varphi)$ can be expressed implicitly in terms of the temperature-dependent macroscopic contact angle $\theta(T)$: $\theta(T_{\rm F}(\varphi)) = \pi/2 - \varphi$. Macroscopically this means that at the filling temperature the liquid–vapor interface is horizontal, that is, perpendicular to the wedge symmetry plane. If the fluid is at bulk liquid–vapor coexistence and $T_{\rm F}(\varphi)$ is approached from below, one encounters the filling transition which—similar to the wetting transition—can be either first order or critical. If one approaches the bulk gas–liquid coexistence curve from the vapor side along an isotherm with $T_{\rm F}(\varphi) < T < T_{\rm C}$, one observes complete filling for $\delta\mu \to 0$. Critical and complete filling are characterized by power laws so that $\bar{\ell}_{\rm w} \sim (T_{\rm F}(\varphi) - T)^{-\beta_{\rm F}}$ for critical filling and $\bar{\ell}_{\rm w} \sim \delta\mu^{-\beta_{\rm F}^{\rm co}}$

for complete filling. Within mean field theory one finds $\beta_F = 0$, that is, $\bar{\ell}_w \sim -\ln((T_F(\varphi) - T)/T_F(\varphi))$, and $\beta_F^{co} = 1$ for short-ranged forces and $\beta_F = 1/p$ and $\beta_F^{co} = 1$ for long-ranged forces (see Eq. 3.4). The effect of fluctuations on the critical exponents depends on the dimensionality of the system. For example, critical filling of a wedge in $d = 2$ is characterized by the universal exponent $\beta_F = 1$. This differs from corresponding mean field predictions obtained by minimizing an effective Hamiltonian that renders a nonuniversal value of this exponent. In the case of complete filling both mean field theory and theories including fluctuations predict the universal value $\beta_F^{co} = 1$, that is, the same for all types of forces and wedge dimensionalities.[129,133–135] It turns out that the midpoint interface height $\bar{\ell}_w(\theta, \varphi)$ is related to the thickness $\bar{\ell}_\pi(\theta) \equiv \bar{\ell}_w(\theta, \pi/2)$ of the wetting layer in the corresponding planar case if the temperature dependence is expressed via the contact angle θ. For shallow wedges, that is, for $\varphi \to \pi/2$, one has $\bar{\ell}_w(\theta, \varphi) = \bar{\ell}_\pi(\theta + \varphi - \pi/2)$.[136] This property called covariance extends also to probability distributions. The probability distribution function $P_w(\ell, \theta, \varphi)$ of the midpoint interface height in the wedge is related to the probability distribution $P_\pi(\ell, \theta)$ in the planar case via $P_w(\ell, \theta, \varphi) = P_\pi(\ell, \theta + \varphi - \pi/2)$.[136]

It is an interesting question whether for $\delta\mu = 0$, the orders of the filling and of the wetting transitions are related if the wedge undergoing filling and the planar substrate undergoing wetting are composed of the same material. In $d = 3$ critical wetting implies critical filling for obtuse wedges, however, it can change to first-order when the wedge becomes more acute.[137] Similarly one can observe critical filling even in the case that the planar substrate undergoes first-order wetting.[138] A first-order wetting transition implies critical filling if the filling temperature $T_F(\varphi)$ is below the spinodal temperature associated with first-order wetting. An additional effect of fluctuations is that the divergence of the interface roughness $\xi_\perp \sim (T_F(\varphi) - T)^{-\nu_\perp}$ in the wedge is governed by a universal exponent $\nu_\perp = 1/4$.[138,139] Since critical filling would have to occur below the first-order wetting transition temperature, one should recall that so far above the triple point, no first-order wetting transition on a solid surface has been observed experimentally.

Thus critical filling associated with first-order wetting seems to be difficult to come by.

However, a similar discussion for the cone geometry[48] shows that in this case the occurrence of critical filling requires the fulfillment of conditions that are much less restrictive than those for critical wetting of planar substrates. Critical wetting requires the fine-tuning of the Hamaker constant a in Eq. 3.5, while in the cone geometry the condition for critical filling amounts to having a negative value of the line tension coefficient $\tau_{\rm lvs}$ (Eq. 3.11). Thus proceeding from a planar substrate via the wedge geometry to the cone geometry weakens the necessary conditions for critical wetting or filling to occur. In the case of complete filling one observes a covariance, which relates the midpoint heights $\bar{\ell}_{\rm w}(\delta\mu, \varphi)$ and $\bar{\ell}_{\rm c}(\delta\mu, \varphi)$ in a wedge and in a cone, respectively, where the angle 2φ denotes the cone opening angle, $\bar{\ell}_{\rm c}(\delta\mu, \pi/2 - \varphi) = \bar{\ell}_{\rm w}(\delta\mu/2, \pi/2 - \varphi)$,[140] omitting the temperature variable.

It is instructive to consider complete filling in *generalized* wedges,[141,142] the shape of which is described by $s(x) = L\,|x/L|^{\gamma}$. The parameter $\gamma \in [0, \infty]$ interpolates between the planar substrate for $\gamma = 0$, the simple wedge for $\gamma = 1$, and the capillary slit for $\gamma = \infty$. In the limit $\delta\mu \to \infty$ the midpoint height $\bar{\ell}_{\rm gw} \sim \delta\mu^{-\beta_{\rm gw}^{\rm co}(\gamma)}$ diverges algebraically with a geometry-dependent exponent: for $p = 2$ (see Eq. 3.4) one has $\beta_{\rm gw}^{\rm co} = 1/3$ for $\gamma < 1/2$, $\beta_{\rm gw}^{\rm co} = \gamma/(2-\gamma)$ for $1/2 < \gamma < 1$, and $\beta_{\rm gw}^{\rm co} = \gamma$ for $\gamma > 1$. The parabolic wedge with $\gamma = 2$ represents a special case in which no meniscus forms unless $\delta\mu$ falls below a certain threshold upon approaching the bulk coexistence curve. This property is predicted by applying the Laplace equation to this particular substrate geometry.

For a macroscopicly large slit of width L, capillary condensation is a first-order transition. The shift of the liquid–vapor coexistence curve $P = P_{1/L}(T)$ in the capillary relative to the bulk pressure $P = P_0(T)$ is described by the Kelvin equation $\delta P = P_0(T) - P_{1/L}(T) \sim 1/L$.[25] In the case of capped capillaries with a finite depth D and width L, capillary condensation becomes a *continuous* phase transition.[143] There is a power law with corresponding critical exponents characterizing the unbinding of the meniscus from the bottom of the capped capillary if the pressure P approaches the capillary coexistence line, that is, $\Delta P = (P_{1/L}(T) - P) \to 0$. For suitable thermodynamic conditions there is also the opposite phenomenon of desorption, in which the meniscus unbinds from the open top of the capillary and approaches the bottom. It turns out that the values of the relevant critical exponents are different in these two cases. For systems governed by dispersion forces, one has for the mean meniscus height in the middle $< \ell > \sim (\Delta P)^{-1/4}$ for adsorption and $(D - < \ell >) \sim (\Delta P)^{-1/3}$ for desorption. Due to the presence of a capping wall at one end of this capillary with a finite depth and its absence at the other open end, there is a relation between the phase behavior in such a capillary[143] and the one in an infinite capillary slit with opposing boundary conditions.[38,39,144,145]

The next level of enriching the liquid morphology on a structured substrate consists of considering complete wetting of periodic arrays of grooves and pits.[146,147] The considered grooves have either rectangular or parabolic cross sections, and similarly, the pits are of cylindrical or parabolic shape. The midpoint interface heights in grooves $\bar{\ell}_g$ and pits $\bar{\ell}_p$ depend algebraically on the undersaturation: $\bar{\ell}_{p,g} \sim \delta\mu^{-\beta_{p,g}}$, where their additional dependences on geometrical parameters characterizing the arrays of grooves and pits, like the groove or pit width, depth, and their separation, have been omitted. Similar to the wedge–cone covariance, for these structures one observes a pit–groove covariance $\bar{\ell}_p(\delta\mu) = \bar{\ell}_g(\delta\mu/2)$ of the midpoint heights. However, this covariance does not extend to the interface height laterally averaged over the whole cavity. In addition, for these structures one observes four distinct scaling regimes corresponding to four different ranges of undersaturation $\delta\mu$. These regimes are determined by the geometry of the substrate, that is, the depth, width, and shape of the cavities, such that the exponent $\beta_{p,g}$ has, in general, different values in each of these ranges. Among them, asymptotically in the limit $\delta\mu \to 0$ one recovers the value $\beta_{p,g} = 1/3$, which is the same as for complete wetting of planar substrate governed by dispersion forces.

Combined experimental[148,149] and theoretical studies[149] of a periodically grooved silicon substrate wetted by perfluoromethylcyclohexane have confirmed the occurrence of the aforementioned wetting regimes as a function of the undersaturation $\delta\mu$. The grooves probed by X-ray scattering have a nearly rectangular cross section with a width of 16 nm at half of the groove height, which equals 39.4 nm; the side walls exhibit a slope of 79.7°. The corresponding theoretical analysis is based on the effective Hamiltonian given by Eq. 3.3, with the effective interface potential $\omega(\ell)$ taking into account the long-ranged character of the substrate–fluid and fluid–fluid interactions. The three wetting regimes correspond to distinct ranges of $\delta\mu$ encountered upon approaching the bulk liquid–vapor coexistence curve. They are the so-called prefilling regime with a thin wetting film ($\bar{\ell} < 1$ nm) uniformly covering the substrate, the capillary condensation regime in which the film thickness in the grooves substantially increases and a concave meniscus is observed, and the postfilling regime in which the grooves are filled with liquid the meniscus of which is located at the upper edge of the cavity.[150]

Analogous to fluids on substrates with chemical wettability patterns, nonvolatile fluids on topographically structured substrates exhibit a variety of morphological, surface tension–driven phase transitions.[151] For example, at straight topographic steps one can observe a transition from elongated, cigar-shaped droplets for small volumes to blob-shaped ones for large volumes;[152] in rectangular grooves small droplets are located at one of the steps forming the groove, while large droplets fill the hole channel.[95] An axially homogeneous rivulet in a wedge of opening angle 2ϕ is linearly unstable with respect to breakup into droplets (pearling) if $\theta_{eq} > \pi/2-\phi$ (see Ref. [153]). In combination with electrowetting, this pearling has been exploited for transporting liquids in grooves.[154,155] However, on the nanoscale these morphological phase transitions have not yet been studied.

Another interesting problem related to wetting of topographically structured substrates is the experimentally observed strong hydrophobicity of surfaces covered by hairs.[156–159] Different mechanisms have been invoked with the purpose of explaining the large contact angle of water drops residing on such surfaces. In particular, the phenomenon of bundling of hairs has been discussed, taking into account their elasticity and capillarity. The relative importance of various mechanisms for the observed hydrophobicity is currently under discussion.[157,160,161]

3.3 Nonequilibrium

3.3.1 *Theoretical Description*

The microscopic description of inhomogeneous condensed matter in thermal equilibrium, for example, in terms of DFT, is based on the solid ground of statistical mechanics.[162,163] For corresponding nonequilibrium systems there is no comparable, truly rigorous, microscopic theory available. A proof of existence of a dynamical DFT for classical systems is available.[164] However, as in the static case, the proof is not constructive, and an established dynamical extension of the equilibrium DFT is only available for overdamped Brownian particles.[165–169] In contrast to overdamped Brownian systems, ideal noninteracting systems are singular in the sense that they cannot relax to equilibrium and thus they cannot be used as a starting point for the development of a dynamic theory. Nevertheless, first steps in the development of a dynamic DFT for simple fluids have been reported in the literature.[170,171]

Apart from these, one has to either resort to simulation methods, in particular to MD, or follow a top-down approach by augmenting macroscopic hydrodynamic equations such that they can be used down to the scale at which the molecular structure becomes relevant.

3.3.1.1 Mesoscopic hydrodynamics

In the following, for simplicity we restrict the discussion to incompressible nonvolatile Newtonian fluids that are well described by the Navier-Stokes equations for the flow velocity $\mathbf{u}(\mathbf{r}, t)$ and the pressure $p(\mathbf{r}, t)$. Nanofluidic systems are in general too small to support strong temperature gradients; accordingly the equation of motion for the energy density does not play a role. For two-phase systems such as wetting films or droplets, one has an additional equation of motion for the interface Γ_{lv}; for wetting films and droplets with small contact angles, it is often parameterized in terms of the so-called Monge representation by the local film thickness $z = f(x, y, t)$. At small scales the resulting free boundary value problem has to take into account the influence of the long range of the intermolecular interactions, which is captured by the effective interface potential ω, the thermal fluctuations, and the hydrodynamic slip.

Thermal fluctuations can be included into the hydrodynamic equations via a Gaussian fluctuating stress tensor $\mathbf{S}(\mathbf{r}, t)$.[172] The stochastic version of the Navier-Stokes equation then reads

$$\varrho_l \left[\frac{\partial \mathbf{u}}{\partial t} + (\mathbf{u} \cdot \nabla)\,\mathbf{u} \right] = -\nabla p + \eta\,\nabla^2 \mathbf{u} + \nabla \cdot \mathbf{S}, \tag{3.16}$$

where η is the fluid viscosity. The mean value of the Gaussian fluctuating stress tensor is zero, $\langle \mathbf{S}(\mathbf{r}, t) \rangle = 0$, and its correlator is determined by the fluctuation-dissipation theorem:

$$\langle S_{\mathrm{ij}}(\mathbf{r}, t)\, S_{\mathrm{lm}}(\mathbf{r}', t') \rangle = 2\,\eta\, k_{\mathrm{B}} T\, \delta(\mathbf{r}-\mathbf{r}')\, \delta(t-t')\, (\delta_{\mathrm{il}}\delta_{\mathrm{jm}}+\delta_{\mathrm{im}}\delta_{\mathrm{jl}}). \tag{3.17}$$

For nanofluidic systems the left-hand side of Eq. 3.16 can be neglected because in general the Reynolds number is tiny. This leads to the linear Stokes equation

$$\nabla p = \eta\, \nabla^2 \mathbf{u} + \nabla \cdot \mathbf{S}. \tag{3.18}$$

For incompressible fluids the conservation law for mass reduces to the condition[a]

$$\nabla \cdot \mathbf{u} = 0. \tag{3.19}$$

Polymeric liquids play an important role in nanofluidics. However, they tend to be non-Newtonian. In this respect Eq. 3.16 can be generalized to take into account viscoelasticity;[174] however, this is beyond the scope of the present discussion.

Together with the disjoining pressure Π, the noise tensor $\mathbf{S}$ appears also in the boundary conditions for the pressure p and the viscous stress tensor $T_{\mathrm{ij}} = \eta \left(\frac{\partial u_{\mathrm{i}}}{\partial r_{\mathrm{j}}} + \frac{\partial u_{\mathrm{j}}}{\partial r_{\mathrm{i}}} \right)$ at the liquid–vapor interface:[175–177]

$$[(\mathbf{T} + \mathbf{S}) \cdot \hat{\mathbf{n}}_{\mathrm{lv}} - (2\,\sigma_{\mathrm{lv}}\, H + p + \Pi)\, \hat{\mathbf{n}}_{\mathrm{lv}}]|_{\Gamma_{\mathrm{lv}}} = 0, \tag{3.20}$$

[a]The incompressibility condition imposes unphysical velocity correlations at short times $|t - t'| < |\mathbf{r} - \mathbf{r}'|/c$, with c as the sound velocity.[173] However, on the micron scale and in particular on the nanoscale, the time scale above which these artifacts are irrelevant is of the order of $10^{-9} \ldots 10^{-12}$ s (using the speed of sound in air $c_{\mathrm{air}} = 340$ m/s or water $c_{\mathrm{water}} = 1,700$ m/s, respectively), that is, beyond the time resolution of most experiments.

with the unit vector $\hat{\mathbf{n}}_{\rm lv}$ normal to the liquid–vapor interface pointing into the vapor phase. H denotes the local mean curvature of the liquid–vapor interface. DFT, from which the disjoining pressure appearing in Eq. 3.20 can be derived, is a theory for ensemble-averaged densities, that is, it should already contain all effects of thermal fluctuations. Therefore one expects that it should not appear in an equation together with a stochastic noise term. However, although in principle fluctuations are included in DFT, all actually available, physically relevant density functionals in three spatial dimensions are of the mean field type (in particular the effective interface Hamiltonians in Eqs. 3.2 and 3.3), and therefore they do not contain important aspects of fluctuations. Pragmatically, one could use effective interface potentials, which are renormalized by the capillary wavelike fluctuations of the interface.[3]

Hydrodynamic slip is usually parameterized in terms of the Navier-slip length b.[178] The underlying idea is that the fluid velocity right at the substrate surface $\Gamma_{\rm ls}$ is proportional to the shear stress in the corresponding direction. For impermeable substrates the component of the flow field normal to $\Gamma_{\rm ls}$ vanishes, that is,

$$\mathbf{u}|_{\Gamma_{\rm ls}} \cdot \hat{\mathbf{n}}_{\rm ls} = 0, \tag{3.21}$$

where the substrate surface normal vector $\hat{\mathbf{n}}_{\rm ls}$ points into the fluid, and the slip boundary condition for the flow velocity in a direction $\hat{\mathbf{p}}_{\rm ls}$ (with $|\hat{\mathbf{p}}_{\rm ls}| = 1$) parallel to $\Gamma_{\rm ls}$ reads

$$\mathbf{u}|_{\Gamma_{\rm ls}} \cdot \hat{\mathbf{p}}_{\rm ls} = \frac{b}{\eta}\, \hat{\mathbf{p}}_{\rm ls} \cdot \mathbf{T}|_{\Gamma_{\rm ls}} \cdot \hat{\mathbf{n}}_{\rm ls}. \tag{3.22}$$

For Newtonian fluids the slip flow velocity is therefore given by the shear rate parallel to the substrate surface times the slip length b.

The dynamics of the liquid–vapor interface is determined by mass conservation. For incompressible and nonvolatile fluids the velocity $v\, \hat{\mathbf{n}}_{\rm lv}$ of the interface in the normal direction is given by the corresponding component of the flow velocity:

$$v = \hat{\mathbf{n}}_{\rm lv} \cdot \mathbf{u}|_{\Gamma_{\rm lv}}. \tag{3.23}$$

Even without taking into account the disjoining pressure, the thermal fluctuations, and the hydrodynamic slip, Eqs. 3.16–3.23 represent a rather complex nonlinear free boundary value problem. Accordingly, solving these mesoscopic hydrodynamic equations is a challenging task. In this context, most progress has been made in the so-called thin-film limit, in which the ratio of the mean film thickness ℓ and the typical lateral length scale λ of the system under consideration is a small number. The solution of the above mesoscopic hydrodynamic equations can be expanded in terms of powers of this parameter.[179] The results of this analysis crucially depend on the value of the slip length b as compared with the ones for ℓ and λ:[180] for $b < \ell$, $b \approx \lambda$, and $b \gg \lambda$ there is a weak-slip, an intermediate-slip, and a strong-slip regime. The qualitative differences of these regimes lead to qualitative differences in the dynamics of thin films. So far, a thin-film model including thermal fluctuations has been derived from Eqs. 3.16–3.23 only for $b = 0$, that is, without slip;[175,176] a phenomenological approach leads to the same result.[181] Up to now efforts to derive systematically a stochastic thin-film equation for finite b have failed, which could imply that the slip boundary condition (3.22) might have to be reconsidered. However, within the phenomenological approach proposed in Ref. [181], it should be possible to develop stochastic thin-film models with slip in all three aforementioned regimes.

Beyond the thin-film approximation, solutions can be found only numerically. For example, boundary integral methods have been used to study the influence of long-ranged forces on two-dimensional nanodroplets.[182–184] However, the presently available numerical codes for solving the Stokes or Navier-Stokes equations are not capable of addressing the full mesoscopic hydrodynamic equations, in particular, including thermal fluctuations. The tracking of the moving interface in three-dimensional systems presents a particular challenge. A promising method that avoids interface tracking by modeling the dynamics of the local number density in phase space (from which the density can be calculated by integrating over the velocity degrees of freedom) is provided by lattice Boltzmann simulations for nonideal systems;[185–189] recently it became possible to include to a certain extent thermal fluctuations into this method.[190] Another method, similar in spirit, is the multiparticle-collision dynamics, also called stochastic rotator dynamics.[191,192] Within this approach the underlying simplified MD fulfills all conservation laws, and therefore, on a large scale compared with the built-in mean free path, the system is well described by stochastic hydrodynamic equations. Within this method fluid–fluid interfaces[193] and the dynamics of capillary waves[194] have been studied. It is straightforward to incorporate a hydrodynamic slip into these two methods,[195,196] but up to now the effective interface potential as the other essential ingredient for the description of nanoscale systems has neither been included into the lattice Boltzmann method nor been included into the multiparticle-collision dynamics.

3.3.1.2 Molecular dynamics

With the only exception of liquid helium, quantum effects are de facto negligible in liquids and gases because the thermal wavelength $\lambda = h/\sqrt{2\pi\, m\, k_B\, T}$ (with the Planck constant h, the Boltzmann constant k_B, the molecular mass m, and the temperature T) is smaller than the average distance of the molecules. Therefore, one can describe the MD of the molecules in a classical fluid by Newton's equations of motion

$$m_n \frac{d^2\mathbf{r}_n}{dt^2} = \mathbf{F}_n(\mathbf{r}_1, \mathbf{r}_2, \ldots, \mathbf{r}_N), \quad n = 1, \ldots, N. \tag{3.24}$$

Here $\mathbf{r}_n(t)$ denotes the position of molecule n with mass m_n at time t, and $\mathbf{F}_n$ is the force exerted on this molecule by all other molecules in the fluid as well as by the molecules forming potential confinements, such as channel walls or substrates. These equations of motion are coupled ordinary differential equations, which in MD simulations are solved numerically.[197,198]

The main challenge for MD simulations is the large number of degrees of freedom. For each molecule and in each numerical time step, the interactions with all other molecules have to be calculated, which requires in the order of $N\,(N-1)/2$ force calculations in total. Even on modern supercomputers this would limit the simulations to a rather small number of molecules. Therefore, in such simulations, often the range of intermolecular interactions is truncated such that the molecules interact only with the small number of neighboring molecules. This implies that the number of force calculations in each numerical time step is proportional to N rather than N^2. However, as a consequence the specific effects due to the actual long range of the intermolecular interactions (which requires large system sizes, too) are lost. Capturing these by MD simulations is computationally very demanding.

The second limitation of MD simulations concerns the time scale. The intermolecular potentials and the molecular mass define the time scale for the chaotic molecular motion. In most actual fluids this time scale is of the order of nanoseconds, and time steps in an MD simulation must resolve this time scale. However, in general, one is interested in phenomena that occur on the scale of seconds or even longer. Currently there are significant efforts to overcome these difficulties by multiscale modeling via suitably bridging the various time and length scales appearing in the systems.[199] However, for driven systems one can reduce this gap in time scales by applying a strong drive in the simulations. If the Reynolds number remains small and if it turns out a posteriori that nonlinearities are not significant, the results can be extrapolated to a weaker drive (see, e.g., Refs. [119] and [120]).

By their very nature MD simulations have to conserve the total energy. This poses a problem for simulations of driven systems. The drive feeds energy into the system, which is dissipated into heat. Experiments are thermally coupled to the environment, which keeps the temperature under control. In MD simulations this environment either is not taken into account at all or is composed of only a small number of molecules. This is aggravated by the need to use unrealistically strong driving forces in order to reduce the gap in time scales between the molecular chaos and the center of mass motion of the driven fluid, as mentioned before. The heating problem can be overcome by introducing a so-called thermostat into the numerical scheme. It keeps the temperature constant by shaping the velocity distribution of the particles, that is, by coupling them to an external heat bath of a given temperature.

To date MD is the most versatile and best-developed simulation method for nonequilibrium systems in which the discrete size of the molecules is important. As a bonus the smaller the size of the system under consideration, the less problematic are most of the challenging issues discussed earlier.

3.3.2 *Homogeneous Substrates*

Spreading of droplets and dewetting of thin films are the two paradigmatic cases of dynamical wetting phenomena on homogeneous surfaces.

3.3.2.1 Spreading of droplets

The surface of a spherical droplet brought into contact with a substrate surface, for example, with the help of a pipette, has a local contact angle with the substrate, which is typically larger than θ_{eq}. As a consequence the interfacial forces acting on the three-phase contact line are not in mechanical equilibrium and the droplet spreads until it reaches its mechanical equilibrium shape. Understanding its moving contact line is a long-standing problem in hydrodynamics because the stress and viscous dissipation in the wedge-shaped liquid volume next to the contact line diverge.[11,179,200,201] Taken literally, as a consequence the contact line should not be able to move,[202] although everyday experience tells that it does.[203] The key to understanding the dynamics of a three-phase contact line is the nanofluidics within the contact line.[15] On the nanoscale the aforementioned singularity is regularized by both the finite slip length[204,205] and the finite thickness of the wetting film at the substrate–vapor interface, as shown in Fig. 3.6. Indeed, a violation of the no-slip boundary condition in the vicinity of moving contact lines has been observed in MD simulations[201,206,207] as well as a precursor film of a finite thickness on which the three-phase contact line can glide.[205]

In a spreading experiment, a droplet is deposited on a surface such that the initial contact angle is larger than θ_{eq}. To a good approximation slowly spreading droplets keep their spherical cap shape throughout the whole spreading process, such that for geometrical reasons the time-dependent contact angle $\theta(t)$ is determined by the droplet base radius $R(t)$ and the fixed droplet volume V:

$$V = \frac{\pi\, R^3(t)}{\sin^3\theta(t)}\,[1-\cos\theta(t)]^2\,[2+\cos\theta(t)] \approx \frac{\pi}{4}\,R^3(t)\,\theta(t), \tag{3.25}$$

where the latter relation holds for $\theta(t) \ll 1$. Assuming a power law dependence of the contact line speed on the contact angle, $dR/dt \sim [\theta(t)]^m$, in the limit of small contact angles one obtains $dR/dt \sim [V/R^3(t)]^m$ and therefore $R(t) \sim [V^m t]^{1/(3m+1)}$ (see section IV in Ref. [11]). Depending on the dissipation mechanism, there are different spreading power laws. If viscous dissipation dominates one has $m = 3$ and therefore $R(t) \sim t^{1/10}$ or $\theta(t) \sim t^{-3/10}$ known as Tanner's law.[208] Dissipation in molecular jumps as a dissipation channel was discussed by Blake,[209] leading to $m = 2$. Accordingly, this type of spreading is faster than in the viscous case, that is, $R(t) \sim t^{1/7}$ and $\theta(t) \sim t^{-3/7}$. More recently there are indications that viscous dissipation dominates at the initial stages of spreading, while molecular kinetic dissipation takes over at later stages.[205,210]

On a completely wetting substrate with $\theta_{eq} = 0$, one observes a so-called precursor film with a thickness of a few Ångstroms spreading ahead of the three-phase contact line. In experiments[211] as well as in MD simulations,[205] the radius of the precursor film has been found to increase much faster, that is, $\sim\sqrt{t}$, than that of the spreading drop. The latter growth law is born out also by kinetic Monte Carlo simulations of spreading liquid monolayers.[212,213] The growth rate $\sim\sqrt{t}$ is consistent both with diffusive growth as well as with hydrodynamic flow in an ultrathin film. Therefore one cannot discern between these two possible underlying dynamics on the basis of the growth law alone.

3.3.2.2 Dewetting of thin films

The dewetting of thin films on partially wetting substrates is much better understood than the spreading of droplets, mainly because during the most crucial phase of the dewetting process—the onset of film rupture—there is no three-phase contact line present yet.[214,215] Moreover, for many aspects of dewetting, the dynamics of the three-phase contact line plays a subdominant role. The following discussion of dewetting is focused on nonvolatile fluids. A further detailed analysis of the stability of thin films is provided in chapter 4, and technological applications of dewetting are discussed in chapter 14.

Depending on its actual thickness ℓ, a thin film on a substrate can be stable, metastable, or linearly unstable. The stability of a thin film is determined by the second variation of the effective interface Hamiltonian in Eq. 3.3. In Fourier space, $f(\mathbf{R}) = \int d^2k\, f(\mathbf{k}) \exp(-i\,\mathbf{k}\cdot\mathbf{R})$, its second variation is diagonal:

$$\frac{\delta^2\mathcal{H}}{\delta f(\mathbf{k})\,\delta f(\mathbf{k}')} = \left[\sigma_{\mathrm{lv}}\,|\mathbf{k}|^2 + \omega''(\ell)\right]\,\delta(\mathbf{k}-\mathbf{k}'). \tag{3.26}$$

For $\omega''(\ell) < 0$ there is a band of modes with $0 < k < \sqrt{-\omega''(\ell)/\sigma_{\mathrm{lv}}}$ for which the second variation is negative so that the film is linearly unstable. Accordingly infinitesimal perturbations (e.g., generated by thermal fluctuations) grow exponentially, which is called spinodal dewetting. For $\omega''(\ell) > 0$ the second variation is positive definite and the film is linearly stable. If, however, ℓ is not the minimum of $\omega(\ell)$, this film is metastable. If the equilibrium wetting film thickness $\bar{\ell}$ is smaller than ℓ, the free energy of the system can be reduced by collecting all the fluid into a large drop, which is connected to a wetting film of a thickness $\approx \bar{\ell}$. But in order to arrive at this state, one has to overcome an energy barrier, which can be calculated from the effective interface Hamiltonians in Eqs. 3.2 and 3.3.[216,217] In general, this barrier increases with the film thickness. Depending on the mechanism for providing the free energy needed to overcome this barrier—thermal fluctuations or heterogeneities present on the substrate (e.g., dirt or roughness) or in the film (e.g., inclusions or bubbles)—the process is called homogeneous or heterogeneous nucleation of holes, respectively. Experimentally it is extremely difficult to observe pure homogeneous nucleation because either the time scale for the formation of holes due to thermal fluctuations is too large or the films are so thin that they are extremely sensitive to the presence of heterogeneities. While spinodal dewetting and nucleation can be discerned due to the resulting dewetting patterns,[218] it is difficult do distinguish homogeneous and heterogeneous nucleation because in both cases the positions of the emerging holes are random and uncorrelated. Sometimes nucleation sites can be identified by repeating a dewetting experiment on the same substrate. A hole that appears repeatedly on the same spot indicates that there is a defect underneath.

Spinodal dewetting has been studied mostly for the case of the negligible slip. But the dewetting behavior with a weak slip (with the slip length b of the order of the film thickness or smaller) and an intermediate slip (with b of the order of the characteristic lateral length scale of the dewetting process) is rather similar; only the time scales differ. In order to observe spinodal dewetting in a well-controlled experiment, one has to prepare an almost perfectly flat film in an immobilized state such that the starting time of the dewetting process can be defined. This can be achieved, for example, by spin-coating a solution of polymers in a volatile solvent. Since the initial conditions for these films are such that they are rather flat and very thin, one can resort to the thin-film limit of the mesoscopic hydrodynamic equations in order to analyze spinodal dewetting. Linearizing the corresponding deterministic thin-film equation around the initial flat film with thickness ℓ renders the temporal evolution of the amplitude of a small perturbation $\delta f(\mathbf{k}, t)$ with wavenumber $\mathbf{k}$,

$$\delta f(\mathbf{k}, t) = \delta f(\mathbf{k}, 0)\, \exp[\Omega(|\mathbf{k}|)\, t] \tag{3.27}$$

where the dispersion relation, that is, the inverse time constant, is given by

$$\Omega(k) = -\frac{M(\ell)}{\eta}\, k^2\, \left[\omega''(\ell) + \sigma_{\mathrm{lv}}\, k^2\right]. \tag{3.28}$$

Apart from the mobility $M(\ell) > 0$ the dispersion relations in the no-slip or weak-slip ($M(\ell) = \ell^3/3 + b\,\ell^2$ with the slip length b) and in the intermediate-slip regime ($M(\ell) = b\,\ell^2$) are equal.[219] For linearly unstable films with $\omega''(\ell) < 0$, the dispersion relation has a maximum $\frac{M(\ell)}{\eta}\,[\omega''(\ell)]^2/(4\,\sigma_{\mathrm{lv}})$ at $|\mathbf{k}| = Q = \sqrt{-\omega''(\ell)/(2\,\sigma_{\mathrm{lv}})}$, and it is positive for $0 < k < \sqrt{2}\,Q$ and negative for $k > \sqrt{2}\,Q$. For stable films with $\omega''(\ell) > 0$, the dispersion relation $\Omega(k)$ is negative for all $k > 0$. The marginal stability of the $k = 0$ mode, $\Omega(k = 0) = 0$, is a consequence of the conserved dynamics.

Concerning the theoretical description, thermal fluctuations have been taken into account on the level of the thin-film equation for the case of a negligible slip.[175,176,181] Within the validity of linear perturbation theory, the resulting power spectrum $S(k, t)$ of the thickness fluctuations of the film is given by

$$\begin{aligned} S(k, t) &= \left\langle |\delta f(k, t)|^2 \right\rangle \\ &= S_0(k)\, e^{2\,\Omega(k)\,t} + \frac{k_{\mathrm{B}}\, T\, M(\ell)}{\eta}\, \frac{k^2}{\Omega(k)}\, \left[e^{2\,\Omega(k)\,t} - 1\right], \end{aligned} \tag{3.29}$$

with $S_0(k)$ as the power spectrum of the initial film roughness. For sufficiently smooth initial power spectra (i.e., without a pronounced peak as a function of k), the deterministic part of the power spectrum (i.e., the first term in Eq. 3.29) has its maximum for all t at the maximum of $\Omega(k)$, that is, at $k = Q$. The position Q of the maximum of $S(k, t)$ renders directly the curvature of $\omega''(\ell) = -2\,\sigma_{\text{lv}}\,Q^2$. Measuring Q for many unstable film thicknesses allows one to determine $\omega(\ell)$ by numerical integration.[220,221] The integration constants are determined by the position $\bar{\ell}$ and the depth $\omega(\bar{\ell})$ of the minimum of $\omega(\ell)$, which have to be determined independently by measuring the wetting film thickness between the droplets after the dewetting process is finished and by measuring the equilibrium contact angle of the droplets θ_{eq}, respectively (see Eq. 3.9).

A peak in a power spectrum that grows exponentially in time but appears at a fixed wavenumber is often taken as an indication that the linearized theory is still valid, whereas peaks that change their positions in time are in general associated with nonlinearities. However, for unstable films with $\omega''(\ell) < 0$ even for structureless initial conditions $S_0(k) = $ const, the maximum of the power spectrum in Eq. 3.29 shifts in time from larger to smaller values of k and it approaches Q from above. This shift toward a larger wavelength can be interpreted as noise-induced coarsening. The reason for this is that initially thermal fluctuations generate mostly fluctuations with short wavelengths and only later the instability associated with longer wavelengths corresponding to the maximum of $\Omega(k)$ sets in. Recently, it has been possible to observe this behavior for spinodal dewetting of thin polymer films.[222] However, numerical studies have also demonstrated that later stages of dewetting, when holes in the film start to coalesce, are hardly influenced by thermal fluctuations.[176] Spinodal dewetting of unstable thin liquid films is a process in which nanoscale properties and phenomena such as the specific shape of the effective interface potential and thermal fluctuations, respectively, influence the resulting structures on much larger scales because the maximum of the power spectrum typically appears at micron length scales.

Beyond the linear regime, thin-film equations can be solved only numerically. But in particular the vanishing of the mobility factor

$M(\ell)$ (i.e., $M(\ell \to 0) \to 0$) poses numerical problems. By using algorithms, which preserve the nonnegativity of $f(x, y, t)$,[223,224] it has been possible to demonstrate,[225] that spinodal dewetting patterns can be described quantitatively by thin-film equations.

The dispersion relation of a thin liquid film in the strong-slip regime with $b \gg 2\pi/Q$ has a form different from Eq. 3.28, that is, the influence of the hydrodynamic slip is not captured by a modified mobility factor $M(\ell)$. Linearizing the corresponding thin-film equation around a flat film of thickness ℓ yields[226]

$$\Omega(k) = -\frac{b\,\ell^2\,k^2\,\left[\omega''(\ell) + \sigma_{\mathrm{lv}}\,k^2\right]}{\eta\,\left(1 + 4\,b\,\ell\,k^2\right)}. \tag{3.30}$$

In the case of linearly unstable films with $\omega''(\ell) < 0$ the strong-slip dispersion relation is positive for $0 < k < \sqrt{2}\,Q$, that is, the stability of a mode is independent of the slip length in all slip regimes. Note that Eq. 3.30 does reduce to Eq. 3.28 with the intermediate-slip mobility coefficient $M(\ell) = b\,\ell^2$ for $b\,\ell\,k^2 \ll 1$. But the position Q of the maximum of $\Omega(k)$ decreases upon increasing the dimensionless slip length $B = 4\,b\,\ell\,Q^2$. The power spectrum of a film in the strong-slip regime is also given by Eq. 3.29 but with the dispersion relation given by Eq. 3.30. As in the case of a weak and intermediate slip, it develops a peak at the position Q of the maximum of $\Omega(k)$. But this position also depends on the slip length. Therefore, the effective interface potential for films with strong slip cannot be determined from the power spectrum (as described earlier) without knowledge of the slip length.[219]

In the later stages of dewetting, the slip influences the dynamics and structure of dewetting patterns, in particular of metastable films. During their dewetting process one initially observes separated holes, which grow in time. The liquid from inside the holes accumulates in a rim around the hole, which therefore grows in time. This growth rate depends on the dissipation mechanism.[227] On substrates without slip, neither the driving force for the rim motion nor the dissipation within the contact line depends on the hole size, such that the hole radius R grows linearly as a function of time, that is, $R \sim t$. In the strong-slip regime, however, the dissipation occurs within the liquid–substrate interface, such that the overall dissipation rate grows with the width of the rim. In this regime one expects $R \sim t^{2/3}$. A combined model, which includes both dissipation mechanisms, yields an implicit relation between R and t, which quantitatively describes the dewetting rate of polymer films with large slip.[228]

The rim around the holes is not stable, and the resulting structures depend on the slip length. On substrates with a weak slip one observes a dimple between the rim and the remainder of the film, which can act as a nucleation site for satellite holes.[229,230] On substrates with strong slip, these dimples are absent[231-233] and the formation of satellite holes is suppressed.[a] Dewetting rims around holes are unstable with respect to pearling.[236,237] This instability (related to the Plateau-Rayleigh instability of fluid cylinders) is a general feature of liquid ridges.[68,238] Recently it has been proposed that hydrodynamic slip not only increases the growth rate of the rim instability, but in the presence of slip, the droplets emerging out of the rim develop protrusions toward the hole.[180]

3.3.3 *Structured Substrates*

As discussed in section 3.2.2 structured substrates exhibit a wide spectrum of equilibrium wetting phenomena. However, the surface structure changes not only equilibrium wetting phenomena but also the dynamics of thin films, rivulets, and droplets. Random heterogeneities, such as roughness or dirt, have significant influence on both the wetting properties[239] and the dynamics of moving contact lines.[240,241] In particular they can lead to contact angle hysteresis (see Ref. [15] and references therein) so that the local contact angle of an advancing contact line is larger than a threshold value θ_a (which, in general, is larger than θ_{eq}) and the local contact angle of a receding contact line is smaller than another threshold value $\theta_r \leq \theta_{eq}$. However, here we focus on regularly structured substrates. We do so because on the one hand experiments involving substrates with a designed structure can be controlled better and on the other hand the theoretical understanding of nanofluidic phenomena on regularly structured substrates is more advanced.

[a]There were speculations that viscoelasticity would prevent the formation of these dimples.[234] However, it has been shown that this is not the case.[235]

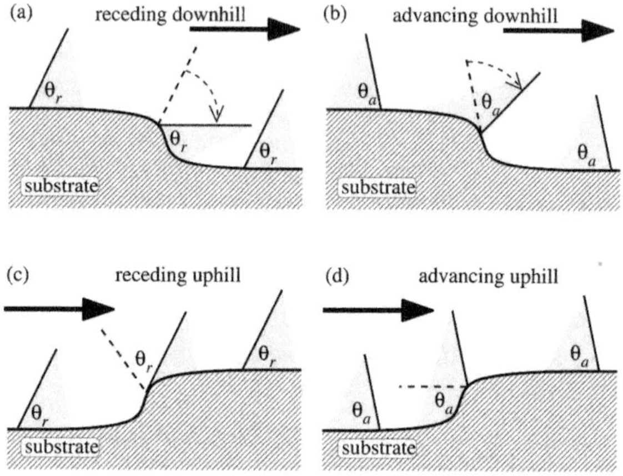

Figure 3.8 Macroscopic picture of a receding (a and c) and an advancing (b and d) three-phase contact line moving across a topographic step in downhill (a and b) and in uphill (c and d) direction (the direction of motion is indicated by the arrow). In this example, at the steep part of the step the slope of the liquid–vapor interface near a three-phase contact line moving in the downhill direction has to decrease from positive (dashed line) to almost zero in the receding case (a), and it has to increase significantly from a negative value (dashed line) to a positive value in the advancing case (b) in order to overcome the step. Keeping the slope of the liquid–vapor interface constant as the three-phase contact line moves in the uphill direction across the step results in a local contact angle smaller than θ_r (dashed line in c) in the receding case and larger than θ_a (dashed line in d) in the advancing case. Thus in the uphill direction the step accelerates the motion of the three-phase contact line because it enhances the imbalance of the driving forces acting on the three-phase contact line.

3.3.3.1 Topographically structured substrates

The discussion of moving contact lines in section 3.3.2.1 has been concerned with homogeneous substrates. If, however, a moving contact line encounters a surface heterogeneity, such as a topographic step, it can be pinned.[11,242] The local contact angle of the liquid–vapor interface with the substrate is measured with respect to the actual substrate surface (see Fig. 3.8). As a consequence the local slope of the liquid–vapor interface changes significantly as the three-phase contact line passes the step. If, for example, a receding three-phase contact line approaches a sufficiently steep step from above, there the slope of the substrate surface decreases such that the local contact angle would increase to a value above θ_r if the local slope of the liquid–vapor interface would not decrease (dashed line in Fig. 3.8a). In the case of an advancing three-phase contact line approaching a sufficiently steep step from above, the local contact angle would decrease to a value smaller than θ_a if the local slope of the liquid–vapor interface would not increase (dashed line in Fig. 3.8b). Depending on the boundary conditions for the liquid–vapor interface far from the three-phase contact line, changing the local slope of the liquid–vapor interface might not be possible such that the three-phase contact line is pinned by the step, or it requires significant transport of fluid, which results in pinning for some time. In the case of a three-phase contact line approaching a step from below, the change in slope of the substrate promotes the motion of the three-phase contact line. The local contact angle increases in the case of a three-phase contact line advancing uphill, and it decreases in the case of a three-phase contact line receding uphill. As a consequence, the imbalance of the interfacial forces acting on the three-phase contact line is enhanced. Within this macroscopic picture the relevant parameter for pinning is the slope of the steepest part of the step. For downhill receding contact lines, experiments have, however, shown that a step has to have a certain minimum height in order to be able to pin a moving contact line.[243] For the polymeric liquid used in these experiments, this minimum height depends on the molecular mass. The mechanisms for this phenomenon are not yet fully understood; nonetheless, the following considerations shed some light on it. Within the framework of DFT in the sharp interface limit (leading to Eq. 3.8) as well as within the mesoscopic hydrodynamic equations discussed in section 3.3.1.1, the statics and dynamics of the free liquid–vapor interface are determined by the disjoining pressure Π, which introduces the microscopic length scale $\bar{\ell}$. Figure 3.9 shows the disjoining pressure $\Pi(x, y)$ in the vicinity of topographic steps of height (a) $2\,\bar{\ell}$ and (b) $0.5\,\bar{\ell}$ which is calculated along the lines proposed in Ref. [111] assuming Lennard-Jones-type interactions. The step in the substrate has a vertical part in both cases, but the contour line $\Pi(x, y) = 0$ (to a good approximation given by the upper perimeter of the white area above the black substrate)

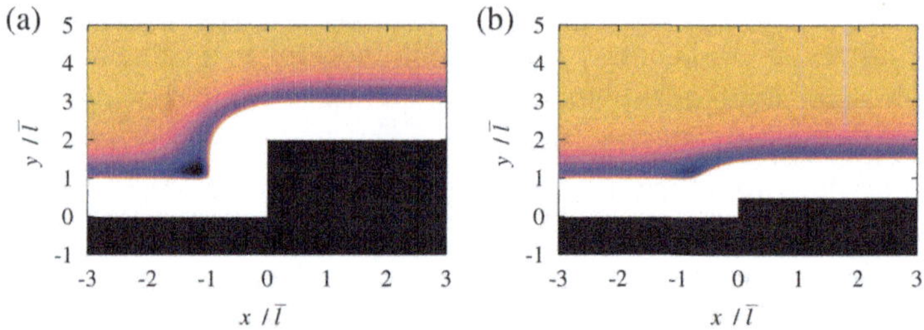

Figure 3.9 False color image of the disjoining pressure $\Pi(x, y)$ in the vicinity of topographic steps (substrate in black) (a) of height $2\bar{\ell}$ and (b) of height $0.5\bar{\ell}$ calculated along the lines proposed in Ref. [111] assuming Lennard-Jones-type intermolecular interactions. Dark colors correspond to negative values of Π and white to very large values of Π. To a good approximation the upper part of the perimeter of the white area coincides with the contour line $\Pi(x, y) = 0$.

becomes rather flat for step heights smaller than $\bar{\ell}$. It is plausible that for distances from the substrate larger than $\bar{\ell}$, the disjoining pressure of a nonvertical step looks rather similar to the one shown in Fig. 3.9b. And in this picture the liquid–vapor interface hardly gets closer to the substrate than the contour line $\Pi(x, y) = 0$. Thus a small vertical step may act as a nonvertical one and thus lose its potential for pinning the liquid–vapor interface. The interplay between the length scale $\bar{\ell}$ in the disjoining pressure and the step height could turn out to be a mechanism explaining the aforementioned experimental findings.

A closer analysis of the disjoining pressure in the vicinity of a topographic step (such as the example shown in Fig. 3.9) shows that the disjoining pressure $\Pi(x, y)$ (or the corresponding effective interface potential $\omega(x, y)$) converges algebraically to the disjoining pressure of an infinite and homogeneous substrate as the distance from the step is increased. This follows from the power law decay of the dispersion forces. As a consequence a nanodroplet in the vicinity of a topographic step experiences a lateral force.[182,244,245] The sign of the force depends on the sign of the leading term of the disjoining pressure for large distances from the substrate, that is, the Hamaker constant. Therefore the direction of motion of the nanodroplets does not depend on whether they start from the top or from the base of the step. They move in the step-down direction for negative and in the step-up direction for positive Hamaker constants.

This migration of nanodroplets has not yet been observed experimentally. Apart from preparing the substrate the main experimental challenges are to position a nanodroplet at a certain distance of a few nanometers and to monitor its motion without influencing its dynamics. A promising technique for positioning nanodroplets is based on using AFMs as pens.[246,247] Experimentally it is much easier to grow droplets from an aerosol or from a vapor phase rather than to deposit them at a specific location. There is experimental evidence that water nanodroplets preferentially condense onto terrace steps of vicinal surfaces.[248] However, this data does not allow one to determine whether the droplets reside on the top terrace, reside on the bottom terrace, or span the step. This example shows that condensation on and evaporation from nanostructured substrates represent promising research topics.

3.3.3.2 Chemically structured substrates

From a macroscopic point of view, chemical surface heterogeneities can both move and stop droplets. While contact lines tend to be pinned at any kind of heterogeneity,[249] by varying continuously the chemical composition of coatings one can prepare surfaces with a smooth chemical gradient, leading to a position-dependent equilibrium contact angle $\theta_{eq}(x, y)$. On such surfaces droplets move toward regions with smaller contact angles.[250–252] The forces acting on the droplets can be strong enough to drive droplets even uphill.[253] Pinning at chemical heterogeneities and the ability to move droplets on surfaces with a chemical gradient both follow from a lower free energy of the droplets on surfaces characterized by a smaller equilibrium contact angle θ_{eq}. Therefore, moving a three-phase contact line onto a surface patch that has a larger local equilibrium contact angle costs free energy and thus requires to overcome a restoring force. If, however, a drop spans the chemical step between two parts of the surface that have a different equilibrium contact angle, it will move toward the side with the lower θ_{eq}. But if the initial (nonequilibrium) droplet shape has a rather small height so that its initial contact angle is much smaller than the equilibrium contact angle on either side of the substrate, it can happen that during its early contraction process, the droplet ends up completely on the less wettable side of the step and stays there[254]—at least on the macroscopic scale.

On the nanoscale the droplets respond to the presence of the chemical step due to the long range of the intermolecular interactions, which is large on this scale. As a result, they migrate, and in an analogy to the behavior of droplets near topographic steps, the direction of motion is determined by the difference of the Hamaker constants of the substrates on the two sides of the chemical step. The direction of motion of the droplet is determined by the sign of the difference of the Hamaker constants: it moves toward the side with the bigger Hamaker constant.[183,245] Since the equilibrium contact angle is determined by the interplay of the long-ranged and short-ranged parts of the intermolecular potentials, it is possible that the droplet moves toward the less wettable side, that is, into the unexpected direction. This means that there can be a free-energy barrier at the chemical step, which prevents the droplet from reaching the absolute minimum of its free energy, which is determined by the well of the effective interface potential and not by the Hamaker constant alone.

The behavior of nanodroplets spanning the chemical step also differs qualitatively from the macroscopically expected one. Within the macroscopic picture sketched before, in this situation the driving force is the difference between the equilibrium contact angles on both sides. But the force on microscopic droplets is given by the integral of the disjoining pressure over the droplet surface.[184] In the limit of large drops one recovers the macroscopic behavior with a local equilibrium contact angle given by Eq. 3.9 in terms of a laterally varying effective interface potential $\omega(x, y, \bar{\ell})$. However, for nanodroplets the disjoining pressure at the droplet apex is nonzero so that the driving force is not given by the difference of the local contact angles at the two sides of the step. Accordingly, even the sign of the force can change with the droplet size.[184] This might have implications for the adsorption and coarsening dynamics of droplets near chemical boundaries, which is strongly influenced by the migration of droplets across chemical steps.[255] Although this has not yet been discussed in detail, based on the same argument chemical gradients can be expected to act in directions that are opposite for macroscopic and microscopic droplets.

3.4 Outlook

Wetting phenomena provide a wide field of research at the nexus of physics, chemistry, and engineering. They pose numerous experimental as well as theoretical challenges with a plethora of possible applications. In order to fully understand wetting phenomena, one has to address their microscopic basis. But nanoscale wetting phenomena provide not just the basis for understanding macroscopic systems. With the progressive miniaturization of microfluidic devices, pushed by an increasing demand from the live sciences, nanofluidics in general and nanoscale wetting phenomena in particular find more and more applications.

The main experimental challenge in this field is to probe very soft systems with sufficient spatial and temporal resolution, while minimizing the influence of the probe on the system. In this context, as described in several chapters of this book, AFM is one of the most widely used methods for imaging, followed by ellipsometry and scattering techniques. On the theoretical side one has to distinguish between equilibrium and nonequilibrium systems. In thermal equilibrium, in addition to the effective interface Hamiltonian approach, one has at one's disposal theories based on rigorous statistical physics, in particular DFT. Most commonly used density functionals are of mean field character (i.e., fluctuations are insufficiently taken into account), which becomes most apparent when studying capillary fluctuations on the nanoscale. A second challenge is presented by polar fluids and electrolytes, which are particularly relevant for biological applications. In addition to discussing more complex fluids one is led to consider substrates that display structures of increasing complexity of both chemical and topographical nature. The analysis of wetting of such substrates and of the corresponding morphological phase transitions often require new theoretical tools as compared to those used for describing planar and homogeneous substrates. In particular further miniaturization of structured substrates and the resulting rapidly varying interface morphologies of the adsorbed fluid will enhance the role of interface fluctuations that need to be included into the analysis of such systems.

The main theoretical challenge for nonequilibrium systems is to bridge the gap in length and time scales between the molecular motion and the collective motion of the fluid, such as the translation of droplets. In top-down approaches macroscopic hydrodynamics is combined with equilibrium statistical physics. The resulting mesoscopic hydrodynamic equations partially include the effects of boundary slip, thermal fluctuations, and the long range of molecular interactions. So far, bottom-up approaches for nonequilibrium systems are only available for a small class of systems with purely diffusive dynamics. For the other liquids one has to resort to numerical simulations.

Among wetting phenomena the ones on homogeneous substrates have been studied most intensively and are understood best. However, even in these systems there are still important open questions concerning, for example, the intrinsic dynamics of moving three-phase contact lines and the dynamics of the adjoining precursor films, in particular for complex fluids such as polymer melts and liquid crystals (see chapter 5).[256–258] The interplay oforientational degrees of freedom and the fluid dynamics in the three-phase contact line can lead to anisotropic spreading.[259] At the three-phase contact line hydrodynamic theories break down. But even the interplay of thermal fluctuations and hydrodynamic slip (or viscoelasticity) in thin liquid films has not been explored. And in many systems hydrodynamic slip itself is barely understood, in particular for complex fluids such as suspensions and polymer melts. Its origin lies in the structure and dynamics of the fluid-substrate interface and therefore is again outside the domain of validity of hydrodynamic theories. The microscopic mechanism of pinning by defects is still awaiting a thorough quantitative understanding.

Chemically patterned substrates are proposed as a means to not only guide liquids but also control their flow in microfluidic devices.[260] While in technological applications micron-sized channels are used, all biological cells use a large number of nanochannels to exchange ions with the extracellular medium. These ion channels are highly sophisticated nanofluidic devices. While the selectivity to a certain type of ions is due to a combination of steric effects and electrostatics, gating, that is, the process of opening and closing the channels, has been recently related to capillary evaporation, that is, the formation of a vapor or gas bubble inside the channel, which blocks the ion exchange.[261] This model for gating provides new insight into how narcotics work.[262]

Wetting phenomena on the nanoscale are also important in less obvious situations such as protein folding. In their native state, proteins are folded such that their hydrophobic amino acids are located in the interior of the molecule and thus avoid contact with the surrounding water.[263] Therefore capillary forces influence protein folding. This has been phrased only recently in terms of an entropy argument.[264] The combination of integralgeometric methods and advanced DFT allows one to calculate efficiently the contribution of the solvent to the total free-energy balance of the protein in a specific configuration.[265-269] This is a first necessary step in order to understand the folding dynamics of proteins in solution.

In summary, understanding wetting phenomena on the nanoscale is key to the miniaturization of microfluidic systems down to the nanoscale as well as a prerequisite for further progress in many active areas in biology and even medicine.

References

1. E. M. Blokhuis and B. Widom, Wetting, *Curr. Opin. Colloid Interface Sci.*, **1**, 424–429 (1996).
2. T. Young, An essay on the cohesion of fluids, *Philos. Trans. R. Soc. Lond.*, **95**, 65–87 (1805).
3. S. Dietrich, Wetting phenomena. In eds., C. Domb and J. L. Lebowitz, *Phase Transitions and Critical Phenomena*. Vol. 12. London: Academic Press. 1–218 (1988).

4. E. Lauga, M. P. Brenner, and H. A. Stone, The no-slip boundary condition: a review. In eds., C. Tropea, A. L. Yarin, and J. F. Foss, *Springer Handbook of Experimental Fluid Mechanics, Part C*. Berlin: Springer. 1219–1240 (2007).
5. L. Bocquet and J.-L. Barrat, Hydrodynamic boundary conditions, correlation functions, and Kubo relations for confined fluids, *Phys. Rev. E*, **49**, 3079–3092 (1994).
6. S. Richardson, On the no-slip boundary condition, *J. Fluid Mech.*, **59**, 707–719 (1973).
7. R. Mukhopadhyay, What does nanofluidics have to offer? *Anal. Chem.*, **78**, 7379–7382 (2006).
8. J. C. T. Eijkel and A. van den Berg, Nanofluidics: what is it and what can we expect from it? *Microfluid. Nanofluid.*, **1**, 249–267, (2005).
9. P. Mitchell, Microfluidics—downsizing large-scale biology, *Nature Biotech.*, **19**, 717–721 (2001).
10. T. Thorsen, S. J. Maerkl, and S. R. Quake, Microfluidic large scale integration, *Science*, **298**, 580–584 (2002).
11. P. G. de Gennes, Wetting: statistics and dynamics, *Rev. Mod. Phys.*, **57**, 827–860 (1985).
12. R. Evans, Microscopic theories of simple fluids and their interfaces. In eds., J. Charvolin, J. F. Joanny, and J. Zinn-Justin, *Liquids at Interfaces, Les Houches Summer School Session XLVIII*. Amsterdam: North-Holland. 1–98 (1990). ISBN: 0-444-88450-5
13. M. Schick, An introduction to wetting phenomena. In eds., J. Charvolin, J. F. Joanny, and J. Zinn-Justin, *Liquids at Interfaces, Les Houches Summer School Session XLVIII*. Amsterdam: North-Holland. 415–497 (1990). ISBN: 0-444-88450-5
14. M. Rauscher and S. Dietrich, Wetting phenomena in nanofluidics, *Ann. Rev. Mater. Res.*, **38**, 143–172 (2008).
15. D. Bonn, J. Eggers, J. Indekeu, J. Meunier, and E. Rolley, Wetting and spreading, *Rev. Mod. Phys.*, **81**, 739–805 (2009).
16. K. Ragil, J. Meunier, D. Broseta, J. Indekeu, and D. Bonn, Experimental observation of critical wetting, *Phys. Rev. Lett.*, **77**, 1532–1535 (1996).
17. D. Ross, D. Bonn, and J. Meunier, Observation of short-range critical wetting, *Nature*, **400**, 737–739 (1999).
18. J. E. Rutledge and P. Taborek, Prewetting phase diagram of ^{4}He on cesium, *Phys. Rev. Lett.*, **69**, 937–940 (1992).

19. P. Taborek and J. E. Rutledge, Tuning the wetting transition: prewetting and superfluidity of ^{4}He on thin cesium substrates, *Phys. Rev. Lett.*, **71**, 263–266 (1993).

20. H. Kellay, D. Bonn, and J. Meunier, Prewetting in a binary liquid mixture, *Phys. Rev. Lett.*, **71**, 2607–2610 (1993).

21. G. Mistura, H. C. Lee, and M. H. W. Chan, Hydrogen adsorption on alkali metal surfaces: wetting, prewetting and triple-point wetting, *J. Low Temp. Phys.*, **96**, 221–244 (1994).

22. D. Nicolaides and R. Evans, Nature of the prewetting critical point, *Phys. Rev. Lett.*, **63**, 778–781 (1989).

23. T. Pompe and S. Herminghaus, Three-phase contact line energetics from nanoscale liquid surface topographies, *Phys. Rev. Lett.*, **85**, 1930–1933 (2000).

24. T. Pompe, Line tension behavior of a first-order wetting system, *Phys. Rev. Lett.*, **89**, 076102 (2002).

25. J. S. Rowlinson and B. Widom, *Molecular Theory of Capillarity*. Mineola, NY: Dover (2002). ISBN: 0-486-42544-4

26. A. K. Rusanov, A. I. Shchekin, and D. V. Tatyanenko, The line tension and the generalized Young equation: the choice of dividing surface, *Colloids Surf. A*, **250**, 263–268 (2004).

27. A. Checco, P. Guenoun, and J. Daillant, Nonlinear dependence of the contact angle of nanodroplets on contact line curvature, *Phys. Rev. Lett.*, **91**, 186101 (2003).

28. A. Checco, H. Schollmeyer, J. Daillant, P. Guenoun, and R. Boukherroub, Nanoscale wettability of self-assembled monolayers investigated by noncontact atomic force microscopy, *Langmuir*, **22**, 116–126, (2006).

29. L. Schimmele, M. Napiórkowski, and S. Dietrich, Conceptual aspects of line tensions, *J. Chem. Phys.*, **127**, 164715 (2007)

30. A. Mortensen, Viewpoint set no. 46: triple lines, *Scripta Mater.*, **62**, 887–972 (2010).

31. M. E. Fisher and A. J. Jin, Effective potentials, constraints, and critical wetting theory, *Phys. Rev. B*, **44**, 1430–1433 (1991).

32. K. Rejmer and M Napiórkowski, Curvature contributions to the capillary-wave Hamiltonian for a pinned interface, *Phys. Rev. E*, **53**, 881–895 (1996).

33. A. O. Parry, C. Rascón, N. R. Bernardino, and J. M. Romero-Enrique, Derivation of a non-local interfacial Hamiltonian for short-ranged wetting: I. double-parabola approximation, *J. Phys.: Condens. Matter*, **18**, 6433–6452 (2006).

34. A. O. Parry, C. Rascón, N. R. Bernardino, and J. M. Romero-Enrique, The critical wetting saga: how to draw the correct conclusions, *J. Phys.: Condens. Matter*, **20**, 494234 (2008).

35. N. R. Bernardino, A. O. Parry, C. Rascón, and J. M. Romero-Enrique, Derivation of a non-local interfacial model for 3D wetting in an external field, *J. Phys.: Condens. Matter*, **21**, 465105 (2009).

36. K. Binder, D. P. Landau, and D. M. Kroll, Critical wetting with short-range forces: is meanfield theory valid? *Phys. Rev. Lett.*, **56**, 2272–2275 (1986).

37. K. Binder and D. P. Landau, Wetting and layering in the nearest-neighbor simple-cubic Ising lattice: a Monte Carlo investigation, *Phys. Rev. B*, **37**, 1745–1765 (1988).

38. D. P. Binder, K. Landau and A. M. Ferrenberg, Character of the phase transition in thin Ising films with competing walls, *Phys. Rev. Lett.*, **74**, 298–301 (1995).

39. K. Binder, D. P. Landau, and A. M. Ferrenberg, Thin Ising films with competing walls: a Monte Carlo study, *Phys. Rev. E*, **51**, 2823–2838 (1995).

40. E. Brézin, B. I. Halperin, and S. Leibler, Critical wetting in three dimensions, *Phys. Rev. Lett.*, **50**, 1387–1390 (1983).

41. E. Brézin, B. I. Halperin, and S. Leibler, Critical wetting: the domain of validity of mean field theory, *J. Phys. (Paris)*, **44**, 775–783 (1983).

42. D. S. Fisher and D. A. Huse, Wetting transitions: a functional renormalization-group approach, *Phys. Rev. B*, **32**, 247–256 (1985).

43. G. Gompper and D. M. Kroll, Critical behaviour of effective interface models for wetting in three dimensions, *EPL*, **5**, 49–53, (1988).

44. C. J. Boulter, Interface Hamiltonian with a position-dependent stiffness: a nonlinear functional renormalization group study, *Phys. Rev. E*, **57**, 2062–2072 (1998).

45. S. Dietrich and M. Napiórkowski, Analytic results for wetting transitions in the presence of van der Waals tails, *Phys. Rev. A*, **43**, 1861–1885 (1991).

46. G. Forgacs, R. Lipowsky, and T. Nieuwenhuizen, The behavior of interfaces in ordered and disordered systems. In eds., C. Domb and J. L. Lebowitz, *Phase Transitions and Critical Phenomena*. Vol. 14. London: Academic Press. 135–363 (1991). ISBN: 0-12-220314-3

47. M. E. Fisher, Interface wandering in adsorbed and bulk phases, pure and impure, *J. Chem. Soc.: Faraday Trans. 2*, **82**, 1569–1603, (1986).

48. A. O. Parry, A. J. Wood, and C. Rascón, Wedge filling, cone filling and the strong-fluctuation regime, *J. Phys.: Condens. Matter*, **13**, 4591–4613 (2001).

49. B. V. Derjaguin and N. V. Churaev, Properties of water layers adjacent to interfaces. In ed., C. A. Croxton, *Fluid Interfacial Phenomena*. Chichester: Wiley. 663–738 (1986).

50. A. Frumkin, On the phenomena of wetting and sticking of bubbles, *Zh. Fiz. Khim.*, **12**, 337–345 (1938).

51. J. R. Henderson, Statistical mechanics of the disjoining pressure of a planar film, *Phys. Rev. E*, **72**, 051602 (2005).

52. R. Lipowsky, Upper critical dimension for complete wetting in systems with long-range forces, *Phys. Rev. Lett.*, **52**, 1429–1432, (1984).

53. S. Dietrich, P. Nightingale, and M. Schick, Applicability of meanfield theory to wetting in binary-liquid mixtures, *Phys. Rev. B*, **32**, 3182–3185 (1985).

54. D. M. Kroll, R. Lipowsky, and R. K. P. Zia, Universality classes for critical wetting, *Phys. Rev. B*, **32**, 1862–1865 (1985).

55. R. Lipowsky and M. E. Fisher, Scaling regimes and functional renormalization for wetting transitions, *Phys. Rev. B*, **36**, 2126–2141 (1987).

56. A. O. Parry, C. Rascón, N. R. Bernardino, and J. M. Romero-Enrique, Derivation of a non-local interfacial Hamiltonian for short-ranged wetting: II. general diagrammatic structure, *J. Phys.: Condens. Matter*, **19**, 416105 (2007).

57. A. O. Parry and C. Rascón, The trouble with critical wetting, *J. Low Temp. Phys.*, **157**, 149–173 (2009).

58. A. O. Parry and C. Rascón, Local scale invariance for wetting and confined interfaces, *J. Phys. A: Math. Gen.*, **43**, 125002 (2010).

59. A. Checco, O. Gang, and B. M. Ocko, Liquid nanostripes, *Phys. Rev. Lett.*, **96**, 056104 (2006).

60. R. C. Tolman, The effect of droplet size on surface tension, *J. Chem. Phys.*, **17**, 333–337 (1949).

61. L. Schimmele and S. Dietrich, Line tension and the shape of nanodroplets, *Eur. Phys. J. E*, **30**, 427–430 (2009).

62. J. A. De Feijter and A. Vrij, I. Transition regions, line tensions and contact angles in soap films, *J. Electroanal. Chem.*, **37**, 9–22 (1972).

63. J. O. Indekeu, Line tension near the wetting transition: results from an interface displacement model, *Physica A*, **183**, 439–461 (1992).

64. T. Getta and S. Dietrich, Line tension between fluid phases and a substrate, *Phys. Rev. E*, **57**, 655–671 (1998).
65. C. Bauer and S. Dietrich, Quantitative study of laterally inhomogeneous wetting films, *Eur. Phys. J. B*, **10**, 767–779 (1999).
66. R. Rosso and E. G. Virga, General stability criterion for wetting, *Phys. Rev. E*, **68**, 012601 (2003).
67. R. Rosso and E. G. Virga, Sign of line tension in liquid bridge stability, *Phys. Rev. E*, **70**, 031603 (2004).
68. M. Brinkmann, J. Kierfeld, and R. Lipowsky, Stability of liquid channels or filaments in the presence of line tension, *J. Phys.: Condens. Matter*, **17**, 2349–2364 (2005).
69. S. Mechkov, G. Oshanin, M. Rauscher, M. Brinkmann, A. M. Cazabat, and S. Dietrich, Contact line stability of ridges and drops, *EPL*, **80**, 66002 (2007).
70. P. Jakubczyk and M. Napiórkowski, The influence of droplet size on line tension, *J. Phys.: Condens. Matter*, **16**, 6917–6928, (2004).
71. P. Jakubczyk, M. Napiórkowski, and A. O. Parry, Point tension in adsorption at a chemically inhomogeneous substrate in two dimensions, *Phys. Rev. E*, **74**, 031608 (2006).
72. P. Lenz, C. Bechinger, C. Schäfle, P. Leiderer, and R. Lipowsky, Perforated wetting layers from periodic patterns of lyophobic surface domains, *Langmuir*, **17**, 7814–7822 (2001).
73. R. Lipowsky, P. Lenz, and P. Swain, Wetting and dewetting of structured and imprinted surfaces, *Colloids Surf. A*, **161**, 3–22 (2000).
74. R. Lipowsky, Morphological wetting transitions at chemically structured surfaces, *Curr. Opin. Colloid Interface Sci.*, **6**, 40–48 (2001).
75. P. Lenz, W. Fenzl, and R. Lipowsky, Wetting of ring-shaped surface domains, *EPL*, **53**, 618–624 (2001).
76. R. J. Jackman, D. C. Duffy, E. Ostuni, N. D. Willmore, and G. M. Whitesides, Fabricating large arrays of microwells with arbitrary dimensions and filling them using discontinuous dewetting, *Anal. Chem.*, **70**, 2280–2287 (1998).
77. S. Dietrich, Fluids in contact with structured substrates. In eds., C. Caccamo, J. P. Hansen, and G. Stell, *New Approaches to Old and New Problems in Liquid State Theory*. Vol. C529. *NATO-ASI Series C*. Dordrecht: Kluwer. 197–244 (1999).
78. M. P. Gelfand and R. Lipowsky, Wetting on cylinders and spheres, *Phys. Rev. B*, **36**, 8725–8735 (1987).

79. R. Hołyst and A. Poniewierski, Wetting on a spherical surface, *Phys. Rev. B*, **36**, 5628–5630 (1987).
80. T. Bieker and S. Dietrich, Wetting of curved surfaces, *Physica A*, **252**, 85–137 (1998).
81. C. Bauer and S. Dietrich, Shapes, contact angles, and line tensions of droplets on cylinders, *Phys. Rev. E*, **62**, 2428–2438 (2000).
82. C. Bauer, T. Bieker, and S. Dietrich, Wetting-induced effective interaction potential between spherical particles, *Phys. Rev. E*, **62**, 5324–5338 (2000).
83. H. B. Eral, J. de Ruiter, R. de Ruiter, J. M. Oh, C. Semprebon, M. Brinkman, and F. Mugele, Drops on functional fibers: from barrels to clamshells and back, *Soft Matter*, **7**, 5138–5143 (2011).
84. A. Tuteja, W. Choi, M. Ma, J. M. Mabry, S. A. Mazzella, G. C. Rutledge, G. H. McKinley, and R. E. Cohen, Designing superoleophobic surfaces, *Science*, **318**, 1618–1622 (2007).
85. A. Tuteja, W. Choi, J. M. Mabry, G. H. McKinley, and R. E. Cohen, Robust omniphobic surfaces, *PNAS*, **105**, 18200–18205 (2008).
86. P. Jakubczyk and M. Napiórkowski, Adsorption in a nonsymmetric wedge, *Phys. Rev. E*, **66**, 041107 (2002).
87. P. Abgrall and N. T. Nguyen, Nanofluidic devices and their applications, *Anal. Chem.*, **80**, 2326–2341 (2008).
88. D. Li, *Encyclopedia of Microfluidics and Nanofluidics*. New York: Springer (2008). ISBN: 978-0-387-49000-7
89. S. Prakash, A. Piruska, E. N. Gatimu, P. W. Bohn, J. V. Sweedler, and M. A. Shannon, Nanofluidics: systems and applications, *IEEE Sens. J.*, **8**, 441–450 (2008).
90. M. L. Kovarik and S. C. Jacobson, Nanofluidics in lab-on-a-chip devices, *Anal. Chem.*, **81**, 7133–7140 (2009).
91. P. Lenz and R. Lipowsky, Morphological transitions of wetting layers on structured surfaces, *Phys. Rev. Lett.*, **80**, 1920–1923 (1998).
92. H. Gau, S. Herminghaus, P. Lenz, and R. Lipowsky, Liquid morphologies on structured surfaces: from microchannels to microchips, *Science*, **283**, 46–49 (1999).
93. K. Rejmer and M. Napiórkowski, Adsorption on a periodically corrugated substrate, *Phys. Rev. E*, **62**, 588–596 (2000).
94. M. Brinkmann, J. Kierfeld, and R. Lipowsky, A general stability criterion for droplets on structured substrates, *J. Phys. A: Math. Gen.*, **37**, 11547–11573 (2004).

95. M. Seemann, R. Brinkmann, F. F. Kramer, E. J. Lange, and R. Lipowsky, Wetting morphologies at microstructured surfaces, *PNAS*, **102**, 1848–1852 (2005).
96. R. N. Wenzel, Resistance of solid surfaces to wetting by water, *Ind. Eng. Chem.*, **28**, 988–994 (1936).
97. R. N. Wenzel, Surface roughness and contact angle, *J. Phys. Colloid Chem.*, **53**, 1466–1467 (1949).
98. A. B. D. Cassie and S. Baxter, Wettability of porous surfaces, *Trans. Faraday Soc.*, **40**, 546–551 (1944).
99. P. S. Swain and R. Lipowsky, Contact angles on heterogeneous surfaces: a new look at Cassie's and Wenzel's laws, *Langmuir*, **14**, 6772–6780 (1998).
100. G. McHale, Cassie and Wenzel: were they really so wrong? *Langmuir*, **23**, 8200–8205 (2007).
101. E. M. Grzelak and J. R. Errington, Nanoscale limit to the applicability of Wenzel's equation, *Langmuir*, **26**, 13297–13304 (2010).
102. M. Brinkmann and R. Lipowsky, Wetting morphologies on substrates with striped surface domains, *J. Appl. Phys.*, **92**, 4296–4306 (2002).
103. G. M. Whitesides and A. D. Stroock, Flexible methods for microfluidics, *Phys. Today*, **54**, 42–48 (2001).
104. W. Koch, S. Dietrich, and M. Napiórkowski, Morphology and line tension of liquid films adsorbed on chemically structured substrates, *Phys. Rev. E*, **51**, 3300–3317 (1995).
105. C. Bauer and S. Dietrich, Wetting films on chemically heterogeneous substrates, *Phys. Rev. E*, **60**, 6919–6941 (1999).
106. C. Bauer, S. Dietrich, and A. O. Parry, Morphological phase transitions of thin fluid films on chemically structured substrates, *EPL*, **47**, 474–480 (1999).
107. C. Bauer and S. Dietrich, Phase diagram for morphological transitions of wetting films on chemically structured substrates, *Phys. Rev. E*, **61**, 1664–1669 (2000).
108. R. Evans, Density functionals in the theory of nonuniform fluids. In ed., D. Henderson, *Fundamentals of Inhomogeneous Fluids*. New York: Marcel Dekker. 85–173 (1992).
109. J. Wu, Density functional theory for chemical engineering: from capillarity to soft materials, *AIChE J.*, **52**, 1169–1193 (2006).
110. R. Roth, Fundamental measure theory for hard-sphere mixtures: a review, *J. Phys.: Condens. Matter*, **22**, 063102 (2010).

111. M. O. Robbins, D. Andelman, and J.-F. Joanny, Thin liquid films on rough or heterogeneous solids, *Phys. Rev. A*, **43**, 4344–4354 (1991).

112. M. Krech, *The Casimir effect in critical systems*. Singapore: World Scientific (1994). ISBN: 981-02-1845-1

113. H. W. Diehl, Field-theoretical approach to critical behaviour at surfaces. In eds., C. Domb and J. L. Lebowitz, *Phase Transitions and Critical Phenomena*. Vol. 10. London: Academic Press. 75–267 (1986).

114. M. Sprenger, F. Schlesener, and S. Dietrich, Critical adsorption at chemically structured substrates, *Phys. Rev. E*, **71**, 056125 (2007).

115. R. Lipowsky, M. Brinkmann, R. Dimova, T. Franke, J. Kierfeld, and X. Zhand, Droplets, bubbles, and vesicles at chemically structured surfaces, *J. Phys.: Condens. Matter*, **17**, S537–S558 (2005).

116. F. Dörfler, *Statics and Dynamics of Simple Fluids on Chemically Patterned Substrates*, PhD thesis. Universität Stuttgart (2010).

117. J. M. Davis and S. M. Troian, Generalized linear stability of noninertial coating flows over topographical features, *Phys. Fluids*, **17**, 072103 (2005).

118. S. Mechkov, M. Rauscher, and S. Dietrich, Stability of liquid ridges on chemical micro- and nanostripes, *Phys. Rev. E*, **77**, 061605 (2008).

119. J. Koplik, T. S. Lo, M. Rauscher, and S. Dietrich, Pearling instability of nanoscale fluid flow confined to a chemical channel, *Phys. Fluids*, **18**, 032104 (2006).

120. M. Rauscher, S. Dietrich, and J. Koplik, Shear flow pumping in open microfluidic systems, *Phys. Rev. Lett.*, **98**, 224504 (2007).

121. A. A. Darhuber, S. M. Troian, S. M. Miller, and S. Wagner, Morphology of liquid microstructures on chemically patterned surfaces, *J. Appl. Phys.*, **87**, 7768–7775 (2000).

122. C. Schäfle, *Morphologie, Verdampfung und Kondensation von Flüssigkeiten auf benetzungsstrukturierten Oberflächen*, PhD thesis. Universität Konstanz (2002).

123. F. Porcheron, P. A. Monson, and M. Schoen, Wetting of rings on a nanopatterned surface: a lattice model study, *Phys. Rev. E*, **73**, 041603 (2006).

124. C. Schäfle, M. Brinkmann, C. Bechinger, P. Leiderer, and R. Lipowsky, Morphological wetting transitions at ring shaped surface domains, *Langmuir*, **26**, 11878–11885 (2010).

125. P. Concus and R. Finn, On the behavior of a capillary surface in a wedge, *PNAS*, **63**, 292–299 (1969).

126. Y. Pomeau, Wetting in a corner and related questions, *J. Colloid Interface Sci.*, **113**, 5–11 (1986).
127. E. H. Hauge, Macroscopic theory of wetting in a wedge, *Phys. Rev. A*, **46**, 4994–4998 (1992).
128. M. Napiórkowski, W. Koch, and S. Dietrich, Wedge wetting by van der Waals fluids, *Phys. Rev. A*, **45**, 5760–5770 (1992).
129. K. Rejmer, S. Dietrich, and M. Napiórkowski, Filling transition for a wedge, *Phys. Rev. E*, **60**, 4027–4042 (1999).
130. E. S. Hernández, F. Ancilotto, M. Barranco, R. Mayol, and M. Pi, Condensation of helium in wedges. In eds., H. Reinholz, G. Roepke, and M. de Llano, *Condensed Matter Thoeries*. Vol. 22. Singapore: World Scientific. 25–34 (2007). ISBN: 978-9812709585
131. M. D. Re, S. Grubišić, and F. Ancilotto, Monte Carlo simulations of argon adsorption in nanoscopic linear channels, *Phys. Rev. B*, **81**, 205427 (2010).
132. M. Schoen and S. Dietrich, Structure of a hard-sphere fluid in hard wedges, *Phys. Rev. E*, **56**, 499–510 (1997).
133. A. O. Parry, C. Rascón, and A. J. Wood, Universality for 2D wedge wetting, *Phys. Rev. Lett.*, **83**, 5535–5538 (1999).
134. C. Rascón and A. O. Parry, Geometry dependent critical exponents at complete wetting, *J. Chem. Phys.*, **112**, 5175–5180 (2000).
135. L. Bruschi, A. Carlin, and G. Mistura, Wetting on a geometrically structured substrate, *J. Chem. Phys.*, **115**, 6200–6203 (2001).
136. A. O. Parry, M. J. Greenall, and A. J. Wood, Wedge covariance for two-dimensional filling and wetting, *J. Phys.: Condens. Matter*, **14**, 1169–1198 (2002).
137. N. R. Bernardino, A. O. Parry, and J. M. Romero-Enrique, *J. Phys.: Condens. Matter*, **24**, 182202 (2012).
138. A. O. Parry, C. Rascón, and A. J. Wood, Critical effects at 3D wedge wetting, *Phys. Rev. Lett.*, **85**, 345–348 (2000).
139. A. Bednorz and M. Napiórkowski, Interfacial fluctuations near the critical filling transition, *Phys. Rev. E*, **63**, 031602 (2001).
140. C. Rascón and A. O. Parry, Covariance for cone and wedge complete filling, *Phys. Rev. Lett.*, **94**, 096103 (2005).
141. C. Rascón and A. O. Parry, Geometry-dominated fluid adsorption on sculpted solid substrates, *Nature*, **407**, 986–989 (2000).
142. L. Bruschi, A. Carlin, and G. Mistura, Complete wetting on a linear wedge, *Phys. Rev. Lett.*, **89**, 166101 (2002).

143. A. O. Parry, C. Rascón, N. B. Wilding, and R. Evans, Condensation in a capped capillary is a continuous critical phenomenon, *Phys. Rev. Lett.*, **98**, 226101 (2007).

144. A. O. Parry and R. Evans, Influence of wetting on phase equilibria: a novel mechanism for critical-point shifts in films, *Phys. Rev. Lett.*, **64**, 439–442 (1990).

145. A. O. Parry and R. Evans, Novel phase behaviour of a confined fluid or Ising magnet, *Physica A*, **181**, 250–296 (1992).

146. M. Tasinkevych and S. Dietrich, Complete wetting of nanosculptured substrates, *Phys. Rev. Lett.*, **97**, 106102 (2006).

147. M. Tasinkevych and S. Dietrich, Complete wetting of pits and grooves, *Eur. Phys. J. E*, **23**, 117–128 (2007).

148. O. Gang, K. J. Alvine, M. Fukuto, P. S. Pershan, C. T. Black, and B. M. Ocko, Liquids on topologically nanopatterned surfaces, *Phys. Rev. Lett.*, **95**, 217801 (2005).

149. T. Hofmann, M. Tasinkevych, A. Checco, E. Dobisz, S. Dietrich, and B. M. Ocko, Wetting of nanopatterned grooved surfaces, *Phys. Rev. Lett.*, **104**, 106102 (2010).

150. A. Checco, T. Hofmann, E. DiMasi, C. T. Black, and B. M. Ocko, Morphology of air nanobubbles trapped at hydophobic nanopatterened surfaces, *Nano Lett.*, **10**, 1354–1358 (2010).

151. S. Herminghaus, M. Brinkmann, and R. Seemann, Wetting and dewetting of complex surface geometries, *Annu. Rev. Mater. Res.*, **38**, 101–121 (2008).

152. M. Brinkmann and R. Blossey, Blobs, channels and "cigars": morphologies of liquids at a step, *Eur. Phys. J. E*, **14**, 79–89 (2004).

153. K. Khare, S. Herminghaus, J.-C. Baret, B. M. Law, M. Brinkmann, and R. Seemann, Switching liquid morphologies on linear grooves, *Langmuir*, **23**, 12997–13006 (2007).

154. J.-C. Baret, M. Decré, S. Herminghaus, and R. Seemann, Electroactuation of fluid using topographical wetting transitions, *Langmuir*, **21**, 12218–12221 (2005).

155. J.-C. Baret, M. M. J. Decré, S. Herminghaus, and R. Seemann, Transport dynamics in open microfluidic grooves, *Langmuir*, **23**, 5200–5204 (2007).

156. A. Otten and S. Herminghaus, How plants keep dry: a physicist's point of view, *Langmuir*, **20**, 2405–2408 (2004).

157. F. Chiodi, B. Roman, and J. Bico, Piercing and interface with a brush: collaborative stiffening, *EPL*, **90**, 44006 (2010).

158. J. Fan and Y. Zhao, Freezing a water droplet on an aligned Si nanorod array substrate, *Nanotechnology*, **19**, 155707 (2008).
159. J. Fan and Y. Zhao, Nanocarpet effect induced superhydrophobicity, *Langmuir*, **26**, 8245–8250 (2010).
160. N. R. Bernardino, V. Blickle, and S. Dietrich, Wetting of surfaces covered by elastic hairs, *Langmuir*, **26**, 7233–7241 (2010).
161. M. J. Blow and J. M. Yeomans, Superphobicity on hairy surfaces. arXiv:1009.5322v1 (2010).
162. R. Evans, The nature of the liquid–vapour interface and other topics in the statistical mechanics of non-uniform, classical fluids, *Adv. Phys.*, **28**, 143–200 (1979).
163. J.-P. Hansen and I. R. McDonald, *Theory of Simple Liquids*. 2nd ed. London: Academic Press (1990). ISBN: 0-12-323852-8
164. G. K.-L. Chan and R. Finken, Time-dependent density functional theory of classical fluids, *Phys. Rev. Lett.*, **94**, 183001 (2005).
165. U. M. B. Marconi and P. Tarazona, Dynamic density functional theory of fluids, *J. Chem. Phys.*, **110**, 8032–8044 (1999).
166. U. M. B. Marconi and P. Tarazona, Dynamic density functional theory of fluids, *J. Phys.: Condens. Matter*, **12**, A413–A418 (2000).
167. A. J. Archer and M. Rauscher, Dynamical density functional theory for interacting brownian particles: stochastic or deterministic? *J. Phys. A: Math. Gen.*, **37**, 9325–9333 (2004).
168. P. Español and H. Löwen, Derivation of dynamical density functional theory using the projection operator technique, *J. Chem. Phys.*, **131**, 244101 (2009).
169. M. Rauscher, DDFT for Brownian particles and hydrodynamics, *J. Phys.: Condens. Matter*, **22**, 364109 (2010).
170. A. J. Archer, Dynamical density functional theory for dense atomic liquids, *J. Phys.: Condens. Matter*, **18**, 5617–5628 (2006).
171. U. M. B. Marconi and S. Melchionna, Kinetic theory of correlated fluids: from dynamic density functional to lattice Boltzmann methods, *J. Chem. Phys.*, **131**, 014105 (2009).
172. L. D. Landau and E. M. Lifshitz, *Fluid Mechanics*. Vol. 6, 2nd ed. *Course of Theoretical Physics*. Amsterdam: Elsevier Butterworth-Heinemann (2005).
173. V. M. Zaĭtsev and M. I. Shliomis, Hydrodynamic fluctuations near the convection threshold, *Soviet Phys. JETP*, **32**, 866–870 (1971).
174. R. B. Bird, R. C. Armstrong, and O. Hassager, *Dynamics of Polymeric Fluids*. Vol. 1. New York: John Wiley & Sons (1977).

175. K. Mecke and M. Rauscher, On thermal fluctuations in thin film flow, *J. Phys.: Condens. Matter*, **17**, S3515–S3522 (2005).

176. G. Grün, K. R. Mecke, and M. Rauscher, Thin-film flow influenced by thermal noise, *J. Stat. Phys.*, **122**, 1261–1291 (2006).

177. K. Mecke, K. Falk, and M. Rauscher, *Dynamics of nanoscopic capillary waves*. In eds., G. Radons, B. Rumpf, and H. G. Schuster, *Nonlinear Dynamics of Nanosystems*, Berlin: Wiley-VCH. 121–142 (2010). ISBN: 978-3-527-40791-0

178. C. L. M. H. Navier, Mémoire sur les lois du mouvement des fluides, *Mémoire de l'Académie Royale des Sciences de lÍnstitut de France*, **6**, 389–440 (1823).

179. A. Oron, S. H. Davis, and S. G. Bankoff, Long-scale evolution of thin liquid films, *Rev. Mod. Phys.*, **69**, 931–980 (1997).

180. A. Münch and B. Wagner, Contact-line instability of dewetting thin films, *Physica D*, **209**, 178–190 (2005).

181. B. Davidovitch, E. Moro, and H. A. Stone, Spreading of thin films assisted by thermal fluctuations, *Phys. Rev. Lett.*, **95**, 244505 (2005).

182. A. Moosavi, M. Rauscher, and S. Dietrich, Motion of nanodroplets near edges and wedges, *Phys. Rev. Lett.*, **97**, 236101 (2006).

183. A. Moosavi, M. Rauscher, and S. Dietrich, Motion of nanodroplets near chemical heterogeneities, *Langmuir*, **24**, 734–742 (2008).

184. A. Moosavi, M. Rauscher, and S. Dietrich, Size dependent motion of nanodroplets on chemical steps, *J. Chem. Phys.*, **129**, 044706 (2008).

185. X. Shan and H. Chen, Lattice Boltzmann model for simulating flows with multiple phases and components, *Phys. Rev. E*, **47**, 1815–1819 (1993).

186. X. Shan and H. Chen, Simulation of nonideal gases and liquid-gas phase transitions by the lattice Boltzmann equation, *Phys. Rev. E*, **49**, 2941–2948 (1994).

187. M. R. Swift, W. R. Osborn, and J. M. Yeomans, Lattice Boltzmann simulation of nonideal fluids, *Phys. Rev. Lett.*, **75**, 830–833 (1995).

188. M. R. Swift, E. Orlandini, W. R. Osborn, and J. M. Yeomans, Lattice Boltzmann simulations of liquid gas and binary fluid systems, *Phys. Rev. E*, **54**, 5041–5052 (1996).

189. A. Dupuis and J. M. Yeomans, Mesoscopic modeling of droplets on topologically patterned substrates. In eds., M. Bubak, G. D. van Albada, P. M. A. Sloot, and J. J. Dongarra, *Computational Science—ICCS 2004*. Vol. 3039. *Lecture Notes in Computer Science*. Berlin: Springer. 556–563 (2004).

190. M. Gross, R. Adhikari, M. E. Cates, and F. Varnik, Thermal fluctuations in the lattice Boltzmann method for non-ideal fluids. arXiv:1009.0254.

191. A. Malevanets and R. Kapral, Mesoscopic model for solvent dynamics, *J. Chem. Phys.*, **110**, 8605–8613 (1999).

192. A. Malevanets and R. Kapral, Solute molecular dynamics in a mesoscale solvent, *J. Chem. Phys.*, **112**, 7260–7269 (2000).

193. E. Tüzel, G. Pan, T. Ihle, and D. M. Kroll, Mesoscopic model for the fluctuating hydrodynamics of binary and ternary mixtures, *EPL*, **80**, 40010 (2007).

194. E. Tüzel, G. Pan, and D. M. Kroll, Dynamics of thermally driven capillary waves for two-dimensional droplets, *J. Chem. Phys.*, **132**, 174701 (2010).

195. J. Harting, C. Kunert, and H. J. Herrmann, Lattice Boltzmann simulations of apparent slip in hydrophobic microchannels, *EPL*, **75**, 328–334 (2006).

196. J. K. Whitmer and E. Luijten, Fluid-solid boundary conditions for multiparticle collision dynamics, *J. Phys.: Condens. Matter*, **22**, 104106 (2010).

197. D. Frenkel and B. Smit, *Understanding Molecular Simulation*. 2nd ed. San Diego: Academic Press (2002).

198. J. De Coninck and T. D. Blake, Wetting and molecular dynamics simulations of simple liquids, *Ann. Rev. Mater. Res.*, **38**, 1–22 (2008).

199. J. J. de Pablo and W. A. Curtin, Multiscale modeling in advanced materials research: challenges, novel methods, and emerging applications, *MRS Bull.*, **32**, 905–911 (2007).

200. L. M. Pismen and B. Y. Rubinstein, Kinetic slip condition, van der Waals forces, and dynamic contact angle, *Langmuir*, **17**, 5265–5270 (2001).

201. T. Qian, X.-P. Wang, and P. Sheng, Power-law slip profile of the moving contact line in two-phase immiscible flows, *Phys. Rev. Lett.*, **93**, 094501 (2004).

202. E. B. Dussan V., On the spreading of liquids on solid surfaces: static and dynamic contact lines, *Ann. Rev. Fluid Mech.*, **11**, 371–400 (1979).

203. J. Ralston, M. Popescu, and R. Sedev, Dynamics of wetting from an experimental point of view, *Ann. Rev. Mater. Res.*, **38**, 23–43 (2008).

204. M. Voué and J. De Coninck, Spreading and wetting at the microscopic scale: recent developments and perspectives, *Acta Mater.*, **48**, 4405–4417 (2000).

205. J. De Coninck, M. J. de Ruijter, and M. Voué, Dynamics of wetting, *Curr. Opin. Colloid Interface Sci.*, **6**, 49–53 (2001).

206. J. Koplik, J. R. Banavar, and J. F. Willemsen, Molecular dynamics of Poiseuille flow and moving contact lines, *Phys. Rev. Lett.*, **60**, 1282–1285 (1988).

207. J. Koplik, J. R. Banavar, and J. F. Willemsen, Molecular dynamics of fluid flow at solid surfaces, *Phys. Fluids A: Fluid Dyn.*, **1**, 781–794 (1989).

208. L. H. Tanner, The spreading of silicone oil drops on horizontal surfaces, *J. Phys. D: Appl. Phys.*, **12**, 1473–1484 (1979).

209. T. D. Blake and J. M. Haynes, Kinetics of liquid/liquid displacement, *J. Colloid Interface Sci.*, **30**, 421–423 (1969).

210. M. J. de Ruijter, M. Charlot, M. Voué, and J. De Coninck, Experimental evidence of several time scales in drop spreading, *Langmuir*, **16**, 2363–2368 (2000).

211. F. Heslot, N. Fraysse, and A. M. Cazabat, Molecular layering in the spreading of wetting liquid drops, *Nature*, **338**, 640–642 (1989).

212. M. N. Popescu and S. Dietrich, Spreading of liquid monolayers: from kinetic Monte Carlo simulations to continuum limit. In eds., H. Emmerich, B. Nestler, and M. Schreckenberg, *Interface and Transport Dynamics*. Heidelberg: Springer. 202–207 (2003). ISBN: 3-540-40367-1

213. M. N. Popescu and S. Dietrich, Model for spreading of liquid monolayers, *Phys. Rev. E*, **69**, 061602 (2004).

214. R. Blossey, Thin film rupture and polymer flow, *Phys. Chem. Chem. Phys.*, **10**, 5177–5183 (2000).

215. M. Rauscher and S. Dietrich, Nanofluidics of thin liquid films. In ed., K. D. Sattler, *Handbook of Nanophysics*. Vol. I: *Principles and Methods*. Boca Raton: CRC Press (2010).

216. R. Bausch, R. Blossey, and M. A. Burschka, Critical nuclei for wetting and dewetting, *J. Phys. A: Math. Gen.*, **27**, 1405–1406 (1994).

217. G. Foltin, R. Bausch, and R. Blossey, Critical holes in undercooled wetting layers, *J. Phys. A: Math. Gen.*, **30**, 2937–2946 (1997).

218. K. Jacobs, R. Seemann, and K. Mecke, Dynamics of structure formation in thin films: a special spatial analysis. In eds., K. R. Mecke and D. Stoyan, *Statistical Physics and Spatial Statistics*. Vol. 554. *Lecture Notes in Physics*, Berlin: Springer. 72–91 (2000). ISBN: 3-540-67750-X.

219. M. Rauscher, R. Blossey, A. Münch, and B. Wagner, Spinodal dewetting of thin films with large interfacial slip: implications from the dispersion relation, *Langmuir*, **24**, 12290–12294 (2008).

220. R. Seemann, S. Herminghaus, and K. Jacobs, Dewetting patterns and molecular forces: a reconciliation, *Phys. Rev. Lett.*, **86**, 5534–5537 (2001).

221. R. Seemann, S. Herminghaus, and K. Jacobs, Gaining control of pattern formation of dewetting liquid films, *J. Phys.: Condens. Matter*, **13**, 4925–4938 (2001).

222. R. Fetzer, M. Rauscher, R. Seemann, K. Jacobs, and K. Mecke, Thermal noise influences fluid flow in thin films during spinodal dewetting, *Phys. Rev. Lett.*, **99**, 114503 (2007).

223. G. Grün and M. Rumpf, Nonnegativity preserving convergent schemes for the thin film equation, *Numer. Math.*, **87**, 113–152 (2000).

224. J. Becker and G. Grün, The thin-film equation: recent advances and some new perspectives, *J. Phys.: Condens. Matter*, **17**, S291–S307 (2005).

225. J. Becker, G. Grün, R. Seemann, H. Mantz, K. Jacobs, K. R. Mecke, and R. Blossey, Complex dewetting scenarios captured by thin film models, *Nat. Mater.*, **2**, 59–63 (2003).

226. K. Kargupta, A. Sharma, and R. Khanna, Instability, dynamics, and morphology of thin slipping films, *Langmuir*, **20**, 244–253 (2004).

227. F. Brochard-Wyart, H. H. P.-G. de Gennes, and C. Redon, Wetting and slippage of polymer melts on semi-ideal surfaces, *Langmuir*, **10**, 1566–1572 (1994).

228. R. Fetzer and K. Jacobs, Slippage of Newtonian liquids: influence on the dynamics of dewetting thin films, *Langmuir*, **23**, 11617–11622 (2007).

229. C. Neto, K. Jacobs, R. Seemann, R. Blossey, J. Becker, and G. Grün, Satellite hole formation during dewetting: experiment and simulation, *J. Phys.: Condens. Matter*, **15**, 3355–3366 (2003).

230. C. Neto, K. Jacobs, R. Seemann, R. Blossey, J. Becker, and G. Grün, Correlated dewetting patterns in thin polystyrene films, *J. Phys.: Condens. Matter*, **15**, S421–S426 (2003).

231. R. Fetzer, K. Jacobs, A. Münch, B. Wagner, and T. P. Witelski, New slip regimes and the shape of dewetting thin liquid films, *Phys. Rev. Lett.*, **95**, 127801 (2005).

232. R. Fetzer, M. Rauscher, A. Münch, B. A. Wagner, and K. Jacobs, Slip-controlled thin film dynamics, *EPL*, **75**, 638–644 (2006).

233. R. Fetzer, A. Münch, B. Wagner, M. Rauscher, and K. Jacobs, Quantifying hydrodynamic slip: a comprehensive analysis of dewetting profiles, *Langmuir*, **23**, 10559–10566 (2007).

234. S. Herminghaus, R. Seemann, and K. Jacobs, Generic morphologies of viscoelastic dewetting fronts, *Phys. Rev. Lett.*, **89**, 056101 (2002).

235. M. Rauscher, A. Münch, B. Wagner, and R. Blossey, A thin-film equation for viscoelastic liquids of Jeffreys type, *Eur. Phys. J. E*, **17**, 373–379 (2005).

236. A. Sharma and G. Reiter, Instability of thin polymer films on coated substrates: rupture, dewetting, and drop formation, *J. Colloid Interface Sci.*, **178**, 383–399 (1996).

237. S. Herminghaus, R. Seemann, D. Podzimek, and K. Jacobs, Strukturbildung und Dynamik in makromolekularen Filmen, *Nachrichten a. d. Chemie*, **49**, 1398–1404 (2001).

238. S. H. Davis, Moving contact line and rivulet instabilities. Part 1: the static rivulet, *J. Fluid Mech.*, **98**, 225–242 (1980).

239. R. R. Netz and D. Andelman, Roughness-induced wetting, *Phys. Rev. E*, **55**, 687–700 (1997).

240. D. Quéré, Wetting and roughness, *Ann. Rev. Mater. Res.*, **38**, 71–99 (2008).

241. R. Fetzer and J. Ralston, Influence of nanoroughness on contact line motion, *J. Phys. Chem. A*, **114**, 12675–12680 (2010).

242. D. Quéré, Non-sticking drops, *Rep. Prog. Phys.*, **68**, 2495–2532 (2005).

243. T. Ondarçuhu and A. Piednoir, Pinning of a contact line on nanometric steps during the dewetting of a terraced substrate, *Nano Lett.*, **5**, 1744–1750 (2005).

244. A. Moosavi, M. Rauscher, and S. Dietrich, Dynamics of nanodroplets on topographically structured substrates, *J. Phys.: Condens. Matter*, **21**, 464120 (2009).

245. M. Rauscher and S. Dietrich, Nano-droplets on structured substrates, *Soft Matter*, **5**, 2997–3001 (2009).

246. M. Yang, P. E. Sheehan, W. P. King, and L. J. Whitman, Direct writing of a conducting polymer with molecular-level control of physical dimensions and orientation, *J. Am. Chem. Soc.*, **128**, 6774–6775 (2006).

247. A. Fang, E. Dujardin, and T. Ondarçuhu, Control of droplet size in liquid nanodispensing, *Nano Lett.*, **6**, 2368–2374 (2006).

248. J. Hu, R. W. Carpick, M. Salmeron, and X.-D. Xiao, Imaging and manipulation of nanometer-size liquid droplets by scanning polarization force microscopy, *J. Vac. Sci. Tech. B*, **14**, 1341–1343 (1996).

249. E. Raphaël and P. G. de Gennes, Dynamics of wetting with nonideal surfaces. The single defect problem, *J. Chem. Phys.*, **90**, 7577–7584 (1989).

250. H. P. Greenspan, On the motion of a small viscous droplet that wets a surface, *J. Fluid Mech.*, **84**, 125–143 (1978).

251. N. Moumen, R. S. Subramanian, and J. B. McLaughlin, Experiments on the motion of drops on a horizontal solid surface due to wettability gradients, *Langmuir*, **22**, 2682–2690 (2006).
252. L. M. Pismen and U. Thiele, Asymptotic theory for a moving droplet driven by a wettability gradient, *Phys. Fluids*, **18**, 042104 (2006).
253. M. K. Chaudhury and G. M. Whitesides, How to make water run uphill, *Science*, **256**, 1539–1541 (1992).
254. T. Ondarçuhu and E. Raphaël, Étalement d'un ruban liquide à cheval entre deux substrats solides différents, *C. R. Acad. Sci. Paris, Sèrie II*, **314**, 453–456 (1992).
255. J. Jéopoldès and D. G. Bucknall, Coalescence of droplets on chemical boundaries, *EPL*, **72**, 597–603 (2005).
256. R. Lucht and C. Bahr, Surface plasmon resonance study of the spreading of a liquid-crystal smectic-A droplet on a gold substrate, *Phys. Rev. Lett.*, **85**, 4080–4083 (2000).
257. C. Poulard and A.-M. Cazabat, The spontaneous spreading of nematic liquid crystals, *Langmuir*, **21**, 6270–6276 (2005).
258. C. Poulard, G. Guéna, A.-M. Cazabat, A. Boudaoud, and M. Ben Amar, Rescaling the dynamics of evaporating drops, *Langmuir*, **21**, 8226–8233 (2005).
259. A. J. McDonald and S. Hanna, Atomistic computer simulations of terraced wetting of model 8CB molecules at crystal surfaces, *Mol. Cryst. Liq. Cryst.* **413**, 135–144 (2004).
260. B. Zhao, J. S. Moore, and D. J. Beebe, Principles of surface-directed liquid flow in microfluidic channels, *Anal. Chem.*, **74**, 4259–4268 (2002).
261. R. Roth and M. Kroll, Capillary evaporation in pores, *J. Phys.: Condens. Matter*, **18**, 6517–6530 (2006).
262. R. Roth, D. Gillespie, W. Nonner, and R. E. Eisenberg, Bubbles, gating, and anesthetics in ion channels, *Biophys. J.*, **94**, 4282–4298 (2008).
263. K. A. Dill, Dominant forces in protein folding, *Biochemistry*, **29**, 7133–7155 (1990).
264. Y. Snir and R. D. Kamien, Entropically driven helix formation, *Science.* **307**, 1067–1067 (2005).
265. R. Roth, Y. Harano, and M. Kinoshita, Morphometric approach to the solvation free energy of complex molecules, *Phys. Rev. Lett.*, **97**, 078101 (2006).
266. Y. Harano, R. Roth, and M. Kinoshita, On the energetics of protein folding in aqueous solution, *Chem. Phys. Lett.*, **432**, 275–280 (2006).

267. Y. Harano, R. Roth, Y. Sugita, M. Ikeguchi, and M. Kinoshita, Physical basis for characterizing native structures of proteins, *Chem. Phys. Lett.*, **437**, 112–116 (2007).

268. H. Hansen-Goos, R. Roth, K. Mecke, and S. Dietrich, Solvation of proteins: linking thermodynamics to geometry, *Phys. Rev. Lett.*, **99**, 128101 (2007).

269. T. Yoshidome, K. Oda, Y. Harano, R. Roth, Y. Sugita, M. Ikeguchi, and M. Kinoshita, Free-energy function based on an all-atom model for proteins, *Proteins: Struct. Funct. Bioinf.*, **77**, 950–961 (2009).

Chapter 4

Stability of Thin Liquid Films

Günter Reiter
Physikalisches Institut, Universität Freiburg, Hermann Herder Str. 3, 79104 Freiburg, Germany
Freiburg Institute for Advanced Studies, Albertstraße 19, Albert-Ludwigs-Universität Freiburg, 79104 Freiburg, Germany
guenter.reiter@physik.uni-freiburg.de

Although weak but long-ranged intermolecular interactions like dispersion forces are present everywhere, we often do not consider them as important. However, when symmetry is broken at interfaces, the strength of these forces becomes clearly relevant, and due to their long-range nature they may have drastic consequences on the stability of thin films. We demonstrate that a force (per unit area) of less than 1 Pa is capable of inducing morphological changes in up to 100 nm thick fluid films, even if molecules are chemically attached to the substrate. We present experimental results on the temporal evolution of different morphologies of unstable thin liquid polymer films caused by destabilizing intermolecular forces. Four distinct stages of pattern evolution have been identified: (A) amplification of surface fluctuations, (B) breakup of the film and formation of holes, (C) growth and coalescence of holes, and (D) droplet formation and ripening. Besides a linear dependence on film viscosity and surface tension, the time τ for film rupture varied

Nanoscale Liquid Interfaces: Wetting, Patterning, and Force Microscopy at the Molecular Scale
Edited by Thierry Ondarçuhu and Jean-Pierre Aimé

ISBN 978-981-4316-45-3 (Hardcover), 978-981-4364-48-5 (eBook)
www.panstanford.com

significantly with film thickness h ($\tau \sim h^5$), as expected by theory. In addition, the influence of the bounding media is investigated by covering such films with different bounding liquids, thus varying the interfacial tension by two orders of magnitude. Clearly, an increase in compatibility between the film and the bounding liquid results in a faster amplification of an instability with a shorter characteristic wavelength. Inversely, based on the characteristic parameters describing the instability, we are able to determine extremely small interfacial tensions with a remarkable accuracy. These results demonstrate the relevance of long-range forces, even at amazingly large separations, with implications in many fields, ranging from physical properties of thin films to biology.

4.1 Introduction

Liquids are characterized by, besides many other features, the fact that an applied force will cause the continuous displacement of molecules: the interactions between the molecules are weaker than the external force acting on them. A net force acting on molecules can also be generated by the system itself. If molecules of a liquid are close to an interface, these molecules are exposed to other types of molecules exhibiting different interaction forces. This asymmetry in forces thus causes a net force acting on these molecules, which either pulls or pushes them in the direction normal to the interface. The same reasoning can be extended to molecules confined between two interfaces in close proximity like in the case of thin liquid films of thickness h. If h gets significantly thinner than one micron, intermolecular forces start to govern the system.[1-16] The two bounding interfaces know about their mutual existence, even at separations much larger than molecular dimensions, via long-range interactions. It is of eminent importance to be able to understand and predict what consequences can arise from such interactions across a film. Often it is assumed that dispersion forces can be neglected at distances of about 100 nm or more. However, as long-range forces like van der Waals (vdW) forces affect not only their nearest neighbors but a much larger volume, they are able to interact with all molecules of this film. It is practically impossible that the sum of all contributions of vdW forces originating from interactions with the bounding media (i.e., substrate and bounding fluid) and the molecules of the film are perfectly balanced. Consequently, a net force will act on the molecules of the film. For the geometry of a thin film between two parallel interfaces, we may distinguish two antagonistic possibilities. In the case of net repulsive forces between the two interfaces, the film will try to get thicker and the interfaces will try to move apart. In the opposite case, when the sum of all forces on the molecules of the film yields a net attraction between the two interfaces, the film would like to get thinner. As a consequence of such net attraction, thin liquid films are unstable and develop various morphological patterns.[8–38]

The stability of thin fluid films is essential in many technological applications like in coatings (paints) or in microelectronic devices (insulating layers). Thus, in these cases one would like to avoid spontaneous pattern formation and the breakup of such films. In thin films, due to the large surface-to-volume ratio, interfacial properties become increasingly important, especially for thicknesses h below one micron.[1–17] As a first attempt to avoid such instabilities one may think of modifying the substrate by chemical surface treatments in order to increase the interactions between the film and the substrate. Short-range interactions, corresponding to forces acting mainly on contact between molecules, determine to a large extent interfacial tensions.[1] Thus, surface treatments and surface modifications consequently allow to improve wettability (= a liquid likes to cover a substrate) and to increase adhesion or to reduce friction. However, such modifications may be insufficient to compete with effects arising from long-range interactions, even if these forces are weak. At a distance of 100 nm these forces (per unit area) are already at least four to five orders of magnitude weaker than atmospheric pressure. Short- and long-range forces may vary independently and even may have different signs[1,9,10] as they originate from different types of interactions, for example, hydrogen bonds and dispersion forces, respectively.

Instabilities of thin liquid films have been investigated for many decades. The general contours of the mechanics and thermodynamics of thin-film instability have emerged during the last 50 years.[3,8–17,20,39–41] On the theoretical side, Sheludko and Vrij[17–19]

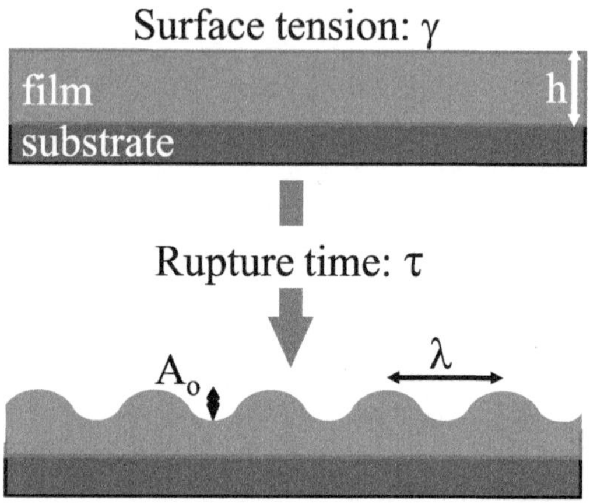

Figure 4.1 Schematic drawing indicating the process of destabilization of a thin liquid film of thickness h by long-range forces. The surface tension of the film tries to keep it flat. The competition of surface tension γ and long-range forces generates an undulation of wavelength λ with an amplitude A_0. After a time τ $A_0 \approx h$ and the film ruptures. The volume of the film separating media 1 and 2 is conserved. Thus, local thinning has to be compensated by thickening at some other place.

have already proposed a long time ago the possibility of an intrinsic instability induced by long-range vdW forces across the film. Initial surface fluctuations at the free surface of an initially rather smooth thin film get amplified spontaneously whenever the second derivative of the excess intermolecular free energy (per unit area) with respect to the local film thickness is negative and their wavelength is larger than a critical wavelength. This amplification eventually leads to the dewetting of the underlying substrate in the form of dry patches.[8,10,17,20–22,26–33,36–38,42,43] In Fig. 4.1, the process is schematically presented. The theory to explain the intrinsic amplification of surface fluctuations and resulting dewetting is firmly in place now.[3,8,10–17,20,21] Real-time direct observation of the initial amplification was facilitated by the introduction of sophisticated measuring techniques such as atomic force microscopy (AFM).[27–29,33,34,44–46]

It is not always clear what kind of interactions are responsible for the instability. One reason for ambiguities is that similar morphologies, for example, wavelike patterns or circular holes, may have evolved from an intrinsic instability or from a (heterogeneous) nucleation process. Different morphologies cannot always be related to different processes that had generated them. Undulation patterns were often taken as signs for an intrinsic instability, called spinodal dewetting, while the formation of isolated holes was thought to reflect an instability induced by defects or heterogeneities.[30,43,47] However, it was shown[28] that spinodal dewetting can result in different patterns if the long-range vdW forces are coupled with various short-range interactions. In turn, based on the morphologies visible on a micrometer scale, together with knowledge about film thicknesses, one may extract the form of the interaction profile on a nanometer scale. Thus, in accordance with three-dimensional computer simulations,[21] it is possible to infer the intermolecular interactions acting in their films just by using a microscope.[48]

Here, we aim to present a comprehensive picture of the film self-destruction process, including the initial amplification and subsequent dewetting, in order to provide a detailed understanding the various stages of thin-film breakup and morphology evolution. In particular, we consider fundamental and unifying scientific aspects of thin liquid films on solid substrates. Some of the central questions that will be treated are:

(1) What happens under the influence of antagonistic (repulsive and attractive) short- and long-range surface interactions?
(2) What is the consequence of changing interfacial tension?
(3) How does the film morphology evolve during different stages of film self-destruction?
(4) What kind of morphological patterns can arise?

4.2 Theoretical Considerations

The "excess" interfacial free energy ΔG (per unit area) of an interface includes a finite thickness domain next to the geometric interface due to long-range interactions, leading to Gibbs's concept of the interfacial tension when integrated over a semi-infinite domain.[3] There are as many (additive) components of the excess free energy as the variety of the underlying intermolecular interactions.[1–16] Most prominent of these are apolar vdW interactions, polar acid-base interactions and interactions of entropic origin, including hydration repulsion and hydrophobic attraction (for aqueous media on hydrophilic and hydrophobic surfaces, respectively), electrostatic double-layer interactions (for charged surfaces in ionic solutions), and more complex shorter-range attraction/repulsion engendered by the entropic effects in polymers close to a confining wall or due to chain adsorption/grafting.[49–55]

Here, we consider a fairly general excess intermolecular interaction free energy (ΔG) composed of antagonistic (attractive/repulsive) long- and (relatively) short-range interactions, which decay with the local thickness, h.[5,6,9,10]

$$\Delta G \sim A/12h^2 + S^{\mathrm{P}} e^{-h/l} \tag{4.1}$$

Here, A is the Hamaker constant, S^{P} is the polar component of the spreading coefficient representing short-range polar interactions, and l is the decay length characterizing the range of these interactions. For unstable systems, there are three universal classes of the form of the free energy:[9,10] type I systems where both long-range apolar and short-range polar forces are attractive ($A > 0$, $S^{\mathrm{P}} < 0$); type II systems, where a long-range attraction ($A > 0$) combines with a shorter-range repulsion ($S^{\mathrm{P}} > 0$); and type IV systems where a long-range repulsion ($A < 0$) combines with a shorter-range attraction ($S^{\mathrm{P}} < 0$). Type III systems (both long- and short-range interactions are repulsive) are stable.

It is well known from colloid science that two particles, labeled 1 and 2, immersed in a fluid, designated as 3, may either attract or repel each other, even over distances (h) of many tens of nanometers. Such long-range interaction may be due to apolar vdW forces, which can be described by a product of two terms.[1–16,56,57] The first term is accounting for the energetic properties of the different materials, expressed by an effective Hamaker constant A_{132}. The second term describes the geometry of the system.[1,5,10,56,57] A_{132} is determined by the individual Hamaker constants (A_{ii}) of the three media and may be related to a good approximation to the dispersion (apolar) part of the individual surface tensions (γ_i^D),[1,5,6,10,57] which are more readily accessible than Hamaker constants.

$$A_{132} = \left(\sqrt{A_{11}} - \sqrt{A_{33}}\right)\left(\sqrt{A_{22}} - \sqrt{A_{33}}\right) \tag{4.2}$$
$$\approx 24\delta_0 \left(\sqrt{\gamma_1^D} - \sqrt{\gamma_3^D}\right)\left(\sqrt{\gamma_2^D} - \sqrt{\gamma_3^D}\right)$$

δ_0 is an atomic cutoff distance (0.158 nm[10]) arising from the short-range Born repulsion. Assuming now that particles 1 and 2 are infinitely large, one ends up with a thin film of medium 3 separating media 1 and 2. Such a change in geometry does not affect A_{132}.

If A_{132} is negative, 1 and 2 repel each other, seeking to make the film infinitely thick. A positive effective Hamaker constant represents attraction between 1 and 2, trying to remove the film completely. Positive A_{132} also means that a thin film is unstable. By thinning the film locally one reduces the free energy more than it costs, due to mass conservation, to increase the film thickness some place else. Mathematically this means that the second derivative of ΔG with respect to h has to be negative for the film to be unstable. Such attraction then can lead to "roughening" of the film. The interfacial tension between the film and the bounding medium γ_{fb}, however, will create a Laplace pressure,[9,10,17–19,21] which will smoothen out undulations, preferentially the ones with a small local radius of curvature. The competition between attractive long-range forces and Laplace pressure thus determines a critical wavelength beyond which fluctuations will grow increasingly more rapidly. This amplification eventually leads to the breakup and subsequent dewetting of the underlying substrate in the form of dry patches.[8–16,21,22,26–28,31–33,36,39–42,48,58] The processes involved in spinodal dewetting, that is, the destruction of thin liquid films, are summarized in Fig. 4.2.

In an approximative theory, the early stages of growth of the amplitude in time can be described by an exponential function. However, it has to be noted that the force acting on the film is varying with film thickness and cannot be represented by the value of the mean thickness if the thickness variations get large. The growth of depressions will accelerate with progressing depth of the depressions. Eventually, the depressions hit the substrate and, on nonwettable substrates, rupture the film, which, in turn, will initiate a dewetting process. Due to such nonlinear behavior, caused by the

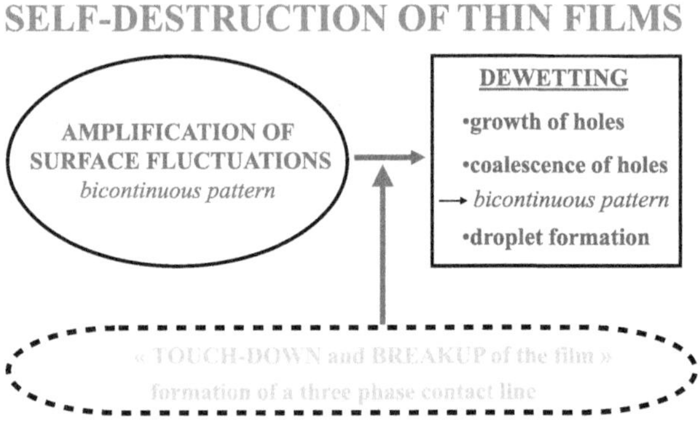

Figure 4.2 Scheme summarizing the various stages in the temporal evolution of an unstable liquid film. One can clearly separate the regimes of amplification of surface fluctuations from the subsequent dewetting stages by the time the three-phase contact line is formed—the "touchdown."

strongly increasing attraction as the thickness gets smaller, most of the time is needed to create depressions of a certain percentage of the film thickness.

As mentioned before, the interfacial tension (γ_{fb}) at the film–bounding fluid interface is one key parameter in the spatial and temporal evolution of this instability. Qualitatively, γ_{fb} plays a purely stabilizing role as it resists the deformation of the film surface. Thus, the higher γ_{fb} is, the more difficult it is to rupture the film.

This situation before the breakup of the film has to be contrasted to the subsequent dewetting. There, the interplay of three interfacial tensions (substrate-bounding fluid, substrate film, and film-bounding fluid), which determine the contact angle at the three-phase contact line,[4] affects the morphological evolution of the film, for example, the growth of dry patches. The dewetting velocity is proportional to γ_{fb}. Quantitatively, linear theory as well as a dimensional analysis of the nonlinear governing equations predict a linear dependence of length and timescales of the instability on the interfacial tension.[8,10–16,59]

It has to be emphasized that varying γ_{fb} by changing the bounding liquid affects, at the same time, the strength of the long-range interactions across the film.[57] Here, we avoided such a correlation by keeping the bounding liquid the same and simply added a very tiny amount of surfactant molecules to this liquid. These surfactant molecules adsorbed at the interface between the film and the bounding liquid and thereby reduced γ_{fb} without significantly changing the long-range interactions.

It is important to note that spinodal dewetting[8,11,11–17,62,63] does not imply that the surface instability and consequent dewetting can occur only in the form of a bicontinuous structure composed of liquid ridges or "hills" and "valleys." The thin-film experiments presented here demonstrate that a rich variety of morphological patterns can evolve.[26–28,30–32,34–36,43,48,64–66]

4.3 Experimental Details

4.3.1 *Why Polymers?*

We chose an experimental system that provided a high control over the forces acting in the thin-film system. In this context, polymers have several advantages. Besides being a nonvolatile fluid of variable viscosity, properties of polymers are often influenced by entropic effects, which also provide a possibility to model and control interfacial interactions.[49–51,53–55] Such effects allow to create repulsive and attractive surfaces, even for the same molecules. That means that the wettability of a polymer-coated substrate depends on the number of conformational states the substrate-attached polymers can realize. At first, it may seem surprising that polymers can dewet a layer of identical but grafted molecules (= polymer brush).[52,55,67–70] However, the reduced number of possible conformational arrangements of grafted polymers makes this grafted layer repulsive for the penetration by free polymers and thus is responsible for such autophobic behavior. However, if one adds long but otherwise identical polymers to such a brush (i.e., creates a bimodal brush), wettability can be reestablished.[70] Consequently, a liquid drop of such a polymer melt will spread out on a bimodal brush of the same polymer but will contract (dewet) on a densely grafted monomodal brush of the same molecules. The long molecules will act as connectors across the nonwettable interface. Already a rather low number of connectors (the connector molecules do not touch each other) may prevent dewetting. Moreover, the dense brush of short polymers underneath the connectors serves to avoid contacts between the chemically different silicon substrate and free polymers. The so generated effective substrate-film interface is between the surface of the brush and the free molecules.[51-55,69] Bimodal brushes have the advantage that they allow to modify short-range interactions at the substrate-film interface without introducing chemical differences between the substrate and the film. Consequently, while interfacial properties can be varied and tuned, A_{132} stays constant. We note, and will prove it by the subsequent experiments, that a wettable bimodal brush surface does not necessarily imply that such a thin film is stable if A_{132} is positive.

4.3.2 *Special Features of the Polymer-Coated Substrate*

Following the above concept, we used silicon wafers coated with mono- and bimodal polydimethylsiloxane (PDMS) brushes as substrates. These wafers were subsequently covered by spin-coating with a thin film of chemically identical PDMS molecules. At room temperature, PDMS is a liquid. In Fig. 4.3 we present a schematic drawing of the samples used. PDMS has the advantage that we can perform experiments at room temperature and that we can change the bounding medium for the PDMS film by putting incompatible liquids like water or ethylene glycol (EG) (extremely bad solvents for PDMS) on top of the polymer.

Several types of coatings were used:

(1) monolayers of reactive, end-functionalized (Si-H terminated) short- or long-chain PDMS (polymer brushes): molecular weight, $M_w = 8.8$ kg/mol and $M_w = 78$ kg/mol, respectively; polydispersity, $I < 1.1$
(2) irreversibly adsorbed PDMS layers representing "pseudo brushes" of PDMS containing a dense part of strongly adsorbed segments and a loose part of dangling ends and loops: $M_w = 38$ kg/mol, $I = 2.6$; $M_w = 80$ kg/mol, $I = 2.1$, respectively.

In addition, we also used bimodal brushes by sequentially grafting chains of two different molecular weights onto the wafers. First, a monomodal dense brush was established from compar-

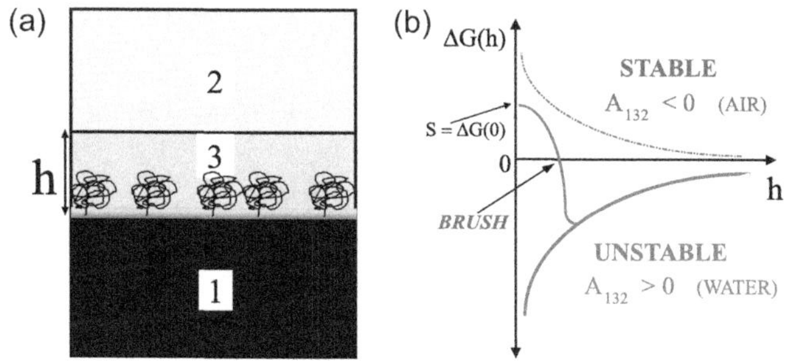

Figure 4.3 (a) Due to long-range interactions a force is acting on a thin film of liquid 3 on substrate 1 and bounded by the surrounding (deformable) medium 2. Attractive forces try to reduce the distance h, while repulsive forces try to increase it. The film of free PDMS molecules is deposited on a bimodal PDMS brush. There, the long-chain PDMS molecules act as connectors that are well intermixed with the free molecules. The thickness h corresponds to the whole PDMS layer, including grafted and free molecules. b) Variations of the excess free energy per unit area (ΔG) with film thickness (h) for the investigated system. The full and dash-dotted line represent the case when the surrounding medium are water and air (positive and negative A_{132}), respectively. Due to the grafted PDMS molecules, a layer proportional to the size of these molecules has to remain on the substrate. This is indicated by a positive spreading coefficient ($S = \Delta G(h = 0)$) and positive values of ΔG at small thicknesses. As a consequence, we obtain a minimum in ΔG at a distance of about the size of the grafted molecules.

atively short Si-H monofunctionalized PDMS chains of uniform length ($M_w = 8,800$ g/mol, polydispersity $I < 1.1$). Due to the high grafting density ($\Sigma = 0.44$ chains/nm^2), the chains were preferentially oriented and stretched in the direction normal to the silicon surface. This caused an autophobic behavior, indicating only little interpenetration (about 1 nm) between the melt and the brush.[49,50,52,55,68–70] Bimodal brushes resulted from a monomodal brush onto which we have grafted a few much longer PDMS chains ($M_w = 121,800$ g/mol, $I < 1.1$), which acted as connectors.[70] The thicknesses of the monomodal 78 k and 8.8 k brush layers were found to be about 15 nm and 6 nm, respectively, by ellipsometric measurements.[32] Adding connectors (i.e., forming bimodal brushes) increased the thickness of the second coating to about 12 nm. For the 38 k and 80 k adsorbed layers, thicknesses of 6 nm and 9 nm were measured, respectively.

The brushes were covered by spin-coating with thin PDMS films of various molecular weights (M_w ranging from 3 to 308 kg/mol, corresponding to viscosities ranging from 10 to 1,000 Pa.s) of variable thickness ($h < 200$ nm). Ellipsometry was used to measure the thickness of the PDMS film. In order to initiate the instability, the surrounding air was replaced by some liquid. Mainly, the PDMS films were embedded in water or an aqueous surfactant solution. This system is a suitable candidate to study the self-destruction of thin polymer films as it provides excellent control over kinetics and intermolecular interactions.[32,35,58,71]

The evolution of the instability was followed at room temperature and in real time by optical microscopy. Removing water resulted in respreading of PDMS. We want to emphasize that PDMS films on bimodal brushes were stable in air. No signs of dewetting or any characteristic patterns have been observed. This has to be contrasted to films on monomodal brushes. There one could initiate dewetting by defects. Putting such films under water led also to dewetting. However, when water was removed dewetting continued and absolutely no respreading was observed.

4.3.3 *Hamaker Constants*

PDMS has a lower Hamaker constant than silicon.[57] The bounding medium air has a Hamaker constant equal to zero. Consequently, A_{132} is negative, and a thin film of PDMS (in air) is stable (upper broken line in Fig. 4.3). The situation changes dramatically if medium 2 is a liquid that has a higher Hamaker constant than PDMS. When we replaced air by putting water on top of the PDMS film, we changed the sign of A_{132} (Fig. 4.3, lower full line). $\gamma_{\text{pdms-air}} \approx 20.7$ mJ/m^2.[60] Water does not wet PDMS. It does not dissolve PDMS either as it is an extremely bad solvent for PDMS. The interfacial tensions between water and grafted or free PDMS molecules, respectively, are roughly the same ($\gamma_{\text{pdms-water}} \approx 38.5$ mJ/m^2.[72]) The increased interfacial tension $\gamma_{\text{pdms-water}}$ with respect to $\gamma_{\text{pdms-air}}$ tends to smoothen the PDMS film. However, long-range forces can destabilize the film. For the present system, the final state is expected to be droplets on a thin wetting film. In theoretical treatments, this situation has been called pseudo-partial wetting.[9] It is evident that a layer of PDMS has to remain because we have attached PDMS molecules chemically to the silicon wafer. In addition, due to the connector molecules this brush is wettable by PDMS, expressed by a positive spreading coefficient $S = \Delta G(h = 0)$. This situation is indicated in Fig. 4.3, resulting in a minimum at a final value of h.

The system was stable in air under the influence of stabilizing short *and* long-range forces corresponding to wettable coatings and the effective Hamaker constant for the system silicon-PDMS-air being negative.[32,35,58,71] The stability of the PDMS layer in air allowed easy handling of the samples, and the instability could be initiated by replacing air with water as the effective Hamaker constant then becomes positive.[32,35,58,71]

4.3.4 *Exchanging the Bounding Fluid Medium*

In actual experiments the instability was initiated by putting a small drop (~10 μl of liquid onto the film. Dilute aqueous solutions of different surfactants provided the possibility of changing interfacial tensions between the PDMS film and the bounding medium. It should be stressed that as predicted by theory, changing the interfacial tension between the film and the bounding medium by adding surfactants changed only the length and time scales of the morphological evolution of the film, without bringing any qualitative change in different stages of evolution as compared to films under pure water (PW).[58]

In total, four different bounding liquids that have a widely different degree of compatibility with PDMS (as expressed by the interfacial tension between the liquid and PDMS, $\gamma_{\text{pdms-liquid}}$) were used for experiments: PW, EG, aqueous solutions of surfactant L77 (polyalkyleneoxide-modified heptamethyltrisiloxane) and a short ($M_w = 3$ k) water-soluble diblock (80/20) copolymer with a small polyethylene oxide (PEO) block, that is, PDMS-PEO (PP). The concentrations of aqueous solutions were about five times the critical micellar concentration; in absolute terms the concentration was 0.00025 v/v for L77 and 0.01 v/v for PP, respectively.

In particular, aqueous solutions of surfactant L77 were used[32,35,58,71] in place of PW to have convenient time and length scales for the breakup of rather thick films (about 100 nm). Such

Table 4.1 Interfacial tensions and contact angles for the different bounding media used in experiments

Liquid (b)	$\gamma_{\text{liquid-air}}$ [mN/m]	θ [degrees]	$\gamma_{\text{pdms-liquid}}$ [mN/m]
PW	72.8 ± 0.2	104 ± 1	$38.4 \pm 3\%$
EG	38.3 ± 0.2	86 ± 1	$18.1 \pm 5\%$
PP	23.3 ± 0.2	57 ± 2	$8.1 \pm 7\%$
L77	21.5 ± 0.1	17 ± 1	$0.3 \pm 33\%$

thick films provided sufficient optical contrast, especially during the early stages, allowing for good statistical analysis. L77 reduced the interfacial tension by about 100 times to a very low value of about 0.3 mN/m compared to 38.5 mN/m, which is the value for PW.[73] For a film of given thickness, the lowering of the interfacial tension led to a faster amplification of surface waves of smaller length scales.

The interfacial tension for different liquids, PW, EG, L77, and PP, was estimated by applying the Young's equation[4] to a liquid drop on the PDMS film in air.

$$\gamma_{\text{pdms-liquid}} = \gamma_{\text{pdms-air}} - \gamma_{\text{liquid-air}} \cos\theta \tag{4.3}$$

where θ is the equilibrium contact angle of the drop and the γ_{i-j} are the interfacial tensions at the interface between species i and j. θ was measured using a contact angle goniometer, and the surface tension of the liquids and PDMS was measured using the Wilhelmy plate apparatus. $\gamma_{\text{pdms-air}}$ was taken from literature as 20.8 ± 0.1 mN/m.[60] The results of the interfacial tensions are presented in Table 4.1.

4.4 Experimental Investigations of Self-Destruction of Thin Liquid Films

4.4.1 *Evolution of Film Morphology*

Figure 4.4 shows a time series of optical micrographs representing the typical spatial and temporal evolution of the morphological pattern during the self-destruction and dewetting of a PDMS film. This visual information is complemented by presentation of

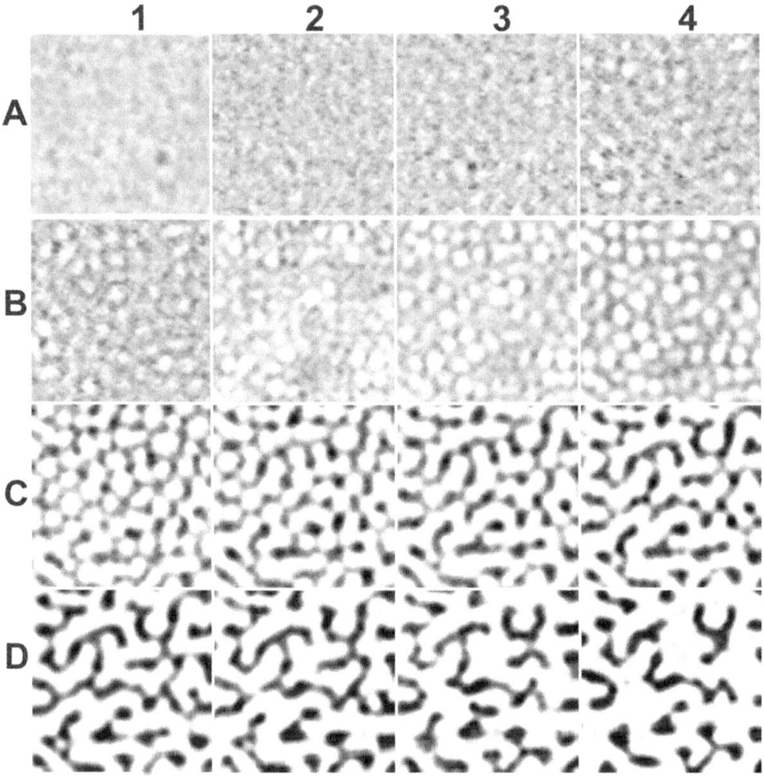

Figure 4.4 Optical micrographs showing the evolution of morphological patterns during the self-destruction of a 85 nm thick and highly viscous (1,000 Pa.s) PDMS film on a silicon wafer coated with a 15 nm thick monolayer of end-grafted PDMS 78k PDMS, embedded in an aqueous L77 solution. The levels of gray indicate the local thickness of the film, with darker regions representing thicker portions. Different micrographs (from left to right and top to bottom, from A1 to D4) correspond to times (in s) 230, 245, 260, 275, 300, 315, 330, 360, 420, 435, 500, 570, 660, 840, 1,110, and 1,680, respectively. The area shown in each image is 25 × 25 μm^2.

structural measures of these patterns in Fig. 4.5. The characteristic wavelength of the patterns was determined by taking the two-dimensional Fourier transform of the pictures. The resulting ring-shaped pattern indicated isotropy (no lateral direction is preferred), as expected from theory. Radially averaging the data resulted in a clear peak with the position of its maximum ($= q_{max}$) determining

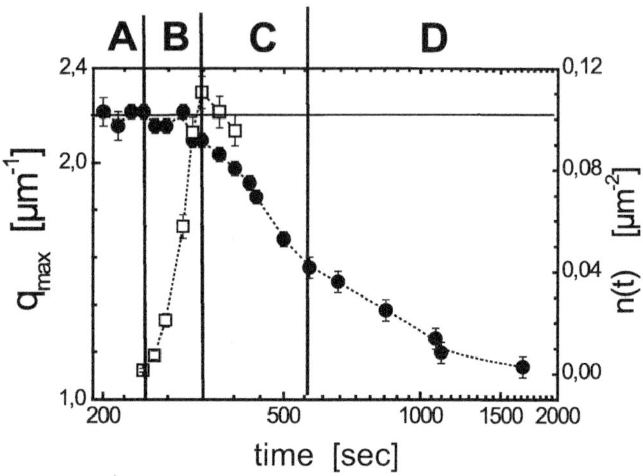

Figure 4.5 The evolution of the position of the peak (q_{max}, filled circles) of the curves shown in Fig. 4.4 and number of isolated circular holes (n, open squares) with time.

the characteristic wavelength, $\lambda = 2\pi/q_{max}$. The variation of q_{max} with time is presented in Fig. 4.5 to show the evolution of the dominant wavelength of the pattern. Figure 4.5 also includes the number of identifiable holes ($n(t)$) as a function of time.

Figure 4.4 allows to identify four different time regimes, *viz.*, A, B, C, and D, in the evolution of the pattern in self-destruction of thin polymer films. An undulating bicontinuous pattern (image A1 of Fig. 4.4) emerged from an initially rather smooth film. This low amplitude pattern amplified and became more pronounced (images A2 to A4 of Fig. 4.4) as evolution continues. The pattern observed at the beginning is thus due to the amplification of surface undulations (capillary waves, small thickness variations due to the spincoating process, etc.) and resulted in the formation of "valleys" and "ridges." This increase in amplitude is reflected by an increase in prominence of the peak at q_{max} during this period. It should be noted that the dominant wavelength stays constant during the first phase of evolution (region A in Fig. 4.5).

While the lateral regularity of the structure is maintained during this initial period, the amplification is nonhomogeneous with different depressions thinning at different rates as indicated by different levels of gray at different depressions (image A4 of Fig. 4.4). This feature is expected due to small difference in the local thickness, reflecting the initial randomly rough surface. The local rate of thinning at these thicknesses is mostly governed by the long-range interactions whose strength increases nonlinearly as the film thins (e.g., the strength of omnipresent vdW forces varies inversely as the third power of film thickness), whereby any small difference in the local thickness results in self-reinforcing vastly different rates of thinning.

It can be noted that most of the amplification has occurred in the last 30 seconds of this stage, which is only about 10% of the total duration of the stage. This also confirms the theoretical predictions[11,12,14–16,39–41] that the amplification is slow in the beginning, whereby the surface fluctuations get arranged on a dominant wavelength, but later enters a nonlinear explosive phase of growth, consistent with an exponential increase of the amplitude.

This amplification of surface fluctuations is followed by the process of "pseudo dewetting"[a] initiated by the "touchdown" of the depressions at the minimum thickness (determined by the grafted layer) to form isolated circular holes (white portions), with the relatively deeper depressions touching earlier (images B1 and B2 in Fig. 4.4) and others following later (image B3 in Fig. 4.4). The variation of the number of holes with time ($n(t)$) is also presented in Fig. 4.5 to highlight the existence of a finite time window for the formation of holes.

The circular holes formed in regime B continue to expand laterally and quickly lose their axi-symmetric shape due to coalescence with other growing neighbors (image C1 of Fig. 4.4). The repeated coalescence of holes results in long pseudo-dewetted channels, which together with the long ridges of polymers give the appearance of a secondary bicontinuous pattern (images C2 to C4 of Fig. 4.4). The long cylindrical ridges of this structure eventually undergo fragmentation (series D of Fig. 4.4) due to Rayleigh instability of long cylinders and break up into small droplets, which become increasingly spherical as time progresses (image D4 of Fig. 4.4). Finally, the structure resembles an array of spherical droplets

---

[a]It is not "true" dewetting because the layer of grafted molecules always remains and thus cannot dewet.

arranged on an intervening thin film of grafted polymers. Ripening of these droplets occurs by material transport via the intervening film until a thermodynamic equilibrium state of a single drop is reached, a scenario similar to the collapse of a smaller bubble, if connected to a bigger bubble, which is driven mainly by Laplace pressure.

In the light of the above presentation we conclude that the self-destruction of thin polymer films is a combination of two serial processes, amplification of initial surface fluctuations and subsequent dewetting of the underlying substrate (see Fig. 4.2). There is a finite time window during which the holes are formed, and both processes overlap. The evolution of the morphological pattern during the whole process can be summarized as follows:

A. Initial largely uncorrelated surface fluctuations organize into a shallow bicontinuous pattern on a distinct characteristic wavelength, which is consistent with linear theory. The pattern **amplifies** without change in its length scale. The amplification is not homogeneous, and some depressions are amplified faster than others.
B. The depressions of the bicontinuous pattern develop into isolated circular holes, which **break** the film. The propagation of the nonhomogeneity in amplification results in different depressions developing into holes at different times. The length scale of the pattern still remains the same, resulting in a one-to-one correspondence between the depressions and the holes.
C. The holes expand laterally and **coalesce** to form a second bicontinuous pattern, which coarsens in time. When compared to the first bicontinuous pattern, this pattern has a much larger amplitude and larger characteristic wavelength, which continuously increases in time.
D. The long and high ridges of this pattern fragment into a collection of spherical **droplets** arranged on an almost-equilibrium-thickness film. These drops may ripen to the final state of a single drop or may remain in time, depending on whether they are interconnected.

Of all these stages only stage A is an exclusive feature of the self-destruction of the film. All other stages (B, C, and D) appear also during nucleation-induced destruction of the film. As similar-

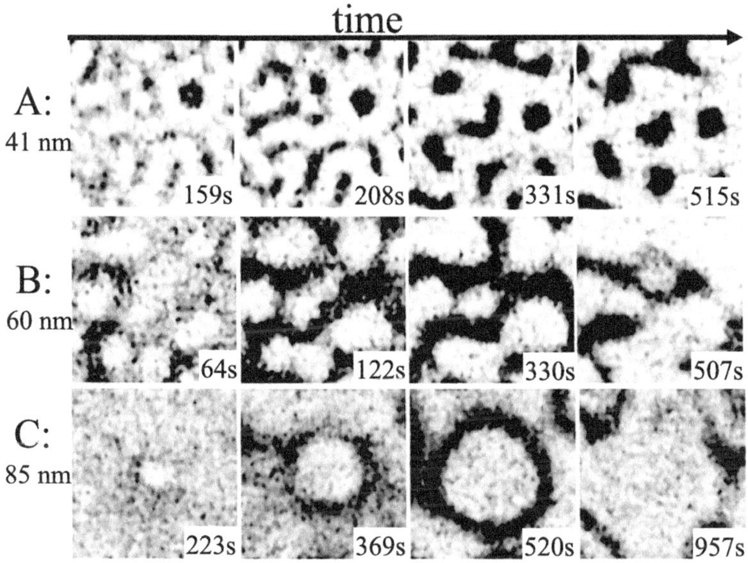

Figure 4.6 Optical micrographs showing the influence of film thickness (A: 41 nm; B: 60 nm; C: 85 nm) on the evolution of morphological patterns during the self-destruction of a highly viscous (1,000 Pa.s) PDMS film on a silicon wafer coated with a 15 nm thick monolayer of end-grafted PDMS 78 k PDMS, embedded in water. The levels of gray indicate the local thickness of the film, with darker regions representing thicker portions. Micrographs were taken after different times as indicated in the images. The area shown in each image is 50 × 50 μm^2.

looking bicontinuous patterns are generated during stages A and C, they can be distinguished only by a time-varying wavelength. Thus, a systematic study of the temporal evolution of the pattern is necessary to identify the various stages and to be sure that the film is self destructing.

4.4.2 *Thickness Dependence*

Figure 4.6 demonstrates that the thickness of the film has a significant influence on the characteristic wavelength of the instability. It can be recognized that for the thickest films investigated, the instability started by the formation of isolated holes, which grew in time and led to the formation of a rim similar to "true" dewetting.[22,30,64,74] However, here we had a wettable substrate as a consequence of the comparatively low grafting density of the polymer brush layer.

A systematic study on the dependence on h larger than the molecular size of the PDMS coil was performed by varying the total thickness of PDMS between 30 nm and 110 nm, deposited onto wettable bimodal brushes.[32] Taking Fourier transforms of the morphological patterns at a stage before the onset of coalescence and ripening yielded the dependence of $q_{\max}$ on h. Almost the same information was obtained in an independent way by simply counting the droplets visible on the micrographs. Differences were mainly observed for the thinnest films studied where significant coalescence has occurred. The results are shown in Fig. 4.7. For $h > 40$ nm the dependence on h is most likely determined by the long-range vdW forces as the slopes on a double logarithmic plot are consistent with theory.[8,11,13,14,17,21] For $h < 40$ nm the results deviate strongly from the power law behavior. For films thinner than 30 nm we never observed the formation of droplets. This may indicate that such films were stable. The value of 30 nm compares well to the thickness of the brush (7 nm) plus the size of the connector molecules (end-to-end distance: 20 nm). Thus, one may assume that the thickness of the remaining wetting layer underneath the droplets is somewhere between 20 nm and 30 nm.

As shown in Figs. 4.4 and 4.5, the time τ of film rupture can be rather precisely determined. Based on this procedure we have determined τ for different film thicknesses h, observing a strong increase in τ with h.[35] Data from films on various PDMS coatings are presented in Fig. 4.8. Furthermore, we used the linear relationship between τ and viscosity η and γ, as predicted by theory,[8,10,11,13,14,17,21] to superpose data for systems differing in η and γ.

$$\tau \sim \eta . \gamma . h^5 \tag{4.4}$$

Although there is significant scatter in the individual data sets, the overall presentation is consistent with $\tau \sim h^5$, as indicated by the full line. In accordance with theory, for lower γ values the absolute value of $q_{\max}$ was larger, while the value of τ was smaller. We thus

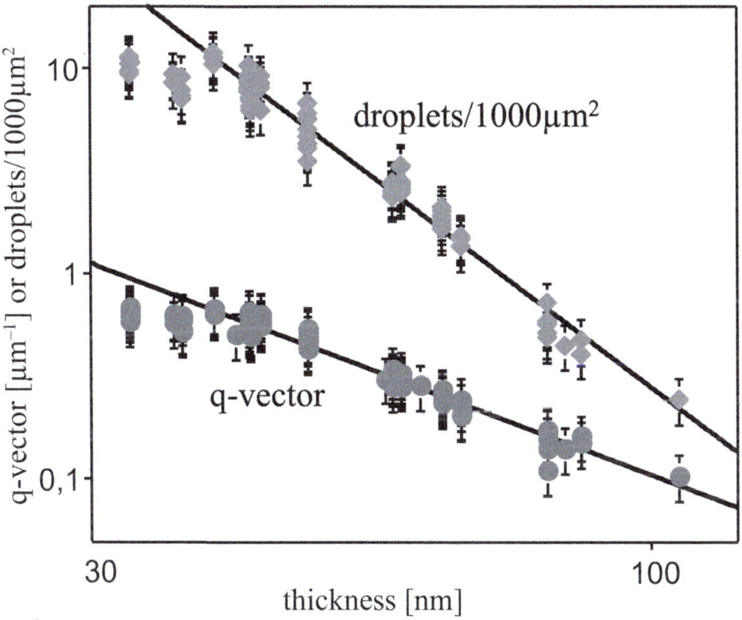

Figure 4.7 Characteristic q-vector as a function of the initial film thickness h on logarithmic scales. $q = 2\pi/d$ (d is the characteristic distance between droplets) is obtained from the position of the maximum of radially averaged two-dimensional Fourier transforms of optical micrographs from systems, as described in Fig. 4.3. Micrographs were taken at a stage where droplets have already been formed. These droplets have also been counted. Their number N per reference area ($= 1,000\ \mu m^2$) is displayed. N and q are related by $N = 1,000q^2/4\pi^2$. The straight lines originate from fitting $q \sim h^\alpha$ and $N \sim h^\beta$ to data points for $h > 40$ nm. We obtained $\alpha = -1.97 \pm 0.05$ and $\beta = -3.95 \pm 0.07$, in accordance with what is expected for vdW interactions.

could superpose all data for $t < \tau$ by normalizing q by $q_{\max}$ and t by τ.

In contrast, at later stages ($t > \tau$) quantitatively different trends for water and surfactant systems were observed, even in the normalized presentation.[35] Hole coalescence, droplet formation, and ripening proceeded much faster when γ was lower as for aqueous L77 surfactant solution as a bounding medium. In Fig. 4.9, we indicate by multiplying the normalized time by the factor 10 that for the low values of γ of the L77 surfactant solution the process

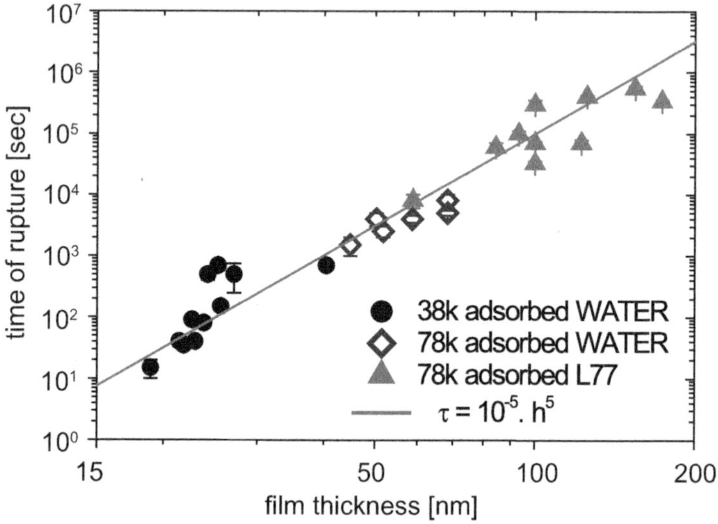

Figure 4.8 Time τ of film rupture as a function of film thickness h. Circles are for PDMS (1,000 Pa.s) films on PDMS 38 k adsorbed layers covered by water, and diamonds and triangles are for PDMS (viscosity varying between 10 and 1,000 Pa.s) films on PDMS 78 k grafted layers covered by water and L77 surfactant solution, respectively. All data is normalized to $\eta = 1{,}000$ Pa.s and $\gamma = 38.4$ mN/m. The line represents $\tau = 10^{-5}$ [sec/nm^5] $\cdot\, h^5$ (the units of τ and h are s and nm, respectively).

was about a factor of 10 faster. One possible explanation is based on the contact angle (θ) at the three-phase contact line (coated substrate - PDMS film - surrounding liquid), which is significantly higher for the surfactant solution—$\cos\theta = (\gamma_{\mathrm{CS-SL}} - \gamma_{\mathrm{CS-PF}})/\gamma_{\mathrm{PF-SL}}$. The interfacial tension between the coating and the film ($\gamma_{\mathrm{CS-PF}}$) is not yet measured for the particular combinations investigated. In comparison with an analogous system,[52] we estimate it to be rather small, about 0.1 mN/m. The interfacial tensions between the coating and the surrounding liquid ($\gamma_{\mathrm{CS-SL}}$) and between the film and the surrounding liquid ($\gamma_{\mathrm{PF-SL}}$) are about the same: 0.3 mN/m and 38.4 mN/m for the surfactant solution and PW, respectively (see Table 4.1). Based on these values and assumptions, we expect for these two cases $\theta \sim 50^{\circ}$ and $\theta \sim 5^{\circ}$, respectively. Thus, significantly

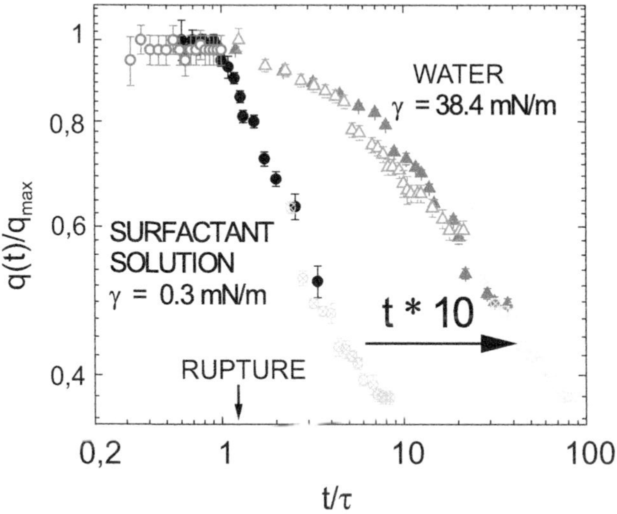

Figure 4.9 Temporal evolution of the characteristic wave-vector $q(t)$ (normalized by its value (q_{max}) at times before film rupture) vs. time (normalized by the time of rupture τ). All films were on 78k grafted PDMS layers. Circles and triangles represent films under L77 surfactant solution and water, respectively. The film thickness and viscosity are 59 nm (10 Pa.S) and 35 nm (1 Pa.S) for open and full triangles and 125 nm (100 Pa.S), 100 nm (1000 Pa.S), and 84 nm (100 Pa.S) for open, full, and crossed circles, respectively. The last data set has also been replotted at 10*t.

different driving forces were pushing at the three-phase contact line for $t > \tau$, resulting in widely different dewetting velocities.

The late stages of film evolution are affected by interfacial tensions represented by the contact angle. In addition, we demonstrate that besides static parameters like θ, dissipation at the film-coating interface can also affect the temporal evolution of the thin film pattern. For that purpose, we have looked at the growth of an isolated hole in PDMS films under water, deposited on various PDMS coatings differing mainly by their molecular weight and conformational properties and thus providing different friction at the coating-film interface. Assuming that dewetting proceeded at a velocity that was constant in time, we obtained a change by more than one order of magnitude in the value of the dewetting velocity V when the molecular weight of the grafted molecules changed from

8.8 kg/mol to 78 kg/mol. Consequently, besides θ also interfacial properties (in particular, friction at the interface) affect V, which, in turn, is the parameter controlling hole coalescence and droplet ripening.

4.4.3 *The Influence of the Interface between Film and Bounding Medium*

Figure 4.10 presents optical micrographs of the decomposition of a 60 nm thick film under different liquids to highlight the widely different "strengths" of the instability as the interfacial tension is changed. Like in Fig. 4.4, for all the bounding media selected here, the evolution follows the characteristic path of spinodal dewetting of thin films.[21,27,28,34–36,48,58] Though, the evolution is qualitatively similar for the three liquids, the time and length scales of the evolution are very different. A lowering of the interfacial tension by a factor of about 100 leads to an about 100-fold increase in the number of holes. By "holes" we mean isolated depressions that have reached the maximum depth. These holes then grow in size laterally with time. It is quite clear now that the instability is the strongest in the L77 Series (lowest interfacial tension between the film and the bounding liquid γ_{fb}) and is the weakest in series A (PW, highest γ_{fb}). As the results are shown for the same film thickness and the bounding liquids differ by only a small concentration of the surfactant, it is reasonable to assume that the excess intermolecular force field is the same for the three cases. Thus, the increased compatibility, or reduced γ_{fb}, results in enhanced instability, that is, a faster and more dense breakup of the film.

Having confirmed the qualitative trends of the effect caused by increased compatibility, we now proceed to quantify it. In order to isolate the effect of interfacial tension, the length scales at similar stages (not absolute times) of evolution have to be compared. Note that the coalescence of holes (i.e., a reduction of $N(t)$) increases the length scale. However, the **maximum** number of holes (depressions) per unit area (N_{max}) will not have such add-on effects. As expected,[21] N_{max} is found to be directly related to the characteristic spinodal wavelength (λ). Thus, N_{max} can be attributed to changes in interfacial tension only.

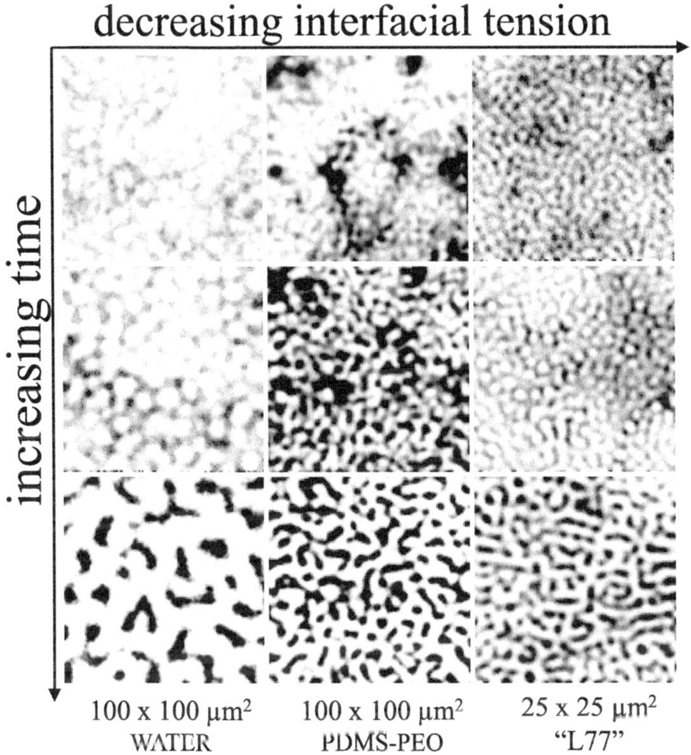

Figure 4.10 Changes in the length and time scales of the breakup of a 60 nm thick PDMS film under bounding liquids that have a varying level of compatibility with PDMS. The interfacial tensions change from 38.4 ± 1.2 mN/m in series PW to 8.1 ± 0.6 mN/m in series PP and then finally to 0.3 ± 0.1 mN/m in series L77 (polyalkyleneoxide modified heptamethyltrisiloxane). The increasing levels of gray denote thicker portions. The actual times in seconds for series PW, PP, and L77 are (25, 65, and 200), (15, 45, and 100), and (5, 10, and 15), respectively. A rather smooth surface of the film evolves into a collection of circular depressions (top frames). These shallow depressions develop into circular holes (middle frames), which grow laterally to coalesce with other growing holes and form long, cylindrical ridges (bottom frames), which break up into small droplets later.

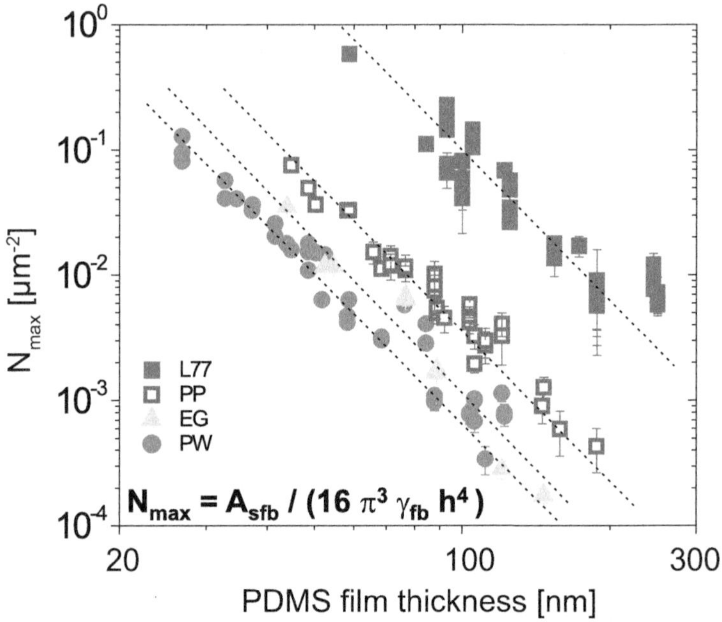

Figure 4.11 The variation of the maximum number density of holes with film thicknesses for different bounding liquids. Results are presented for the four liquids that have a varying level of compatibility with PDMS: PW (open circles), PP (open squares), L77 (filled squares), and EG (filled triangles). The dotted lines indicate a slope of −4, consistent with theory (see text).

In addition, for $t < \tau$ Fig. 4.5 also shows that λ stays rather constant, as expected in spinodal dewetting of thin films,[21,27,28,36] before N_{max} is reached at time τ: $q(t < \tau) = 2\pi/\lambda = 2\pi(N_{max})^{1/2}$.

Figure 4.11 presents the variation of N_{max} with the PDMS layer thickness (h) for different interfacial tensions. Based on a multitude of similar experiments, the following general observations can be drawn: N_{max} increased rapidly with the decreasing film thickness. For any given thickness, L77 and PW resulted in the maximum and minimum densities of holes, respectively. The results for PP are intermediate, and those of EG are very similar to those of water. The relative densities of holes among different liquids for any given h are largely invariant of h. There is an inverse power law dependence of N_{max} on h, as shown by the dotted straight lines, which adequately represent the data on a double logarithmic plot. It is also clear that the exponent of the power law is similar for different liquids. We averaged these results, to obtain a higher accuracy, by fitting straight lines of fixed slope -4 to the data for individual liquids. A slope of -4 is very similar to that given by the best-fitted values for the different liquids, which are -4 ± 0.3, -3.7 ± 0.1, -4.4 ± 0.5, and -3.6 ± 0.4 for the cases of L77, PP, PW, and EG, respectively. It is worthwhile to recall at this point that a slope of -4 strongly suggests that the instability is engendered by long-range vdW interactions. According to the theoretical model for instability engendered by vdW forces, N_{max} is given by.[8,10–17,20,21,59]

$$N_{max} = A_{sfb}/(16\pi^3\gamma_{fb}h^4) = C\,h^{-4} \tag{4.5}$$

where A_{sfb} is the effective Hamaker constant for the system silicon-PDMS-liquid. Equation 4.5 clearly shows that N_{max} is directly proportional to A_{sfb} and inversely proportional to γ_{fb} and h^4. In Table 4.2 we have normalized γ_{fb} and N_{max} with respect to the values for PW to highlight the relative enhancement of the instability. We find that for PW, PP, and L77 the relative hole density is inversely proportional to the relative interfacial tensions. The large error bar in the relative interfacial tension in the case of L77 is a propagation of uncertainty in the initial measurement of γ_{fb}.

The deviation of data for EG from this inverse dependence is expected as the excess intermolecular force field strength has to be different from that in the case of aqueous solutions. To further prove this, we calculated the Hamaker constants by using Eq. 4.5 and the measured values of γ_{fb} for different liquids. Table 4.2 demonstrates that the addition of a few surfactant molecules results in the same Hamaker constant.[27,28,61,71]

Table 4.2 Results obtained from the analysis of Fig. 4.11

Liquid (b)	$\gamma_{pdms-PW}/\gamma_{pdms-b}$	$(N_{max})_b/(N_{max})_{PW}$	A_{sfb} [10^{18} J]
PW	1	1	$1.8 \pm 10\%$
EG	$2.1 \pm 10\%$	$1.4 \pm 25\%$	$1.2 \pm 15\%$
PP	$4.7 \pm 10\%$	$4.6 \pm 15\%$	$1.8 \pm 10\%$
L77	$128 \pm 40\%$	$96 \pm 20\%$	$1.4 \pm 40\%$

As an additional feature of these experiments we present an approach to estimate the interfacial tension. The standard methods suffer from large uncertainties when the interfacial tensions are very low (refer to the L77 case). We can apply the theoretical model of spinodal decomposition to back-calculate the interfacial tension from the characteristic length *and* time scales of the instability.

To highlight the robustness of the method we apply it for a general form of intermolecular potential, which can have *any arbitrary* dependence on the film thickness. For a *general* form of potential, N_{max} and τ are given by[10,21,59]

$$N_{max} = P/(8\pi^2 \gamma_{fb}) \tag{4.6}$$

$$\tau \approx 12\pi^2 \eta \gamma_{fb}/(h^3 P^2) \tag{4.7}$$

P is the force per unit volume at the initial film thickness, and η is the viscosity of the film liquid (PDMS). Please note that Eq. 4.5 is a special form of Eq. 4.6 for the case of vdW interactions.

After eliminating P from Eqs. 4.6 and 4.7, one obtains

$$Y = 3\eta/(16\pi^4 N_{max}^2 \tau) = \gamma_{fb} h^3 \tag{4.8}$$

As N_{max}, τ, and h were measured independently, γ_{fb} can be calculated using Eq. 4.8. From the variation of the parameter Y with h for L77 yields γ_{fb} to be 0.38 ± 0.05 mN/m.

Furthermore, using relevant values for the above experiment [$h = 100 \pm 1$ nm, $\gamma = 0.4 \pm 0.1$ mN/m, $\eta = 1,000$ Pa.s, $q_{max} = 2.2 \pm 0.03\ \mu\text{m}^{-1}$ and $\tau = 300 \pm 15$ seconds], we also can *independently* obtain values for the force P per unit volume. Using Eqs. 4.6 and 4.7 we get the following similar values of the force, P: $3.8 \pm 0.08 \times 10^{-15}$ N.m^{-3} and $4 \pm 0.4 \times 10^{-15}$ N.m^{-3}, respectively. The larger error bar in the second value is a propagation of the error in the value of τ, which reflects the existence of a finite time window for formation of holes, that is, the uncertainty in determining τ. This represents a strong support for the validity of the theory of self-destruction of thin liquid films.[3,8,10–17,20,21]

4.5 Concluding Remarks

The influence of ubiquitous long-range dispersion forces at separation distances up to about 100 nm can be drastic, although these forces are, at such comparatively large distances, already rather weak. This is widely neglected because directly at a surface or an interface, dispersion forces are mostly weak compared to specific interactions of short range. For example, long-range vdW forces (per unit area) at a distance of 100 nm amount to less than 1 Pa. Nonetheless, they are capable of rupturing 100 nm films, even if these films are "glued" to the substrate by specific short-range interactions. In fact, experiments show that the destabilizing forces can be sufficiently strong to even deform thin liquid films confined between two solid walls.[75–77] The experiments presented here and also others[28,34,38,48] show that it is not possible to avoid such instabilities by a simple surface treatment or surface modification of the substrate. Improving adhesion between the substrate and the film by grafting polymers onto the substrate or by adding a thin layer of adhesion promoter will only stabilize films that are sufficiently thin, that is, about the thickness of this additional layer. Thus, we may be forced to reconsider the influence of long-range forces, especially at large separations, in many fields, ranging from physical properties of thin films (there, processes like phase transitions or phase separation differ from the behavior in bulk) to biology (cell adhesion or deformation).

The presented experimental results strongly corroborate the theoretical relationship between the interfacial tension γ_{fb} and the strength of instability in thin liquid films. Compared to the reference interface of PDMS-water, a lowering of interfacial tension results in a much faster instability of reduced wavelength. For very low values of γ_{fb}, as in the case of L77, even several-hundred-nanometer-thick films could be ruptured in seconds. These results also demonstrated that an unstable thin liquid film passes through several stages until it finally reaches equilibrium. The amplification stage is reasonably accurately described by various theoretical works and has been predicted in a quite correct way already by Vrij.[17] Both film viscosity and the tension at the film-surrounding liquid interface affect linearly the time needed for rupturing such films. The late stages of pattern formation, however, are dominated by the three-phase contact angle and friction at the substrate-film interface.

The observed variation of the q-vector with h can be directly related to the variation of the intermolecular forces with h. In that sense, such experiments allow to determine the intermolecular-forces-versus-distance profile for such films.[48] As direct force measurements using techniques like AFM do not allow to measure these forces *within* a liquid film, the stability analysis presented here for varying film thickness provides a valuable alternative.[44–46] In advanced technologies, dimensions below the micrometer have become increasingly more important. This is exactly the range for which our experiments prove the importance and relevance of long-range intermolecular interactions. Even if one sometimes may have the impression that such "nanodevices" are stable, this impression may be wrong as apparent stability may only be the result of slow kinetics. Thus, we are confident that experiments of this type will allow to shed more light on intermolecular forces and interfacial properties relevant for many areas as diverse as biology and materials science.

While we have presented results only for unstable thin liquid films, it should not be ignored that similar instability processes occur also in (visco)-elastic[13,78–81] or even solid films.[82–88] We also did not discuss related problems of the consequences of density gradients[89] or the influence of additives.[90] As an alternative to long-range vdW forces, destabilization may also be induced by an electric field.[37,91–93] These concepts can also be extended to bilayer systems.[93–96] For a review of technological applications of dewetting, see chapter 14.

The process of spinodal dewetting is an intrinsic phenomenon that can hardly be avoided without changing the system. Thus, in view of various applications involving thin films or, in a more general sense, two parallel interfaces separated by some medium, the understanding and control of the relevant molecular interactions are highly desirable. So, if one cannot avoid such instability, can it be of potential use? One may consider to use this creative power of intermolecular forces to produce regular patterns, for example, by employing some lateral confinement in competition with the thickness-dependent characteristic wavelength of the instability.[97]

Consequently, I am expecting that efforts in these directions will be intensified, and not just because of the aesthetic aspects of the evolving patterns.

Acknowledgments

Fruitful discussions with Ashutosh Sharma, Rajesh Khanna, Francoise Brochard, Elie Raphael, and Ullrich Steiner are gratefully acknowledged. I am indebted to Philippe Auroy for providing us with end-functionalized PDMS molecules. Early parts of this work were supported by the Indo-French Centre for the Promotion of Advanced Research/Centre Franco-Indien Pour la Promotion de la Recherche.

References

1. J. N. Israelachvili, *Intermolecular and Surface Forces*. 3rd ed. London: Academic Press (2010).
2. I. B. Ivanov, *Thin Liquid Films: Fundamentals and Applications*. New York: Marcel Dekker (1988).
3. B. V. Derjaguin, *Theory of Stability of Thin Films and Colloids*. New York: Consultants Bureau/Plenum Press (1989).
4. P. G. de Gennes, Wetting: statics and dynamics, *Rev. Mod. Phys.*, **57**, 827 (1985).
5. C. J. Van Oss, M. K. Chaudhury, and R. J. Good, Interfacial Lifshitz–van der Waals and polar interactions in macroscopic systems, *Chem. Rev.*, **88**, 927 (1988).
6. C. J. Van Oss, Acid/base interracial interactions in aqueous media, *Colloid Surf. A*, **78**, 1 (1993).
7. G. H. Findenegg and S. Herminghaus, Wetting: statics and dynamics, *Curr. Opin. Colloid Interface Sci.*, **2**, 301 (1997).
8. F. Brochard-Wyart and J. Daillant, Drying of solids wetted by thin liquid films, *Can. J. Phys.*, **68**, 1084 (1990).
9. F. Brochard-Wyart, J.-M. di Meglio, D. Quéré, and P.-G. de Gennes, Spreading of nonvolatile liquids in a continuum picture, *Langmuir*, **7**, 335 (1991).

10. A. Sharma, Equilibrium contact angles and film thicknesses in the apolar and polar systems—role of intermolecular interactions in coexistence of drops with thin films, *Langmuir*, **9**, 861, 3580 (1993).

11. A. Oron, S. Davis, and S. G. Bankoff, Long-scale evolution of thin liquid films, *Rev. Mod. Phys.*, **69**, 931 (1997).

12. J. A. Diez, and L. Kondic, On the breakup of fluid films of finite and infinite extent, *Phys. Fluids*, **19**, 072107 (2007).

13. R. Blossey, Thin film rupture and polymer flow, *Phys. Chem. Chem. Phys.*, **10**, 5177 (2008).

14. R. V. Crastera and O. K. Matar, Dynamics and stability of thin liquid films, *Rev. Mod. Phys.*, **81**, 1131 (2009).

15. U. Thiele, Thin film evolution equations from (evaporating) dewetting liquid layers to epitaxial growth, *J. Phys.: Condens. Matter*, **22**, 084019 (2010).

16. R. Khanna, N. K. Agnihotri, M. Vashishtha, A. Sharma, P. J. Jaiswal, and S. Puri, Kinetics of spinodal phase separation in unstable thin liquid films, *Phys. Rev. E*, **82**, 011601 (2010).

17. A. Vrij, Possible mechanism for spontaneous rupture of thin free liquid films, *Discuss. Faraday Soc.*, **42**, 23 (1966).

18. A. Scheludko, Thin liquid films, *Adv. Colloid Interface Sci.*, **1**, 391 (1967).

19. A. Vrij and J.Th.G. Overbeek, Rupture of thin liquid films due to spontaneous fluctuations in thickness, *J. Am. Chem. Soc.*, **90**, 3074 (1968).

20. M. B. Williams and S. H. Davis, Non-linear theory of film rupture, *J. Colloid Interface Sci.*, **90**, 220 (1982).

21. A. Sharma and R. Khanna, Pattern formation in unstable thin liquid films, *Phys. Rev. Lett.*, **81**, 3463 (1998).

22. G. Reiter, Dewetting of thin polymer films, *Phys. Rev. Lett.*, **68**, 75 (1992).

23. F. Brochard-Wyart, P. Martin, and C. Redon, Liquid-liquid dewetting, *Langmuir*, **9**, 3682 (1993).

24. F. Brochard-Wyart, P.-G. de Gennes, H. Hervert, and C. Redon, Wetting and slippage of polymer melts on semi-ideal surfaces, *Langmuir*, **10**, 1566 (1994).

25. A. Sharma and G. Reiter, Instability of thin polymer films on coated substrates: rupture, dewetting, and drop formation, *J. Colloid Interface Sci.*, **178**, 383 (1996).

26. M. Sferazza, M. Heppenstall-Butler, R. Cubitt, and R. A. L. Jones, Interfacial instability driven by dispersive forces: the early stages of

spinodal dewetting of a thin polymer film on a polymer substrate, *Phys. Rev. Lett.*, **81**, 5173 (1998).

27. R. Xie, A. Karim, J. F. Douglas, C. C. Han, and R. A. Weiss, Spinodal dewetting of thin polymer films, *Phys. Rev. Lett.*, **81**, 1251 (1998).
28. S. Herminghaus, K. Jacobs, K. Mecke, J. Bischof, A. Fery, Md. Ibn-Elhaj, and S. Schlagowski, Spinodal dewetting in liquid crystal and liquid metal films, *Science*, **282**, 916 (1998).
29. T. G. Stange, D. F. Evans, and W. A. Hendrikson, Nucleation and growth of defects leading to dewetting of thin polymer films, *Langmuir*, **13**, 4459 (1997).
30. K. Jacobs, S. Herminghaus, and K. R. Mecke, Thin liquid polymer films rupture via defects, *Langmuir*, **14**, 965 (1998).
31. U. Thiele, M. Mertig, and W. Pompe, Dewetting of an evaporating thin liquid film: heterogeneous nucleation and surface instability, *Phys. Rev. Lett.*, **80**, 2869 (1998).
32. G. Reiter, A. Sharma, A. Casoli, M.-O. David, R. Khanna, and P. Auroy, Destabilising effect of long-range forces in thin liquid films on wettable substrates, *Europhys. Lett.*, **46**, 512 (1999); Thin film instability induced by long-range forces, *Langmuir*, **15**, 2551 (1999).
33. R. A. Segalman and P. F. Green, Dynamics of rims and the onset of spinodal dewetting at liquid/liquid interfaces, *Macromolecules*, **32**, 801 (1999).
34. J. Becker, G. Grün, R. Seemann, H. Mantz, K. Jacobs, K. Mecke, and R. Blossey, Complex dewetting scenarios captured by thin-film models, *Nat. Mater.*, **2**, 59 (2003).
35. G. Reiter, R. Khanna, and A. Sharma, Self-destruction and dewetting of thin polymer films: the role of interfacial tensions, *J. Phys.: Condens. Matter*, **15**, S331 (2003).
36. G. Reiter, Thin-film pattern formation—the artistic side of intermolecular forces, *Science*, **282**, 888 (1998).
37. S. Herminghaus, Dynamical instability of thin liquid films between conducting media, *Phys. Rev. Lett.*, **83**, 2359 (1999).
38. R. Yerushalmi-Rozen, T. Kerle, and J. Klein, Alternative dewetting pathways of thin liquid films, *Science*, **285**, 1254 (1999).
39. K. Kargupta, R. Konnur, and A. Sharma, Spontaneous dewetting and ordered patterns in evaporating thin liquid films on homogeneous and heterogeneous substrates, *Langmuir*, **17**, 1294 (2001).

40. K. Kargupta and A. Sharma, Creation of ordered patterns by dewetting of thin films on homogeneous and heterogeneous substrates, *J. Colloid Interface Sci.*, **245**, 99 (2002).
41. A. L. Bertozzi, G. Grün, and T. P. Witelski, Dewetting films: bifurcations and concentrations, *Nonlinearity*, **14**, 1569 (2001).
42. A. M. Higgins and R. A. L. Jones, Anisotropic spinodal dewetting as a route to self-assembly of patterned surfaces, *Nature*, **404**, 476 (2000).
43. J. Bischof, D. Scherer, S. Herminghaus, and O. Leiderer, Dewetting modes of thin metallic films: nucleation of holes and spinodal dewetting, *Phys. Rev. Lett.*, **77**, 1536 (1996).
44. Y. J. Wang, C.-H. Lam, X. Zhang, and O. K. C. Tsui, Stability of polymer films as a 2D system, *Eur. Phys. J. Spec. Top.*, **141**, 181 (2007).
45. O. K. C. Tsui, Y. J. Wang, F. K. Lee, C.-H. Lam, and Z. Yang, Equilibrium pathway of spin-coated polymer films, *Macromolecules*, **41**, 1465 (2008).
46. Z. Yang, Y. Fujii, F. K. Lee, C.-H. Lam, and O. K. C. Tsui, Glass transition dynamics and surface layer mobility in unentangled polystyrene films, *Science*, **328**, 1676 (2010).
47. I. Karapanagiotis and W. W. Gerberich, Particle nucleation for dewetting initiation of thin polymer films, *J. Appl. Polym. Sci.*, **98**, 138 (2005).
48. R. Seemann, S. Herminhaus, and K. Jacobs, Dewetting patterns and molecular forces: a reconciliation, *Phys. Rev. Lett.*, **86**, 5534 (2001).
49. L. Leibler, A. Ajdari, A. Mourran, G. Coulon, and D. Chatenay, Wetting of grafted polymer surfaces by compatible chains. In eds., A. Teramoto, M. Kobayashi, and T. Norisuje, *Ordering in Macromolecular Systems.* Berlin: Springer-Verlag. 301–311 (1994).
50. K. Shull, Wetting autophobicity of polymer melts, *Faraday Discuss.*, **98**, 203 (1994).
51. J. I. Martin, Z.-G. Wang, and M. Schick, Effects of polymer brush self-assembly on spreading and thin film stability, *Langmuir*, **12**, 4950 (1996).
52. G. Reiter and R. Khanna, Kinetics of autophobic dewetting of polymer films, *Phys. Rev. Lett.*, **85**, 2753; *Langmuir*, **16**, 6351 (2000).
53. M. W. Matsen, Scaling behavior of a brush-homopolymer interface in the limit of high grafting density, *J. Chem. Phys.*, **122**, 144904 (2005).
54. C. Pastorino, K. Binder, and M. Müller, Static and dynamic properties of the interface between a polymer brush and a melt of identical chains, *J. Chem. Phys.*, **124**, 064902 (2006).

55. X. Zhang, F. K. Lee, and O. K. C. Tsui, Wettability of end-grafted polymer brush by chemically identical polymer films, *Macromolecules*, **41**, 8148 (2008).

56. H. C. Hamaker, The London–van der Waals attraction between spherical particles, *Physica*, **4**, 1058 (1937).

57. J. Visser, On Hamaker constants: a comparison between Hamaker constants and Lifshitz–van der Waals constants, *Adv. Colloid Interface Sci.*, **3**, 331 (1972).

58. G. Reiter, R. Khanna, and A. Sharma, Enhanced instability in thin liquid films by improved compatibility, *Phys. Rev. Lett.*, **85**, 1432 (2000).

59. R. Khanna and A. Sharma, Pattern formation in spontaneous dewetting of thin apolar films, *J. Colloid Interface Sci.*, **195**, 42 (1997).

60. B. B. Sauer and G. T. Dee, Molecular weight and temperature dependence of polymer surface tension—comparison of experiment with theory, *Macromolecules*, **24**, 2124 (1991).

61. W. Zhao, M. H. Rafailovich, J. Sokolov, L. J. Fetters, R. Plano, M. K. Sanyal, S. K. Sinha, and B. B. Sauer, Wetting properties of thin liquid polyethylene propylene films, *Phys. Rev. Lett.*, **70**, 1453 (1993).

62. A. Sharma and E. J. Ruckenstein, The role of lipid abnormalities, aqueous and mucus deficiencies in the tear film breakup, and the implications for tear substitutes and contact lens tolerance, *Colloid Interface Sci.*, **111**, 8 (1986).

63. V. S. Mitlin, Dewetting of solid surface—analogy with spinodal decomposition, *J. Colloid Interface Sci.*, **156**, 491 (1993).

64. G. Reiter, Unstable thin polymer films—rupture and dewetting processes, *Langmuir*, **9**, 1344 (1993).

65. J. M. Guerra, M. Srinivasarao, and R. S. Stein, Photon tunneling microscopy of polymeric surfaces, *Science*, **262**, 1395 (1993).

66. M.-O. David, G. Reiter, T. Sitthai, and J. Schultz, Deformation of a glassy polymer film by long-range intermolecular forces, *Langmuir*, **14**, 5667 (1998).

67. Y. Liu, M. H. Rafailovich, J. Sokolov, S. A. Schwarz, X. Zhong, A. Eisenberg, E. J. Kramer, B. B. Sauer, and S. Satija, Wetting behavior of homopolymer films on chemically similar block copolymer surfaces, *Phys. Rev. Lett.*, **73**, 440 (1994).

68. G. Reiter, P. Auroy, and L. Auvray, Instabilities of thin polymer films on layers of chemically identical grafted molecules, *Macromolecules*, **29**, 2150 (1996).

69. J. H. Maas, G. J. Fleer, F. A. M. Leermakers, and M. A. Cohen Stuart, Wetting of a polymer brush by a chemically identical polymer melt: phase diagram and film stability, *Langmuir*, **18**, 8871 (2002).
70. G. Reiter, J. Schultz, P. Auroy, and L. Auvray, Improving adhesion via connector polymers to stabilize non-wetting liquid films, *Europhys. Lett.*, **33**, 29 (1996).
71. G. Reiter, A. Sharma, R. Khanna, A. Casoli, and M. David, The strength of long-range forces across thin liquid films, *J. Colloid Interface Sci.*, **214**, 126 (1999).
72. V. Bergeron and D. Langevin, Monolayer films of poly(dimethylsiloxane) on aqueous surfactant solutions, *Macromolecules*, **29**, 306 (1996).
73. The interfacial tension was estimated on the basis of surface tension measurements by the Wilhelmy Plate apparatus, the contact angle goniometer, and Young's equation.
74. C. Redon, F. Brochard-Wyart, and F. Rondelez, Dynamics of dewetting, *Phys. Rev. Lett.*, **66**, 715 (1991).
75. K. Dalnoki-Veress, B. G. Nickel, and J. R. Dutcher, Dispersion-driven morphology of mechanically confined polymer films, *Phys. Rev. Lett.*, **82**, 1486 (1999).
76. G. A. Buxton and N. Clarke, Stress-guided self-assembly in Dutcher films, *Phys. Rev. E*, **73**, 041801 (2006).
77. G. A. Buxton, Ultrathin, dynamic and groovy: the physics of Dutcher films, *Annu. Rep. Prog. Chem. C*, **105**, 40 (2009)
78. S. A. Safran and J. Klein, Surface instability of viscoelastic thin films, *J. Phys. II (Fr.)*, **3**, 749 (1993).
79. M. Rauscher, A. Münch, B. Wagner, and R. Blossey, A thin-film equation for viscoelastic liquids of Jeffreys type, *Eur. Phys. J. E*, **17**, 373 (2005).
80. S.-Q. Huang and X.-Q. Feng, Spinodal surface instability of soft elastic thin films, *Acta Mech. Sin.*, **24**, 289 (2008).
81. X.-H. Pan, S.-Q. Huang, S.-W. Yu, and X.-Q. Feng, Interfacial slippage effect on the surface instability of a thin elastic film under van der Waals force, *J. Phys. D: Appl. Phys.*, **42**, 055302 (2009).
82. D. J. Srolovitz and S. A. Safran, Capillary instabilities in thin films. 1. Energetics and Capillary instabilities in thin films. 2. Kinetics, *J. Appl. Phys.*, **60**, 247 and 255 (1986).
83. E. Jiran and C. V. Thompson, Capillary instabilities in thin, continuous films, *Thin Solid Films*, **208**, 23 (1992).
84. E. Jiran and C. V. Thompson, Capillary instabilities in thin films, *J. Electr. Mater.*, **11**, 1153 (1990).

85. J. Ye and C. V. Thompson, Mechanisms of complex morphological evolution during solid-state dewetting of single-crystal nickel thin films, *Appl. Phys. Lett.*, **97**, 071904 (2010).

86. O. Pierre-Louis, A. Chame, and Y. Saito, Dewetting of ultrathin solid films, *Phys. Rev. Lett.*, **103**, 195501 (2009).

87. H. Krishna, R. Sachan, J. Strader, C. Favazza, M. Khenner, and R. Kalyanaraman, Thickness-dependent spontaneous dewetting morphology of ultrathin Ag films, *Nanotechnology*, **21**, 155601 (2010).

88. N. Shirato, H. Krishna, and R. Kalyanaraman, Thermodynamic model for the dewetting instability in ultrathin films, *J. Appl. Phys.*, **108**, 024313 (2010).

89. A. Sharma and J. Mittal, Instability of thin liquid films by density variations: a new mechanism that mimics spinodal dewetting, *Phys. Rev. Lett.*, **89**, 186101 (2002).

90. R. Yerushalmi-Rozen and J. Klein, Stabilization of nonwetting thin liquid films on a solid substrate by polymeric additives, *Langmuir*, **11**, 2806 (1995).

91. N. Arun, A. Sharma, P. S. G. Pattader, I. Banerjee, H. M. Dixit, and K. S. Narayan, Electric-field-induced patterns in soft viscoelastic films: from long waves of viscous liquids to short waves of elastic solids, *Phys. Rev. Lett.*, **102**, 254502 (2009).

92. D. R. Barbero and U. Steiner, Nonequilibrium polymer rheology in spin cast films, *Phys. Rev. Lett.*, **102**, 248303 (2009).

93. D. Bandyopadhyay, A. Sharma, U. Thiele, and P. D. S. Reddy, Electric-field-induced interfacial instabilities and morphologies of thin viscous and elastic bilayers, *Langmuir*, **25**, 9108 (2009).

94. D. Bandyopadhyay and A. Sharma, Nonlinear instabilities and pathways of rupture in thin liquid bilayers, *J. Chem. Phys.*, **125**, 054711 (2006).

95. D. Bandyopadhyay, A. Sharma, S. W. Joo, and S. Qian, Self-organized micropatterning of thin viscous bilayers under microgravity, *Microgravity Sci. Technol.*, **22**, 273 (2010).

96. A. Pototsky, M. Bestehorn, D. Merkt, and U. Thiele, Alternative pathways of dewetting for a thin liquid two-layer film, *Phys. Rev. E*, **70**, 025201(R) (2004).

97. R. Mukherjee, D. Bandyopadhyay, and A. Sharma, Control of morphology in pattern directed dewetting of thin polymer films, *Soft Matter*, **4**, 2086 (2008).

Chapter 5

Thin Films in Complete Wetting and the Specific Case of Nematic Liquid Crystals

A. M. Cazabat,[a,b,*] U. Delabre,[a,b] C. Richard,[a,b] and Y. Yip Cheung Sang[a,b]

[a] *Université Pierre et Marie Curie, Paris, France*
[b] *Laboratoire de Physique Statistique de l'ENS, UMR 8550 CNRS, 24 rue Lhomond, 75231 Paris Cedex 05, France*
*anne-marie.cazabat@lps.ens.fr

5.1 Introduction

Complete wetting, at least if mutual miscibility of the contacting phases is discarded, seems to be a simple situation, expectedly accounted for by general laws. For instance, there is an actual separation of scales during the spontaneous spreading of a simple, pure liquid on a smooth solid substrate under inert atmosphere.[1,2] Therefore, the macroscopic dynamics is only weakly dependent on the specificities of the liquid–substrate interaction, which control the thin precursor film ahead. In contrast, the separation of scales in the dynamics does not exist on liquid substrates,[3,4] nor if the spreading fluid is a liquid crystal, due to the long-range orientational order.[5,6]

Nanoscale Liquid Interfaces: Wetting, Patterning, and Force Microscopy at the Molecular Scale
Edited by Thierry Ondarçuhu and Jean-Pierre Aimé

ISBN 978-981-4316-45-3 (Hardcover), 978-981-4364-48-5 (eBook)
www.panstanford.com

In the case of simple, pure liquids and smooth substrates, mesoscopic precursor films, that is, films whose thickness is much larger than molecular sizes, are also controlled by general laws,[1,2,7-10] while specific behavior shows up at molecular scale. In the case of liquid crystals, specific behavior is present whatever the scale. How and why such behavior is observed is the subject of the present contribution, with specific focus on the nanoscopic thickness scale. At least in the case of liquid crystals, many questions are still waiting for answers.

5.2 Thin Films in Complete Wetting: Generalities

5.2.1 *Tanner's Law*

The macroscopic contact angle θ of a simple, pure, nonvolatile wetting liquid wedge advancing with a small constant velocity U on a smooth solid substrate (Fig. 5.1) is given by the so-called Tanner's law.[1,2-17] A simple form reads:

$$\theta^3 \approx K_T \frac{\eta U}{\gamma} = K_T Ca \tag{5.1}$$

Here, η and γ are, respectively, the viscosity and the surface tension of the liquid, Ca is the capillary number, and K_T is approximately

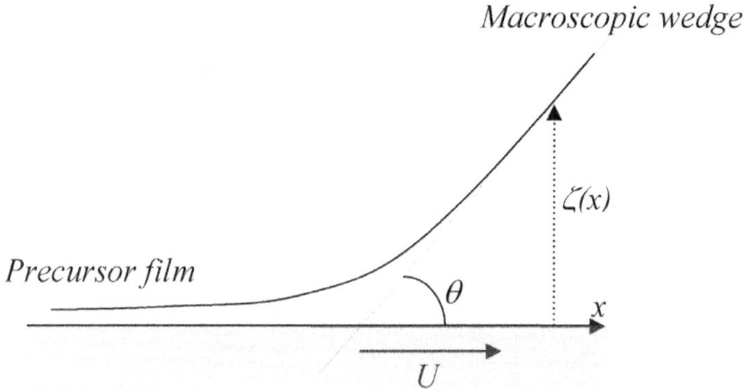

Figure 5.1 Complete wetting. Stationary, advancing liquid edge drawn in a moving frame. Here, the solid substrate moves to the right with a constant velocity U.

constant. More precisely, in the framework of de Gennes's analysis,[1] which holds at macroscopic and mesoscopic scales, the local thickness ζ is a solution of the equation:

$$U = \frac{\zeta^2}{3\eta}\frac{\partial}{\partial x}\left[\gamma\frac{\partial^2\zeta}{\partial x^2} + \Pi(\zeta)\right] \tag{5.2}$$

Equation 5.2 results from the stationary, linearized Navier–Stokes equation in the lubrication approximation with a no-slip condition on the solid substrate, complemented by a disjoining pressure term $\Pi(\zeta)$ in the mesoscopic range. The occurrence of the disjoining pressure[7] characterizes a "thin" film: the chemical potential of the molecules depends on the film thickness ζ, which means that the film is neither a three dimensional (3D) nor a two-dimensional (2D) phase. As mentioned before, thin films are mesoscopic if ζ is large compared to molecular sizes, they are treated as continuous media, and they obey general formulae. Molecular films are obviously quite specific.

In the thick macroscopic part, the profile $\zeta(x)$ obeys asymptotically (Fig. 5.1):

$$\zeta(x) \approx x\left(9\,Ca\,\mathrm{Log}\frac{x}{x_1}\right)^{1/3} \qquad x >> x_1 \tag{5.3}$$

In the same limit, the slope of the interface is

$$\left(\frac{\partial\zeta}{\partial x}\right)^3 \approx 9\,Ca\,\mathrm{Log}\,\frac{x}{x_1} \qquad x >> x_1 \tag{5.4}$$

Here, x_1 is a crossover toward the precursor film. In complete wetting, and for a mesoscopic precursor controlled by van der Waals interaction, the disjoining pressure Π is positive and is written as:

$$\Pi_{vdW}(\zeta) = \frac{A}{6\pi\zeta^3} \tag{5.5}$$

Here, A is the absolute value of the (negative) Hamaker constant. Then the crossover reads:[1]

$$x_1 \approx 2.5\;a\;(Ca)^{-2/3} \qquad a = \sqrt{\frac{A}{6\pi\,\gamma}} \tag{5.6}$$

The length a belongs to the molecular scale. Usually, the ratio x/x_1 is of the order 10^6; therefore K_T in Tanner's law is around 120. The liquid–solid interaction enters in the interface profile only through a logarithmic term. Following de Gennes's picture, the unbalanced macroscopic capillary force is dissipated in the macroscopic wedge of angle θ, while the initial spreading parameter is dissipated in the precursor. Neither line tension nor dissipation at the contact line[2,18–20] is considered.

Similar approaches, where the crossover x_1 appears as a slip length,[10,13,16,17] have been developed. We shall keep to the Gennes's presentation in the following, as in the complete wetting case the precursor film plays an actual role.

The hydrodynamic model, with a no-slip condition on the solid substrate, and supplemented by van der Waals interaction in the precursor film, accounts for macroscopic and mesoscopic scales and provides a reference base for finer, now more specific analyses. In very thin films, the no-slip condition is not valid. Moreover, short-range contributions come into play in the disjoining pressure.

In the case of liquid crystal films, both short-range and long-range contributions must be added to the van der Waals term, and the occurrence of ordered phases makes the problem more complex. Let us introduce the main characteristics of the long-range contribution due to nematic elasticity.

5.2.2 *Elementary Picture of a Nematic Film*[21,22]

Molecules in a nematic liquid crystal tend to be parallel to a common axis, which for uniaxial nematics (the sole considered here) is a symmetry axis. Opposite directions are equivalent; therefore the orientational order is quadrupolar, characterized by the nematic director $\overleftrightarrow{n}$. For sake of simplicity, the unit vector $\vec{n}$ parallel to the director is often used (formulae are invariant if $\vec{n} \to -\vec{n}$).

A consequence of long-range orientational order is the occurrence of an elastic contribution to the free energy of a distorted nematic film. Without an applied external field, distortion is induced by antagonist anchorings at the film interfaces. In a hybrid aligned nematic (HAN)[23] film, the director rotates by an angle $\Delta\theta$ inside a plane normal to the interfaces (Fig. 5.2). Let us ignore for the moment the anisotropy of the elastic response. If K_{el} is the common value of elastic constants, the contribution to the disjoining pressure

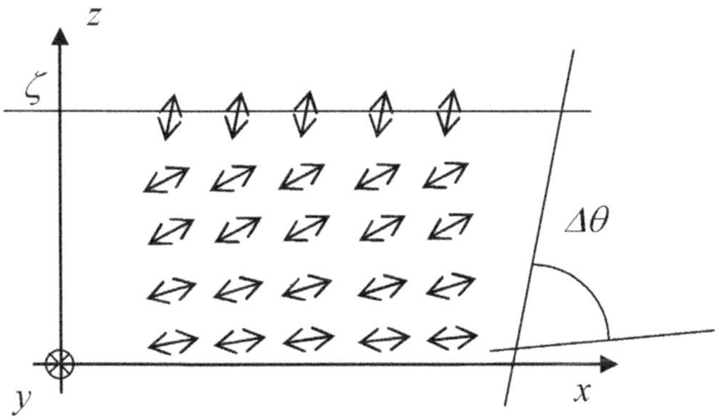

Figure 5.2 HAN structure (mesoscopic scale). The system is invariant with respect to x and y; the only variable is z. Here, the director is normal to y and rotates from bottom to top inside any given y plane. The preferred anchoring is parallel to x in the plane of the substrate ($z = 0$) and parallel to z at the free interface ($z = \zeta$). If $\zeta \to \infty$, $\Delta\theta \to \pi/2$. If $\zeta \leq d_{\mathrm{HAN}}$, $\Delta\theta = 0$, the stronger anchoring imposes a common orientation over the film. If the anchoring is stronger at the substrate, the director is everywhere parallel to x (planar film). If the anchoring is stronger at the free interface, the director is everywhere parallel to z (homeotropic film).

reads:

$$\Pi_{\mathrm{elastic}}(\zeta) = K_{\mathrm{el}} \frac{(\Delta\theta)^2}{2\zeta^2} \tag{5.7}$$

The HAN structure has a minimum thickness d_{HAN}, below which the cost of elastic energy is too large: the orientation in the film becomes homogeneous, imposed by the stronger anchoring.[24] At increasing thickness, the elastic contribution decreases slower (ζ^{-2}) than the van der Waals one (ζ^{-3}). No well-cut separation between the macroscopic wedge and the microscopic precursor can be defined any more.

Consequences are remarkable, both on the shape and on the dynamics of a macroscopic nematic droplet spreading spontaneously on a solid substrate.[5,6] When spreading proceeds, the initial spherical cap profile evolves progressively toward a characteristic bell shape, while the dynamics is accelerated with respect to the one of a simple spreading liquid. Tanner's law is no longer obeyed. The elastic contribution, significant at the drop edge, can be formally treated as some macroscopic line tension.[2,25]

However, the most intriguing features occur at nanoscopic scale, where structural interactions, anchoring, and elastic response compete to provide a complex landscape.

Before starting a detailed presentation of these contributions in the specific case of thin nematic films, let us proceed further with a brief overview of the techniques used in our research wetting team, with emphasis put on the nematic liquid crystals case. The field of wetting expanded significantly in the eighties, before the blossoming of nanoscopic probes. Interferometry,[26] ellipsometry,[27] and X-ray reflectivity were used to provide information. The specificity of these techniques is the large aspect ratio in the investigated scales: the thickness resolution is much better than the spatial resolution along the substrate. Our films or drops are quite flat. The relevant information is contained in the characteristic thicknesses and in the lateral extension of the films. The exact shape of the film edge is usually not accessible due to the poor spatial resolution.

5.3 Techniques and Investigated Systems

The spontaneous spreading of thin microdroplets on solid substrates is slow and can be conveniently monitored using spatially resolved ellipsometry. On the other hand, extended thin wetting films can be spin-coated on solids or deposited on liquid substrates and thereafter studied under a polarized microscope for nanometric thicknesses or a Brewster angle microscope for molecular ones. X-ray reflectometry provides complementary information.

5.3.1 *Brief Overview of Our Experimental Techniques*

5.3.1.1 Microdroplets: spatially resolved ellipsometry

Nonvolatile wetting microdroplets spreading on oxidized silicon wafers have been studied using spatially resolved ellipsometry.[28,29] The phase-modulated ellipsometer in Collège de France worked at the Brewster angle and a fixed wavelength, using the red line 632.8 nm of a short He-Ne laser in the TEM00 mode, which ensures high optical quality of the light beam. The well-defined Gaussian beam is focused on the sample, improving the spatial resolution. The contrast index between the underlying silicon, on the one hand, and silica plus film, on the other hand, is strong, allowing a thickness resolution of the order of 10^{-2} nm over the laser spot. The size of the laser spot on the substrate was progressively decreased from approximately 200 μm × 1 mm in the first experiments[28,29] to 30 μm × 120 μm.[30] As the lateral extension of the droplets is millimetric, then the corresponding spatial resolution is 30 μm.

While precursor films at the edge of a macroscopic spreading drop are controlled by the macroscopic edge velocity (stationary, or "adiabatic," films, in de Gennes's terminology[1]), no bulk reservoir exists for microscopic droplets. The dynamics is slow but not stationary, controlled by the gradient of the disjoining pressure.[7] At mesoscopic scale, and for slowly varying profiles, the local conservation of film thickness, written here in the simpler one-dimensional (1D) case, reads:

$$\frac{\partial \zeta(x,t)}{\partial t} + \frac{\partial}{\partial x}[\zeta(x,t)U(\zeta,t)] = 0 \tag{5.8a}$$

where

$$U(\zeta,t) = \frac{\zeta(x,t)^2}{3\eta}\frac{\partial}{\partial x}\left[\gamma\frac{\partial^2\zeta(x,t)}{\partial x^2} + \Pi(\zeta(x,t))\right] \tag{5.8b}$$

The velocity $U(\zeta,t)$ in Eq. 5.8b is merely the velocity of Eq. 5.2. It results from the stationary Navier-Stokes equation written at a given time t. This is allowed because the characteristic times for the building-up of velocity profiles of the Navier-Stokes equation are much shorter than the ones of the spreading dynamics.

In the flat part of the film, where the curvature term is negligible, Eq. 5.8a can be understood as a diffusion-like equation:

$$\frac{\partial \zeta}{\partial t} - \frac{\partial}{\partial x}\left[D(\zeta)\frac{\partial \zeta}{\partial x}\right] = 0 \tag{5.9}$$

where the thickness-dependent diffusion coefficient is given by[7]

$$D(\zeta) = -\frac{\zeta^3}{3\eta}\frac{d\Pi}{d\zeta} \tag{5.10a}$$

This means that in Eq. 5.9 the profile $\zeta(x,t)$ depends on the variable x^2/t and not independently on x and t; these films are referred to as "diffusive films." In the mesoscopic range of thickness, and for disjoining pressure controlled by van der Waals

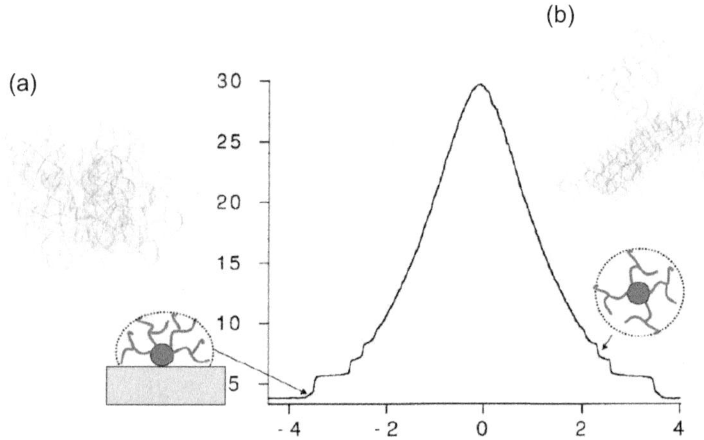

Figure 5.3 Ellipsometric profile of a TK microdroplet on a bare, oxidized silicon wafer. Vertical scale, nm; horizontal scale, mm; baseline, silica layer; lateral resolution, 30 μm; mesoscopic scale, above 10 nm, smooth profile, controlled by van der Waals interaction (Eqs. 5.9 and 5.10b); molecular scale, below 10 nm, stepped profile. In the present case, the first layer is thinner and smoother than the next ones; the conformation of the molecules changes in contact with hydrophilic silica. The molecular structures are calculated ones. *Abbreviation*: TK, tetrakis-2-ethyl hexoxysilane.

interaction, a good agreement is obtained between calculated Hamaker constants and the ones deduced from the widening of droplets profiles with time (Fig. 5.3). Here, Eq. 5.10a takes the simple form:

$$D(\zeta) = \frac{A}{6\pi\eta\zeta} \tag{5.10b}$$

In Eq. 5.10a, the factor $\zeta^3/3\eta$ corresponds to a Poiseuille flow, with no slip at the solid. Therefore, the validity of Eq. 5.10a is restricted to mesoscopic films. However, for molecular ones, the proportionality between $D(\zeta)$ and $d\Pi/d\zeta$ is retained. Note that as the droplet shape $\zeta(x, t)$ evolves slowly in time, $D(\zeta)\ \partial\zeta/\partial x$ evolves slowly in space; therefore the profile $\zeta(x, t)$ gives qualitative information on the shape of the disjoining pressure:[31] ellipsometric droplets profiles are not merely striking pictures. Let us comment further on short-range contributions.

5.3.1.2 Stepped ellipsometric profiles and structural interaction

Surface-induced layering corresponds to oscillations of the disjoining pressure in the molecular range of thickness[32] (see also chapter 1). Well-known saw-tooth force profiles are obtained with surface forces apparatus (SFA) setups.[32] In the case of microdroplets, where the thickness is free to adjust, an unstable part in the disjoining pressure behaves just like the usual pressure at three dimensions: a Maxwell construction leads to a horizontal part in $\Pi(\zeta)$, corresponding to the coexistence of films with different thickness (see chapter 3).

Stepped profiles result if the liquid is nonvolatile, not only at 3D, but also at 2D, where the number of neighbors is less.[29,30]

Then a thickness range where $\Pi(\zeta)$ is constant corresponds to a steep part in the droplet profile. The exact profile of the steps edges is not accessible using ellipsometry, because of the poor spatial resolution. This is why Eqs. 5.2 and 5.8a, where the interface slope is assumed to be small, are enough for ellipsometric studies. Although the spatial resolution is poor, the high-thickness resolution provides information about interactions and sometimes about molecular configurations in the film.

Let us take a few examples:

The thickness profiles of microdroplets of polydimethylsiloxane (PDMS) oligomers deposited on oxidized silicon wafers exhibit equal thickness steps, each 0.7 nm thick (Fig. 5.4). The number of visible steps, their lengths, and their growth dynamics depend on the surface chemistry of the wafers.[33] The common step thickness is the transverse size of the PDMS chain, which means that each step corresponds to a compact flat monolayer of chains, that is, a 2D nonvolatile liquid. Information on the disjoining pressure, and on the friction coefficients between successive layers, can be extracted.[30,31,34–36] The dynamics of adsorption of silanol ends when silanol-terminated PDMS (PDMS-OH) is deposited on the wafer can also be monitored in real time.[30,35]

With TK, the picture is somewhat different.[36] Depending on the substrate cleaning, either all the steps have equal thickness, around 1 nm, which is the diameter of the molecule in a spherical conformation, or the first step is thinner (0.8 nm) and the next

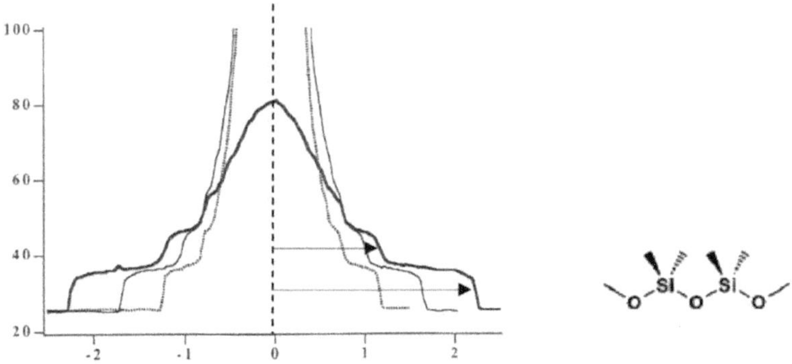

Figure 5.4 Ellipsometric profiles of a PDMS microdroplet. Right: The PDMS chain whose viscosity = 420 cP (M_P = 16 kg/mole), polydispersity index, I_P = 1.20 (fractionated sample), and surface tension, γ = 21.1 mN/m. Substrate: Silicon wafer grafted with hexamethyldisilazane; critical surface tension close to 21.5 mN/m. Vertical scale, Angström; horizontal scale, mm; baseline, silica layer; lateral resolution, 30 μm. Profiles recorded at 18 h, 44 h, and 95 h after drop deposition. Each step is a flat, compact monolayer of PDMS chains at 2D (right). (Adapted from N. Fraysse[37]).

ones thicker (1.3 nm). The latter case occurs on a high-energy substrate (Fig. 5.3), where the molecules in the first layer change their conformation due to the strong interaction with the substrate. Chemical models account satisfactorily for such a behavior.[36]

For trimethyl-terminated PDMS as well as for TK, steps correspond to superimposed compact monolayers, except very close to the wetting transition. There, a single, thicker step is observed, referred to as a "pancake" since de Gennes's analysis.[1]

5.3.1.3 Pancakes

Oscillations in the disjoining pressure may also be due to the structure of the substrate. Stratified substrates, with a weakly wettable thin surface layer, may lead to such a behavior and ultimately to a wetting transition.[34] The negative contribution to the disjoining pressure produces, after Maxwell construction, a vertical part in the drop profile, whose height increases when the wetting transition is approached (Fig. 5.5). The final state of a nonvolatile

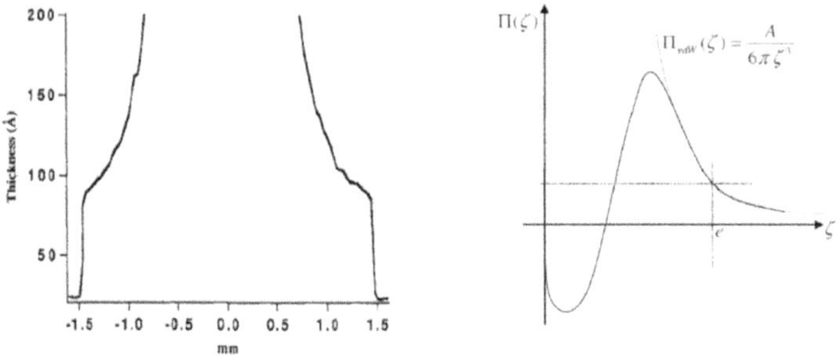

Figure 5.5 A typical pancake. A PDMS microdroplet, whose viscosity = 5,600 cP (M_P = 52 kg/mole), I_P = 1.08 (fractionated sample), and γ = 21.37 mN/m. Substrate: Silicon wafer grafted with hexamethyldisilazane; critical surface tension close to 21.5 mN/m. Pancake thickness 7.2 nm, corresponding to de Gennes's formula (Eq. 5.11a) if $\gamma_C = 21.43$ mN/m, with the Maxwell construction shown on the right. Initial spreading parameter, 0.06 mN/m; baseline, silica layer; lateral resolution, 30 μm. Profile recorded 26 days after drop deposition.[34,37]

liquid on such a substrate is a flat pancake with steep edges, whose thickness depends on the initial spreading parameter.[1,34]

The pancake thickness is given by a simple formula in the mesoscopic range, that is, very close to the wetting transition. The molecular film on the substrate has a negligible thickness, and the mesoscopic film is controlled by van der Waals interaction. Then the pancake thickness is written as:

$$e = \sqrt{\frac{A}{4\pi S_0}} \tag{5.11a}$$

$$S_0 = \gamma_{SG} - \gamma_{SL} - \gamma = \gamma_c - \gamma \int_0^{\infty} \Pi(\zeta) d\zeta \tag{5.11b}$$

Here, S_0 is the initial spreading parameter, γ_{SG} the solid–gas interfacial tension (the liquid is nonvolatile), γ_{SL} the solid–liquid interfacial tension, and γ_C the critical surface tension.[38]

Approaching a wetting transition by monitoring the substrate wettability is not easy, as getting mesoscopic pancakes requires[1] the spreading parameter S_0 to be of the order of 10^{-5} N/m, while interfacial and critical surface tension values are in the range 10^{-2}–10^{-1} N/m.

Let us now introduce our studies on nematic liquid crystals in reference to the discussion presented before.

Considering the previous examples, the behavior of nematic liquid crystal microdroplets or thin films is expected to be very rich. Surface-induced layering (see also chapter 1), which in the present case corresponds to a smectic order, is well known to occur at the free interface, which means that structural interactions play an important role in the disjoining pressure. Moreover, in the specific case of antagonist anchoring at interfaces considered here, the positive elastic contribution plays at a longer range than the van der Waals one. Surface layering and nematic elasticity depend on temperature. The droplet shape, or the observed thicknesses of deposited films, reflects this complex behavior.

5.3.1.4 Ellipsometric profiles of nematic microdroplets

Thickness profiles of microdroplets of the very common 5CB liquid crystal on bare, oxidized silicon wafers appear on Fig. 5.6.[39] The nematic–isotropic (NI) transition temperature T_{NI} is close to 35°C. The scenario is complex, both at the molecular and at mesoscopic scale.

A remarkable feature is the vertical part on the thickness profiles, whose upper boundary (UB) and lower boundary (LB) are fixed at a given temperature, whatever the drop volume and time. On a given profile, the part above the UB is smooth. Steps are visible below the LB in a narrow range of temperature.[40] This corresponds to a complex shape of the disjoining pressure: a Maxwell construction leads to a horizontal part between the LB and the UB, defining a forbidden range of thickness. Below the LB, steps, if any, reveal surface-induced layering and therefore a smectic-like structure.

Far below T_{NI} (Fig. 5.6a), ellipsometric measurements agree with X-ray reflectivity to show that the LB is a trilayer of molecules.[41] The trilayer is built as the superposition of a monolayer with ferroelectric order (the molecules are parallel to each other with the same orientation) and a bilayer of molecules with head-to-foot orientation.

Closer to the transition, the number of bilayers on top of the monolayer increases (Fig. 5.6b,c), and then the profile becomes

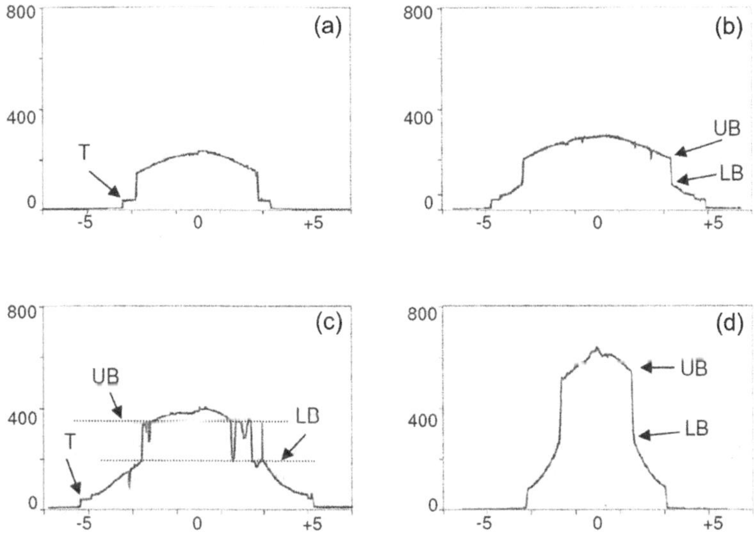

Figure 5.6 5CB microdroplets on bare, oxidized silicon wafers.[39] Vertical scale, Angström; horizontal scale, mm; lateral resolution, 20 μm; NI transition temperature, 35°C. The UB and LB are, respectively, the upper and lower boundaries of the vertical part in the ellipsometric profile; T is the trilayer. (a) 18°C, 18 h after drop deposition; (b) 28°C, 24 h after drop deposition; (c) 31°C, 30 h after drop deposition; and (d) 33°C, 10 h after drop deposition.

smooth again (Fig. 5.6d). Even if the exact number of observable bilayers is not known, considering the poor lateral resolution of the experiment, the smoothing out of the profile close to the NI transition seems to be an actual fact.

The height of the vertical part in the profile increases when the NI transition is approached, like in a wetting transition, which is actually the case: the isotropic phase above T_{NI} does not wet the wafers.[40] What is specific here is that both LB and UB increase close to T_{NI} (Fig. 5.7).

The droplet dynamics is also similar to the one observed with "pancakes." At intermediate times, the thicker part of the drop becomes flat at thickness UB. Later, this "pancake" becomes more and more narrow (sometimes breaks into narrower pancakes with the same thickness; see profile c) and finally disappears, giving

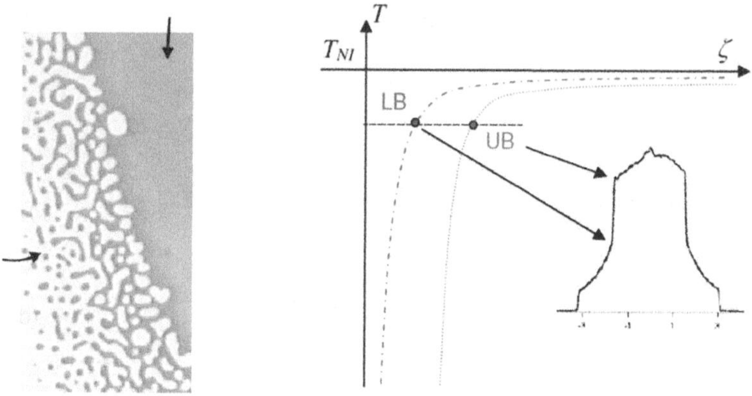

Figure 5.7 Left: Edge of a spin-coated wafer, where a weak interface slope is present. The average thickness of the 5CB film changes slowly from the UB (bottom, left) to the LB (top, right). The ratio of the area covered by UB and LB films, respectively, changes accordingly.[43] Right: Schematic "phase diagram" as a function of temperature. At a given temperature, the forbidden range of thickness is between the LB and the UB.[44] It corresponds to the vertical part in the microdroplet profile.

birth to a new pancake of thickness LB. Later, the edges of the new pancake widen very slowly.

5.3.1.5 Extended nematic films

If the nematic liquid crystal is spin-coated on a silicon wafer, films whose average deposited thickness is larger than the UB or smaller than the LB are stable with an actual thickness equal to that average thickness. For intermediate average thickness, late time coexistence is observed under a microscope between two films with respective thicknesses UB and LB, just like during 3D phase separation at the common liquid–gas transition, or in binary mixtures (Fig. 5.7)[42,43] (see also chapter 3).

More precisely, the average deposited thickness is measured at short times using ellipsometry, before "phase separation," and then the ratio of the areas covered by each film at late times allows calculating their thickness, if mass conservation is assumed. The calculated values agree with the values of the UB and LB provided by ellipsometric profiles of microdroplets. Precise thickness measurements have been performed using X-ray reflectometry[44] and confirm the point. We shall come back to this X-ray experiment in section 5.4.

One might expect the situation to be the same if the nematic liquid crystal is spread on water or glycerol in the complete wetting case. Here, polarized microscopy, or Brewster angle microscopy,[45–46] is a convenient method. In our setup, the films are observed under a microscope (Polyvar MET-Reichert-Jung) using either a black-and-white camera (CoolSnap, Photometrics) or a color one (EXI aqua, Q imaging). The Newton scale of colors is used to measure film thicknesses ζ larger than 50 nm typically. For thinner films, the calibrated gray levels of the black-and-white camera are used. Image processing provides high sensitivity, as we are able to image the boundary between a compact monolayer and a trilayer of molecules. Brewster angle microscopy[45,47] would be required to image the edge of the monolayer.

The situation in liquids has strong similarities with the one in solid substrates but is more complex. Anticipating on the experimental results reported later, we recover the forbidden range of thickness, with values of the LB and UB quite similar to the ones found on wafers, and the wetting transition.

In contrast, metastability is the rule. This means that several thicknesses may coexist both above the UB and below the LB (see Fig. 5.8). Note that we use "metastability" in a sense that is not the common one in phase transitions: no thickness is observed inside the forbidden range (LB, UB). But we observe simultaneously several thickness values, while they should correspond to different values of the free energy.

We always assume in the analysis that true equilibrium is achieved between and only between the films with thicknesses LB and UB, respectively. Such a "metastable" behavior was previously observed with smectic films on water.[46,48,49] Note that the smectic elasticity differs from the nematic one.

Another striking observation is the occurrence of long-wavelength stripe patterns in films thicker than the UB ($\zeta \geq$ UB), Fig. 5.8., while such patterns are not observed on wafers.

It is clear that a proper description of the nematic elasticity is now required to write correctly the free energy of these films, explain the origin of the patterns, and explain why they show up

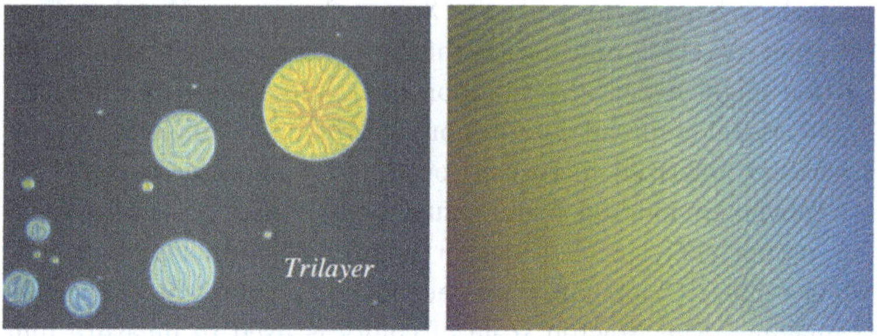

Figure 5.8 Nematic 8CB deposited on a liquid substrate far from NI and SmN transitions (left) on water and (right) on glycerol. Image size 0.18 mm × 0.13 mm. On glycerol, extended films with or without thickness gradients are the rule. In the present example, the local thickness decreases from bottom left to top right. On water, circular, striped domains of various sizes and thicknesses coexist on a homogeneous continuous phase, which is a trilayer of molecules. The Newton scale of colors allows measuring the film thickness. It is currently accepted that thickness modulations for a given domain are negligible; the changes in color are due to changes in the effective index induced by variable local orientation.

on water and glycerol but not on wafers (without entering into details, neither a forbidden range of thicknesses nor striped patterns exist in smectic films). However, the presence on both types of substrates of the same forbidden range of thickness, with quite similar values of the LB and UB, suggests that the film structures should be comparable. In fact, we shall see later that instability patterns are present also on wafers but with short wavelengths and are therefore nonobservable under a microscope. Note that they might be detectable with nanoscopic probes.

Before going into experimental details, let us present briefly the parameters required to describe thin nematic films. Comprehensive analyses are available in reference books.[21,22] Here we restrict the presentation to parameters directly useful for the interpretation of experiments. References to more elaborated approaches are provided in the text.

Substrates are isotropic ones, oxidized silicon wafers, water, or glycerol. The liquid crystals are investigated in the nematic range of temperature, where the bulk phase is a uniaxial nematic.

5.3.2 *The Physics of Thin Nematic Films*

5.3.2.1 Nematic order at mesoscopic scale

At mesoscopic, continuous scale, the nematic order is characterized by the local orientation $\vec{n}\,(\vec{r})$ and the quality of the order. We restrict ourselves to uniaxial nematics. The parameters are introduced in the frame of the Landau–de Gennes analysis,[22] which is convenient close to the NI transition, using the pedagogical presentation by Sheng and Priestley.[50] In the local coordinate system where $\vec{n}(\vec{r})$ coincides with the z-axis,[50,51] the local order parameter is given by:

$$\overleftrightarrow{Q}(\vec{r}) = S(\vec{r}) \begin{pmatrix} -1/2 & 0 & 0 \\ 0 & -1/2 & 0 \\ 0 & 0 & 1 \end{pmatrix} \tag{5.12a}$$

$$S(\vec{r}) = \frac{1}{2}\left\langle 3\cos^2\theta - 1 \right\rangle \tag{5.12b}$$

The brackets correspond to the average of the orientations of all molecules with respect to the z-axis, at point $\vec{r}$. For arbitrary orientation of $\vec{n}(\vec{r})$ relative to the external coordinate system, the components of $\overleftrightarrow{Q}(\vec{r})$ can be expressed as:

$$Q_{ij}(\vec{r}) = \frac{1}{2}S(\vec{r})[3n_i(\vec{r})n_j(\vec{r})\delta_{ij}] \tag{5.13}$$

where $i, j = 1, 2, 3$ denote the components along the three orthogonal axes of the Cartesian coordinate system and δ_{ij} is the Kronecker symbol.

The free energy density of a nematic contains powers of the order parameter and spatial derivatives.[52–56] The NI transition is weakly first order, which allows writing the first group of terms as:

$$f_1 = f_0(T) + a_1(T - T^*)S^2(\vec{r}) + a_2S^3(\vec{r}) + a_3S^4(\vec{r}) \tag{5.14}$$

where T^* is the temperature below which the isotropic phase is unstable, and is slightly less than the NI transition temperature T_{NI}. Considering the spatial derivatives, the general case is quite complex.

5.3.2.2 About variations of nematic order

Specific situations can be defined:

If $S(\vec{r}) = S$ is uniform, the spatial derivatives play only on $\vec{n}(\vec{r})$ and the corresponding terms are the usual ones associated with nematic elasticity[50,51] (see later).

If $\vec{n}(\vec{r}) = \vec{n}$ is uniform, no elasticity comes into play; however, the value of $S(\vec{r})$ changes in the vicinity of interfaces[57–65] (see also S. Jarvis, this issue). Noticeably, the nematic order $S(\vec{r})$ increases at interfaces with homeotropic anchoring or nondegenerate planar anchoring (rubbed substrates). It decreases at interfaces with planar degenerate anchoring, that is, oxidized silicon wafers and liquid substrates. This can be understood as "preferential wetting" of the interface by either the nematic (if the local order increases) or the isotropic phase (if the local order decreases) or as a "surface-phase transition."[52–56,63]

The consequences are serious for thin films, as the two interfaces interplay.[66] Noticeably, there is a shift $\Delta T \propto \zeta^{-1}$ of the transition temperature at low film thickness ζ. For identical anchorings, the shifts are moderate, a fraction of a degree: nondegenerate anchorings increase the transition temperature, while degenerate planar anchorings reduce it.[52,56,60]

With antagonist anchorings, both $\vec{n}(\vec{r})$ and $S(\vec{r})$ are variable. Without entering into details, the net result for mesoscopic films is a decrease of the transition temperature $\Delta T \propto \zeta^{-2}$, significantly larger than in the previous case.[56] An experimental confirmation is given by Wittebrood[67] in a nondegenerate planar/homeotropic configuration.

Shifts in transition temperature are studied at a fixed film thickness. The presence of a free interface allows our systems to prefer coexistence between two thicknesses, LB and UB. However, we must be aware that the ingredients of preferential wetting, confinement, and capillary condensation are present in the global evolution of the LB and UB with temperature.

5.3.2.3 Elasticity and anchoring

Let us forget preferential wetting for the moment and introduce the usual formulae for nematic elasticity and anchoring energy in the framework of the continuous description:

— The complete equations, where both $\vec{n}(\vec{r})$ and $S(\vec{r})$ are variable, are very complicated.[50] Nevertheless, in the presence of antagonist anchorings, one may reasonably assume that the main contribution to the free energy comes from the spatial derivatives of $\vec{n}(\vec{r})$ and treat $S(\vec{r})$ as being a constant S. Then the spatial gradients in the free energy appear as the usual nematic elasticity:[68]

$$F = \iiint f_{\mathrm{Frank}} dv + \iint \Big\lfloor f_s + K_{13}\vec{k} \cdot \vec{n} div\vec{n} - (K_2 + K_{24})\vec{k} \cdot (\vec{n} div\vec{n} + \vec{n} \wedge \overrightarrow{rot\vec{n}}) \Big\rfloor d\sigma \qquad (5.15a)$$

The bulk elastic energy density is written as:[69]

$$f_{\mathrm{Frank}} = \frac{1}{2}K_{11}(div\vec{n})^2 + \frac{1}{2}K_{22}(\vec{n} \cdot \overrightarrow{rot\vec{n}})^2 + \frac{1}{2}K_{33}(\vec{n} \wedge \overrightarrow{rot\vec{n}})^2 \qquad (5.15b)$$

The three (positive) elastic constants K_{11} (splay), K_{22} (twist), and K_{33} (bend) are associated to the three principal deformations. In the surface term, f_S is the contribution of the two anchorings, $\vec{k}$ is the unit vector normal to the surface and directed outward, K_{13} is the splay-bend constant, and K_{24} is the saddle-splay constant. The two last surface terms play only for thin films; the mere existence of the splay-bend constant K_{13} is a matter of debate.[70-74] In the framework of Landau–de Gennes analysis, $K_{11} - K_{33}$ and the elastic constants[50] are proportional to S^2.

— In a continuous description, anchoring terms introduce boundary conditions at interfaces. On isotropic substrates like oxidized wafers, water, or glycerol and for flat films (without thickness gradient), the only condition is on the preferred orientation with respect to the surface normal, that is, the polar angle θ. In the cases considered here, the preferred orientation at the free interface is along the normal ($\theta = 0$, homeotropic anchoring). The preferred orientation at the substrate is degenerate, in the plane of the substrate ($\theta = \pi/2$, planar anchoring). In the following, the subscript H will be used for the free interface and P for the substrate. At mesoscopic scale, the anchoring free energy per unit area can be written as:[75]

$$f_s = \frac{1}{2} W_H \sin^2 \theta_H + \frac{1}{2} W_P \cos^2 \theta_P \tag{5.15c}$$

At the free interface, W_H is the homeotropic anchoring energy and θ_H the actual angle between $\vec{n}$ and the normal to the film. On the substrate, W_P is the planar anchoring energy and θ_P the actual angle between $\vec{n}$ and the normal.

The anchoring energy decreases close to the NI transition.[76–78]

— Considering the range of investigated thicknesses (see Fig. 5.6), the continuous description will be relevant for $\zeta \geq$ UB, even if the thinnest films far from the NI transition are somewhat at the verge. For $\zeta \leq$ LB, the situation is more complex. On the molecular side, monolayers and trilayers are well characterized; this is the case far from the NI transition. Closer to the transition, the description of these films is still a matter of debate. Noticeably, a general shape of the free energy, allowing us to describe both thinner and thicker films (see also G. Reiter, this issue) and therefore to account for the forbidden range of thickness (LB, UB), is not yet available.

Let us now discuss more specifically experiments on extended nematic films, either spin-coated on silicon wafers or deposited on water or glycerol. Usually, a solution of the liquid crystal in hexane is deposited, allowing a fast spreading on the substrate, followed by a rapid evaporation of hexane. The first experiments with 5CB on wafers reported before have been completed by a systematic investigation of the series of n-cyanobiphenyls on water and glycerol in the nematic range. The aim of the study is to define a common scheme of behavior for the *n*CB series on the three isotropic substrates, as a base for the theoretical analysis. Specific deviations from the common scheme must also be understood.

5.4 Experimental Studies of Extended Nematic Films

5.4.1 *Investigated Liquid Crystals: The Series of n-Cyanobiphenyls*

5.4.1.1 General properties

The molecules of n-cyanobiphenyl liquid crystals, schematized on Fig. 5.9, are known to be relatively stable. They form dimers in bulk.[79]

As previously said, the preferred anchoring is homeotropic at the free interface and planar on silica, water, and glycerol. Transition temperatures available in the literature are, for the usual compounds:

$$\begin{aligned}
T_{\text{solid–nematic}} &= T_{\text{SN}} = 24^\circ\text{C for 5CB, } 14.5^\circ\text{C for 6CB, } 30^\circ\text{C for 7CB}\\
T_{\text{solid–smectic}} &= T_{\text{SSm}} = 21.5^\circ\text{C for 8CB}\\
T_{\text{smectic–nematic}} &= T_{\text{SmN}} = 33.5^\circ\text{C for 8CB}\\
T_{\text{nematic–isotropic}} &= T_{\text{NI}} = 35.3^\circ\text{C for 5CB, } 29^\circ\text{C for 6CB, } 42.8^\circ\text{C for}\\
&\qquad 7\text{CB, } 40.5^\circ\text{C for 8CB}
\end{aligned}$$

While purity is not critical in most bulk studies, where a mere shift of the NI transition temperature is often the sole consequence of the presence of impurities, the problem is severe as soon as interfaces are concerned.[80] We found it necessary to use liquid

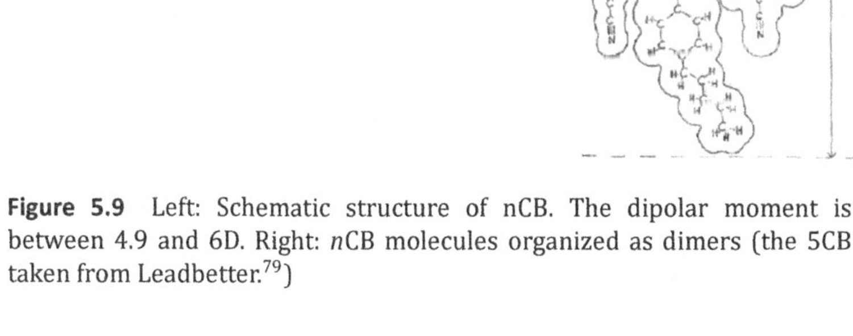

Figure 5.9 Left: Schematic structure of nCB. The dipolar moment is between 4.9 and 6D. Right: *n*CB molecules organized as dimers (the 5CB taken from Leadbetter.[79])

crystal samples, purity >99%, purchased from Frinton. These "pure" samples wet silicon wafers, water, and glycerol in the nematic range of temperature, although 5CB and 7CB are clearly at the verge on water substrates. The isotropic phase does not wet, at least on silicon wafers and water, which means that the NI transition coincides with a wetting transition, as already mentioned.[40,41,81]

Note that experiments above room temperature are difficult with liquid substrates, due to possible convection in the liquid, especially with water, which is less viscous than glycerol. On glycerol, two experimental difficulties are met: first, glycerol is hygroscopic and must be kept in dry atmosphere, and second, the solubility of glycerol molecules inside the liquid crystal is not negligible, especially for short cyanobiphenyls, and increases with increasing temperature.[81]

5.4.1.2 Observed behavior: similarities and differences

Let us list them briefly:

Similarities: For all investigated *n*CB (from 5 to 8), and for the three substrates, there is a forbidden range of thickness (LB–UB) in the bulk nematic range of temperatures. Far below the NI transition, the LB is a trilayer of molecules (3–4 nm), while the UB is around 20–30 nm.

When approaching the NI transition, both LB and UB increase; they diverge at the transition.

Partial similarities: For all investigated *n*CB, and for the two liquid substrates, metastable states are observed. Several films with different thicknesses $\geq$ UB may be in contact with the LB film (see Fig. 5.8). The LB film may be in contact with thinner films. Contact between these thinner films and films $\geq$ UB is not observed during significant time scales.

For all investigated *n*CB, and for the two liquid substrates, stripe patterns are observed in a well-defined range of thickness between the UB and an upper threshold D (0.5–1 μm). At a given temperature, the wavelength λ depends only on the film thickness ζ. – $\lambda(\zeta)$ increases faster than linearly and is typically 100 times larger than ζ. A typical example is given in Fig. 5.10.

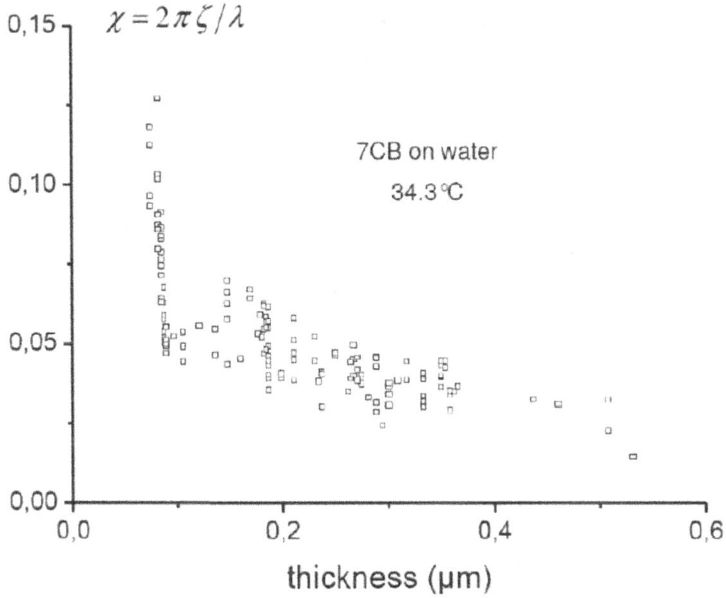

Figure 5.10 The parameter χ is the proper one to characterize stripe patterns. Long-wavelength patterns correspond to $\chi << 1$, which allows specific approximations in the theoretical analyses. The dispersion of the points is quite typical of experiments on a heated water substrate, which are especially critical with 7CB, due to wettability problems. Above 0.5 μm, the usual defects of nematic phases replace stripes. The threshold UB has not been reached in the present experiment.

Although stripe patterns are controlled by a competition between elasticity and anchoring, which both decrease strongly close to the NI transition, the main effect on $\lambda(\zeta)$ is the increase of the UB and seemingly a decrease of D. The curve $\lambda(\zeta)$ itself depends only weakly on temperature.

Differences: On wafers, only the two thicknesses LB and UB are observed (see Fig. 5.7). No stripe is visible under the microscope on the UB film.

On water, flat, striped, circular domains are observed (see Fig. 5.8). If two domains with different thickness merge, the thickness gradient relaxes rapidly (seconds), and the final domain is flat and circular.

On glycerol, extended striped films are observed, with (see Fig. 5.8) or without thickness gradients. $\lambda(\zeta)$ does not depend on the presence of gradients; ζ is merely the local thickness. The relaxation of these gradients is extremely slow (day). One might invoke the high viscosity of (dry) glycerol at room temperature (1.5 Pa.s, i.e., 1,500 times larger than the one of water), but the trend persists with 7 and 8CB, that is, around 40°C, where the viscosity is less (0.4 Pa.s). We believe that the persistence of thickness gradients in glycerol is related to anchoring energy values and not merely to hydrodynamic time scales (see next).

5.4.1.3 Elastic constants and anchoring energies

Measured values of the Frank elastic constants K_{11}, K_{22}, and K_{33} and their temperature dependence are available in the literature.[82-84] The common order of magnitude far from T_{NI} is around 10^{-11} N. K_{11} and K_{33} are close (they are equal in the Landau–de Gennes analysis). For the 5CB, $K_{33} \approx 1.3K_{11}$, while $K_{22} \approx 0.5K_{11}$. Similar trends are found for the other compounds. Note that the nematic range is relatively narrow in the 8CB and that K_{33} diverges at the (second order) SmN transition.

Considering the surface elasticity, the situation is more complex, as direct measurements are not possible. As previously mentioned, the mere existence of K_{13} in a continuous description is still a matter of debate. In 5CB, NMR studies[85] suggest that $-0.6K_{11} \leq K_{24} \leq 0.6K_{11}$.

The value of the anchoring energy W_{H} at the 5CB free interface is around 10^{-5} J.m^{-2} at room temperature.[86] It is currently accepted that:

$$W_{\mathrm{Pglycerol}}, W_{\mathrm{Pwater}} > W_{\mathrm{H}} > W_{\mathrm{Psilica}} \tag{5.15d}$$

even if the measured values on silica are considerably scattered.[77,78,87] This means that the homeotropic anchoring at the free interface is stronger than the planar one on silica, while this is the opposite on liquid substrates. We must keep in mind that the reference formula of Eq. 5.15c is written at mesoscopic scale and that the physical processes globally referred to as "anchoring" depend on the scale.[88,89]

5.4.2 *"Phase Diagrams" and Coexisting Phases*

5.4.2.1 Phase diagrams

Figure 5.7 shows a schematic picture of the phase diagram of 5CB on a silicon wafer. The high-quality diagram in Fig. 5.11 has been obtained using X-ray reflectometry.[44,90] An interesting point in the data analysis is the value of free-surface roughness required to fit the data.

For isotropic films above T_{NI}, and LB films below T_{NI}, the roughness is the same, typically 0.45 nm. This is an expected value, considering the contribution of thermal fluctuations at a 5CB free interface. Below T_{NI}, for UB films, and for uniform nematic films at larger thickness, the roughness is again the same but significantly larger than in the previous case, around 0.75 nm. One may assume that defects on the solid substrate are more efficiently coupled to the free-interface shape when the order parameter is larger.[44,90] Another explanation is the presence of short-wavelength stripe

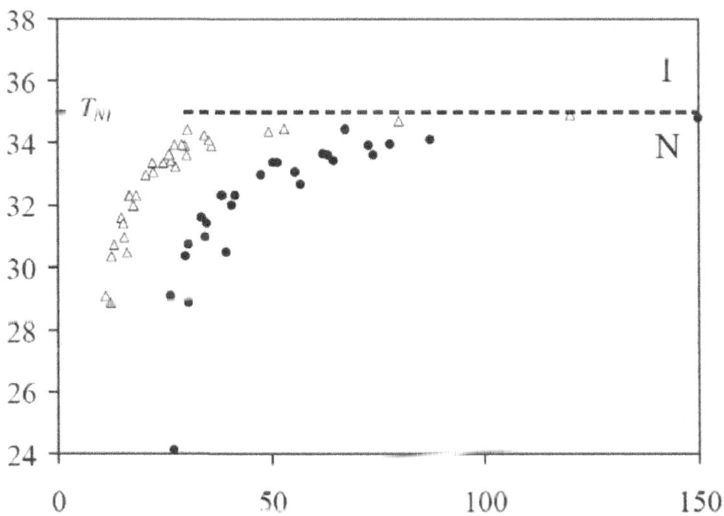

Figure 5.11 Measured phase diagram of 5CB on an oxidized silicon wafer using X-ray reflectivity (adapted from Van Effenterre[90]). Vertical scale, temperature °C; horizontal scale, thickness nm. Triangles: LB values; full circles, UB values.

patterns, with $\lambda \approx \zeta$. As we shall see in the following, the occurrence of such stripe patterns on silica is theoretically predicted. Nanoscopic probes might provide an answer to this open question.

The phase diagram of the nCB on water is globally compatible with the previous one but with some specific features and a much larger scattering of the points, due to the hydrodynamic mobility of the substrate, and the simultaneous presence of thicknesses $\zeta \geq$ UB and $\zeta \leq$ LB. Actually, the UB is defined as the thinnest striped film. Even if many experiments with decreasing average thickness are performed, we cannot prove that the thinnest possible film, that is, the UB, has actually been observed.

Considering the LB, the smooth variation obtained on silica (see Figs. 5.6, 5.7, and 5.10) is not recovered. The most reliable system is 6CB on water, where T_{NI} is the lowest (29.5±0.1°C in pure samples). The trilayer, around 3 nm thick, is observed till 1.5°C below T_{NI}; then the LB jumps suddenly to mesoscopic values, around 40 nm at 28°C and 60 nm at 29°C. Closer to the transition, thermal fluctuations cause the system to oscillate continuously between striped films and the LB because the temperature width of the forbidden range at given thickness tends to zero.

On glycerol, the diffusion of glycerol molecules inside the liquid crystal does not allow concluding unambiguously on the LB side; therefore we did not try to explore the vicinity of the NI transition.

5.4.2.2 Coexisting phases

In the thicker films $\zeta \geq$ UB, the simple HAN structure introduced in Fig. 5.2 is deeply altered by oscillations of the nematic director out of the initial xz plane, giving rise to stripe patterns. The continuous description of nematic films with antagonist anchoring, Eq. 5.15a–c are used. Considerable theoretical work has been devoted to this problem.[23,73,86,91–95] The difficulty is that Eq. 5.15a–c are complex and that the oscillations of the director are large, typically $\pm\ \pi/4$ out of the plane, which makes any linear analysis questionable.[92,94] Long-wavelength stripes $\lambda >> \zeta$ are predicted if the planar anchoring is larger; this is the case on water and glycerol.

Short-wavelength stripes $\lambda \approx \zeta$ are predicted if the homeotropic anchoring is larger; this is the case on silica.

Stripes observed under a microscope correspond to changes in the local orientation of the director, inducing changes in the local effective index in normal incidence. We do not know if the local thickness of the film is also modulated, as surface scattering using a laser beam does not allow separating thickness and optical index contributions. As previously mentioned, nanoscopic probes could provide an answer in the case of silicon wafers, where the characteristic length scales belong to the proper range.

The structure of films $\zeta \leq$ LB is still largely unknown. Far from the NI transition, the LB is a trilayer. Very close to the transition, it might become like an isotropic phase with a homogeneous, weak value of the order parameter.[90] In between, we merely know that smectic order is induced by the free interface and that the degenerate planar anchoring on the substrate decreases the local order parameter.[61,78,96] No birefringence is observed. This excludes a homogeneous planar structure but does not distinguish between homeotropic and isotropic structures.

Therefore, it is wise to restrict the conclusion to a few accepted points:

(a) The occurrence of antagonist anchorings is the main origin of the specific behavior of the films.
(b) The free energy of striped films is considerably reduced with respect to the one of the HAN state. While a linear perturbation analysis accounts acceptably for $\lambda(\zeta)$ in the intermediate range of thickness[86] (see Fig. 5.10), the HAN state is not the proper reference to discuss UB values. Noticeably, the minimum thickness d_{HAN} of the HAN state, below which the orientation of the director becomes homogeneous,[24] is around 0.35 μm for 5CB on glycerol,[94] much larger than the UB.
(c) Except possibly very close to T_{NI}, UB values can rather be estimated as the threshold thickness between a homogeneous film (fully planar on liquid substrates) and a striped one.[81,93,97]
(d) No general expression of the free energy of films $\zeta \leq$ LB is available. A fortiori, no formula covering the whole range of thicknesses and allowing to deduce LB and UB values from some "common tangent" construction is available.

5.4.3 *Line Tension of 6CB Films on Water*

Up to now, only flat films have been considered. Although their precise structure is not resolved in our experiments, film edges and the associated line tension contribute to the free energy and deserve further study. The results are significantly different depending on the distance from the NI transition, with a sudden change of behavior 1.5°C below T_{NI}.

5.4.3.1 Line tension far from T_{NI}

We have investigated line tensions in the 6CB/water system. For $T \leq 27$°C, the striped films are in contact with the trilayer, which, at low coverage, is in contact with the monolayer.[99] Both static measurements of 2D curvature and Laplace pressure (Fig. 5.12), and dynamic ones, where the relaxation dynamics of coalescing films is recorded, have been performed.

In the latter case, one must find two domains with approximately the same thickness and the same radius, largely surrounded by the outer phase (Fig. 5.13). At late times, the relaxation toward a circular shape with final radius R_F results from the competition between the

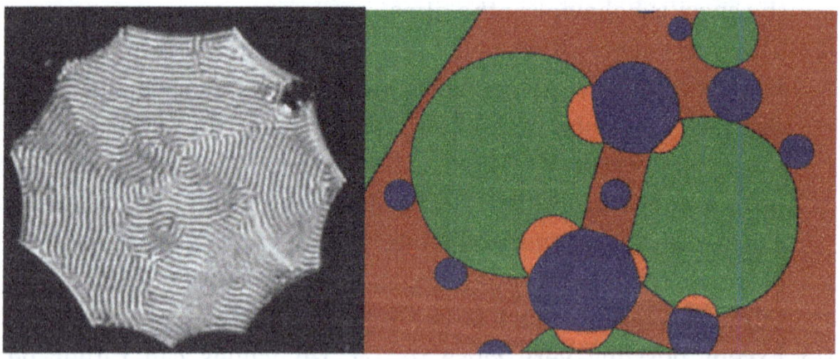

Figure 5.12 Observations at low coverage. Left: A funny picture of a striped 6CB domain on water (approximate size 0.16 mm × 0.16 mm). Each curved part in the perimeter is in contact with a domain of trilayer, which is itself in contact with the monolayer. Image processing would allow observing the edge of the trilayer. Right: Reconstruction of a typical configuration (blue: striped domains; red: trilayer; green: monolayer). (Adapted from Delabre.[98])

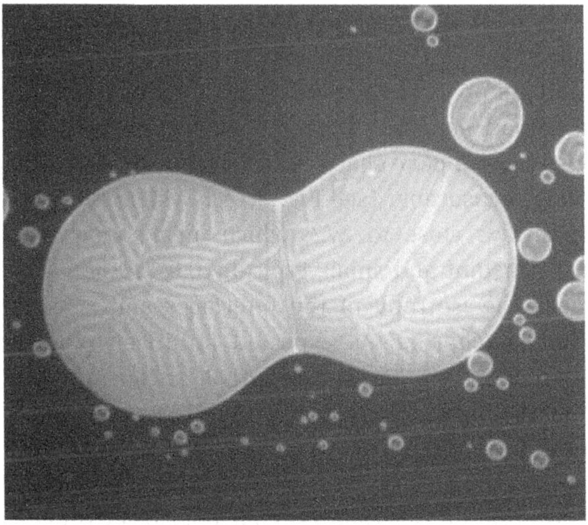

Figure 5.13 6CB on water far below T_{NI}. Coalescence between two striped domains with the same size and thickness. The continuous surrounding phase is a trilayer of molecules (from Delabre.[98])

driving term, that is, the line tension τ and the viscous dissipation in the liquid substrate.

The relaxation is an exponential one,[100] with a characteristic time T_{rel}.

$$T_{rel} = \frac{5\pi}{16}\frac{\eta_b R_F^2}{\tau} \tag{5.15e}$$

Here, η_b is the bulk viscosity of the liquid substrate.

The measured value of the line tension τ_{TM} between the trilayer and the monolayer is[99] 11±3 pN, in good agreement with the one measured with 8CB on water in the smectic range.[47,49] This is expected, as the nature of the bulk phase does not play at that scale.

The line tension $\tau_{NT}(\zeta)$ between nematic striped domains and trilayer increases linearly with ζ up to 0.2 μm typically, with a slope 0.7 mN/m, $\tau_{NT}(\zeta) \approx 120$ pN for $\zeta = 0.2$ μm, and then saturates[99] around 200 pN.

In comparison, the line tension between smectic domains of 8CB on water and the trilayer has been investigated by Mann *et al.*[49] Here the slope is larger, 3.3 mN/m, that is, 11 pN per added bilayer, and well accounted for by the folding of the smectic layers at the edge of the domain.[101]

Experiments provide some information on the structure of the edges of striped nematic domains. In contrast with the smectic case, where they appear sharp under an optical microscope, here a width w is measurable (see Fig. 5.14). Systematic studies show that w is always close to $\lambda/2$, which means that the slope of the edge is constant and small, typically 0.5°. Although weak, this slope breaks the isotropy of the interfaces and is equivalent to an azimuthal anchoring. As expected intuitively, and confirmed by calculations,[23,81,93,102] an azimuthal anchoring reduces the excursion of the director out of the initial HAN plane and efficiently restrains the formation of stripes. Therefore the edge of a domain behaves like a HAN structure below d_{HAN}, that is, it becomes planar. Actually, observation between crossed polarizers proves that the edge has a planar structure (Fig. 5.14). Interestingly, the measured line tensions agree well with a very primitive model, assuming that the excess energy at the edge is mainly due to the anchoring energy of the planar area:[99]

$$\tau_{\text{NT}}(\zeta) \approx w(\zeta) W_{\text{H}} \approx \frac{\lambda(\zeta)}{2} W_{\text{H}} \tag{5.15f}$$

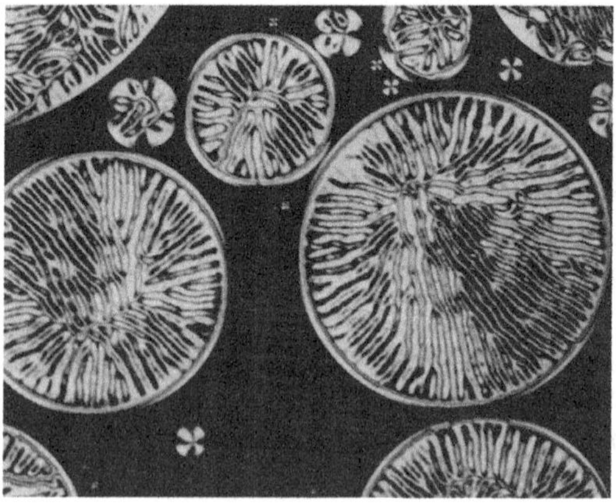

Figure 5.14 6CB on water far below the NI transition. All the domains here have the same color, that is, the same thickness. Between crossed polarizers, the edge of the striped domains shows a planar structure (crossed extinctions). The domains smaller than the wavelength have a planar radial structure with a defect in the center.

One might wonder why on glycerol stripes exist in the presence of thickness gradients. A first answer is that the gradients observed on glycerol (Fig. 5.8) are at least 10 times less than the previous ones. A second answer is that as the azimuthal anchoring associated to a given gradient scales as the inverse of the planar anchoring W_P, W_P should be significantly larger on glycerol than on water.[98,99] This requires experimental confirmation.

5.4.3.2 Line tension close to T_{NI}

As previously mentioned, the picture changes 1.5°C below T_{NI}. Several configurations show up, some of them with a short lifetime. Long-life configurations correspond to striped films in contact with a mesoscopic LB film, which is itself in contact with a monolayer. Surprisingly, the trilayer appears as a transient structure when it is in contact with the mesoscopic LB film. At very low coverage, the coexistence between the monolayer and the trilayer is observable. In that case, the measured line tension τ_{TM} is 13±2 pN, similar to the one far from T_{NI}. This agrees with the fact that the nature of the bulk phases does not play at that scale.

Measuring the line tensions $\tau_{NLB}(\zeta)$ between striped nematic films and the LB and $\tau_{LBM}(\zeta)$ between the LB and the monolayer is quite difficult, considering temperature fluctuations and mobility of the liquid substrate. Noticeably, we are not able to make the difference between measurements done at different temperatures in the range $T_{NI} - T \leq 1.5$°C.

$\tau_{NLB}(\zeta)$ behaves qualitatively as $\tau_{NT}(\zeta)$ far from T_{NI} but is 10 times smaller (see Fig. 5.15). Unexpectedly, the width of edges of the striped domains is still $w \approx \lambda/2$ (see Fig. 5.16), that is, it does not change significantly. The primitive model above (Eq. 5.18) would therefore suggest that W_H is 10 times less than far from T_{NI}. This is plausible, considering the temperature variations of W_H close to T_{NI} at the free surface of 4-methoxybenzylidene-4'-butylaniline (MBBA)[76] but not proved. Moreover, this model ignores the nature of the LB phase in contact with the striped nematic.

In the present case, the thickness of the LB film is between 40 and 70 nm, and the edge in contact with the monolayer is sharp (see Fig. 5.16). The line tension $\tau_{LBM}(\zeta)$ value is $\tau_{LBM}(\zeta) = 37\pm8$ pN, whatever the thickness of the LB, considering the scattering of the points.

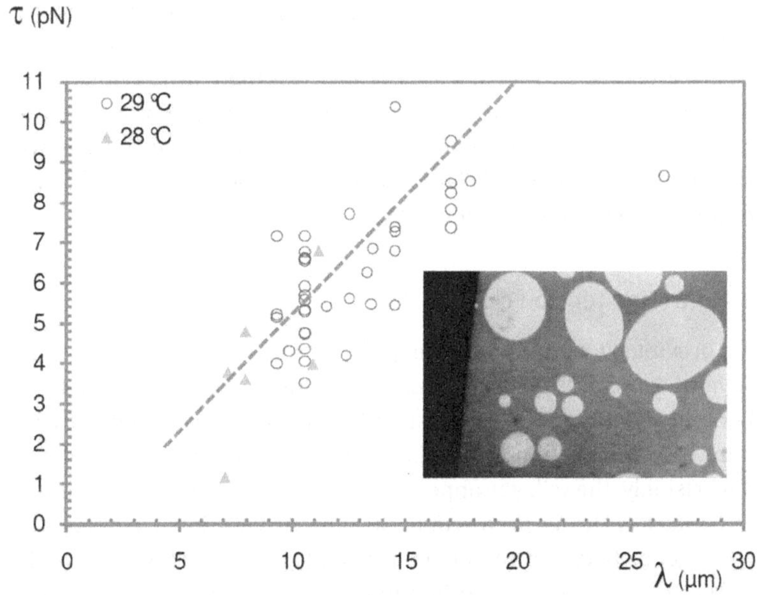

Figure 5.15 6CB on water close to the NI transition ($T_{NI} = 29.5 \pm 0.1°C$). Line tension between striped nematic domains and an LB film. Inset: On the right, the striped domains are in contact with the LB film; on the left, the LB film is in contact with the monolayer. (Adapted from Delabre.[81])

As for the LB structure, no birefringence is measurable; therefore a planar structure is discarded. As the trilayer becomes unstable, this suggests that a homeotropic configuration is not plausible. It is therefore tempting to refer to the Landau–de Gennes analysis[56,67] and assume that the structure of the LB is close to isotropic, with a low value of the order parameter, as expected close to a disordering substrate. This is quite probable but does not help to write equations explicitly.

5.5 Summary and Conclusion

Complete wetting situations generate films whose thickness meets nanoscopic scale. Extended thin films and almost flat microdroplets are usual investigated systems. They have in common a large aspect

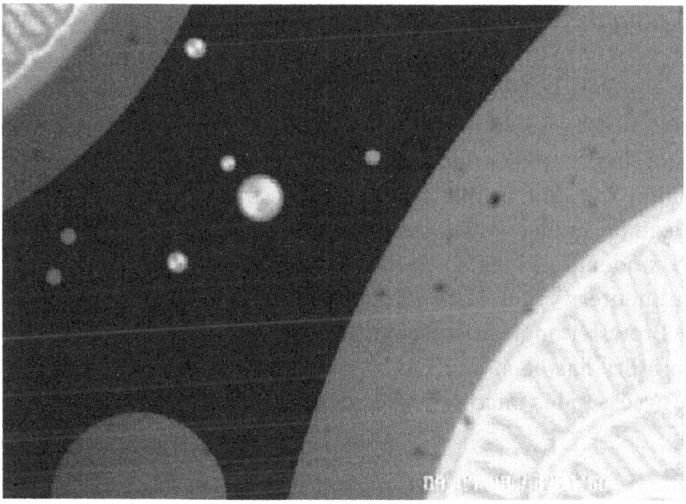

Figure 5.16 Same situation as the inset before, with a larger magnification. The edge and stripe patterns of the striped domains (top left and bottom right) are well visible. The width is half the wavelength, just like on Fig. 5.14, 5 far from T_{NI}. The edge of the LB film appears sharp. Size of the picture: approx. 0.07 mm × 0.10 mm.

ratio in the length scales. This allowed using spatially resolved ellipsometry, interferometry, or optical microscopy to investigate such systems before the blossoming of nanoscopic probes.

The case of pure, isotropic, nonvolatile liquids spreading on solid substrates is well characterized and understood in terms of the various short-range and long-range contributions to the disjoining pressure. Wetting transitions take an easy place in the general description. Going from mesoscopic to microscopic scale does not generate specific difficulties. Conformational changes of molecules in contact with the substrate can even be analyzed.

The occurrence of long-range order in a pure, nonvolatile liquid crystal does no longer allow to separate length scales so easily. In the case of thin nematic films, the mere structure of the phases if often under debate, and the coincidence between the wetting transition and the NI transition is not yet fully understood. Here, a proper imaging of the film's free surface and edge on silicon wafers using nanoscopic probes might provide fruitful information.

Acknowledgments

Our experimental work on wetting was initiated in Collège de France by Pierre-Gilles de Gennes. Working there was invaluable. François Heslot built up the ellipsometer, which during years provided striking thin-film profiles. The liquid crystal venture started in Collège de France with Marie-Pierre Valignat and proceeded with PhD students, Sébastien Bardon, François Vandenbrouck, Damien Van Effenterre, and, later, Christophe Poulard and Geoffroy Guéna. The Brewster angle microscope at Ecole Normale Supérieure (ENS) laboratory has been built up by Jacques Meunier, who also gave us free access to his microscopes and high-sensitivity cameras.

References

1. P. G. de Gennes, Wetting, statics and dynamics, *Rev. Mod. Phys.*, **57**, 827 (1985).
2. S. Dietrich, M. Rauscher, and M. Napiorkowski, "Wetting phenomena on the nanometer scale," in *Nanoscale Liquid Interfaces: Wetting, Patterning and Force Microscopy at the Molecular Scale* (this issue).
3. J. F. Joanny, Wetting of a liquid substrate, *PhysicoChem. Hydrodyn.*, **9**, 183 (1987).
4. J. G. E. M. Fraaije and A. M. Cazabat, Dynamics of spreading on a liquid substrate, *J. Colloid Interface Sci.*, **133**, 452 (1989).
5. C. Poulard and A. M. Cazabat, Spontaneous spreading of nematic liquid crystals, *Langmuir*, **21**, 6270 (2005).
6. C. Poulard, M. Voué, J. De Coninck, and A. M. Cazabat, Spreading of nematic liquid crystals on hydrophobic substrates, *Colloids Surf.* **A282**, 240 (2006).
7. B. V. Derjaguin, Z. M. Zorin, N. V. Churaev, and V. A. Shishin, *Wetting, Spreading and Adhesion*, ed., J. F. Padday., London: Academic Press (1977).
8. B. V. Derjaguin, N. V. Churaev, and V. M. Muller, *Surface Forces*. New York and London: Consultant Bureau (1987).
9. G. F. Teletzke, H. T. Davis, and L. E. Scriven, Wetting hydrodynamics, *Rev. Phys. Appl.*, **23**, 989 (1988).

10. N. V. Churaev, Wetting films and wetting, *Rev. Phys. Appl.*, **23**, 975 (1988).
11. L. H. Tanner, The spreading of silicone drops on horizontal surfaces, *J. Phys. D*, **12**, 1473 (1979).
12. V. E. B. Dussan, On the spreading of liquids on solid surfaces: static and dynamic contact lines, *Annu. Rev. Fluid Mech.*, **11**, 371 (1979).
13. R. G. Cox, The dynamics of the spreading of liquids on a solid surface. Part 1. Viscous flow, *J. Fluid Mech.*, **168**, 169 (1986).
14. V. E. B. Dussan, E. Ramé, and S. Garoff, On identifying the appropriate boundary conditions at a moving contact line: an experimental investigation, *J. Fluid Mech.*, **230**, 97 (1991).
15. J. A. Marsh, S. Garoff, and E. Dussan, Dynamic contact angles and hydrodynamics near a moving contact line, *Phys. Rev. Lett.*, **70**, 2778 (1993).
16. O. V. Voinov, Meniscus and precursor film in steady-state wetting, *J. Colloid Interface Sci.*, **201**, 127 (1998).
17. O. Vinogradova, "Wetting, roughness and hydrodynamic slip," in *Nanoscale Liquid Interfaces: Wetting, Patterning and Force Microscopy at the Molecular Scale* (this issue).
18. T. D. Blake, The physics of the moving contact line, *J. Colloid Interface Sci.*, **299**, 1 (2006).
19. S. Mechkov, G. Oshanin, M. Rauscher, A. M. Cazabat, M. Brinkmann, and S. Dietrich, Contact line stability of ridges and drops, *Europhys. Lett.*, **80**, 66002 (2007).
20. S. Mechkov, G. Oshanin, and A. M. Cazabat, Post-Tanner stages of droplet spreading: the energy balance approach revisited, *J. Phys. Condens. Matter*, **21**, 464131 (2009).
21. S. Chandrasekar, *Liquid Crystals*, 2nd ed. Cambridge University Press (1992).
22. P. G. de Gennes and J. Prost, *The Physics of Liquid Crystals*. 2nd ed. Oxford: Clarendon Press (1993).
23. A. Sparavigna, L. Komitov, B. Stebler, and A. Strigazzi, Static splay-stripes in a hybrid aligned nematic layer, *Mol. Cryst. Liq. Cryst.*, **207**, 265 (1991).
24. G. Barbero and R. Barberi, Critical thickness of a hybrid aligned nematic liquid crystal cell, *J. Phys.*, **44**, 609 (1983).
25. S. Mechkov, G. Oshanin, and A. M. Cazabat, Post-Tanner spreading of nematic droplets, *J. Phys. Condens. Mat.*, **21**, 464134 (2009).

26. Z. M. Zorin, D. Platikanov, and T. Kolarov, The transition region between aqueous wetting films on quartz and the adjacent meniscus, *Colloids Surf.*, **22**, 147 (1987).
27. L. Léger, M. Erman, A. M. Guinet-Picard, D. Aussérré, and C. Strazielle, Precursor film profiles of spreading liquid drops, *Phys. Rev. Lett.*, **60**, 2390 (1988).
28. F. Heslot, A. M. Cazabat, and P. Levinson, Dynamics of spreading of tiny drops, *Phys. Rev. Lett.*, **62**, 1286 (1989).
29. F. Heslot, N. Fraysse, and A. M. Cazabat, Layering phenomena in the spreading of tiny drops, *Nature*, **338**, 640 (1989).
30. N. Fraysse, M. P. Valignat, A. M. Cazabat, F. Heslot, and P. Levinson, The spreading of layered microdroplets, *J. Colloid Interface Sci.*, **158**, 27 (1993).
31. P. G. de Gennes and A. M. Cazabat, Etalement d'une goutte stratifiée incompressible, *C.R.A.S. Paris, Série II*, **310**, 1601 (1990).
32. J. N. Israelachvili, *Intermolecular and Surface Forces*. London: Academic Press (1985).
33. F. Heslot, A. M. Cazabat, P. Levinson, and N. Fraysse, Wetting at the microscopic scale; influence of surface energy, *Phys. Rev. Lett.*, **65**, 599 (1990).
34. M. P. Valignat, N. Fraysse, and A. M. Cazabat, Multilayered "pancakes" of non volatile liquids close to a wetting transition, *Langmuir*, **9**, 3255 (1993).
35. M. P. Valignat, N. Fraysse, A. M. Cazabat, P. Levinson, F Heslot, and M. Cazabat, Some specific aspects of the behavior of ultrathin films, *Colloids Surf. A*, **83**, 193 (1994).
36. S. Villette, M. P. Valignat, A. M. Cazabat, L. Jullien, and F. Tiberg, Wetting on the molecular scale and the role of water. A case study of wetting of hydrophilic silica surfaces, *Langmuir*, **12**, 825 (1996).
37. N. Fraysse, Etude ellipsométrique de films de mouillage ultra-minces à l'équilibre (adsorption en phase vapeur) et en situation dynamique (étalement). PhD Thesis. Paris: Université Pierre et Marie Curie (1991).
38. W. Zisman, Relation of the equilibrium contact angle to liquid and solid constitutions. In *Contact Angle, Wettability and Adhesion, Chemical Series*. Vol. 43. Washington: ACS. 1 (1964).
39. S. Bardon, Organisation des cristaux liquides au voisinage d'une surface solide. PhD Thesis. Paris: Université Pierre et Marie Curie (1999).

40. F. Vandenbrouck, S. Bardon, M. P. Valignat, and A. M. Cazabat, Wetting transition and divergence of the extrapolation length near the nematic-isotropic transition, *Phys. Rev. Lett.*, **81**, 610 (1998).

41. S. Bardon, R. Ober, M. P. Valignat, F. Vandenbrouck, A. M. Cazabat, and J. Daillant, Organization of cyanobiphenyl liquid crystal molecules in prewetting films spreading on silicon wafers, *Phys. Rev. E*, **59**, 6808 (1999).

42. F. Vandenbrouck, M. P. Valignat, and A. M. Cazabat, Thin nematic films: metastability and spinodal dewetting, *Phys. Rev. Lett.*, **82**, 2693 (1999).

43. F. Vandenbrouck, Films minces de cristaux liquides. PhD Thesis. Paris: Université Pierre et Marie Curie. Available online (2001).

44. D. Van Effenterre, R. Ober, M. P. Valignat, and A. M. Cazabat, Binary separation in very thin nematic films: thickness and phase coexistence, *Phys. Rev. Lett.*, **87**, 125701 (2001).

45. S. Hénon and J. Meunier, Microscope at the Brewster angle: direct observation of first-order phase transition in monolayers, *Rev. Sci. Inst.*, **62**, 936 (1991).

46. M. N. G. De Mul and J. A. Mann, Jr., Determination of the thickness and optical properties of a Langmuir film from the domain morphology by Brewster angle microscopy, *Langmuir*, **14**, 2455 (1998).

47. J. Laüger, C. R. Robertson, C. W. Frank, and G. Fuller, Deformation and relaxation processes of mono- and bilayer domains of liquid crystalline Langmuir films on water, *Langmuir*, **12**, 5630 (1996).

48. M. N. G. De Mul and J. A. Mann, Jr., Multilayer formation in thin films of thermotropic liquid crystals at the air-water interface, *Langmuir*, **10**, 2311 (1994).

49. L. Zou, J. Wang, P. Basnet, and E. K. Mann, Line tension and structure of smectic liquid-crystal multilayers at the air-water interface, *Phys. Rev. E* **76**, 031602 (2007).

50. P. Sheng and E. B. Priestley, Chapter 10. In eds., E. B. Priestley, P. Wojtowicz, and P. Sheng, *Introduction to Liquid Crystals*. New York and London: Plenum Press (1974).

51. T. J. Sluckin and A. Ponierwierski, Chapter 5. In ed., C. A. Croxton, *Fluid Interfacial Phenomena*. Chichester: Wiley (1986).

52. P. Sheng, Phase transition in surface-aligned nematic films, *Phys. Rev. Lett.*, **37**, 1059 (1976).

53. P. Sheng, Boundary-layer phase transition in nematic liquid crystals, *Phys. Rev. A*, **26**, 1610 (1982).

54. T. J. Sluckin and A. Ponierwierski, Novel surface phase transition in nematic liquid crystals: wetting and the Kosterlitz-Thouless transition, *Phys. Rev. Lett.*, **55**, 2907 (1985).

55. A. K. Sen and D. E. Sullivan, Landau-de Gennes theory of wetting and orientational transitions at a nematic liquid-substrate interface, *Phys. Rev. A* **35**, 1391 (1987).

56. T. J. Sluckin and A. Ponierwierski, Wetting and capillary condensation in liquid crystal systems, *Mol. Cryst. Liq. Cryst.*, **179**, 349 (1990).

57. K. Miyano, Surface ordering of a liquid crystal in the isotropic phase, *J. Chem. Phys.*, **71**, 4108 (1979).

58. K. Miyano, Wall-induced pretransitional birefringence: a new tool to study boundary aligning forces in liquid crystals, *Phys. Rev. Lett.*, **43**, 52 (1979).

59. J. Tarczon and K. Miyano, Surface-induced ordering of a liquid crystal in the isotropic phase. II. 4-methoxybenzylidene-4'-n-butylaniline (MBBA), *J. Chem. Phys.*, **73**, 1994 (1980).

60. H. Yokohama, Nematic-isotropic transition in bounded thin films, *J. Chem. Soc. Faraday Trans. 2*, **84**, 1023 (1988).

61. T. Moses and Y. R. Shen, Pretransitional surface ordering and disordering of a liquid crystal, *Phys. Rev. Lett.*, **67**, 2033 (1991).

62. M. B. Feller, W. Chen, and Y. R. Shen, Investigation of surface-induced alignment of liquid crystal molecules by optical second-harmonic generation, *Phys. Rev. A*, **43**, 6778 (1991).

63. P. Sheng, B-Z. Li, M. Zhou, T. Moses, and Y. R. Shen, Disordered-surface-layer transition in nematic liquid crystals, *Phys. Rev. A*, **46**, 946 (1992).

64. X. Zhuang, L. Marrucci, and Y. R. Shen, Surface-monolayer-induced bulk alignment of liquid crystals, *Phys. Rev. Lett.*, **73**, 1513 (1994).

65. T. Qian, X. Zhuang, and Y. R. Shen, Surface-monolayer-induced bulk alignment of liquid crystals: from nematic to smectic-A phase, *Phys. Rev. E*, **59**, 1873 (1999).

66. D. Mattia and H. Leese, "Wetting in carbon inorganic and organic nanotubes and nanochannels," in *Nanoscale Liquid Interfaces: Wetting, Patterning and Force Microscopy at the Molecular Scale* (this issue).

67. M. Wittebrood, D. Luijendijk, S. Stallinga, T. Rasing, and I. Muševi, Thickness-dependent phase transition in thin nematic films, *Phys. Rev. E*, **54**, 5232 (1996).

68. J. Nehring and A. Saupe, On the elastic theory of uniaxial liquid crystals, *J. Chem. Phys.*, **54**, 337 (1971).

69. F. C. Frank, On the theory of liquid crystals, *Discuss. Faraday Soc.*, **25**, 19 (1958).
70. C. Oldano and G. Barbero, An ab initio analysis of the second-order elasticity effect on nematic configurations, *Phys. Lett. A*, **110**, 213 (1985).
71. C. Oldano and G. Barbero, Derivative-dependent surface-energy terms in nematic liquid crystals, *Nuovo Cimento D*, **6**, 479 (1985).
72. G. Barbero and C. Oldano, On the K_{13}-dependent energy term in nematic liquid crystals, *Mol. Cryst. Liq. Cryst.*, **168**, 1 (1989).
73. V. M. Pergamenshchik, Surfacelike-elasticity-induced spontaneous twist deformations and long-wavelength stripe patterns in a hybrid nematic layer, *Phys. Rev. E*, **47**, 1881 (1993).
74. V. M. Pergamenshchik, K_{13} term and effective boundary condition for the nematic director, *Phys. Rev. E*, **58**, R16 (1998).
75. A. Rapini and M. Papoular, Distorsion d'une lamelle nématique sous champ magnétique, conditions d'ancrage aux parois, *J. Phys. Colloques*, **30**(C4), 54 (1969).
76. P. Chiarelli, S. Faetti, and L. Fronzoni, Critical behaviour of the anchoring energy of the director at the free surface of a nematic liquid crystal, *Phys. Lett.*, **101A**, 31 (1984).
77. S. Faetti, M. Gatti, V. Palleschi, and T. Sluckin, Almost critical behaviour of the anchoring energy at the interface between a nematic liquid crystal and a SiO substrate, *Phys. Rev. Lett.*, **55**, 1681 (1985).
78. H. Yokoyama, S. Kobayashi, and H. Kamei, Temperature dependence of the anchoring strength at a nematic liquid crystal-evaporated SiO interface, *J. Appl. Phys.*, **61**, 4501 (1987).
79. A. J. Leadbetter, R. M. Richardson, and C. N. Colling, The structure of a number of nematogens, *J. Phys. Colloques*, **36**(C1), 37 (1975).
80. U. Delabre, C. Richard, and A. M. Cazabat, Thin nematic films on liquid substrates, *J. Phys. Chem. B*, **113**, 3647 (2009). U. Delabre, C. Richard, and A. M. Cazabat, Some specificities of wetting by cyanobiphenyls liquid crystals, *J. Phys: Condens. Matter*, **21**, 464129 (2009).
81. U. Delabre, C. Richard, Y. Yip Cheung Sang and A. M. Cazabat, Thin liquid crystal films in the nematic range of temperatures, *Langmuir*, **26** 13368 (2010).
82. P. Karat and N. Madhusudana, Elastic and optical properties of some 4'-n-alkyl-4-cyanobiphenyls, part II, *Mol. Cryst. Liq. Cryst.*, **40**, 239 (1977).
83. J. Bunning, T. E. Faber, and P. Sherrell, The Frank constants of nematic 5CB at atmospheric pressure, *J. Phys. Fr.*, **42**, 1175 (1981).

84. M. Bradshaw, E. Raynes, J. Bunning, and T. E. Faber, The Frank constants of some nematic liquid crystals, *J. Phys. Fr.*, **46**, 1513 (1985).
85. D. W. Allender, G. P. Crawford, and J. W. Doane, Determination of the liquid-crystal surface elastic constant K_{24}, *Phys. Rev. Lett.*, **67**, 1442 (1991).
86. O. D. Lavrentovich and V. M. Pergamenshchik, Stripe domain phase of a thin nematic film and the K_{13} divergence term, *Phys. Rev. Lett.*, **73**, 979 (1994).
87. H. A. Van Sprang and R. G. Aartsen, Torsional Anchoring of 5CB and 5PCH on various substrates, *Mol. Cryst. Liq. Cryst.*, **123**, 355 (1985).
88. J.-B. Fournier and P. Galatola, Coarse-grained surface energies and temperature-induced anchoring transitions in nematic liquid crystals, *Phys. Rev. Lett.*, **82**, 4859 (1999).
89. J.-B. Fournier and P. Galatola, Effective anchoring and scaling in nematic liquid crystals, *Eur. Phys. J. E: Soft Matter Biol. Phys.*, **2**, 59 (2000).
90. D. Van Effenterre, Instabilités d'un film mince de nématique. PhD Thesis. Paris : Université Pierre et Marie Curie. Available online (2002).
91. O. D. Lavrentovich and V. M. Pergamenshchik, Periodic domain structures in thin hybrid nematic layers, *Mol. Cryst. Liq. Cryst.*, **179**, 125 (1990).
92. O. D. Lavrentovich and V. M. Pergamenshchik, Patterns in thin liquid crystal films and the divergence ("surfacelike") elasticity, *Int. J. Mod. Phys. B*, **9**, 2389 (1995).
93. A. Sparavigna, O. D. Lavrentovich, and A. Strigazzi, Periodic stripe domains and hybrid-alignment regime in nematic liquid crystals: threshold analysis, *Phys. Rev. E*, **49**, 1344 (1994).
94. A. Sparavigna, O. D. Lavrentovich, and A. Strigazzi, Magnetic field effect on periodic stripe domains in nematic liquid crystals, *Phys. Rev. E*, **51**, 792 (1995).
95. A. L. Alexe-Ionescu, G. Barbero, and I. Lelidis, Periodic deformations in nematic liquid crystals, *Phys. Rev. E*, **66**, 061705 (2002).
96. P. Guyot-Sionnest and Y. R. Shen, Bulk contribution in surface second-harmonic generation, *Phys. Rev. B*, **38**, 7985 (1988).
97. D. Krzyzanski and G. Derfel, Structure of periodic deformations in hybrid aligned nematic layers, *Phys. Rev. E*, **63**, 021702 (2001).
98. U Delabre, Films nématiques minces sur substrats liquides. PhD Thesis. Paris: Université Pierre et Marie Curie. Available online. (2009).

99. U. Delabre, C. Richard, J. Meunier, and A. M. Cazabat, Line tension of nematic pancakes at the air-water interface: static and dynamic measurements, *Eur. Phys. Lett.*, **83**, 66004 (2008); U. Delabre, C. Richard, G. Guéna, J. Meunier, and A. M. Cazabat, Nematic pancakes revisited, *Langmuir*, **24**, 3998 (2008).

100. H. Stone and H. M. McConnell, Hydrodynamics of quantized shape transitions of lipid domains, *Proc. R. Soc. Lond. A*, **448**, 97 (1995).

101. R. Holyst and P. Oswald, Dislocations in uniaxial lamellar phases of liquid crystals, polymers and amphiphilic systems, *Int. J. Mod. Phys. B*, **9**, 1515 (1995).

102. O. D. Lavrentovich, Geometrical anchoring at an inclined surface of a liquid crystal, *Phys. Rev. A*, **46**, R722 (1992).

Chapter 6

Droplet Morphology at Nanometer Scale

Antonio Checco
Condensed Matter Physics and Materials Science Department,
Brookhaven National Laboratory, Upton, New York 11973, USA
checco@bnl.gov

6.1 Introduction

Understanding the wetting behavior of nanometer-scale drops or "nanodrops" on a solid substrate is an important scientific problem with implications in many areas of technology. From a theoretical standpoint, nanodrops are not expected to obey the macroscopic laws of capillarity because their size is comparable with the finite range of intermolecular interactions. Consequently many of the familiar aspects of droplets as seen in everyday life do not necessarily apply to the "nanoworld." This has profound implications for many important phenomena such as surface corrosion and cloud formation. It is also relevant to many emerging technologies—such as microfluidics and inkjet printing—aiming at creating and manipulating smaller and smaller amounts of liquids on a substrate. For these reasons, understanding wetting phenomena on small scales has become a compelling problem in condensed matter

Nanoscale Liquid Interfaces: Wetting, Patterning, and Force Microscopy at the Molecular Scale
Edited by Thierry Ondarçuhu and Jean-Pierre Aimé

ISBN 978-981-4316-45-3 (Hardcover), 978-981-4364-48-5 (eBook)
www.panstanford.com

physics. This field of research has greatly benefited from the development of nonconventional microscopy techniques, such as scanning electron, tunneling, and atomic force microscopy (AFM). In particular, because of its ability to image nonconductive materials under a variety of environmental conditions, AFM has become the instrument of choice for imaging the topography of liquid interfaces with lateral resolution in the nanometer range. In this chapter, we trace a historical time line of advances in high-resolution imaging of liquid drops by AFM, providing a review of the various imaging modes and their advantages and limitations. We also illustrate the details of the nanodroplet morphology on homogenous as well as inhomogenous substrates, as revealed by AFM, and we discuss them in the context of current theories. Finally we show how these achievements have contributed to advancing our understanding of wetting phenomena at the nanoscale.

6.2 Imaging of Nanoscale Drops by AFM: Time Line of Achievements

The atomic force microscope was initially conceived as a tool for mapping the topography of a nonconductive material with atomic resolution.[1] At the heart of the instrument is a cantilever beam with a sharp tip (probe) at its end that is used to trace the specimen surface. When the tip is brought in contact with the surface, it experiences a force that causes the cantilever to bend. Because the cantilever behaves as a tiny spring, measuring the cantilever deflection provides a measure of the force acting on the tip. Piezoelectric elements are used to scan the tip along the $x - y$ plane parallel to the specimen surface, whereas another piezoelectric element coupled to a feedback loop varies the z position of the tip in order to maintain constant the cantilever deflection. Provided that the tip and the specimen are made of nondeformable materials, a topographical image of the sample surface can be obtained by recording the x, y, and z coordinates of the tip during the scan. The inventors of the atomic force microscope readily pointed out that the sensitivity of the microscope could be improved if the tip was oscillated in close proximity to the sample surface[1] rather than being in physical contact with it. A year later Martin *et al.*[2] demonstrated the use of a vibrating cantilever for profiling and force mapping with a lateral resolution of ≈10 nm. This technique, now referred to as "amplitude modulation AFM" (AM-AFM), consists in oscillating the cantilever near its natural frequency and measuring the oscillation amplitude with optical or electrical methods. The amplitude of vibration depends on the gradient of the long-range van der Waals force acting between the tip and the sample.[2] Therefore, a three-dimensional map of constant surface potential can be obtained by scanning the tip over the surface, while maintaining constant the cantilever amplitude (also called "set-point amplitude"). In the case of a chemically homogeneous sample, this map provides a picture of the surface topography convoluted with the shape of the tip. The best lateral resolution is on the order of the tip–sample distance when the latter is in the nanometer range.[3] This method minimizes the sample perturbation by the atomic force microscope tip, thus enabling the imaging of soft and liquid surfaces. It was originally dubbed "noncontact" (NC)-AFM, although this nomenclature is now usually associated with atomic resolution, frequency modulation (FM)-AFM in ultrahigh vacuum. In this chapter we refer to NC-AFM as AM-AFM, where the tip experiences exclusively long-range attractive forces.

Mate *et al.*[4] were able to image the free surface of a thin film of liquid polymer using NC-AFM, but early attempts to image liquid droplets proved difficult. Indeed, in order to steadily trace the profile of a drop, the AFM feedback system has to continuously compensate for changes in the sample height due to the curvature of the liquid surface. However, because the tip and the surface are separated by only a few nanometers, an accidental contact might occur during the scanning. In this case the liquid wets the probe apex, forming a capillary meniscus around it. This happens very quickly because capillary waves can propagate much faster than the cantilever motion. Indeed, it can be estimated based on capillary wave theory[5] that in the case of low-viscosity liquids such as water, the meniscus forms on a time scale of about 10 ns, which is much shorter than the typical period of the cantilever oscillation (≈10 μs). The meniscus exerts on the probe a capillary force that largely exceeds the van der Waals interaction between the tip and the liquid.[6] As a result the tip is "captured" by the liquid and the cantilever oscillation is heavily damped. Moreover, the contact between the probe and the sample causes deformation of the liquid surface and tip contamination. All these effects are detrimental to the quality and stability of the imaging process.

To circumvent these issues, in the early 1990s various groups experimented novel imaging modes and techniques based on AM-AFM. Zhong and Inniss[8] used AM-AFM with cantilever amplitudes in the range 20–100 nm, significantly larger than used in the traditional noncontact mode (3–15 nm) to allow the tip to escape the capillary attraction by the surface fluid layer. This expedient promotes a stable oscillating mode, referred to as "intermittent contact" (IC) or "tapping," where the tip touches the sample surface at each oscillation cycle, therefore experiencing not only long-range forces but also contact, adhesion, and capillary forces. The IC mode is well suited to image soft and viscoelastic materials such as polymer films. Later, Tamayo and Garcia[9] showed that droplets of highly viscous liquids could also be imaged with this technique. Simultaneously to the topography signal, they also measured the phase lag, referred to as the "phase contrast," of the vibrating cantilever with respect to the external excitation. The phase contrast was larger on those regions of the substrate occupied by the drops. Remarkably, Herminghaus *et al.*[7] successfully used IC-AFM to image small drops of aqueous salt solutions (see Fig. 6.1a) and hexaethylene glycol[10] deposited, respectively, on mica and a chemically patterned silicon surface.

These results were surprising because they demonstrated the possibility to image liquid drops in the IC mode. As reported earlier by Tamayo and Garcia, the droplets were found to exhibit a large phase contrast ($\approx 40^\circ$, see Fig. 6.1b) and an apparent topography dependent on the set-point amplitude. The most reliable images where obtained using a set-point amplitude up to 20% smaller than the cantilever free amplitude. Lower set-point amplitudes lead to either unstable or distorted profiles. In the late 1990s, progress in the understanding of the cantilever dynamics in AM-AFM allowed one to relate the phase signal to the amount of energy dissipated by the cantilever in the interaction with the sample.[11,12] The large phase contrast observed in IC imaging of liquids originates from energy dissipation caused by either the viscoelastic properties of the

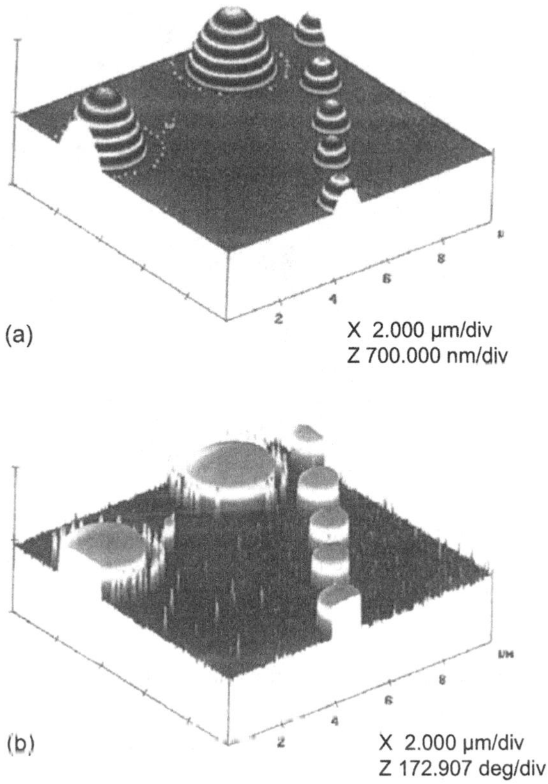

Figure 6.1 IC-AFM topography (a) and phase (b) image of droplets of an aqueous solution of phosphorus pentoxide on mica (reprinted with permission from Ref. 7, copyright 1997, Elsevier).

liquid or the nonconservative capillary forces acting on the atomic force microscope tip.[13,14] However, the good lateral resolution of the resulting AFM images (≈10 nm) suggests that the meniscus extension is very limited. It is still not clearly understood why in certain instances the meniscus formation is somehow inhibited, whereas in other cases the capillary forces are so strong as to prevent reliable imaging.

Simultaneously to the development of the IC-AM imaging mode, a variant of the NC-AM mode that uses long-range electrostatic forces—rather than van der Waals forces—was introduced by Hu

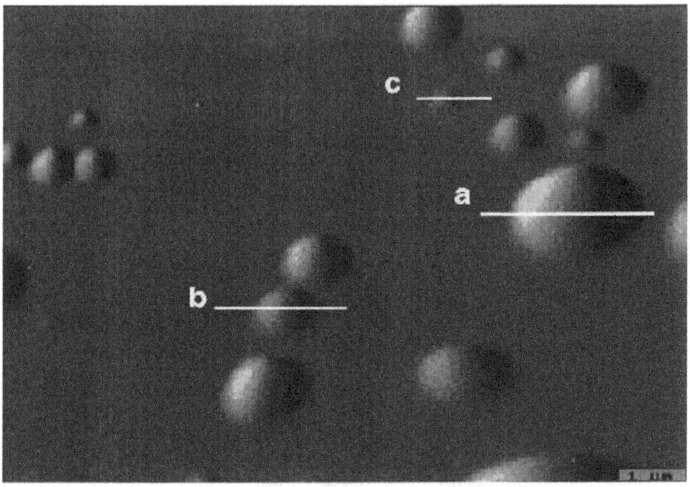

Figure 6.2 Electrostatic NC-AFM image of sulfuric acid droplets on mica (reprinted with permission from Ref. 15, copyright 1998, American Chemical Society).

et al.[16] specifically to image liquid surfaces. These authors applied an electric field between the tip and the liquid surface through biasing of the tip. The resulting long-range attractive polarization force can be used to image the sample at a distance larger than conventional NC-AFM, thereby eliminating the occurrence of capillary instabilities. However, this is achieved at the expense of the lateral resolution, which is on the order of the tip-sample distance or tip radius ≈20–40 nm. The apparent height of the liquid samples may also depend on differences among the dielectric constants of the liquid and the substrate. The same group[15] later used this technique to image nanodroplets of sulfuric acid on a mica surface (see Fig. 6.2). This was one of the first quantitative studies of droplet morphology at the submicron length scale.

In the late 1990s and the beginning of the 2000s, other groups initiated novel efforts to image liquid drops by the original NC-AM mode, relying only on van der Waals forces. Luna *et al.* were able to image water droplets condensed on graphite,[18] mica, and gold[19] at various relative humidities. Although these droplets actually appeared as very flat (10–20 molecules thick) islands, these studies

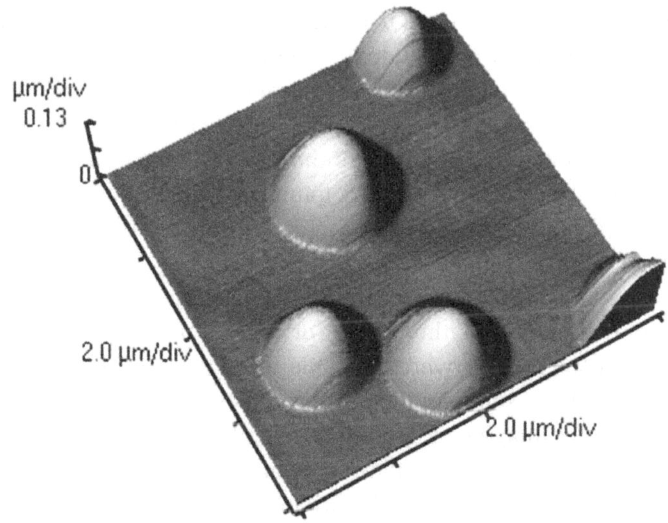

Figure 6.3 NC-AFM picture of decane drops on OTS-coated silicon (reprinted with permission from Ref. 17, copyright 2006, American Chemical Society).

demonstrated that conventional NC-AM could rival its electrostatic variant to image thin liquid films and droplets. Another achievement in this direction came with the work of Checco *et al.*[20] These authors focused their efforts on the imaging of simple apolar liquids such as short-length alkanes. Due to their volatility, submicron-sized alkane droplets can only be stabilized under a controlled environment where the alkane vapor pressure is controlled precisely. For this purpose, Checco *et al.* developed a miniature environmental cell containing the atomic force microscope tip and the substrate.[17] The cell was saturated with vapors of alkanes, and droplets were nucleated on the substrate—a silicon surface passivated with a monolayer of octadecyltrichlorosilane (OTS)—and imaged *in situ* using NC-AFM (Fig. 6.3).

After nucleation, the droplet size increased up to a few microns but could be subsequently reduced by slowly releasing the vapor pressure inside the cell. With this method, the morphology of droplets resting on a variety on homogeneous[20] as well as inhomogeneous[17] surfaces could be studied as a function of the drop radius ranging from $\approx$50 nm to a few microns. The authors reported that small cantilever amplitudes ($\approx$10 nm) and tips with curvature radii $\approx$10–30 nm were required to achieve stable imaging in the NC regime. The absence of tip–sample contact was also indicated by the negligible phase contrast of the droplet compared to the dry substrate.[20] However, the authors found that the droplets could be easily displaced as a result of an accidental contact between the tip and the liquid.[17]

By the late 2000s, various groups reported imaging of droplets of various liquids by either the IC or the NC mode. In particular, Zhang and Ducker[21,22] were also able to image emulsion droplets condensed at the solid–liquid interface. These results demonstrate nowadays that imaging of liquid interfaces by AFM can be achieved routinely. Although the NC mode is the method of choice because of its intrinsically small sample disturbance, the IC mode can also be used, provided that the sample deformation is minimized through appropriate choice of the imaging parameters such as the free and set-point amplitude. In all cases it is important to record the phase signal simultaneously to the topography in order to quantify the energy dissipated in the tip–sample interaction and the amount of sample deformation.

6.3 Nanodroplet Morphology on Homogeneous and Inhomogeneous Substrates

6.3.1 *Homogeneous Substrates*

The majority of experimental studies of droplet morphology at nanometer scale have focused on substrates that are as close as possible to the ideal case of atomically smooth and chemically homogenous surfaces.

In this case the comparison between experiment and theory is more straightforward, and any deviation from the macroscopic wetting behavior due to microscopic effects can be readily discerned through the analysis of the droplet profile. Among the most studied substrates are freshly cleaved mica[15,23] and chemically passivated silicon.[17,20,22,24] The surface of freshly cleaved mica is atomically

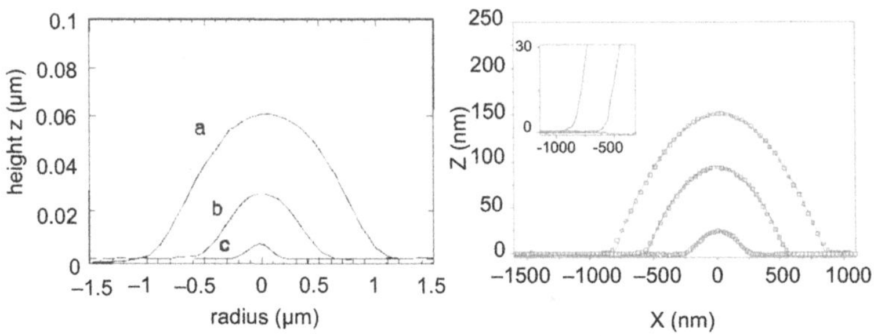

Figure 6.4 (Left) Electrostatic NC-AFM profiles of surfulric acid droplets on mica (reprinted with permission from Ref. 15, copyright 1998, American Chemical Society). (Right) NC-AFM profiles of dodecane droplets on OTS-coated silicon (reprinted with permission from Ref. 17, copyright 2006, American Chemical Society).

smooth but electrostatically charged and hence prone to contamination by airborne chemical species. Silicon surfaces bearing a native oxide layer are smooth (root mean square [rms] roughness $\approx$0.3 nm) and can be passivated with organic monolayers[25,26] to tailor their surface energy for wetting studies. However, highly homogenous monolayers only can be obtained through well-controlled preparation methods.[25] Droplets of both polar (sulfuric acid, glycerol, hexaethylene glycol) and apolar (octane, nonane, decane, dodecane) liquids were studied. The experiments show that independent of the liquid used, droplets of size ranging from a few microns to hundreds of nanometers have a spherical cap shape as expected for homogeneous substrates. However, the liquid profile near the contact line region often appears smooth (see Fig. 6.4a,b). Therefore, the apparent droplet contact angle is usually estimated by fitting an arc to the entire AFM profile with exclusion of the region near the contact line (see Fig. 6.4b). The experimental results clearly show that the apparent contact angle depends on the droplets' size. In the majority of cases, the angle is found to decrease with the droplet radius, as illustrated in Fig. 6.5, for droplets of alkanes on OTS-coated silicon. In a few other cases, the opposite behavior is observed. Both these findings—the size dependence of the contact angle and the profile smoothening near the contact line—can be

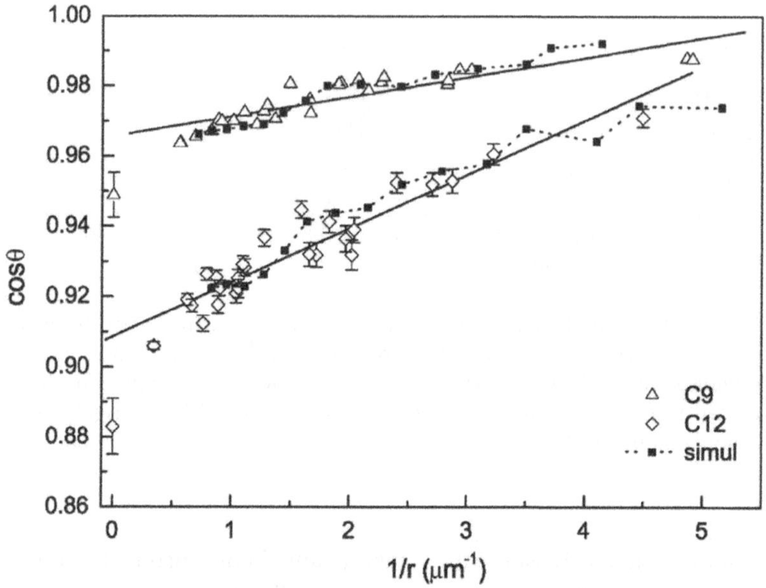

Figure 6.5 Cosine of the contact angle vs. contact line curvature for nonane (open triangles) and dodecane (open diamonds) droplets on OTS-coated silicon (reprinted with permission from Ref. 17, copyright 2006, American Chemical Society).

interpreted as nanoscale effects due to the finite range of the intermolecular interactions.

In principle, the analysis of the droplet profile near the triple line provides quantitative information on the shape of the "effective interface potential" or the excess free energy of a thin wetting film due to intermolecular interactions (see also chapter 3). In turn, the so-called "line tension"—or the energy density of the triple line—can be estimated by integration of the effective interface potential. Alternatively, the line tension can be inferred from the size dependence of the droplet contact angle according to the "modified Young equation"[27]

$$\cos\theta_r = \cos\theta_\infty - \frac{\tau}{\gamma r}, \tag{6.1}$$

which relates the contact angle θ_r of a drop of finite radius r to the macroscopic contact angle θ_∞, the liquid surface tension γ, and the line tension τ. However, the first approach for measuring τ is less accurate because the finite lateral resolution of the instrument may also contribute to the smoothing of the liquid profile. This convolution artifact—present both in NC and IC imaging modes—cannot be easily quantified and subtracted from the data. On the other hand, the contact angle measurements are less sensitive to convolution effects because the droplets studied by AFM typically exhibit a low contact angle ($\approx 20°$) and a large curvature radius of the liquid–vapor interface (≈ 1 μm). For this reason the analysis of the size dependence of the contact angle is considered the most reliable method to measure τ. Nonetheless, Pompe and Herminghaus[28] have shown that the two approaches can provide consistent results.

In experiments measuring the size dependence of the droplet contact angle, the line tension coefficient is obtained by fitting a straight line to a set of contact angle versus contact line curvature measurements, as shown in Fig. 6.5 for droplets of nonane and dodecane on an OTS-terminated silicon surface.[17] The slope of the curve provides the ratio $-\tau/\gamma$. In this case the fit renders a negative line tension in the 10^{-10} N range. A few results obtained for various liquids and substrates are summarized in Table 6.1.

Both negative and positive values of τ laying in the range 10^{-11}–10^{-10} N have been reported. Theory predicts values of line tension of both signs in the range 10^{-12}–10^{-11} N (see also chapter 3).

Despite the reasonable agreement between experiments and theory, there is no general consensus on the interpretation of the

Table 6.1 Summary of AFM measurements of line tension of droplets of various liquids

Liquid	Substrate	Line tension	Imaging mode	Reference
Sulfuric acid	Mica	$\approx -10^{-10}$ N	NC (electrostatic)	15
Nonane	OTS-coated silicon	$-1.3(0.1)*10^{-10}$ N	NC	20
Decane	OTS-coated silicon	$-2.0(0.1)*10^{-10}$ N	NC	20
Dodecane	OTS-coated silicon	$-3.8(0.2)*10^{-10}$ N	NC	20
Water	Silicon	$7.0(0.3)*10^{-10}$ N	IC	28
Hexaethylene glycol	Perfluorodimethyl chlorosilane-coated Si	$-3(1)*10^{-10}$ N	IC	28
Hexaethylene glycol	Silicon	$-1.0(0.3)*10^{-10}$ N	IC	28
Decane (in diluted ethanol)	OTS-coated silicon	$+8*10^{-11}$ N	IC	22

size dependence of the contact angle data. For example, Checco *et al.*[20,17] have pointed out that the experimental relationship between $\cos\theta$ and $1/r$ is not always perfectly linear and that the macroscopic contact angle extrapolated from the fit does not necessarily agree with the measured value as expected (see Fig. 6.5). Moreover, the same authors have measured line tension moduli about two orders of magnitude larger than the theoretical value calculated specifically for their well-defined system.

For these reasons, Checco *et al.* have postulated that the observed size dependence of the contact angle may not originate solely from line tension effects but also from nanoscale chemical inhomogeneity of the sustrate. This hypothesis is particularly applicable to the experiments of Checco *et al.* where the drops condense on the substrate from the vapor, therefore nucleating preferentially on the nanoscale chemical "defects" of the solid. In this situation, a small drop wetting a single surface defect exhibits a contact angle smaller than a larger drop covering an area of the susbtrate containing only a few isolated defects. This may explain, at least partially, the observed decrease of contact angle with the drop size. Although the technique used to deposit the droplets varies from one experiment to the other, the issue of chemical inhomogeneity of the substrate is a highly relevant one. Indeed, the modified Young equation holds only for perfectly homogeneous surfaces, whereas actual substrates always exhibit defects, while also being contaminated by atmospheric chemical species during the course of the experiment.

The interpretation of the size dependence of the contact angle of nanodroplets in terms of a modified Young equation has also been questioned from the theoretical standpoint. Recently, Dietrich *et al.*[29] have shown that a mere fit of the experimental data with the modified Young equation does not render the intrinsic line tension but rather a combination of the line tension, the Tolman length, and the stiffness coefficients of the line (see also chapter 3). It is therefore legitimate to claim that AFM studies of the morphology of nanodroplets have established an upper bound of 10^{-10}N for the modulus of the line tension, close to theoretical expectations. However, further advances in both theoretical and experimental methods are required before being able to extract intrinsic values of line tension from the experimental data.

6.3.2 *Inhomogeneous Substrates*

In recent years, there have been a growing number of AFM studies on the morphology of nanodroplets wetting inhomogeneous substrates. On the one hand, this interest stems from the fact that all real surfaces are not ideal but rather exhibit physical (roughness) or chemical inhomogeneity with random spatial distribution, which may affect the wetting behavior of small drops in ways that are not yet well understood. On the other hand, advances in nano-lithography techniques now allow one to fabricate surfaces with controlled topography and chemical composition at the nanoscale. These nanopatterned surfaces provide a means to confine and manipulate extremely small amounts of liquid within well-defined regions of a substrate for specific applications such as microfluidics. Moreover, they enable fundamental studies of wetting behavior at the nanoscale.

Seemann *et al.*[30] have studied the morphology of droplets wetting microstructured surfaces. Their results are described accurately by classical capillary theory. Checco *et al.*[17] have shown that both random nanoscale roughness and chemical defects have a strong influence on the shape of nanodrops. Substrates with (rms) roughness of only a few nanometers were prepared by thermal evaporation in vacuum of a thin layer of gold on a silicon wafer. Due to the limited surface diffusion of the gold atoms, the metallic layer exhibits a polycrystalline structure with grain size $\approx$40 nm, as shown in Fig. 6.6a. Droplets of dodecane condensed on the grainy gold surface passivated with a monolayer of dodecanethiols exhibit an irregular shape, as shown in Fig. 6.6b. Moreover, the contact line of the droplets was not found to easily slide over the solid if the droplet volume were slightly increased (decreased) by condensation (evaporation). These results clearly indicate that even a small amount of surface roughness may cause the wetting behavior to deviate from the quasi ideal case discussed in the previous section.

Similar results were obtained in a study of dodecane droplets wetting a chemically inhomogeneous substrate made of an incomplete OTS monolayer on oxidized silicon (see Fig. 6.7a). In this case the droplet contact line is quite irregular (see Fig. 6.7c) even

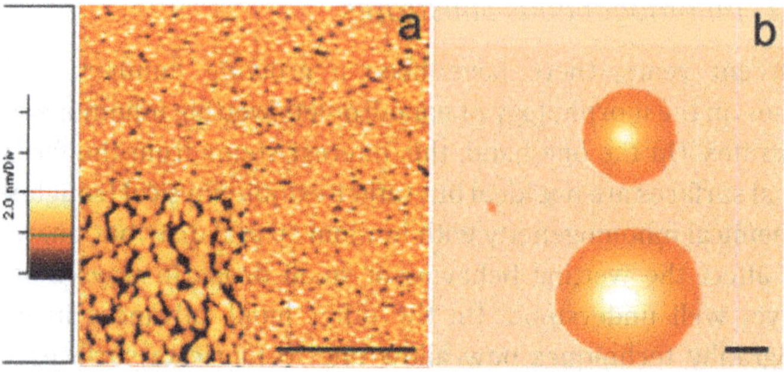

Figure 6.6 (a) IC-AFM topography of a gold monolayer thermally evaporated on an oxidized silicon wafer and then passivated with dodecanethiols. Inset shows a 400 nm wide detail of the same surface imaged using scanning tunneling microscopy. (b) NC-AFM images of dodecane droplets condensed on the surface shown in (a). For all pictures, the lateral scale bar denotes 1 μm (reprinted with permission from Ref. 17, copyright 2006, American Chemical Society).

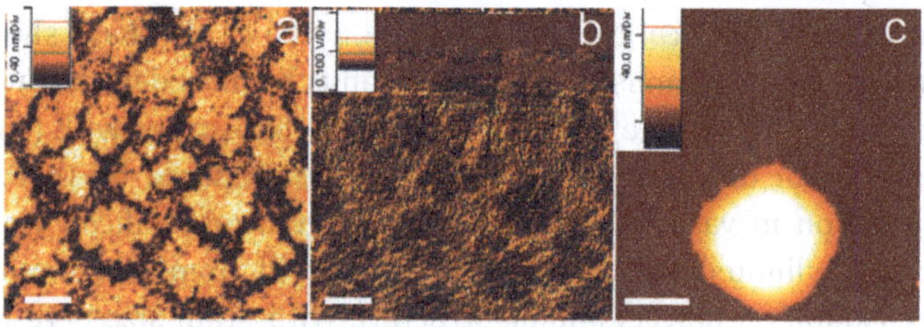

Figure 6.7 (a) IC-AFM image of a partially grown OTS monolayer, (b) the corresponding phase image, and (c) an NC-AFM picture of a dodecane droplet condensed on the partial monolayer. For all pictures, the lateral scale bar denotes 1 μm. (Adapted with permission from Ref. 17, copyright 2006, American Chemical Society.)

though the surface roughness is small (<1 nm rms). The contact line roughness is induced by the different wettability among the densest regions of the monolayer (bright fractal island in Fig. 6.7a) and to the less dense region in between them. This chemical inhomogeneity is apparent in the phase contrast image shown in Fig. 6.7b.

Other studies have aimed at controlling the inhomogeneity of the substrate to probe the fundamental properties of nanodroplets. Herminghaus *et al.*[10] have patterned a silicon surface with

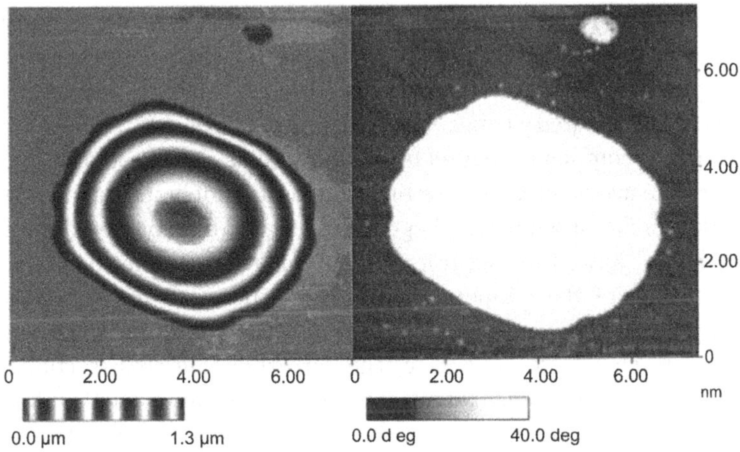

Figure 6.8 IC-AFM topography (left side) and phase image (right side) of a droplet of diethylene glycol on a silicon wafer patterned with alternating hydrophilic and hydrophilic stripes, respectively, 250 and 550 nm wide. In the phase image the hydrophilic regions of the substrate show a higher phase shift as compared to the hydrophobic regions (reprinted with permission from Ref. 10, copyright 1998, American Chemical Society).

alternating hydrophobic and hydrophilic lines using microcontact printing.[31] Droplets of various liquids deposited on the patterned surface were found to have a shape elongated in the direction of the stripes, whereas the contact line crossing the stripes was largely distorted by the wettability gradient (see Fig. 6.8a). The largest contact line curvature (either positive or negative) was measured in the middle of the stripes and could be controlled by varying the width of the stripes. In addition, the contact angle at the point of largest curvature could be measured from the AFM images. Pompe and Herminghaus[28] used this method to study the dependence of the contact angle on the contact line curvature. They have interpreted their results according to the modified Young equation and measured line tension values similar to those obtained by other groups using the drop size dependence of the contact angle (see Table 6.1).

Checco *et al.* have studied the morphology of ethanol[32,33] and octane[34] droplets condensed on chemically nanopatterned surfaces made by electro-oxidative nanolithography.[26] The chemical patterns were composed of alternating carboxylic acid-terminated stripes and methyl-terminated stripes. In contrast to the nonwetting methyl-terminated stripes, the carboxylic stripes (80 to 300 nm wide) are completely wetted by organic liquids such as ethanol and therefore act as templates for the condensation of liquid nanodrops with cylindrical symmetry. A specially designed environmental AFM chamber allowed for accurate measurements of droplet height as a function of the temperature offset ΔT between the substrate and an ethanol reservoir. NC-AFM profiles of the drops at various ΔTs are shown in Fig. 6.9a–c. The corresponding cross-sectional profiles, shown in Fig. 6.9d, illustrate that the droplet's contact line is pinned to the boundaries of the wettable stripe independent of the droplet volume. The maximum droplet thickness is about 10 nm, thin enough for the long-range intermolecular forces to cause noticeable deviation from the macroscopic droplet morphology. This effect is clearly visible for the largest drops whose profile is flat at small heights (<5 nm) but becomes curved as the drop volume increases.

A quantitative analysis of the effect of long-range forces on the liquid morphology can be carried out by plotting the droplet height as a function of the stripe width w and the temperature ΔT. This plot is shown in Fig. 6.10a. Three different regimes can be distinguished. For a small, positive ΔT the droplet height increases gradually with w and then "saturates" at large enough droplet widths. This behavior is easily understood by considering that in the undersaturated regime ($\Delta T > 0$), when the stripe width largely exceeds the drop height, the droplet resembles a uniform liquid film. Close to saturation and for the range of w studied, the drop height scales with w according to a power law with an exponent $\approx 1/2$. As detailed next, this scaling exponent is expected theoretically for nanodroplets of liquids interacting with purely dispersive potentials. In the oversaturated regime ($\Delta T < 0$) the droplet height increases faster than a power law with exponent $1/2$, particularly for a large w. These findings indicate that van der Waals forces become less and less important as the droplet thickness increases and eventually surface tension alone dictates the droplet shape (supposing the effects of gravity are negligible).

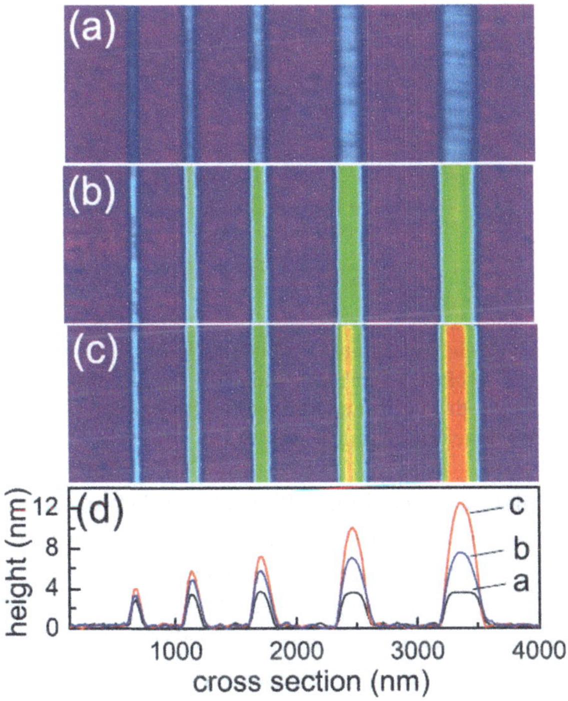

Figure 6.9 (a–c) NC-AFM images (topography, false-color height scale) showing the condensation patterns of ethanol onto hydrophilic nanostripes for a substrate–reservoir temperature differential equal to (a) 71, (b) 0, and (c) −26 mK. (d) Corresponding cross-sectional liquid profiles (reprinted with permission from Ref. 33, copyright 2008, American Physical Society).

These results are described quantitatively by the "augmented Young equation" model of droplet morphology[35]:

$$\gamma K(x) = \rho \Delta \mu + \frac{\partial \omega(x, l(x))}{\partial l} \tag{6.2}$$

This equation relates the droplet capillary pressure (left-hand side where $K(x)$ is the local curvature of the liquid surface at a specific coordinate x normal to the striped pattern) to the chemical potential difference $\Delta\mu(\propto \Delta T)$ between the substrate and the liquid reservoir (at $T = T_0$) and the partial derivative of the effective interface potential ω with respect to the liquid thickness $l(x)$ (often referred to as the disjoining pressure, see also chapter 3).

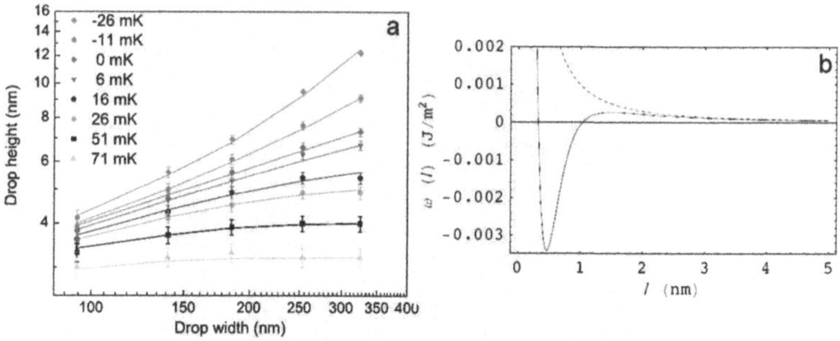

Figure 6.10 (a) Height of ethanol droplets vs. width of the wettable stripe for various ΔT (colored symbols) and (b) effective interface potential used to model the droplet shape. (Adapted with permission from Ref. 33, copyright 2008, American Physical Society.)

The droplet shape is obtained by solving the augmented Young equation with an appropriate analytical form of ω derived from density functional theory with intermolecular interactions modeled by pairwise Lennard-Jones potentials (see Fig. 6.10b). Because ω scales as l^{-2} for thick films, theory predicts the height of droplets to scale with the width according to a power law with exponent 1/2 at saturation, as found in the experiments. Moreover, the model describes accurately the droplet shape off-coexistence (solid lines in Fig. 6.10a).

In addition to the static measurements described before, time-resolved AFM measurements have been recently used to study the out-of-equilibrium behavior of droplets on chemically nanopatterned surfaces. Checco and Ocko[33] have demonstrated the ability to study the evaporation dynamics of ethanol droplets wetting chemical nanostripes by recording the droplet profiles by NC-AFM with temporal resolution ≈ 0.1 seconds. In a different set of experiments, by using the same technique, Checco has studied the morphology of squalane rivulets spreading along wettable nanostripes of widths ranging from 200 to 650 nm.[36] The results indicate that the profile of the spreading nanorivulets is truncated near the contact line, as illustrated in Fig. 6.11. The truncation thickness $\approx$2.7 nm as well as the weak dependence of the liquid profile on the spreading velocity are in excellent agreement with classical mesoscale hydrodynamic theory.[37] However, the truncation region is not as sharp as predicted

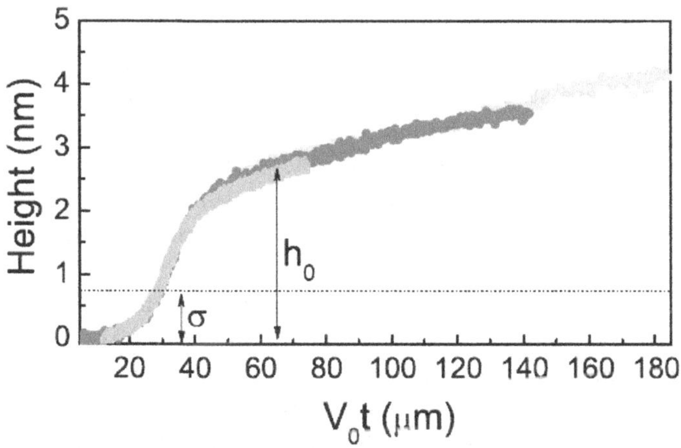

Figure 6.11 NC-AFM profile of a squalane rivulet spreading along a 650 nm wide wettable stripe. The profile near the contact line is truncated at a height $h_0 \approx 2.7$ nm (adapted with permission from Ref. 36, copyright 2009, American Physical Society).

by the model. This discrepancy could originate, at least partially, from the contact line roughness in the plane of the stripe, which reflects a slight chemical inhomogeneity of the patterned substrate. Nevertheless, the latter effect is small, and it is more likely that molecular diffusion and thermal fluctuations of the contact line—which are not accounted for in the theory—are responsible for the line broadening. Similar to the case of static wetting, these findings indicate that long-range forces have a profund effect on the morphology of spreading nanoliquids.

6.4 Summary and Outlook

In the past two decades AM-AFM has emerged as a powerful tool to study the morphology of liquid drops with nanometer resolution. Various imaging modes have been developed, and the conditions and parameters necessary to ensure reliable, nonperturbative imaging have been identified. The science of wetting phenomena on the nanoscale has greatly benefited from these advancements. Extensive AFM studies of the morphology of nanoscale droplets on a variety of homogenous substrates proved instrumental to solve a long-standing controversy about the magnitude and sign of the line tension. These studies have established an upper bound of 10^{-10} N for the modulus of the line tension, which is very close to the range of values predicted theoretically. Moreover, it has been found that a small chemical inhomogeneity of the substrate can mimic the effects of the line tension, thus complicating the measurement of this elusive physical quantity.

On the other hand, advances in patterning techniques have allowed the control of the spatial distribution of the substrate inhomogeneity at the nanoscale, so finding several applications in fundamental studies of nanowetting. In particular, chemical nanopatterns were used to induce nanoscale deformations of the droplet contact line in a controlled fashion, which led to an alternative method to estimate the line tension. Chemical patterns made of wettable regions embedded in a nonwettable substrate were used as templates for the condensation of nanodroplets with lateral dimensions as small as 50 nm and with heights less than 15 nm. Detailed analysis of nanodroplet shape indicates that van der Waals forces dictate the nanoliquid morphology consistent with the augmented Young equation with the intermolecular interactions modeled by Lennard-Jones pairwise potentials. In a time-resolved study, the morphology of nonvolatile nanorivulets spreading along wettable nanostripes was investigated, revealing that the liquid profile is truncated near the microscopic contact line in good qualitative agreement with classical mesoscale hydrodynamics.

These remarkable achievements have enriched our understanding of wetting on small scales, while also raising new questions, hopefully to be addressed in the near future, thanks to the increase in sensitivity and resolution of AFM techniques. In particular, the development of high-speed AFM imaging may enable the study of non-equilibrium nanowetting phenomena, a still largely unexplored area.

References

1. G. Binnig, C. Quate, and C. Gerber, Atomic force microscope, *Phys. Rev. Lett.*, **56**, 930 (1986).

2. Y. Martin, C. Williams, and H. Wickramasinghe, Atomic force microscope—force mapping and profiling on a sub 100-Å scale, *J. Appl. Phys.*, **61**, 4723 (1987).
3. U. Hartmann, van der Waals interactions between sharp probes and flat sample surfaces, *Phys. Rev. B*, **43**, 2404 (1991).
4. M. Mate, M. R. Lorenz, and V. Novotny, Atomic force microscopy of polymeric liquid films, *J. Chem. Phys.*, **90**, 7550 (1989).
5. D. Langevin, ed., *Light Scattering by Liquid Surfaces and Complementary Techniques*. New York: Marcel Dekker (1992).
6. U. Hartmann, Intermolecular and surface forces in noncontact scanning force microscopy, *Ultramicroscopy*, **42**, 59 (1992).
7. S. Herminghaus, A. Fery, and D. Reim, Imaging of droplets of aqueous solutions by tapping-mode scanning force microscopy, *Ultramicroscopy*, **69**, 211 (1997).
8. Q. Zhong and D. Inniss, Fractured polymer/silica fiber surface studied by tapping mode atomic force microscopy, *Surf. Sci. Lett.*, **290**, 688 (1993).
9. J. Tamayo and R. Garcia, Deformation, contact time, and phase contrast in tapping mode scanning force microscopy, *Langmuir*, **12**, 4430 (1996).
10. T. Pompe, A. Fery, and S. Herminghaus, Imaging liquid structures on inhomogeneous surfaces by scanning force microscopy, *Langmuir*, **14**, 2585 (1998).
11. J. P. Cleveland, B. Anczykowski, A. E. Schmid, and V. B. Elings, Energy dissipation in tapping-mode atomic force microscopy, *Appl. Phys. Lett.*, **72**, 2613 (1998).
12. R. Garcia, R. Magerle, and R. Perez, Nanoscale compositional mapping with gentle forces, *Nat. Mater.*, **6**, 405 (2007).
13. L. Zitzler, S. Herminghaus, and F. Mugele, Capillary forces in tapping mode atomic force microscopy, *Phys. Rev. B*, **66**, 155436 (2002).
14. N. Hashemi and R. Montazami, Basins of attraction of tapping mode atomic force microscopy with capillary force interactions, *Appl. Phys. Lett.*, **94**, 251902 (2009).
15. F. Rieutord and M. Salmeron, Wetting properties at the submicrometer scale: a scanning polarization force microscopy study, *J. Phys. Chem. B*, **102**, 3941 (1998).
16. J. Hu, X.-D. Xiao, and M. Salmeron, Scanning polarization force microscopy: a technique for imaging liquids and weakly adsorbed layers, *Appl. Phys. Lett.*, **67**, 476 (1995).

17. A. Checco, H. Schollmeyer, J. Daillant, P. Guenoun, and R. Boukherroub, Nanoscale wettability of self-assembled monolayers investigated by noncontact atomic force microscopy, *Langmuir*, **22**, 116 (2006).
18. M. Luna, J. Colchero, and A. M. Baro, Study of water droplets and films on graphite by noncontact scanning force microscopy, *J. Phys. Chem. B*, **103**, 9576 (1999).
19. M. Luna, J. Colchero, A. Gil, J. Gomez-Herrero, and A. M. Baro, Application of non-contact scanning force microscopy to the study of water adsorption on graphite, gold and mica, *Appl. Surf. Sci.*, **157**, 393 (2000).
20. A. Checco, P. Guenoun, and J. Daillant, Nonlinear dependence of the contact angle of nanodroplets on contact line curvature, *Phys. Rev. Lett.*, **91**, 186101 (2003).
21. H. X. Zhang and W. Ducker, Formation of interfacial nanodroplets through changes in solvent quality, *Langmuir*, **23**, 12478 (2007).
22. H. X. Zhang and W. Ducker, Interfacial oil droplets, *Langmuir*, **24**, 110 (2008).
23. L. Xu and M. Salmeron, Scanning polarization force microscopy study of the condensation and wetting properties of glycerol on mica, *J. Phys. Chem. B*, **102**, 7210 (1998).
24. F. Mugele, T. Becker, R. Nikopoulos, M. Kohonen, and S. Herminghaus, Liquid microstructures at solid interfaces, *J. Adhes. Sci. Technol.*, **16**, 951 (2002).
25. A. Ulman, Formation and structure of self-assembled monolayers, *Chem. Rev.*, **96**, 1533 (1996).
26. R. Maoz, S. R. Cohen, and J. Sagiv, Nanoelectrochemical patterning of monolayer surfaces: toward spatially defined self-assembly of nanostructures, *Adv. Mater.*, **11**, 55 (1999).
27. J. L. Boruvka and A. Neumann, Generalization of the classical theory of capillarity, *J. Chem. Phys.*, **66**, 5464 (1977).
28. T. Pompe and S. Herminghaus, Three-phase contact line energetics from nanoscale liquid surface topographies, *Phys. Rev. Lett.*, **85**, 1930 (2000).
29. S. L. Schimmele, M. Napiórkowski, and S. Dietrich, Conceptual aspects of line tensions, *J. Chem. Phys.*, **127**, 164715 (2007).
30. R. Seemann, M. Brinkmann, E. J. Kramer, F. F. Lange, and R. Lipowsky, Wetting morphologies at microstructured surfaces, *Proc. Natl. Acad. Sci. USA*, **102**, 1848 (2005).
31. J. L. Wilbur, A. Kumar, H. A. Biebuyck, E. Kim, and G. M. Whitesides, Microcontact printing of self-assembled monolayers: applications in microfabrication, *Nanotechnology*, **7**, 452 (1996).

32. A. Checco, O. Gang, and B. M. Ocko, Liquid nanostripes, *Phys. Rev. Lett.*, **96**, 056104 (2006).
33. A. Checco and B. M. Ocko, Morphology and phase behavior of ethanol nanodrops condensed on chemically patterned surfaces, *Phys. Rev. E*, **77**, 061601 (2008).
34. A. Checco, Y. Cai, O. Gang, and B. M. Ocko, High resolution non-contact AFM imaging of liquids condensed onto chemically nanopatterned surfaces, *Ultramicroscopy*, **106**, 703 (2006).
35. J. B. Sweeney, T. Davis, L. E. Scriven, and J. A. Zasadzinski, Equilibrium thin films on rough surfaces. 1. Capillary and disjoining effects, *Langmuir*, **9**, 1551 (1993).
36. A. Checco, Liquid spreading under nanoscale confinement, *Phys. Rev. Lett.*, **102**, 106103 (2009).
37. D. Bonn, J. Eggers, J. Indekeu, J. Meunier, and E. Rolley, Wetting and spreading, *Rev. Mod. Phys.*, **81**, 739 (2009).

Chapter 7

Surface Bubbles in Micro- and Nanofluidics

Bram Borkent and Detlef Lohse
Physics of Fluids Group, Department of Science and Technology, MESA+ Institute, and JM Burgers Center for Fluid Dynamics, University of Twente, 7500AE Enschede, The Netherlands
d.lohse@utwente.nl

In this book chapter we review and summarize recent findings of the Twente Physics of Fluids group on controlled cavitation of bubbles on micro- and nanostructured surfaces[1–5] and in addition give some overview on the Twente findings on surface nanobubbles.[6–9] Gas accumulation at liquid-solid interfaces can occur in the form of bubbles, for instance, at a surface defect. The resulting micro- and nanoscale air pockets can act as nucleation sites in shockwave-induced cavitation experiments, leading to the random nature of cavitation, both in space and in time. However, by controlling the size and position of the defects, Bremond *et al.*[2,3] succeeded in making cavitation perfectly reproducible both in time and in space. This technique made it possible to study the dynamics of individual bubbles, bubble pairs, and bubble clusters emerging from the defects in an extremely reproducible fashion. In addition, by further miniaturizing the pits down to diameters of 100 nm, Borkent *et al.*[4] could measure the pressure at which these bubbles start to nucleate,

Nanoscale Liquid Interfaces: Wetting, Patterning, and Force Microscopy at the Molecular Scale
Edited by Thierry Ondarçuhu and Jean-Pierre Aimé

ISBN 978-981-4316-45-3 (Hardcover), 978-981-4364-48-5 (eBook)
www.panstanford.com

giving results in perfect agreement with the theory by Atchley and Prosperetti.[10] In contrast, so-called surface nanobubbles, though comparible in size and volume to the bubbles in the nanopits, do not act as nucleation sites in cavitation experiments.[5] Finally, we report on some typical properties of surface nanobubbles[6] and report that they can be created in a reproducible and controllable way by electrolysis.[8] That suggests that these surface nanobubbles are in dynamic equilibrium, that is, gas in- and outflux are balanced. Such dynamic stabilization has also been suggested for standard surface nanobubbles.[11]

7.1 Introduction

Water in contact with (in particular hydrophobic) solid surfaces can contain submicroscopic air bubbles at the solid interface. The bubbles are stabilized by surface defects and/or impurities, and generally they are not visible with the naked eye. An example is provided by the daily-life situation of freshly drawn tap water in a glass beaker. Since the water is slightly oversaturated once it comes out of the pipe system, the water gets rid of its excess gas through diffusion, and bubbles—which were initially not visible—suddenly appear at the glass interface. These growing surface bubbles originated on much smaller length scales, where they started from microscopic gas bubbles trapped in tiny surface imperfections. In general, surface bubbles may become important in micro- and nanofluidic environments, since—depending on the shape, stability, and number densities of the bubbles—they may alter hydrodynamic boundary conditions, form nucleation sites in boiling and cavitating liquids, or affect chemical surface reactions. A detailed understanding of surface bubbles on these small length scales will therefore be of great potential value for various applications like lab-on-a-chip devices, inkjet printing, megasonic cleaning, boiling, and sonochemistry.

7.2 Controlled Surface Bubble Cavitation

Cavitation is a well-known phenomenon that occurs undesirably around ship propellers, around rudders, and in water turbines.[13]

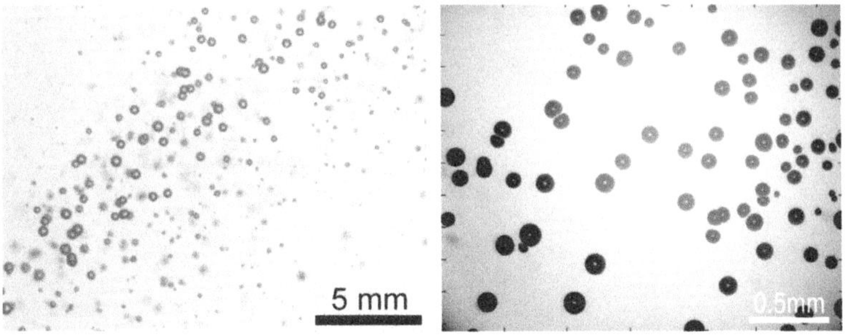

Figure 7.1 Cloud cavitation at particles in suspension (left, taken from[12]) and randomly distributed cavitating bubbles on a surface (right).

The noise emitted by the collapsing bubbles can be a nuisance, while the destructive nature of aspherical bubble collapses is even more problematic, since it can lead to surface erosion. The destructive power of cavitation bubbles, however, can also be beneficial, for example, for cleaning jewelry, glasses, or semiconductor surfaces or a snapping shrimp killing prey animal through a cavitating bubble.[14] In all these cases, the rapid growth of vapor bubbles occurs due to a local stress in the liquid, which can be applied either by hydrodynamic or by acoustic means.[13] One of the main issues in cavitation research is, Where do the bubbles originate exactly? It has been postulated that the bubbles are nucleated from microscopic gas pockets, trapped at surface defects,[15] such as walls or suspended particles. Indeed, in many practical applications surfaces are not completely smooth and the water is not entirely free from tiny particles, and as a result cavitation clouds of all kinds can occur[1,12,16,17] (see Fig. 7.1). How can one control these cavitation nuclei? Once the bubbles emerge from their nucleation sites, another issue arises too: how to describe the bubble–bubble dynamics? For sure the bubbles will interact with each other, but how can one measure and describe this quantitatively?

7.2.1 *Experimental Setup*

A way to tackle both problems in one experiment is by controlling the surface defects down to the microscopic level and by positioning them very precisely on a two-dimensional surface

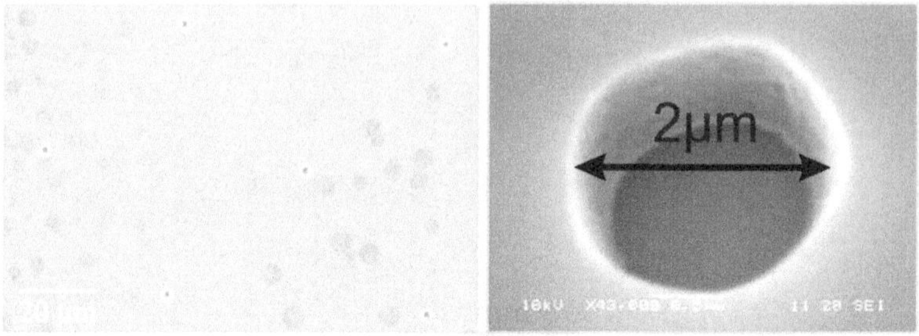

Figure 7.2 Microholes positioned on a silicon surface, shown on two different scales (figure taken from.[2])

(Bremond *et al.*[1-3]) Standard lithography techniques allow to etch cylindrical pits of uniform sizes (typically 2, μm) into pieces of silicon and to put the pits on predefined locations (Fig. 7.2). After coating the surface with a monolayer of hydrophobic molecules, the pits will entrap air once the surface is submerged in water.

To nucleate bubbles Bremond *et al.* used a commercially available lithotripter, which consists of a focused array of piezoelectric elements at the bottom of a liquid bath, which is filled with gas-saturated ultrapure MilliQ water (Fig. 7.3). This technique made it possible to generate shockwaves in water, which typically look like the pulse depicted in Fig. 7.3. The pressure first rose, and after a few μs it droped and became negative down to values of ~ -15 bar or more. After some small reflections the pressure got restored to its atmospheric value. The bubbles could be recorded in reflection mode through either high-speed video imaging or simply single snapshots. Since the resulting bubble dynamics turned out to be extremely reproducible, Bremond *et al.* chose the latter option and recorded images with a high resolution charge-coupled device (CCD) camera. By shifting the trigger of the shockwave source in repeated experiments, they could obtain high-resolution stroboscopic videos, from which the bubble sizes could be extracted.

7.2.2 *Single Bubble*

The simplest scenario to study is the case of a single bubble. Figure 7.4 depicts the resulting images both from top and side views

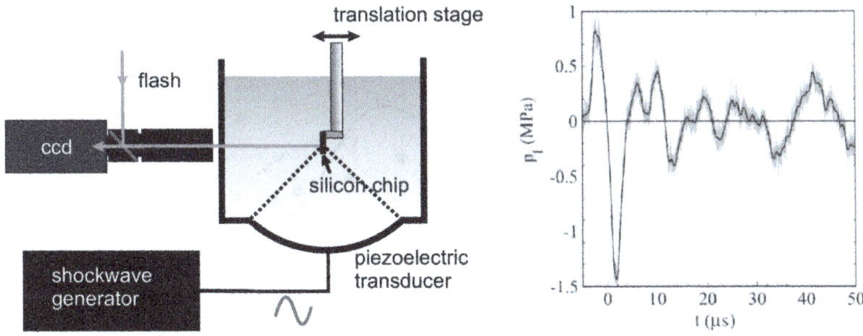

Figure 7.3 Experimental setup (left) and employed pressure wave (right) employed in the controlled cavitation experiments of Bremond *et al.*[2]

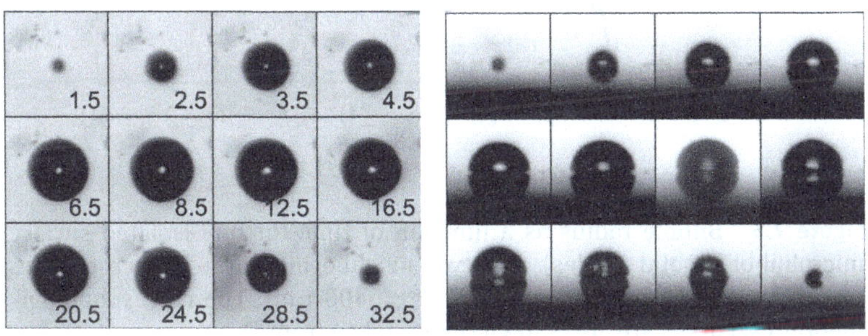

Figure 7.4 Top and side views of the dynamics of a single cavitating surface microbubble. The numbers represent the time (in microseconds) after inception. At maximum, the bubble is around 150 μm in radius (figure taken from.[3])

for a single bubble subjected to a minimum negative pressure of $p_m = -2$ MPa. The bubble grows rapidly in the first microseconds and then collapses at around 32.5 μs. The bubble radius R can be extracted from the top-view images. The side view shows that the bubble is essentially hemispherical in shape during the majority of its lifetime. The hemispherical bubble together with its hemispherical image bubbles can be interpreted as a spherical bubble (neglecting viscous effects directly at the wall). The situation is thus equivalent to that of a spherically symmetric bubble, and the standard Rayleigh-Plesset equation can be applied:

$$R\ddot{R} + \frac{3}{2}\dot{R}^2 = \frac{1}{\rho}\left(p_v + p_g - p_\infty - \frac{2\sigma}{R} - 4\mu\frac{\dot{R}}{R}\right) \qquad (7.1)$$

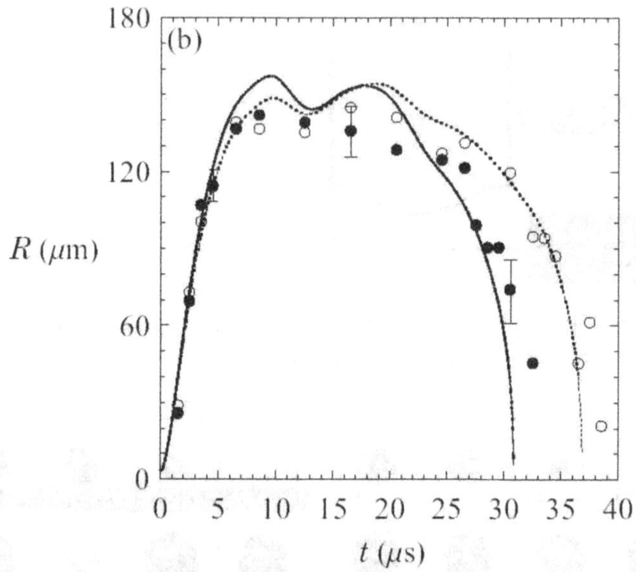

Figure 7.5 Bubble radius as a function of time for one isolated surface microbubble (solid circles) and two neighboring surface microbubbles (open circles; the initial distance is $d = 400$ μm). The correspondingly theoretical curves are also shown (figure taken from.[3])

Here p_v and p_g are the pressure of the gas and vapor in the bubble, p_∞ the liquid pressure far away, and σ and μ the liquid surface tension and dynamic viscosity, respectively. Integration of Eq. 7.1 using the measured pressure profile $p_\infty(t)$ gives a theoretical prediction of the radial bubble dynamics. Comparing this result with the experimental observations gives a perfect match, as shown in Fig. 7.5.

7.2.3 *Two Bubbles*

In the case of two bubbles the situation becomes more complex since the bubbles will interact with each other through acoustic coupling. Typical experimental results of two bubbles subjected to $p_m = -2$ MPa, but with different interdistances d, are shown in Fig. 7.6. In the left case ($d = 400$ μm), both bubbles remain essentially spherical during their whole lifetime (so-called "weak

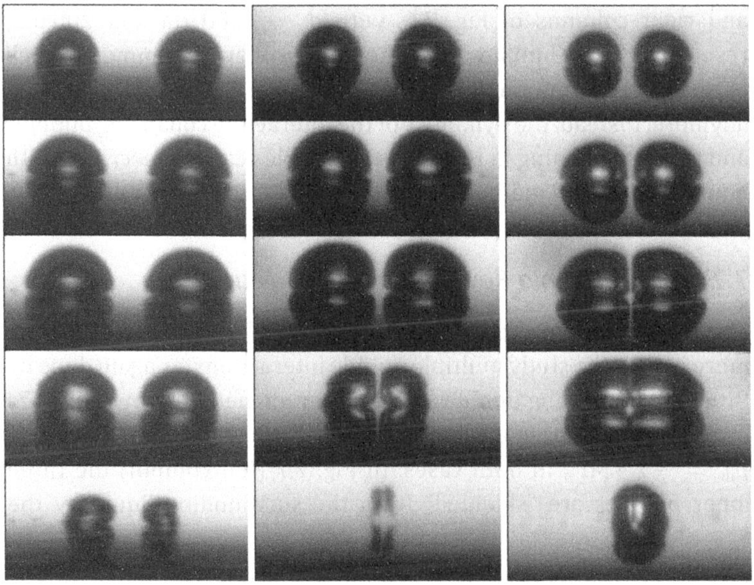

Figure 7.6 Dynamics of the interaction of two surface microbubbles with an initial distance of $d = 400$ μm, $d = 300$ μm, and $d = 200$ μm, respectively (left to right). At maximum, the bubbles are about 150 μm in radius (figure taken from.[3])

interaction regime"). However, the stage of bubble collapse happens a few microseconds later than in the single-bubble case, thanks to mutual shielding of the bubbles. This effect can be embodied into the Rayleigh-Plesset equation by adding the sound emission of the neighboring bubble, leading to

$$R\ddot{R} + \frac{3}{2}\dot{R}^2 = \frac{1}{\rho}\left(p_v + p_g - p_\infty - \rho\frac{R}{d}(R\ddot{R} + 2\dot{R}^2) - \frac{2\sigma}{R} - 4\mu\frac{\dot{R}}{R}\right). \tag{7.2}$$

Indeed, the $R(t)$ curve resulting from this extended Rayleigh-Plesset equation (7.2) leads to a delayed bubble collapse for the two-bubble case, see Fig. 7.5, in good quantitative agreement with the measurements, despite the fact that the bubbles feel an anisotropic pressure field around their surface, especially in the last stage of their collapse. When the bubbles come closer to each other (middle and right columns of Fig. 7.6, with $d = 300\,\mu\text{m}$ and $200\,\mu\text{m}$, respectively), the bubbles loose their hemispherical symmetry much earlier ("strong interaction regime"), and Eq. 7.2 cannot be applied anymore. Instead, the dynamics of the interacting collapsing bubbles and the film thinning in between the bubbles can be described with boundary integral methods.[3]

7.2.4 *Bubbles in a Row and a Bubble Matrix*

The next step is to extend the number of bubbles to three or even more in order to study multiple bubble interactions, as again done by Bremond *et al.*[3] The bubbles were either aligned in a row (Fig. 7.7) or ordered in a hexagonal structure (Fig. 7.8), with $d = 200\ \mu\text{m}$ and $p_{\text{m}} = -1.4\,\text{MPa}$ in both cases. In Fig. 7.7 (left column) the three inner bubbles are "shielded" from the surrounding liquid by the

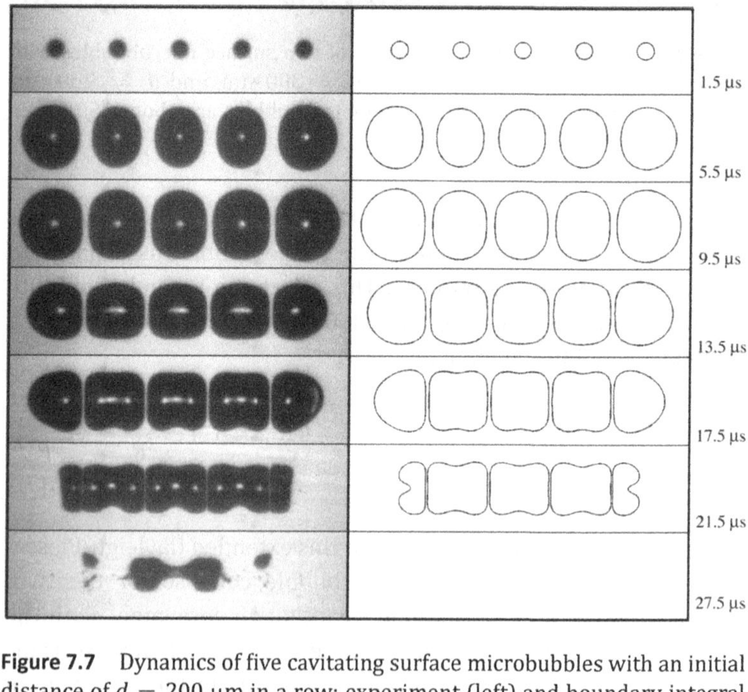

Figure 7.7 Dynamics of five cavitating surface microbubbles with an initial distance of $d = 200\ \mu\text{m}$ in a row: experiment (left) and boundary integral simulation (right) (figure taken from.[3])

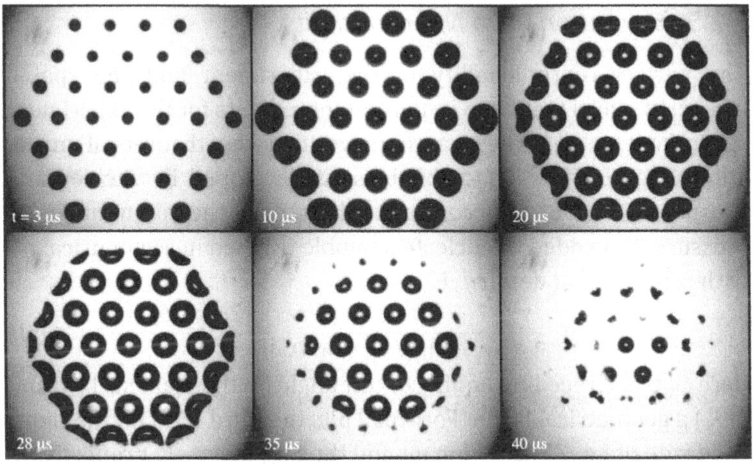

Figure 7.8 Dynamics of cavitating surface microbubbles with an initial distance of $d = 200$ μm (figure taken from.[2])

outer bubbles and collapse much later. Since the bubbles strongly deform, their dynamics cannot be described by Eq. 7.2. Fortunately, two plains of symmetry are present (wall and z-axis), which makes it possible to employ an axi-symmetrical boundary integral code to describe the bubble dynamics. The features of the numerical scheme were based on a method developed by Oğuz and Prosperetti[18] and were validated for the single- and two-bubble cases. The result of the simulations (Fig. 7.7, right column) could be directly compared with experimental top views (left column) with nice agreement between the two. The screening effect, as visible in the previous example, is even more pronounced for the bubbles located in an hexagonal array (Fig. 7.8). The center bubble is now shielded from all sides and clearly collapses at last. In addition, wall-normal jets directed toward the center of the bubble cluster become nicely visible during the collapse phases of the outer bubbles. These jets develop due to an asymmetry in the liquid phase surrounding the bubble. They are similar to wall-normal jets for single bubbles close to a wall. Since the bubbles in this example stayed hemispherical during the majority of their lifetimes, Eq. 7.2 could be applied with good success.[2]

7.2.5 *Cavitation from Nanosized Pits*

The perfect reproducibility of the experiment and the spatial control of nuclei were two important ingredients for the experiment of Borkent *et al.*[4] These authors verified the theoretical nucleation threshold of surface bubbles, as developed by Atchley and Prosperetti in 1989.[10] According to this theory, the minimum liquid pressure p_{m} needed to nucleate a bubble trapped in a cylindrical pit with radius r_{c} and depth d_{c} is

$$p_{\mathrm{m}} = p_{\mathrm{v}} + \frac{3\,p_{g,0}}{3 + (r_{\mathrm{c}}/d_{\mathrm{c}})g(\theta_{\mathrm{R}})/\sin^3\theta_{\mathrm{R}}} - \frac{2\sigma\sin\theta_{\mathrm{R}}}{r_{\mathrm{c}}}. \qquad (7.3)$$

Here θ_{R} is the receding liquid contact angle on the plain surface and $g(\theta_{\mathrm{R}})$ a geometrical factor. For small pits the third term on the right-hand side of Eq. 7.3 is dominant and thus p_{m} inversely proportional to r_{c}. Therefore one needs very small cavities to measure large differences in p_{m}. For this purpose, Borkent *et al.* have produced three samples with square arrays containing uniformly sized pits with diameters of 900 nm, 500 nm, and 100 nm, respectively (Fig. 7.9).

For each array, Borkent *et al.* decreased the minimum pressure of the shockwave applied to the liquid, until all bubbles of the array were nucleated (as observed by the camera). The bubble patterns are shown in Fig. 7.10, and the resulting nucleation phase diagram is shown in Fig. 7.11. In this plot, the line is the theoretical prediction according to Eq. 7.3. It perfectly matches the experimental results. The results of Borkent *et al.*[4] were the first demonstration that nucleation of surface bubbles down to nanometer length scales can be fully controlled and quantitatively understood.

Figure 7.9 SEM picture of nanopits with diameters of 900 nm, 500 nm, and 100 nm, respectively (figure taken from[4]) *Abbreviation*: SEM, scanning electron microscopy.

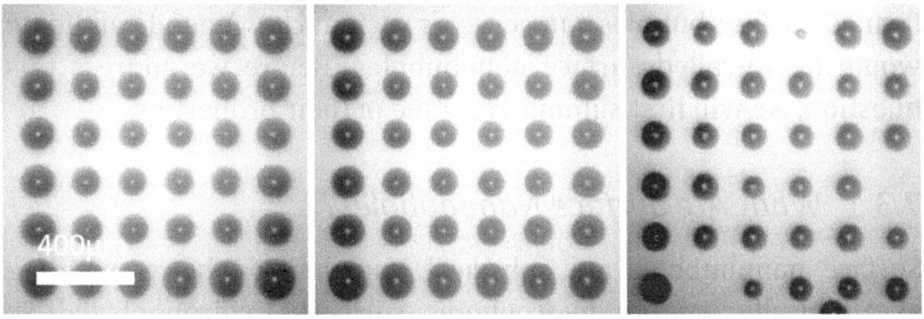

Figure 7.10 Cavitation bubbles nucleated from cylindrical pits with radius $r_c = 495$ nm (left), $r_c = 246$ nm (middle), and $r_c = 53$ nm (right) and depth $d_c = 2r_c$ for a pressure pulse with $p_m = -3.2$ MPa (figure taken from.[4])

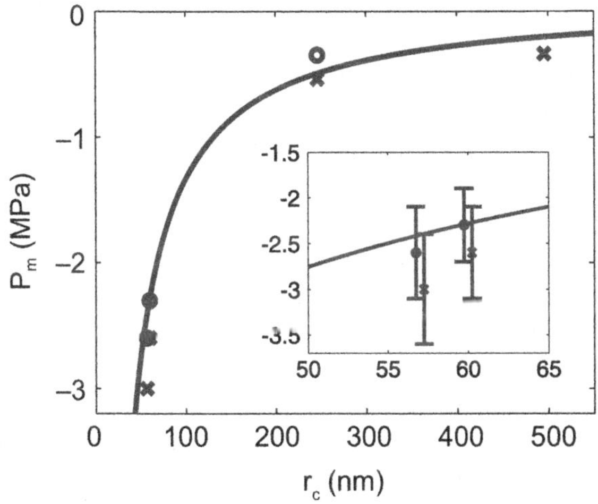

Figure 7.11 Nucleation threshold as a function of the pit radius for both theory (line) and experiment (symbols: crosses—nucleation; circles no nucleation). The inset shows a zoom-in with error bars. For visibility, overlapping points are shifted ±0.25 nm with respect to each other (figure taken from.[4])

7.3 Surface Nanobubbles

In this review we up to now dealt with surface micro- and nanobubbles in artificially produced surface pits. These bubbles served as cavitation nuclei. One wonders whether also naturally formed surface bubbles can serve as such nuclei. A good candidate would be so-called surface nanobubbles. Before we address this question, we will first summarize some of their properties.

7.3.1 *What Are Surface Nanobubbles?*

Surface nanobubbles are spherical, cap-shaped objects residing at atomically smooth hydrophobic surfaces (hydrophobized Si, graphite, etc.)[19,20] The features are typically ~10 nm in height and ~100 nm in diameter, and they are able to merge together,[21,22] disappear in degassed water,[23] and reappear when the water is gas oversaturated.[5-7,24-26] Hence, they are believed to consist of gas. As they are not visible with normal light, atomic force microscopy (AFM) in tapping mode is employed to image the bubbles. Here, the oscillations of a vibrating cantilever (with a sharp tip) are measured and are kept constant with a feedback mechanism. Once the tip of the cantilever comes close to a surface, the amplitude of the vibration becomes smaller and the system tracks the surface structure. In this way, topography images of a solid–liquid interface can be obtained. With this technique, surface nanobubbles have been imaged successfully, and it is found that they are surprisingly stable. Since the bubbles have a radius of curvature R_c on the order of ~100 nm, the Laplace excess pressure $2\sigma/R_c$ is several bars and should lead to rapid dissolution of the bubbles within milliseconds.[27,28] This is not observed: in experiments the bubbles are stable for hours, or even days.[26]

One example of a series of AFM images is shown in Fig. 7.12. Here, the temperature of a water droplet is increased before deposition on a piece of hydrophobized silicon wafer.[6] The topography of the

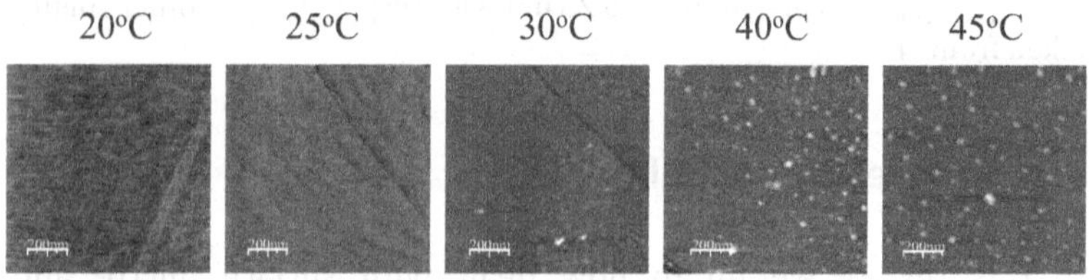

Figure 7.12 Surface nanobubble formation on a silanated Si surface with increasing temperature (figure taken from.[6])

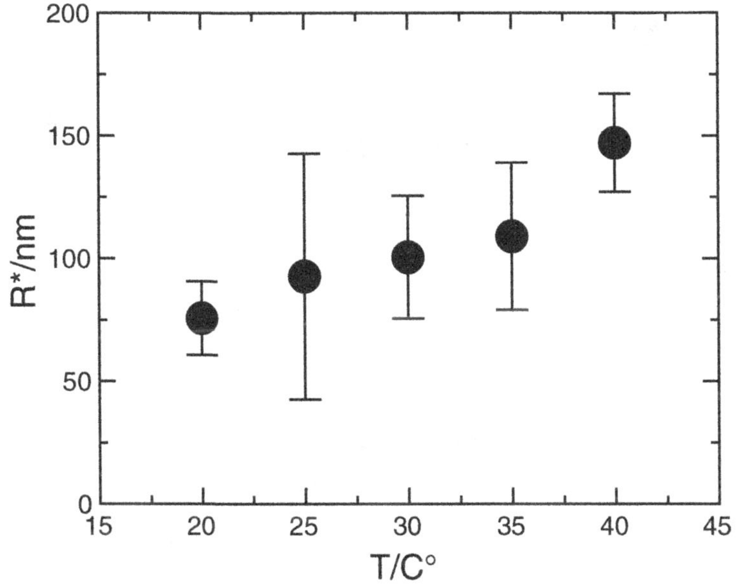

Figure 7.13 Preferred surface nanobubble radius of the nanobubbles of Fig. 7.12. The error bar represents one standard deviation.

solid–liquid interface shows a rapid increase in the number density of nanobubbles for increasing temperature. This observation can be understood as follows: The gas solubility of water decreases with increasing temperature. If the water is heated quickly, the excess amount of gas does not have enough time to escape to the environment. As a result, the water becomes slightly oversaturated with gas, which stimulates the formation of interfacial air bubbles. The bubbles become larger as well. A statistical analysis on the radius of the nanobubles present in Fig. 7.12 and in a number of repeated experiments (not shown) shows that the nanobubble radius increases as a function of temperature (see Fig. 7.13).

7.3.2 *Surface Nanobubbles Subjected to Extreme Stresses*

As surface nanobubbles have sizes similar to the nanopits used in the previous experiment, Borkent *et al.*[4] asked whether it is possible to nucleate surface nanobubbles to visible sizes. According

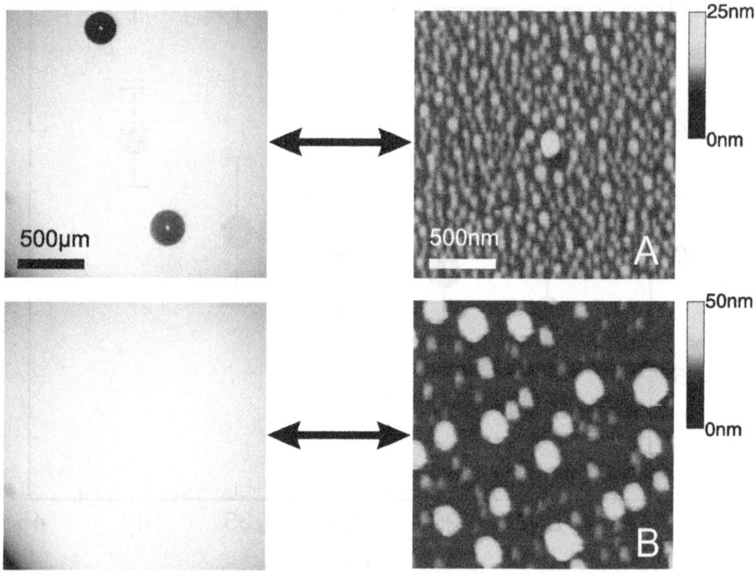

Figure 7.14 In spite of a high nanobubble concentration (right), there is hardly any cavitation activity (left). Note that the scale on the left is larger than that on the right by a factor 1,000 (figure taken from.[5])

to Eq. 7.3, or alternatively Eq. 7.1, this should be possible. If one could nucleate these nanobubbles, this would disclose a new type of nucleation sites, while in addition providing a relatively easy method to check whether surface nanobubbles are present on a surface. In the experiment Borkent *et al.* used hydrophobized silicon samples without any surface defects. First, they checked with AFM whether surface nanobubbles were densely present on these surfaces, after having submerged the sample in clean water (Fig. 7.14, top right). However, when Borkent *et al.* subjected the same sample to a negative pressure, which was 10 times stronger than what would be needed according to Eq. 7.3, they saw hardly any cavitation (Fig. 7.14, top left). Only occasionally one or two bubbles were visible, presumably from a small bubble trapped by some surface contamination (note the difference in scales between the left and right pictures). By applying forced oversaturation (flushing ethanol away with water), larger surface nanobubbles could be created in a reproducible way. Strikingly, this did not help at all: not a single nanobubble was nucleated to visible size in the cavitation experiment (Fig. 7.14, down left). To check whether the nanobubbles had been removed by the strong shockwave, Borkent *et al.* took the very same surface, kept it immersed in water, and measured the solid–liquid interface with AFM. Remarkably, they found that the surface was still covered with hundreds of large surface nanobubbles (Fig. 7.14, down right)! The experiments by Borkent *et al.* have thus revealed that surface nanobubbles are not suitable nucleation sites in shockwave-induced cavitation experiments, in contrast to nanobubbles trapped in pits of similar sizes (see Fig. 7.15). Remarkably, surface nanobubbles are able to survive huge pressure sweeps. Borkent *et al.* denoted this property as "superstability" of surface nanobubbles.

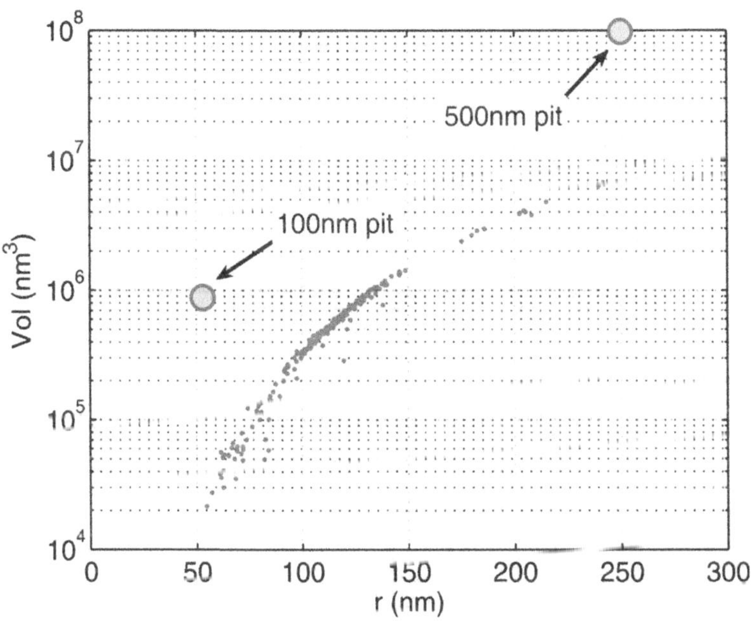

Figure 7.15 Volume vs. radius of artifical (green) and natural (blue) surface nanobubbles. While the former do cavitate for a large enough pressure reduction, the latter do not, though they are of a similar size and volume.

7.3.3 *Preference in Size*

The AFM pictures shown in Fig. 7.14 show dense populations of surface nanobubbles of roughly similar sizes within each picture (cases A and B, respectively). Therefore one may ask whether nanobubbles have a preferred radius. The answer was found by Borkent *et al.*[9] through a statistical analysis of the nanobubble diameters. The result is shown in Fig. 7.16. Indeed, Borkent *et al.* observed a single preferred size in case A, while in case B two maxima showed up. Notice that in case B the liquid was oversaturated temporally to stimulate nanobubble formation. Possibly, this led to the formation of the larger set of nanobubbles (right peak). The left peak in Fig. 7.16 is representing small nanobubbles, which were still present in between the larger ones and were similar in size to those in case A. Presumably, this set of

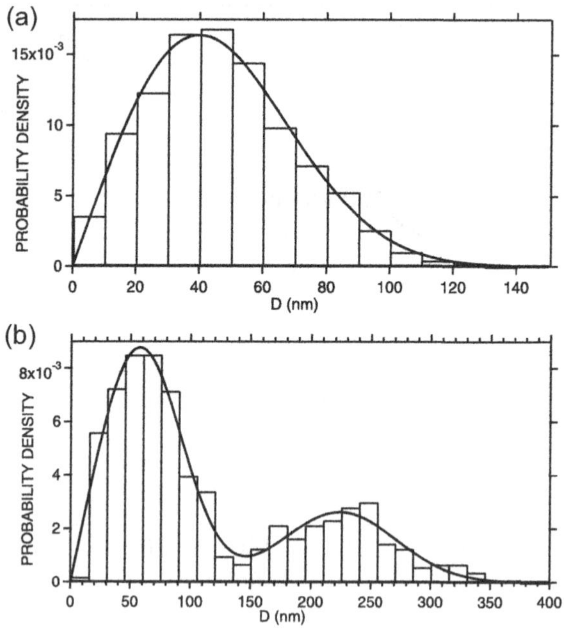

Figure 7.16 Probability distribution functions for the nanobubble diameter for two different experimental procedures for the nanobubble creation (see text) (figure taken from.[9])

smaller bubbles formed once the oversaturation had disappeared. The results of Borkent *et al.*[9] suggest that, indeed, nanobubbles in dense populations have a tendency toward an equilibrium radius, which increases when the liquid becomes oversaturated.

7.3.4 *Electrolysis: A Controlled and Reproducible Way to Create Surface Nanobubbles*

A well-known method to produce gas at solid–liquid interfaces is through electrolysis. What happens here at the nanoscale? Yang *et al.*[8] could monitor the growth of gaseous surface nanobubbles in real time using AFM,[8] when the sample surface acted as of the hydrogen-producing electrode. A typical result of that experiment, at 1 V, is shown in Fig. 7.17.

Although the bubble grew in time, after some tens of seconds a stable situation set in and the bubble maintained a constant shape. This observation was reproduced by a further experiment, in which Yang *et al.*[8] measured the electrical current as a function

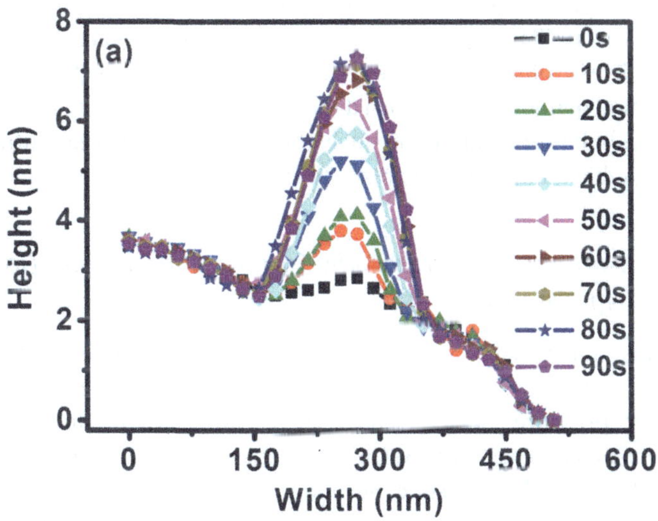

Figure 7.17 Time evolution of the cross-section height profile of an electrolytically generated hydrogen surface nanobubble (1V) (figure taken from.[8])

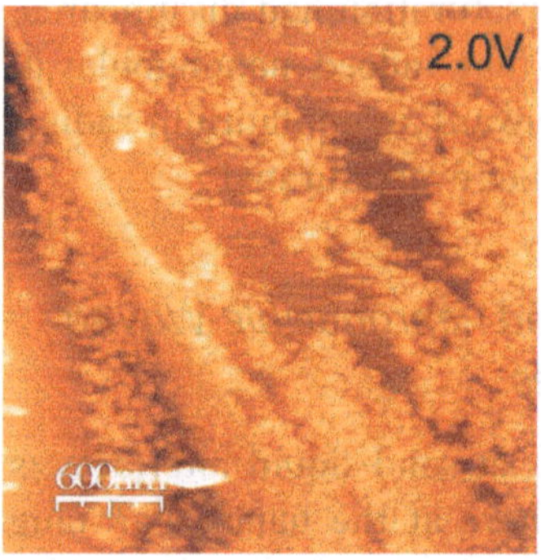

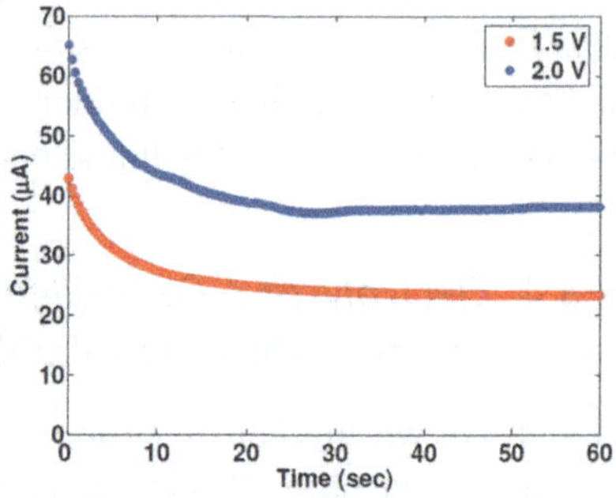

Figure 7.18 Electrolytically generated hydrogen surface nanobubbles (left) and the time evolution of the current toward that situation at 1.5 V and 2.0 V (right) (figures taken from.[8])

of time (Fig. 7.18) at constant voltage: in the first ~20s the current decreased exponentially, but then it saturated and became constant. One would have expected that surface nanobubbles would constantly accumulate gas on the electrode surfaces and therefore grow without bounds, especially since the current is nonzero. However, Yang *et al.*'s AFM images showed stationary nanobubbles of certain sizes. These observations therefore suggest that electrolytically generated nanobubbles are in a dynamic equilibrium state: gas flows into the bubble, but simultaneously this is balanced by a gas outflow, presumably thanks to the Laplace pressure.

7.3.5 *Another Conjecture Why Surface Nanobubbles Do Not Dissolve?*

Brenner and Lohse[11] suggested that such a dynamic equilibrium may also stabilize standard nanobubbles. On the one hand the Laplace pressure $2\sigma/R_c$ creates an overpressure in the bubble, which leads to a gas outflux. On the other hand, a gas influx, similar to that in the electrolysis experiments, could compensate for the outflux. It is known from experiments, but also from molecular dynamics simulations,[29] that hydrophobic surfaces attract gas molecules. Brenner and Lohse[11] suggested that these accumulated gas molecules are able to enter the nanobubble at the contact

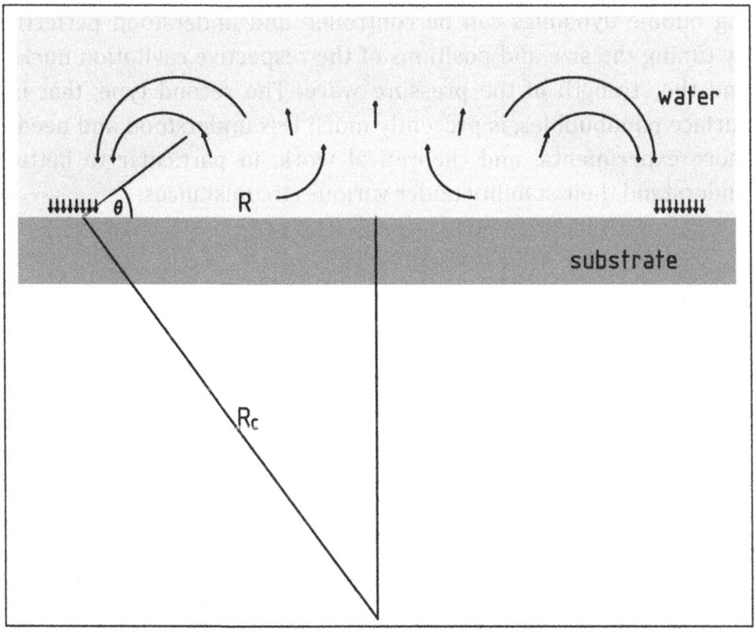

Figure 7.19 Sketch of the dynamic equilibrium of surface nanobubbles (figure taken from.[11])

line, resulting in a gas influx (see the sketch in Fig. 7.19), though it is unclear what drives this mechanism in the long term. They calculated the effective gas in- and outflux and found that these effects are of a similar order of magnitude and could balance each other at a particular stable equilibrium radius R^*. Brenner and Lohse's theory would therefore indeed explain the observation that nanobubbles show a preferred radius (Figs. 7.13 and 7.16), but various other bubble stabilization mechanisms may be possible, too.

7.4 Conclusions

In summary, this overview article discussed two types of micro- and nanoscopic surface bubbles. The first type—gas entrapped in controlled surface defects—serves as nucleation sites in cavitation experiments. For these bubbles cavitation inception and the resulting bubble dynamics can be controlled and understood perfectly by tuning the size and positions of the respective cavitation nuclei and the strength of the pressure wave. The second type, that is, surface nanobubbles, is presently much less understood and needs more experimental and theoretical work, in particular to better understand their stability under various circumstances.

Acknowledgments

Most of the results presented here were obtained by PhD students and postdocs associated with the Twente Physics of Fluids group. In particular, we thank Nicolas Bremond for his pioneering work on controlled cavitation from micropits and Shangjiong Yang for his work on nanobubbles. Furthermore, we thank Manish Arora, Michael Brenner, G. Le Caër, Stefan Dammer, Benjamin Dollet, Stephan Gekle, Stefan Kooij, Claus-Dieter Ohl, Bene Poelsema, Andrea Prosperetti, Holger Schönherr, Peichun Tsai, Julius Vancso, and Harold Zandvliet, who all contributed significantly to the research results presented here. We also thank the Foundation for Fundamental Research on Matter (FOM) and Stichting voor de Technische Wetenschappen (STW) for financial support.

References

1. N. Bremond, M. Arora, C. D. Ohl, and D. Lohse, Cavitation on surfaces, *J. Phys.: Condens. Matter*, **17**, S3603 (2005).
2. N. Bremond, M. Arora, C. D. Ohl, and D. Lohse, Controlled multibubble surface cavitation, *Phys. Rev. Lett.*, **96**, 224501 (2006).
3. N. Bremond, M. Arora, S. M. Dammer, and D. Lohse, Interaction of cavitation bubbles on a wall, *Phys. Fluids*, **18**, 121505 (2006).
4. B. M. Borkent, S. Gekle, A. Prosperetti, and D. Lohse, Nucleation threshold and deactivation mechanisms of nanoscopic cavitation nuclei, *Phys. Fluids*, **21**(10), 102003 (2009).
5. B. M. Borkent, S. M. Dammer, H. Schönherr, G. J. Vancso, and D. Lohse, Superstability of surface nanobubbles, *Phys. Rev. Lett.*, **98**, 204502 (2007).

6. S. Yang, S. M. Dammer, N. Bremond, H. J. W. Zandvliet, E. S. Kooij, and D. Lohse, Characterization of nanobubbles on hydrophobic surfaces in water, *Langmuir*, **23**, 7072–7077 (2007).
7. S. Yang, E. S. Kooij, B. Poelsema, D. Lohse, and H. J. W. Zandvliet, Correlation between geometry and nanobubble distribution on hopg surface, *EPL*, **81**, 64006 (2008).
8. S. Yang, P. Tsai, E. S. Kooij, A. Prosperetti, H. J. W. Zandvliet, and D. Lohse, Electrolytically generated nanobubbles on highly orientated pyrolytic graphite surface, *Langmuir*, **25**, 1466–1474 (2009).
9. B. Borkent, H. Schönherr, G. L. Caër, B. Dollet, and D. Lohse, Preferred sizes and ordering in surface nanobubble populations, *Phys. Rev. E*, **80** (3), 036315 (2009).
10. A. A. Atchley and A. Prosperetti, The crevice model of bubble nucleation, *J. Acoust. Soc. Am.*, **86**(3), 1065–1084 (1989).
11. M. P. Brenner and D. Lohse, Dynamic equilibrium mechanism for surface nanobubble stabilization, *Phys. Rev. Lett.*, **101**, 214505 (2008).
12. B. M. Borkent, M. Arora, and C. D. Ohl, Reproducible cavitation activity in water-particle suspensions, *J. Acoust. Soc. Am.*, **121**, 1406–1412 (2007).
13. C. E. Brennen, *Cavitation and Bubble Dynamics*. New York: Oxford University Press (1995).
14. M. Versluis, A. v. d. Heydt, B. Schmitz, and D. Lohse, How snapping shrimp snap: through cavitating bubbles, *Science*, **209**, 2114–2117 (2000).
15. E. N. Harvey, D. K. Barnes, W. D. McElroy, A. H. Whiteley, D. C. Pease, and K. W. Cooper, Bubble formation in animals, *J. Cell. Comp. Physiol.*, **24**, 1–22 (1944).
16. M. Arora, L. Junge, and C. D. Ohl, Cavitation cluster dynamics in shock-wave lithotripsy: part 1. Free field, *Ultrasound Med. Biol.*, **31**(6), 827–839 (2005).
17. M. Arora, C. D. Ohl, and D. Lohse, Effect of nuclei concentration on cavitation cluster dynamics, *J. Acoust. Soc. Am.*, **121**, 3432–3436 (2007).
18. H. Oguz and A. Prosperetti, Bubble entrainment by the impact of drops on liquid surfaces, *J. Fluid Mech.*, **219**, 143–179 (1990).
19. S. Lou, Z. Ouyang, Y. Zhang, X. Li, J. Hu, M. Li, and F. Yang, Nanobubbles on solid surface imaged by atomic force microscopy, *J. Vac. Sci. Technol. B*, **18**, 2573–2575 (2000).
20. N. Ishida, T. Inoue, M. Miyahara, and K. Higashitani, Nano bubbles on a hydrophobic surface in water observed by tapping-mode atomic force microscopy, *Langmuir*, **16**, 6377–6380 (2000).

21. A. C. Simonsen, P. L. Hansen, and B. Klösgen, Nanobubbles give evidence of incomplete wetting at a hydrophobic interface, *J. Colloid Interface Sci.*, **273**(1), 291–299 (2004).

22. L. Zhang, Y. Zhang, X. Zhang, Z. Li, G. Shen, M. Ye, C. Fan, H. Fang, and J. Hu, Electrochemically controlled formation and growth of hydrogen nanobubbles, *Langmuir*, **22**, 8109–8113 (2006).

23. X. H. Zhang, G. Li, N. Maeda, and J. Hu, Removal of induced nanobubbles from water/graphite interfaces by partial degassing, *Langmuir*, **22**, 9238–9243 (2006).

24. X. H. Zhang, N. Maeda, and V. S. J. Craig, Physical properties of nanobubbles on hydrophobic surfaces in water and aqueous solutions, *Langmuir*, **22**, 5025–5035 (2006).

25. X. H. Zhang, A. Khan, and W. A. Ducker, A nanoscale gas state, *Phys. Rev. Lett.*, **98**, 136101 (2007).

26. X. H. Zhang, A. Quinn, and W. A. Ducker, Nanobubbles at the interface between water and a hydrophobic solid, *Langmuir*, **24**, 4756–4764 (2008).

27. P. S. Epstein and M. S. Plesset, On the stability of gas bubbles in liquid-gas solutions, *J. Chem. Phys.*, **18**, 1505–1509 (1950).

28. S. Ljunggren and J. C. Eriksson, The lifetime of a colloid-sized gas bubble in water and the cause of the hydrophobic attraction, *Phys. Rev. E*, **129–130**, 151–155 (1997).

29. S. M. Dammer and D. Lohse, Gas enrichment at liquid-wall interfaces, *Phys. Rev. Lett.*, **96**, 206101 (2006).

Chapter 8

Use of Ultrasonic Shear Waves for the Study of Liquid Interfaces

Diethelm Johannsmann* and Anne Finger
Institute of Physical Chemistry, Clausthal University of Technology, Arnold Sommerfeld-Str. 4, D-38678 Clausthal-Zellerfeld, Germany
*johansmannn@pc.tu-clausthal.de

Acoustic shear waves and shear-wave resonators, in particular, have a long tradition in interfacial sensing. Typically one infers the thickness and softness of an adsorbate layer from the shifts of resonant frequency and bandwidth. Laterally heterogeneous samples (vesicles, cells, adsorbed particles) can be modeled numerically. The shifts of frequency and bandwidth are proportional to the stress–velocity ratio (the "load impedance") at the interface, and this load impedance can be calculated by (for instance) the finite element method (FEM).

This chapter is, to a large part, concerned with the extension of this formalism to (nano-)bubbles. Simple-minded planar models of interfacial slippage predict that nanobubbles should look like a film with an apparent negative thickness where the negative thickness is identified with the "slip length". A frequency increase upon bubble formation (corresponding to a negative apparent film thickness) has

Nanoscale Liquid Interfaces: Wetting, Patterning, and Force Microscopy at the Molecular Scale
Edited by Thierry Ondarçuhu and Jean-Pierre Aimé

ISBN 978-981-4316-45-3 (Hardcover), 978-981-4364-48-5 (eBook)
www.panstanford.com

repeatedly been observed. However, one can also observe negative frequency shifts upon gas evolution at the resonator surface.

This situation can be traced back to the fact that nanobubbles at a solid surface constitute a laterally heterogeneous sample. Nanobubbles are rather stiff objects due to surface tension. Even though they have a low density, bubbles increase the roughness of the surface and may lead to a negative frequency shift (positive apparent film thickness, negative apparent slip length) because the bubbles drag the adjacent liquid along in their motion. A second effect of importance is bubble-induced acoustic streaming: Due to the nonlinear term in the Navier-Stokes equation, a bubble sets off a stationary flow. This convective flux increases the rate of gas dissolution into the liquid and therefore may prevent bubble formation at voltages where nanobubbles would otherwise evolve.

8.1 Introduction

Acoustic resonators are well-known tools in gravimetric sensing.[1–3] Although there is a family of related instruments (including, for instance, torsional resonators,[4] Love wave devices,[5] and FBARs,[6]) the workhorse of the field still is quartz crystal microbalance (QCM).[7] QCM is a very simple device (Fig. 8.1). It consists of a disk of a piezoelectric material (typically quartz) coated with electrodes. Upon application of an alternating current (AC) voltage, the crystal undergoes a shear vibration. When the frequency of the driving voltage coincides with one of the acoustic resonance frequencies of the disk, the vibration amplitude strongly increases. At the same time, the current passed into the electrodes increases accordingly. For that reason, the acoustic resonance is easily probed by electrical means. The quality factor (Q-factor—frequency divided by bandwidth) well exceeds 1,000 even in liquids. The resonances are very sharp, and the accuracy, by which the resonance frequency is determined, is correspondingly good. The use of quartz crystals as clocks and sensors builds on the sharpness of the resonances. In the context of liquid sensing, the *tangential* motion of the crystal surface is of equal importance. Tangential motion gives rise to *shear* waves, and these decay within a distance from the surface of less than a

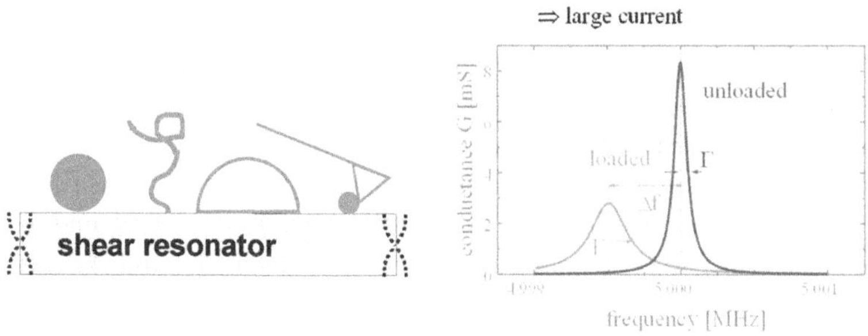

Figure 8.1 Sketch of QCM. The figure on the left is not to scale. The crystal thickness is around 300 μm. The sample, on the other hand, typically has a thickness of well below a micron. Right: Conductance spectrum as obtained in impedance analysis. These measurements may be carried out on different harmonics. The ring-down technique (QCM-D) yields equivalent parameters,[11] where the "dissipation" is given as $D = 2\Gamma/f$. Resonance frequency (f) and resonance bandwidth (Γ) are derived by fitting resonance curves to the experimental conductance spectra. The presence of the sample changes both f and Γ. In the modeling process one tries to reproduce the experimental values of Δf and $\Delta\Gamma$.

micron. In contrast to conventional ultrasonic instruments (which employ longitudinal sound), shear-wave resonators are *surface specific*.

In the past decades, QCM has extensively been adapted to applications in the liquid phase.[8,9] In the liquid, one usually analyzes the resonance bandwidth, $\Delta\Gamma$[10] (or equivalently, dissipation, $\Delta D = 2\Delta\Gamma/f$,[11]) in addition to the frequency shift, Δf, and one does so on a few different overtones.[12] Comparing Δf and $\Delta\Gamma$ as determined on the different overtones, one obtains a detailed picture of the sample. In particular, its softness may be derived.[13]

For planar layers, analytic equations predicting Δf and $\Delta\Gamma$ from the model parameters (layer thickness, storage modulus, loss modulus) exist.[14–17] Mathematical details of modeling have been provided in a recent book chapter.[18] The models basically amount to a prediction of the acoustic reflectivity of the resonator t sample interface. For planar layers, this reflectivity is calculated in essentially the same way as in optics. The role of the refractive index is taken by the material's acoustic impedance, $Z_{ac} = (\rho G)^{1/2}$.

While the planar models have been successful and are widely applied, many samples of practical interest are laterally heterogeneous. Examples are biological cells[19] or vesicles.[20] Of course one can always try to describe such samples by an equivalent planar layer. Importantly, this approach is unsatisfying in the sense that the equivalent layer system may easily miss essential aspects of the sample's structure and dynamics.[21] However, Δf and $\Delta\Gamma$ can be predicted even for heterogeneous samples if one goes through the effort of modeling with the **f**inite **e**lement **m**ethod (FEM).[22] We have used the commercial software package supplied by COMSOL. Such calculations always entail certain simplifications. The most important shortcoming of the current FEM code is the fact that it only covers two-dimensional (2D) geometries. Still, one can back up interpretations of experimental findings with such calculations. The assumption here is that the behavior of a three-dimensional (3D) object (like a hemisphere) is similar to the behavior of the corresponding 2D object (which is the infinite hemicylinder in this case).

This chapter is concerned with the study of *liquid* interfaces. Liquid interfaces would mostly mean droplets,[22,23] vesicles,[24] or bubbles.[25] Bubbles and "nanobubbles," in particular, have been studied extensively in the context of slippage[26] and the long-range hydrophobic attraction.[27] Today they are studied by numerous groups independently of these specific applications.[28,29] This chapter only covers a rather narrow subtopic of this research field. We show in the second part that nanobubbles look like stiff objects when investigated with ultrasonic shear waves.

The chapter is organized as follows. Section 8.2 provides a short reminder of what acoustic shear waves can and cannot do. Shear waves have distinct advantages (compared to other surface analytical techniques like optical reflectometry or atomic force microscopy [AFM]), but there are also some caveats to be kept in mind. Section 8.3 briefly summarizes some predictions from simple planar models of slip. An experimental result, which stands as an example for an experience in the authors' laboratory, is presented in section 8.4. Section 8.5 provides the results from FEM calculations. Section 8.6 discusses nonlinear phenomena and acoustic streaming, in particular.

8.2 Peculiarities of Acoustic Shear Waves

8.2.1 *Shear Waves Do Not Propagate in Liquids*

When discussing the benefits of acoustic waves close to surfaces, it is essential to not confuse MHz shear waves with conventional ultrasound. Conventional ultrasound employs compressional (longitudinal) waves. Compressional waves *do* propagate in liquids. For that reason, they may be reflected from the opposing wall of the cell, which complicates the experiment. Shear waves in liquids decay within less than a micron, and the shear-wave resonator therefore only sees the portion of the sample immediately adjacent to the surface (Fig. 8.2). For a 5 MHz bulk acoustic wave (BAW) resonator in water, the penetration depth of the wave, δ, is 250 nm. Admittedly, even shear-wave resonators sometimes *do* emit a small component of compressional waves.[30] These are sources of artifacts.

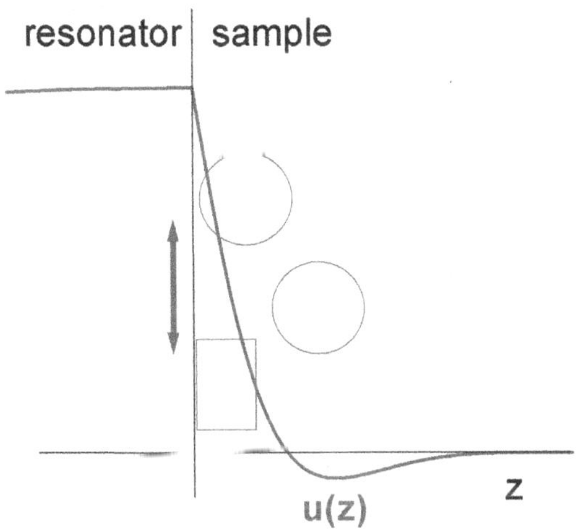

Figure 8.2 Tangential displacement close to the resonator surface ($u(z)$, thick blue line). Any object inside the tail of the shear wave will deviate the flow in one way or another. For a given geometry, the flow field can be calculated numerically. The shifts of frequency and bandwidth are proportional to the (area-averaged) ratio of tangential stress and tangential velocity at the resonator surface, where the latter quantities are to be understood as complex amplitudes of oscillatory variables

The compressional component is caused by the nonideal shape of the resonator (energy trapping, in particular).

8.2.2 *The Amplitude of Oscillation Is Small*

Typical amplitudes of oscillation are below 1 nm. The shear strain is much below 1%, the Reynolds number is below 0.01, and nonlinear effects are weak. However, nonlinearities are not strictly absent, unless one goes to very small drive powers. Effects of varying amplitude are a way to gain additional insights into the dynamics of the sample. We come back to nonlinearities in section 8.6.

8.2.3 *QCM Averages over the Active Area*

Given the recent emphasis on nanoscopic objects, one may ask whether QCM can detect a single molecule (or, to be more modest, a single colloidal particle). With respect to gravimetric measurements in vacuum, single-molecule sensitivity appears possible with specialized devices.[31] With standard devices in liquids, measurements on individual nanobubbles, nanoparticles, cells, vesicles, etc., are prohibitively difficult. One may even question whether such measurements are meaningful. After all, averaging has its virtues. When sensing the concentration of analytes, there is not much point in looking for single molecules. It is the concentration that matters, and the concentration is the (average) number of molecules (or particles) per unit volume. Determination of the corresponding number of adsorbed molecules per unit area is a perfectly logical procedure.

Typically, QCM averages over the active area, A_q, which is between 3 and 30 mm^2. Actually, there is an amplitude distribution with an approximately Gaussian shape, which introduces a statistical weight (large in the center, small at the rim).

8.2.4 *Even Dilute Soft Samples Are Easily Detected*

In principle, there is a close analogy between QCM and surface plasmon resonance (SPR) spectroscopy. Both types of resonators couple to the environment by rapidly decaying waves, and the mathematics is indeed very similar. The acoustic impedance, Z_{ac}, is the acoustic analog of the refractive index, n. In contrast to the refractive index (generating the contrast in SPR spectroscopy, variations amount to 20% at most), the acoustic impedance may vary over a decade or more. Somewhat loosely speaking, QCM compares the stiffness of any object within its view to the stiffness of the liquid. If the sample contains two different objects, where one is, for example, 10 times as rigid as water and the second is 100 times as rigid as water, QCM cannot tell the difference. Both objects scatter the acoustic wave efficiently. QCM efficiently differentiates between "soft" and "very soft" objects. On the contrary, a very dilute polymer layer is hardly visible in SPR spectroscopy due to its small refractive index increment, Δn. Such dilute layers are only detected in QCM. For a more detailed discussion, the reader is referred to Ref. 32.

8.2.5 *QCM Operates at MHz Frequencies*

The fact that the oscillations occur in the MHz range has a number of consequences. First, the shear modulus of the adsorbate layer as determined with QCM must not naively be compared to the low frequency (<100 Hz) modulus as determined with standard rheological equipment. Soft materials have viscoelastic dispersion, and the modulus therefore depends on frequency. Second, the transport of momentum across the liquid is more efficient at high frequencies than in the usual stationary flow. The loss modulus, G'', of a Newtonian liquid is given as $G'' = \omega\eta$, with ω the angular frequency and η the viscosity. Inserting $\eta \approx 10^{-3}$ Pa.s and $\omega = 2\pi \times 5$ MHz, one finds $G'' = 30$ kPa. Water does not flow easily at 5 MHz. Finally, *inertia* sometimes plays a role not known in low-frequency experiments. This effect is particularly important when measuring in the dry. A 100 μm sphere contacting the crystal across a point contact is clamped in space by inertia. It does not take part in the crystal's movement. The "inertia clamp" is a very convenient aspect of QCM experiments.

Inertia is the QCM experimenter's best friend. In many ways, QCM experiments come down to the measurement of an (oscillatory) force. One may even spell out "QCM" as "quartz crystal micro- (force) balance." In contrast to AFM, force calibration is close to trivial in

QCM experiments because the forces of inertia are known rather precisely. There is no need to calibrate a cantilever.[33] The QCM experimenter usually balances the force of interest (such as the force transmitted across some kind of contact) against inertia. In this respect, we follow a time-honored recipe first put forward by Sir Isaac Newton in the late seventeenth century. Newton inferred the strength of the gravitational interaction between the sun and the planets from the orbital period. Today we study microparticles rather than planets, and for that reason we have to hurry up.

8.2.6 *There Are Electric and Dielectric Artifacts*

Anyone who is not particularly interested in (high-frequency) electrochemistry is strongly advised to ground the front electrode well before starting a QCM experiment. Otherwise the electric boundary conditions will affect Δf and $\Delta\Gamma$ via piezoelectric stiffening.[34] Such effects are left aside in the discussion next.

8.2.7 *The Vibration May Induce Stationary Flow*

As emphasized before, QCM basically operates in the domain of linear acoustics. Nonlinear effects have, however, been observed. For instance, Edvardsson *et al.* found that the adsorption of particles to the QCM surface was prevented by exciting the crystal with a voltage of 10 V (typical is 100 mV).[35] This effect can be explained with acoustic streaming.[36,37]

8.3 Simple Planar Models of Slip

The term “slip” has acquired different meanings. It can mean “solid-like sliding” in the sense of friction and tribology.[38] While such phenomena can definitely be investigated with shear-wave resonators,[39] they are outside the scope of this chapter. Second, slip may denote interfacial shear thinning in complex liquids.[40] The well-known everyday example is toothpaste. When squeezed out of the tube, toothpaste experiences “plug flow” because the large stress at the wall induces structural transformations inside the paste (such as an alignment of polymer chains along the direction of shear), which make the boundary less viscous than the bulk. Acoustic shear waves are not well suited to study such effects. Complex fluids usually are so viscous that the shear wave is highly overdamped. A second problem with slip in complex fluids is the large stress required to induce structural changes at the interface. Also, the transformation in the contact zone needs time. Chain alignment is slower than the oscillation period of QCM.

Third, slip can imply interfacial sliding between a simple liquid and a solid[41,42] (see chapter 2). Generally speaking, slip in this sense is the exception rather than the rule. The interactions between small molecules and a solid wall are expected to be at least as strong as the interactions between the molecules in the bulk. The effective viscosity at the interface therefore should be similar to (or larger than) the bulk viscosity. Still, there is experimental evidence in favor of slip even in simple liquids. Acoustic shear waves should be a well-suited tool of investigation.[43-46]

This chapter is limited to simple liquids. We do, however, adopt a bit of an open-minded view of the term "interfacial layer." Whether or not this layer consists of nanobubbles depends on the definition. The following text uses the term "nanobubble," but this usage should not imply that these are just small bubbles. Nanobubbles are expected to behave peculiarly.[28] Line tension should play a role.[47] The Laplace pressure can be many atmospheres. The vertical scale of surface roughness compares to bubble size. Understanding nanobubbles comes down to understanding a significant portion of slip (see also chapter 7 for a general description of nanobubbles).

In the one-dimensional (1D) continuum model, a slipping layer is a layer close to the surface with a viscosity lower than that of the bulk. Decreased viscosity implies increased the shear gradient dv/dz (v is the tangential velocity). The slip length, b_S, (see Fig. 8.3) is defined by extrapolating the linear portion of the flow profile to the plane of zero shear. Should this plane be located below the interface, the distance between the two is called slip length.[48] Evidently, the near-surface viscosity might also be increased. The reason might be surface roughness, the presence of adsorbates, or an increased density. The distance between the surface and the plane of zero shear then is the "hydrodynamic thickness" of the respective layer.

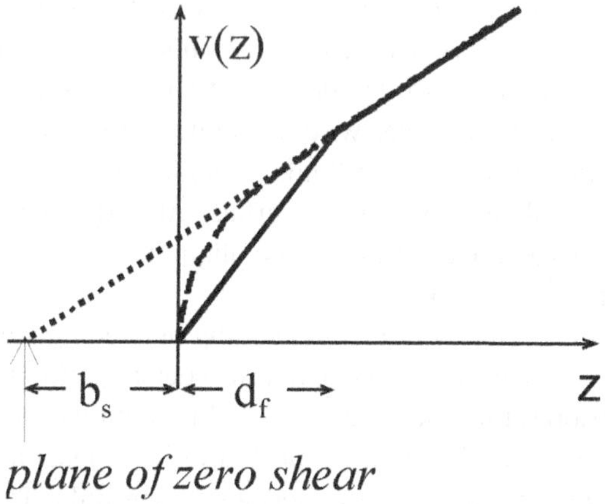

Figure 8.3 Flow profile, $v(z)$, above a solid surface with slip. The viscosity $\eta(z)$ close to the surface is lower than in the bulk. The shear gradient, dv/dz, is increased correspondingly. Solid line: The viscosity is reduced inside a hypothetical discrete layer of thickness d_f. The slip length, b_S, is the distance from the surface to the extrapolated plane of zero shear (dotted). Dashed line: The viscosity gradually decreases at the interface.

The slip length can be viewed as an apparent negative hydrodynamic thickness.

One can show that the slip length, b_S, is given by

$$b_S = \left[\frac{\eta_{liq}}{\eta_f} - 1\right] d_f \tag{8.1}$$

where η_f is the viscosity inside the film and d_f is the thickness of the layer with reduced viscosity (solid line in Fig. 8.3). Since the above relation is linear in d_f, it also holds for continuous profiles $\eta(z)$ (dashed line in Fig. 8.3) in an integral sense:

$$b_S = \int_0^\infty \left[\frac{\eta_{liq}}{\eta(z)} - 1\right] dz \tag{8.2}$$

The term "slip length" does not imply a statement about the physical origin of slippage. It only rephrases the phenomenon.

For a given profile of the viscosity, $\eta(z)$, one can straightforwardly calculate Δf and $\Delta\Gamma$. For the calculation of the displacement, $u(z)$, start from the wave equation:

$$\frac{du(z)}{dz} = -ik(z)u$$

$$k(z) = \frac{\omega}{c(z)} = \frac{\omega\sqrt{\rho}}{\sqrt{G(z)}} = \frac{\omega\sqrt{\rho}}{\sqrt{i\omega\eta(z)}} \qquad (8.3)$$

$k = k' - ik''$ is the wave vector, c is the speed of sound, G is the shear modulus, ρ is the density, and η is the viscosity. Fix $u(z = z_{max})$ and $du/dz(z = z_{max})$ as unity. z_{max} is some location outside the range of decreased viscosity. These choices amount to picking some arbitrary amplitude and phase of the acoustic wave. Solve the wave equation numerically (Eq. 8.3, an ordinary differential equation with known initial values) and compute the stress at the surface as

$$\sigma = \left(\eta\frac{d\dot{u}}{dz}\right)\bigg|_{z=0} \qquad (8.4)$$

Finally divide the stress, σ, by the velocity, $\dot{u}(z = 0) = i\omega u(z = 0)$, to derive the load impedance Z_L. Clearly, the ratio of stress and velocity (both complex amplitudes) does not depend on the choice of amplitude and phase at $z = z_{max}$. Δf and $\Delta\Gamma$ are calculated from the stress–velocity ratio at the surface by the small-load approximation (SLA).[18] The latter states that the complex frequency shift $\Delta f^* = \Delta f + i\Delta\Gamma$ is proportional to the stress–velocity ratio at the resonator surface:

$$\frac{\Delta f^*}{f_F} = \frac{i}{\pi Z_q} Z_L = \frac{i}{\pi Z_q}\left\langle\frac{\sigma}{\dot{u}}\right\rangle \qquad (8.5)$$

f_F is the frequency of the fundamental, $Z_q = 8.8\times 10^6$ kg/(m^{2}s) is the acoustic impedance of AT-cut quartz, Z_L is the load impedance, σ is the tangential stress, and $\dot{u}$ is the tangential velocity.

The above formalism works for arbitrary profiles of the viscosity, $\eta(z)$. However, it does not algebraically relate Δf and $\Delta\Gamma$ to the slip length, b_S. There is a simple relation between Δf and b_S, if the slipping layer is much thinner than the wavelength of sound ("long-wavelength approximation"). Treat the slipping layer as an acoustic multilayer, where each layer, i, is characterized by its thickness, d_i, its acoustic impedance, Z_i, and its wave vector, k_i. Since $k_i d_i << 1$ for all layers, trigonometric functions in $k_i d_i$ can be linearized, whenever they occur.[49] Making these approximations, one finds[18]

$$\frac{\Delta f^*}{f_F} \approx -\frac{-\omega}{\pi Z_q} \sum_i \frac{Z_i^2 - Z_{\text{bulk}}^2}{Z_i^2} \rho_i d_i$$

$$\approx -\frac{-\omega}{\pi Z_q} \int_0^\infty \left[\frac{Z_2(z) - Z_{\text{bulk}}^2}{Z_2(z)} \right] \rho(z) \mathrm{d}z$$

$$\approx -\frac{\rho\omega}{\pi Z_q} \int_0^\infty \left[\frac{G(z) - G_{\text{bulk}}}{G(z)} \right] \mathrm{d}z \qquad (8.6)$$

Constant density, ρ, was assumed in the last step. Δf as derived in this way is proportional to the slip length. To see this, write:

$$\frac{\Delta f}{f_F} = -\frac{2f}{Z_q} \int_0^\infty \left(1 - \frac{\mathrm{i}\omega\rho_{\text{bulk}}\eta_{\text{bulk}}}{\mathrm{i}\omega\rho(z)\eta(z)}\right) \rho(z)\mathrm{d}z \approx \frac{2f}{Z_q}\rho \int_0^\infty \left(\frac{\eta_{\text{bulk}}}{\eta(z)} - 1\right) \mathrm{d}z$$

$$= \frac{2f}{Z_q}\rho b_s \qquad (8.7)$$

where $\rho(z) \approx \rho_{\text{bulk}} \approx \rho$ has been used. To the experimentalist, the slip looks like a negative Sauerbrey mass, where the slip length is equal to the negative Sauerbrey thickness.

8.4 Electrochemically Produced Bubbles May Induce Negative Frequency Shift

The standard picture of slip predicts an increase of frequency upon bubble formation. This has repeatedly been observed.[50-54] Tsionski *et al.* went through a particularly careful study, which yielded a negative result: these authors did not see a frequency shift under conditions where they expected the presence of nanobubbles on the basis of their knowledge of the electrochemistry going on in the cell.[55]

In the authors' laboratory a negative frequency shift upon gas was also observed.[56] This effect is most pronounced when the bubbles are small. A set of experimental results is shown in Fig. 8.4. This data was acquired with an electrochemical QCM (EQCM[57]) in water. In EQCM, the front electrode of the resonator at the same

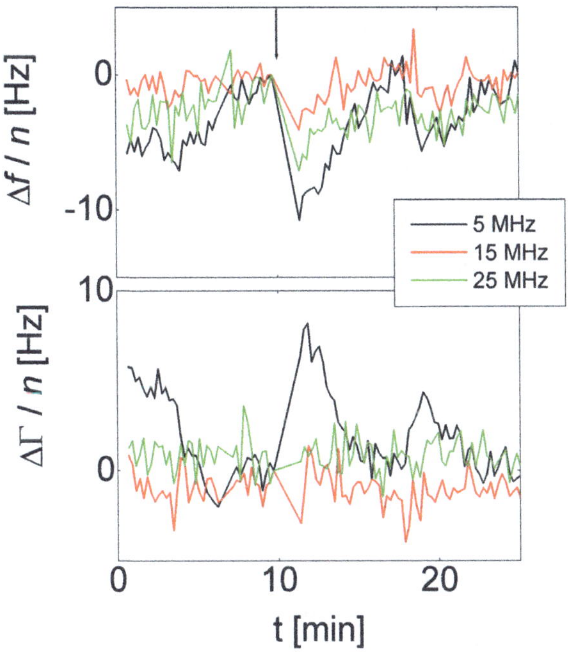

Figure 8.4 Frequency shift (Δf, top) and shift of bandwidth ($\Delta\Gamma$, bottom, both normalized to overtone order) observed upon producing small bubbles by electrochemical means. After the potential is set to -1 V vs. SCE ($t = 10$ min, vertical arrow), a thin layer of hydrogen evolves at the gold surface. The layer is visually observable via a slight coloration. Individual bubbles cannot be discerned by the naked eye. The voltage is turned off after 0.5 s, and the bubbles subsequently dissolve. Importantly, Δf is negative, as expected for the deposition of solid objects. A laterally continuous slipping layer would have induced $\Delta f > 0$. The same would have happened with macroscopic bubbles. Bubble formation here is accompanied by in increase in bandwidth, $\Delta\Gamma$, which occurs because the bubbles deviate the liquid flow. *Abbreviation*: SCE, standard calomel electrode.

time is the working electrode of an electrochemical cell. Employing EQCM (for instance, during an electrodeposition process), one can correlate charge and deposited mass.

At $t = 10$ minutes, the voltage at the working electrode was lowered by -1 volt, which induced hydrogen evolution. The voltage was shut off after 0.5 seconds. Had the voltage been applied for a longer time, macroscopic bubbles would have evolved. Due to the short duration of the pulse, the bubbles remained small and stuck to the electrode surface. One visually observes a change in color, but the individual bubbles were not discernable to the naked eye. With time the coloration disappeared, and Δf and $\Delta\Gamma$ returned to their original values. As Fig. 8.4 shows, the bubble-induced frequency shift is *negative*. Also, there is an increase in bandwidth. Such a behavior would typically be interpreted as evidence for the presence of solid (nano-)particles. The explanation for this finding starts from the notion that nanobubbles may be viewed as stiff objects due to Laplace pressure. Bubbles can *increase the coupled mass* because they speed up the movement of adjacent liquid relative to the situation on a flat surface. Bubbles in this sense create surface roughness, which may increase the moving mass even though the bubbles themselves have zero density.

Next, we put this argument into a more quantitative form. We argue that the effect of surface tension can be captured by assigning stiffness equal to the Laplace pressure to the interior of the bubble. Consider the deformed bubble depicted in Fig. 8.5. The curvature of the interface has decreased on the left-hand side, while it has increased on the right. In the limit of small amplitude, shear transforms a spherical bubble into an ellipsoid.

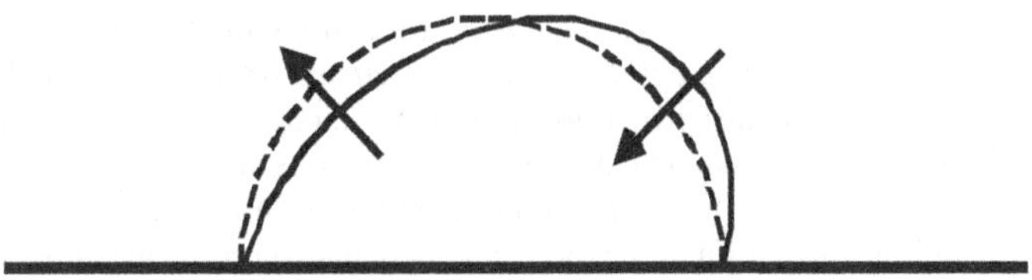

Figure 8.5 Illustration of the forces connected to bubble deformation. Arrows indicate deformation-induced pressure. Upon shearing a hemispherical bubble, a Laplace pressure of magnitude $\gamma \times \Delta(1/r)$ develops at the deformed surfaces. Here $1/r$ is the curvature and $\Delta(1/r)$ is the shear-induced change in curvature. For a given shear deformation, the change in curvature is proportional to the curvature in the undeformed state. The Laplace stress will significantly deviate the flow whenever γ/r is comparable the shear stiffness of the liquid, which is $\omega\eta$. With $\eta \approx 10^{-3}$ Pa.s, $\omega \approx 2\pi$ 5 MHz, and $\gamma \approx 72 \times 10^{-3}$ N/m, one finds that shear-induced Laplace pressure matters when the radius is less than about a micron.

The changes in principle curvature can be estimated as

$$\Delta\left(\frac{1}{r}\right) \sim \frac{1}{r_0}\left.\frac{\mathrm{d}u}{\mathrm{d}z}\right|_{z=0} \tag{8.8}$$

$1/r_0$ is the curvature of the undeformed bubble, and $\mathrm{d}u/\mathrm{d}z$ is the shear rate. Due to deformation, the bubble exerts a normal stress onto the neighboring liquid. Disregarding the tensorial nature of stress, its magnitude can be approximated as

$$\Delta\sigma \sim \gamma\Delta\left(\frac{1}{r}\right) \sim \frac{\gamma}{r_0}\cdot\left.\frac{\mathrm{d}u}{\mathrm{d}z}\right|_{z=0} \tag{8.9}$$

The stress is proportional to the shear gradient. A similar stress would be exerted by an elastic body of stiffness γ/r_0.

8.5 Numerical Calculations

Central to the view outlined the previous section is the notion that slipping layers will in most practical situations be laterally heterogeneous.[58] For a start, we treat the slipping layer as an assembly of nanobubbles and analyze this geometry by means of a finite element analysis.

Δf and $\Delta\Gamma$ can be predicted even for heterogeneous systems making use of SLA.[22] The latter states that the complex frequency shift $\Delta f^* = \Delta f + \mathrm{i}\Delta\Gamma$ is proportional to the area-averaged tangential stress at the resonator surface (Eq. 8.5).

$$\frac{\Delta f^*}{f_\mathrm{F}} = \frac{i}{\pi Z_q} Z_L \tag{8.10}$$

Ref. 18 further elaborates on the meaning of the load impedance and the electromechanical analogy. SLA is applicable to a wide range of samples. That certainly includes the Sauerbrey film. For the Sauerbrey film, the stress is $m_\mathrm{f}(-\omega^2 u_0)$, where m_f is the mass per unit are, u_0 is the oscillation amplitude, and $-\omega^2 u_0$ is acceleration. The speed, $\dot{u}$, is given as $\mathrm{i}\omega u_0$. Inserting these relations, one finds the Sauerbrey equation $(\Delta f = -2 f_\mathrm{F} m_\mathrm{f}/Z_{q},)^1$ recovered.

Since Δf^* depends linearly on the load, SLA can be applied in an average sense.[59] One may replace the stress, σ, by an area-averaged stress, $\langle\sigma\rangle$. Assume that the sample is a complex material, such as a layer of biological cells, a sand pile, a froth, an assembly of vesicles, or a droplet. If the average stress at the resonator surface can be calculated in one way or another, a quantitative analysis of the QCM experiment is in reach. In the following we build on the fact the Δf and $\Delta\Gamma$ are proportional to stress and that the stress can be calculated numerically.

We made use of the incompressible Navier-Stokes module of the multiphysics software package supplied by COMSOL (COMSOL GmbH, Göttingen, Germany). The bubble was treated as a domain with a density of 1 kg/m^3 and viscosity of 10^{-6} Pa.s. These are the values for air. In order to emulate surface tension, we introduce a very thin elastic shell at the interface between air and liquid. The shell is chosen so thin ($\approx$ 1 nm) that the bending stiffness of the membrane is negligible compared to its elongational stiffness. Since bending stiffness scales as the thickness cubed and elongational stiffness scales as thickness, such a situation can always be achieved. More specifically, the thickness, d_S, must obey

$$\frac{1}{24} d_S^2 \Delta \left(\frac{1}{r}\right)^2 << \varepsilon \tag{8.11}$$

where $\Delta(1/r)$ is the flow-induced curvature and ε is the flow-induced elongation of the membrane. The Young's modulus of the membrane, E, was chosen such that the following relation holds:

$$\frac{E}{1-\nu^2} d_S = \gamma \tag{8.12}$$

Here, ν is the Poisson number, $E/(1-\nu^2)$ is the biaxial modulus, and γ is the surface tension. E is related to the viscosity via $\eta = -\mathrm{i}\eta'' = -\mathrm{i}G/\omega = -\mathrm{i}E/(3\omega)$.

Evidently, this model entails simplifications. First, we only compute the properties of a 2D object. We study hemicylinders rather than hemispheres. A second somewhat subtle difference between the model and a real liquid surface has to do with local or global expansion and contraction. If an elastic membrane if forced to contract at some spot and to expand by the same amount somewhere else, the membrane responds with a restoring force. A liquid surface would not do this. Liquid surfaces only respond to a change in *total* area. This is a caveat to be kept in mind.

Because the software can only handle incompressible media, the bubble needs to be treated as incompressible. With the geometry investigated here, compressibility is not a problem because the flow pattern does not induce a volume change of the bubble. The bubble is situated in the center of a symmetric cell, and any contraction occurring on the left is compensated by a corresponding expansion on the right. Incompressibility has to be taken care of by the choice of the geometry. With regard to the details of the FEM methodology, we refer the reader to Ref. 22. Here, we proceed directly to the results.

Figure 8.6 shows two raw outputs of the FEM calculation. The geometry is described in the figure caption. The difference between the two graphs is in the size of the bubble. The radii of the bubbles

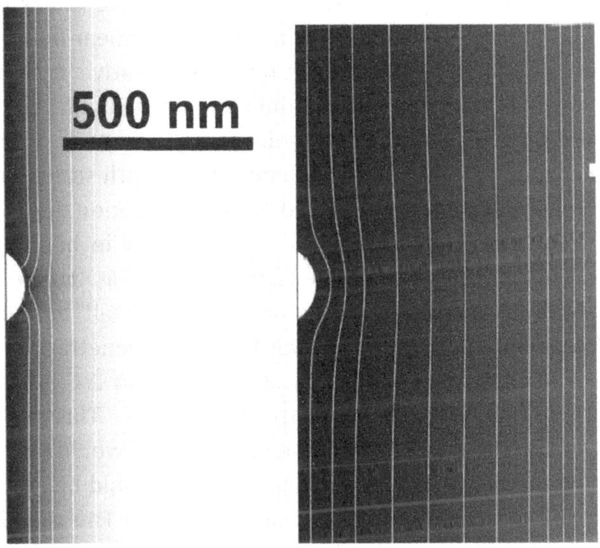

Figure 8.6 Results of the FEM calculation. The resonator surface is located on the left. The bubble is shown in white. Stream lines denote the flow of the liquid around the surface. The gray scale encodes the tangential component of the flow, where dark is the speed of the substrate and light denotes a liquid at rest in laboratory frame. The radius of the bubble is 100 nm (left) and 10 nm (right). The scale differs between the two images by a factor of 10. The decay depth of the shear wave is $\delta \approx 250$ nm in both cases. In the image on the left, the bubble is comparable in size to the penetration depth of the shear wave. The right image shows a bubble smaller than δ.

are 100 nm and 10 nm, respectively. The depth of penetration can be inferred from the gray scale, which displays the tangential component of the velocity of the liquid. In both cases, the depth of penetration is 250 nm. The lines in white are the stream lines. The stream lines are strikingly different in the two cases. For the large bubble, the air–liquid interface is only weakly curved. Note that the inverse radius of the undeformed bubble enters as a prefactor in Eq. 8.9. The interface is readily deformed, and the stream lines at the bubble surface are almost parallel to the surface. Actually, the stream lines are even curved towards the bubble.

This situation is reversed for the small bubble. This bubble resists deformation due to its high Laplace pressure, and the liquid therefore deviates to the other side, moving around the bubble, rather than deforming it. These two graphs capture the essence of the peculiar behavior of nanobubbles in acoustic shear fields.

Figure 8.7 makes this argument more quantitative. It shows Δf and $\Delta\Gamma$ for one fixed geometry (bubble radius of 50 nm) but varied surface tension, γ, and overtone order, n. The overtone order here sets the spatial scale because the penetration depth scales as $n^{-1/2}$. When the bubble is small compared to the penetration depth (small n), the frequency shift is negative and the shift in bandwidth is positive. This corresponds to the right panel in Fig. 8.6 (solid-like object). As the overtone order becomes larger, a "small" bubble turns into a "large" bubble because the yardstick (the penetration depth) decreases in size. At a large n, Δf is positive and $\Delta\Gamma$ is negative, as would be expected for macroscopic bubbles. This particular calculation does *not* encompass a range where Δf would be positive and $\Delta\Gamma$ would be close to zero. Such a finding would be associated with slip in the sense of a planar slipping layer. The calculations suggest that bubbles (small or large) never appear as slipping layers in the sense of Eq. 8.7 and Fig. 8.3.

8.6 Bubble-Induced Acoustic Streaming

The FEM calculations were carried out at an amplitude where nonlinear effects can be neglected. Actual oscillation amplitudes in experiment, on the other hand, are between 0.1 and 1 nm, and there

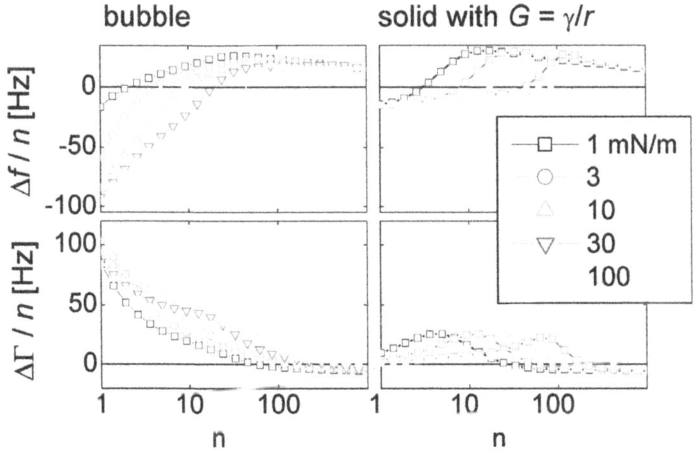

Figure 8.7 Normalized shifts of frequency ($\Delta f/n$) and bandwidth ($\Delta\Gamma/n$) vs. overtone order, n, for the geometry shown in Fig. 8.6. The bubble radius was 50 nm in this particular calculation. The different symbols correspond to different air–water surface tensions, as indicated in the legend. At low overtone orders, ($n \approx 1$) the penetration depth is larger than the bubble size. In this limit, Δf is negative, while $\Delta\Gamma$ is positive. Such a behavior is expected for a solid object (right in Fig. 8.6). In the limit of high overtones, the bubble is larger than the penetration depth (left in Fig. 8.6). The shifts of frequency and bandwidth reverse sign. This is the behavior expected for a slipping interface. The diagram on the right-hand side shows the corresponding shifts of frequency and bandwidth calculated with a model, where the bubble has been replaced by an object of the same size, shape, and density, the shear modulus of which was chosen as γ/r_0. While the details differ, the qualitative behavior is similar.

is the possibility of slight nonlinearities. Because the nonlinearities are small, one can do a perturbation analysis with the amplitude as the small parameter and break off the perturbation after the second order. In this way one only considers effects that are at most quadratic in amplitude. Again, the nonlinear effects are small, but they can be observed neverthcless because forces that depend quadratically on a sinusoidal excitation with frequency ω occur at either $\omega = 0$ ("rectification," "acoustic streaming") or at 2 ω ("second-harmonic generation").[60,61] Both effects are fairly easily observed because they occur against a zero background.

For acoustic waves in incompressible liquids with constant density, the source term of acoustic streaming is the Reynolds force density, f_R, given as

$$f_{R,i} = \rho \sum_j \frac{\partial}{\partial x_j} \langle v_i v_j \rangle = \rho \sum_j \left\langle v_j \frac{\partial v_i}{\partial x_j} \right\rangle = \frac{\rho}{2} \sum_j \left(v_{j,0}^* \frac{\partial v_{i,0}}{\partial x_j} \right) \tag{8.13}$$

v is the instantaneous speed, the index 0 denotes an amplitude of an oscillatory quantity, the indices i,j denote directions in space, the star denotes the complex conjugate and angle brackets denote time averaging. The first relation on the left makes use of incompressibility ($\Sigma dv_j/dx_j = 0$). The second relation makes use of the fact that the time average of $\cos^2(\omega t)$ is 1/2.

The Reynolds force can be extracted from the FEM calculation and be inserted as a source term into a second FEM calculation, again using the incompressible Navier-Stokes model but applied to stationary motion. One finds that particles located inside the evanescent tail of a shear wave are repelled from the surface. This result matches the experimental finding by Edvardsson.[35] Bubbles attached to the surface, on the contrary, induce a flow *toward* the bubble surface. Figure 8.8 shows the result for a hemicylinder with a radius of 10 nm. The crystal surface is located at the left. The bubble is shown in white. The stream lines indicate *stationary* acoustic streaming (as opposed to the streamlines in Fig. 8.6, which show the MHz motion of the liquid). The gray scale is the normal component of the flow, where the scale extends from −15 nm/s (dark) to +5 nm/s (bright). The amplitude of oscillation here is 0.1 nm. Shear-induced acoustic streaming of this magnitude is by no means a "large amplitude effect" in the sense that it would be completely negligible at the amplitudes typically encountered in practice. The speed of flow scales with the square of the amplitude and the speed certainly is small when compared to the action of normal pumps. On the other hand, the flow occurs right at the surface. As a consequence, mass transport (and the dissolution of gas into the liquid, in particular) is more efficient than in a quiescent liquid. This phenomenon is known from sono-electrochemistry,[62,63] although sono-electrochemistry usually is carried out with externals sources of ultrasound.

Figure 8.8 Oscillation-induced stationary flow of liquid around a bubble. See the main text for the geometry. Acoustic streaming is absent in pure shear flow because the flow direction and the gradient direction are perpendicular. However, the bubble deviates the flow, which creates a time-averaged net force. The stationary flow increases the rate of dissolution of the bubble in the liquid phase, which presumably is part of the reason why nanobubbles are less often observed in EQCM experiments than expected. We believe that shear-induced acoustic streaming is a second reason—in addition to Laplace pressure and apparently "solid" nanobubbles—why nanobubbles are less commonly observed in EQCM experiments than one might think.

8.7 Conclusions

Nanobubbles located on the surface of a shear-wave resonator behave peculiarly. First, they may look like rigid objects due to Laplace pressure. They may decrease the frequency and increase the bandwidth and thereby seemingly contradict the simple-minded models of slip. Also, they act as a source of acoustic streaming. Since acoustic streaming increases mass transport close to the interface, the bubbles thereby speed up their own dissolution.

This work highlights the peculiar behavior of soft matter in acoustic shear fields. It also shows how FEM calculations help in the interpretation of experimental findings. The authors are convinced that acoustic shear waves and the related phenomena are of broad relevance in other fields of soft matter research, such as dynamic AFM and cantilever-based sensing.

References

1. G. Sauerbrey, Verwendung von Schwingquarzen zur Wagung Dunner Schichten und zur Mikrowagung, *Z. Phys.*, **155**, 206–222 (1959).
2. A. Arnau, *Piezoelectric Transducers and Applications*. Heidelberg: Springer (2004).
3. C. Steinem and A. Janshoff, eds., *Piezoeletric Sensors*. Springer (2006).
4. T. M. Stokich, D. R. Radtke, C. C. White, and J. L. Schrag, An instrument for precise measurement of viscoelastic properties of low-viscosity dilute macromolecular solutions at frequencies from 20 to 500 kHz, *J. Rheol.*, **38**, 1195–1210 (1994).
5. E. Gizeli, N. J. Goddard, C. R. Lowe, and A. C. Stevenson, A love plate biosensor utilizing a polymer layer, *Sens. Actuat. B—Chem.*, **6**, 131–137 (1992).
6. H. Zhang and E. S. Kim, Micromachined acoustic resonant mass sensor, *J. Microelectromech. S.*, **14**, 699–706 (2005).
7. D. Johannsmann, Viscoelastic, mechanical, and dielectric measurements on complex samples with the quartz crystal microbalance, *Phys. Chem. Chem. Phys.*, **10**, 4516–4534 (2008).
8. S. Bruckenstein and M. Shay, Experimental aspects of use of the quartz crystal microbalance in solution, *Electrochim. Acta*, **30**, 1295–1300 (1985).
9. M. D. Ward and D. A. Buttry, In-situ interfacial mass detection with piezoelectric transducers, *Science*, **249**, 1000–1007 (1990).
10. Γ is the *half* bandwidth at half height (HBHW).
11. M. Rodahl, F. Hook, A. Krozer, P. Brzezinski, and B. Kasemo, Quartz-crystal microbalance setup for frequency and Q-factor measurements in gaseous and liquid environments, *Rev. Sci. Instrum.*, **66**, 3924–3930 (1995).
12. R. Beck, U. Pittermann, and K. G. Weil, Impedance analysis of quartz oscillators, contacted on one side with a liquid, *Ber. Bunsen Phys. Chem.*, **92**, 1363–1368 (1988).
13. D. Johannsmann, Viscoelastic analysis of organic thin films on quartz resonators, *Macromol. Chem. Phys.*, **200**, 501–516 (1999).

14. T. Nakamoto, and T. Moriizumi, A theory of a quartz crystal microbalance based upon a mason equivalent-circuit, *Jpn. J. Appl. Phys. 1*, **29**, 963–969 (1990).

15. H. L. Bandey, S. J. Martin, R. W. Cernosek, and A. R. Hillman, Modeling the responses of thickness-shear mode resonators under various loading conditions, *Anal. Chem.*, **71**, 2205–2214 (1999).

16. R. Lucklum, C. Behling, and P. Hauptmann, Role of mass accumulation and viscoelastic film properties for the response of acoustic-wave based chemical sensors, *Anal. Chem.*, **71**, 2488–2496 (1999).

17. E. Benes, Improved quartz crystal microbalance technique, *J. Appl. Phys.*, **56**, 608–626 (1984).

18. D. Johannsmann, Studies of viscoelasticity and micromechanics with the QCM. In ed., C. Steinem, A. J., *Piezoelctric Sensors*. Heidelberg: Springer (2006).

19. J. Wegener, A. Janshoff, H. J. Galla, Cell adhesion monitoring using a quartz crystal microbalance: comparative analysis of different mammalian cell lines, *Eur. Biophys. J. Biophy.*, **28**, 26–37 (1998).

20. C. A. Keller, and B. Kasemo, Surface specific kinetics of lipid vesicle adsorption measured with a quartz crystal microbalance, *Biophys. J.*, **75**, 1397–1402 (1998).

21. E. Tellechea, D. Johannsmann, N. F. Steinmetz, R. P. Richter, and I. Reviakine, Model-independent analysis of QCM data on colloidal particle adsorption, *Langmuir*, **25**, 5177–5184 (2009).

22. D. Johannsmann, I. Reviakine, E. Rojas, and M. Gallego, Effect of sample heterogeneity on the interpretation of QCM(-D) data: comparison of combined quartz crystal microbalance/atomic force microscopy measurements with finite element method modeling, *Anal. Chem.*, **80**, 8891–8899 (2008).

23. Z. X. Lin, and M. D. Ward, Determination of contact angles and surface tensions with the quartz crystal microbalance, *Anal. Chem.*, **68**, 1285–1291 (1996).

24. L. McKenna, M. I. Newton, G. McHale, R. Lucklum, and J. Schroeder, Compressional acoustic wave generation in microdroplets of water in contact with quartz crystal resonators, *J. Appl. Phys.*, **89**, 676–680 (2001).

25. I. Reviakine, F. F. Rossetti, A. N. Morozov, and M. Textor, Investigating the properties of supported vesicular layers on titanium dioxide by quartz crystal microbalance with dissipation measurements, *J. Chem. Phys.*, **122**, 204711 (2005).

26. X. H. Zhang, Quartz crystal microbalance study of the interfacial nanobubbles, *Phys. Chem. Chem. Phys.*, **10**, 6842–6848 (2008).
27. C. Neto, D. R. Evans, E. Bonaccurso, H. J. Butt, and V. S. J. Craig, Boundary slip in Newtonian liquids: a review of experimental studies, *Rep. Prog. Phys.*, **68**, 2859–2897 (2005).
28. J. N. Israelachvili and R. M. Pashley, Measurement of the hydrophobic interaction between 2 hydrophobic surfaces in aqueous-electrolyte solutions, *J. Colloid Interface Sci.*, **98**, 500–514 (1994).
29. V. S. J. Craig, Very small bubbles at surfaces-the nanobubble puzzle, *Soft Matter*, **7**, 40–48 (2011).
30. B. Borkent, S. Dammler, H. Schonherr, G. Vansco, and D. Lohse, Superstability of surface nanobubbles, *Phys. Rev. Lett.* **98**, 204502 (2007).
31. Z. X. Lin and M. D. Ward, The role of longitudinal-waves in quartz-crystal microbalance applications in liquids, *Anal. Chem.*, **67**, 685–693 (1995).
32. Y. T. Yang, C. Callegari, X. L. Feng, K. L. Ekinci, and M. L. Roukes, Zeptogram-scale nanomechanical mass sensing, *Nano Lett.*, **6**, 583–586 (2006).
33. M. A. Plunkett, Z. H. Wang, M. W. Rutland, and D. Johannsmann, Adsorption of pNIPAM layers on hydrophobic gold surfaces, measured in situ by QCM and SPR, *Langmuir*, **19**, 6837–6844 (2003).
34. J. L. Hutter and J. Bechhoefer, Calibration of atomic-force microscope tips, *Rev. Sci. Instrum.*, **64**, 1868–1873 (1993).
35. D. Johannsmann, W. Bucking, B. Bode, and J. Petri, Simple frequency-based sensing of viscosity and dielectric properties of a liquid using acoustic resonators, *IEEE T. Ultrason. Ferr.*, **57**, 677–683 (2010).
36. M. Edvardsson, M. Rodahl, and F. Hook, Investigation of binding event perturbations caused by elevated QCM-D oscillation amplitude, *Analyst*, **131**, 822–828 (2006).
37. T. M. Squires and S. R. Quake, Microfluidics: fluid physics at the nanoliter scale, *Rev. Mod. Phys.*, **77**, 977–1026 (2005).
38. J. Lighthill, Acoustic streaming, *J. Sound Vib.*, **61**, 391–418 (1978).
39. F. P. Bowden and D. Tabor, Friction lubrication and wear—a survey of work during last decade, *Brit. J. Appl. Phys.*, **17**, 1521 (1966).
40. S. Berg and D. Johannsmann, High speed microtribology with quartz crystal resonators, *Phys. Rev. Lett.*, **91** (2003).
41. C. L. Tucker and P. Moldenaers, Microstructural evolution in polymer blends, *Annu. Rev. Fluid Mech.*, **34**, 177–210 (2002).

42. O. I. Vinogradova, Slippage of water over hydrophobic surfaces, *Int. J. Miner. Process.*, **56**, 31–60 (1999).

43. P. A. Thompson and S. M. Troian, A general boundary condition for liquid flow at solid surfaces, *Nature*, **389**, 360–362 (1997).

44. F. Ferrante, A. L. Kipling, and M. Thompson, Molecular slip at the solid-liquid interface of an acoustic-wave sensor, *J. Appl. Phys.*, **76**, 3448–3462 (1994).

45. G. McHale, R. Lucklum, M. I. Newton, and J. A. Cowen, Influence of viscoelasticity and interfacial slip on acoustic wave sensors, *J. Appl. Phys.*, **88**, 7304–7312 (2000).

46. S. J. Martin, G. C. Frye, A. J. Ricco, and S. D. Senturia, Effect of surface-roughness on the response of thickness-shear mode resonators in liquids, *Anal. Chem.*, **65**, 2910–2922 (1993).

47. J. S.Ellis and G. L. Hayward, Interfacial slip on a transverse-shear mode acoustic wave device, *J. Appl. Phys.*, **94**, 7856–7867 (2003).

48. J. Buehrle, S. Herminghaus, and F. Mugele, Impact of line tension on the equilibrium shape of liquid droplets on patterned substrates, *Langmuir*, **18**, 9771–9777 (2002).

49. D. C. Tretheway and C. D. Meinhart, Apparent fluid slip at hydrophobic microchannel walls, *Phys. Fluids*, **14**, L9–L12 (2002).

50. To be precise, the requirement $k_i d_i << 1$ implies that the viscosity does not drop drastically. Otherwise, the wave vector will become large and $k_i d_i$ is no longer less than unity.

51. Z. L. Li, Z. J. Niu, T. H. Wu, H. D. Nie, and X. M. Xiao, In situ monitoring of potential oscillations in the reduction of IO3- by electrochemical quartz crystal microbalance, *Electrochem. Commun.*, **5**, 297–300 (2003).

52. F. B. Li, A. R. Hillman, S. D. Lubetkin, and D. J. Roberts, Electrochemical quartz crystal microbalance studies of potentiodynamic electrolysis of aqueous chloride solution—surface processes and evolution of H_2 and Cl_2 gas-bubbles, *J. Electroanal. Chem.*, **335**, 345–362 (1992).

53. X. H. Zhang, Formation of interfacial nanodroplets through changes in solvent quality, *Phys. Chem. Chem. Phys.*, **10**, 6842–6848 (2008).

54. S. T. Lou, Z. Q. Ouyang, Y. Zhang, X. J. Li, J. Hu, M. Q. Li, F. J. Yang, Nanobubbles on solid surface imaged by atomic force microscopy, *J. Vac. Sci. Technol. B*, **18**, 2573–2575 (2000).

55. G. M. Liu, Z. H. Wu, V. S. J Craig, Cleaning of protein-coated surfaces using nanobubbles: an investigation using a quartz crystal microbalance, *J. Phys. Chem. C*, **112**, 16748–16753 (2008).

56. V. Tsionsky, A. Kaverin, L. Daikhin, G. Katz, and E. Gileadi, Physical chemistry an experimental verification of the possible influence of gas nano-bubbles on the response of an electrochemical quartz crystal microbalance, *Chem. Phys.*, **7**, 1830–1835 (2005).
57. A. R. Hillman, D. C. Loveday, M. J. Swann, S. Bruckenstein, and C. P. Wilde, Analytical applications of the electrochemical quartz crystal microbalance, *ACS Sym. Ser.*, **487**, 150–163 (1992).
58. B. Y. Du, E. Goubaidoulline, and D. Johannsmann, Effects of laterally heterogeneous slip on the resonance properties of quartz crystals immersed in liquids, *Langmuir*, **20**, 10617–10624 (2004).
59. A. Laschitsch and D. Johannsmann, High frequency tribological investigations on quartz resonator surfaces, *J. Appl. Phys.*, **85**, 3759–3765 (1999).
60. K. Wondraczek, A. Bund, and D. Johannsmann, Acoustic second harmonic generation from rough surfaces under shear excitation in liquids, *Langmuir*, **20**, 10346–10350(2004).
61. S. Wehner, K. Wondraczek, D. Johannsmann, and A. Bund, Roughness-induced acoustic second-harmonic generation during electrochemical metal deposition on the quartz-crystal microbalance, *Langmuir*, **20**, 2356–2360 (2004).
62. R. G. Compton, J. C. Eklund, and F. Marken, Sonoelectrochemical processes: a review, *Electroanalysis*, **9**, 509–522 (1997).
63. O. Schneider, S. Matic, and C. Argirusis, Application of the electrochemical quartz crystal microbalance technique to copper sonoelectrochemistry—part 1. Sulfate-based electrolytes. *Electrochim. Acta*, **53**, 5485–5495 (2008).

Chapter 9

Nanomeniscus Mechanical Properties

Jean-Pierre Aimé,[a,*] Thierry Ondarçuhu,[b,] Rodolphe Boisgard,[c] Laure Fabié,[b] Mathieu Delmas,[b] and Charlotte Fouché[d]**

[a]*CBMN, B8 Avenue Geoffroy Saint Hilaire, 33402 Talence Cedex, France*
[b]*CEMES CNRS, 29 rue Jeanne Marvig, 31055 Toulouse Cedex, France*
[c]*LOMA, 351 Cours de la libération, 33405 Talence Cedex, France*
[d]*Agence Nationale de la Recherche, 7 rue Watt, 75013 Paris, France*
*jp.aime@cbmn.u-bordeaux.fr and **thierry.ondarcuhu@cemes.fr

9.1 Overview and Motivations

Water is an everyday life liquid playing a key role in many domains: it is the unique and unavoidable environment necessary for biological activities, and it dramatically changes the properties of nonhomogeneous materials such as sand and microporous systems. When the gas phase is confined at micro- and nanometer scales, it may condense into a liquid at vapor pressure lower than the saturation vapor pressure. Capillary condensation[1,2] has several consequences such as reinforcing sandcastle stability. Capillary condensation may also be a key force acting on atomic force microscope and scanning tunneling microscope (STM) tips, in which humidity becomes an important parameter on the tip surface adhesion property and may be detrimental for imaging or may lead to artifacts.

Nanoscale Liquid Interfaces: Wetting, Patterning, and Force Microscopy at the Molecular Scale
Edited by Thierry Ondarçuhu and Jean-Pierre Aimé

ISBN 978-981-4316-45-3 (Hardcover), 978-981-4364-48-5 (eBook)
www.panstanford.com

Understanding the formation, stability, size, and shape of liquid bridges between two proximal surfaces is also crucial to describe the behavior of gas and liquid in porous materials or else to have an insight into the ultimate resolution one can achieve when using dip-pen lithography,[3] tip-based oxidation lithography[4] (see chapter 13), or liquid nanodispensing.[5] (NADIS; see chapter 12) techniques. In all of these cases, surface and interface effects require special attention as they are dominant parameters. Indeed, when scaling down, capillary forces, which scale as r (the lateral dimension of the meniscus), become predominant compared to the surface (r^2) of volumic forces (r^3). Capillary forces can therefore be used as a mechanism for soft manipulation of objects at the micro- or nanoscale. They, for example, provide an efficient gripping mechanism[6] and can be used to align[7] (molecular combing; see chapter 16), assemble[8,9] (capillary assembly; see chapter 15), or deform[10–12] (capillary origami) objects down to nanometer scale.

The dynamics of confined liquids is also of prior importance in many situations. In biology, for example, many chemical reactions that sustain metabolic and anabolic activity are performed in confined environments. A precise description of flow in confined geometry is therefore essential and requires a good knowledge of boundary conditions, which play a central role.

We face new fundamental and technical challenges where the domain of the nanofluidic is at the first place. The use of atomic force microscopes with specially designed tips such as nanoneedles, carbon nanotubes, or NADIS at the tip apex are very much appropriate to elucidate the physical behavior of a tiny volume of liquid such as small drops, liquid bridges, and nanomeniscus. In this chapter, we give a comprehensive overview of static and dynamic AFM studies that should provide new insights into the mechanical properties of small quantities of liquid.

9.2 Theoretical Background

In this section, we give a short overview of the models and methods used to compute the capillary force exerted by a liquid meniscus bridging two solid surfaces. We describe the main practical cases of studies, in particular those encountered in AFM experiments, such as the examples given in the section 9.3.

9.2.1 *Static Capillary Force*

We describe here the capillary force resulting from a liquid meniscus bridging two solid surfaces. We make the following assumptions:

- The gravity effects are neglected. This is valid for AFM experiments where the characteristic lengths are much smaller than the capillary length defined as $\kappa^{-1} = \sqrt{\gamma/\rho g}$, where γ is the liquid surface tension, ρ is its volumic mass, and g the acceleration. For usual liquids K^{-1} is of the order of 1–3 mm.
- We also assume that the influences of van der Waals forces, electrostatic forces, or chemical forces can be neglected. Very thin films or prewetting films where van der Waals forces may play an important role are not considered.
- All the methods presented later assume matter as a continuous medium, which is valid down to dimensions of the order of a few nanometers. For smaller sizes, the discrete nature of matter can be accounted for using molecular dynamics simulations. This issue is discussed in section 9.3.1.3.

9.2.1.1 Boundary conditions and constraints

Let us consider a liquid meniscus bridging two solid surfaces such as a plane substrate and a conical tip as in the case of AFM experiments (Fig. 9.1a). In order to describe the system, it is necessary to determine the boundary conditions and the constraints applied to the meniscus.

The boundary conditions determine the contact between the liquid and the solid and are mainly governed by the wetting properties of the solid surfaces. Let us note R_{tip} (R_{surf}) and θ_{tip} (θ_{surf}), the radius of the contact line (CL) and the contact angle on the tip and on the surface, respectively. If the surface does not show any contact angle hysteresis (CAH), the Young equation defines the equilibrium contact angles θ_{tip} and θ_{surf} that should be

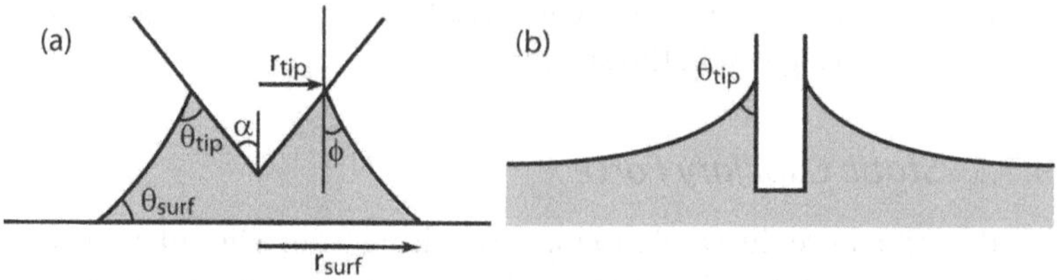

Figure 9.1 Sketches of the systems considered. (a) A meniscus bridging the gap between a conical tip and a flat surface; (b) an unbounded meniscus created by dipping a cylinder in a liquid bath.

verified at each point of the CL. This is essentially valid with clean, rather hydrophobic surfaces. If one considers tip retraction on more hydrophilic surfaces, the receding contact angle may be equal to zero, which means that the CL remains fixed, leading to a meniscus with given R_{tip} and R_{surf} values. For the case of an unbounded meniscus obtained when a tip is dipped in a liquid bath (Fig. 9.1b), the R_{surf} value is defined by the capillary length K^{-1}.

The constraints in the liquid meniscus are mainly of two types. If one considers low volatile liquids or experiments with short characteristic times compared to the evaporation time, the volume of the meniscus must remain constant. On the contrary, with volatile liquids, the volume is not constant anymore and the pressure inside the meniscus is imposed instead. This is the case of capillary condensation.[1] Menisci connected to a reservoir or tips dipped into a large liquid bath (Fig. 9.1b) also fall into this category.

Examples of the influence of boundary conditions and constraints on capillary forces are reported in detail in section 9.3.1.1.

9.2.1.2 General approaches

Once boundary conditions and constraints are defined, the capillary force can be calculated. Two main approaches have been developed to compute the capillary force between the two objects:[6]

- In the so-called energetic approach, one first computes the total energy of the system, including surface and bulk energies,[13]

$$W = \gamma_{LV} A_{LV} + \gamma_{SV} A_{SV} + \gamma_{SL} A_{SL} - p_V V_V - p_L V_L \quad (9.1)$$

where A_{SL} (γ_{SL}), A_{SV} (γ_{SV}), and A_{LV} (γ_{LV}) denote the solid–liquid, solid–vapor, and liquid–vapor surface area (surface tension), respectively, and V_V (V_L) is the gas (liquid) volume. The capillary force is then computed as the derivative of the energy with respect to the distance z between the tip and the surface, that is,

$$F = -\frac{dW}{dz}. \tag{9.2}$$

- Within the framework of a geometrical approach, the force is calculated from the shape of the meniscus. In simple axisymmetrical geometries, it has two components:

The Laplace force results from the pressure difference between the meniscus and the surrounding medium $\Delta P = P_{in} - P_{out}$. It is related to the total mean curvature of the meniscus through the Laplace equation. The Laplace force therefore reads as $F_L = \pi R_{tip}^2 \Delta P$.

The tension force is due to the surface tension force applying on the CL, that is, $F_T = 2\pi R_{tip}\gamma . \cos\phi$.

The total capillary force can therefore be written as

$$F = F_L + F_T = \pi R_{tip}^2 \cdot \Delta P + 2\pi R_{tip}^{\gamma} \cdot \cos\phi. \tag{9.3}$$

It can be demonstrated that both approaches are equivalent.[14]

9.2.1.3 Meniscus shape

Both methods described in the previous section require the determination of the meniscus shape. We first describe the case of a finite meniscus bridging two surfaces and then the case of an unbounded meniscus obtained when dipping a tip in a liquid bath.

9.2.1.3.1 Finite meniscus

Given the assumptions mentioned before, the pressure inside the meniscus is constant and related to the meniscus curvature by the Laplace pressure. In the axisymmetric situation, the interface shape $r(z)$ thus verifies:

$$\Delta P = \gamma \left(\frac{r''}{\left(1 + r'^2\right)^{3/2}} + \frac{1}{r\left(1 + r'^2\right)^{1/2}} \right) \tag{9.4}$$

where $r' = dr/dz$ and $r'' = d^2r/dz^2$. When the pressure is given, as, for example, in the case of capillary condensation, Eq. 9.4 allows computing the profile of the meniscus. In the particular case where $\Delta P = 0$, this leads to the well-documented problem of minimal surfaces, which can be solved analytically and lead to solutions in the form of a catenoid (surface formed by the revolution of a catenary).

In order to facilitate the calculation of capillary forces, several approximations on the meniscus shape have been proposed. They are mainly applied for experimental conditions where the radius of curvature of the meniscus interface is much smaller than the radius of curvature of the solid surfaces. This is relevant for the surface force apparatus where the surface has centimetric radius, while the meniscus is typically tens of hundreds of nanometers. The most used approximation is the toroidal approximation assuming the liquid interface has a circular profile. Obviously, such a meniscus does not exhibit a constant curvature. Nevertheless, this approximation gave good results, in particular for small contact angles, and is therefore widespread (see Ref. 15 for its application in various geometries and section 9.3.1.1 for an example of its application in atomic force microscopy [AFM]). In the case of capillary condensation between a plane and a sphere with a large radius of curvature R, in contact, the tension term of the capillary force is negligible and the Laplace term leads to the simple formula $F = 4\pi R\gamma \cos\theta$.[2] A parabolic profile is also sometimes used to eliminate some numerical difficulties[16] inherent in circle approximation.

For a liquid bridge with a given volume, the resolution of the meniscus shape is not trivial. A numerical resolution was proposed using a double-iterative method. It can handle any surface shapes and boundary conditions and compute the capillary force from the geometrical method.[17]

In the case of surfaces with nonaxisymmetric morphologies, which may be relevant with pyramidal atomic force microscope tips, Eq. 9.4 is no longer valid. A numerical solution can be obtained using the Surface Evolver software,[18] which minimizes the total energy of the system with given boundary conditions and constraint (pressure or volume). The capillary force can be extracted by a numerical derivation according to Eq. 9.2, that is, using an energetic approach. The advantage of this method, which is rather simple to implement, is that it applies to any situation as shown on section 9.3.1.1 in the case of nanodispensing atomic force microscope tips.

9.2.1.3.2 Unbounded meniscus

The theoretical study involving a cylindrical geometry is of practical interest as it deals with the coating of fibers, an important domain for industrial application. Therefore, there have been several attempts to describe the shape of an unbounded meniscus formed when a rod of radius R is dipped in a liquid bath (Fig. 9.1b). Contrary to the case of a meniscus on a planar wall,[19] there is no general analytical solution in the cylindrical geometry. In the domain where $R/\kappa^{-1} << 1$, solving the Laplace equation without gravity leads to the profile of the catenary. However its expression does not allow determining the meniscus height h. This value can be deduced from the volume of the meniscus[20] or by asymptotic matching with an outer solution where gravity comes into play.[21] In the latter case, the following expression is found

$$h = R\cos\theta\left[\ln\left(\frac{4\kappa^{-1}}{R\,(1+\sin\theta)}\right) - 0,57\right], \tag{9.5}$$

where θ is the wetting contact angle and 0.57 the Euler constant. In the case where $b < 0.3K^{-1}$, Eq. 9.5 reproduces the numerical results with a relative error smaller than 1%. Note that the expression of h deduced from the meniscus volume[20] only differs from Eq. 9.5 by the Euler constant, which is negligible as soon as small fibers diameters R are considered.

In the case of nanometric fibers the use of a large scale cutoff length of millimetric dimension (K^{-1}) is questionable. The lateral extension of the meniscus might be smaller than K^{-1}. The meniscus size may also be limited in the case of a finite meniscus when a liquid bridges a tip and a surface or by dynamic effects. We therefore anticipate that Eq. 9.5 could be used to describe all situations by replacing K^{-1} by the appropriate large cutoff length L corresponding to the effective lateral extension of the meniscus. The study of meniscus spring constants in various situations can therefore be of interest to determine experimentally the L value, which is a recurrent problem in wetting science (see sections 9.3.2.1 and 9.3.2.2)

9.2.1.4 A nanomeniscus modeled as an equivalent mechanical oscillator: meniscus spring constant

In many applications, it will be useful to describe the nanomeniscus as an equivalent mechanical oscillator and thus to derive a meniscus spring constant. Here, we propose to determine this value in the simple case of a CL pinned on a fiber dipped in a liquid bath. Since the pressure inside the meniscus is zero, the capillary force is simply given by the tension term:

$$F = 2\pi R\gamma \cos\theta \tag{9.6}$$

The associated spring constant that describes the elastic restoring force in response to a weak vertical displacement of the CL is:

$$k = -2\pi R\gamma \sin\theta \frac{d\theta}{dh} \tag{9.7}$$

For a forced vertical displacement along the z-axis, the corresponding variation of the contact wetting angle θ is given by the derivative of Eq. 9.5:

$$\frac{dh}{d\theta} = -R\sin\theta \left[\ln\left(\frac{4\kappa^{-1}}{R(1+\sin\theta)}\right) - 0.57\right] - R\frac{\cos^2\theta}{1+\sin\theta} \tag{9.8}$$

Combining Eqs. 9.7 and 9.8 gives the spring constant:

$$k = \frac{2\pi\gamma\sin\theta}{\sin\theta\left[\ln\left(\frac{4\kappa^{-1}}{R(1+\sin\theta)}\right) - 0.57\right] + \frac{\cos^2\theta}{1+\sin\theta}} \tag{9.9}$$

Equation 9.9 is quite cumbersome and does not help to analyze the experimental data. In particular, the measurement of the spring constant k should allow deducing the contact angle θ (see section 9.3.2.2). A first-order expansion of Eq. 9.9 gives a much tractable equation, still accurate enough to be used, in particular for small fibers, and small contact angles (Fig. 9.2):

$$k \approx \frac{2\pi\gamma}{C + \frac{1-\theta}{\theta}} + O\left(\theta^2\right) \tag{9.10}$$

with

$$C = \ln\left(\frac{4\kappa^{-1}}{R(1+\sin\theta)}\right) - 0.57 \tag{9.11}$$

The important parameter in the expressions given above is the ratio of the fiber diameter to the capillary length. However, in several situations the validity of K^{-1} as a large scale cutoff length is questionable and has not been clearly established. In section 9.3,

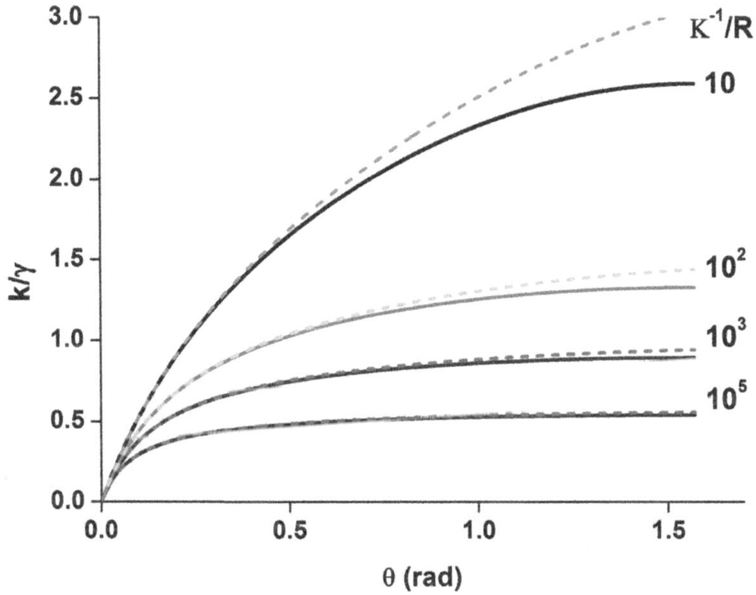

Figure 9.2 Computation of the spring constant as a function of the wetting angle. Comparison between Eq. 9.9 (solid lines) and first-order expansion Eq. 9.10 (dashed lines) for four values of the K^{-1}/R ratio. If one takes K^{-1} = 2 mm, these values correspond to fiber radii of 20 nm, 2 μm, 20 μm, and 200 μm from bottom to top.

we show that different experimental values of meniscus stiffness can be interpreted using characteristic lengths other than K^{-1}.

9.2.1.5 Contact line spring constant

It is interesting to compare the elasticity of a meniscus with another elasticity involved in wetting processes: the CL spring constant. It was shown experimentally that the elasticity energy of a CL is not Hookean. By studying the relaxation of a CL deformed periodically, Ondarçuhu and Veyssié[22] verified that this energy has a nonclassical expression given by:[23]

$$E_{\mathrm{el}} = \frac{1}{2}\gamma\theta^2 \int_{-\infty}^{+\infty} \frac{dq}{2\pi} |q|\,|u(q)|^2 \tag{9.12}$$

where q is the wavelength of the deformation, $u(q)$ the Fourier transform of the CL shape $u(x)$, and L a macroscopic characteristic length. The q dependence of E_{el} (instead of the q^2 term in the case of a Hookean system) comes from the fact that a deformation of the CL induces a deformation of the liquid meniscus, which extends over a distance of the order of q^{-1}. Despite its nonclassical energy, if one considers a small force applied locally on the CL over a distance d (e.g., by pinning on a defect), the force recovers a Hookean behavior with a spring constant:

$$k = \frac{\pi \gamma \sin^2 \theta}{\ln\left(\frac{L}{d}\right)} \tag{9.13}$$

This expression is different from the one derived for a meniscus (Eq. 9.9). However, it gives the same order of magnitude and includes the same dependence with $\ln(L/d)$ parameter, which again raises the issue of the transversal characteristic length to include in these expressions. The noticeable difference is given by the dependence in θ that varies between Eq. 9.9 and Eq. 9.13.

9.2.2 *Dynamic Capillary Force*

In the previous section we described static capillary forces. Viscous forces may also bring an additional contribution that will be readily measured with AFM dynamical modes. Therefore, in the framework of AFM studies, it is of interest to evaluate the effect of the liquid viscosity. We distinguish two cases corresponding to different configurations discussed in the next section of this chapter. We first evaluate the viscous force compared to the capillary one and then the effect of the boundary condition.

9.2.2.1 Competing contributions: viscous force vs. capillary force in a meniscus

If we consider a tip oscillating in the vicinity of a surface with a liquid bridge between the tip and the surface, we can estimate the force due to viscous losses during the oscillation and compare it with the capillary force. In the typical case of an oscillating tip with an amplitude $A = 10$ nm and frequency $\omega = 2\pi\nu = 2\pi 10^5$ cycles/s, the loss force scales as $F_{\mathrm{loss}} = \eta r A \omega$ so that the ratio with the capillary force is $F_{\mathrm{loss}}/F_{\mathrm{cap}} = \eta A \nu/\gamma \cos\theta$. In the case of water, this ratio is of the order of 10^{-4}, showing that viscous force is a negligible contribution compared to capillary forces. The dissipation associated with the nonconservative character of the capillary force resulting from the fact that meniscus forms and breaks at each oscillation ($E_{adh} \approx F_{cap}A$) is therefore predominant over the viscous dissipation ($E_{loss} \approx F_{loss}\ A$) as already pointed.[24] When the meniscus breaks at each oscillation, dynamical AFM modes acting on a meniscus bridging a tip and a surface can therefore be described using the static description of liquid meniscus (see section 9.3.1.3).

9.2.2.2 Viscous dissipation in a nonbreaking nanomeniscus: a boundary condition problem

In the case of a nanoneedle oscillating at a liquid interface,[25,26] since capillary force is reversible, viscous dissipations become predominant and thus play a significant role. Indeed, contrary to the previous case, there is not any more the large energy dissipation due to the creation and rupture of a meniscus. The viscous dissipation occurs in the volume of liquid dragged by the needle oscillation, whose lateral extension δ corresponds to the penetration depth of the fluid motion in the liquid.[19] δ is given by

$$\delta = \sqrt{\eta/\rho\omega}. \tag{9.14}$$

In the particular case of a nanoneedle at the end of a typical oscillating atomic force microscope cantilever ($f = 260$ kHz), this evanescent length is of about $\delta \approx 10^{-6}$ m that can be larger than the meniscus size.

Let us consider first an oscillating flat plane inducing a shear motion in a semi-infinite thick liquid (see Fig. 9.3a), the dissipated energy per period is:[19]

$$\langle E_{diss} \rangle_T = \pi M_\delta \frac{v_0^2}{2} \tag{9.15}$$

where $M_\delta = \rho S \delta$ is the mass of liquid of thickness δ surrounding the plate and S the contact area between the plate and the liquid.

We now consider the case of interest reported in Fig. 9.3b of a nanomeniscus where the film thickness d is much smaller than δ ($d << \delta$). With an oscillating velocity forcing $u(t) = u_0 e^{i\omega t}$ (with $u_0 = \omega A$), the boundary conditions are $v(x = 0) = u(t)$ at the

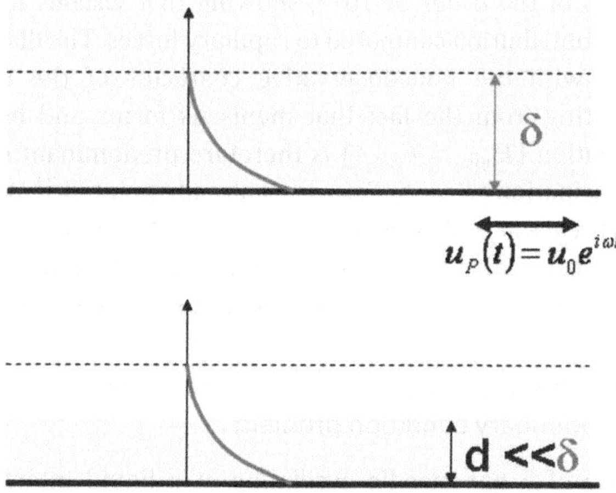

Figure 9.3 Influence of the film thickness on the dissipative force acting on a flat oscillating surface. The velocity field created in the liquid is the product of a cosine function and an exponential decay. The decreasing length δ gives the spatial extension of the velocity field.

liquid nanoneedle interface with no slip effect of water molecules on the needle and at the air–liquid interface $\eta\frac{\partial v}{\partial x}(x = d) = 0$, ensuring the continuity of the shear constraint. The velocity profile in the film is $v(x, t) = u_0 e^{i\omega t}\left[\frac{\cos k(d-x)}{\cos kd}\right]$ and the hydrodynamic viscous force $F = \eta\frac{\partial v}{\partial x}S = \eta v S\mathrm{Re}[k \tan kd]$, with $k = \frac{1+i}{\delta}$ and *Re* the real part of the product in the brackets. The loss of energy per period in the thin liquid film is given by:

$$\langle E_{\text{diss}}\rangle_T = T\,\langle F v\rangle = \frac{T}{2}v_0^2\eta S\mathrm{Re}\,[k \tan kd]\,. \tag{9.16}$$

Since $d << \delta$, Eq. 9.16 can be expanded as a function of the ratio $\frac{d}{\delta}$, giving:

$$\langle E_{\text{diss}}\rangle_T = T\,\langle F v\rangle \approx 4\frac{T}{6}v_0^2\eta S\frac{d^3}{\delta^4}. \tag{9.17}$$

Using $T\omega = 2\pi; 1/\delta^2 = \rho\omega/2\eta$; $m_d = \rho S d$, the equation is rewritten in a more readable form:

$$\langle E_{\text{diss}}\rangle_T \approx \frac{2}{3}\pi\left(\frac{d}{\delta}\right)^2 m_d v_0^2 \tag{9.18}$$

These calculations show that the evanescent length δ and its relative value compared to meniscus thickness can play a major role in dynamic oscillating AFM modes. This point is discussed in section 9.3.2.2.

9.2.3 *Characteristic Lengths*

In the situations described in the present chapter, there are several lengths that may characterize the static or dynamical properties of a meniscus:

(i) The capillary length, of about millimeter size for water, involving gravity. It is not relevant when one considers the nanomeniscus between a tip and a surface but may play an important role for an unbounded meniscus.
(ii) For oscillating tips, the evanescent length δ can play a major role. The evanescent length gives the extension of the velocity field in the liquid generated by tip oscillation and thus should have an effect on the mechanical properties of the meniscus. δ is given by the relationship $\delta = \sqrt{\eta/\rho\omega}$, which for water is within the micrometer range.
(iii) The radius R_{tip} of the tip or the needle is the important length of the problem, as it governs the local curvature of the meniscus.
(iv) In the case of a meniscus bridging a tip and a surface, the radius R_{surf} of the wetted part of the substrate is also relevant.

It is of great importance to determine the relevant characteristic lengths depending on the experimental situations. This is an open question. In the following, we bring some indications on this issue.

9.3 Experimental Results

In the following sections, we report on AFM experimental studies. The examples given next were chosen to illustrate the large variety of situations encountered and interpreted using the theoretical background exposed earlier. In the first part, we focus on capillary forces occurring when approaching a tip from a solid surface. It comprises the standard AFM configuration as well as a nonconventional use, such as nanodispensing methods.[5] In the second part, we discuss recent experiments where a nanoneedle is used to probe a liquid interface.

9.3.1 *Capillary Forces in AFM on Surfaces*

In ambient conditions, when an atomic force microscope tip is brought in contact with a surface, a liquid meniscus may form due to the presence of a liquid layer adsorbed on the substrate or to capillary condensation. This meniscus leads to large adhesion forces, which may be detrimental for sample imaging, in contact or dynamic modes. The tip can also be used to dispense liquid on the surface as detailed in chapter 12. This latter case leads to different situations, which are important to understand in order to optimize the deposition process. In the following, we report examples of capillary force measurements in static contact mode. We then mention numerical simulations aiming at computing the smallest size that a nanomeniscus can have as a function of the relative humidity (RH) and of the size of the tip apex. Such numerical simulations are of interest as they set the limiting meniscus size below which a modeling of the liquid behavior with a continuous liquid medium approximation is no longer hold. Finally, we show that the understanding of the effect of liquid bridges in tapping mode might be of importance for the interpretation of AFM images.

9.3.1.1 Static measurement of forces

In this section we report on force measurements performed in contact mode. Since the tip velocity is small, it is generally considered that the meniscus is always at equilibrium: dynamic effects are therefore not considered. We present two examples emphasizing the role of boundary conditions and constraints in the capillary force.

9.3.1.1.1 Capillary adhesion forces: influence of relative humidity

The influence of RH on the magnitude of adhesion force between a tip and a surface is an important issue in AFM. The commonly observed behavior is that the capillary force exhibits a maximum for a certain humidity value, which varies from experiment to experiment.[27-29] Recently, Köber *et al.* investigated the influence of tip sizes and aspect ratios (characterized by scanning electron microscopy [SEM]) on the pull-off force.[30] The experimental results show that for tips with a large radius (typically above 15 nm), a

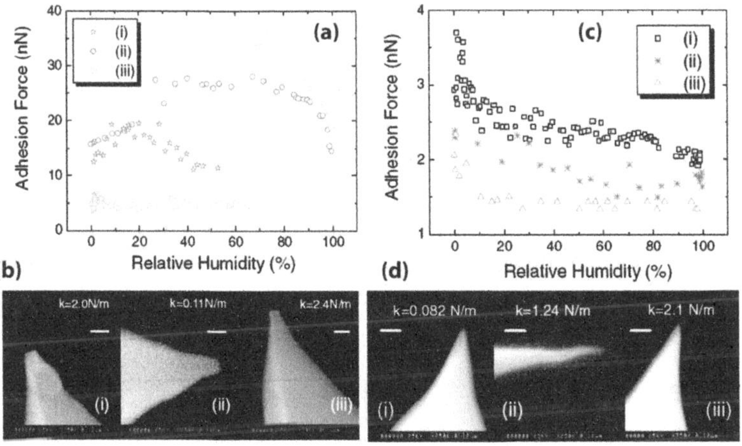

Figure 9.4 Left: Common maximum behavior. (a) Adhesion force vs. RH curves for three different tips on a flat mica surface. (i) Initially sharp Si tip (nominal radius <7 nm), normal force limited to 50 nN; (ii) Si 3 N 4 tip of nominal tip radius 15 nm; no normal load limitations; (iii) special case with initially sharp tip and normal force limited to 18 nN; at about RH 65% the tip crashed against the sample (the load remained 18 nN). (b) SEM images showing the three tips after they had been used to measure the adhesion force (scale bars: 60 nm). The SEM images of the tips taken after the experiments show larger tip radii than the nominal values. Right: Decreasing adhesion force. (a) Adhesion force measured between a sharp Si tip and a flat mica sample as a function of the RH for three different tips. Nominal tip radii are <7 nm for (i) and (iii) and about 2 nm for (ii). The experiments were performed with a maximum applied load of 10 nN. (b) SEM images showing the three tips after they had been used to measure the adhesion force (scale bars: 60 nm). Within the resolution the nominal tip radius was preserved after all three experiments. (Reprinted figure with permission from Ref. 30, copyright 2010, Wiley.)

maximum shows up as a function of RH (Fig. 9.4a,b) in agreement with previous experiments. Nevertheless, tips with a small enough size, exhibit an adhesion force that decreases monotonously as the vapor saturation increases (Fig. 9.4c,d). At first sight, this latter result appears surprising since an increase of the vapor saturation would lead to an increase of the size of the liquid bridge, in turn, increasing the magnitude of the capillary force. The authors interpret the nonintuitive results as an effect of the precise shape of the tip apex and therefore of the boundary conditions of the meniscus on the tip.

The precise dependence of the meniscus shape on RH values can be derived from an energetic approach by considering the grand canonical potential that contains bulk and surface contributions.[13] The meniscus shape is obtained by a minimization procedure, whereas the force is obtained as the derivative of the potential with respect to the tip–surface separation. The results reported in Fig. 9.5 give the overall evolution of the adhesion force for different

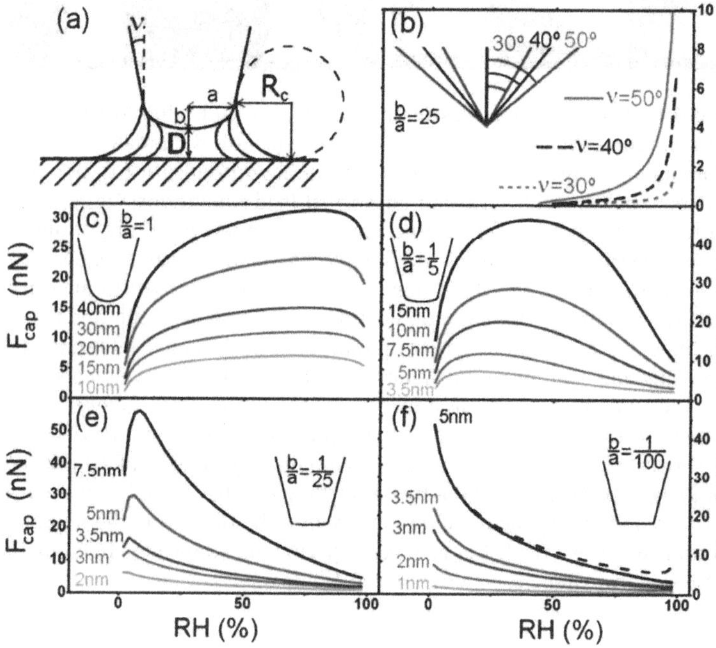

Figure 9.5 Capillary forces vs. geometry. (a) Sketch of the modeled tip geometry and water meniscus. (b–f) Calculated adhesion force vs. humidity for different values of the tip apex form factor b/a (b/a = 25, 1, 0.2, 0.04, and 0.01, respectively) and the ellipsoid transverse semiaxis a. In (b) the results for a quasi-conical tip (b/a = 25, a = 5 nm) are shown for different aperture angles ν. In (c–f) the solid lines are results for tips with $\nu = 10^\circ$ and different a values. In (f) the dashed line corresponds to a quasi-truncated tip with a = 5 nm and $\nu = 30^\circ$. (Reprinted figure with permission from Ref. 30, copyright 2010, Wiley.)

tip geometries and sizes as a function of RH. It shows that a large variety of behaviors can be obtained depending on the tip shape, from increase of the adhesion force with increasing RH to a decrease and a maximum.

An interpretation of the origin of these behaviors can be obtained by a geometrical approach. If one considers the so-called toroidal approximation for which the meniscus profile is assumed circular and if one neglects the tension force, the capillary force can simply be written as a Laplace term:

$$F \approx P_{\text{Laplace}} \pi r_{\text{tip}}^2 \approx \frac{\gamma}{r_{\text{K}}} \pi r_{\text{tip}}^2 \tag{9.19}$$

where r_{K} is the Kelvin radius and r_{tip} the radius of the wetted part of the tip. As RH increases, the liquid condensation is favored and r_{K} increases proportionally to $-1/\ln(RH)$.[1] Nevertheless, one also needs to match the boundary conditions on the tip and the surface, that is, zero contact angle, since the tip and the surface are considered perfectly hydrophilic. In the case of a conical tip (Fig. 9.5b), r_{tip} increases linearly with r_{K}, leading to a continuous increase of the force, which corresponds to the intuitive idea. In the general case of a rounded tip, r_{tip} increases with r_{K} at low RH but then does not evolve so much when the CL lies on the corner of the tip. Since r_{K} continues to increase with RH, the total capillary force decreases. In the case of a truncated cone (Fig. 9.5f), the CL is immediately pinned on the sharp tip edge.[31] (r_{tip} remains constant), leading to a continuous decrease of the force due to the increase of r_{K}, as observed with small tips.

This simple explanation illustrates how boundary conditions play a crucial role in the capillary force and may lead to nonintuitive observations. This is true for tip dimensions that are of the order of the Kelvin radius, typically below 10 nm. If no special care is taken, atomic force microscope tips are rather rounded with a radius of the order of 10 nm or more, which leads to the maximum of force observed as a function of RH. The position of this maximum strongly depends on the precise shape, explaining the large range of reported values. In their experiments, Köber *et al.* used sufficiently small forces to preserve the tip shape down to nanometric dimensions, which was the condition to observe the continuous decrease of the force.

The use of macroscopic equations, Laplace and Kelvin equations, appears quite convincing to quantify the variation of the adhesion force as the result of change of the geometrical parameters. However, several questions rise:

- One is the determination of the relationship between the equilibrium size as determined with the minimization procedure and the corresponding radius of curvature of the meniscus close to the instability.
- Another one stands with an accurate knowledge of the tip–surface distance at which the tip pulls off from the surface. The minimization of the grand potential is calculated at a fixed tip–surface distance that, indeed, varies as a function of the RH.
- A later question, a key one for very small sizes down to the nanometer, that is, only containing a few water molecules, is the ability of macroscopic theory to describe the mechanical properties of the meniscus.

9.3.1.1.2 Capillary forces during nanodispensing

Atomic force microscope tips can also be used to deposit matter on a surface at nanometer scale. The reference example is dip-pen nanolithography where molecules deposited on the tip diffuse to the surface when in contact and define patterns with dimensions as small as 15 nm.[3] An extension is the so-called NADIS technique, for which a nanochannel drilled at the tip apex is used to deposit liquid from a reservoir defined on the top of the cantilever[5] (for a review of these methods refer to chapter 12). Besides its applications for nanopatterning, the NADIS method is a unique method to study liquids at submicron scale and, more particularly, liquid meniscus with controlled dimensions. The capillary force curve experienced by the tip during the deposition and measured with AFM is a good way to assess calculation methods and finally to get more insight into the liquid transfer mechanism from the reservoir to the surface.[32]

We only consider here the retraction of the tip. During this phase the tip pulls a liquid meniscus, which finally breaks. Interestingly, the shape of the curves obtained with NADIS tips is greatly influenced

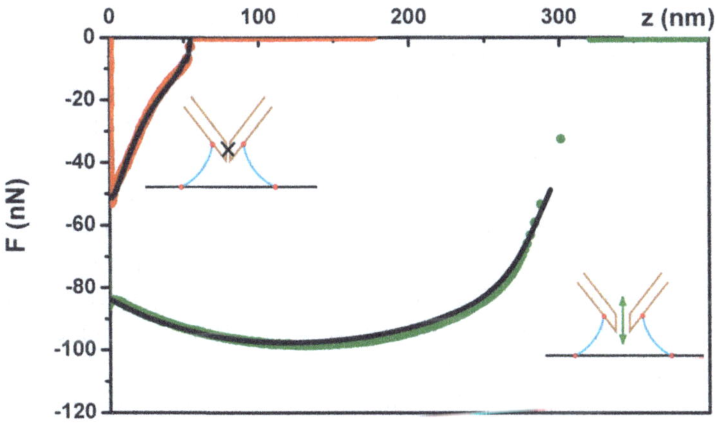

Figure 9.6 Experimental (dots) and simulated force curves (lines) in the case of a NADIS tip with an aperture of 280 nm (in green) and 35 nm (in red). The insets are sketches of the constraints used in the modeling, constant pressure for the large aperture and constant volume for the small one.

by the experimental conditions, as illustrated in Fig. 9.6, gathering the two main observed behaviors when changing aperture size for a hydrophilic tip on a hydrophilic surface. Force curves recorded during a deposit with a tip with a 280 nm large aperture (in green) present a minimum. The magnitude of the (attractive) force slowly increases and then decreases before the snap-off. Curves obtained with a tip with very small aperture (35 nm) show a totally different behavior (in red). The force magnitude continuously decreases, and the concavity of the curve is opposite from the one of the previous example. The functionalization of the tip and/or the surface in order to make them hydrophobic leads to different behaviors, which are discussed in detail in Ref. 32.

In order to extract information from these curves we performed simulations using the Surface Evolver software,[18] which allows to compute the equilibrium shape of a liquid body in many different situations, even nonaxisymmetric geometry. We considered a liquid bridging a flat plane (the surface) and a conical surface (the tip), as represented in Fig. 9.1a. We used as boundary conditions on the surface a circular CL with a fixed R_{surf} radius corresponding to the molecules deposit observed in the experiment and measured by AFM. Since the surface is rather hydrophilic, we considered that the CL on the surface did not recede during the retraction process, which is coherent with the fact that the spots have uniform thickness. With a hydrophilic tip we also used a fixed CL with an R_{tip} radius. As for the constraints, we can impose either a fixed volume or a fixed pressure for the liquid. For every tip–surface separation, we deduced the force exerted by the liquid by computing the energy change associated with a small change of the tip height (energetic approach). Force curves were obtained by incrementing the tip height from zero to the point where rupture occurred (no convergence of the optimization).

We discuss here only the influence of the channel aperture since it gives a nice illustration of the importance of constraints in the meniscus. As reported in Fig. 9.6, the experimental force curves can be fitted fairly well using this method with appropriate boundary conditions and constraints.

- For large apertures, the calculations were performed with, as a constraint, a fixed pressure inside the liquid, given by Laplace pressure in the reservoir droplet. In these cases, it is assumed that the liquid can flow inside the channel to maintain a constant pressure in the liquid at all times. The smooth minimum observed in the experimental force curves is well reproduced with, as only an adjustable parameter, the R_{tip} value, which corresponds to the radius of the wet part of the tip.
- For small apertures, we infer that the channel is too narrow and prevents the liquid to flow from the reservoir during withdrawal, leading to a liquid bridge with constant volume. The computations using such a constraint fit again very well the experimental curve, leading to a change in concavity compared to the constant pressure curves.

The good agreement between calculated and experimental force curves, obtained with R_{tip} as the only adjustable parameter, validates the force calculation method. We find R_{tip} values about twice the aperture diameter, which is compatible with a rather hydrophilic tip. The flow in the nanochannel may thus limit the amount of dispensed

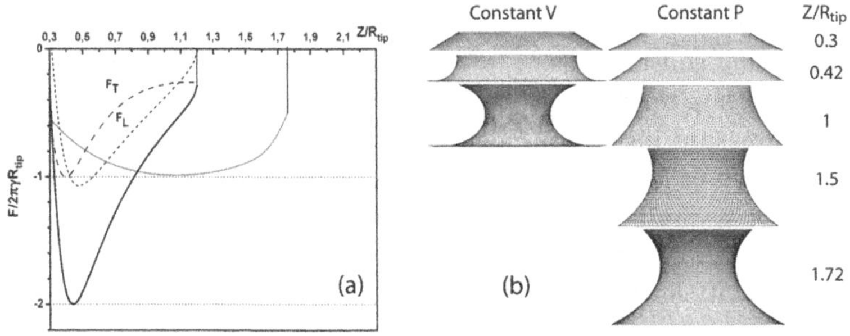

Figure 9.7 (a) Normalized force curves calculated for constant pressure constraint (gray solid line) and constant volume (black solid line). In the latter case, the Laplace (dotted line) and tension (dashed line) components are also represented. (b) Shapes of menisci calculated with Surface Evolver in both conditions for several tip heights.

liquid, which is validated by estimations of flow rate in the channel using a Poiseuille equation.[32]

The modelization allows computing the Laplace and tension terms of the total capillary force (see Fig. 9.7a for normalized force curves).

In the case of the large apertures, the normalized Laplace force is 0.05, showing that its contribution is negligible. The capillary force is therefore mainly due to the tension term. The maximum magnitude is obtained when the liquid interface is vertical on the tip ($\phi = 0$), which leads to a total normalized force equal to -1, as observed in Fig. 9.7. As for small apertures, the Laplace pressure is no more negligible and plays an important role. A negative pressure[33] builds in the meniscus and then decreases to zero (see chapter 11 for a discussion on negative pressures). The magnitude of the Laplace force is comparable to the one due to tension and is even predominant during a large part of the separation process. In this normalized representation, it is worth noting that the liquid bridge breaks faster in the constant volume regime than in the constant pressure regime, as evidenced by the snapshots reported in Fig. 9.7b.

Interestingly, the curves presented above allow an estimation of the meniscus spring constant k by differentiating the $F(z)$ curves. Due to the nonlinear nature of these curves, the spring constant changes with z. In particular it vanishes at the maximum of the $F(z)$

curves. If we focus on the maximum value of $k(z)$ curves, we estimate the spring constants as $k_P/\gamma \approx 12$ and $k_V/\gamma \approx 18$ for constant pressure and constant volume curves, respectively. Since, in the first case pressure inside the meniscus is negligible, we can compare the values of k_P with the ones deduced from Eq. 9.10. The large value of k_P can be obtained using a ratio of characteristic lengths of the order of two, which is coherent with the R_{surf}/R_{tip} used in the simulations. Equation 9.10 therefore gives a proper estimation of the spring constant of a meniscus by using the actual extension of the meniscus as a large cutoff length. Experiments and calculations are underway in order to precisely define the role of meniscus size on its spring constant. In the case of constant volume experiments, the pressure term brings an additional contribution compared to constant pressure, leading to larger spring constants.

This example clearly evidences the importance of constraints on the capillary forces. In particular, the detachment of a tip with a small aperture leads to larger normalized forces, which may be important to consider on soft substrates.

9.3.1.2 Numerical simulations of minimal meniscus size

In order to discuss issues raised before about the validity of macroscopic descriptions at nanometer scale, we use some numerical results dedicated to the description of nanomeniscus properties. Using numerical methods (lattice gas model and mean field density functional theories) Jang *et al.*[34] studied the influence of RH on the pull-off force of a tip on a surface. They reproduced the behavior with a maximum observed with highly hydrophilic rounded tips surfaces (see section 9.3.1.1) and proposed other scenarios for different wetting conditions.

Another issue of particular interest is the determination of the smallest size of a meniscus as a function of the tip geometry and the RH,[35] which is relevant for the experimental results described before. Whatever the process involved, with a surface exhibiting an elastic-like behavior, the force at which the tip detaches from a surface, that is, the adhesion force measured, occurs when the tip reaches a critical contact area that corresponds to the smallest contact area between the tip and the meniscus. Therefore, the

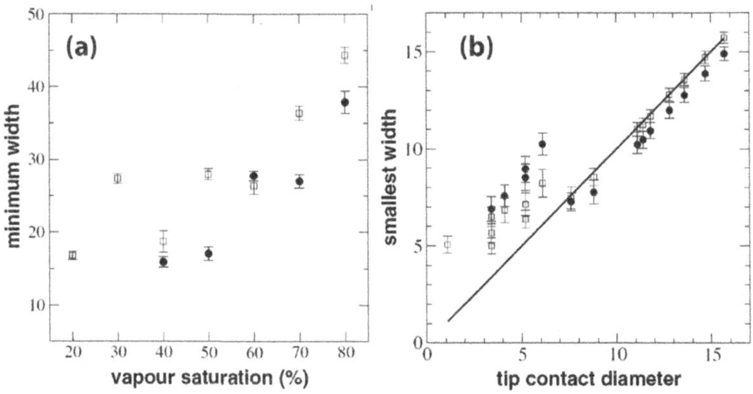

Figure 9.8 (a) Minimum width of stable liquid meniscus vs. vapor saturation. Liquid menisci with width fluctuations less than 10% are considered. At vapor saturations ranging from 10% to 80%, the minimum width is determined for a given saturation by varying the tip–substrate distance. The minimum width is then plotted as a function of saturation for the partially wetting (circles) and completely wetting (squares) tips. The standard deviation in the width is drawn as an error bar in the figure. (b) Smallest meniscus widths for various tip contact diameters. The smallest possible width for a fixed tip geometry is determined by examining stable liquid menisci that form at saturations ranging from 10% to 80%. The smallest width is then plotted as a function of the tip contact diameter for the partially (circles) and completely (squares) wetting tips (the standard deviation in the width drawn as an error bar). The line refers to the case where the smallest width is given by the tip contact diameter. Multiple points for a given contact diameter correspond to tips with different sharpness (aspect ratio). (Reprinted figure with permission from Ref. 35, copyright 2004, American Physical Society.)

determination of the smallest size of the meniscus for a given tip–surface distance and RH can be readily compared with the measured variation of the adhesion force.

The numerical results show that the smallest size of the meniscus does not exhibit a monotonous, increasing, variation as a function of the vapor saturation (Fig. 9.8a).

The variation of the meniscus minimum width shows accidents suggesting a more complicate evolution of the contribution of the capillary force as a function of RH. There is a balance between the feeding of the meniscus with increasing vapor saturation—thus an increase of the meniscus size and the distance at which the instability occurs. As a result, the smallest sizes of the meniscus might reach values that do not increase with the vapor saturation.

A second result of interest seeks for the relationship between the smallest width of the meniscus and the tip geometrical parameters. As displayed in Fig. 9.8b, the smallest width of the liquid bridge follows a linear relationship down to about a value of 8 nm of the tip contact diameter. Below that threshold value, a plateau-like behavior is obtained with a limiting value of the smallest meniscus width of about 5 nm for a tip contact diameter as low as 1 nm. Such a numerical result suggests that for contact diameters smaller than 8 nm, the adhesion force should be nearly a constant independently of the tip geometry and size. In other words, the numerical results suggest that while giving a general evolution in agreement with the experimental data, the calculation using macroscopic theory could be questionable below a contact diameter less than 8 nm.

9.3.1.3 Capillary forces in tapping mode

The most commonly used mode for AFM imaging is the so-called tapping (or intermittent contact) mode. In this configuration, the cantilever is vibrated close to its resonance frequency and the amplitude and phase lag of the oscillation relative to the driving signal are measured with a lock-in amplifier. The amplitude signal is used in the close loop to maintain a given setpoint, while the phase shift signal is linked to energy dissipation processes (adhesion, viscoelastic properties, etc.) of the sample.

It is known that for an asymmetric back-and-forth motion of an oscillating tip, dissipation occurs during an oscillation cycle.[36] Such an asymmetric motion can be due to the adhesion force when the tip touches the surface that gives a backward oscillating path different from the forward path. Therefore the adhesion leads to a loss of energy either in tapping or noncontact AFM mode.[37] When operated in an ambient condition, a nanomeniscus may form between the tip and the surface at each oscillation and modify the imaging process and mechanical information, especially with soft objects. This is more detrimental than in contact mode, for which the presence of the meniscus is less problematic.

The influence of capillary forces in tapping mode was investigated through the measurements of amplitude A and phase φ curves as a function of the tip–surface distance.[24,38] While these two values are directly available with the AFM instruments, another important quantity, that is, the average dissipated power, can be deduced using:[39]

$$P_{\text{dis}} = \frac{k\omega}{2Q}\left(A_0 A \sin\varphi - A^2 \frac{\omega}{\omega_0}\right) \tag{9.20}$$

where A_0 is the free oscillation amplitude, ω the oscillation pulsation, ω_0 the resonance frequency of the cantilever, k the cantilever spring constant, and Q the quality factor.

Because the variations of the amplitude and of the phase depend both on the conservative and dissipative parts, the tapping mode is a more complex mode to use compared to the frequency modulation (FM) mode.[40] In particular, the tapping oscillating behavior depends on the initial experimental condition chosen, oscillation amplitude A, and value of the frequency forcing ω.

For instance, for a cantilever oscillating at its resonance frequency, below an amplitude threshold, the oscillator may only experience the attractive regime at proximity of the surface (curves A in Fig. 9.8). The amplitude and phase signals are superimposed for approach and retraction curves, while when the amplitude is increased above this threshold, which depends on the RH and hydrophilicity of the tip, a hysteresis shows up both in amplitude and phase signals. At given tip–sample distances, the system switches from attractive to repulsive interaction.[38]

These results can be interpreted by adding to the standard differential equation describing the cantilever dynamics in presence of van der Waals and repulsive forces, a capillary force applying during the retraction process.[24] Using the circle approximation for the meniscus profile, one gets a precise picture of the process that reproduces the experimental observations, validating the influence of the formation and rupture of a liquid bridge at each oscillation.

More recently, Sahagun *et al.*[41] extended these investigations to the dissipated energy, which is a crucial issue if one intends to use the dissipation as characteristics of the wetting properties of the sample and therefore map its hydrophobicity. This is of particular interest for biological applications for which the recognition of

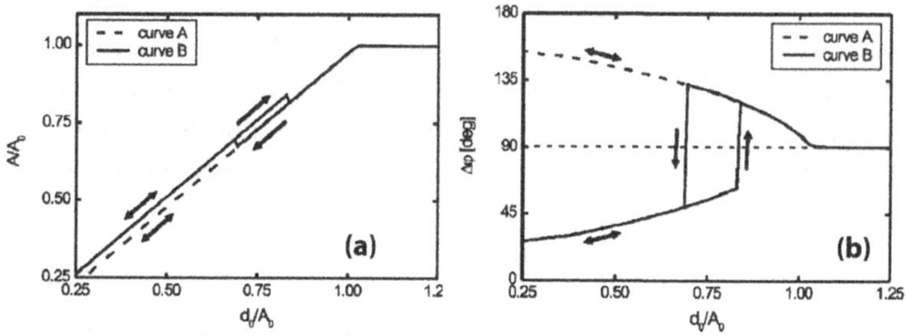

Figure 9.9 Simulated amplitude (a) and phase (b) distance curves for two different free oscillation amplitudes $A_0 = 20$ nm (curve A) and $A_0 = 21$ nm (curve B). While curves A ($A_0 < A_c$) show no jump to the repulsive regime and therefore no hysteresis, the retracting part of curve B ($A_0 > A_c$) remains significantly longer in the repulsive regime than the approaching part, leading to a hysteresis cycle. (Reprinted figure with permission from Ref. 24, copyright 2002, American Physical Society.)

species is usually based on their hydrophilic or hydrophobic nature. The dissipated energy per cycle is given by:

$$U_{\text{dis}} \approx \pi r_K R_{\text{tip}} \gamma (\cos\theta_{\text{tip}} + \cos\theta_{\text{surf}} - a_0/r_K)^2 \qquad (9.21)$$

related to the sample hydrophobicity through θ_{surf}.

Figure 9.10 shows the influence of the free amplitude on (a) the phase and (b) the dissipated power as a function of A/A_0.

In this example, the red and blue curves are obtained on the attractive regime as deduced from phase curves. The dissipation curves clearly show that in this regime, the dissipation is not constant as given by Eq. 9.21. The elastic properties of both the cantilever and the substrate come into play, making an interpretation of a contrast impossible. A hydrophobicity mapping using the dissipation signal is therefore only possible in the repulsive regime (black curves) for which the dissipation jumps to the value given by Eq. 9.21 independently on the tip–surface distance.

This example illustrates the importance of nanomenisci in the interpretation of AFM images. It is only through a precise understanding of the physical mechanisms involved that such images can be used with confidence. It also shows that when used in appropriate conditions, the AFM signal contains information that may be used to map sample properties. However, for quantitative

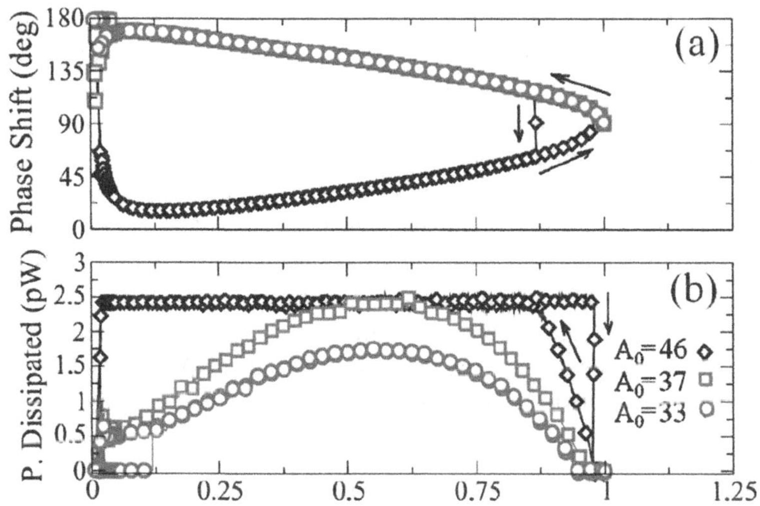

Figure 9.10 (a) Phase and (b) power dissipated vs. normalized amplitude A/A_0 (reprinted figure with permission from Ref. 41, copyright 2007, American Physical Society).

analysis with dynamical modes, it is preferable to use the FM method.

Note that the understanding of the formation of nanomenisci in oscillating modes is also of importance in the framework of nanopatterning. As demonstrated in chapter 13, the presence of water at the tip apex can be used to perform local surface oxidation through the application of an electrical potential. In this case, the influence of the electric field on the stability of menisci has to be taken into account.

9.3.2 *Wetting Conical Tip, Nanoneedle and Carbon Nanotube at Liquid–Air Interfaces*

Because the study of liquid interfaces by scanning probe microscopy (SPM) requires using specific tips usually milled with FIB or else using high-aspect-ratio tips grown at the apex, micro- and nanofluidics using SPM has started only recently. When a standard pyramidal tip is put in contact with a liquid surface, a large meniscus is created and is detrimental to any measurement. Noncontact modes have been developed in order to image the shape of droplets or bubbles. However, if one wants to probe the liquid interface without inducing fast and large tip wetting involving huge capillary forces, it is required that we use much smaller tips, which was made possible only recently by the use of nanoneedles or carbon nanotube tips. Nevertheless, the use of such a method depends on the precise description of the wetting properties of these objects. In the following, we first describe an experiment probing the statics of this mechanism using carbon nanotips, which allows testing the pinning of the CL on nanometric defects. The next section is devoted to original experiments using dynamic AFM modes, which give information about dissipation effects at nanometer scale.

9.3.2.1 Static wetting of carbon nanocones: contact angle hysteresis at nanometer scale

When a liquid droplet is deposited on a real surface, it forms a contact angle whose value is not unique, as expected from the Young equation. It can take a whole range of values bounded by the so-called advancing θ_{av} and receding θ_{rec} contact angles.[42,43] This phenomenon, known as contact angle hysteresis (CAH) is due to topographical or chemical surface defects that pin the CL. Despite many experiments, a precise correlation between defects size and repartition and CAH is still lacking. In order to study this issue, we performed experiments allowing a direct study of the elementary event at the origin of CAH, that is, the pinning on individual nanometric defects.[44] The idea of the experiment is to study the wetting of surfaces with nanometric dimension as, for example, a carbon nanocone.[45] The one-dimensional structure of this surface will be favorable to probe the surface defects one by one. With that aim the carbon nanocones were attached to an atomic force microscope cantilever and dipped in a liquid interface (Fig. 9.11a). The resulting force is measured by the cantilever deflection in a nanometer-scale version of the Wilhelmy balance technique.[46,47]

An example of a force curve is reported in Fig. 9.11b, in the case of a conical tip with a 120 nm long nanotube protruding at its extremity, dipped in heptadecane. After contact with the liquid interface, a short plateau corresponding to the cylindrical nanotube

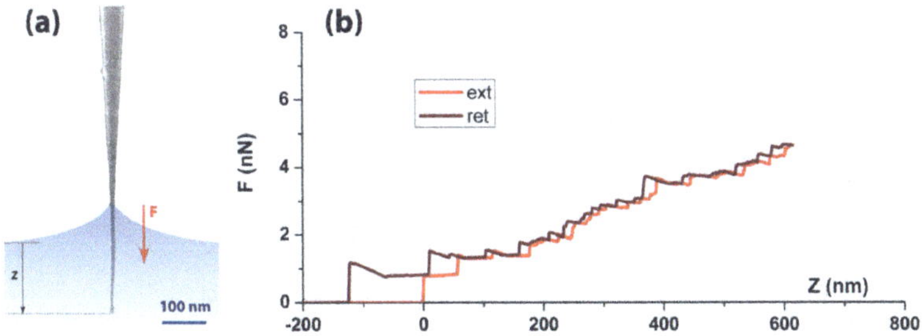

Figure 9.11 (a) Sketch of the experiment. A carbon nanotip is dipped in a liquid interface. The capillary force F measured by AFM is recorded as a function of the tip displacement Z during extension and retraction of the piezoelectric element, corresponding to advancing and receding CLs, respectively; (b) example of a force curve measured with a nanocone tip with a 120 nm long nanotube protruding at its extremity. The extension and retraction curves are represented showing localized hysteresis cycles.

is observed. The force then increases as soon as the CL reaches the conical part of the needle. When tip motion is reversed, the force curve follows approximately the same path, both on the conical and nanotube portions. An increase of the force is always observed at the end of the curve for negative Z values. It is characteristic of the meniscus stretching when the CL is pinned at the tip extremity before snap-off.[48] The main features of the force curve can therefore be interpreted by comparing with the needle shape. The curve reported in Fig. 9.11b is rather different from a typical experiment with a Wilhelmy balance for which advancing and receding curves are clearly separated, allowing a determination of advancing and receding contact angles. In the present case, advancing and receding curves are superimposed on portions of the nanoneedle surface and hysteresis cycles only appear locally. We attribute these cycles to the pinning of the CL on individual (or a small number of) defects of the tip surface. In the geometry we used, the CL is only a few tens of nanometers in length. The surface can thus be perfectly defined at this length scale, leading to portions with no hysteresis. The other advantage is that this short CL can only interact with one or a small number of defects at once. Zooming on parts of the force curves allows thus to study pinning on individual defects as described next.

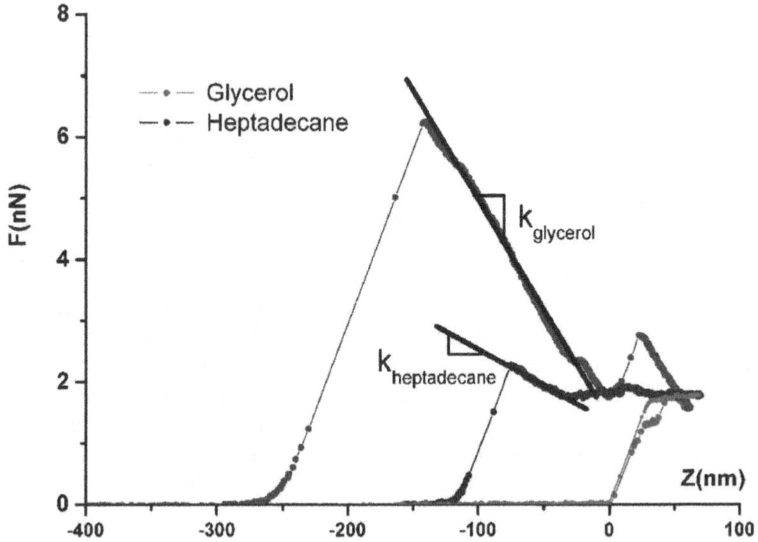

Figure 9.12 Force curves $F(d)$ obtained by dipping the extremity of a carbon tip in heptadecane (red curves) and glycerol (blue curves). The increase of the capillary force for negative d values corresponds to the stretching of the liquid meniscus pinned on the tip extremity before it breaks.[46,47] The slopes of these curves give a direct measurement of meniscus spring constant k.

The last part of the separation process is important in the framework of the elasticity of meniscus since it corresponds to the stretching of the meniscus. The slope of the curve is a direct determination of the meniscus spring constant. In Fig. 9.12 is reported this part for two different liquids (heptadecane and glycerol). The spring constant values determined with several tips for each liquid are $k_{\text{heptadecane}} = 8\pm 3$ mN/m and $k_{\text{glycerol}} = 18\pm 3$ mN/m.

For both liquids we find that $k/\gamma = 0,3\pm 0.1$. Interestingly these values are more than 10 times smaller than the one obtained with nanomeniscus between a tip and a surface (see section 9.3.1.1). A comparison with Eq. 9.10 giving the meniscus spring constant provides a way to estimate the macroscopic characteristic length of the meniscus. Due to the fact that this parameter is involved in a logarithmic term and given the uncertainties in the experimental determinations, a precise determination is difficult. Nevertheless, the small measured k values indicate that this length is large (at least in the order of 1 mm) and therefore comparable with the capillary length (see Fig. 9.2).

The elastic energy stored in the deformation of the meniscus is an important issue if one considers its pinning on defects. An interesting situation is shown on Fig. 9.13, which corresponds to a zoom on a force curve (tip motion amplitude 100 nm). Two qualitatively different behaviors are evidenced. On the right part of the curve ($z > 50$ nm), a modulation of the force amplitude gives rise to a hysteresis cycle between the advancing and receding curves. On the contrary, on the left part, a smaller modulation of the force without any noticeable difference between both curves is observed. This latter case demonstrates that not every surface defect does necessarily induce a CAH. Some defects induce a local change in the contact angle whose related force is fully reversible.

Since it was established early that the CAH originates in the pinning of the CL on surface defects, the single-defect situation attracted much attention. A pioneer paper by Joanny and de Gennes[23] described the case as a balance between the elastic response of the CL and the force due to a local heterogeneity of the surface. Two cases can be distinguished depending on the relative values of the line spring constant and the gradient of defect force:

(i) In the weak heterogeneity regime, for any tip position, only one equilibrium position of the CL exists. The contact angle is thus locally modified but in a fully reversible way.
(ii) On the contrary, for strong heterogeneities, three equilibrium points may exist: the equilibrium point follows two different branches, depending on the direction of motion of the CL, leading to a hysteresis cycle.

Figure 9.13b shows the expected force curves for two successive Gaussian defects with different forces. Depending on the defect strength, either a reversible modulation (as in the left part of the curve in Fig. 9.13b) or a hysteresis cycle (as in the right part of the curve in Fig. 9.13b) is observed. The two defects of Fig. 9.13a therefore correspond to weak and strong defects, respectively,

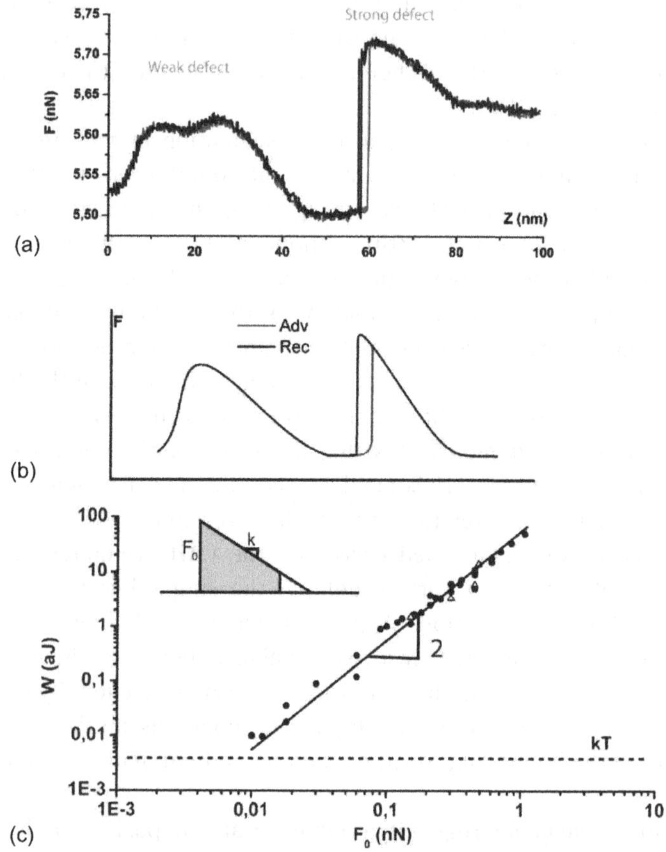

Figure 9.13 (a) (top) Zoom on a portion of the force curve where the CL interact with two defects with different natures; (bottom) calculated force curve for two different Gaussian defects. (b) Plot of the dissipated energy W as a function of the defect force F_0; inset: schematic representation of the pinning on a strong defect, in the simplified model. k is the spring constant of the meniscus (or CL). (Adapted from Ref. 44.)

according to Joanny-de Gennes's definition. To our knowledge, this is the first direct experimental evidence of the weak regime on an individual defect.

In order to quantitatively investigate the strong defect regime, which is the most frequent situation, we proposed a simplified model for the pinning on a single defect. Considering the case of a defect with a small dimension and sharp edges where F_0 and a are the force and lateral extension of the defect, respectively, an $F(z)$ curve with a triangular shape is expected (Fig. 9.13c, inset). The dissipated energy per defect is given by the area of the cycle shaded gray, which is written as $W \approx \frac{F_0^2}{2k}$ in the case of a single localized defect ($a << D = F_0/k$), with k as the meniscus (or CL) spring constant. Both the pinning force F_0 and the dissipated energy W were measured experimentally as the jump height and the area of the local hysteresis cycle, respectively, for a whole set of different events attributed to individual defects, for various tips dipped in heptadecane. The results plotted in Fig. 9.13c evidence a quadratic relationship between W and F_0 over four orders of magnitude in energy, compatible with the expression given above. A value of the meniscus spring constant $k = 9\pm\ 2$ mN/m is obtained in very good agreement with the value deduced from the meniscus stretching at the end of the separation process. The single defect case is therefore quantitatively described by a simple model. Interestingly, this description assuming Hookean elasticity accounts for pinning forces as small as 10 pN and dissipated energies in the range of 10^{-20} J. For such values close to the thermal energy $kT = 410^{-21}$ J, thermal fluctuations may play an important role.

In order to evaluate the sizes of the defects responsible of the wetting hysteresis, we considered a topographical defect consisting of a bump with lateral size a and a maximum slope φ.[50] Despite the fact that the value of φ is unknown, we estimated that the points reported on Fig. 9.13 correspond to defects sizes ranging from tens of nanometers down to 2 nm, which is compatible with electron microscopy observation of the nanoneedles.[45] We also determined a minimal defect size necessary to obtain hysteresis, by comparing the dissipated energy on the defect to kT. This led to a value comparable with the size of liquid molecules, compatible with previous results,[51] which explains why it is so challenging to obtain hysteresis-free surfaces.

9.3.2.2 Dynamics

The dynamical AFM method allows to access simultaneously both conservative and nonconservative interactions.[40] When liquid motions are involved in a small volume of liquid, it provides the information on the processes of dissipation. On the other hand, little is known on the influence of the velocity fields in a confined volume of liquid. Therefore, by using nanoneedles and a forcing with oscillation amplitude of few tens of nanometers, one can investigate how a velocity profile extends in a confined volume and the corresponding contribution of the viscous properties on its mechanical properties.

9.3.2.2.1 Frequency modulation AFM dynamical method

A nanoneedle carved by a focused ion beam (FIB) is approached gently to the air–liquid interface with a step motor. When the tip touches the surface, it oscillates in the liquid and the FM-AFM mode is used to record the changes in the oscillating properties of the cantilever.[25,26] With the FM-AFM mode, the resonance frequency shift, measuring the conservative force gradient, and the damping coefficient, measuring the dissipative force, are simultaneously recorded.[40]

9.3.2.2.2 Experimental results

A typical sequence of dipping events in water is displayed in Fig. 9.14. The liquid is approached with a step motor toward the nanoneedle. When the tip is far from the surface, that is, more than a few nanometers, there is no interaction between the needle and the air–liquid interface and thus no frequency shift (phase 1). As soon as the tip touches the liquid, the elastic response of the meniscus induces a positive shift of the resonance frequency (phase 2). Because of the water evaporation, the average contact angle decreases with time leading to a decrease of the frequency shift (phase 3). The phase 4 is identical to phase 1 with no significant interaction between the tip and the liquid.

Resonance frequency and dissipation are simultaneously recorded (Fig. 9.15) and show totally different behaviors. The frequency shift continuously decreases as the liquid level recedes due to evaporation while the dissipation shows a more complex variation. The damping first decreases as the interface moves downward; then a sharp turn follows with a noticeable diverging-like increase, before a breakdown, which brings back the dissipation to zero when the tip goes off.

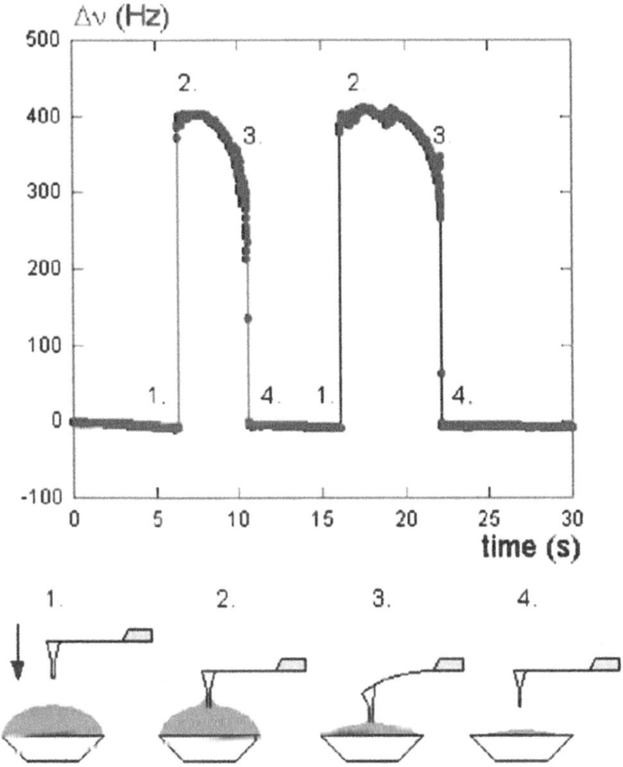

Figure 9.14 Wetting processes of a nanoneedle at the air–liquid interface. As soon as the tip touches the interface, an abrupt increase of the resonance frequency happens. The positive resonance frequency shift is given by the elastic response of the nanomeniscus. Note that due to the evaporation of the water, the typical duration time of one dipping event is 3–5 s. (Reprinted figure with permission from Ref. 25, copyright 2004, American Chemical Society.)

The nanomeniscus spring constant k_{men} is obtained from the experimental resonant frequency shift using $\frac{\Delta\nu}{\nu} \approx \frac{1}{2}\frac{k_{\text{men}}}{k_c}$, where k_c is the cantilever spring constant. From several experiments in water, the values deduced from the first part of the curves, where $\Delta\nu$ is slowly varying, are between 0.7 and 1.1 N/m, which correspond to $k/\gamma = 1.2 \pm 0.2$. This value is much larger than the one resulting from static experiments (see section 9.3.2.1). According to Eq. 9.10, an increase of the meniscus spring constant may result from a

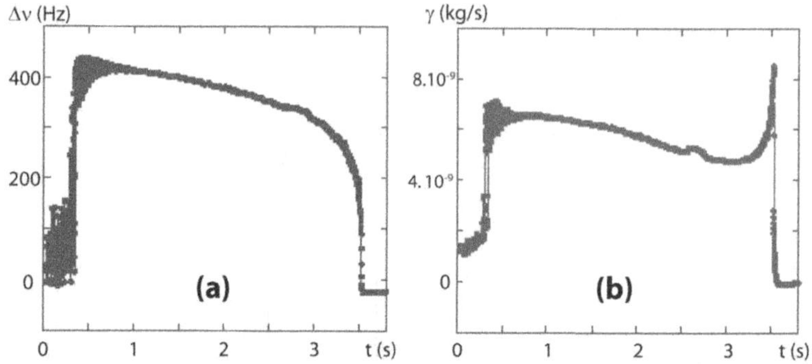

Figure 9.15 Variations of (a) the resonance frequency shift and (b) the damping coefficient.

decrease of the lateral extension of the meniscus with respect to the capillary length used in the equation. Indeed, the experimental values deduced from dynamic experiments can be obtained from Eq. 9.10 using a characteristic length between 300 nm and 1.5 μm, comparable with the evanescent δ. The length δ determines a radial domain around the rod out of which the velocity field induced by the nanorod motion vanishes (see section 9.2.2), so it is reasonable to consider that the meniscus extension involved in its dynamics is of the order of δ. For a liquid, it is not surprising that an induced velocity field confined in a small volume can have an influence on its mechanical properties. Nevertheless, an accurate and quantitative analysis has still to be done and more experiments are needed to confirm this point. In the following, we use Eq. 9.10, replacing K^{-1} by δ. However, an accurate measurement of the lateral extension of the nanomeniscus is very challenging. This limits the description to a qualitative analysis.

From the dissipation data, an appropriate calibration gives the absolute value of the damping coefficient associated with hydrodynamic loss of energy in the nanomeniscus. With the FM-AFM method, the dissipation measured is proportional to $D \propto \beta A$, where β is the damping coefficient and A the oscillation amplitude at the resonance. The damping coefficient of the cantilever is $\beta_0 = k_C/\omega_0 Q$, where k_C is the cantilever stiffness, w_0 the cantilever pulsation at the resonance, and Q the quality factor. When the tip is dipped in the liquid, the total dissipation parameter is given by a linear combination of two contributions $D \propto D_0 + D_{\text{men}}$ with the dissipation of the cantilever $D_0 \propto \beta_0 A$, a simple algebraic equation gives the associated damping coefficient of the meniscus β_{men}:

$$\beta_{\text{men}} = \beta_0 \left(\frac{D}{D_0} - 1 \right) \quad (9.22)$$

Since a goal of interest is to identify the dissipation processes at work in the meniscus, the experimental data gives the opportunity to establish a relationship between the variation of the damping coefficient and the relative change of the meniscus shape. Once the meniscus extension is fixed as discussed above, the variation of the resonant frequency shift, and in turn the variation of the spring constant, gives the variation of the meniscus shape, as deduced from Eq. 9.10. The contact angle θ is thus given by:

$$\theta = \frac{k_{\text{men}}}{2\pi\gamma - k_{\text{men}}(C - 1)} \quad (9.23)$$

A variation of the contact angle θ at the triple line calculated with Eq. 9.23 is shown in Fig. 9.16a. In the inset of Fig. 9.16a the simultaneous variation of the damping coefficient is displayed.

When one plots the dissipation as a function of the contact angle θ, for various oscillation amplitudes, two different behaviors are

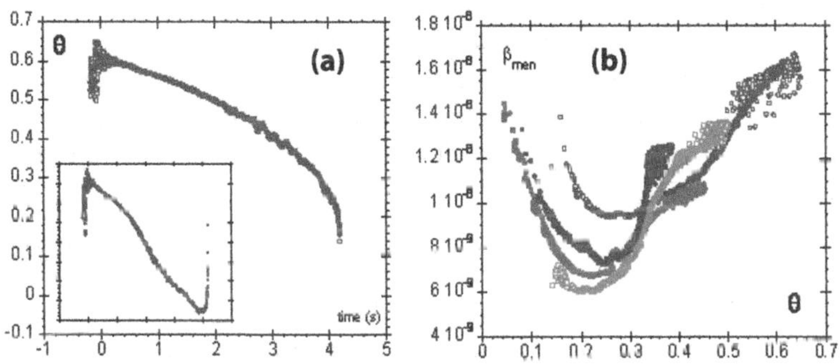

Figure 9.16 (a) Variation of the wetting angle θ calculated from the resonant frequency shift variation with Eq. 9.23. In the inset is shown the corresponding variation of the damping coefficient; (b) experimental damping coefficient deduced from damping experimental data and Eq. 9.22 as a function of the wetting calculated from the resonant frequency shift data with Eq. 9.23. Four oscillation amplitudes are used: 8, 12, 16 and 20 nm.

evidenced (Fig. 9.16b). The nonmonotonous variation is observed with a minima located at angular values lying between 0.2 and 0.3 radians. In the following we give an analysis of these two regimes.

9.3.2.2.3 Influence of the shape of meniscus

For a harmonic oscillator the dissipated energy is given by $< E_{\text{diss}} >_T = \pi\beta\omega A^2$. In section 9.2.2.2, we reported in Eq. 9.18 the dissipated energy for an oscillating tip in a liquid with a meniscus smaller than the evanescent length δ, a situation corresponding to the experimental configuration. Leaving aside the numerical factors, it gives the associated damping coefficient β_{men} that expresses the loss of energy due to the motion of the liquid in the meniscus:

$$\beta_{\text{men}} \approx \left(\frac{d}{\delta}\right)^2 m_d\omega \tag{9.24}$$

Note that δ^2 scaling as the viscosity η, Eq. 9.24 leads the nonintuitive result that the amount of energy loss decreases with the viscosity. That happens for a film of thickness smaller than the length δ and can be understood as the fact that the velocity gradient scaling as v/δ, the larger the viscosity, the smaller the velocity gradient within the thin film at a given thickness $d << \delta$.

We first investigate the capability of Eq. 9.24 to explain the decrease of the amount of the energy loss with the thinning of the meniscus, that is, for contact angles larger than 0.2–0.3 radians. To do so, we need to estimate the amount of liquid m_{d} trapped in the meniscus that contributes to the energy loss. Expressing the thickness d as a function of the meniscus angle $d \approx \theta h$, change of the mass m_{d} can be readily evaluated and Eq. 9.14 is rewritten in a more usable form:

$$\beta_{\text{men}}/\theta^3 h^4 \approx \left(\frac{1}{\delta}\right)^2 2\pi R\omega = Cst \tag{9.25}$$

Equation 9.25 normalizes the damping coefficient with the nanomeniscus geometry taking into account the thickness effect as described with Eq. 9.24. It predicts a plateau behavior with a height that should not depend on the oscillation amplitude. In Fig. 9.17 the normalized dissipation $\beta_{\text{men}}/\theta^3 h^4$ is reported for the experimental curves of Fig. 9.16.

For contact angles larger than 0.3 radians, all curves collapse on a master curve which tends toward a plateau. Therefore a thickness

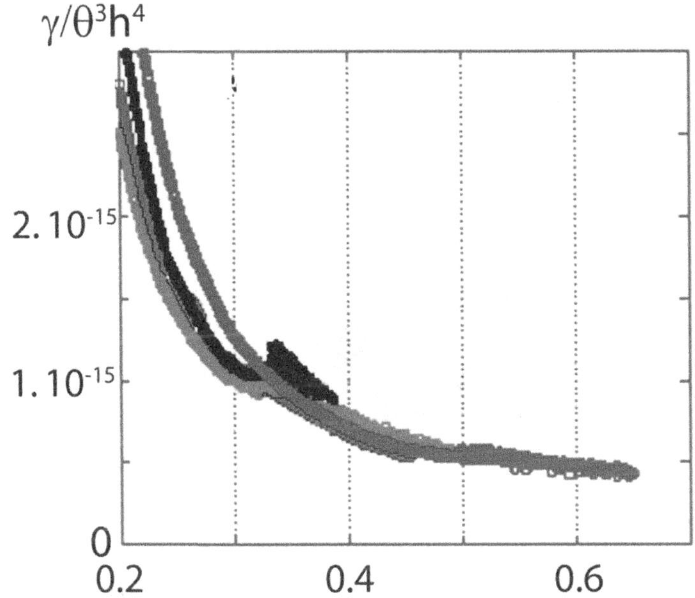

Figure 9.17 Plot of the experimental data shown in Fig. 9.16 divided by $\theta^3 h^4$.

effect as described in Fig. 9.3 might certainly be involved in the decrease of the damping but is not able to account for the damping evolution within a significant range of contact angle values. In order to emphasize the thinning effect, we computed the damping coefficient assuming that it is proportional to the mass of fluid involved in the oscillating motion without taking into account the thinning effect. The volume of liquid involved in the oscillating motion is $V \approx 2\pi Rhz = 2\pi R\theta h^2$, so for a variation of the damping only governed by the variation of the volume of fluid trapped in the meniscus, the following expression must be verified:

$$\frac{\beta}{\theta h^2} \approx Cst \tag{9.26}$$

The damping coefficient normalized to the volume of fluid is displayed in Fig. 9.18.

This normalization procedure, taking only into account the volume of liquid involved in the oscillating motion, is unable to remove the nonmonotonous variation of the damping coefficient.

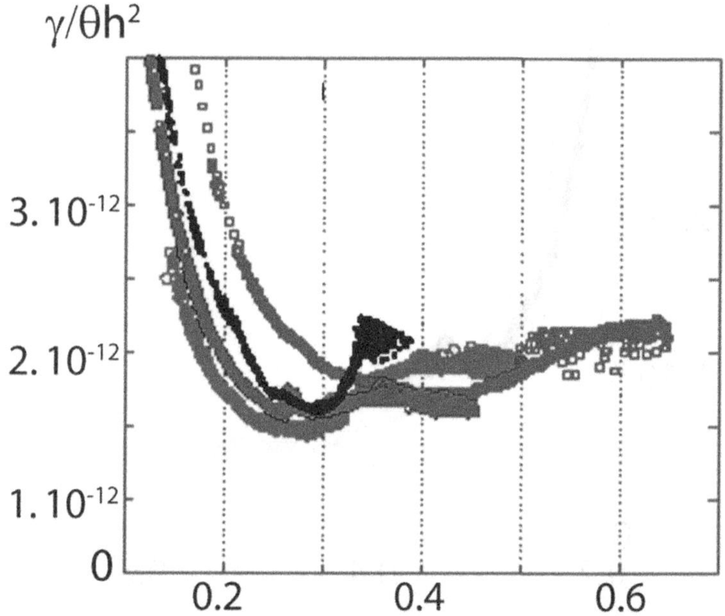

Figure 9.18 Plot of the experimental data shown in Fig. 9.16 divided by θh^2, corresponding to a normalization to the volume of fluid inside the meniscus.

This result supports the fact that the thickness of the meniscus, with values smaller than the evanescent length, is relevant to explain the decrease of the damping coefficient when the meniscus start to recede.

The increase of the dissipation associated with a diverging behavior at the end of the process, when the meniscus becomes thinner and finally breaks, is not described by the model. Recent experiments showed that the diverging behavior is also observed for nonevaporating liquids. Therefore, for ultrathin liquid films the noticeable dissipation increase before the meniscus burns out cannot be a consequence of the water evaporation as mentioned previously for a water nanomeniscus.[26] There is no immediate analysis that may help to understand the diverging behavior before the meniscus burns out. A few routes can be envisioned, one to focus on the boundary conditions at the air liquid interface and another to consider the effect of a very thin liquid, surrounding the cylinder, held by the disjunction pressure.

9.3.2.2.4 Case of a sliding triple line and influence of surface wettability

The previous results were devoted to study of dynamical properties of a meniscus with a pinned triple line and of a thinning effect that exhibits the influence of the boundary condition at the air–liquid interface. In this part, we examine the influence of a sliding triple CL on the dynamical properties of the nanomeniscus; in particular the modeling with an associate spring constant related to nanomeniscus shape might become questionable. Also, the influence of the wetting ability of the liquid on the nanorod is investigated.

The experiment is done with a cantilever on which an FIB was used to grow three tungsten carbon nanoneedles (see Fig. 9.19). The fabrication of a cantilever bearing three nanoneedles was performed by Mariolle in CEA-LETI (Grenoble, France). It was designed to ensure a very similar dipping process in order to compare the mechanical response of each tip.

The tips were dipped in two different solutions of peptide. One peptide was selected through a phage display technique to recognize

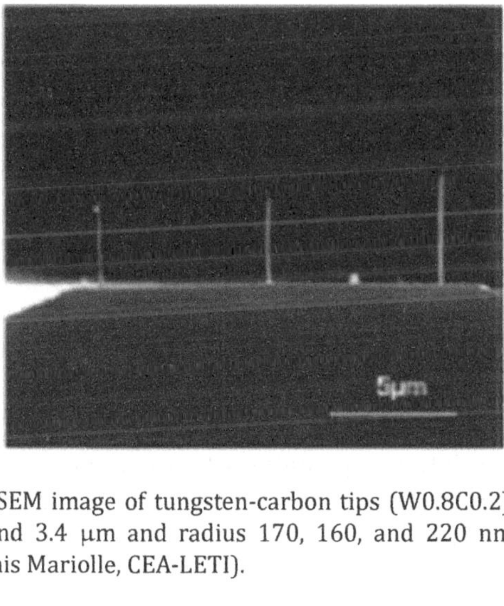

Figure 9.19 SEM image of tungsten-carbon tips (W0.8C0.2) of length 2.5 μm, 3 μm, and 3.4 μm and radius 170, 160, and 220 nm, respectively (courtesy Denis Mariolle, CEA-LETI).

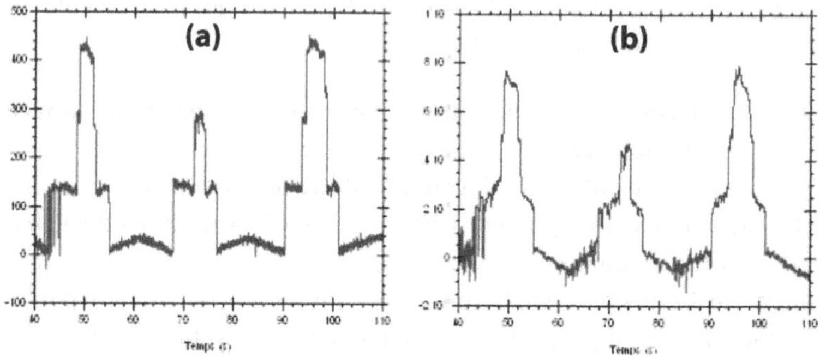

Figure 9.20 Left: Successive positive resonant frequency shifts. Right: Successive corresponding damping coefficients. Note that in the middle only two events were measured.

selective carbon species,[52] hereafter called peptide-carbon solution, and the second peptide specifically interacted with silica,[53] hereafter called peptide-silica solution. Because the tips contain carbon atom aggregates, it is expected that peptides act as a surfactant, inducing a smaller wetting angle when the tips are dipped in the peptide-carbon solution.

Typical variations of the resonance frequency shifts and damping coefficients are displayed in Fig. 9.20. Each dipping event shows three marked steps corresponding to the successive dipping of the tips. A well-defined plateau of resonant frequency shifts and damping appears first during the wetting process and then when the liquid recedes. In spite of a distance between the tips larger than 5 μm, a distance that should minimize the hydrodynamic interaction between tips, the second and third steps are not as well defined as the first one. Therefore, one cannot completely rule out that hydrodynamic interaction between tips may occur when the second and third dipping events occur. Still, the three events are well separated.

In Fig. 9.21 is displayed a comparison between the dipping process in the solution of peptide-carbon and peptide-silica. Unambiguously, the resonant frequency shift is smaller, about twice less, in the peptide-carbon solution than that observed when dipping in the peptide-silica solution. Therefore the wetting contact angle is

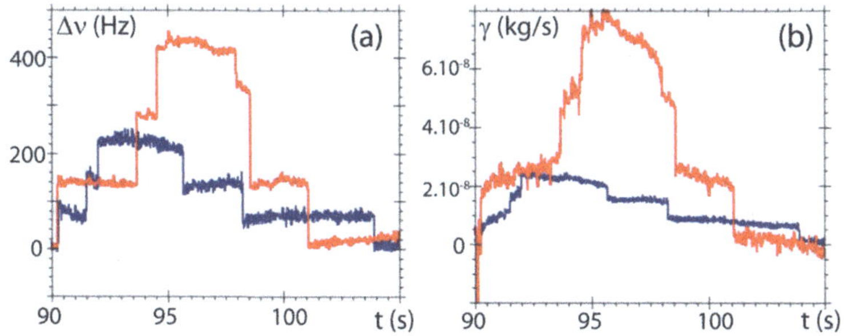

Figure 9.21 Left: Comparison between the dipping events in the peptide-carbon solution (blue) and the peptide-silica solution. The positive frequency shift is smaller in the peptide-carbon solution, indicating a smaller wetting angle. Right: Corresponding variation of the damping coefficient showing a lower dissipation in the peptide-carbon solution.

smaller with a solution able to wet the nanoneedles more efficiently. The amount of dissipation for each event is also much smaller.

Here again, several well-defined steps are observed. Contrary to what was observed with the silicon tip dipped in water, there is no noticeable variation of the resonant frequency shift during the evaporation of the solution. As discussed before, a constant resonance frequency shift gives a constant value of the associate spring constant, and based on Eq. 9.9 or 9.10, this absence of variation of the spring constant means that the contact angle remains a constant value. Therefore, the CL is not pinned but slides along the rod when the liquid evaporates. In other words, the nanomeniscus keeps the same shape over the whole receding motion of the liquid.

Since the CL slides, the elastic restoring force of the nanomeniscus against the nanorod oscillation forcing should be much smaller than the one observed with a pinned line. In particular, for the limiting case of a complete free sliding CL, we do not expect to have any restoring force at all. For each tip, the meniscus slides slowly as each resonant frequency shift step remains over few seconds. Therefore, at the time scale of the second, one should not see an effect of the force, or a very weak effect, if any. However, the observed resonant frequency shift means that the nanomeniscus does show a mechanical response. As the forcing applied at frequencies of hundreds of kiloHertz, one can understand the mechanical behavior as follows: the sliding process is slow enough so that at the time scale of the oscillation period, a few microseconds, the CL remains pined, thus allowing the spring constant of the meniscus to be measured. Therefore, in what follow, as was done for the case of pinned CLs, we shall use the frequency shift as a measurement of the meniscus shape.

In Fig. 9.22 is displayed the damping coefficient as a function of the resonant frequency shift for the three events. Due to fluctuations of the damping and of the wetting angle, clouds of experimental data are obtained (Fig. 9.22). The dipping in the peptide-carbon solution gives a resonant frequency shift centered at 75 Hz, with values varying between 50 and 100 Hz; the damping coefficient is centered at 10^{-8} kg/s. For the peptide-silica solution, the resonant frequency is centered at 130 Hz, with a smaller width of frequency fluctuations varying between 120 and 150 Hz. The damping shows a larger domain of fluctuations centered at a value slightly below $4\ 10^{-8}$ kg/s.

Whatever the solution used, these results strongly support that the heights of the plateau, both for frequency shifts and damping, are nearly identical. The dispersion of the experimental data corresponds to fluctuation of the heights of the plateau, either on the resonant frequency shift or on the damping parameter. Keeping in mind that the largest difference in radius is between 160 nm and 220 nm, the data is not precise enough to allow the extraction of the influence of the value of the radius on the behavior of the meniscus. However, the two solutions show marked differences correlated to their ability to wet the nanorod that we can readily analyzed in greater detail.

In the peptide-silica solution, the resonant frequency shift is twice as large as the one measured in the peptide-carbon solution. In addition, the damping coefficient is about 3.5 times larger in the peptide-silica solution than in the peptide-carbon solution. As the accuracy of the experimental data does not allow seeing differences on the values of the plateau between wetting and receding of the liquid surface, the data treatment is based on a resonant frequency shift that remains nearly constant over the whole process.

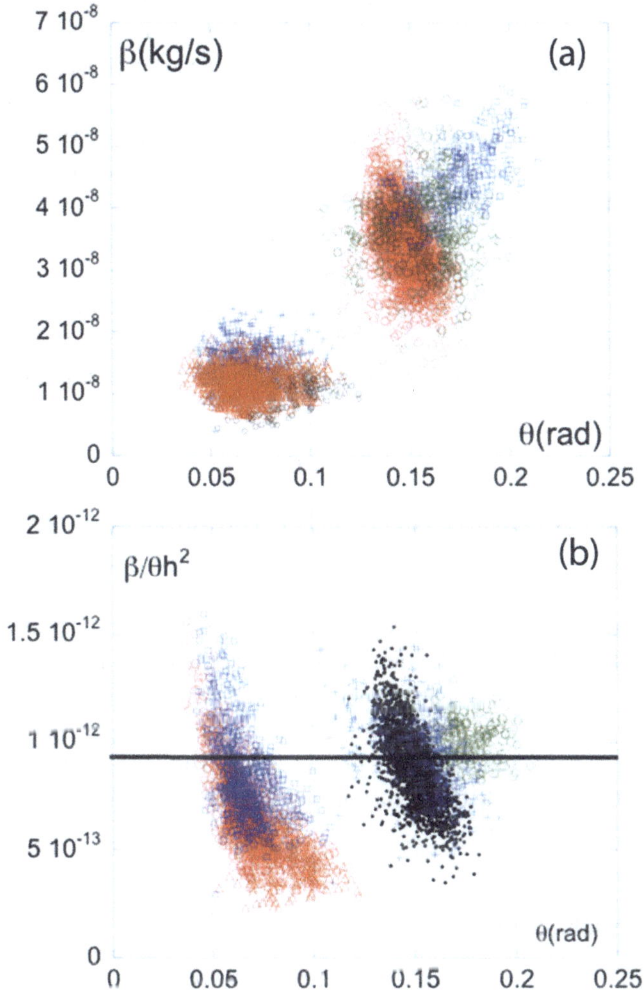

Figure 9.22 Experimental data showing the variation of the damping coefficient as a function of the calculated contact angle deduced from the frequency shift (see text) for the three damping events in the two solutions. The clouds of experimental data gather the height of the steps observed for the three tips in the peptide-carbon solution—left clouds, first tip (orange), second tip (black), and third tip (blue); in the peptide-silica solution—right clouds, first tip (red), second tip (green), and third tip (blue).

Because the resonant frequency shift is a measurement of the meniscus shape, we also assume the size of the nanomeniscus remains nearly constant over the whole process. Based on this assumption, we can establish a relationship between the amount of dissipation observed and the shape of the meniscus. With a sliding three-phase CL, we might expect the amount of dissipation is mostly related to the mass of fluid involved in the oscillating motion. In a sense, it is a case much simpler to analyze than the one in which a thinning process takes place.

Taking into account the volume of liquid involved in the oscillating motion for a sliding contact angle, keeping a constant volume of liquid in the meniscus, the following expression must be relevant:

$$\frac{\beta}{\theta h^2} \approx Cst$$

In Fig. 9.22 is shown a comparison between the damping coefficient of the two solutions as measured and the one normalized to the size of the meniscus.

Whatever the tip involved, there is a large difference, about a factor three, between the damping coefficients in the peptide-silica solution and the ones in the peptide-carbon solution. When the damping coefficients are divided by the volume of the meniscus, the damping coefficients become nearly equal. The large difference in dissipation is mostly the result of the difference of the contact angle at the three-phase CL.

9.4 Conclusions

In this chapter, we reported on the mechanical properties of nanomenisci investigated using AFM techniques. The fabrication of nonconventional tips (nanoneedles, nanotube tips, NADIS tips), combined with the variety of modes available with this technique, allows to perform controlled measurements at nanometer scale and opens new routes toward a better understanding of what nanofluidics is. Despite recent advances on both static and dynamical aspects, several important issues remain to be clarified. Among them we can cite the following:

- Continuum theory vs. molecular theory (lattice gas model and mean field): The capability to determine the domain of validity of the macroscopic equations to describe the behavior of nanodrops and nanomenisci is of primary importance. Continuum modeling allows to end with analytical predictive equations or at least to provide comprehensive pictures of meniscus and drop properties. This is the case in most of the experiments presented earlier, in particular those addressing specific questions such as instability and rupture of menisci, sustainability and strength of liquid nanobridges, or pinning of a CL on individual nanometric defects. They show nicely that the use of Poiseuille, Laplace formula, and surface tension effects are enough to explain the experimental results down to nanometer scale. However, despite the many good agreements obtained when using continuum theory, several questions remain. The key one is to know if it makes sense to keep macroscopic theory to describe the mechanical properties of a small collection of molecules. Numerical results suggest that for contact diameters smaller than 8 nm, the adhesion force should nearly be a constant independent of the tip geometry and size. In other words, while giving a general evolution in agreement with the experimental data, the calculation using macroscopic theory could be questionable below contact diameters lesser than a few nanometers. Experiments using nanotube tips as the ultimate probes may allow to answer this question.
- Relevant length scales: Most of models of wetting and capillarity involving liquid menisci include characteristic lengths. At a small scale, in hydrodynamic description, a cutoff length is introduced to remove singularity at the CL. In molecular kinematics theory, the issue of the length of activated jumps is crucial. At a large scale, one can cite the extension of the meniscus around a fiber. In the case of nanoscale fibers, the use of a capillary length is questionable. In this chapter we showed that the study of the meniscus spring constant may be a good way to address these questions. We have shown that the confinement of

the meniscus leads to much larger meniscus stiffness, which can be interpreted by a decrease of the large-scale cutoff length. The dynamical stiffness may also result from a limitation of the meniscus to the distance δ over which the velocity field vanishes. These primary results could be studied thoroughly using nanofabrication techniques to obtain menisci with controlled sizes.

- Static vs. dynamic experiments—influence on the mechanical properties of the velocity field in a confined volume: The investigation of the mechanical properties of a nanomeniscus shows significant differences that depend on whether static or dynamic experiments are performed. We mentioned that the use of the evanescent length δ right be appropriate to describe the spring constant of a dynamical meniscus. There is no rigorous mathematical proof allowing replacing the capillary length by the evanescent length. However, the dissipation behavior, as a function of the meniscus thickness, does support such a proposition. The decrease of the dissipation can be readily understood as a consequence of a meniscus thickness smaller than the evanescent length. The velocity profile generated by the cylinder's oscillating motion that expands over the evanescent length is frustrated when the thickness is much smaller. Following this line, one may consider that for very thin films, the molecules of the fluid move without significant velocity variation between them, thus canceling the viscous dissipation. The comparison between static and dynamic experiments shows that the velocity field plays a role and does modify the mechanical response of a nanoscale volume of liquid. As a matter, it is not that much surprising for a liquid.
- Dissipation processes at the contact line and the role of surface defects: The description of the dissipation processes occurring at a moving CL is of primary importance in the framework of wetting dynamics.[42,43] In the hydrodynamic model (de Gennes and Cox-Voinov models), the dissipation is considered in the liquid wedge of length between microscopic and macroscopic cutoff lengths, while in

kinematic theories, it comes from the elementary events at the CL when a liquid molecule jumps the next adsorption site. In addition to these contributions, the pinning on defects, which leads to CL hysteresis, brings an additional contribution, which is still poorly understood. Despite the fact that the theories can fit experimental data, there are many unknown mechanisms, which could be resolved if one is able to probe the spreading dynamics, at nanometer scale, at the CL. We have shown that a nanotube tip dipped in a liquid interface is a very favorable configuration, mainly because of the very short CL involved. In static mode, we studied the pinning on individual defects down to nanometer scale. These elementary events may account for the large jump lengths sometimes reported in molecular kinetics theory.[54] As suggested by Prevost et al.,[55] the pinning of the CL on defects may be at the origin of both the CAH and dynamics. It is therefore important to extend our results to dynamic processes[56] whose investigation should be possible by using dynamic AFM modes.

Acknowledgments

We thank Denis Mariolle, François Bertin, and Amal Chabli from CEA-LETI, Grenoble (France), for financial support by TRAIN^2 (Trans Pyrénées action, Interreg IV B SUDOE program) is acknowledged, fruitful discussions and FIB milling of nanoneedles on cantilevers.

References

1. A. Adamson and A. Gast, *Physical Chemistry of Surfaces*. 6th ed. New York: John Wiley and Sons (1997).
2. J. Israelachvili, *Intermolecular and Surface Forces*. New York: Academic Press (1992).
3. R. D. Piner, J. Zhu, F. Xu, S. H. Hong, and C. A. Mirkin, "Dip-pen" nanolithography, *Science*, **283**, 661–663 (1999).

4. R. Garcia, M. Calleja, and H. Rohrer, Patterning of silicon surfaces with noncontact atomic force microscopy: field-induced formation of nanometer-size water bridges, *J. Appl. Phys.*, **86**, 1898–1903 (1999).
5. A. P. Fang, E. Dujardin, and T. Ondarçuhu, Control of droplet size in liquid nanodispensing, *Nano Lett.*, **6**, 2368–2374 (2006).
6. P. Lambert, *Capillary Forces in Microassembly*. New York: Springer (2010).
7. A. Bensimon, A. Simon, A. Chiffaudel, V. Croquette, F. Heslot, and D. Bensimon, Alignment and sensitive detection of DNA by a moving interface, *Science*, **265**, 2096–2098 (1994).
8. L. Malaquin, T. Kraus, H. Schmid, E. Delamarche, and H. Wolf, Controlled particle placement through convective and capillary assembly, *Langmuir*, **23**, 11513–11521 (2007).
9. H. G. Duan and K. K. Berggren, Directed self-assembly at the 10 nm scale by using capillary force-induced nanocohesion, *Nano Lett.*, **10**, 3710–3716 (2010).
10. C. Py, P. Reverdy, L. Doppler, J. Bico, B. Roman, and C. N. Baroud, Capillary origami: spontaneous wrapping of a droplet with an elastic sheet, *Phys. Rev. Lett.*, **98** (2007).
11. J. H. Cho, A. Azam, and D. H. Gracias, Three dimensional nanofabrication using surface forces, *Langmuir*, **26**, 16534–16539 (2010).
12. J. W. van Honschoten, J. W. Berenschot, T. Ondarcuhu, R. G. P. Sanders, J. Sundaram, M. Elwenspoek, and N. R. Tas, Elastocapillary fabrication of three-dimensional microstructures, *Appl. Phys. Lett.*, **97** (2010).
13. F. Restagno, L. Bocquet, and T. Biben, Metastability and nucleation in capillary condensation, *Phys. Rev. Lett.*, **84**, 2433–2436 (2000).
14. P. Lambert, A. Chau, A. Delchambre, and S. Regnier, Comparison between two capillary forces models, *Langmuir*, **24**, 3157–3163 (2008).
15. H. J. Butt and M. Kappl, Normal capillary forces, *Adv. Colloid Interface Sci.*, **146**, 48–60 (2009).
16. X. Pepin, D. Rossetti, S. M. Iveson, and S. J. R. Simons, Modeling the evolution and rupture of pendular liquid bridges in the presence of large wetting hysteresis, *J. Colloid Interface Sci.*, **232**, 289–297 (2000).
17. P. Lambert and A. Delchambre, Parameters ruling capillary forces at the submillimetric scale, *Langmuir*, **21**, 9537–9543 (2005).
18. K. Brakke, The surface evolver software, *Exp. Math.*, **1**, 141–165 (1992).
19. L. Landau and E. Lifshitz, *Course of Theoretical Physics: Fluid Mechanics*. Oxford: Elsevier (1987).

20. C. Clanet and D. Quere, Onset of menisci, *J. Fluid Mech.*, **460**, 131–149 (2002).

21. D. F. James, Meniscus on outside of a small circular-cylinder, *J. Fluid Mech.*, **63**, 657–664 (1974).

22. T. Ondarçuhu and M. Veyssié, Relaxation modes of the contact line of a liquid spreading on a surface, *Nature*, **352**, 418–420 (1991).

23. J. F. Joanny and P. G. de Gennes, A model for contact-angle hysteresis, *J. Chem. Phys.*, **81**, 552–562 (1984).

24. L. Zitzler, S. Herminghaus, and F. Mugele, Capillary forces in tapping mode atomic force microscopy, *Phys. Rev. B*, **66** 155436 (2002).

25. C. Jai, J. P. Aime, D. Mariolle, R. Boisgard, and F. Bertin, Wetting an oscillating nanoneedle to image an air-liquid interface at the nanometer scale: dynamical behavior of a nanomeniscus, *Nano Lett.*, **6**, 2554–2560 (2006).

26. C. Jai, J. P. Aimé, and R. Boisgard, Dynamical behavior of an evaporating nanomeniscus: a boundary condition problem at the local scale, *EPL*, **81**, 34003 (2008).

27. D. B. Asay and S. H. Kim, Effects of adsorbed water layer structure on adhesion force of silicon oxide nanoasperity contact in humid ambient, *J. Chem. Phys.*, **124** (2006).

28. X. D. Xiao and L. M. Qian, Investigation of humidity-dependent capillary force, *Langmuir*, **16**, 8153–8158 (2000).

29. L. Xu, A. Lio, J. Hu, D. F. Ogletree, and M. Salmeron, Wetting and capillary phenomena of water on mica, *J. Phys. Chem. B*, **102**, 540–548 (1998).

30. M. Kober, E. Sahagun, P. Garcia-Mochales, F. Briones, M. Luna, and J. J. Saenz, Nanogeometry matters: unexpected decrease of capillary adhesion forces with increasing relative humidity, *Small*, **6**, 2725–2730 (2010).

31. J. F. Oliver, C. Huh, and S. G. Mason, Resistance to spreading of liquids by sharp edges, *J. Colloid Interface Sci.*, **59**, 568–581 (1977).

32. L. Fabie, H. Durou, and T. Ondarçuhu, Capillary forces during liquid nanodispensing, *Langmuir*, **26**, 1870–1878 (2010).

33. N. R. Tas, P. Mela, T. Kramer, J. W. Berenschot, and A. van den Berg, Capillarity induced negative pressure of water plugs in nanochannels, *Nano Lett.*, **3**, 1537–1540 (2003).

34. J. K. Jang, G. C. Schatz, and M. A. Ratner, Capillary force in atomic force microscopy, *J. Chem. Phys.*, **120**, 1157–1160 (2004).

35. J. Y. Jang, G. C. Schatz, and M. A. Ratner, How narrow can a meniscus be? *Phys. Rev. Lett.*, **92** (2004).

36. U. Durig, Interaction sensing in dynamic force microscopy, *New J. Phys.*, **2**, 5 (2000).
37. S. Marsaudon, C. Bernard, C. Nguyen, A. M. Bonnot, J. P. Aimé, and R. Boisgard, carbon nanotubes as spm tips: understanding cnt tips mechanical properties and imaging. In eds., B. Bushan and H. Fuchs, *Applied Scanning Probe Methods.* New York: Springer. 137–183 (2009).
38. B. Anczykowski, D. Kruger, and H. Fuchs, Cantilever dynamics in quasinoncontact force microscopy: spectroscopic aspects, *Phys. Rev. B*, **53**, 15485–15488 (1996).
39. J. P. Cleveland, B. Anczykowski, A. E. Schmid, and V. B. Elings, Energy dissipation in tapping-mode atomic force microscopy, *Appl. Phys. Lett.*, **72**, 2613–2615 (1998).
40. R. Garcia and R. Perez, Dynamic atomic force microscopy methods, *Surf. Sci. Rep.*, **47**, 197–301 (2002).
41. E. Sahagun, P. Garcia-Mochales, G. M. Sacha, and J. J. Saenz, Energy dissipation due to capillary interactions: hydrophobicity maps in force microscopy, *Phys. Rev. Lett.*, **98**, 085504 (2007).
42. P. G. de Gennes, Wetting—statics and dynamics, *Rev. Mod. Phys.*, **57**, 827–863 (1985).
43. D. Bonn, J. Eggers, J. Indekeu, J. Meunier, and E. Rolley, Wetting and spreading, *Rev. Mod. Phys.*, **81**, 739–805 (2009).
44. M. Delmas, M. Monthioux, and T. Ondarçuhu, Contact angle hysteresis at the nanometer scale, *Phys. Rev. Lett.*, **106**, 136102 (2011).
45. H. Allouche and M. Monthioux, Chemical vapor deposition of pyrolytic carbon on carbon nanotubes. Part 2. Texture and structure, *Carbon*, **43**, 1265–1278 (2005).
46. A. H. Barber, S. R. Cohen, and H. D. Wagner, Static and dynamic wetting measurements of single carbon nanotubes, *Phys. Rev. Lett.*, **92**, 186103 (2004).
47. M. M. Yazdanpanah, M. Hosseini, S. Pabba, S. M. Berry, V. V. Dobrokhotov, A. Safir, R. S. Keynton, and R. W. Cohn, Micro-Wilhelmy and related liquid property measurements using constant-diameter nanoneedle-tipped atomic force microscope probes, *Langmuir*, **24**, 13753–13764 (2008).
48. X. M. Xiong, S. O. Guo, Z. L. Xu, P. Sheng, and P. Tong, Development of an atomic-force-microscope-based hanging-fiber rheometer for interfacial microrheology, *Phys. Rev. E*, **80**, 061604 (2009).
49. S. M. M. Ramos, E. Charlaix, A. Benyagoub, and M. Toulemonde, Wetting on nanorough surfaces, *Phys. Rev. E*, **67**, 031604 (2003).

50. T. Ondarçuhu and A. Piednoir, Pinning of a contact line on nanometric steps during the dewetting of a terraced substrate, *Nano Lett.*, **5**, 1744–1750 (2005).

51. M. J. Pender, L. A. Sowards, J. D. Hartgerink, M. O. Stone, and R. R. Naik, Peptide-mediated formation of single-wall carbon nanotube composites, *Nano Lett.*, **6**, 40–44 (2006).

52. R. R. Naik, L. L. Brott, S. J. Clarson, and M. O. Stone, Silica-precipitating peptides isolated from a combinatorial phage display peptide library, *J. Nanosci. Nanotechnol.*, **2**, 95–100 (2002).

53. T. D. Blake, The physics of moving wetting lines, *J. Colloid Interface Sci.*, **299**, 1–13 (2006).

54. A. Prevost, E. Rolley, and C. Guthmann, Thermally activated motion of the contact line of a liquid He-4 meniscus on a cesium substrate, *Phys. Rev. Lett.*, **83**, 348–351 (1999).

55. E. Raphaël and P. G. de Gennes, Dynamics of wetting with nonideal surfaces—the single defect problem, *J. Chem. Phys.*, **90**, 7577–7584 (1989).

Chapter 10

Wetting in Carbon Inorganic and Organic Nanotubes and Nanochannels

Hannah Leese and Davide Mattia*
Chemical Engineering, University of Bath, Claverton Down, Bath, BA27AY, United Kingdom
*d.mattia@bath.ac.uk

10.1 Introduction

How a liquid wets a surface continues to fascinate scientists 200 years after Thomas Young laid down the foundation of this branch of science with his seminal equation.[1] While fundamental debates about the exact nature of the forces involved in the wetting process are still ongoing (as can be seen in other chapters in this book—chapter 3), the possibility of manipulating and directly observing liquid behavior under nanometer-scale confinement has opened an exciting new chapter in the study of wettability. This renewed interest has coincided with the availability of carbon nanotubes (CNTs), which constitute an ideal self-contained platform for wettability studies. These studies have been later extended to other kinds of nanotubular or fabricated nanochannel structures, as they have become available.

Nanoscale Liquid Interfaces: Wetting, Patterning, and Force Microscopy at the Molecular Scale
Edited by Thierry Ondarçuhu and Jean-Pierre Aimé

ISBN 978-981-4316-45-3 (Hardcover), 978-981-4364-48-5 (eBook)
www.panstanford.com

This review will first briefly reintroduce the fundamental equations associated with wetting phenomena and how they are affected by nanoscale confinement and subsequently review both modeling and experimental results of wetting in and outside nanotubes and nanochannels.

In 1805 Young established a relationship between the contact angle of a drop of liquid and the interfacial tensions at the three-phase contact line between a solid, a liquid, and its vapor at equilibrium (Eq. 10.1). J. Willard Gibbs in 1928 then derived the contact angle θ, from thermodynamic quantities, the surface free energies for the three interfaces.[2] Substantial refinements in terms of the nature of the forces involved in the wetting process have been made since then, and interested readers can consult fundamental texts on surface forces.[3,4]

$$\cos\theta = \frac{\gamma_{\mathrm{SV}} - \gamma_{\mathrm{SL}}}{\gamma_{\mathrm{LV}}} \tag{10.1}$$

where γ_{SV}, γ_{SL} and γ_{LV} represent the interfacial surface tensions of the solid–vapor, solid–liquid, and liquid–vapor interfaces, respectively.

The constraints for the substrate characteristics of Young's equation have led to the development of several modified versions to account for surface roughness and topographical or chemical inhomogeneities. Two models are commonly used to evaluate the effect of these surface characteristics on contact angles,[5] the Wenzel[6,7] model (Eq. 10.2) and the Cassie-Baxter[8] model (Eq. 10.3). The Wenzel model accounts for surface roughness by introducing a proportional factor between the measured contact angle, θ', and the thermodynamic one derived by Young for a flat, smooth surface, θ

$$\cos\theta' = r\left(\frac{\gamma_{\mathrm{SV}} - \gamma_{SL}}{\gamma_{\mathrm{LV}}}\right) = r\ \cos\theta. \tag{10.2}$$

The surface roughness, r, is defined as the ratio of the actual surface area to the horizontal projection of the surface area.[6] In the Cassie-Baxter model, the interaction of the fluid with the air trapped in the gaps of rough (or porous) surfaces is considered alongside the liquid–solid interaction. The effect is strongest in the case of water, which forms a contact angle of 180° with air. The Cassie-Baxter model introduces a surface fraction parameter, f, which represents

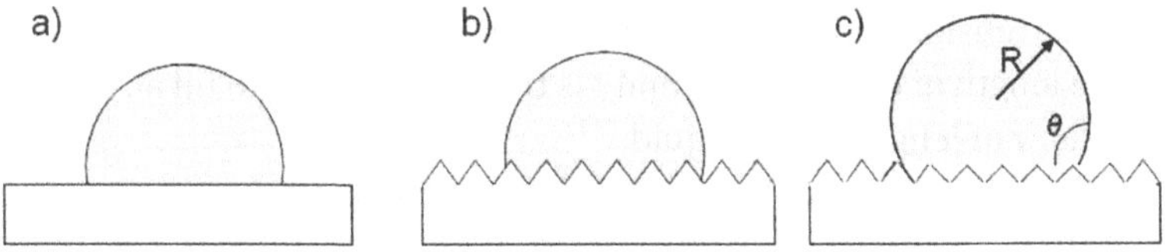

Figure 10.1 Schematic of a drop partially wetting (a) a smooth, defect-free solid (Young's equation), (b) a homogenous, rough surface (Wenzel's equation), and (c) a heterogeneous, rough surface (the Cassie-Baxter equation).

the unit area of the wetted surface:

$$\cos\theta = f\ \cos\theta + (1 - f)\ \cos\ 180^\circ = f\ \cos\theta + f - 1 \quad (10.3)$$

A schematic of a drop partially wetting a solid surface in the case of a solid in the Young, Wenzel, and Cassie-Baxter states is shown in Figs. 10.1a, 10.1b, and 10.1c, respectively.

Similarly to wetting, the flow of liquids through nanotubes has renewed interest in the area of capillary flow. The fundamental equations used to describe liquid flow in capillaries and tubular structures are the Young-Laplace equation (Eq. 10.4), the Hagen-Poiseuille equation (Eq. 10.5), and the Washburn equation (Eq. 10.6). The first describes the capillary pressure difference across the interface between two fluids (such as at the air–water interface). The second describes the well-developed, laminar flow of an incompressible fluid in a capillary under pressure alone. A simplified version of the third is used to describe the filling of an empty capillary by capillary pressure only, as a function of the wettability of the liquid:

$$p = \frac{2\gamma_{LV}}{R}\ \text{Air/water Interface} \quad (10.4)$$

$$Q = \frac{\pi \Delta P R^4}{8\mu L} \quad (10.5)$$

$$L^2 = \left(\frac{R\gamma_{LV}\cos\theta}{2\eta}\right) t \quad (10.6)$$

where R is the radius of the capillary, ΔP is the pressure across the liquid–vapor interface, Q is the flow rate, η is the viscosity, γ_{LV} is the surface tension at the liquid–vapor interface, θ is the contact angle, L is the length of the capillary, and t is the time elapsed to fill an empty capillary of length L by a liquid.

The Hagen-Poiseuille equation reported above assumes a no-slip condition at the capillary wall, that is, the molecules in the fluid layer in contact with the capillary wall have zero velocity. It should be noted that this is an approximation and there is no reason why particles at the wall should not have a finite velocity. The nature and very existence of slip is an intensely debated topic and is covered in detail elsewhere in this book. Several fluid flow experiments in nanometer channels showing huge slip lengths will be described later in this chapter. In these experiments, the slip length λ has been calculated via the Hagen-Poiseuille equation with a slip wall boundary condition:

$$Q_{\text{avg}} = \frac{\pi \Delta P R^4}{8\eta L}\left(1 + \frac{4\lambda}{R}\right) \tag{10.7}$$

A diagram of the no-slip/slip boundary condition can be seen in Fig. 10.2.

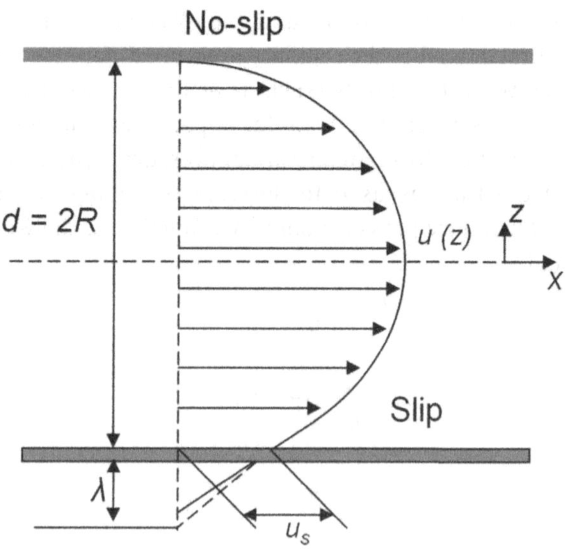

Figure 10.2 Shape of a meniscus inside a nanochannel representing slip and no slip.

It is important to note that although surface tension and a contact angle are not explicitly found in the Hagen-Poiseuille equation, there is substantial evidence that the wettability of the channel wall has an effect on slip.[9,10,11] Several models linking the slip length to the tube wall and liquid characteristics have been suggested,[12,13] though there is not yet enough experimental evidence to convincingly back up these models. Indeed, the interpretation of large slip values observed in certain experiments or theoretical models is complicated by the potential influence of a large number of parameters, such as the surface chemistry and structure of the channel, the type of fluid confined within the channel (i.e., polar or nonpolar, its viscosity), and the height or diameter of the channel. A study that considered what influences slip looked at how the height of a photoresist-fabricated channel (varying from 2.7 μm–40 nm) affected the slip length and whether different fluids would have different slip lengths at these scales.[14] Four different fluids were investigated to evaluate how the slip observed experimentally compared with the theoretical model (Eq. 10.7). Results for all channel heights were consistent with theoretical predictions. Interestingly, the experimental values for slip in 42 nm tall photoresist channels for water was also well predicted by the model (the values of slip length were not given, although an increase in flow rate was recorded), but the slip lengths for the other three fluids (hexane, hexadecane, and decane) deviated from the predicted values. The experimental values of slip length for hexadecane were found to be $\approx$30 nm, for decane $\approx$14 nm, and for hexane $\approx$9 nm. These values need greater consideration, in that the fluid–wall interaction could play a major role in the slip length, so the composition of the walls as well as the type of fluid all have to be considered accordingly. A similar result was observed for the flow of both polar and nonpolar liquids in CNTs.[15]

The effect has been attributed to ordering effects of the liquid molecules at the wall. The slip boundary condition for the Navier–Stokes equation considers the wall surface as a whole, as do the Wenzel and Cassie-Baxter models for wetting. Several molecular dynamics (MD) simulations have looked at local changes of the wall surface chemistry and structure, leading to local increases in the slip length, resulting in a macroscopic larger slip length.[16,17,18,19,20]

The effect of gravity on wetting can be analyzed by computing the capillary length, κ^{-1}, by equating the Laplace pressure and the hydrostatic pressure for a given liquid:

$$\kappa^{-1} = \sqrt{\gamma/\rho g} \quad (10.8)$$

where g is equal to 9.81 m.s^{-2}. Therefore when the radius, r, of the capillary is close to or less than the capillary length, gravity is negligible and surface forces dominate.[21] For water, the capillary length is $\approx$3 mm.

10.2 Wetting at the Nanoscale

From an applications point of view, the macroscopic behavior of a surface is what matters, for example, being hydrophilic or hydrophobic, as well as the possibility of switching between these two behaviors.[22-25] Understanding how individual nanostructures, perhaps composed of different materials, are wet at nanometer scale is the key to obtaining surfaces with controlled wetting behavior.[26] As recalled in the introduction, the appearance of CNTs in the early 1990s led almost immediately to investigation of the wetting behavior of liquids under nanoscale confinement. In more recent years the attention has shifted to studying the flow of liquids through nanochannels and nanotubes both for theoretical and practical applications.[9,10] Preliminary experimental results have shown that below a certain size threshold, tentatively around 10 nm, the continuum model of fluid mechanics no longer applies and up to five orders of magnitude higher than expected flow velocities have been observed (based on Eq. 10.5).[15,27] Later experiments inside slightly larger CNTs with different surface structures have shown, though, much smaller enhancements of the fluid velocity.[28] Despite their differences, all these studies point to the fact that at this scale, intermolecular forces of the fluid and the molecular interactions at the fluid–solid interface have much stronger effects than at the macroscale. As a consequence, efforts are now being made to investigate the effect of the wettability of nanotube and nanochannel walls on the flow of liquids through them.

Several publications have used MD to investigate the effect of the wettability of nanotube and nanochannel walls on the flow of liquids through them. As in the macroscopic case, the effect of surface roughness and inhomogeneities has been extensively simulated. A first paper looked at the capillary filling by wetting fluids in nanochannels with both smooth and roughened wall structures. Lattice-Boltzmann calculations were used in parallel to MD simulation and showed the formation of a thin precursor film moving ahead of the main capillary front (Fig. 10.3). Simulations suggested that the meniscus profile under nanoscale confinement consistently deformed from the ideal circular meniscus.[29] This distortion has been attributed to the interaction with the boundary surface (the wall) and the disjoining pressure induced by van der Waals interactions between the fluid and the solid.

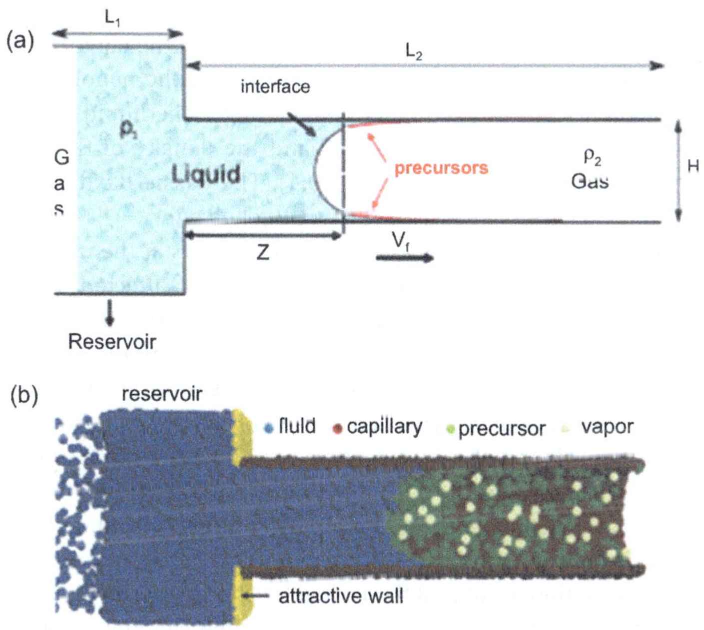

Figure 10.3 (a) Schematic of the geometry used to simulate capillary imbibition of a wetting fluid in smooth and rough capillaries; (b) snapshot of MD simulation showing fluid imbibition and the formation of thin precursor films ahead of the capillary front. (with permissions from ref.[29])

The same authors also investigated the effect of hydrophilic and hydrophobic rough channel walls on flow and found that the flow rate was higher for rough hydrophilic than smooth hydrophobic surfaces.[29] This result runs contrary to the general belief that hydrophilicity and roughness both reduce the flow velocity by increasing drag[12] (see Fig. 10.5). The higher flow velocity has been attributed to the presence of the precursor film, which would shield the wetting field from the rough surface. It is necessary to note here that the authors consider "hydrophobic" to be a surface wet by a liquid forming a contact angle of 30° to 40°. MD simulations of flow through nanotubes and nanochannels typically use much higher values of the contact angle for hydrophobic surfaces.[30–33] In addition, the roughness was simulated by adding a single rectangular obstacle inside the channel whose height can be varied to observe the effect of roughness on the meniscus.

On the other hand, MD simulations of the wetting of liquid argon droplets in 10 nm platinum channels showed that the contact angle increased from 36° $\approx$ 150° when the surface of the nanochannels was roughened.[34] For the higher contact angle, the liquid filled all the gaps in the nanostructured wall and the density of the liquid decreased near the surface[34] (Fig. 10.4). The Cassie-Baxter model (Eq. 10.3) was used to describe the wetting behavior.

Another MD study considered how a fluid wets at the nanoscale nanopatterned surfaces and how the flow is affected by the hydrophobicity of a surface and the density of the fluid.[35] The simulation considered a dense (with Reynolds number, Re $\approx$3) and a rarefied fluid (Re $\approx$8) nanoscale flow within attractive and repulsive nanochannels. The channel walls were built from two [001] lattice planes of a face-centered cubic lattice, with the left side of the channel being the attractive end (hydrophilic) and the right side of the channel being the repulsive half (hydrophobic). For the dense fluid interacting with the attractive half of the channel, the nanoscale flow exhibited the Poiseuille character, whereas when the dense fluid was in the repulsive half of the channel, plug flow was observed and, surprisingly, the peak velocity was less than that of the peak velocity of the Poiseuille nanoscale flow. This result appears to be in contrast with the extremely high fluxes observed in CNTs with diameters <10 nm[15,27] whose surface can be considered

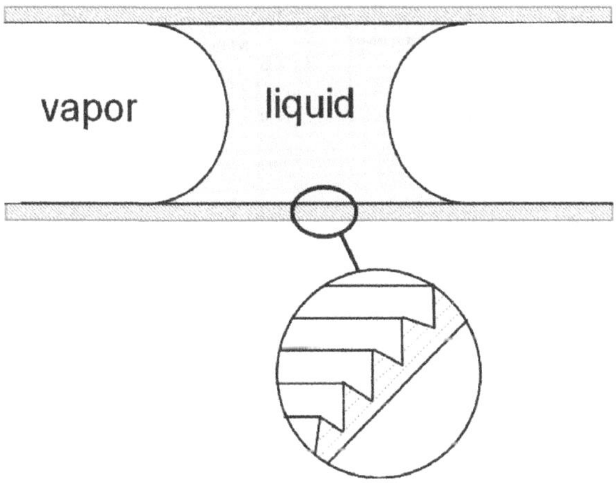

Figure 10.4 Vapor and liquid inside the nanochannel (meniscus shown) and a magnified view of the nanostructured nanochannel.

relatively hydrophobic, although, as discussed elsewhere,[9] water forms a contact angle on graphite in the 80°–86° range—below the 90° threshold between hydrophilic and hydrophobic surfaces.[36]

In the case of the rarefied nanoscale flow, the plug flow velocity was higher than the peak velocity of the Poiseuille flow, in line with the experimental results mentioned above. An explanation of the Poiseuille character observed for the attractive channel walls (to the fluid) is that the wall offers a mechanism for energy dissipation, resulting in a lower flow velocity at the channel walls, whereas with the repulsive walls there is less fluid–wall interaction and therefore less occurrence of energy dissipation. This is in agreement with models relating fluid flow to contact angle and surface roughness.[12,13,37] A similar result has been also obtained simulating water flow in a CNT with Lennard-Jones (LJ) parameters mimicking a hydrophilic surface.[30]

A broader study compared the effect of different surface chemistry and roughness on water flow (and thus wettability) in a CNT, a boron nitride nanotube (BNNT), a silicon nanotube, and a roughened nanotube.[30] Figure 10.5 expresses the axial velocity profiles of each nanotube. The simulations found that there was a plug-like velocity flow profile with a "jump" in velocity at the solid–

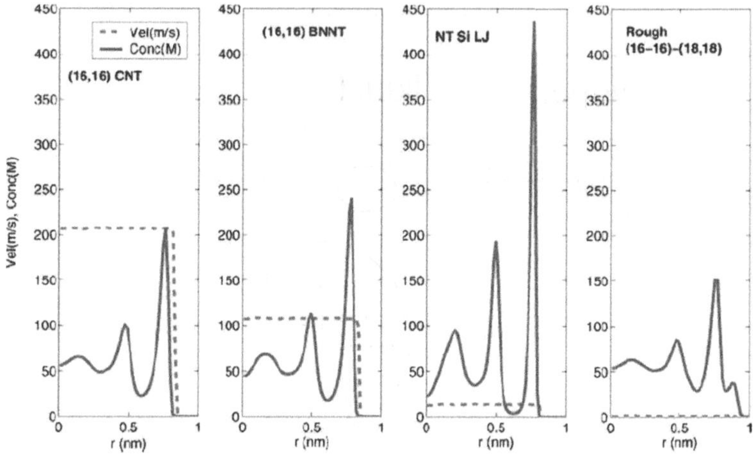

Figure 10.5 Water concentration and axial velocity profiles as a function of the radial distance toward the wall from the center of the tube at $r = 0$ nm. Only one-half of each tube cross section is shown. From the left to the right are (i) a (16, 16) CNT, (ii) a (16, 16) BNNT, (iii) a (16, 16) nanotube with Si LJ, and (iv) a rough tube with alternating sections of (16,16) and (18, 18) tubes. For the first three cases, even though the tubes have smooth surfaces, the magnitude of the flow velocities show marked differences. The bulk concentration is 55 M. (with permissions from ref.[30])

fluid interface. This was observed with all the nanotubes except the rough structured nanotube, and the largest enhancement was observed within the (16, 16) CNT, for which a hydrophobic structure is assumed. As will be discussed in section 10.3.1, this is a debatable assumption and the choice of specific values for the LJ interaction potentials can dramatically alter the result of MD simulations.[9] The velocity "jump" has been attributed to nanobubbles forming on hydrophobic surfaces, resulting in ballistic flow profiles.[38] Another study, though, used X-ray reflectivity to investigate the interaction of water with a smooth hydrophobic surface and observed the formation of a low density depletion layer (of about one water molecule thickness) at the wall but no nanobubbles segregating at the surface.[39] MD simulations of dewetting of films on smooth surfaces found that not only the mobility of the liquid molecules adjacent to the solid increases with decreasing solid–liquid interactions, but also the mobility adjacent to the free surface of the film is higher than in the bulk so that the effective viscosity of the film decreases with thicknesses below 50.[40–43]

MD simulations of the wetting of hydrophobic substrates with surface inhomogeneities displayed a similar behavior as hydrophobic rough surfaces, with high contact angles ($>130°$ for fluid comprised of "small-chain" molecules).[35] Upon application of a lateral push, the droplet detached from the surface, akin to the lotus effect but at the nanoscale. This was not observed with the simulations with a homogenous surface.

Nonequilibrium molecular dynamics (NEMD) simulations have been carried out to observe the effect on the slip boundary condition[12] in wetting at the nanoscale:[44] Simulations showed that for high contact angles ($>140°$), the slip length can be as high as 30 molecular diameters ($\approx$5 nm for mercury on glass), whereas it is negligible for low contact angles.[44]

10.2.1 *Liquids under Nanoscale Confinement*

One of the most debated aspects of the behavior of liquids under nanoscale confinement is their phase. Most results on this topic have been obtained via MD simulations due to the complexity of obtaining direct experimental evidence.[9,10] Numerous MD results suggest that water would form hydrogen bond networks at room temperature and physiological conditions,[45] as well as at high pressure,[46] giving rise to complex geometries from square to heptagonal ice nanotubes (Fig. 10.6).[47,48]

The few experimental results available appear to be in agreement with the above picture: Pressure-induced formation of water "wires" in zeolite channels has been observed,[49] and helical chains of an aqueous solution in CNTs were also observed by transmission electron microscopy (TEM).[50] The transition from liquid to an amorphous bilayer above room temperature in 1 nm wide hydrophobic pores has also been simulated.[51] On the other hand, MD results have been reported with water molecules forming an ordered layer in the region near the wall and showing no order along the axis of the nanotube, both in hydrophilic and in hydrophobic channels.[52] The addition of CO_2 was shown to induce layering along the walls of a

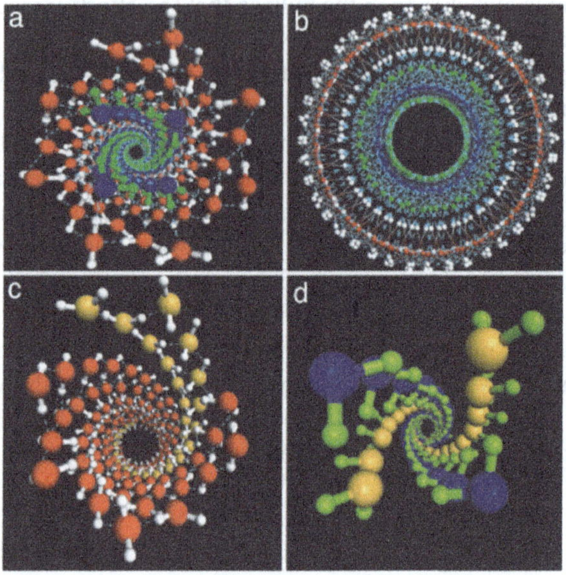

Figure 10.6 Snapshots of quenched molecular coordinates of the helical nanoice form in a (17, 0) CNT at 4 GPa axial pressure. (a) Top view of the double-walled nanoice helix in the axial direction. Water molecules in the outer wall are in red-white, whereas those in the inner wall are in blue-green (the blue dashed lines denote hydrogen bonds). (b) Projected top view in the axial direction. Due to helicity, the projected top view shows ring-like outer and inner walls. (c) The outer wall: An octuple-stranded helix where two strands (gold) are proton donors and two (blue) are proton acceptors to molecules of the outer. (with permissions from ref.[46])

closed, single-walled nanotube (SWNT) of water molecules with the establishing of liquid–gas equilibrium.[53]

A study of the transport properties of water in smaller tubes with diameters ranging from 1.1 to 2.1 nm suggested strong anisotropy, with axial diffusivity and viscosity being much larger that those in the radial direction, leading water molecules to assume an ordered, helical structure inside a (10,10) nanotube.[54]

X-ray diffraction (XRD) studies showed that a substantial amount of water adsorbed in SWNTs bundles, with a liquid–solid transition at 235 K. Simulation of XRD patterns suggested that water adsorbed primarily inside the tubes rather than in the interstitial channels of the bundle.[55] Comparison of the vibrational and rotational spectra of water confined in SWNTs and in bulk was performed, showing a significant shift of the absorption bands as a function of the pore diameter in which water is confined.[56,57] Simulations also suggest the presence of a new peak around 3,650 cm^{-1} attributed to the splitting of the symmetric and antisymmetric components of the stretching vibration of liquid water.

The findings presented above all imply that below a certain size threshold, liquids will modify their phase and/or flow characteristics due to nanoscale confinement. The critical dimension below which this occurs is currently debated.[10,58] Fundamental studies on surface forces under confinement offer some clues as to where to place such a threshold.[3,59] When a polar liquid comes in contact with a hydrophobic surface, the system dries up because of the energy cost of maintaining a hydrogen bond network close to a large hydrophobic surface below a critical dimension (or distance). The energy involved in the drying process is proportional to the surface area through the surface tension. This energy is balanced by the favorable bulk free energy proportional to the difference of the chemical potentials of the liquid, μ_l, and gas, μ_g, phases. So the critical dimension for breakdown of the system is.[60]

$$D_c = \frac{2\gamma_{LV}}{\left(n_l \left|\mu_l - \mu_g\right|\right)} \tag{10.9}$$

where n_l is the average number of liquid molecules per unit volume and D has the dimension of a length. For water molecules between two flat, hydrophobic surfaces at room temperature and atmospheric pressure, $D_C \sim 100$ nm.[60] Below this threshold, water molecules tend to form a depletion layer near the surface, due to the breakdown of the hydrogen bond network, and eventually dry up. This model conforms to the formation of a depletion layer near the wall of a hydrophobic CNT of water molecules flowing through it (compare Fig. 10.5). It is not clear, though, whether the formation of a depletion layer would also occur in nanotubes with diameters up to the value of $D_c \approx 100$ nm reported previously, nor if there is a difference between water in contact with a flat or a curved surface.

Experiments on capillary filling of molten metals in 0.6–1.2 nm channels in zeolites show a much smaller threshold closer to 1 nm.[61] This value was obtained considering that interphase transition layers between the liquid touching the wall and the bulk along the axis of the channel have typical thickness, $h \sim 0.15$ nm, comparable with the diameter of the capillary, D. As a result, measuring the effective diameter available for filling, $d = D - 2h$, and the critical pressure, P_c, meant that it was possible to determine the effective surface tension.[61]

$$\gamma_c = P_c \frac{d}{4} \tag{10.10}$$

This value differed from the "macroscopic" surface tension for each metal, and the difference has been attributed to the effect of the presence of the interface transition layers:

$$\gamma_c = P_c \frac{d}{4} + P_c \frac{\delta_c}{2} \tag{10.11}$$

where $\delta_c \approx h \sim 0.15$ nm was obtained experimentally. This value is about one order of magnitude larger than the so-called Tolman length,[62,63] which accounts for the effects on surface tension of curvature of the liquid–solid–vapor interface with a size comparable to that of characteristic atomic dimensions. Therefore, the authors suggest that the influence of curvature alone on the wetting properties is negligible compared to the effect of the interphase layer pressure.[61]

From the mechanical point of view, a liquid confined to a few molecules in thickness will exhibit a response both to compression and to tangential shear typical of a solid.[64] For a dodecane droplet between two mica surfaces, this happens for thicknesses ≤ 5 nm, with the molecules aligned preferentially parallel to the boundaries. When considering a flow of dodecane through mica surfaces, viscosity measurements showed that the relaxation time in confinement is much slower than in bulk liquid; thus the liquid reacted as a solid to mechanical solicitations.

10.3 Wetting of Liquids Inside Carbon Nanotubes

As discussed earlier, infiltration of molten metals and salts was used to prove that CNTs were indeed hollow, soon after their discovery in the early 1990s.[65–67] Renewed interest in the wetting and filling behavior of CNTs with low-molecular-weight fluids, such as water, occurred in the early 2000s with the first direct images of water confined inside CNTs (Fig. 10.7). Readers interested in a historical review of the first wetting experiments in CNTs can consult two well-known reviews on this topic.[9,10]

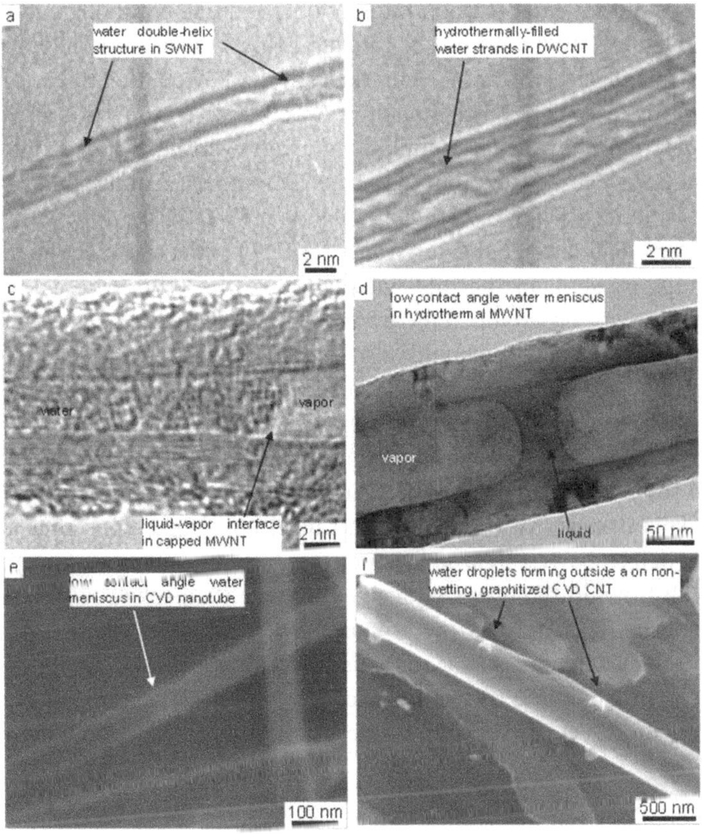

Figure 10.7 TEM and SEM images of water inside CNTs. (a, b) TEM micrographs of double-helix water strands in (a) an SWNT and (b) a DWNT. (c) TEM micrograph of the interface between water molecules and vapor in a capped MWNT. (d) TEM micrograph of a low-contact-angle liquid meniscus between 2 vapor bubbles in a hydrothermal MWNT. (e) A water-filled CVD MWNT in the environmental SEM. (f) Water droplets forming outside a graphitized CVD nanotube in the environmental SEM. (with permissions from ref.[50]) *Abbreviations*: DWNT, double-walled nanotube; MWNT, multiwalled nanotube.

MD simulations have so far played a vital role in understanding wetting properties of individual tubes and CNT arrays or membranes. A key— and much debated—aspect of MD simulations is the choice of the water–carbon interaction potential. A seminal study has correlated the wetting of water on graphite with different values of the water–carbon interaction potential, obtaining contact angle values varying from 0° to over 140°.[31] Clearly not all values have a physical meaning, but this study showed the risks (and rewards) associated with modeling. A well-known example is a 2001 paper that showed MD simulations of nonwetting water droplets inside a CNT.[32] A few months earlier an experimental paper was published, showing a water plug inside a CNT with a low contact angle.[68] The latter behavior was a result of hydrophilic functional groups present on the internal surface of the tube, whereas the former was based on a hydrophobic behavior of the graphitic wall of the CNT. It should also be noted that graphite is not actually hydrophobic, with a macroscopic contact angle below 90°.[9] Interested readers are referred to an extensive review considering the scope of computational simulations of the thermal, mechanical, and transport properties of CNTs using a wide range of fluids.[69]

This first group of results has raised novel questions that are now being answered through experiments and simulations (see Fig. 10.7): How do the water molecules arrange themselves within the nanotube, and what is their phase? What happens to the water molecules when the surface chemistry of the carbon wall is changed (from hydrophobic to hydrophilic)? Do all these aspects have an effect on the flow of water inside small (sub-10 nm in diameter) CNTs?

10.3.1 *Effect of Wall Structure and Surface Chemistry*

Many of the discrepancies between numerical and experimental results concerning liquid behavior inside CNTs discussed above can be attributed to the presence of surface defects or functionalities on the tube walls, or to the structure of the tube walls.

Systematic analysis of the effects of surface chemistry and structure on the wettability of CNTs has shown that it is possible to control their wettability from hydrophobic to hydrophilic. An example is the hydrophilization of CNTs under boiling sodium hydroxide,[70] with the formation of oxygen-containing functional groups on the carbon surface.[71] X-ray photoelectron spectroscopy (XPS) analysis of nanotubes fabricated following the same procedure shows a strong peak for oxygen and none for sodium; Fourier transform infrared (FTIR) spectroscopy analysis on the same sample also indicated the presence of peaks at 1,580 and 1,680 cm^{-1}, which were attributed to oxygen-containing groups.[72] It should be noted that after these results were first published, a similar treatment with NaOH of carbon films produced by physical vapor deposition yielded the same result in terms of contact angle reduction.[73] The authors, though, reported that successive washes in water brought the contact angle back up to its initial value, therefore suggesting that the increase in hydrophilicity observed with the NaOH treatment is not due to formation of covalent C-O bonds but rather to the formation of salts on the surface of the films. These contact angle measurements can also be used to calculate the polar and dispersion components of the surface tension.[74,75] The results show that the treatment in NaOH does not alter the structure of the carbon but changes substantially its surface chemistry (Fig. 10.8).[9] Readers interested in further details on functionalization and chemical modification techniques of CNTs are referred to a comprehensive review on the topic.[76]

Similar experiments on the contact angle of water on CNTs films annealed at temperatures ranging from 670°C to 2,000°C in high vacuum have shown an increase from 40° to 80° due to graphitization of the carbon structure.[77]

The surface chemistry and structure of nanotubes play a crucial role in their use as reinforcement for polymer matrix composites, determining the effectiveness of load transfer between the matrix and the CNTs.[78,79] The strong affinity of different types of polymers for CNTs has been reported without additional modifications of the CNT/polymer matrix,[80,81] whereas other studies have found that small changes of the matrix chemistry result in a greater adhesion between the CNTs and the polymer.[82] An example of the latter case is given by contact angle measurements of solidified polyvinylidene fluoride (PVDF) nanodroplets on individual CNTs inside an environmental scanning electron microscope (ESEM)

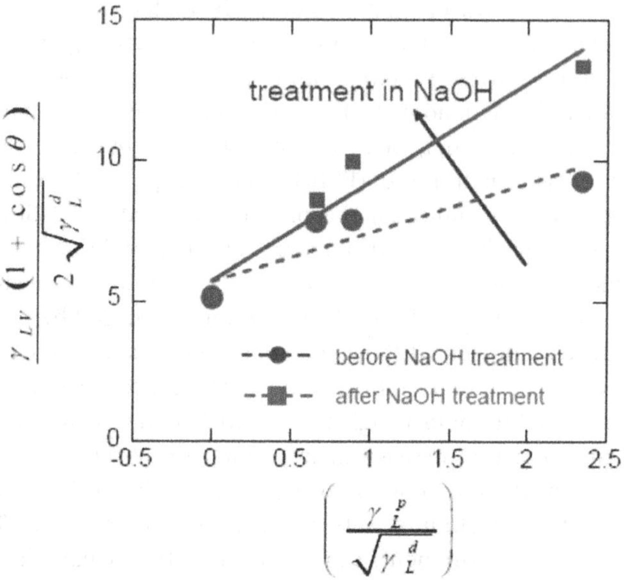

Figure 10.8 Change of the surface structure, γ_S^d, and surface chemistry, γ_S^p, components of the surface tension of CVD CNTs as a result of NaOH treatment. It should be noted that both curves have a common origin for $x = 0$, x being the abscissa. (with permissions from ref.[9])

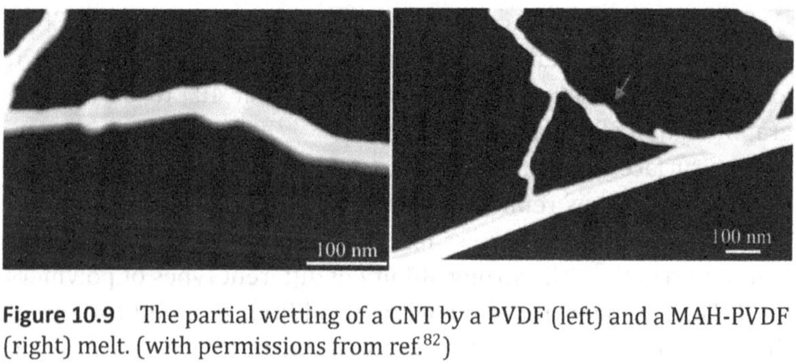

Figure 10.9 The partial wetting of a CNT by a PVDF (left) and a MAH-PVDF (right) melt. (with permissions from ref.[82])

(Fig. 10.9). The addition of a small amount (ranging from 0.5 ppm to 1.25 ppm) of maleic anhydride (MAH) improved the wettability of the polymer onto the CNTs.

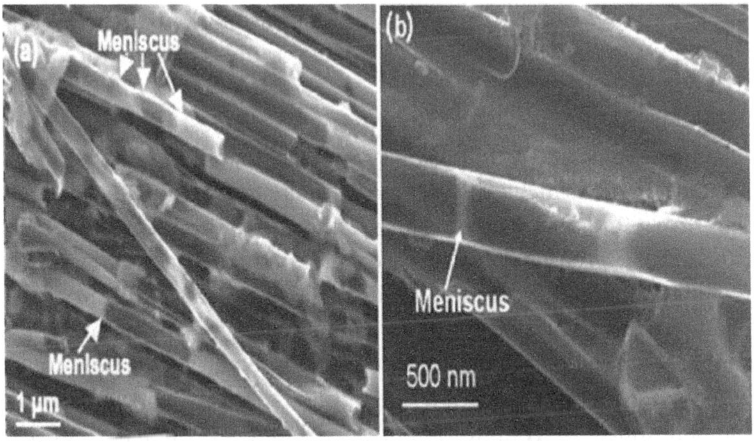

Figure 10.10 (a) ESEM image of LCs inside an agglomerate of CNTs. (b) ESEM image of menisci inside an individual CNT. The lighter phase is the LC plug, while the darker phase is the empty part of the CNT, which is transparent to the electron beam. (with permissions from ref.[83])

Recent studies have also investigated the wetting and flow of liquid crystals (LCs) inside CNTs,[83] showing that the LCs readily filled hydrophilized CNTs, with the formation of menisci with low contact angles (Fig. 10.10). Raman spectroscopy was used to confirm that no chemical reaction occurred between the LCs and the nanotube and that the results of the interactions of the materials were primarily physical. The net interfacial energy of the LCs and CNTs was calculated by considering their respective surface tensions ($\gamma_{LC} = 30\text{–}40$ mN m^{-1} and $\gamma_{CNT} = 41$ mN m^{-1}, the latter value taken from Ref. 70):

$$\gamma_{LC-CNT} = \gamma_{LC} + \gamma_{CNT} - 2\sqrt{(\gamma_{LC}\gamma_{CNT})} - \left(\sqrt{\gamma_{LC}} - \sqrt{\gamma_{CNT}}\right)^2 \tag{10.12}$$

Considering these values, the net interfacial energy was quite low, even when considering the higher value of γ_{LC}, and indicates that the LCs can freely flow inside the CNTs, and therefore there is complete wetting of the tubes by the LCs.[83] It should be noted that the surface tension value used in the above calculations is higher than values in the 25–30 mN m^{-1} range obtained using atomic force microscopy measurements.[84-86] This discrepancy has been attributed to the turbostratic nature of the graphitic structure of the CNTs used in Ref. 83 compared to a more graphitized structure used in the latter case.

NEMD simulations were carried out to investigate the effect of temperature and wall elasticity on capillary filling in SWNTs.[87] Simulations in armchair SWNTs with diameters ranging from D = 0.81 to 2.69 nm showed that increasing the thermal energy inside the tubes decreases the pressure barrier to capillary filling. The flexibility of nanotubular structures can also have a significant effect on the wetting and flow of fluids inside them. MD simulations showed that gases and liquids can be pumped at higher flow rates when oscillations or vibrations are applied to the external surface of the nanotube.[88–90] A limitation of these models is the length of the system compared with the diameter. In fact, in this study the length of the nanotube was only six times that of the diameter of the tube, whereas typical aspect ratios for commercial SWNTs start at about 1,000 up several orders of magnitude.

10.3.2 *Effect of Diameter*

To examine the role of tube diameter on their wettability, MD simulations of filling of SWNTs with different diameters—(6,6), (8,8), and (10,10), cfr[91] for SWNT nomenclature—with liquid isotopes were performed to investigate confinement effects.[57] Spectral densities of oxygen, hydrogen, and deuterium were computed along the tube length. Results showed a shift in the vibration frequencies of light and heavy water molecules from their bulk values, with the strongest deviation observed for the smallest tube. Subsequent simulations[92–95] and some experimental data[96–98] have supported these findings.

ESEM has been successfully used to observe the formation of water plugs inside CNTs[99,100] from condensation of the vapor present in the microscope chamber. It was found that for a fixed temperature, as the water vapor pressure was increased, water first formed a thin film on the walls of the CNT, which subsequently thickened to form a plug (Fig. 10.11).

Interestingly, the onset of water condensation inside the CNTs occurred at a lower than expected vapor pressure (5.6 Torr instead

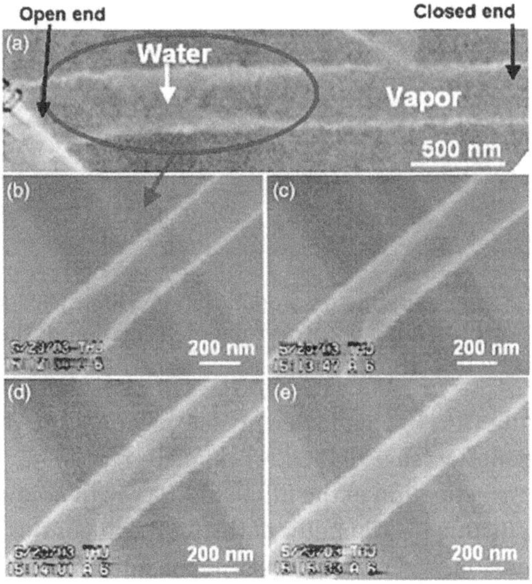

Figure 10.11 (a) High-magnification ESEM image of a liquid plug inside a CNT. (b–e) ESEM images showing how, upon an increase in water vapor pressure, the water begins to form the plug seen in (a). (with permissions from ref.[100])

of 6.2 Torr, at 4°C). A possible explanation could be the effect of the tube curvature on the vapor pressure (cfr. Eq. 10.4). Further analysis of the above results showed that the apparent diameter of the nanotubes varied as a function of the partial pressure in the ESEM chamber:[96] Upon formation of a liquid plug inside the CNT at 5.6 Torr the CNT diameter decreased from when it was empty. As the pressure was increased from 5.6 Torr to 6.7 Torr the CNT diameter increased again to return to the minimum value when the pressure was returned to 5.6 Torr (Fig. 10.12). The increase in diameter was explained by considering Lévy's classical analysis of pipe buckling and including the Kelvin theory of capillary condensation. The initial decrease in diameter was attributed to the formation of the thin liquid film formation and the elasticity of the nanotube below the saturation vapor pressure. The thickness of the thin liquid film is related to the disjoining pressure necessary to maintain the thickness of the film when the pressure is decreasing

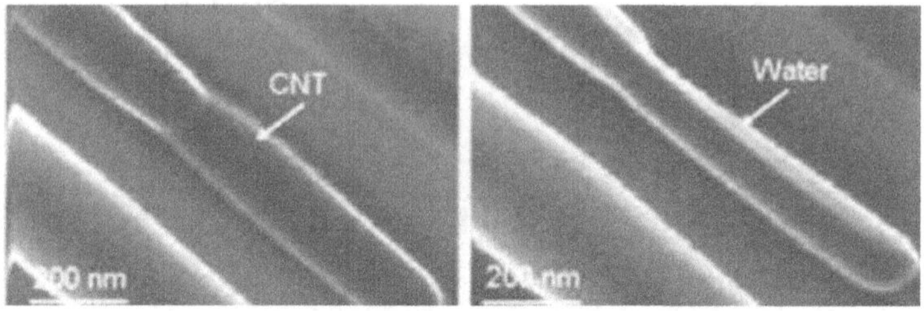

Figure 10.12 (Left) ESEM before condensation. (Right) ESEM after condensation vapor pressure 5.6 Torr; visible deformation of the CNT. (with permissions from ref.[96])

within the tube.[101,102] This lower pressure causes the deformation and subsequent decrease of the CNT diameter.

A recent investigation of the filling of CNTs with nonpolar liquids found that reducing the diameter of the CNT reduces the solid liquid interfacial tension.[103] From theoretical considerations, the authors calculated the surface tension of the CNT to be 50 mN m^{-1}, close to experimental values found in literature (28–45 mN m^{-1}).[70,86] A transition between Poiseuille laminar flow and ballistic flow inside a CNT below a $\approx$7 nm diameter threshold was also obtained from theoretical calculations.

With their particular structure, CNTs have been the ideal test bed to directly investigate liquid behavior under confinement. Some of these findings are now being tested for applications, in particular in the area of water filtration and desalination.[104]

10.4 Wetting in Metal Oxide and Ceramic Nanochannels

While CNTs have been the first materials whose wetting behavior has been extensively investigated, the more recent availability of nanochannels of different compositions has renewed the interest in wetting at nanometer scale.

The wetting behavior of water in hydrophilic silica nanochannels with an approximate height of 100 nm produced in a polysilicon wafer using the sacrificial layer method was investigated.[105] The water meniscus had an abnormal shape (see Fig. 10.13c,d), and there was a significant pressure drop. A model, based on the Laplace

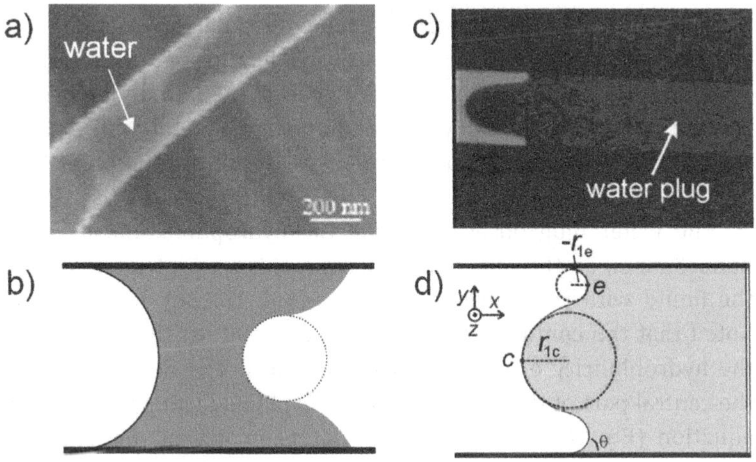

Figure 10.13 (a) ESEM micrograph and (b) schematic representation of a water plug inside an individual CNT.[99] (c) Top view of a water plug in a silica rectangular nanochannel (scale bar: 10 μm; height of channel: 100 nm). (d) A schematic of the meniscus curvature. (with permissions from ref.[105])

equation (Eq. 10.4), was used to predict the negative pressure by considering the tensile capillary pressure (from the change in surface free energy by the wetting of the walls). The equation predicted a capillary pressure of 13 bar (from values of $h = 108$ nm and $\theta = 18°$), which was close to that of the experimental results. Overall this gave a negative pressure of -12 bar (ambient pressure ≈ 1 bar).[106]

A comparison of the previous images with those for water plugs inside CNTs (cfr. Fig. 10.11) shows a substantial difference in terms of the shape of the meniscus, which is most probably driven by the different surface chemistry of the nanotubes. In the case of the CNT, asymmetric menisci shapes have been observed (Fig. 10.13a), with a regular meniscus shape at the open side of the tube (left in Fig. 10.13a) and an irregular meniscus deformed inward at the close end of the nanotube (right in Fig. 10.13b).[99] This asymmetry has been attributed to the difference in pressure between the open and closed ends of the CNT. It should be noted that after the system stabilized, the right-hand-side meniscus formed a more regular shape. The menisci shapes described above have been observed also elsewhere.[100,107] In particular, the same irregular meniscus shape was obtained upon electron beam–assisted heating of the liquid plug inside the CNT. After 120 seconds of heating the liquid plug broke, leading to the formation of two thick liquid films on the wall, akin to the effect of reducing the pressure below the saturation point (cfr. Fig. 10.11).

The water plug observed inside the hydrophilic silicon oxide nanochannel, on the other hand, has a meniscus protruding out of the liquid with curvature at the walls (Fig. 10.13c). It should be noted that the contact angle remains below 90° at the wall, given the hydrophilicity of the silica channel walls. The deformation of the central part of the meniscus can be explained using the Laplace equation (Eq. 10.4) and the particular geometry of the channel. Experiments in silica channels coated with a hydrophobic organic layer showed the presence of a concave meniscus in opposition to the convex ones observed for the hydrophilic case (Fig. 10.13).[108] These results were confirmed by investigating the wetting of a nanosized silicon grating ($\approx$16 nm pitch) using small angle X-ray scattering.[109] The X-ray scattering profiles verified that the Young-Laplace equation was valid down to $\approx$10 nm. Below this value hysteresis effects substantially affected the ability to accurately quantify the results. For a detailed description of the influence of channel deformation on the shape and dynamics of spreading, refer to chapter 11.

Capillary filling of Vycor glass with a three-dimensional network of interconnecting silica nanopores ($\approx$10 nm in diameter) by water has recently been observed (Fig. 10.14).[110] Infrared (IR) spectroscopy confirmed the presence of silanol (SiOH) groups on the pore surface, leaving the material hydrophilic with a contact angle close to 0°. When put in contact with water, a monotonic mass increase was observed. The time behavior of the imbibition of water into the porous monolith followed the Lucas–Washburn law (Eq. 10.6). An earlier study also used hydrophilic nanoporous silica (Vycor, mean pore diameter $\approx$7 nm) to study the spontaneous capillary rise of fluids (two long-chain alkanes) at the nanoscale.[111] Also in this case there was no breakdown of the classical hydrodynamic flow of the two liquids in the nanoporous silica.

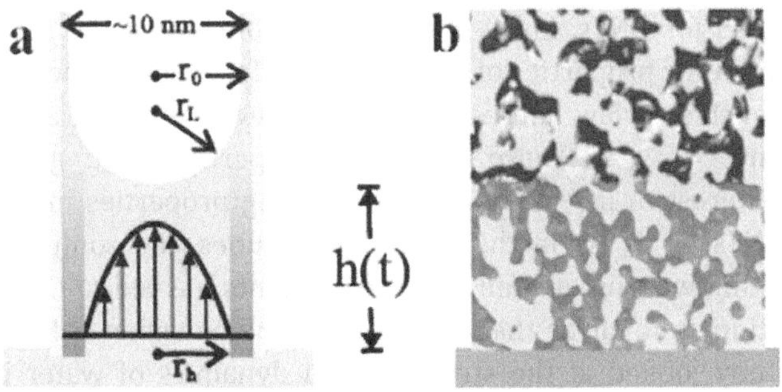

Figure 10.14 (a) Schematic side view of the capillary rise of water in comparison with a ray-tracing illustration of spontaneous water imbibition in Vycor. (b) A liquid column has advanced up to the height $h(t)$. A parabolic fluid velocity profile along with preadsorbed water layers beyond $h(t)$ and a shaded resting boundary layer are sketched for the nanocapillary in panel (a). (with permissions from ref.[110])

The effect of channel diameter (in the range 30–170 nm) on the wettability of silica nanochannels using a range of water/methanol mixtures was investigated.[108] The results showed a transition from a wetting to nonwetting regime when a cosolvent is added, leading to a failure of the Young-Laplace equation to accurately predict the mol fraction where the transition occurs. This occurred for all channel diameters investigated with no substantial difference from the macroscopic case. Surprisingly, the mol fraction of methanol for which channel filling occurred was about 10 times higher than what predicted by the Young-Laplace equation. Certain factors might have contributed to such a high value, such as the fact that the results may have been recorded before equilibration occurred or there may have been air inside the nanotubes when immersed into solution. A higher than expected capillary pressure, though, suggests that the diameter and surface chemistry of the channel are the main contributors to this effect (especially for the 30 nm diameter tube).[112]

BNNTs have a structure similar to CNTs and promise exciting electrical and mechanical properties beyond what can be obtained with CNTs.[113] BNNTs are semiconductors with a tunable band gap of 5.5 eV,[114,115] with high tensile strength[116,117] and high heat conduction[118] BNNTs can be produced as both MWNTs and

SWNTs as for CNTs. Many studies have considered the structure and electronic properties of CNTs and BNNTs, which are beyond the scope of this review and can be found elsewhere in the abundant literature.[113,119–122] These attractive properties have generated some interest in investigating their wetting properties in analogy with CNTs.[123,124] As in the case of nanotubes, the complexity of conducting wetting or fluid flow experiments has led to the use of simulations to analyze the wetting and flow behavior of BNNTs. A first study looked at the structure and dynamics of water inside BNNTs.[123] using density functional theory (DFT) and MD. DFT allows the simulation to capture the unique polarization effects observed with BNNTs to more accurately observe how water is affected by it. Multiple DFT and MD simulations were carried out for both (5,5) BNNTs and (6,6) BNNTs and subsequently compared with CNT simulation results. The results implied that water molecules inside an uncharged (5,5) BNNT align in a single file within the tube. A similar arrangement of the water molecules is observed when the tube is partially charged, whilst making the tube more hydrophilic and slowing down water diffusion due to the interaction of the water molecules and nitrogen atoms of the BNNT.

Another MD study compared the wetting behavior of sub-nanometer BNNTs and CNTs, finding that a (5,5) BNNT of 6.9 Å diameter could still conduct water and that comparable (5,5) CNTs could not.[125] The different regime was attributed to strong van der Waals interaction between the nitride atoms and the water molecules. It was also found that the water properties inside (5,5) BNNTs were very similar to those of water in (6,6) CNTs. To understand the wetting behavior of the BNNTs the nitride atoms were given the same LJ parameters of boron atoms and vice versa. When all the atoms had the LJ parameters of the boron atoms, the nanotube was almost empty, but when all the atoms were given the LJ parameters of nitride atoms, the BNNT was completely filled with water.

Further comparisons of BNNTs and CNTs have focused on applications, in particular water filtration. Recent publications have looked at salt rejection.[126,127] and ion separation,[128] in both cases finding that BNNTs gave better results than CNTs.

To the authors' knowledge there has been no experimental investigation of water inside BNNTs reported in the literature,

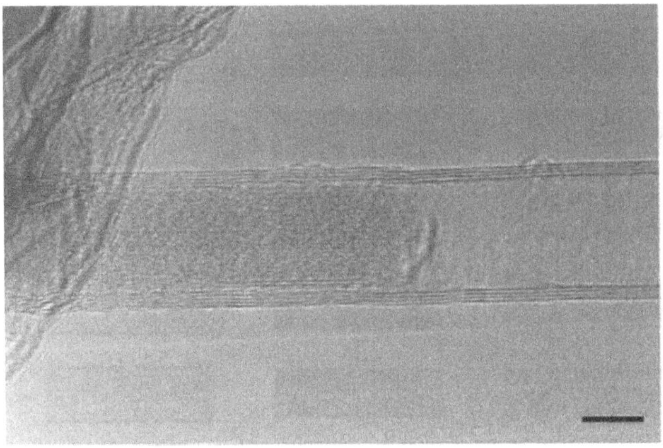

Figure 10.15 High-resolution TEM of amorphous materials contained inside the four-layer BNNT (scale bar: 5 nm). (with permissions from ref.[129])

although in 2001 an amorphous material plug, resulting from the synthesis process, was observed inside a BNNT, forming a nonwetting meniscus (Fig. 10.15).[129]

The surface tension of BNNTs was measured using the Wilhelmy balance method with nonpolar and polar fluids, obtaining a value of 27 $mN\ m^{-1}$, comparable to values obtained for CNTs[124] (see section 10.3.1). Contact angle measurements using different liquids showed an increase in value as a function of the liquid's polarity, with a value of $85 \pm 5°$ water.

The capillary imbibition of a metal halide melt inside BNNTs, molybdenum disulfide (MoS_2) nanotubes, and CNTs was simulated using MD.[130] From the simulation of initial insertion of the PbI_2 melt, it was found that CNTs and BNNTs reacted almost identically, with the formation of a convex meniscus due to the weak wetting of the graphene-like surfaces by the melt. The capillary imbibition of the PbI_2 melt is quite different for the MoS_2 nanotubes; in contrast to the CNT and BNNT a partial adsorption of the Pb (lead) and I (iodine) molecules is observed on both the outer surface and inner surface of the nanotube, resulting in a convex meniscus (Fig. 10.16). The filling of the MoS_2 nanotubes was found to obey the Lucas–Washburn equation, whereas for the CNTs and BNNTs a modification of the expression was necessary. The simulations also showed that

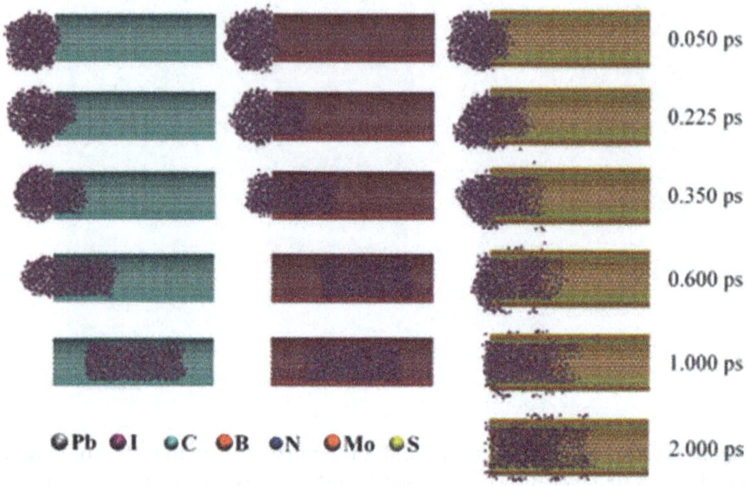

Figure 10.16 Advancing menisci of PbI_2 in three different nanotube materials. (with permissions from ref.[130])

the good wettability in the tubes can cause crystallization of the melt into cylindrical nanostructures, as observed in several other cases.[131,132]

10.5 Wetting in Polymer-Based Nanochannels

The development of lab-on-a-chip devices based, at least partially, on polymers has let to studying the wetting behavior of polymer-based nanochannels, as it can be used to control the flow of liquids in the chip.[133,134] Additionally, polymer coatings have since long been used to alter the wettability of other materials, and this method is now applied to the nanoscale as well.[135–137]

Anodic aluminum oxide (AAO) membranes are a popular choice as templates for nanowires and nanotubes. As such they have been used to fabricate polymer nanochannels, mostly via melt infiltration, which has given the possibility of studying the wetting and flow of polymer melts in the alumina pores. In one such work an enhanced flow rate of poly-caprolactone was observed when mechanical vibration is applied to the AAO during the filling process.[138] It

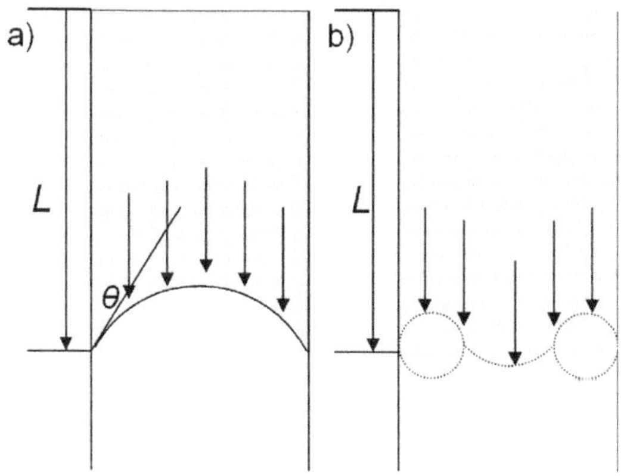

Figure 10.17 Meniscus profile for a nonwetting polymer melt in an AAO under (a) static and (b) vibration conditions. (schematic redrawn from ref.[138])

was found that with increasing vibrations the velocity profile of the melt changed, resulting in an increase in the flow rate. As a results of the vibrations, the flow profile of the polymer melt changed from parabolic to plug flow like (Fig. 10.17), with the dynamic contact angle of the liquid deviating from the equilibrium value, as measured by the Lucas–Washburn equation (Eq. 10.6). This effect was attributed to two factors, one related to the reduction in viscosity of the melt due to the disentanglement of the polymer chains and the other due to an increase in slip length due to the applied vibrations (Fig. 10.17). While the latter case appears to be a reasonable assumption, the former is probably the dominant mechanism. In fact, MD simulations have investigated the used of external vibrational forces to deform CNTs to pump fluids through them,[90] attributing this to friction effects between the fluid molecules and the tube walls. In the case of AAO the pore walls are rigid and further apart, but the authors have suggested that this mechanism will apply to larger space/time scales.

MD and Monte Carlo simulations of the wetting of alumina nanochannels by thermoplastic polymer melts have investigated the effect on wetting of surface functionalization and channel diameter.[135] It was found that the flow rate of the polymer inside the nanochannels was greatly increased when the polymer was heated to its glass transition temperature, at which point the channel was completely wetted. It was also found that the flow rate of the polymer melt through the nanochannel increases with the channel diameter (from 80 nm to 220 nm). This implied that the nanochannels were wetted more rapidly at the larger diameters. The findings were in agreement with filling of CNTs by low-molecular-weight organic liquids,[139] with both experiments following the Washburn equation (Eq 10.6). It was also found that alumina nanochannels treated with alkysilane self-assembled monolayers (SAMs) had a lower flow rate of the polymer melt than the untreated alumina. This can be explained by the decrease in surface free energy associated with the SAMs coating with the contact angle increasing from 3° to 41°.[135] Due to the high surface energy of alumina (for α-alumina 2.04 Jm^{-2}),[140] polymer melts with low surface energy can wet the surface in a partial or complete wetting regime and the wetting driving force p from a high-energy surface can be described using the Laplace equation (Fig. 10.18).

An MD study investigated the wettability of nanotubes lined with polymer brushes on their inner surface.[136,141] The fluid was modeled as molecules interacting through LJ potential, and it was found, as expected, that the wettability of the coated tube was solely a function of the wettability of the polymer brushes[137] and that the imbibition rate increases with increasing length of the brush chains. The simulation also showed that the polymer brushes are distorted by the motion of the meniscus and formed a plug of monomers at the rear of the meniscus, while the polymer chains still untouched by the fluid exhibit little stretching and lie closer to the capillary wall (Fig. 10.19). Simulations for a nonwetting case showed that for a short chain length, there was no meniscus movement, while for long polymer chains no curvature of the meniscus could be identified, with behavior consistent with that of a column of fluid rising through a porous medium.

Wetting experiments in polyurethane acrylate (PUA) rectangular nanochannels (with width ranging from 50 to 800 nm and a constant height of 200 nm) showed negative capillary pressure (as discussed earlier for silicon oxide nanochannels, Fig. 10.13).[142] Several surface chemistry treatments were performed to alter the wettability of the channels. In the case of complete wetting, the formation of a

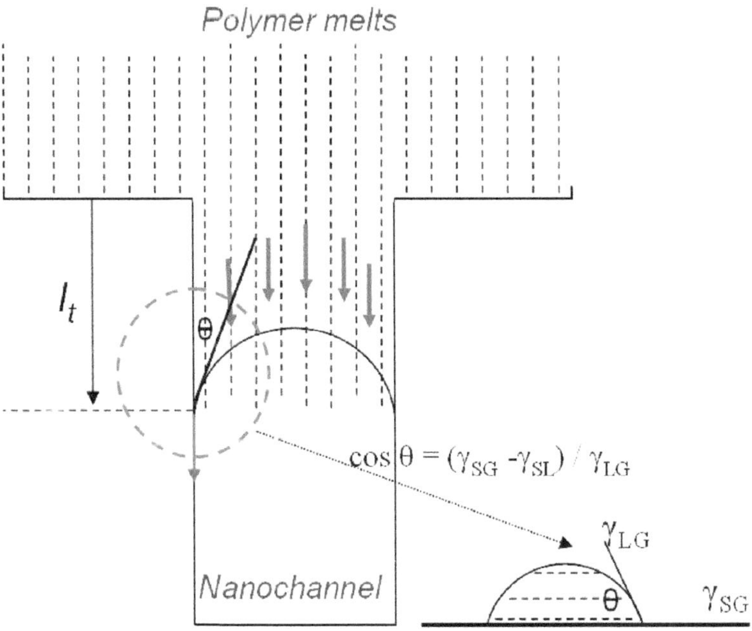

Figure 10.18 Dynamic parameters of nanoflow of polymer melts in nanochannels. (with permissions from ref.[135])

Figure 10.19 Schematic representing the stretching of polymer brushes with a moving meniscus.

precursor film in front of the meniscus was observed, as well as the meniscus having convex curvature (cfr. Fig. 10.3). For a partial wetting case, a complex meniscus shape was observed similar to what has been observed in silicon oxide nanochannels (Fig. 10.13). Interestingly, for the smallest channel investigated, this complex meniscus led to a plug-like meniscus shape due to an overlap between the convex curvatures at the channel walls.

10.6 Conclusions

The combination of unusual properties and promising applications has generated widespread interest in studying wetting inside carbon, inorganic, and organic nanotubes and nanochannels. The results obtained so far have generated more questions than those that have been answered, in particular regarding the effect of nanoscale confinement on important quantities such as surface tension and contact angle. Whether a breakdown of the continuum model—used to derive all of current fluid mechanics models—actually occurs and below which threshold is another question still open to investigation. These exciting topics will ensure that scientific investigation in this area will continue for many years on developing methods to control wetting and flow at nanometer scale and correlating macroscopic quantities (such as an applied pressure to filter water through a nanotube membrane) to the molecular interactions occurring inside the channels. Harnessing these properties will lead to radical advancements in several technological areas, the most promising being, in our opinion, water filtration and desalination.

References

1. T. Young, An essay on the cohesion of fluids, *Philos. Trans. R. Soc. Lond.*, **95**, 65 (1805).
2. J. W. Gibbs, *The Collected Works of J. Willard Gibbs—Volume I. Thermodynamics*. New Haven, CT: Yale University Press. 434 (1928).
3. J. Israelachvili, *Intermolecular & Surface Forces*. 2nd ed. San Diego, CA: Academic Press. 450 (1991).
4. A. W. Adamson and A. P. Gast, *Phys. Chem. of Surfaces*. 6th ed. New York: Wiley-Interscience. 784 (1997).
5. G. McHale, N. J. Shirtcliffe, and M. I. Newton, Contact-angle hysteresis on super-hydrophobic surfaces, *Langmuir*, **20**, 10146 (2004).

6. R. N. Wenzel, Resistance of solid surfaces to wetting by water, *Ind. Eng. Chem.*, **28**, 988 (1936).
7. T. N. Wenzel, Surface roughness and contact angle, *J. Phys. Colloid Chem.*, **53**, 1466 (1949).
8. A. B. D. Cassie and S. Baxter, Wettability of porous surfaces, *Trans. Faraday Soc.*, **40**, 546 (1944).
9. D. Mattia and Y. Gogotsi, Review: static and dynamic behavior of liquids inside carbon nanotubes, *Microfluid. Nanofluid.*, **5**, 289 (2008).
10. M. Whitby and N. Quirke, Fluid flow in carbon nanotubes and nanopipes, *Nat. Nanotechnol.*, **2**, 87 (2007).
11. D. M. Huang, C. Sendner, D. Horinek, R. R. Netz, L. Bocquet, eacute, and ric, Water slippage versus contact angle: a quasiuniversal relationship, *Phys. Rev. Lett.*, **101**, 226101 (2008).
12. E. Lauga, M. P. Brenner, and H. A. Stone, Microfluidics: the no-slip boundary condition. In eds., J. Foss, C. Tropea, and A. Yarin, *Handbook of Experimental Fluid Dynamics*. New York: Springer. 1219 (2005).
13. T. D. Blake, Slip between a liquid and a solid: D.M. Tolstoi's (1952) theory reconsidered, *Colloids Surf.* **47**, 135 (1990).
14. J.-T. Cheng and N. Giordano, Fluid flow through nanometer-scale channels, *Phys. Rev. E*, **65**, 1 (2002).
15. M. Majumder, N. Chopra, R. Andrews, and B. J. Hinds, Nanoscale hydrodynamics: enhanced flow in carbon nanotubes, *Nature*, **438**, 44 (2005)
16. R. S. Voronov, D. V. Papavassiliou, and L. L. Lee, Boundary slip and wetting properties of interfaces: correlation of the contact angle with the slip length, *J. Chem. Phys.*, **124**, 10 (2006).
17. M. Sun and C. Ebner, Molecular-Dynamics study of flow at a fliud-wall interface, *Phys. Rev. Lett.*, **69**, 3491 (1992).
18. P. A. Thompson and M. O. Robbins, Shear-flow near solids—epitaxial order and flow boundary—conditions, *Phys. Rev. A* **41**, 6830 (1990).
19. J. L. Barrat and L. Bocquet, Large slip effect at a nonwetting fluid-solid interface, *Phys. Rev. Lett.*, **82**, 4671 (1999).
20. K. Falk, F. Sedlmeier, L. Joly, R. R. Netz, and L. r. Bocquet, Molecular origin of fast water transport in carbon nanotube membranes: superlubricity versus curvature dependent friction, *Nano Lett.*, **10**, 4067 (2010).
21. P. G. de Gennes, F. Brochard-Wyard, and D. Quere, *Capillarity and Wetting Phenomena: Drops, Bubbles, Pearls, Waves*. New York: Springer. 291 (2004).

22. B. Su, M. Li, Z. Y. Shi, and Q. H. Lu, From superhydrophilic to superhydrophobic: controlling wettability of hydroxide zinc carbonate film on zinc plates, *Langmuir*, **25**, 3640 (2009).
23. G. P. Li, T. Chen, B. Yan, Y. Ma, Z. Zhang, T. Yu, Z. X. Shen, H. Y. Chen, and T. Wu, Tunable wettability in surface-modified ZnO-based hierarchical nanostructures, *Appl. Phys. Lett.*, **92**, 173104 (2008).
24. M. Guo, P. Diao, and S. M. Cai, Highly hydrophilic and superhydrophobic ZnO nanorod array films, *Thin Solid Films*, **515**, 7162 (2007).
25. X. Q. Meng, D. X. Zhao, J. Y. Zhang, D. Z. Shen, Y. M. Lu, L. Dong, Z. Y. Xiao, Y. C. Liu, and X. W. Fan, Wettability conversion on ZnO nanowire arrays surface modified by oxygen plasma treatment and annealing, *Chem. Phys. Lett.*, **413**, 450 (2005).
26. T. S. Wong, A. P. H. Huang, and C. M. Ho, Wetting behaviors of individual nanostructures, *Langmuir*, **25**, 6599 (2009).
27. J. K. Holt, H. G. Park, Y. Wang, M. Stadermann, A. B. Artyukhin, C. P. Grigoropoulos, A. Noy, and O. Bakajin, Fast mass transport through sub-2-nanometer carbon nanotubes, *Science*, **312**, 1034 (2006).
28. M. Whitby, L. Cagnon, M. Thanou, and N. Quirke, Enhanced fluid flow through nanoscale carbon pipes, *Nano Lett.*, **8**, 2632 (2008).
29. S. Chibbaro, L. Biferale, K. Binder, D. Dimitrov, F. Diotallevi, A. Milchev, and S. Succi, Hydrokinetic simulations of nanoscopic precursor films in rough channels, *J. Stat. Mech.—Theory Exp.* (2009).
30. S. Joseph and N. R. Aluru, Why are carbon nanotubes fast transporters of water? *Nano Lett.*, **8**, 452 (2008).
31. T. Werder, J. H. Walther, R. L. Jaffe, T. Halicioglu, and P. Koumoutsakos, On the water-carbon interaction for use in molecular dynamics simulations of graphite and carbon nanotubes, *J. Phys. Chem. B*, **107**, 1345 (2003).
32. T. Werder, J. H. Walther, R. L. Jaffe, T. Halicioglu, F. Noca, and P. Koumoutsakos, Molecular dynamics simulation of contact angles of water droplets in carbon nanotubes, *Nano Lett.*, **1**, 697 (2001).
33. T. Werder, J. H. Walther, and P. Koumoutsakos, *Hydrodynamics of Carbon Nanotubes-Contact Angle and Hydrophobic Hydration. Computational Nanoscience and Nanotechnology*. Weinbergstrasse, Zurich: Institute of Computational Science (2002).
34. B. Y. Cao, M. Chen, and Z. Y. Guo, Liquid flow in surface-nanostructured channels studied by molecular dynamics simulation, *Phys. Rev. E*, **74**, 7 (2006).

35. M. Cieplak, J. Koplik, and J. R. Banavar, Nanoscale fluid flows in the vicinity of patterned surfaces, *Phys. Rev. Lett.*, **96**, 4 (2006).

36. P. G. de Gennes, Wetting: statics and dynamics, *Rev. Mod. Phys.*, **57**, 827 (1985).

37. D. M. Huang, C. Sendner, D. Horinek, R. R. Netz, and L. Bocquet, Water slippage versus contact angle: a quasiuniversal relationship, *Phys. Rev. Lett.*, **101**, 226101 (2008).

38. M. P. Brenner and D. Lohse, Dynamic equilibrium mechanism for surface nanobubble stabilization, *Phys. Rev. Lett.*, **101**, 214505 (2008).

39. A. Poynor, L. Hong, I. K. Robinson, S. Granick, Z. Zhang, and P. A. Fenter, How water meets a hydrophobic surface, *Phys. Rev. Lett.*, **97**, 266101 (2006).

40. F. Brochardwyart and P. G. Degennes, Dynamics of partial wetting, *Adv. Colloid Interface Sci.*, **39**, 1 (1992).

41. P. G. Petrov and J. G. Petrov, A combined molecular-hydrodynamic approach to wetting kinetics, *Langmuir*, **8**, 1762 (1992).

42. T. D. Blake, The physics of moving wetting lines, *J. Colloid Interface Sci.*, **299**, 1 (2006).

43. E. Bertrand, T. D. Blake, V. Ledauphin, G. Ogonowski, J. De Coninck, D. Fornasiero, and J. Ralston, Dynamics of dewetting at the nanoscale using molecular dynamics, *Langmuir*, **23**, 3774 (2007).

44. J. L. Barrat and L. Bocquet, Influence of wetting properties on hydrodynamic boundary conditions at a fluid/solid interface, *Faraday Discuss.*, **112**, 119 (1999).

45. W. H. Noon, K. D. Ausman, R. E. Smalley, and J. Ma, Helical Ice-sheets inside carbon nanotubes in the physiological condition, *Chem. Phys. Lett.*, **355**, 445 (2002).

46. J. Bai, J. Wang, and X. C. Zeng, Multiwalled ice helixes and ice nanotubes, *Proc. Natl. Acad. Sci. USA*, **103**, 19664 (2006).

47. K. Koga, G. T. Gao, H. Tanaka, and X. C. Zeng, Formation of ordered ice nanotubes inside carbon nanotubes, *Nature*, **412**, 802 (2001).

48. K. Koga, G. T. Gao, H. Tanakac, and X. C. Zeng, How does water freeze inside carbon nanotubes? *Physica A*, **314**, 462 (2002).

49. Y. Lee, C. D. Martin, J. B. Parise, J. A. Hriljac, and T. Vogt, Formation and manipulation of confined water wires, *Nano Lett.*, **4**, 619 (2004).

50. D. Mattia, M. P. Rossi, H. Ye, and Y. Gogotsi, *In situ Fluid Studies in Carbon Nanotubes with Diameters Ranging from 1 to 500 nm.* 5th IASME/WSEAS International Conference on Fluid Mechanics and Aerodynamics. Athens, Greece. 297 (2007).

51. K. Koga, H. Tanaka, and X. C. Zeng, First-order transition in confined water between high-density liquid and low-density amorphous phases, *Nature*, **408**, 564 (2000).
52. J. M. D. Leo and J. Maragnon, Confined water in nanotube, *THEOCHEM*, **623**, 159 (2003).
53. J. L. Rivera, C. McCabe, and P. T. Cummings, Layering behavior and axial phase equilibria of pure water and water + carbon dioxide inside single wall carbon nanotubes, *Nano Lett.*, **2**, 1427 (2002).
54. Y. C. Liu and Q. Wang, Transport behavior of water confined in carbon nanotubes, *Phys. Rev. B*, **72**, 085420 (2005).
55. Y. Maniwa, H. Kataura, M. Abe, S. Suzuki, Y. Achiba, H. Kira, and K. Matsuda, Phase transition in confined water inside carbon nanotubes, *J. Phys. Soc. Jpn.*, **71**, 2863 (2002).
56. J. Marti and M. C. Gordillo, Effects of confinement on the vibrational spectra of liquid water adsorbed in carbon nanotubes, *Phys. Rev. B*, **63**, 165430_1 (2001).
57. J. Marti and M. C. Gordillo, Time-dependent properties of liquid water isotopes adsorbed in carbon nanotubes, *J. Chem. Phys.*, **114**, 10486 (2001).
58. A. Noy, H. G. Park, F. Fornasiero, J. K. Holt, C. P. Grigoropoulos, and O. Bakajin, Nanofluidics in carbon nanotubes, *Nanotoday*, **2**, 22 (2007).
59. J. Israelachvili and D. Gourdon, Putting liquids under molecular-scale confinement, *Science*, **292**, 867 (2001).
60. K. Lum, D. Chandler, and J. D. Weeks, Hydrophobicity at small and large length scales, *J. Phys. Chem. B*, **103**, 4570 (1999).
61. V. N. Bogomolov, Capillary effects in ultrathin channels, *Sov. Phys. Tech. Phys.*, **37**, 79 (1992).
62. R. C. Tolman, The superficial density of matter at a liquid-vapor boundary, *J. Chem. Phys.*, **17**, 118 (1949).
63. R. C. Tolman, The effect of droplet size on surface tension, *J. Chem. Phys.*, **17**, 333 (1949).
64. S. Granick, Motions and relaxations of confined liquids, *Science*, **253**, 1374 (1991).
65. E. Dujardin, T. Ebbesen, H. Hiura, and K. Tanigaki, Capillarity and wetting of carbon nanotubes, *Science*, **265**, 1850 (1994).
66. E. Dujardin, T. W. Ebbesen, A. Krishnan, and M. M. J. Treacy, Wetting of single shell carbon nanotubes, *Adv. Mater.*, **10**, 1472 (1998).

67. P. M. Ajayan and S. Iijima, Capillarity-induced filling of carbon nanotubes, *Nature*, **361**, 333 (1993).
68. Y. Gogotsi, J. A. Libera, A. G. Yazicioglu, and C. M. Megaridis, *In situ* multiphase fluid experiments in hydrothermal carbon nanotubes, *Appl. Phys. Lett.*, **79**, 1021 (2001).
69. H. Rafii-Tabar, Computational modelling of thermo-mechanical and transport properties of carbon nanotubes, *Phys. Rep.—Rev. Sec. Phys. Lett.*, **390**, 235 (2004).
70. D. Mattia, H. H. Bau, and Y. Gogotsi, Wetting of CVD carbon films by polar and non-polar liquids and implications for carbon nanopipes, *Langmuir*, **22**, 1789 (2006).
71. E. G. Shafrin and W. A. Zisman, Constitutive relations on the wetting of low energy surfaces and the theory of the retraction method of preparing monolayers, *J. Phys. Chem.*, **64**, 519 (1960).
72. H. Orikasa, N. Inokuma, S. Okubo, O. Kitakami, and T. Kyotani, Template synthesis of water-dispersible carbon nano "test tubes" without any post-treatment, *Chem. Mater.*, **18**, 1036 (2006).
73. A. H. Yan, X. C. Xiao, I. Kulaots, B. W. Sheldon, and R. H. Hurt, Controlling water contact angle on carbon surfaces from 5 degrees to 167 degrees, *Carbon*, **44**, 3116 (2006).
74. F. M. Fowkes, *Dispersion Force Contributions to Surface and Interfacial Tensions, Contact Angles, and Heats of Immersion. Contact Angle, Wettability, and Adhesion* Vol. 43. Los Angeles, CA: American Chemical Society. 99 (1964).
75. D. K. Owens and R. C. Wendt, Estimation of the surface free energy of polymers, *J. Appl. Polym. Sci.*, **13**, 1741 (1969).
76. E. G. Rakov, Chemistry of carbon nanotubes. In ed., Y. Gogotsi, *Nanomaterials Handbook*. Boca Raton, FL: CRC Press. 105 (2006).
77. D. Mattia, M. P. Rossi, B. M. Kim, G. Korneva, H. H. Bau, and Y. Gogotsi, Effect of graphitization on the wettability and electrical conductivity of CVD carbon nanotubes and films, *J. Phys. Chem. B*, **110**, 9850 (2006).
78. J. N. Coleman, U. Khan, and Y. K. Gun'ko, Mechanical reinforcement of polymers using carbon nanotubes, *Adv. Mater.*, **18**, 689 (2006)
79. P. J. F. Harris, Carbon nanotube composites, *Int. Mater. Rev.*, **49**, 31 (2004).
80. T. Arai, Y. Tominaga, S. Asai, and M. Sumita, Study on correlation between physical properties and interfacial characteristics in highly loaded graphite-polymer composites, *J. Polym. Sci. Pt. B—Polym. Phys.*, **43**, 2568 (2005).

81. S. M. Hong, Y. W. Nam, S. S. Hwang, and D. W. Chae, Physical properties of multi-walled carbon nanotube-filled PVDF composites prepared by melt compounding, *Mol. Cryst. Liq. Cryst.*, **464**, 777 (2007).

82. M. Q. Tran, J. T. Cabral, M. S. P. Shaffer, and A. Bismarck, Direct measurement of the wetting behavior of individual carbon nanotubes by polymer melts: the key to carbon nanotube polymer composites, *Nano Lett.*, **8**, 2744 (2008).

83. H. J. Shah, A. K. Fontecchio, M. P. Rossi, D. Mattia, and Y. Gogotsi, Imaging of liquid crystals confined in carbon nanopipes, *Appl. Phys. Lett.*, **89**, 043123 (2006).

84. A. H. Barber, S. R. Cohen, and H. D. Wagner, Static and dynamic wetting measurements of single carbon nanotubes, *Phys. Rev. Lett.*, **92**, 186103 (2004).

85. A. H. Barber, S. R. Cohen, and H. D. Wagner, External and internal wetting of carbon nanotubes with organic liquids, *Phys. Rev. B*, **71**, 115443 (2005).

86. S. Nuriel, L. Liu, A. H. Barber, and H. D. Wagner, Direct measurement of multiwall nanotube surface tension, *Chem. Phys. Lett.*, **404**, 263 (2005).

87. J. B. Zhao, L. Liu, P. J. Culligan, and X. Chen, Thermal effect on the dynamic infiltration of water into single-walled carbon nanotubes, *Phys. Rev. E*, **80** (2009).

88. A. Barreiro, R. Rurali, E. R. Hernandez, J. Moser, T. Pichler, L. Forro, and A. Bachtold, Subnanometer motion of cargoes driven by thermal gradients along carbon nanotubes, *Science*, **320**, 775 (2008).

89. Y. Q. Xue and M. D. Chen, Dynamics of molecules translocating through carbon nanotubes as nanofluidic channels, *Nanotechnology*, **17**, 5216 (2006).

90. Z. Insepov, D. Wolf, and A. Hassanein, Nanopumping using carbon nanotubes, *Nano Lett.*, **6**, 1893 (2006).

91. N. Hamada, S. Sawada, and A. Oshiyama, New one-dimensional conductors—graphitic microtubules, *Phys. Rev. Lett.*, **68**, 1579 (1992).

92. A. Alexiadis and S. Kassinos, Self-diffusivity, hydrogen bonding and density of different water models in carbon nanotubes, *Mol. Simul.*, **34**, 671 (2008).

93. A. Alexiadis and S. Kassinos, Molecular simulation of water in carbon nanotubes, *Chem. Rev.*, **108**, 5014 (2008).

94. Y. Lin, J. Shiomi, S. Maruyama, and G. Amberg, Dielectric relaxation of water inside a single-walled carbon nanotube, *Phys. Rev. B*, **80** (2009).

95. G. Garberoglio, Collective properties of water confined in carbon nanotubes: a computer simulation study, *Eur. Phys. J. E*, **31**, 73 (2010).

96. M. P. Rossi, Y. Gogotsi, and K. G. Kornev, Deformation of carbon nanotubes by exposure to water vapor, *Langmuir*, **25**, 2804 (2009).

97. W. Sekhaneh, M. Kotecha, U. Dettlaff-Weglikowska, and W. S. Veeman, High resolution NMR of water absorbed in single-wall carbon nanotubes, *Chem. Phys. Lett.*, **428**, 143 (2006).

98. M. D. Ellison, A. P. Good, C. S. Kinnaman, and N. E. Padgett, Interaction of water with single-walled carbon nanotubes: reaction and adsorption, *J. Phys. Chem. B*, **109**, 10640 (2005).

99. M. P. Rossi, H. Ye, Y. Gogotsi, S. Babu, P. Ndungu, and J. C. Bradley, Environmental scanning electron microscopy study of water in carbon nanopipes, *Nano Lett.*, **4**, 989 (2004).

100. A. L. Yarin, A. G. Yazicioglu, C. M. Megaridis, M. P. Rossi, and Y. Gogotsi, Theoretical and experimental investigation of aqueous liquids contained in carbon nanotubes, *J. Appl. Phys.*, **97**, 124309 (2005).

101. M. Rauscher and S. Dietrich, Wetting phenomena in nanofluidics, *Annu. Rev. Mater. Res.*, **38**, 143 (2008).

102. J. W. van Honschoten, N. Brunets, and N. R. Tas, Capillarity at the nanoscale, *Chem. Soc. Rev.*, **39**, 1096 (2010).

103. S. S. Rekhviashvili and E. V. Kishtikova, Properties of nonpolar fluids inside a carbon nanotube, *Prot. Met. Phys. Chem. Surf.*, **46**, 55 (2010).

104. T. V. O. Ratto, J. K. Holt, and A. W. Szmodis, *Membranes with Embedded Nanotubes for Selective Permeability*. San Jose, CA: NanOasis (2010).

105. N. R. Tas, P. Mela, T. Kramer, J. W. Berenschot, and A. van den Berg, Capillarity induced negative pressure of water plugs in nanochannels, *Nano Lett.*, **3**, 1537 (2003).

106. C. E. Brennen, *Cavitation and Bubble Dynamics*. Oxford Engineering Science Series 44. Oxford University Press, 282 (1995).

107. A. L. Yarin, A. G. Yazicioglu, and C. M. Megaridis, Thermal stimulation of aqueous volumes contained in carbon nanotubes: experiment and modeling, *Appl. Phys. Lett.*, **86**, 013109 (2005).

108. K. Jayaraman, K. Okamoto, S. J. Son, C. Luckett, A. H. Gopalani, S. B. Lee, and D. S. English, Observing capillarity in hydrophobic silica nanotubes, *J. Am. Chem. Soc.*, **127**, 17385 (2005).

109. T. Hofmann, M. Tasinkevych, A. Checco, E. Dobisz, S. Dietrich, and B. M. Ocko, Wetting of nanopatterned grooved surfaces, *Phys. Rev. Lett.*, **104**, 4 (2010).

110. S. Gruener, T. Hofmann, D. Wallacher, A. V. Kityk, and P. Huber, Capillary rise of water in hydrophilic nanopores, *Phys. Rev. E*, **79**, 4 (2009).
111. P. Huber, S. Gruner, C. Schafer, K. Knorr, and A. V. Kityk, Rheology of liquids in nanopores: a study on the capillary rise of water, n-hexadecane and n-tetracosane in mesoporous silica, *Eur. Phys. J.—Spec. Top.*, **141**, 101 (2007).
112. R. Helmy, Y. Kazakevich, C. Y. Ni, and A. Y. Fadeev, Wetting in hydrophobic nanochannels: a challenge of classical capillarity, *J. Am. Chem. Soc.*, **127**, 12446 (2005).
113. D. Golberg, Y. Bando, C. C. Tang, and C. Y. Zhi, Boron nitride nanotubes, *Adv. Mater.*, **19**, 2413 (2007).
114. K. H. Khoo, M. S. C. Mazzoni, and S. G. Louie, Tuning the electronic properties of boron nitride nanotubes with transverse electric fields: a giant DC Stark effect, *Phys. Rev. B*, **69**, 4 (2004).
115. C. Y. Zhi, Y. Bando, C. C. Tang, and D. Golberg, Engineering of electronic structure of boron-nitride nanotubes by covalent functionalization, *Phys. Rev. B*, **74**, 4 (2006).
116. E. Hernandez, C. Goze, P. Bernier, and A. Rubio, Elastic properties of C and BxCyNz composite nanotubes, *Phys. Rev. Lett.*, **80**, 4502 (1998).
117. K. N. Kudin, G. E. Scuseria, and B. I. Yakobson, C2F, BN, and C nanoshell elasticity from ab initio computations, *Phys. Rev. B*, **64**, 10 (2001).
118. C. W. Chang, A. M. Fennimore, A. Afanasiev, D. Okawa, T. Ikuno, H. Garcia, D. Y. Li, A. Majumdar, and A. Zettl, Isotope effect on the thermal conductivity of boron nitride nanotubes, *Phys. Rev. Lett.*, **97**, 4 (2006).
119. R. Saito, G. Dresselhaus, and M. S. Dresselhaus, *Physical Properties of Carbon Nanotubes*. London: Imperial College. 259 (1998).
120. M. S. Dresselhaus, M. A. Pimenta, P. C. Eklund, and G. Dresselhaus, Raman scattering in fullerenes and related carbon-based materials. In eds., W. H. Weber and R. Merlin, *Raman Scattering in Materials Science*. Berlin: Springer-Verlag. 314 (2000).
121. P. M. Ajayan, Nanotubes from carbon, *Chem. Rev.*, **99**, 1787 (1999).
122. C. Branca, F. Frusteri, V. Magazù, and A. Mangione, Characterization of carbon nanotubes by TEM and infrared spectroscopy, *J. Phys. Chem. B*, **108**, 3469 (2004).
123. C. Y. Won and N. R. Aluru, Structure and dynamics of water confined in a boron nitride nanotube, *J. Phys. Chem. C*, **112**, 1812 (2008).
124. K. Yum and M. F. Yu, Measurement of wetting properties of individual boron nitride nanotubes with the Wilhelmy method using a nanotube-based force sensor, *Nano Lett.*, **6**, 329 (2006).

125. C. Y. Won and N. R. Aluru, Water permeation through a subnanometer boron nitride nanotube, *J. Am. Chem. Soc.*, **129**, 2748 (2007).

126. T. A. Hilder, D. Gordon, and S. H. Chung, Salt rejection and water transport through boron nitride nanotubes, *Small*, **5**, 2183 (2009).

127. M. E. Suk, A. V. Raghunathan, and N. R. Aluru, Fast reverse osmosis using boron nitride and carbon nanotubes, *Appl. Phys. Lett.*, **92**, 133120 (2008).

128. T. A. Hilder, D. Gordon, and S. H. Chung, Boron nitride nanotubes selectively permeable to cations or anions, *Small*, **5**, 2870 (2009).

129. R. Ma, Y. Bando, T. Sato, and K. Kurashima, Thin boron nitride nanotubes with unusual large inner diameters, *Chem. Phys. Lett.*, **350**, 434 (2001).

130. A. N. Enyashin, R. Kreizman, and G. Seifert, Capillary imbibition of PbI2 melt by inorganic and carbon nanotubes, *J. Phys. Chem. C*, **113**, 13664 (2009).

131. R. Kreizman, S. Y. Hong, J. Sloan, R. Popovitz-Biro, A. Albu-Yaron, G. Tobias, B. Ballesteros, B. G. Davis, M. L. H. Green, and R. Tenne, Core-shell PbI2@WS2 inorganic nanotubes from capillary wetting, *Angew. Chem., Int. Ed.*, **48**, 1230 (2009).

132. W. Q. Han, C. W. Chang, and A. Zettl, Encapsulation of one-dimensional potassium halide crystals within BN nanotubes, *Nano Lett.*, **4**, 1355 (2004).

133. S. Ramakrishna, N. Lala, H. Garudadhwaj, R. Ramaseshan, and V. K. Ganesh, *Polymer Nanofibers for Biosensor Applications*. 377 (2007).

134. M. Sokuler and L. A. Gheber, Nano fountain pen manufacture of polymer lenses for nano-biochip applications, *Nano Lett.*, **6**, 848 (2006).

135. K. L. Yung, J. Kong, and Y. Xu, Studies on flow behaviors of polymer melts in nanochannels by wetting actions, *Polymer*, **48**, 7645 (2007).

136. D. I. Dimitrov, A. Milchev, and K. Binder, Polymer brushes in cylindrical pores: simulation versus scaling theory, *J. Chem. Phys.*, **125**, 15 (2006).

137. D. I. Dimitrov, A. Milchev, and K. Binder, Molecular dynamics simulations of capillary rise experiments in nanotubes coated with polymer brushes, *Langmuir*, **24**, 1232 (2008).

138. J. Kong, Y. Xu, K. L. Yung, Y. C. Xie, and L. He, Enhanced polymer melts flow through nanoscale channels under vibration, *J. Phys. Chem. C*, **113**, 624 (2009).

139. B. M. Kim, S. Sinha, and H. H. Bau, Optical microscope study of liquid transport in carbon nanotubes, *Nano Lett.*, **4**, 2203 (2004).

140. S. Blonski and S. H. Garofalini, Molecular dynamics simulations of [alpha]-alumina and [gamma]-alumina surfaces, *Surf. Sci.*, **295**, 263 (1993).

141. D. I. Dimitrov, A. Milchev, and K. Binder, Polymer brushes in solvents of variable quality: Molecular dynamics simulations using explicit solvent, *J. Chem. Phys.*, **127**, 9 (2007).

142. P. Kim, H. Y. Kim, J. K. Kim, G. Reiter, and K. Y. Suh, Multi-curvature liquid meniscus in a nanochannel: evidence of interplay between intermolecular and surface forces, *Lab Chip*, **9**, 3255 (2009).

Chapter 11

Capillary Action in Silicon-Based Nanochannels

Joost van Honschoten, Nataliya Brunets, and Niels Tas*

Transducers Science and Technology Group, MESA+ Institute for Nanotechnology, University of Twente, P.O. Box 217, 7500 AE Enschede, The Netherlands

*n.r.tas@utwente.nl

We commemorate Dr. Joost van Honschoten, who could not witness the completion of this book. He lives on in our thoughts as a passionate scientist and a concerned colleague and friend.

11.1 Introduction

Capillary action, in particular capillary rise, is an old subject in physics literature. Already in 1717 Jurin[1,2] described in detail capillary rise experiments from which he concluded that the height of liquid in a capillary tube is inversely proportional to the diameter of the tube at the location of the liquid meniscus only. Note that at the end of his letter he gave credits to "our incomparable president" (of the Royal Society), which was Isaac Newton, for having described the same principle around the same time in a supplement to the later editions of his "Opticks."[3] For a complete description of the phenomenon of capillarity (relating

Nanoscale Liquid Interfaces: Wetting, Patterning, and Force Microscopy at the Molecular Scale
Edited by Thierry Ondarçuhu and Jean-Pierre Aimé

ISBN 978-981-4316-45-3 (Hardcover), 978-981-4364-48-5 (eBook)
www.panstanford.com

the pressure drop across a meniscus to its curvature), concepts like surface tension, contact angle, and double curvature of the liquid meniscus were needed. These were described and properly related by Young[4] in 1805. A thorough mathematical description of capillarity, including application of what is now known as the Young-Laplace equation to many different capillarity configurations, was finally provided by Laplace[5] in 1806. The extent of his account suggests that he was aware of the main principle for quite some time.

With the development of chemistry, the measurement of the size of atoms and molecules, and the increasing understanding of the forces between molecules (nineteenth and early twentieth centuries), an eminent question rose: Down to which scale does the Young-Laplace equation hold, in particular considering gaps with characteristic diameters in the range of the working distances of the intermolecular forces? Pioneering experiments on capillary action in nanometric-sized channels and gaps have been performed by Israelachvilli[6] (using the surface forces apparatus [SFA]) and Churaev *et al.*[7] (using pulled quartz capillaries). Fisher and Israelachvili[6] produced accurate experimental data that supports the validity of the Young-Laplace equation in nanometric gaps. They used the so-called SFA to measure the pull-off force between cylindrical mica surfaces in a vapor atmosphere, while multiple-beam interferometry allowed the measurement of the mean curvature of the liquid–vapor interface of the condensed liquid. Churaev *et al.*[7] studied the surface tension and viscosity of liquids in microcapillaries down to 40 nm in radius, pulled from high-purity quartz glass. With this radius accurately measured, they found that even in the smallest capillaries the capillary pressure is equal to what can be expected based on the Young-Laplace equation and the macroscopic value of the surface tension. From dynamic measurements they determined the (apparent) viscosity of the liquids in confinement. No changes in the viscosity were found for the apolar liquids CCl_4 and benzene, while for water an increase of 35%–40% in the (apparent) viscosity was found in the smallest capillaries.

Silicon-based nanotechnology offers techniques for precise fabrication of artificial nanopores. Often these channels are planar (in the plane of the substrate wafer), facilitating microscope observation of liquid inside these channels. The combination of well-defined

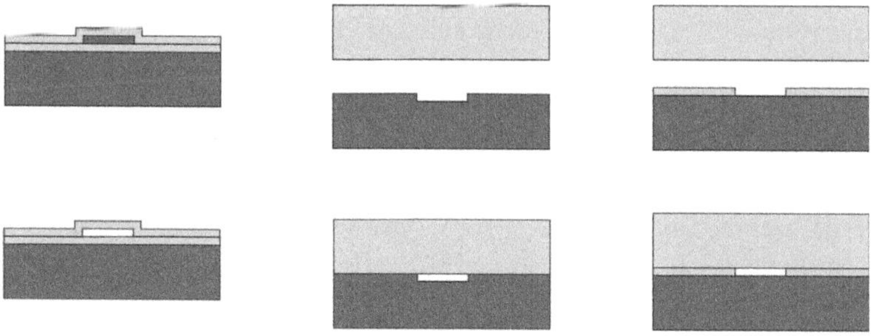

Figure 11.1 The three main techniques for the creation of (1D-confined) nanochannels: sacrificial layer etching (left), bulk etching (middle), and the spacer layer technique (right).

dimensions and observability makes silicon-based nanochannels an ideal model system for fundamental studies in capillarity. Figure 11.1 illustrates the three main techniques for the creation of (one-dimensional [1D]-confined) nanochannels: sacrificial layer etching (left), bulk etching (middle), and the spacer layer technique (right).

The sacrificial layer technique to produce nanochannels is based on the thin-film encapsulation of a strip of sacrificial material and the subsequent selective removal of the sacrificial layer through access holes.[8,9] Typical combinations of materials are silicon oxide as the sacrificial layer and polysilicon as the structural material or polysilicon as the sacrificial layer and silicon oxide or silicon nitride as the structural material. The latter combination provides optically transparent nanochannels. A practical issue in this fabrication scheme is the finite selectivity of the etchant during removal of the sacrificial layer, resulting in a tapered channel height (in the length direction). In the bulk etching technique[10] a trench is slowly and accurately etched into the silicon substrate. Wet anisotropic etching can produce the required smoothness and uniformity.[11] An enclosed channel is formed by bonding of a second wafer to the substrate wafer containing the trench. This second wafer can be a glass wafer, producing an optically transparent channel capping. Using wet etching[11] uniform channel heights typically in the range of 20–100 nm can be produced, with an accuracy of a few nanometers. To define the channel height for even smaller channels, the spacer layer technique[12,13] is particularly suitable. Uniform and well-defined spacer layers of silicon oxide can be created on a silicon substrate by thermal oxidation. Their thickness can be accurately measured before wafer bonding by ellipsometry, creating channels as small as 5 nm in height with a precision better than 1 nm.[14] Note that in both bulk etching and the spacer layer technique, in practice, at least a native oxide layer of 1–2 nm silicon oxide is formed prior to wafer bonding, ensuring complete silica-based channel walls.

11.2 Static Capillarity

11.2.1 *Negative Pressure in Capillary Liquid*[85]

In sufficiently narrow lyophilic capillaries the negative Laplace pressure can exceed the atmospheric pressure (often 1 bar in absolute value), and a negative absolute pressure can develop in the liquid. Liquids can withstand a considerable tension due to the cohesive forces between the molecules. The tensile stress developed when the liquid is stretched is isotropic normal stress; therefore it is justified to refer to it as a negative pressure. Reports about liquids under tension date back as far as the late seventeenth and early eighteenth centuries and originate from some of the greatest scientists of all time, including Huygens[15] and Newton.[16] They addressed the phenomenon that when a tube of a Torricelli barometer is completely filled with mercury and then perfectly purged of air and inverted, the mercury inside sometimes stayed top-full in the tube, suspended to heights as much as 75 inches.[15] This is much higher than the expected 29 inches caused by the weight of the atmosphere. In particular the explanation of Newton is remarkably accurate, referring to the cohesion of the liquid particles to each other and to the glass composing the tube.[16] Tension in water plays a central role in the cohesion-tension theory of sap ascent in trees, as first proposed by Dixon and Joly,[17] and has therefore drawn the interest of plant physiologists. Classical laboratory methods to induce negative pressures in liquids are the Berthelot tube method[18] based on isochoric cooling and the rotating capillary technique introduced by Briggs.[19]

Bulk liquid at negative absolute pressure is superheated, which is a metastable state. It is a metastable state because the sytem can move to a state of lower free energy by creating cavities filled by vapor; however, to create a cavity (vapor bubble) an energy barrier has to be overcome.[20] If the liquid is at a pressure p, then the work required to create a bubble of volume V and surface area A, filled with vapor at a pressure p_v, is given by:[20]

$$W = \gamma A + (p - p_V)V = 4\pi r^2\gamma + \frac{4}{3}\pi r^3(p - p_v) \qquad (11.1)$$

where r is the radius of the bubble. Assuming that the magnitude of p is much larger than p_v, the work required to create the bubble has a maximum $16\pi\gamma^3/3p^2$ for a critical radius:[20]

$$r_{crit} = -2\gamma/p \qquad (11.2)$$

The capillary action in narrow lyophilic tubes, in open communication with the surrounding atmosphere, can easily be stronger in magnitude than the surrounding absolute atmospheric pressure. Figure 11.2 shows a longitudinal section of such a capillary, partially filled with liquid. The absolute pressure in the liquid is given by the sum of the ambient pressure p_0 and the negative Laplace pressure p_{Lap}.

Assume in Fig. 11.2 that we have a thin rectangular capillary (height $h << $ width) Neglecting the meniscus curvature in the lateral plane, the pressure of the liquid phase p_{liq} is given by:

$$p_{liq} = p_0 - (2\gamma\cos\theta)/h \qquad (11.3)$$

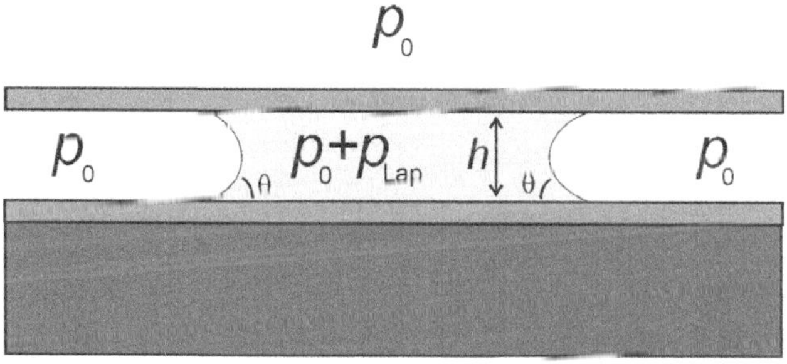

Figure 11.2 Longitudinal section of a partially filled flat, rectangular capillary. The pressure in the liquid is equal to the ambient pressure p_0 plus the negative Laplace pressure p_{Lap}. The channel height is h, and the contact angle of the liquid to the channel walls is represented by θ.

where h is the channel height, γ the surface tension of the liquid, and θ the contact angle of the liquid to the channel walls. If the liquid partially filling the hydrophilic channel is water, capillarity-induced negative pressure will occur when the channel height is below approximately 1 μm. The first notion that capillary action in narrow tubes can induce tension in the liquid was probably by Kelvin when he derived what is now known as the Kelvin equation.[21] Experimental proof of the existence of capillary negative pressure was provided by Wiig and Juhola[22] through the measurement of a reduced density of water condensed in nanoporous activated charcoal, though quantitative interpretation was hampered by the unknown compressibility of water in the negative pressure regime as well as lack of exact knowledge of the pore diameter.

Through pull-off force measurements in an SFA,[6] consistency with the Young-Laplace equation could be confirmed in nanometric gaps; however, in this method the conversion from forces to pressures requires application of the Kelvin equation. A direct indication for the existence of capillary negative pressures was recently found by measuring the elastocapillary deformation of silicon-based nanochannels[23] (see also deformation of synthetic nanochannels in chapter 10). Figure 11.3 shows a microscope image of the typical shape of the meniscus (top view of a flat, rectangular channel). The peculiar curvature of the meniscus in the lateral plane can be explained by a downward bending of the flexible channel cover. The downward bending can be quantified by detailed analysis of the meniscus shape.[23,24]

Together with the known stiffness of the channel cover it was possible to estimate the negative pressure of the liquid to be, depending on the method to analyze the meniscus shape, -17 ± 10 bar[23] or -14 ± 8 bar.[24] Within the limits of the measurement accuracy this is in accordance with Eq. 11.3 (-12 bar).

11.2.2 *Stability of Capillary Liquid*

Liquid plugs like the one shown in Fig. 11.3 are remarkably stable and can (depending on their volume and the speed of evaporation) exist for at least several minutes. The explanation for this observation[23] is that when we substitute Eq. 11.3 in the

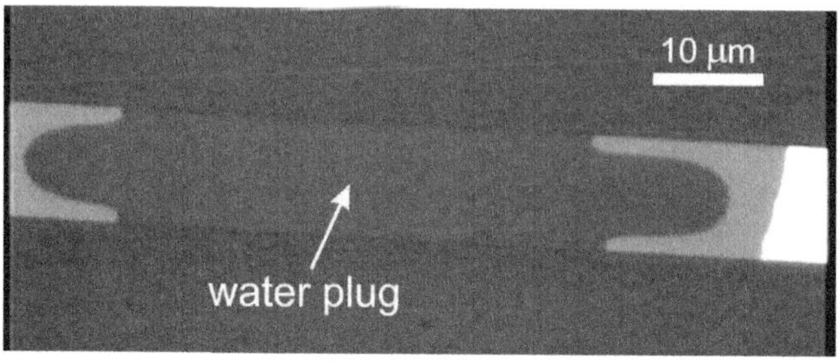

Figure 11.3 Top view of a silicon oxide nanochannel (flat, rectangular cross section) partially filled by water.[23] The channel capping is a silicon oxide thin film, which deforms under the negative pressure load. This deformation can explain the peculiar shape of the meniscus.[23,24] (Reprinted with permission from Ref. 23, copyright 2003, American Chemical Society.)

expression for the critical bubble size (Eq. 11.2), we find (neglecting p_0 with respect to the negative Laplace pressure) $r_{\text{crit}} = h/\cos(\theta)$. This shows that for capillary negative pressure the critical bubble size for cavitation is larger than the size of the channel, so the liquid at negative pressure is in a stable state. Note that we have implicitly considered channels of uniform cross section.

11.2.3 *Elastocapillary Deformation of Nanochannels*

11.2.3.1 Introduction

The capillary pressure scales with $1/D$, where D is the characteristic dimension like channel diameter (circular tubes) or height (flat channels). With $\gamma = 0.072$ N/m for water and $D \approx 1$ μm, a capillary pressure on the order of 1 bar is found; therefore it may be clear that on the nanoscale, capillary forces are extremely large. They can easily lead to the deformation of elastic structures, which is well known in the design, fabrication, and operation of microelectromechanical systems (MEMS).[25–29] As we showed in the previous section, the elastocapillary equilibrium between the surface tension forces of the wetting liquid and the mechanical forces of the deforming nanochannel results in a very peculiar shape of the interfacial meniscus. The explanation for this shape is that the downward bending of the channel capping leads to a nonuniform channel height, implying a nonuniform curvature of the

Figure 11.4 Top view of 80 nm high nanochannels that are partially filled with ethanol. The flexible membranes on top of these membranes are bent downward due to the induced negative pressure in the liquid. This leads to remarkable shapes of the liquid menisci in these channels.[23,24] The peculiar protrusion of the meniscus becomes more prominent for wider channels because of the increasing compliance with increasing channel width.[24] (Reprinted with permission from Ref. 24, copyright 2009, Elsevier.)

meniscus in the plane normal to the observed surface. To obey the constant mean curvature imposed by the Young-Laplace equation everywhere along the meniscus, the curvature in the lateral plane will vary, as observed. As can be seen in Fig. 11.4, this effect becomes more pronounced for more compliant channels. In this section, after introducing some general concepts of elastocapillarity, we will develop an analytical model for the observed curvature of the meniscus.

Let us first have a look at capillary forces in terms of scaling laws. If the characteristic size of a structure is L, capillary forces are proportional to L, while pressure forces (such as wind drag) or elastic forces are proportional to L^2 and body forces (such as gravity) to L^3. When all the dimensions of a given structure are scaled down, capillary forces decrease slower than pressure and body forces and eventually become dominant. As a consequence,

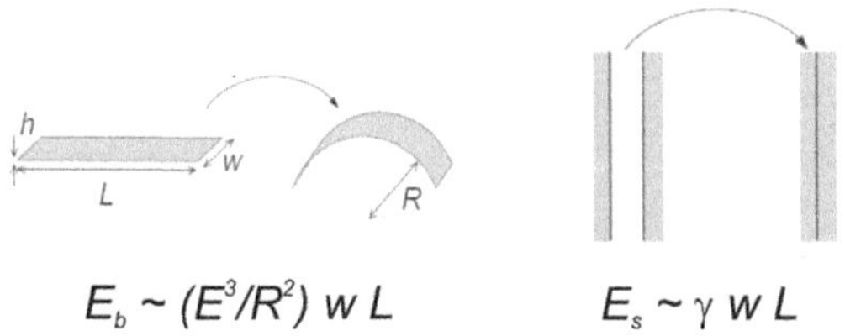

Figure 11.5 Comparison of elasticity and capillarity. The figure on the left shows a thin plate that is bent into a curved shape of radius R; the right figure shows two plates covered by a thin liquid layer, represented by the solid lines. Balancing mechanical bending energy (left) and interfacial energy (right) leads to the elastocapillary length $L_{ec} = \sqrt{Eh^3/\gamma}$.

a liquid droplet smaller than the capillary length $L_c = \sqrt{\gamma/\rho g}$, where γ and ρ are respectively the surface tension and the density of the liquid, adopts a spherical shape when deposited on a surface. In the case of elastocapillarity the capillary forces have to be compared with elastic forces. In this respect we will have to define a characteristic length scale, the elastocapillary length, L_{ec}, present in any problem combining mechanical bending rigidity and surface tension forces. To this purpose, consider an elastic plate (Young's modulus E) of thickness h and area wL that is covered by a liquid (see Fig. 11.5). Bending the plate to a curvature $\kappa = 1/R$ requires a mechanical bending energy of the order $E_b \approx Eh^3\kappa^2 wL$. The surface energy of the area wL covered by a liquid (represented by the dark area in the right picture of Fig. 11.5) is given by γwL, so the comparison of both energies leads to the characteristic elastocapillary length L_{ec}.[30,31]

$$L_{ec} = \sqrt{Eh^3/\gamma} \tag{11.4}$$

L_{ec} sets the typical curvature that capillary forces can induce on a flexible sheet: a structure is significantly deformed by surface tension if its length is large in comparison with L_{ec}. Indeed capillary forces produce a larger torque (larger lever arm) on a larger structure. If a given structure is scaled down homothetically by a factor $\lambda < 1$, L_{ec} is multiplied by a factor $\lambda^{3/2}$ ($L_{ec} \approx h^{3/2}$) and thus decreases faster than the length of the structure. Capillary forces are effectively crucial for smaller structures since the condition $L \approx L_{ec}$ may eventually be reached, which is the case in many situations of microstructures and microcapillaries.

The meaning of the elastocapillary length is well illustrated by the beautiful experiment performed recently by Py *et al.*[30] They used the interaction between elasticity and capillarity to produce three-dimensional (3D) structures. A liquid droplet is deposited on a thin planar sheet, and when the droplet evaporates it is spontaneously wrapped by the membrane. The final encapsulated 3D shape can be controlled by tailoring the initial geometry of the flat membrane. We now see that the critical length below which encapsulation cannot occur is the elastocapillary length L_{ec}.

Examples of elastocapillary forces in everyday life are legend: When a paintbrush is dipped into a pot of paint and withdrawn, surface tension forces cause the individual hairs in the brush to coalesce; the wet fur of a dog coming out of a pool assembles into bunches. The elastocapillary coalescence of a bundle of flexible lamellae (the "hairs" of the brush) when withdrawn from a bath was also studied systematically in 2004 by Bico *et al.*[31] More dramatically, capillarity can cause the collapse of lung airways of premature infants with fatal consequences.[32] As mentioned before, the collapse of fabricated MEMS structures by capillary forces during drying of the structures after a wet-release step[25] is a well-known problem. Different strategies have been developed to reduce or overcome this problem, including freeze drying and supercritical drying of fabricated structures.[27–29] Similarly, arrays of micro- or nanorods tend to form bundles in wet conditions. Cellular or "tepee" structures have indeed been observed with "carpets" of carbon nanotubes,[33–37] ZnO or Si nanowires,[38] or polymeric micropillars.[39,40] Elastocapillary deformation also plays a role in stamp deformation in microcontact printing and influences the quality of the pattern transfer.[41,42]

11.2.3.2 Elastocapillarity in nanocapillaries: the formation of a liquid meniscus in a flexible fluid channel

As we saw in Fig. 11.4, thin flexible membranes of micrometer-sized capillaries can be significantly deformed due to the surface tension forces of the liquid inside.[24] The figure shows silicon oxide channels of several micrometer width and about 80 nm height. These channels are capped by a thin, flexible silicon oxide–silicon nitride (SiO_2/Si_3N_4) bilayer membrane that is transparent so that we can observe from above the liquid menisci inside.

When a partially wetting liquid fills a capillary, the pressure in the liquid is lower than outside. The sum of the two curvatures of the liquid–air interface is negative, and according to the Young-Laplace equation the pressure at the liquid side of the meniscus is negative compared to the gas side. The induced negative pressure difference between the liquid pressure and the pressure outside bends the thin membrane on top of the channels downward. It was observed[23] that the elastocapillary equilibrium between the surface tension of the wetting liquid and the mechanical forces in the capillary results in a very peculiar shape of the interfacial meniscus, visible from the top through the transparent membrane. For increasing deflection of the membrane, the meniscus is seen to protrude along the channel and its curvature changes from concave to convex in the center.

Let us analyze this effect in more detail, following the analysis presented in Ref. 24. As we will show here, it is possible, under some conditions, to deduce an expression for the characteristic shape of the meniscus in the capillary. Besides, the analysis shows us that the protrusion length of the meniscus, which can be measured easily, is an accurate and useful indicator for the membrane deflection.

Figure 11.6 shows a schematic representation of a microcapillary cross section. The capillary shown here is a hydrophilic channel several micrometers in width and a height h of about 80 nm, with a thin ($\approx$180 nm) SiO_2/Si_3N_4 membrane on top. When an array of those capillaries was filled with a fluid, observation using an optical microscope showed the peculiar shapes of the menisci at the fluid–air interface at the front of the liquid plugs (see Fig. 11.4).

To describe this typical shape of a meniscus, we adopt the geometry of Fig. 11.7, in which the channel is directed along the

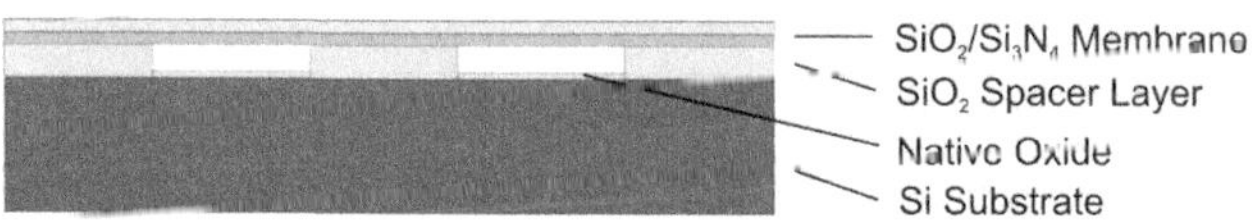

Figure 11.6 Cross-sectional view of a nanochannel structure covered by a SiO_2/Si_3N_4 membrane (reprinted with permission from Ref. 24, copyright 2009, Elsevier).

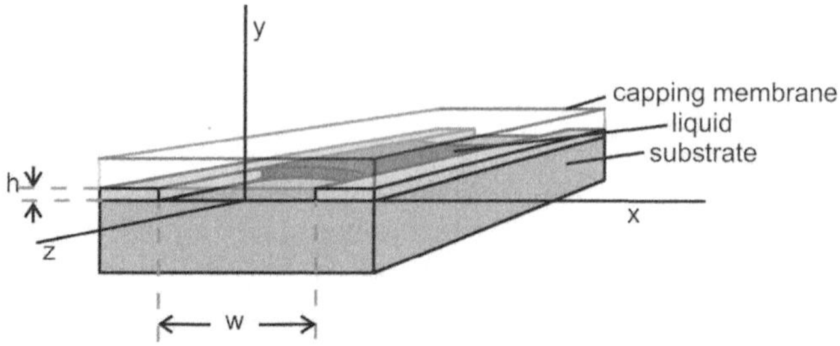

Figure 11.7 Schematic picture of the analyzed geometry. The channel is directed along the z-axis and has a width w in the x-direction and a height h along the y-axis. The channel is capped by a membrane and partially filled with liquid. (Reprinted with permission from Ref. 24, copyright 2009, Elsevier.)

z-axis, with its height in the y-direction and the width along the x-direction. Let the fluid have a contact angle θ and surface tension γ. The curvatures of the surface in the x- and y-directions are $1/R_1$ and $1/R_2$. One can then write the curvatures in terms of the function $z = f(x, y)$ describing the meniscus shape.

Using the notation, the Young-Laplace equation becomes:

$$-\vec{\nabla} \cdot \left(\frac{\vec{\nabla} f}{\sqrt{1 + (\vec{\nabla} f)^2}} \right) = \frac{p^{(l)} - p^{(g)}}{\gamma} \tag{11.5}$$

where the left-hand side is the sum of the curvatures $1/R_1 + 1/R_2$ and $p^{(l)}$ and $p^{(g)}$ represent the pressures in the liquid and in the gas, respectively. Equation 11.5 has to be solved for f, together with the boundary condition that the fluid surface and the channel walls intersect at the contact angle θ.

The rectangular channel is covered by an elastic membrane that is pulled downward due to the negative pressure inside the liquid. This downward deflection can be written as $u(x, z)$. If we make the assumption that, the z-derivatives of u are much smaller than the x-derivatives and that the length of the meniscus is not much larger than the channel width w, then the deflection in the elasticity problem is virtually independent of z and we are left with a quasi one-dimensional problem, the so-called cylindrical bending of a long, rectangular plate. The equation of elasticity for the deflection

$u(x, z)$ becomes:

$$-\kappa \partial_x^4 u + S_x \partial_x^2 u = \begin{cases} 0 & \text{membrane} - \text{air} \\ p^{(l)} - p^{(g)} & \text{membrane} - \text{liquid} \end{cases} \quad (11.6)$$

where $\Delta = \partial_x^2 + \partial_z^2$, $\kappa = Et^3/12(1 - \nu^2)$, S_x denotes the tensile stress in the x-direction, E is the effective Young's modulus, and ν the Poisson ratio of the membrane composite. Besides $u(x, z)$ should fulfill the boundary conditions that read $\partial_x u(x, z) = 0$ and at $x = 0$ and $x = w$. In principle Eqs. 11.5 and 11.6 are coupled via the pressure difference Δp. Besides, the membrane area on which the pressure acts is defined by the shape of the meniscus, which is still unknown. To solve the problem analytically one should therefore make some approximations. Doing so, one can write for the solution $u(x)$ for small deflections a fourth-order polynomial function in x with a maximum deflection U_0 at $x = w/2$ determined by the lateral and tensile forces. In normalized variables $\xi - 1 - 2x/w$ and $U = u/h$, one obtains as a solution of the deflection Eq. 11.6:

$$U(\xi) = U_0(1 - \xi^2)^2 \quad (11.7)$$

where U_0, representing the center deflection at $x = w/2$, is a positive constant proportional to Δp and further only depends on the mechanical properties of the membrane.

The central idea of the approach is now that the sum of the two curvatures, proportional to the Laplace pressure drop across the meniscus, is constant while the radius in the vertical (y)+ direction is determined by the local height of the channel, whereas the radius in the $x - z$ plane can be directly related to the x-derivatives of f, $\partial_x f$ and $\partial_x^2 f$. By integration we can thus obtain the function f as a function of x.

The capping membrane and the bottom of the channel are separated by the local height $h_l(x) = h - u(x)$ at a given z. Since this distance is much smaller than w, the curvature in the y-direction can be calculated locally as the curvature of the meniscus between two flat plates separated by the distance $h_l(x)$:

$$\frac{1}{R_2} = -\frac{2\cos\theta}{h_l(x)} \quad (11.8)$$

In the quasi one-dimensional approach, when the meniscus length is smaller or about equal to the channel width w, the original

Eqs. 1.5 and 1.6 are reduced to the relatively simple one-dimensional equation:

$$-\frac{d}{dx}\left(\frac{\frac{df}{dx}}{\sqrt{1+\left(\frac{df}{dx}\right)^2}}\right)=\frac{\Delta p}{\gamma}+\frac{2\cos\theta}{h(1-U(\xi))}. \tag{11.9}$$

Equation 11.9 includes the unknown function $f(x)$ and the unknown pressure difference Δp between the liquid and the gas. It has to be solved together with the boundary conditions that can be used to define Δp and one of the arbitrary constants. The solution of Eq. 11.9 has been presented in Ref. 24 to be:

$$f(\xi)=\frac{w}{2}\int\frac{d\xi\, P(\xi)}{\sqrt{1-P^2(\xi)}} \tag{11.10}$$

with $P(\xi)$ defined by:

$$P(\xi)=\frac{w\Delta p}{2\gamma}\xi+\frac{w\cos\theta}{4h\xi_1\sqrt{U_0}}\left(\frac{2\xi_1}{\xi_2}\arctan\frac{\xi}{\xi_2}+\ln\frac{\xi_1+\xi}{\xi_1-\xi}\right) \tag{11.11}$$

and $\xi_{1,2}^2 = 1/\sqrt{U_0}\pm 1$. The constant Δp in this equation has to be determined from the nonlinear equation $P(1)=-\cos\theta$. This is done numerically since both $\xi_{1,2}$ and U_0 depend on Δp. Figure 11.8 shows the observed meniscus of an ethanol–air interface in a 5.4 μm

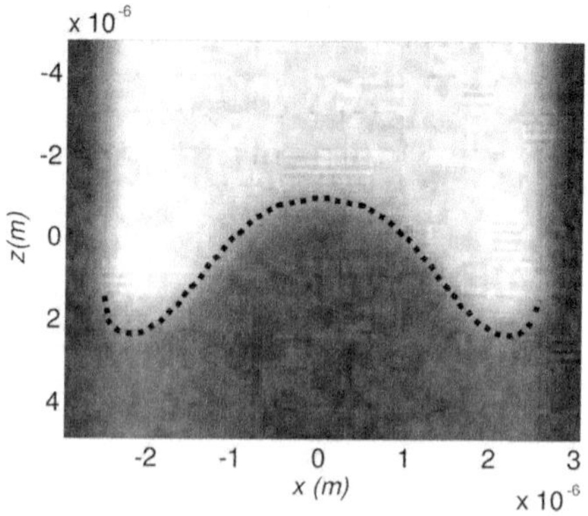

Figure 11.8 Image of an ethanol–air interface in a 5.4 μm wide channel, together with a calculated curve (black squares) according to the analysis of Ref. 24 (reprinted with permission from Ref. 24, copyright 2009, Elsevier).

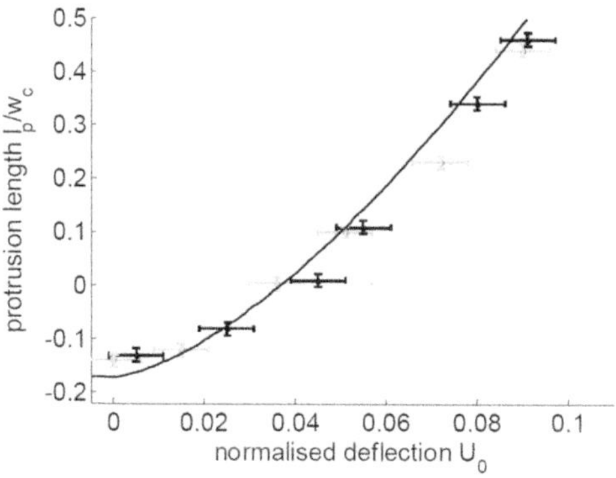

Figure 11.9 Measured normalized protrusion length l_p/w_c of the meniscus (ethanol in 80 nm channels) vs. the normalized deflection of the membrane U_0 as measured by AFM. The solid line shows the model prediction. At $U_0 = 0.1$, the pressure drop over the meniscus amounts to -6.7 bar. The protrusion length l_p has been normalized by the critical width w_c, defined as the channel width for which the center deflection is in the order of the channel height.[24] (Reprinted with permission from Ref. 24, copyright 2009, Elsevier.)

wide rectangular fluid channel together with the obtained solution according to Eqs. 11.10 and 11.11.

The observed elastocapillary interaction has a second interesting consequence. It has been found that the protrusion length of the meniscus (its extension along the channel, in the z-direction) can be directly related to the deflection of the membrane. More specific, there is a one-to-one relation between the deflection of the membrane and the meniscus length (that can be observed easily using a standard optical microscope), and therefore, the value of the pressure in the liquid can be accurately determined. To analyze the protrusion of the meniscus along the channel, one can define the protrusion length, l_p, as the distance between the front of the meniscus and its attachment point at the wall of the channel. The dependence of the measured protrusion length versus the membrane deflection as measured by atomic force microscopy (AFM) is shown in Fig. 11.9. The protrusion length l_p has been normalized by w_c. The solid line represents a model description according to the equations above. As is clearly seen, the protrusion length of the meniscus is a sensitive measure for the deflection of the membrane. This provides us with a new method to quantify the deflection of flexible sheets: the observation of the protrusion length l_p is a very convenient and accurate procedure to determine the deflection of the channel membrane.

In a previous analysis of a water plug in a flexible microchannel by Tas *et al.*,[23] a similar typical meniscus shape was found (Fig. 11.3). By measuring the protrusion length for this experiment and applying the model described above (Eqs. 11.5–11.11), a pressure in the water plug of -14 ± 8 bar is found, which is in relatively good correspondence with the pressure as measured via the local curvatures (-17 ± 10 bar) by the authors of Ref. 23 and the predicted -12 bar (Eq. 11.3).

11.2.4 *Limits to the Macroscopic Theory*[85]

When the characteristic diameter of nanochannels becomes smaller than roughly 10 nm, one has to take care in applying the macroscopic theory of capillarity. There are a number of effects to be considered when downscaling to nanoscopic dimensions: Possible changes in the surface tension, the validity of the macroscopic contact angle, and the possible presence of a thin wetting film changing the effective diameter (see also chapter 3 for a detailed description of these effects). Finally, the extreme strong capillary forces can possibly lead to elastocapillary deformation of the elastic bodies composing the channels. The effect of the wetting film and changes in the contact angle are strongly related and can be theoretically described in the framework of Derjaguin's disjoining pressure theory.[43] Firstly, we briefly address the surface tension of a strongly curved meniscus.

For strongly curved interfaces changes of the surface tension were predicted by Tolman based on Gibbs thermodynamic theory of capillarity. The Gibbs-Tolman-Koenig equation[44] predicts an increase in the surface tension for strong negative curvature, according to:

$$\gamma = \gamma_\infty (1 - \kappa \cdot \Delta\lambda) \tag{11.12}$$

where γ is the surface tension of the curved interface with mean curvature κ, γ_∞ is the surface tension of the flat interface, and $\Delta\lambda$ is the "thickness" of the interfacial region. Scarce experimental data

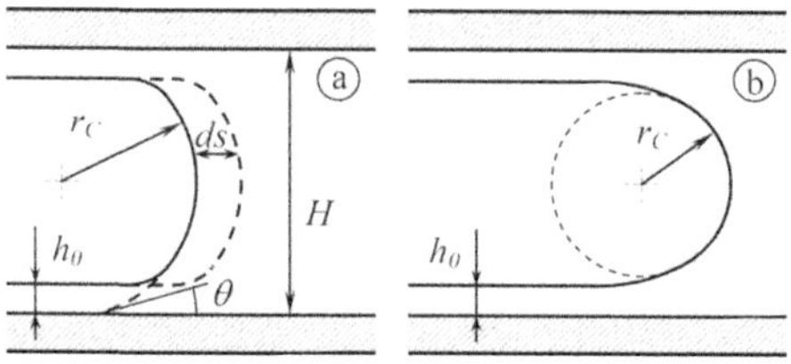

Figure 11.10 The coexistence of a thin wetting film in static equilibrium with a meniscus and a bulk liquid in the capillary. (a) We consider a small displacement of the meniscus δs, while the meniscus itself remains its shape. In the case of partial wetting the radius of the meniscus r_c exceeds half the capillary height, and it is possible to determine the effective contact angle θ needed to determine r_c. (b) In the case of complete wetting $r_c < H/2 - h_0$.

indicates that depending on the liquid, considerable deviations from Eq. 11.12 can be expected, with for water even decreasing values of γ with increasing negative curvature.[45]

In the case of partial or pseudo-partial wetting a wetting film can be in coexistence with a macroscopic amount of liquid.[46] This important situation applies to small capillaries where a thin wetting film covers the capillary walls and is in contact with a meniscus and where the thickness of the film h_0 is not negligible compared with the capillary height H (see Fig. 11.10). If the meniscus is in the contact with the wetting film, one can introduce the contact angle of the liquid phase by extending the curvature of the middle part of meniscus where the disjoining pressure does not act until the surface of the solid phase (Fig. 11.10a).

To describe the static case of the coexistence of the bulk liquid with a wetting film, Frumkin[47] proposed to use the concept of film tension, γ_{film}. This film tension γ_{film} is found by taking the derivative of the film free energy to the covered area under condition of constant volume and including an energy term representing the disjoining pressure effect.[46] Now, however, one should consider the constancy of the liquid volume of the total system, that is, of the film and the liquid in the channel. This was elaborated by Derjaguin,[48] who considered a small isothermal shift of the meniscus (by retaining its shape) over a distance ds, as shown in Fig. 11.10a. This process is accompanied by an exchange of dN moles of liquid between the film and the bulk liquid. During this isothermal process the infinitesimal variation of the energy in equilibrium, dU, is zero, so that an increase of the free energy of the wetting film equals the decrease of the energy of the liquid behind the meniscus. Assuming the molar volume of the liquid in the film and in the bulk to be the same, Derjaguin showed that:

$$-\frac{\partial P(h)}{\partial h} = \frac{\gamma + P(h)}{H/2 - h_0} = \Pi(h) \tag{11.13}$$

where $P(h)$ represents the interfacial interaction energy of the thin film as defined by $P(h) = \int_h^\infty \Pi(z)dz$, in which $\Pi(z)$ is the disjoining pressure as a function of the film thickness. The film thickness h_0 of the film in this analysis can be very small when, in the case of partial wetting, only a few adsorbed monolayers cover the solid surface.

The disjoining pressure in the film must be equal to the capillary pressure so that one can deduce from Eq. 11.13 an expression for the contact angle of the meniscus (Fig. 11.10a) for the partial wetting case:

$$\gamma \cos\theta = \gamma + \int_{h_0}^{\infty} \Pi(h)dh + \Pi(h_0)h_0 \tag{11.14}$$

We see from this equation that the contact angle is only larger than zero ($\cos\theta < 1$) under certain conditions for the shape and range of the disjoining pressure isotherm $\Pi(h)$. In the case of complete wetting (Fig. 11.10b), Eq. 11.14 cannot be applied, and the left-hand side of the equation is replaced by $p_c H/2$, with p_c the capillary pressure (we consider the absolute value in this paragraph). For complete wetting, the isotherm of the disjoining pressure can be expressed in the form $\Pi(h) \approx h^{-n}, n > 2$, as was done by Derjaguin,[49] and one can find the radius r_c of the meniscus via $r_c = \gamma/p_c$. In this way one arrives at an expression for the meniscus radius:

$$r_c = \frac{H}{2} - \frac{n}{n-1}h_0 \tag{11.15}$$

Equation 11.15 reveals that r_c is smaller than $H/2 - h_0$: the effective radius is less than half the capillary height minus the film thickness, and a transition zone between meniscus and film exists.

It has to be pointed out that due to the simplified estimation of the disjoining pressure in form of $\Pi(h) \approx h^{-n}, n > 2$, this homogeneous approach gives very accurate quantitative results only for the channels and pores with an effective diameter larger than 5–7 nm. If proper parameters for the molecular interaction are chosen, then pores with an effective size smaller than 7 nm can be quantitatively described with the nonlocal density functional theory (NLDFT), as was shown by Neimark *et al.*[50] for the case of capillary condensation of a Lennard-Jones fluid in nanopores. This theory gives a good match with Monte Carlo simulations for the pores as small as 2 nm and smoothly approaches Derjaguin's equations applicable for the pores larger than 7 nm.

It has also been shown that the thickness of the wetting film can be deduced from the behavior of the film tension as a function of disjoining pressure (this was done explicitly by, among others, Derjaguin,[51] and in Ref. 52). If a capillary is placed vertically and a meniscus coexists with a wetting film on the sides of the capillary, then the weight of the liquid column is carried by the tension of the wetting film.

In thermodynamic equilibrium, a small vertical displacement of the meniscus together with the liquid column below it, δs, implies that the weight of the liquid column performs an amount of work per unit length of the column that is equal to an increase of the free energy of the film. It is then possible to find the thermodynamic thickness of the wetting film on the channel wall from the known isotherm of the disjoining pressure and the film tension:[51,52]

$$\frac{d\gamma_{\text{film}}}{d\Pi} = h_0 \tag{11.16}$$

11.3 Dynamic Capillarity

11.3.1 *Lucas Washburn Model of Capillary Filling*[85]

The filling dynamics of capillaries driven by the negative Laplace pressure was first modeled by Washburn[53] and Lucas.[54] The filling speed is determined by the surface tension γ of the liquid filling the capillary, its (advancing) contact angle θ to the channel walls, the channel diameter d (or in the case of a rectangular cross section by the width w and the height h), and finally by the viscosity η of the liquid. For sufficiently long and narrow capillaries the flow of the liquid entering the capillary may be considered laminar and fully developed and the hydraulic resistance R = (pressure drop Δp)/(volume flow ϕ) follows from the Poiseuille flow model. Taking the negative of the Laplace pressure for the pressure drop and realizing that the change of the position x of the moving meniscus is related to the volume flow through $dx/dt = \phi/A$ (where A is the cross-sectional area of the channel), the following general expression for the position of the moving meniscus as a function of time results:

$$xdx = \frac{-P_{\text{Lap}}}{R^* \cdot A} dt \tag{11.17}$$

where R^* is the hydraulic resistance per unit length of the liquid plug ($R = xR^*$). As the hydraulic resistance is proportional to the liquid plug length x, it follows that the filling velocity decreases in time. More precisely, after integration the result is $x^2 \approx t$ or $x \approx t^{1/2}$. For example, for a flat, rectangular channel ($w >> h$) by substitution of R^* and p_{Lap} we find:

$$x = \sqrt{\frac{\gamma \cos(\theta) h}{3\eta} \cdot t} \tag{11.18}$$

11.3.2 *Nanochannel-Based Measurements of Apparent and Real Viscosity Effects*[85]

In recent years this expression has been experimentally verified in silicon-based nanochannels with a perfectly defined rectangular cross section, and down to channel heights of 5 nm,[55–60] the qualitative $x = at^{1/2}$ dependence has been confirmed for water and aqueous solutions as well as for some organic liquids. Figure 11.11 shows a photograph of a chip used in such experiments. Liquid can be introduced in the access holes and fills the chip by capillary action, first through the inlet microchannel and then through the nanochannels. Rulers have been etched next to the nanochannels to follow the filling process in time.

Figure 11.12 shows an example of the measured position of the moving meniscus versus time in capillary filling of hydrophilic nanochannels by deionized (DI) water (thin “native” silicon dioxide channel bottom and borosilicate glass channel roof, rectangular

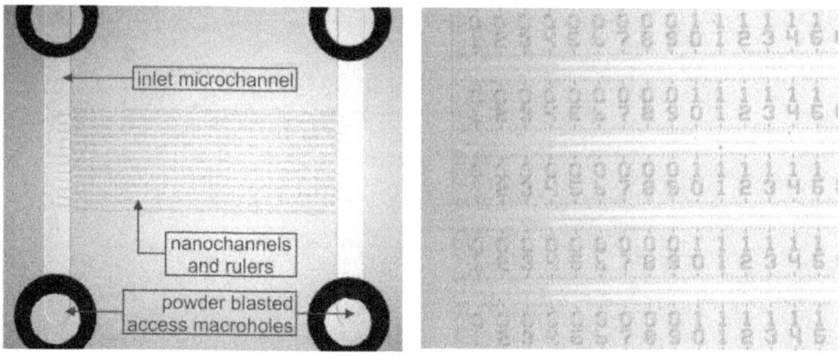

Figure 11.11 Top view of a chip (approx. $1 \times 1\ cm^2$) used in capillary-filling experiments (left). Zoom-in showing a still image of the filling of 30 nm channels (right).

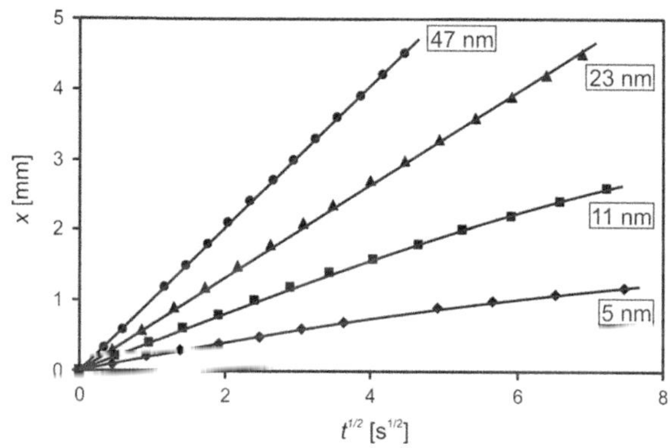

Figure 11.12 Measured position of the moving meniscus as a function of $t^{1/2}$ for filling of 5, 11, 23, and 47 nm deep nanochannels with DI water at temperatures of 22.7, 22.2, 21.6, and 20.8 ± 0.5°C, respectively (reprinted with permission from Ref. 60, copyright 2008, American Institute of Physics).

cross section $h << w$). Note that down to the level of h = 5 nm the moving meniscus can be observed by optical microscopy. For the smallest channel heights, differential interference contrast (DIC)[60] or an embedded interferometer[59] were used to enhance the contrast. In the analysis of the filling dynamics, usually macroscopic contact angles and surface tension values are used, which for small molecules should be correct when channel hydraulic diameters are in the order of 10 nm or more.

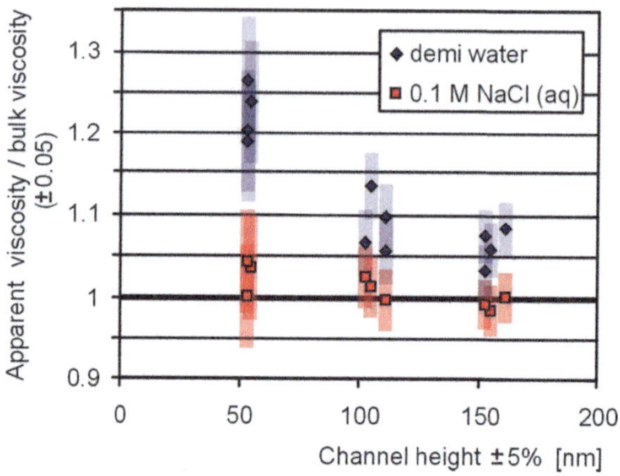

Figure 11.13 Ratio of the apparent viscosity over the bulk viscosity of DI water (blue diamonds) and aqueous NaCl solution (red squares) as derived from capillary-filling experiments (reprinted with permission from Ref. 56, copyright 2004, American Institute of Physics).

While the qualitative agreement with the Washburn model is good, several significant quantitative deviations have been observed. As they all concern slower than expected filling they are usually described by a dimensionless apparent viscosity η_{app}/η larger than 1 (where $\eta_{\text{app}}/\eta = 1$ refers to the situation described by Eq. 11.18). Figure 11.13 shows the results as reported by Tas *et al.*[56] for filling of silica channels (20 nm SiO_2, boroslicate glass channel cover), by DI water and aqueous 0.1 M NaCl solution. A significant slower than expected filling for DI water was attributed to the electroviscous effect, as the values of η_{app}/η returned to one when filling the channels by aqueous NaCl solution. The electroviscous effect is related to the electrical double layer that is formed when the capillary walls in contact with the liquid are charged (see Schoch *et al.*[61] for an extensive review of electrokinetic effects in nanochannels). Due to the flow a net displacement of charge and an induced streaming potential result, which results into a counter electro-osmotic flow in a part of the electrical double layer that is closest to the channel walls[62] (see also chapter 2). A comparison[63] of the observed value of η_{app}/η with the theory of Levine, however, is not straightforward because of deprotonation of the hydroxylated channel walls[64] and possible release of adsorbed ions from these walls during filling, leading to a modified and possibly axially nonuniform ionic composition of the solution filling the channel. In addition, it is likely that the constant potential assumption should be replaced by some form of charge regulation.

An important aspect of capillary filling, especially in channels having sharp inner corners, is the formation of bubbles.[65] In sharp inner corners a strong capillary suction develops, which leads to capillary "fingers" possibly proceeding in front of the main meniscus (see section 11.3.4). The length of these fingers depends on the filling time and the "consumption" at the back by the proceeding main meniscus. As the main meniscus slows down in time the length and volume of the fingers tend to increase and eventually can lead to bubble formation through enclosure of air in front of the main meniscus. It has been suggested[65] that the presence of bubbles can be responsible for the sometimes observed slower than expected filling, as shown in Fig. 11.13. However, the Washburn model is derived for filling without bubbles and should therefore only be applied up to the length where the first bubbles are formed.[56]

11.3.3 *The Viscosity of Water in Nanometer Confinement*[85]

Since the early 1970s there is a notion that water near polar or charged surfaces is somehow ordered or structured, leading to a so-called structural component of the disjoining pressure. The first indications came from interpretation of the disjoining pressure isotherms of wetting films of water on hydrophilic substrates,[66] while unambiguous experimental proof came through application of the so-called SFA,[67–71] showing short-range repulsion forces between charged surfaces brought in close proximity (<5 nm), in addition to the double-layer repulsion and van der Waals attraction as described by Derjaguin and Landau, and Verwey and Overbeek (DLVO) theory. Despite the enormous importance of water as a liquid in nature as well as engineering systems, the exact nature of this hydration force as it is called for water has still not been conclusively determined. The main mechanisms that have been discussed in literature are related to the compression of hydrated ions (ionic surface groups, adsorbed ions, or counterions in the electrical double layer), orientation of water dipoles in the strong electric field near a charged wall, or the hydrogen-bonding nature of water. An important question that emerges is if and how the supposed interfacial "structuring" of water affects its local viscosity. In the early 1970s measurements of the flow velocity of liquid plugs in quartz microcapillaries driven by a disbalance between the capillary pressure and an applied external pressure revealed an elevated viscosity of water up to 40% in capillaries of 0.04 μm.[72] These results could not be reproduced in nanometer-thin films in an SFA. A dynamic SFA technique revealed a viscosity of water equal to the bulk viscosity, even in 2 nm thin films and under high salt concentrations necessary to induce a strong repulsive hydration force between the mica surfaces.[73] A similar conclusion was reached by SFA squeeze-out experiments of thin water films between silica surfaces.[71] More recently, in an SFA where also shear force can be induced and measured, Raviv *et al.*[74] found that the viscosity of demineralized water when confined to films of 3.5 $\pm$ 1 to 0.0 $\pm$ 0.4 nm between mica surfaces remains within a factor three of its bulk value. Similar experiments with aqueous salt solution (in the presence of repulsive hydration forces) show that the bounded water molecules in the hydration layers retain the shear fluidity characteristics of bulk water, with an upper limit estimated for the viscosity of 0.08 Pa·s (which is maximally 80 times the bulk viscosity).[75] Remarkably, recent AFM experiments, measuring directly and simultaneously the normal and lateral forces of a nanosized silicon tip approaching a solid hydrophilic surface, reveal orders-of-magnitude increase in the viscosity of water with respect to bulk water in subnanometer confinement.[76] Capillary filling of silica nanochannels by water and aqueous salt solutions reveals a significant slower than expected filling in channels with heights in the range of 10–50 nm.[58,60] If interpreted as a viscosity effect, the averaged effective viscosity is up 30% higher for 11 nm channels[60] (Fig. 11.14). The red line is a fit to the experimental data assuming a highly viscous layer next to the hydrophilic channel walls. The best fit is found for a thickness of this immobile layer of 0.9 nm at each interface. Filling experiments under different conditions are currently carried out in our lab to test the hypothesis

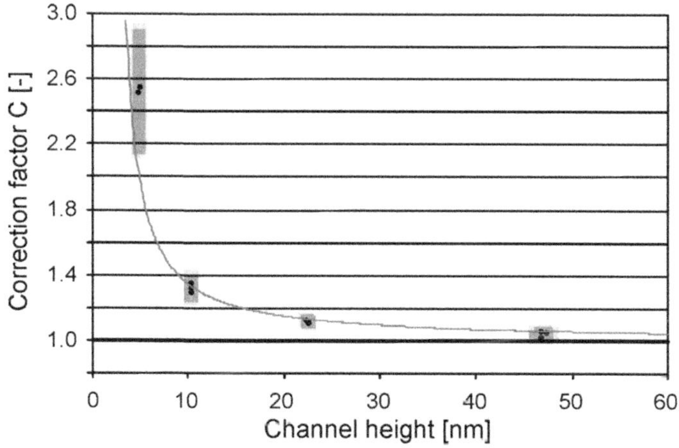

Figure 11.14 Dimensionless correction factor (prefactor of the denomitator in Eq. 11.18, indicating a slower than expected filling of water in flat, rectangular silica nanochannels,[60] as a function of channel height. The red line indicates the effect of a solidified layer next to the channel walls with a thickness of 0.9 nm or four monolayers. (Reprinted with permission from Ref. 60, copyright 2008, American Institute of Physics.)

that it is indeed the enhanced viscosity of hydration layers that causes the observed slower than expected filling. On the basis of all experimental data available, one must conclude that the question if and how hydration forces are connected to viscosity changes of interfacial water has still not been completely answered.

11.3.4 *The Flow of a Wetting Liquid along the Corners of a Noncircular Capillary*

A specific phenomenon related to capillaries having sharp corners, occupied by a wetting liquid, is the occurrence of a laminar flow of the liquid along the corners of the duct. The problem of microcapillaries occupied predominantly by a nonwetting phase and a wetting liquid is encountered in many different situations, such as oil trapping, foam generation, weeping flow, and capillary imbibition.[77] Also in the wetting of powders[78] and in oil recovery, one has to deal with a wetting liquid imbibing the capillary tubes.

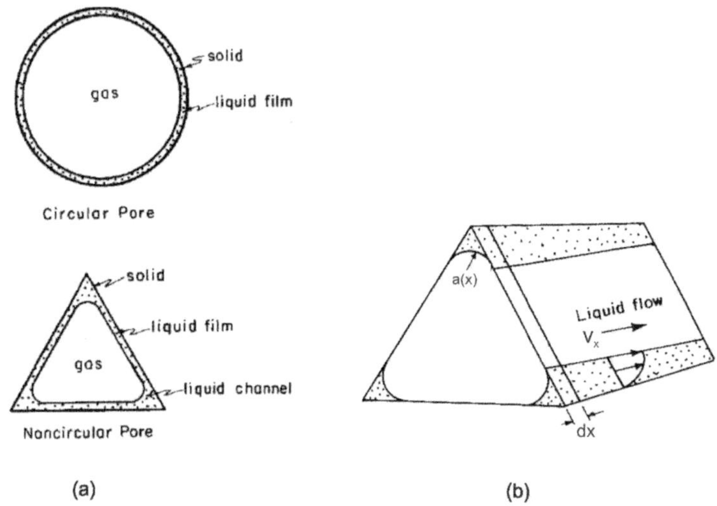

Figure 11.15 (a) Cross-sectional views of a circular and noncircular capillary occupied by a gas and a wetting liquid. (b) Liquid flow along the corners of a capillary tube, directed along the x-axis. The corner flow problem can be solved using the lubrication approximation so that the velocity profile adopts a parabolic shape, as depicted. The gas–liquid interface is a circular arc of radius a, which weakly depends on x. (Reprinted with permission from Ref. 83, copyright 1988, Elsevier.)

If a cylindrical capillary is occupied by a wetting liquid and a gas, a thin film of the liquid is observed on the interior wall of the capillary in the regions occupied by the gas. In many cases of (micro-) capillaries, however, one has to deal with capillary cross sections that are noncircular, characteristic for underground porous media and microfabricated channels. A considerable amount of liquid flow occurs in these cases in the corner regions of the capillary tubes (see Fig. 11.15a).

The problem of two-phase flow through noncircular ducts has received considerable attention in the last decades[79–82] and was studied specifically by Ransohoff and Radke.[83] They addressed the problem of the low Reynolds number wetting liquid flow in a noncircular capillary occupied predominantly by a nonwetting gas phase by presenting a solution in terms of a dimensionless flow resistance, β, depending on corner geometry and fluid parameters.

To understand the main steps of their approach, we consider the geometry depicted in Fig. 11.15b.

It is assumed that the curvature of the capillary in the axial, x-, direction is small compared to the curvature of the gas–liquid interface. Also, the variation of interfacial curvature and surface tension gradients with respect to axial position is assumed to be small. We neglect gravitational forces and consider cases of low Reynolds numbers so that inertial terms may be neglected. The wetting liquid is an incompressible and Newtonian fluid of viscosity μ. A perfectly sharp corner in the capillary cross section can be characterized fully by its interior angle α only. To describe as well a corner that is "rounded," Ransohoff and Radke used the parameter R_0 for the degree of roundedness, which equals zero for perfectly sharp corners, while for $R_0 = 1$ the liquid in the "corner region" only exists as a thin film. Because the gas phase pressure, p_g, is a constant, and the liquid-phase pressure, p_l, varies due to the liquid flow in the x-direction, the mean curvature of the interface, κ_m, must also vary with the axial position, as described by the Young-Laplace equation. Since the curvature of the interface in the axial direction is negligible, the curvature κ_m is simply equal to $1/a(x)$, which does not change along a distance dx (see Fig. 11.15b). To find now the full velocity profile in the cross-sectional plane, we should solve the Navier-Stokes equations for the specific corner geometry, together with the no-slip condition at the solid walls and the continuity equation. In their analysis Ransohoff and Radke adopt a radial coordinate system and the axial velocity of the wetting liquid, v_x, is calculated for a variety of corner geometries using a numerical approach. The velocity v_x is linearly proportional to the pressure gradient, dp_l/dx, and the constant of proportionality depends on the cross-sectional area of the fluid in the corner and the viscosity μ. Hence a dimensionless flow resistance, β, is defined as:

$$\beta = \frac{a^2}{\mu < v_x >}\left(\frac{-dp_l}{dx}\right) \tag{11.19}$$

The average velocity, $< v_x >$, is calculated by dividing the flow rate with the entire area of the corner. This means that if the corner is rounded (i.e., $R_0 > 0$), then the area used for determining $< v_x >$ will include the filled-in region of the corner as well as the flow region. According to the definition of the parameter β in Eq 11.19, we see that it is a constant for a given corner geometry and is independent of the absolute size of the corner.

The flow resistance β could thus be calculated and was presented in Ref. 83 as a function of corner geometry, surface shear viscosity, and the contact angle. The effect of corner geometry and the contact angle is to change the cross-sectional area available for the flow, while the surface shear viscosity affects the boundary condition at the gas–liquid interface. It is found that β increases with increasing half-angle, degree of roundedness, surface shear viscosity, and contact angle.

We can now apply the dimensionless flow resistance to analyze the dynamics of the imbibition of a wetting liquid in the corners of a square capillary tube (that is occupied predominantly by air). If a small amount of wetting liquid is introduced into one end of a dry square capillary, the liquid can imbibe only along the corners of the tube due to capillary pressure gradients. The capillary tube being sealed, the origin meniscus of the liquid does not move during this imbibition. Along the length of the capillary tube, the curvature of the liquid meniscus increases with increasing distance relative to the origin meniscus and consequently the pressure in the liquid decreases in the flow direction. Under this condition, the liquid will flow from the high-pressure to the low-pressure part of the system. The imbibition of a wetting liquid in the corners of a noncircular capillary is the result of capillary pressure gradients. By assuming a very large radius of curvature in the plane parallel to the tube, the shape of the meniscus seen in a cross section normal to the tube axis is a fraction of a circle with radius r, which, as we saw above, is related to the capillary pressure $p_c = p_l - p_g$ and the surface tension γ at any location x and time t by $p_c(x, t) = \gamma / r(x, t)$. Dong and Chatzis[80] considered this situation of imbibition along the corners of square capillaries specifically, starting from the situation as sketched in Fig. 11.16. Let us consider the imbibing film in the rectangular interior corner and define S as the cross-sectional area of the wetting liquid at the location x. With q as the volumetric flow rate, the equation of continuity gives:

$$\frac{\partial q}{\partial x} = -\frac{\partial S}{\partial t} \tag{11.20}$$

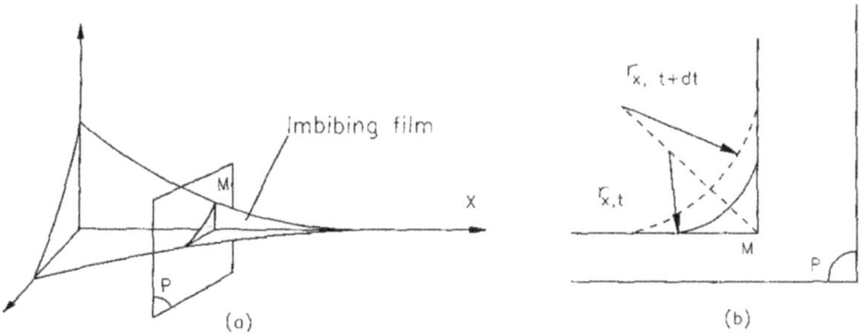

Figure 11.16 (a) Schematic representation (a) and cross-sectional view (b) of an imbibing film of a wetting liquid in the interior corner of a capillary. (Reprinted with permission from Ref. 80, copyright 1995, Elsevier.)

An expression for the corner area S available for fluid flow herein can be found in terms of the contact angle θ and radius of curvature r.

For the calculation of the average flow velocity $< v_x >$, one can use the dimensionless flow parameter β as defined by Ransohoff and Radke:

$$< v_x > = -\frac{r^2(x,t)}{\beta\mu}\frac{\partial p_c}{\partial x} \tag{11.21}$$

which is a function of x and t. The dimensionless flow resistance provides an exact solution to the noncircular conduit flow problem. The average velocity is calculated by dividing the flow rate by the entire cross-sectional area of the corner within the fluid meniscus. The volumetric flow rate q at any location x and time t is therefore:[80]

$$q(x,t) = C\frac{r^4(x,t)}{\beta\mu}\frac{\partial p_c}{\partial x} \tag{11.22}$$

where C is a shape factor that depends on the contact angle and number of the corners in a tube (in Ref. 80 it was calculated explicitly for square capillaries). Inserting now the Young-Laplace relation, $p_c(x,t) = \gamma/r(x,t)$, into Eq. 11.22, one obtains:[80]

$$q(x,t) = C\frac{r^4(x,t)}{\beta\mu}\frac{\partial p_c}{\partial x} \tag{11.23}$$

Together with the continuity equation, Eq. 11.20, this leads to the nonlinear partial differential equation:[80]

$$\frac{\partial S}{\partial t} = B\frac{\partial}{\partial x}\left(S^{1/2}\frac{\partial S}{\partial x}\right) \tag{11.24}$$

with $B = 1/2C^{-1/2}\gamma/\beta\mu$. Dong and Chatzis analyzed this equation in detail and showed that the saturation of the wetting liquid in the main meniscus varies rapidly from 100% (all of the cross-sectional area of the capillary is liquid) to 6% (only the corners are wetted by the imbibing liquid). After the imbibition front advances a certain distance x_0, the shape of the main meniscus can be considered constant ($t = 0$). At position $x = 0$ we then define a constant $S = S_0$. The nonlinear diffusion Eq. 11.24 had been analyzed before in 1983 by Mayer *et al.*,[81] who provided a solution for different boundary conditions equivalent to constant pressure, constant flow rate, and instantaneous deposition of the fluid. In line with the approach of Ref. 81, Dong and Chatzis adopted a solution for constant pressure at the origin meniscus, which can be written in good approximation as:

$$S = S_0\left(1 - \frac{ax}{x_0\sqrt{1+2\nu t}} + \frac{(a-1)x^2}{x_0^2(1+2\nu t)}\right)^2 \qquad (11.25)$$

where a is a parameter that has to be determined later numerically by equating the spread volumes that are calculated for long times and those obtained by numerical integration using the distribution of fluid in the corner, according to Eq. 11.25. The time constant ν is related to the initial shape of the liquid meniscus, proportional to $\gamma/\beta\mu$, and can be calculated by a variational method as Dong and Chatzis elaborated. Assume we introduce an amount of liquid at one end of a square capillary filled with gas. When we seal this end, the liquid can imbibe only along the corners. The front of spreading moves according to:[80]

$$x_f = x_0(1+2\nu t)^{1/2} \qquad (11.26)$$

Equation 11.25 can therefore be written as:[80]

$$S = S_0\left(1 - a\frac{x}{x_f} + (a-1)\left(\frac{x}{x_f}\right)^2\right)^2 \qquad (11.27)$$

For large times, $\nu t >> 1$, Eq. 11.26 shows that the front of spreading follows the classical square root of time behavior. The expression for x_f can thus be written for large times, as done in Ref. 80, as:

$$x_f = \sqrt{2K\frac{\gamma}{\beta\mu}Rt} \qquad (11.28)$$

with R the radius of the liquid meniscus in the corners at the equilibrium capillary pressure at $S = S_0$ and K a constant to be determined numerically.

Along this approach, Dong and Chatzis also deduce an expression for the volumetric flow rate, showing that q is proportional to $(\gamma/\beta\mu)^{1/2}$ and $t^{-1/2}$, which yields after integration an expression for the total volume V imbibed at time t. It is seen that the imbibed volume V is proportional, apart from geometrical and numerical constants, to $(\gamma/\beta\mu)^{1/2}t^{1/2}$. Experimental results of imbibition of a slug of a wetting fluid introduced in a 300 and 500 μm square capillary tube support these expressions. The spreading of different liquids of varying viscosity and surface tension show the proportionality of the imbibition rate with $(\gamma/\mu)^{1/2}$. The imbibition rate decreases greatly when the spreading liquid does not perfectly wet the wall of the channel. This is mainly due to the cross-sectional area available for fluid flow of the wetting liquid in the corners becoming smaller as the contact angle increases. It was also found that the imbibition rate of a wetting fluid square capillary tube decreases with decreasing tube size. Theoretically, the rate is proportional to $d^{1/2}$, where d is the tube size. The latter result is important for the flow of the wetting phase in the corners of the continuum of pore space involving large and small pores in a pore network, such as in the water-flooding method of oil recovery in water-wet media. The mobility of water along the corners of larger pores is very significant and plays an important role in the oil-trapping mechanisms.[84]

The reader may note that the above analysis leads to a singularity at the front location of the liquid column in the corner because of the infinite curvature at the point $x = x_f$. It is important to realize that the usefulness of the solution is restricted to slender fluid columns and slight curvature along the x-axis. Weislogel and Lichter[82] placed the results obtained by Dong and Chatzis in a broader context by developing a very general analytical model of capillary flow in an interior corner of arbitrary geometry and under varying boundary conditions. The capillary channel they consider is placed vertically in an environment of varying gravitational forces. They point out that different types of flow behaviors can be distinguished depending on the boundary conditions for the flow. The similarity solutions developed in Ref. 82 also account explicitly for much of the variation of flow resistance due to the geometry of the capillary, because of the scaling, so that the effects of corner geometry can be separated from those of surface curvature.

11.4 Summary and Conclusions

We have introduced the concept of capillary negative pressure that can occur in submicrometer capillaries. It exists when the negative Laplace pressure exceeds the atmospheric pressure of the gas phase. Flexible nanochannels have been used to measure this negative pressure. An elastocapillary model for deformable nanochannels has been discussed. It describes the peculiar shape of the meniscus in the deformed channel and can be used to calculate the negative pressure from the observed meniscus shape.

For truly nanoscopic capillaries several factors that can lead to deviations from macroscopic theory have been identified. A first step in including these effects in the theory of capillarity has been described in section 11.2.4. It was shown how the capillary pressure can be found, taking into account the presence of a wetting film through the use of Derjaguin's disjoining pressure.

We presented capillary filling as an important experimental technique to determine the dynamic properties of liquids in confinement. In 50–150 nm nanochannels (silicon covered by insulating 20 nm silicon oxide), a significant slower than expected filling with DI water was observed. This deviation disappears when electrolyte solutions are used for the filling. This behavior indicates that an electroviscous effect plays a role in the filling by DI water. In 10–50 nm nanochannels (silicon covered by native oxide only), a significantly slower than expected capillary filling was observed, which is possibly related to an enhanced viscosity of the hydration layers near the polar or charged channel walls. However, different experiments from different groups are not conclusive in answering the question if and how hydration forces are connected to viscosity changes of interfacial water.

An important mode of transport in noncircular capillaries having sharp corners, such as most silicon-based nanochannels, is through so-called corner flow. We summarized some of the key theories on corner flow in predominantly gas-filled capillaries. Note that model discussed based on Ref. 80 assumes that the capillary is sealed after introduction of the wetting liquid. This is an important boundary condition, which prevents the straightforward use of the model for the case of capillary filling of nanochannels, as discussed in sections 11.3.1 and 11.3.2.

References

1. J. Jurin, An account of some experiments shown before the Royal Society; with an enquiry into the cause of the ascent and suspension of water in capillary tubes, *Philos. Trans.*, **30**, 739 (1717).
2. J. Jurin, An account on some new experiments, relating to the action of glass tubes upon water and quicksilver, *Philos. Trans.*, **30**, 1083 (1717).
3. Sir Isaac Newton, *Opticks*. 3rd ed. London (1721).
4. Th. Young, An essay on the cohesion of fluids, *Philos. Trans. R. Soc. Lond.*, **95**, 65 (1805).
5. P. S. Laplace, *Mécanique Céleste*, Supplement to Vol. 4 of the 10th ed. (1806).
6. L. R. Fisher and J. N. Israelachvili, Determination of the capillary pressure in menisci of molecular dimensions, *Chem. Phys. Lett.*, **76**, 325 (1980).
7. N. V. Churaev, V. D. Sobolev, and Z. M. Zorin, Measurement of viscosity of liquids in quartz capillaries. In *Special discussions on Thin Liquid Films and Boundary Layers*. New York: Academic Press. 213 (1971).
8. M. B. Stern, M. W. Geis, and J. E. Curtin, Nanochannel fabrication for chemical sensors, *J. Vac. Sci. Technol. B*, **15**, 2887 (1997).
9. S. W. Turner, A. M. Perez, A. Lopez, and H. G. Craighead, Monolithic nanofluid sieving structures for DNA manipulation, *J. Vac. Sci. Technol. B*, **16**, 3835 (1998).
10. J. Han and H. G. Craighead, Entropic trapping and sieving of long DNA molecules in a nanofluidic channel, *J. Vac. Sci. Technol. A*, **17**, 2142 (1999).
11. J. Haneveld, H. Jansen, E. Berenschot, N. Tas, and M. Elwenspoek, Wet anisotropic etching for fluidic 1D nanochannels, *J. Micromech. Microeng.*, **13**, S62 (2003).

12. J. Haneveld, Nanochannel fabrication and charaterization using bond micromachining. PhD Thesis. University of Twente (2006).
13. F. Persson, L. H. Thamdrup, M. B. L. Mikkelsen, S. E. Jaarlgard, P. Skafte-Pedersen, H. Bruus, and A. Kristensen, Double thermal oxidation scheme for the fabrication of SiO_2 nanochannels, *Nanotechnology*, **18**, 245301 (2007).
14. J. Haneveld, N. R. Tas, N. Brunets, H. V. Jansen, and M. Elwenspoek, Capillary filling of sub-10nm nanochannels, *J. Appl. Phys.*, **104**, 014309 (2008).
15. C. Huygens, *Philos. Trans.*, **7**, 5027 (1672).
16. Sir Isaac Newton, *Opticks*, 3rd ed. London (1721).
17. H. H. Dixon and J. Joly, On the ascent of sap, *Phil. Trans. R. Soc. Lond. B*, **186**, 563 (1895).
18. M. Berthelot, Sur quelques phénomènes de dilatation forcée des liquides, *Ann. Chim. Phys.*, **30**, 232 (1850).
19. L. J. Briggs, Limiting negative pressure of water, *J. Appl. Phys.*, **21**, 721 (1950).
20. J. C. Fisher, The fracture of liquids, *J. Appl. Phys.*, **19**, 1062 (1948).
21. Lord Kelvin, On the equilibrium of vapour at a curved surface of liquid, *Proc. R. Soc. Edinburgh*, **7**, 63 (1870).
22. E. O. Wiig and A. J. Juhola, The adsorption of water vapor on activated charcoal, *J. Am. Chem. Soc.*, **71**, 561 (1949).
23. N. R. Tas, P. Mela, T. Kramer, J. W. Berenschot, and A. van den Berg, Capillarity induced negative pressure of water plugs in nanochannels, *Nano Lett.*, **3**, 1537 (2003).
24. J. W. van Honschoten, M. Escalante, N. R. Tas, and M. Elwenspoek, Formation of liquid mensici in flexible nanochannels, *J. Colloid Interface Sci.*, **329**, 133 (2009).
25. C. H. Mastrangelo and C. H. Hsu., Mechanical stability and adhesion of microstructures under capillary forces—part ii: experiments., *J. Microelectromech. Syst.*, **2**, 33 (1993).
26. T. Tanaka, M. Morigami, and N. Atoda, Mechanism of resist pattern collapse during development process, *Jpn. J. Appl. Phys.*, **32**, 6059 (1993).
27. N. Tas, T. Sonnenberg, H. Jansen, R. Legtenberg, and M. Elwenspoek, Stiction in surface micromachining, *J. Micromech. Microeng.*, **6**, 385 (1996).
28. C. H. Mastrangelo, Adhesion-related failure mechanisms in micromechanical devices, *Tribol. Lett.*, **3**, 223 (1997).
29. R. Maboudian, and R. T. Howe, Critical review: adhesion in surface micromechanical structures, *J. Vac. Sci. Technol. B*, **15**, 1 (1997).

30. C. Py, P. Reverdy, L. Doppler, J. Bico, B. Roman, and C. N. Baroud. Capillary Origami: Spontaneous wrapping of a droplet with an elastic sheet, *Phys. Rev. Lett.*, **98**, 156103 (2007).

31. J. Bico, B. Roman, L. Moulin, and A. Boudaoud, Elastocapillary coalescence in wet hair, *Nature*, **432**, 690 (2004).

32. A. L. Hazel and M. Heil, Surface-tension-induced buckling of liquid-lined elastic tubes: a model for pulmonary airway closure, *Proc. R. Soc. Lond. A*, **461**, 1847 (2005).

33. C. V. Nguyen, L. Delzeit, A. M. Cassell, J. Li, J. Han, and M. Meyyappan, Preparation of nucleic acid functionalized carbon nanotube arrays, *Nano Lett.*, **2**, 1079 (2002).

34. K. K. S. Lau, J. Bico, K. B. K. Teo, M. Chhowalla, G. A. J. Amaratunga, W. I. Milne, G. H. McKinley, and K. K. Gleason, Superhydrophobic carbon nanotube forests, *Nano Lett.*, **3**, 1701 (2003).

35. N. Chakrapani, B. Wei, A. Carrillo, P. M. Ajayan, and R. S. Kane, Capillarity-driven assembly of two-dimensional cellular carbon nanotube foams, *Proc. Natl. Acad. Sci. USA*, **101**, 4009 (2004).

36. H. Liu, S. H. Li, J. Zhai, H. J. Li, Q. S. Zheng, L. Jiang, and D. B. Zhu, Self-assembly of large-scale micropatterns on aligned carbon nanotube films, *Angew. Chem., Int. Ed.*, **43**, 1146 (2004).

37. C. Journet, S. Moulinet, C. Ybert, S. T. Purcell, and L. Bocquet, Contact angle measurements on superhydrophobic carbon nanotube forests: effect of fluid pressure, *Europhys. Lett.*, **71**, 104 (2005).

38. A. Dev and S. Chaudhuri, Uniform large-scale growth of micropatterned arrays of ZnO nanowires synthesized by a surfactant assisted approach, *Nanotechnology*, **18**, 175607 (2007).

39. D. Chandra, J. A. Taylor, and S. Yang, Replica molding of high-aspect-ratio (sub-) micron hydrogel pillar arrays and their stability in air and solvents, *Soft Matter*, **4**, 979 (2008).

40. D. Chandra and S. Yang, Capillary-force-induced clustering of micropillar arrays: is it caused by isolated capillary bridges or by the lateral capillary meniscus interaction force? *Langmuir*, **25**, 10430 (2009).

41. C. Y. Hui, A. Jagota, Y. Y. Lin, and E. J. Kramer, Constraints on microcontact printing imposed by stamp deformation, *Langmuir*, **18**, 1394 (2002).

42. K. G. Sharp, G. S. Blackman, N. J. Glassmaker, A. Jagota, and C. Y. Hui, Effect of stamp deformation on the quality of microcontact printing: theory and experiment, *Langmuir*, **20**, 6430 (2004).

43. B. V. Derjaguin and M. M. Kursakov, Properties of the liquid in thin layers, *Izv. Akad. Nauk SSSR Ser Khilm. NS*, 741 (1936).

44. R. C. Tolman, The effect of droplet size on surface tension, *J. Chem. Phys.*, **17**, 333 (1949).
45. L. R. Fisher and J. N. Israelachvili, Determination of the capillary pressure in menisci of molecular dimensions, *Chem. Phys. Lett.*, **76**, 325 (1980).
46. P.-G. de Gennes, F. Brochard-Wyart, and D. Quéré, *Capillarity and Wetting Phenomena: Drops, Bubbles, Pearls, Waves*. New York: Springer (2004).
47. A. Frumkin, About wetting and adherence phenomena, *J. Phys. Chem. USSR*, **12**, 337 (1938).
48. B. V. Derjaguin, A theory of capillary condensation in the pores of sorbents and of other capillary phenomena taking into account the disjoining action of polymolecular liquid films, *Acta Physicochim. USSR*, **12**, 181–200 (1940).
49. B. V. Derjaguin, N. V. Churaev, and V. M. Muller, *Surface Forces*. New York: Consultants Bureau/Plenum Publishing (1987).
50. A. V. Neimark, P. I. Ravikovitch A. Vishnyakov, Bridging scales from molecular simulations to classical thermodynamics: density functional theory of capillary condensation in nanopores, *J. Phys.: Condens. Matter*, **15**, 347 (2003).
51. B. V. Derjaguin, Determination of tension in wetting film by liquid column capillary ascent. Development of wetting and capillary theories, *Kolloidnyj Zhurnal*, **56**, 47 (1994) (in Russian).
52. V. V. Krotov and A. I. Rusanov, *Physicochemical Hydrodynamics of Capillary Systems*. London: Imperial College Press (1999).
53. E. W. Washburn, The dynamics of capillary flow, *Phys. Rev.*, **17**, 273 (1921).
54. R. Lucas, Ueber das Zeitgesetz des Kapillaren Aufstiegs von Flussigkeiten, *Kolloid Z.*, **23**, 15 (1918).
55. A. Hibara, T. Saito, H.-B. Kim, M. Tokeshi, T. Ooi, M. Nakao, and T. Kitamori, Nanochannels on a fused silica micro-chip and liquid properties investigation by time-resolved fluorescence measurements, *Anal. Chem.*, **74**, 6170 (2002).
56. N. R. Tas, J. Haneveld, H. V. Jansen, M. Elwenspoek, and A. van den Berg, Capillary filling speed of water in nanochannels, *Appl. Phys. Lett.*, **85**, 3274 (2004).
57. A. Han, G. Mondin, N. G. Hegelbach, N. F. de Rooij, and U. Staufer, Filling kinetics of liquids in nanochannels as narrow as 27 nm by capillary force, *J. Colloid Interface Sci.*, **293**, 151 (2006).

58. F. Persson, L. H. Thamdrup, M. B. L. Mikkelsen, S. E. Jaarlgard, P. Skafte-Pedersen, H. Bruus, A. Kristensen, Double thermal oxidation scheme for the fabrication of SiO_2 nanochannels, *Nanotechnology*, **18**, 245301 (2007).

59. K. M. van Delft, J. C. T. Eijkel, D. Mijatovic, T. S. Druzhinina, H. Rathgen, N. R. Tas, A. van den Berg, and F. Mugele, Micromachined Fabry-Pérot interferometer with embedded nanochannels for nanoscale fluid dynamics, *Nano Lett.*, **7**, 345 (2007).

60. J. Haneveld, N. R. Tas, N. Brunets, H. V. Jansen, and M. Elwenspoek, Capillary filling of sub-10 nm nanochannels, *J. Appl. Phys.*, **104**, 014309 (2008).

61. R. B. Schoch, J. Han, and Ph. Renaud, Transport phenomena in nanofluidics, *Rev. Mod. Phys.*, **80**, 839 (2008).

62. S. Levine, J. R. Marriott, and K. Robinson, Theory of electrokinetic flow in a narrow parallel-plate channel, *J. Chem. Soc. Faraday. Trans. 2*, **71**, 1 (1975).

63. N. A. Mortensen and A. Kristensen, Electroviscous effects in capillary filling of nanochannels, *Appl. Phys. Lett.*, **92**, 063110 (2008).

64. K. G. H. Jansen, H. T. Hoang, J. Floris, J. de Vries, N. R. Tas, J. C. T. Eijkel, and Th. Hankemeier, Solution titration by wall deprotonation during capillary filling of silicon oxide nanochannels, *Anal. Chem.*, **80**, 8095 (2008).

65. L. H. Thamdrup, F. Persson, H. Bruus, A. Kristensen, and H. Flyvbjerg, Experimental investigation of bubble formation during capillary filling of SiO_2 nanoslits, *Appl. Phys. Lett.*, **91**, 163505 (2007).

66. B. V. Derjaguin and N. V. Churaev, Structural component of disjoining pressure, *J. Colloid Interface Sci.*, **49**, 249 (1974).

67. J. N. Israelachvili and G. E. Adams, Measurements of the forces between two mica surfaces in aqueous electrolyte solutions in the range of 0 – 100 nm, *J. Chem. Soc., Faraday Trans.*, **74**, 975 (1978).

68. R. M. Pashley, Hydration forces between mica surfaces in aqueous electrolyte solutions, *J. Colloid Interface Sci.*, **80**, 153 (1981).

69. R. M. Pashley, DLVO and hydration forces between mica surfaces in Li^+, Na^+, K^+, and Cs^+ electrolyte solutions: a correlation of double-layer and hydration forces with surface cation exchange properties, *J. Colloid Interface Sci.*, **83**, 531 (1981).

70. R. M. Pashley and J. N. Israelachvili, Molecular layering of water in thin films between mica surfaces and its relation to hydration forces, *J. Colloid Interface Sci.*, **101**, 511 (1984).

71. R. G. Horn, D. T. Smith, and W. Haller, Surface forces and viscosity measured between silica sheets, *Chem. Phys. Lett.*, **162**, 404 (1989).
72. N. V. Churaev, V. D. Sobolev, and Z. M. Zorin, Measurement of viscosity of liquids in quartz capillaries. In *Special Discussion in Thin Liquid Films and Boundary Layers*. New York: Academic Press. 213 (1971).
73. J. N. Israelachvili, Measurement of the viscosity of liquids in very thin films, *J. Colloid Interface Sci.*, **110**, 263 (1986).
74. U. Raviv, P. Laurat, and J. Klein, Fluidity of water confined to subnanometer films, *Nature*, **413**, 51 (2001).
75. U. Raviv and J. Klein, Fluidity of bound hydration layers, *Science*, **297**, 1540 (2002).
76. T.-D. Li, J. Gao, R. Szoszkiewicz, U. Landman, and E. Riedo, Structured and viscous water in subnanometer gaps, *Phys. Rev. B*, **75**, 115415 (2007).
77. R. Lenormand, C. Zarcone, Role of roughness and edges during imbibition in square capillaries, Proceedings of the 59th Annual Technical Meeting of the SPE. SPE Paper 13264. Houston, TX (1984).
78. P. M. Heertjes and W. C. Witvoet, Some aspects of the wetting of powders, *Powder Technol.*, **3**, 339 (1969/1970).
79. A. K. Singhal and W. H. Somerton, Two-phase flow through a non-circular capillary at low Reynolds numbers, Proceedings of the 21st Annual Technical Meeting of the Petroleum Society of CIM. Calgary (1970).
80. M. Dong and I. Chatzis, The imbibition and flow of a wetting liquid along the corners of a square capillary tube, *J. Colloid Interface Sci.*, **172**, 278 (1995).
81. F. J. Mayer, J. F. McGrath, and J. W. Steele, A class of similarity solutions for the nonlinear thermal conduction problem, *J. Phys. A: Math. Gen.*, **16**, 3393 (1983)
82. M. M. Weislogel and S. Lichter, Capillary flow in an interior corner, *J. Fluid Mech.*, **373**, 349 (1998).
83. T. C. Ransohoff and C. J. Radke, Laminar flow of a wetting liquid along the corners of a predominantly gas-occupied noncircular pore, *J. Colloid Interface Sci.*, **121**, 392 (1988).
84. I. Chatzis and F. A. L. Dullien, Dynamic immiscible displacement mechanisms in pore doublets: theory vs. experiment, *J. Colloid Interface Sci.*, **91**, 199 (1983).
85. J. W. van Honschoten, N. Brunets, and N. R. Tas, Capillarity at the nanoscale, *Chem. Soc. Rev.*, **39**, 1096 (2010) (reproduced by permission of the Royal Society of Chemistry).

Chapter 12

Manipulation of Liquid Nanodrops

Erik Dujardin,* Thierry Ondarçuhu, and Laure Fabié**

Nanosciences Group, CEMES-CNRS, 29, rue Jeanne Marvig, 31055 Toulouse Cedex 4, France

*dujardin@cemes.fr and **ondar@cemes.fr

12.1 Introduction

The recent development of nanosciences has brought to light the importance of surface structuration at nanometer scale, from both fundamental and applied viewpoints. In particular, taking advantage of the incredible variety of objects that can be synthesized by chemical methods, the use of individual nano-objects or molecules as a functional part of nanodevices may lead to new effects. The functionalities emerging from this approach bring new paradigms in almost all fields of physical sciences such as electronics, optics, magnetism, and biology. The crucial role of micrometer- and nanometer-scale surface patterning in fundamental phenomena as well as technological processes has been progressively uncovered thanks to the progress of new lithographic methods that enabled

Nanoscale Liquid Interfaces: Wetting, Patterning, and Force Microscopy at the Molecular Scale
Edited by Thierry Ondarçuhu and Jean-Pierre Aimé

ISBN 978-981-4316-45-3 (Hardcover), 978-981-4364-48-5 (eBook)
www.panstanford.com

the modification of substrates over multiple length scales and to the advent of a genuine functional surface engineering. Meanwhile, the cost and still limited spatial resolution of these lithographic techniques prompted the emergence of alternative low-cost or very-high-resolution approaches such as inkjet printing, nanoimprinting, soft lithography, self-assembly, and scanning probe–derived techniques.

However, in the framework of hybrid device fabrication, one technological challenge remains—the immobilization of one or a few molecules or nano-objects on a predefined nanodevice fabricated by standard lithography methods. A technique that would be (i) sufficiently generic to handle a large range of nano-objects, (ii) fast enough to cover a macroscopic surface area, and (iii) highly resolved to manipulate individual molecules is not yet available, but partial solutions to this challenge have appeared in recent years and will be reviewed in this chapter.

Since the molecules or nanoparticles (NPs) of interest are generally produced in the form of solutions and suspensions, the most intuitive method to handle one such nano-object would be to manipulate a small amount of the liquid and to deposit it on the surface as performed, at macroscopic scale, using a pipette. Down-scaling pipettes or pins to submicron scale proved to be a difficult task, strongly limiting the application of this approach to nanosciences. Various methods were then developed to manipulate nanodroplets with the aim to provide a general technique allowing depositing any soluble molecule or NP with high spatial resolution. Atomic force microscopy (AFM) certainly appears to be one of the most appealing techniques for this aim since the sharp atomic force microscope tips reach the size of the droplets or ultimately that of the molecules to be deposited.

In this chapter, we first review the liquid-based patterning tools developed so far before describing in greater detail the liquid nanodispensing (NADIS) technique that allows the direct manipulation of droplets with volumes down to the attoliter (10^{-18} L). Besides its potential as a nanopatterning method, we also show that NADIS is a unique tool to study wetting and capillarity at submicron scale.

12.2 Liquid-Based Patterning Tools

The methods recently developed to manipulate droplets and take advantage of the flexibility of liquid lithography in terms of deposited molecules are reviewed.

The range of feature sizes made accessible by these techniques spans an impressive 12 orders of magnitude in delivered volumes from microliter (10^{-6} L) to attoliter (10^{-18} L), as illustrated in Fig. 12.1. Ink-based patterning can be split into two approaches. Noncontact methods consist in projecting liquid drops onto the surface from a droplet ejector (e.g., inkjet printing). They are reviewed briefly in section 12.2.1. Contact methods transfer molecules, droplets, and larger species thanks to the liquid meniscus formed between a tip or a pen and the substrate (direct scanning probe lithography or micromachined fountain pen techniques). In section 12.2.2, we will successively consider scanning ionic conductance microscopy (SICM), microcantilevers, and nanodispensing technologies. We will also dedicate specific attention to current challenges

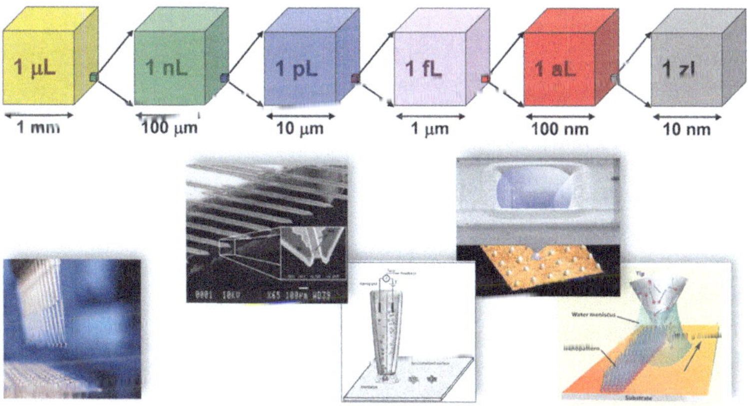

Figure 12.1 Surface nanopatterning techniques by manipulation of liquid droplets spanning five orders of magnitude in diameter from 1 mm to 10 nm. Micropipettes, slit microcantilevers, nanopipettes, NADIS, and DPN (from left to right). (Adapted in part from Refs. 1–3, copyright 2005, 2007, Wiley-VCH; copyright 2006, IOP Publishing.) *Abbreviation*: DPN, dip-pen nanolithography.

such as reservoir and on-board microfluidics, parallelization, and positioning.

12.2.1 *Ink Delivery by Noncontact Methods*

The development of methods enabling to deliver liquid on a surface without any contact has attracted a lot of attention. The main advantage of these methods is their compatibility with any type of substrates, especially those with irregular surfaces that are difficult to pattern in direct contact. Nevertheless, the downscaling of these techniques is a challenging issue. In the following, we briefly review several strategies that have been pursued to produce and deliver ink droplets.

The reference noncontact technique is inkjet printing. This mature technology has been used for several decades in plotters and printers to produce low-cost, high-quality images with high throughput. The principle of inkjet techniques is the ejection of liquid droplets from a chamber through a nozzle. In the "continuous charge and deflect" technology, a continuous jet of liquid droplets is produced and oriented toward the substrate or deviated by an electric field on an ink collector. This method provides high throughput appropriate for industrial processes requiring large area coverage. In the "drop-on-demand" systems, which are more suitable for research activities, two different mechanisms have been developed to expel the drop. The digital input activates either a piezoelectric transducer, which compresses the chamber until the liquid is propelled out of the nozzle, or a heater, which generates a bubble that expels a liquid drop. The ejection mechanism is rather complex and depends on the physical and chemical properties of the nozzle, ink, and substrate. Many parameters therefore need to be optimized for a given application. For example, inkjet heads are usually designed for low-viscosity liquids, and the extension of this technique to more viscous materials such as polymers is not straightforward. In this respect, piezoelectric inkjet printing is more versatile than thermal inkjet, which mainly remains limited to aqueous solutions. Nevertheless, inkjet has been lately used to pattern surfaces with a wide variety of materials[4–6] such as polymers, NPs, sol-gel materials, and metals for many different applications.

Inkjet printing is also quite promising for the deposition of biological compounds since it preserves their functionality and is used for the fabrication of functional bioarrays.[7–9] Another application of this noncontact deposition method is the functionalization of fragile structures such as cantilever-based biosensors.[10]

Despite its potential, inkjet printing cannot accommodate a drastic downscaling. The constraints mainly come from the surface tension, which increases the pressure gradients required to form droplets, and viscous dissipation, which dramatically decreases liquid flows at small scale. As a consequence, the direct printing of features much smaller than 10 μm is rather unlikely.

The resolution of noncontact delivery techniques has been improved recently by using an electric field instead of acoustic or thermal actuation.[11] In the e-jet printing technique, an electric field applied between a pipette and the surface generates a Taylor cone at the tip apex.[12] For sufficiently high applied voltage, the electrostatic stress overcomes surface tension and ejects droplets from the cone apex. This electrospray technique developed in the 1980s is commonly used in ionization mass spectroscopy. In the frame of downscaling spots size, the main advantage of electrospraying is the fact that the generated droplets have a diameter much smaller than the capillary used to generate them. The method is also versatile since many different materials could be patterned (polymers, NPs, nanowires, carbon nanotubes, etc.).[11] Examples of DNA patterns formed by e-jet printing are reported in Fig. 12.2. They show the ability of the method to form patterns in a controlled fashion with an ultimate resolution approaching 100 nm.[13]

In-plane emitters with aperture width as small as 20 nm were also designed using micromachining techniques and can be compatible with parallelization,[14] although these silicon-based sources have not yet been used for surface patterning.

E-jet printing techniques are thus very attractive to pattern surfaces with liquids at submicron scale. Nevertheless, the use of high voltages (typically 1,000 V) requires a complicated setup. The alignment on predefined structures, which is a strong requirement in order to combine with other lithographic techniques, is also a challenging issue with noncontact methods and may limit their application.

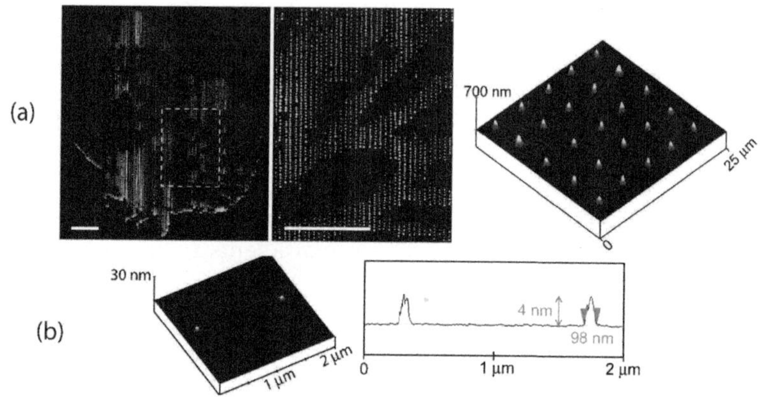

Figure 12.2 Images of patterns of DNA formed by e-jet printing. (a) Fluorescence micrograph of a butterfly pattern printed on a Si wafer using an oligonucleotide suspension and a 2 µm nozzle. Magnified image and AFM image show that the pattern is composed of arrays of dots ≈ 2 µm in diameter. Scale bars are 100 µm. (b) AFM image and height profile of dots with average diameter ≈ 100 nm obtained with a 500 nm nozzle. (Adapted from Ref. 13, copyright 2008, American Chemical Society.)

12.2.2 *Ink Delivery by Contact Methods*

12.2.2.1 Micro- and nanopipettes

Putting a liquid-filled capillary in contact with a surface produces a liquid meniscus, which then spreads onto the substrate, provided the latter is not too hydrophobic. Pipettes have therefore been used for the local deposition of liquids by direct contact on substrates. The downscaling of pipetting techniques by profiling capillaries with micron-scale apertures benefitted from the association with electrochemistry in the confined droplet volume. Scanning ionic conductance microscopy (SICM),[15] which has recently become commercially available from Ionscope Ltd,[16] enables the deposition of a large variety of materials with a spatial resolution of about 1 µm.[17] In its simplest version, SICM consists of a tapered capillary with a micrometer-scale aperture, fitted with an electrode and filled with the electrolyte to be deposited. The capillary is immersed into a solution, in which the counterelectrode is placed and brought near the substrate. The ionic current between the two electrodes varies nonlinearly when the nanopipette capillary altitude with respect to the substrate is modulated below 10 μm.[18] Similarly to standard scanning probe microscopy (SPM) techniques, a feedback loop stabilizing the ion current allows to maintain the tip–substrate distance. Conversely, SICM can be used to measure local ion flux. For example, if the counterelectrode is placed inside a cellular cytoplasm, the SICM capillary can measure the ion flux through the membrane and thus map the spatial distribution of ion channels.[18] In order to exploit the nanopipette for local deposition onto dry substrates, Klenerman *et al.* have modified the capillary by splitting it into two compartments, each containing an electrode.[2] When the double-barreled capillary is far from the substrate, no current exists between the electrodes because no liquid bridge allows for electrolyte transport. Upon touching the substrate, the small meniscus establishes this link and the ion current increases rapidly, therefore making it possible to control the capillary position (Fig. 12.3a) and to deposit materials brought by the ionic flux.

So far, two deposition modes have been explored: when the aqueous meniscus is immersed in an oil environment, submicrometer deposition resolution is reached, while direct deposition on dry substrates leads to micrometer-sized deposits. In both cases, fluorophore-tagged DNA spots were placed within less than 1 μm from each other. The amount of deposited material varies proportionally to the applied tension between the two electrodes. Interestingly, the two compartments of the capillary can be filled with different electrolytes, which will be independently deposited depending on the polarity of the bias voltage applied to the electrodes. Such a unique feature is illustrated by the reproduction of the blue-filtered 75 × 61–pixel image shown in Fig. 12.3b. Two DNA strands tagged with green and red fluorophores are loaded in the two compartments, and the nanopipette is scanned with a 1 μm step. For each green (resp. red) pixel, the "green" (resp. "red") DNA is deposited, and for the yellow pixels, both DNAs are superimposed. The complete patterning is performed in 35 minutes at 2 pixels.s^{-1}, and its fluorescent image is shown in Fig. 12.3c. Significantly, microfabricated equivalents of the capillary nanopipettes have recently been fabricated as arrays of silicon or polymer tubes on substrates.[19] While this proof of principle, as

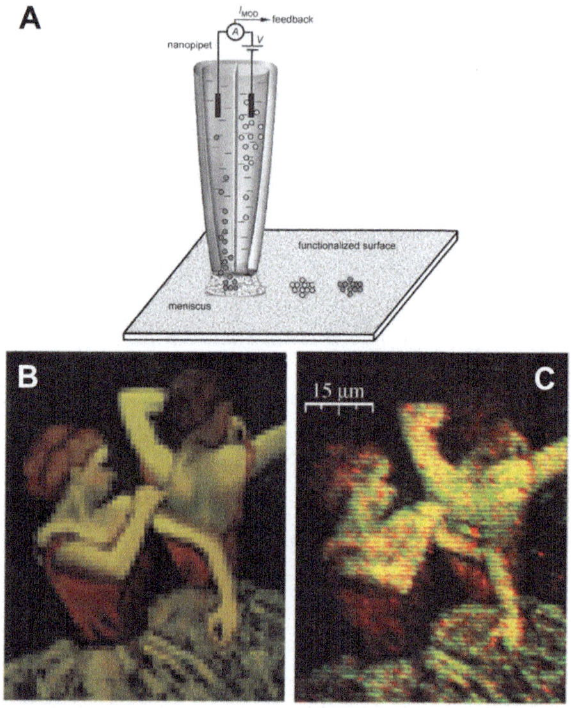

Figure 12.3 Nanopipette technology. (A) Principle of double-barreled nanopipette deposition. The two barrels of the nanopipette are filled with different fluorophore-labeled antibodies for deposition onto a glass surface. The voltage between the two barrels produces the distance-feedback control current and controls the molecular delivery. (B) 75×61-pixel image of the painting *Degas Dancers* by G. Candeloriin, in which the blue channel has been removed. (C) Fluorescence image of the image in (B) printed with 1 µm diameter spots of DNA strands marked wither with Alexa 647 (red) or Rhodamine (green). Yellow pixels are obtained by depositing both fluorophores on the same pixel. Total printing time: 35 min. (Adapted from Ref. 2, copyright 2005, Wiley VCH.)

most early results, is restricted to DNA deposition, the advent of a commercial version of this technique marks the onset of a more varied usage, in particular in proteomics. SICM addresses a number of challenges where direct liquid dispensing at the micrometer scale is required, but the full achievement of the targets of these niche applications will yet require the automation of scanning movements to minimize the stress applied to the rigid pulled or microfabricated pipettes and the development of repositioning routines to enable the aligned deposition on prepatterned substrates.

12.2.2.2 In-plane microcantilevers for liquid delivery

While microspotting with capillaries is only entering now the stage of parallel microfabrication because of the difficulty of reliably manufacturing small-diameter tubing, other liquid-dispensing technologies owe their downscaling performances to the development of microlithographic processes. This is precisely the recent path followed by quill spotting, which relies on the transfer of liquid along a narrow slit perforated in a flat cantilevered tip, a structure which can be easily be implemented in lithographic processes.[20,21] The upper part of the slit can be widened to act as a liquid reservoir, therefore permitting multiple deposits from one liquid load. Moreover, the flat tip design allows the addition of metallic tracks, which can be used for active loading by promoting the spreading of the liquid on the tip surface by electrowetting.[1,22] When standard photolithography processes are used for the microcantilever fabrication, the typical slit width and quill apex thickness are 1–2 μm. This size directly sets the deposit dimensions in the range of 2 to 10 μm, which match the requirements of the main application target, namely the downsizing of optical biochips. Indeed, the resolution of this approach is ultimately limited by the performance of the optical fluorescence detection system for which a 1 μm minimal spot size is sufficient. Nevertheless, we should point out that the extension of the quill approach to nanosized droplets is indeed being explored by profiling the microslit with a nanometric overtip.[23] Moreover, microfabricated 1 μm wide slits were prolonged with 20 nm wide slits milled in the cantilever slab by a focused ion beam (FIB).[14] The cantilever-based design can be straightforwardly parallelized by the fabrication of arrays,[1] which has already been demonstrated and commercialized in a 30-cantilever version[24] With the extra commodity of piezoresistive detection of the cantilever deflection rather than the usual optical approach, individual deflection of cantilevers in an array can be monitored, which paves the way to a fully flexible deposition scheme in which individual droplets can be deposited from an arbitrarily chosen cantilever in the array. The diversity in material deposition is probably one of the main assets of the quill design since liquid solutions are manipulated rather than surface-coated molecular films. Deposition of cells and biomolecules has been demonstrated.[1,22,25,26] Moreover, with a metalized slit, electrochemical reactions can be locally induced upon electrolyte deposition. In this case, the precursors (metallic salt, electropolymerizable monomer) is loaded in the reservoir and deposited as a droplet while the electrode potential is actuated to trigger the reduction (for metals) or oxidation (for polypyrrole) reaction in the meniscus.[1,22]

12.2.2.3 Scanning probe microscopy

Soon after its invention, SPM appeared as an interesting patterning method. The tips used to image a surface down to atomic resolution could thus be used as tools to modify the surface with ultimate resolution. In this context, scanning tunneling microscopy (STM) imaging led to STM lithography as early as 1992, while AFM imaging led to indentation and anodic patterning. Nevertheless, in spite of their initial achievements and subsequent improvements, these early SPM-based nanopatterning methods could merely modify the material of the substrate itself and not add locally extraneous matter (molecules or NPs). In this respect, dip-pen nanolithography (DPN) invented in 1995 by Butt[27] and fully developed by Mirkin in 1999 demonstrated for the first time[28] the possibility to prepare sub-100 nm patterns composed of molecules brought with an atomic force microscope tip, deposited onto the substrate by molecular diffusion on solid surfaces through a liquid meniscus. This technique is a striking benchmark in terms of spatial resolution, versatility, and technological development. Its remaining limitations have however been tackled and sometimes overcome by other probe-based approaches, which rely on the direct and local delivery of materials onto a surface through a liquid meniscus or a droplet.

12.2.2.3.1 Dip-pen lithography: a scientific and technological reference

The method consists in (1) "inking" an atomic force microscope tip in a solution of molecules such as alkanethiols, (2) drying the tip, and (3) bringing it in contact with a gold substrate in order to transfer the molecules to the surface. Molecules diffuse from the tip to the surface with the help of a water meniscus, which naturally condensates at the contact point for relative humidity (RH) exceeding 50%. At the circumference of this liquid bridge, molecules self-assemble and anchor on the gold surface, forming a dense self-assembled monolayer (SAM).

Due to the presence of the water meniscus, the mechanism of molecule transport from the tip to the surface is rather complex and not fully understood.[29] Nevertheless, it has been clearly established that the lateral sizes of the patterns are determined by parameters such as RH, molecule solubility in water, contact time (usually as $t^{1/2}$), writing speed (usually as V^{-1}),[30] and tip functionalization. Interestingly, contact force has no influence on patterns, which is an important point for parallelization of the method.[31] In about 10 years, the influence of these parameters has been intensively studied and optimized on model systems such as thioalkanes (octadecylthiol [ODT], mercaptohexadecaneoic acid [MHA]). It comes out that a careful control of experimental conditions, and more particularly of RH, is required to obtain reproducible patterns. Doing so, the best reported spatial resolution is 15 nm, and the typical spot sizes obtained routinely with DPN are of the order of 100 nm.

During the last decade, a very large range of materials have been patterned using DPN, showing that this technique is technologically relevant in many applications. Systems are now commercially available. Since several reviews have recently listed the applications of DPN,[32–36] we will not describe them in great detail here and discuss only the two main deposition modes developed.

Since thioalkanes and thiolated oligonucleotides have shown to be the best suitable molecules for DPN, most of the applications of DPN used patterns of these molecules as templates to attach the molecules of interest in a second step, often performed *ex situ*. Using specific interactions (chemical or electrostatic, for example) with the functionalized thiols, a wide range of molecules and nano-

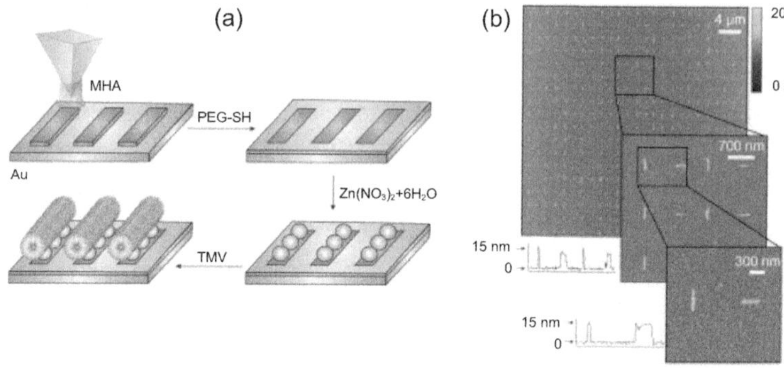

Figure 12.4 (a) Sketch of the selective immobilization of single virus particles on DPN-generated MHA nanotemplates treated with $Zn(NO_3)_2{\cdot}6H_2O$; (b) topography AFM images and height profiles of a perpendicular array of single virus particles ($40 \times 40\ \mu m^2$). (Adapted from Ref. 37, copyright 2005, Wiley VCH.)

objects have been patterned using this bottom-up approach with typical sizes of 100 nm. Example of such an *indirect patterning* is shown in Fig. 12.4, where a template of MHA is patterned by DPN.[37] After passivation of the surface around the patterns with thiolated ethylene glycol, a metal-ion-based coupling approach was used to immobilize single tobacco mosaic virus (TMV) particles on the templates, provided their dimensions are compatible with the ones of the particles. This example of a bottom-up process not only demonstrates the validity of the indirect DPN patterning approach but also shows that the positioning and orientation of individual biological structures will be achieved in a foreseeable future.

Direct patterning has also been demonstrated for a large range of objects, including DNA, polymers, dendrimers, proteins, and NPs. In contrast to indirect patterning which benefits from the well-understood and controlled deposition of thiol molecules and the power of self-assembly methods, direct patterning requires a systematic study and optimization of the transport mechanism of the molecules from the tip to the surface. Specific surface and tip functionalizations are then needed to optimize the transfer of a given molecule.[38] It was shown recently that the use of a porous coating of the atomic force microscope tip increases the amount of ink carried and facilitates the transfer of molecules to the surface.[39] Agarose gel covering the tip was also presented as a "universal" carrier and could extend the field of application of direct patterning.[40] Nevertheless, in the latter situation the matrix gel is deposited alongside the molecular ink and its removal may be problematic. Beyond these rather primitive solutions to increase the amount of ink, technological developments have been performed to design new devices with on-chip microfluidic reservoirs. These solutions are detailed in section 12.2.3.

Despite many applications demonstrated using DPN, there is a need for a more general patterning method that could apply to any soluble molecule or nanoobject, independently of its chemical nature.

12.2.2.3.2 Liquid nanodispensing

In order to combine the spatial resolution obtained with DPN with the flexibility of the liquid-based patterning methods presented earlier, Heinzelmann *et al.* in CSEM (Neuchâtel, Switzerland) developed an AFM-based technique called NADIS.[41] The principle is to use a hollow atomic force microscope tip with a nanochannel drilled at its apex. The aperture is used to transfer liquid from a droplet reservoir deposited on the cantilever to the surface by simple mechanical contact. Cantilevers with channel diameters larger than 1 μm can be prepared using microfabrication techniques,[42] while tips with submicron apertures are obtained using FIB milling of commercial atomic force microscope tips. This latter approach was used to demonstrate ultimate resolution of 35 nm leading to the deposition of spots with diameters of 75 nm.[43] The NADIS technique is therefore approaching the resolution of DPN and may provide an alternative to this method in the case of larger objects such as NPs or proteins, which need to be kept in solution rather than dried on the tip surface. Since NADIS is an interesting technique to manipulate ultrasmall droplets (femto to attoliter), we dedicate section 12.3 to the description of the experimental procedure, the deposition mechanism, and examples of applications in nanopatterning. We also demonstrate that NADIS is a unique tool to perform model experiments with liquids at submicron scale.

12.2.3 *On-Chip Microfluidic Reservoirs*

The future developments of probe-based liquid nanopatterning techniques rely on their capabilities to deliver a large number of small droplets in a minimal time. One major limitation, compared to jet printing and except for SICM, is the finite amount of ink carried by the probes and the cumbersome or time-consuming procedures to reload them. The second limitation is the serial deposition of droplets from a single probe, which dramatically slows down the overall process of complex patterning. This last issue is being addressed by the development of parallel-probe arrays and will be discussed in section 12.2.4. Regarding the probe replenishment, DPN circumvented this problem by adjoining a series of microfabricated wells in which the single or arrayed atomic force microscope tip can be automatically dipped to re-cover the surface with alkanethiols and dried in order to resume patterning.[3,31] A similar strategy was used in the case of slit microcantilevers for which specific microfluidic networks with hydrophobic barriers separating the inking wells were added in order to prevent cross-contamination.[44] While partially solving the issue, this extra procedure is limiting the overall throughput of single-probe systems for large scale patterning and is particularly delicate to implement for massively parallel-probe arrays.[45] A more sophisticated approach consists in fabricating on-chip microfluidic systems comprising a large liquid reservoir on the silicon chip and the required capillaries on the cantilever so that the liquid can be continuously supplied close to the tip apex.[46] On the flat microquill systems, open channels rather than closed capillaries have been implemented so that tip cleaning and solution rinsing could be facilitated.[47] Since the channels are relatively large, these tools are rather insensitive to evaporation issues. However, upon downscaling the probe size, evaporation of the ink solution can very rapidly render the deposition devices unusable; therefore SPM-based approaches have favored closed-channel designs.

Essentially two implementation strategies of this concept were developed and named "nanofountain pen" (NFP).[46,48] In both cases, microchannels with typical sections of 300–500 nm by 2–10 μm were fabricated in the cantilever and its support using sacrificial layer techniques. The channels connect the tip to nanoliter-sized reservoirs (i.e., typically 0.1 mm by side), which can themselves be connected to larger remote reservoirs using standard capillaries.[49,50] The filling of the tip from the reservoir can be performed either passively by capillary force or actively by embedding an evaporation-based pump.[51]

So far, the two systems developed differ by the position of the outlet connecting the channel to the tip. The tips developed by Deladi *et al.* are based on hollow, pyramidal atomic force microscope tips pierced with an aperture at the tip apex using an FIB in a NADIS-like configuration or, alternatively, at the base of the pyramid. The liquid is brought from the reservoir to the aperture by capillarity through the cantilever-borne microfluidic channels (Fig. 12.5a–c).[48]

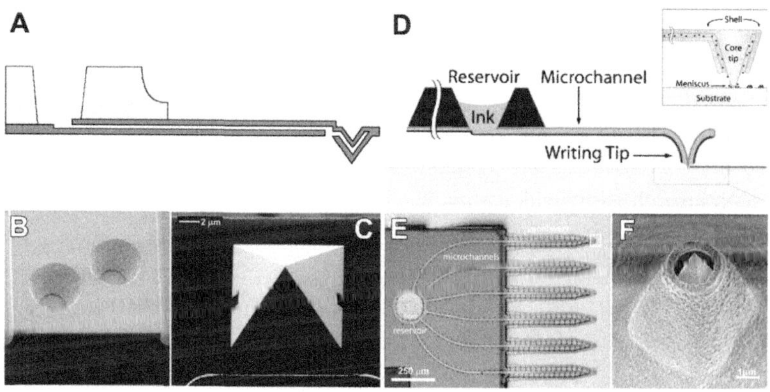

Figure 12.5 Nanofountain pens. (A) Cross-sectional sketch of the AFM-based NFP developed by Deladi *et al.*, showing the chip-borne reservoir, the embedded microchannel, and the apertures at the basis of the pyramidal tip. (B and C) SEM images showing the entire NFP chip with the reservoir and the apertured tip (adapted from Ref. 49, copyright 2005, IOP Publishing). (D) Schematic representation of the NFP developed by Espinosa *et al.*, which shows a volcano tip enlarged in the inset (adapted from Ref. 53, copyright 2009, Wiley VCH). (E) Optical top view of an array of parallel NFP tips with a common reservoir and dedicated microchannels (adapted from Ref. 55, copyright 2008, National Academy of Sciences of the USA). (F) SEM image of the volcano tip in which the microchannel brings the liquid around the central pyramid where it can freely diffuse along the apex toward the substrate (adapted from Ref. 50, copyright 2006, IOP Publishing). *Abbreviation*: SEM, scanning electron microscopy.

When the end of the capillary is away from the tip apex, the ink is let to diffuse from the upper outlet toward the tip apex, where it is transferred onto the substrate following a DPN mechanism. Proper ink feeding along the pyramid side requires a hydrophilic functionalization of the tip outer wall in order to favor the spreading of a liquid film. This film actually limits the resolution of this approach, which is an order of magnitude larger than original DPN (500 and 350 nm wide lines for ODT deposition and chromium etching, respectively).[48] The NFP developed by Espinosa *et al.* is very similar except that the liquid is brought much closer to the tip apex, thanks to a volcano tip design (Fig. 12.5d–f).[46] The external wall of the tip ends in a truncated conical shape at the center of which protrudes the sharp tip formed by the inner wall of the tip. This sharp protrusion is the deposition tip, which touches the substrate upon deposition. The gap separating the two walls constitutes the microfluidic channel, which continuously extends to the reservoir place on the chip. The liquid is therefore driven to the very end of the tip by capillary action, where it forms a liquid–air interface. Solute molecules then diffuse as in DPN or in Deladi's design but over a much shorter distance. Besides achieving constant ink supply, this system presents the advantage to keep the molecules solvated throughout the patterning process, which may be crucial to maintain the activity of biological molecules during their deposition. Since the last step of the deposition at the tip apex is strictly the same as in conventional DPN, the feature sizes are comparable. Indeed, the best resolution reported for the MHA line is of 40 nm line width.[46] Interestingly, this method has proven efficient also for a direct deposition of gold NPs,[52] nanodiamonds,[53] DNA,[54] and proteins[55] with resolutions of the order of 100 nm. It is noteworthy that the microfabrication process of the volcano tip could be parallelized to create arrays of NFP, as described in the next section.

An original outcome of the combination of microfluidics and apertured SPM probes is the development of FluidFM, which exploits the liquid dispensing directly at or close to a sharp atomic force microscope tip to deliver the solution on or through the cytoplasmic membrane of living cells in a liquid environment.[56] The sensitive force feedback inherited from the SPM configuration provides a convenient way to approach the tip to the fluid membrane in a much more controlled and softer way than conventional cell manipulation techniques such as micropipettes and patch clamping. When the aperture is placed at the tip apex, the blunt profile of the tip allows it to land it on the lipidic membrane where liquid can flow, while controlling the applied force and tip movement, therefore avoiding membrane disruption. Alternatively, when the aperture is located on the tip side, the intact sharp tip can be used to surgically pierce through the membrane until the aperture penetrates the endoplasmic space where solutions can be injected. Upon withdrawal of the tip, the membrane heals immediately and perfectly so that the cell survives the injection. Finally, the injected volumes can be much smaller than with other techniques, therefore limiting the stress undergone by the living cells.

12.2.4 *Toward Massive Parallelization*

Scanning probe techniques have always seen their ultimately high spatial precision promises tempered by their intrinsically localized action field and slow and serial operating mode. While the new physics made accessible by the former largely compensated for the limitations of the latter, the issue of speeding up scanning probe processes became essential when they were considered for surface-patterning applications. It is thus not surprising that the development of DPN immediately took into consideration ways to parallelize the patterning produced by a single probe. Inspired by the multicantilever Millipede innovation by IBM.[57] Mirkin *et al.* chose to simultaneously produce multiple copies of an arbitrary motif by an array of DPN tips rather than try to speed up the process itself. This approach proved successful with DPN going from one probe in 1999 to eight probes side by side in 2000 and 250 parallel tips in 2005.[28,31,58] The fabrication of arrays of slit microcantilevers[1] and, to a lesser extent, NADIS probes has also been demonstrated at the prototype level. Similar to parallel DPN, the inking of slit microcantilevers from external reservoirs enabled the parallel deposition of a series of different oligonucleotide solutions with each probe.[1,23] However, the quantitative improvement toward centimeter-scale nanopatterning was demonstrated by extending this to a two-dimensional (2D) array comprising 55,000 tips (500 columns of 100 tips) over 1 cm^2.[45] Quite remarkably, the ensemble of probes simultaneously deposits 55,000 times the same pattern as the Jefferson polygraph would do macroscopically, but each probe preserves the routine resolution performances of a standard DPN tip (60–80 nm spot diameter). Moreover, the entire array is brought in contact with the substrate, after a simple, gravity-driven alignment step, which ensures excellent reproducibility of the pattern (>99%). Figure 12.6a,b illustrates how this massively parallel DPN can produce 55,000 copies of a 5-cent coin pattern containing 8,773 spots of 80 nm each. The total 500 million spots are deposited in 30 minutes (contact time 80 ms, writing speed 60 μm/s) for an overall area of 0.1 cm^2. Alternatively, the entire coverage of 1 cm^2 with 600 nm spots spaced by 2 μm has also been demonstrated on a regular glass substrate with total thickness variation ca. 2.6 $\mu m.cm^{-2}$, instead of a flat single-crystal silicon surface. These achievements upgrade DPN from a strictly local manipulation instrument to a large-scale nanopatterning tool, and the technology is now commercially available.

In expanding from one-dimensional (1D) to 2D arrays of probes, parallel DPN has lost the ability to load each individual tip with an arbitrarily chosen molecular ink. Moreover, arrayed probe devices are limited to the production of multiple, identical copies of a single pattern. These two new challenges define the two foreseeable evolutions of parallel systems toward (i) on-chip microfluidic systems on arrayed probes and (ii) individual probe actuation. The parallelization of NFP probes has been proposed as a first possible route to implement linear or 2D arrays of volcano tips, each supplied by a dedicated microfluidics circuit, therefore allowing for the delivery of specific inks by each tip of the array (Fig. 12.6c).[59] The early demonstration of individual DPN tip actuation was based on the bimorph bending upon electrical heating of an array of gold-coated silicon nitride cantilevers.[60] Finally, one should mention that the recent efforts to fabricate complex probe arrays also offers the perspective of integrating several nanopatterning tips with analytical probing tools. For example, a double triangular cantilever NFP was developed by Deladi, which comprised a regular atomic force microscope imaging tip and a microfluidic-assisted deposition tip.[48] More generally, Wang and Liu designed a linear array of DPN tips

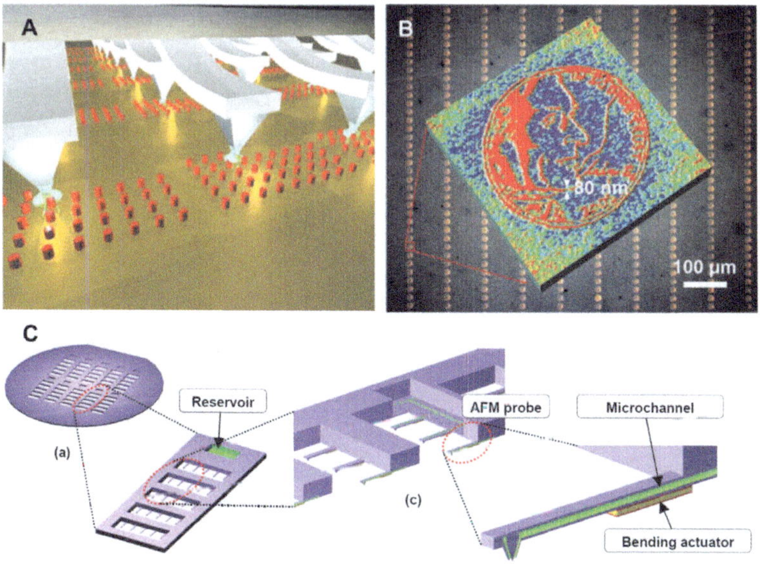

Figure 12.6 Parallel-probe arrays. (A) Artistic view of a 2D array of 55,000 parallel DPN passive tips. (B) Optical micrograph of a representative region of a substrate on which the approx. 55,000 duplicates were generated by the DPN array pictured in (A). Each of the circular features is a miniaturized replica of the face of the 5-cent coin. Inset: High-resolution topographical AFM image of a representative replica (adapted from Ref. 45, copyright 2006, Wiley VCH). (C) Schematic showing the concept of parallel NFPs: (a) arrays of the device on a wafer, (b) view of a unit cell with a feeding reservoir, (c) an embedded microfluidics network, and (d) a cantilever with a dispensing tip, a microchannel, and a bending actuator for vertical motion control (adapted from Ref. 59, copyright 2003, Society for Experimental Mechanics, Inc.).

with mixed lithography (DPN or scanning probe contact printing) and imaging (atomic or lateral force microscopy) capabilities.[61]

Initially motivated by the intrinsic low speed and highly local characteristics of scanning probe lithographic techniques, the paradigm of parallelization of nanopatterning tip devices has recently brought these techniques from exquisite laboratory instruments to technologically viable nanopatterning tools, which not only combine nanometer scale resolution and macroscopic-scale surface structuring but also combine material versatility and *in situ* structural and chemical analysis. It is therefore conceivable that the realm of SPM-based nanopatterning techniques will go beyond local surface modification and become performing tools for combinatorial synthetic and analytical chemistry as well as biomolecular and medical diagnosis.[62]

12.2.5 *Positioning*

The challenge of building multiscale architectures with nanometer-scale feature size supposes that molecular ink or nanodroplets can be positioned with a similar accuracy at an arbitrarily chosen location on a prestructured substrate, for example, a nanogap in molecular electronics applications. While stitching successive steps of downscaling lithography techniques is well mastered down to a few tens of nanometers, the preservation of surface registry during multiple steps of matter deposition by nanopatterning techniques remains challenging. Among the methods presented above, the SPM-based ones can provide highly resolved surface images in a nondestructive process. Therefore, a favorite strategy consists in preimaging the targeted patterning area in order to provide an accurate local map of the prestructured surface and then proceed to the deposition. Interestingly, the same functionalized DPN tip can be used for both generating and imaging nanostructures by simply changing the scan velocity. Indeed, DPN writing is typically performed at tip velocities smaller than 1 $\mu m.s^{-1}$, but no ink transfer occurs at velocities larger than 5 $\mu m.s^{-1}$, while such high speeds are suitable for lateral mode force microscopy. It has been proposed that large scanning velocities may disrupt the water meniscus between the tip and the substrate, thus interrupting the molecular diffusion from the tip to the surface through the meniscus. Nevertheless, since undetectable amounts of ink are likely to be transferred to the surface during imaging and may contaminate the pattern, more secure alignment procedures involve the addition of marks produced by DPN close to the zone of interest.[28] Thanks to the possibility of *in situ* characterization of the surface with the writing tip, patterns with different inks are aligned using serial or parallel procedures with a precision estimated of the order of 5 nm. The high accuracy of this alignment method was illustrated by the study of the diffusion of an MHA spot produced by DPN on a substrate in the presence of a pure ODT or a mixed ODT/MHA pattern acting as a diffusion barrier.[31] Similarly, the narrow gap between two predefined nanoelectrodes was locally functionalized with oligonucleotides deposited by DPN in order to subsequently direct the attachment of single NPs in the nanojunction by base pair recognition for electrical characterization.[63] The success of this hybrid top-down and bottom-up approach relied on the sub-10 nm positioning accuracy of the DPN patterning.

In the case of liquid-dispensing methods using apertured atomic force microscope tips, such as NFP or NADIS, the permanent presence of a large liquid meniscus at the tip apex prevents the surface imaging with the writing tip. Nanopositioning procedures therefore require the addition of a supplementary tip dedicated to surface imaging. Two different options have been demonstrated. For the NADIS technique, the deposition tip was incorporated on an independent holder under a commercial AFM setup dedicated to imaging. A nanopositioning table was inserted under the sample stage in order to move and position it with nanometric precision under the deposition or imaging tips.[64] The full procedure involves a referencing deposition step during which spots are patterned near the targeted zone. Then AFM imaging of the reference spots together with the location of interest provides the displacement coordinates to be applied to the sample in order to perform the actual patterning at the desired place. The positioning accuracy of this approach was demonstrated to be better than 100 nm.[65] This vectorial referencing approach could be fully integrated into the double triangular cantilever NFP described earlier.[48] The innermost tip is used for imaging the substrate before deposition, while the outer cantilever, which is fed ink by a closed channel, is used for writing process. Since the distance between the two tips is fixed and known, the displacement needed to perform the deposition after the reference image acquisition is straightforward.

12.2.6 *Summary*

With the recent advent of a panel of technologies able to manipulate liquid droplets with decreasing probe size, ranging from standard micropipettes to NADIS tips and DPN, it is now conceivable to arbitrarily pattern and functionalize a substrate from 1 mm (1 μL) down to a few tens of nanometers (100 zL), as introduced in Fig. 12.1. At smaller scales, (10–0.1 nm), one can anticipate that surface patterning by addition of matter will not involve a liquid carrier but rely on molecular self-assembly, which could nevertheless occur in a manipulated droplet. One remarkable trend, in recent literature, is that all these probe-based techniques converge toward common challenges (parallelization, embedded microfluidics, individual probe actuation), which allow them to scale up from local patterning with individual probes to simultaneous macroscopic-scale patterning. Therefore the techniques described in this section, which have proven their potential as laboratory tools, have reached a stage of maturity required to become relevant complementary technologies alongside standard lithography or nanoimprint processes.

12.3 Liquid Nanodispensing

In this section, we describe the NADIS technique, which allows a direct manipulation of ultrasmall liquid quantities (Fig. 12.7a). We detail the experimental procedures used for tip fabrication and droplet deposition and review the main parameters controlling the size of the spot. We then show that besides its potential as a nanopatterning method, the NADIS technique is a unique tool to study capillary mechanisms at submicron scale.

12.3.1 *Tip Fabrication and Deposition Process*

NADIS tip fabrication comprises two technological steps. The first one is FIB milling of the aperture at the apex of commercially available silicon nitride atomic force microscope tips. The second step consists in controlling the tip surface energy by functionalizing the intact gold film with a SAM. Both steps are briefly described here, and full details can be found in Ref. 43.

NADIS tips are fabricated by FIB nanofabrication of a small aperture on commercially available, silicon nitride, double-sided, gold-coated atomic force microscope tips with a hollow, square-pyramidal shape. Apertures with a diameter larger than 200 nm are milled through the apex from the inner side. However, the rapid degradation of the aperture quality when its aspect ratio (tip wall thickness:diameter) exceeds 4:1 does not allow to extend this approach below 200 nm apertures. For smaller apertures, a preliminary thinning process is performed in order to reduce the tip wall thickness around the apex from 800 nm to 80–100 nm. (Fig. 12.7b). For a hole size between 200 and 100 nm, the aperture itself can then be drilled through from the inner side, leaving the outer surface of the NADIS tip intact (Fig. 12.7c2). The centering of the aperture on the tip apex is crucial to ensure efficient liquid transfer from the tip to the substrate. The two-step protocol ensures a satisfactory centering precision for aperture size down to 100 nm, as shown in Fig. 12.7d, where the NADIS tip is viewed from the outer (tip) side. However, the sharp atomic force microscope tip apex profile and the residual misalignment between the inner and outer apexes frequently result in off-centered apertures for sub-100 nm holes. Consequently the milling of ultimately small apertures is performed from the outer side to allow a precise positioning, as illustrated in Fig. 12.7b2'. With this approach, one can reach the smallest attainable aperture diameter, which is about 25–30 nm, as shown in Fig. 12.7e. Finally, FIB exposure is also used to modify the NADIS tip surface properties. Indeed, when the outer surface of the silicon nitride tips, which is coated with 30 nm of gold, is left intact further chemical functionalization by thiol chemistry can be performed (Fig. 12.7f). Alternatively, since the sputtering rates of gold and silicon nitride are significantly different, the gold layer can be removed over a desired area around the aperture, thus exposing the underlying nitride (Fig. 12.7g). SixNy is intrinsically hydrophilic upon oxidation in ambient conditions, but it can also be chemically functionalized orthogonally to gold by using silane and siloxane chemistry.

Surface chemistry is used to functionalize tip and substrate surfaces with SAMs in order to spatially control their wetting properties. All surface treatments are operated in gaseous phase in a homebuilt vacuum line. Thermally grown SiO_2 on silicon, 100 nm thick, is used as a substrate, the surface energy of which is tuned

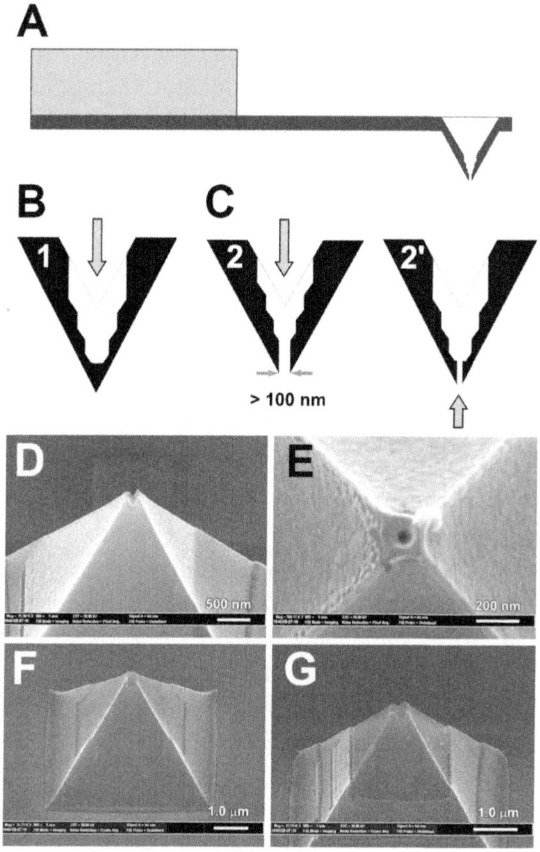

Figure 12.7 FIB fabrication of NADIS tips. (A) Cross-sectional sketch of a NADIS tip showing the hollow pyramid reservoir and the aperture perforated at the tip apex. (B and C) Schematics of the two-step milling protocol, namely, (B) thinning from the inner side and (C) aperture drilling. Apertures larger than 100 nm in diameter are drilled from the inner side (2), while smaller apertures have to be perforated from the tip side (2'). (D) Side view of a 120 × 70 nm aperture. (E) Axial view of a centered 35 nm round aperture. (F) NADIS tip with an intact gold coating on the outer surface (hole size 280 nm). (G) NADIS tip with a silicon nitride area exposed by locally removing the gold coating (hole size 215 nm). (Adapted from Ref. 43, copyright 2006, American Chemical Society.)

by treatments with 1H,1H,2H,2H-perfluorodecyltrichlorosilane, (aminopropyl)triethoxysilane (APTES), or octadecyltrichlorosilane (OTS). The surface treatment of the gold-coated NADIS tips is achieved in a similar way but in an atmosphere of dodecanethiol. These surface treatments allow varying surface properties to a large extent, from hydrophilic silica to moderately hydrophobic amine-covered and highly hydrophobic fluorinated surfaces.

All droplet deposition experiments are performed with glycerol-based liquids loaded in the NADIS reservoir with a microinjector controlled by a micromanipulator. Owing to its very low volatility (boiling point 290°C; vapor pressure 25°C, $<10^{-6}$ atm), glycerol can be stored under ambient atmosphere in the open reservoir of the NADIS tips. To characterize the dimensions and spacing of the deposited droplets and droplet arrays, we typically dissolve a highly stable ruthenium polypyridine complex[13] into a 9:1 glycerol:deionized water mixture at a concentration of 16 mM. The ruthenium complex acts as a marker for deposited droplets after complete evaporation of the solvent, and the spots left on the surface are imaged by tapping-mode AFM with a silicon probe (Veeco). A multimode PicoForce atomic force microscope (Veeco) is operated in force curve mode to transfer the liquid by contact of the loaded tip with the surface.

Arrays of droplets are obtained by repeating the contact at different locations on the surface following an automated procedure (Fig. 12.8a). The force exerted on the tip is recorded during the deposition process. A typical force curve obtained during the contact of a loaded tip with a given point of the surface is shown in Fig. 12.8b. It markedly differs from standard force curves obtained with an unmodified atomic force microscope tip or an unloaded NADIS tip. The specific features observed for liquid-loaded NADIS tips are attributed to the evolution of a liquid bridge between the tip and the surface. Indeed, during the approach phase, a rapid jump to a negative tip deflection is observed when the tip touches the surface and is bent downward by an attractive capillary force (from Fig. 12.8a to Fig. 12.8b). The cantilever bends upward until it reaches a preset threshold value (Fig. 12.8c). This contact force is maintained constant during a given period defined as "surface delay." During the retraction phase, a prolonged adhesion arises from the capillary

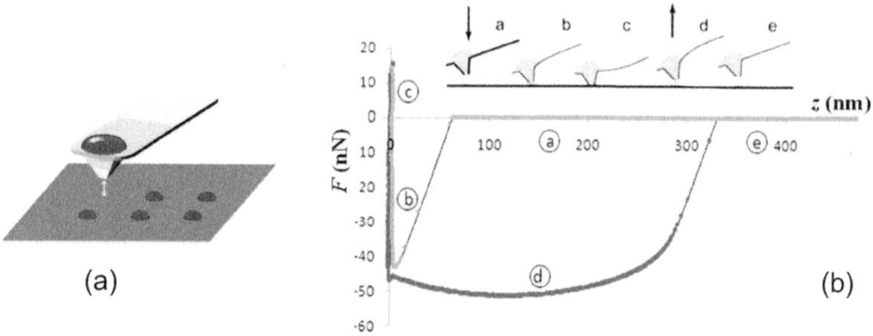

Figure 12.8 (a) Sketch of the spot deposition process with a NADIS tip; (b) force curve for a loaded cantilever probe with a 400 nm aperture at tip apex. The schematic labeling shows the probe–surface interaction: (a) approach, (b) jump to contact, (c) contact with the surface until a given applied force, (d) adhesion in the presence of a liquid meniscus, and (e) breaking of the liquid meniscus and pulling off of the tip.

force exerted by the liquid meniscus and results in a complex force curve shape (Fig. 12.8d). The liquid bridge eventually breaks abruptly, and the tip is pulled off the surface (Fig. 12.8e).

Such curves provide useful information concerning the deposition process, as will be detailed in section 12.3.3. Several parameters can be optimized in the deposition process in order to control the droplets size for a given tip. The impact of these dynamical parameters is discussed in the next section after the description of the influence of chemical properties of substrate and tip surfaces.

12.3.2 *Parameters Controlling Droplet Size*

The mechanism of liquid transfer from a tip to the surface is rather complex and involves a large number of parameters, which are related to liquid properties, tip and surface geometrical, or chemical aspects or to the deposition procedure itself. In the following sections the influence of these parameters is reviewed.

The spots of molecules left on the surface were imaged in tapping-mode AFM. Nevertheless, the size of the spots strongly depends on the mechanism of evaporation of the solvent[66] and does not necessary correspond to the size of the deposited droplets. In the case of rather hydrophilic surfaces or highly concentrated solutions, evaporation takes place in a constant radius mode. Consequently, the spots of molecules left on the surface after solvent evaporation coincide with the initial droplet size. On the contrary, for hydrophobic surfaces and dilute solutions, the droplet shrinks during evaporation leading to a spot with a much smaller size. The diameter of the dried spots no longer reflects the initial droplet diameter. In such case, linear arrays of droplets with diminishing pitches from line to line were deposited by NADIS. The droplet diameter can then be indirectly estimated as the smallest distance between two noncoalescing droplets.[41] Alternatively, force curves measured during deposition can provide the height of the deposited droplet from which the deposited volume can be determined.[43]

In the following section, we use highly concentrated glycerol solutions of a ruthenium complex, which give thick spots reflecting the size of the deposited droplets and facilitating AFM imaging.

12.3.2.1 Substrate surface properties

Substrate wettability plays a predominant role in setting the amount of deposited liquid. During the tip–substrate contact, the more hydrophilic the surface, the more the liquid spreads out. For example, the same hydrophilic NADIS tip with a 400 nm aperture can deposit droplets of several microns on a freshly clean SiO_2 surface, while it leads to the formation of sub-100 nm spots on a highly hydrophobic fluorinated surface.[43] However, in the latter case, the droplet size is less uniform, even leading to missing droplets.

Although fluorinated surfaces would provide a means to achieve droplets much smaller than the tip aperture,[67,68] their inertness, which prevents favorable molecule–surface interactions, makes them less attractive for practical applications. We therefore focused on the parameters provided by the deposition tool (tip geometry and functionalization) and method (contact time, applied force, etc.) in order to develop a precise control of the deposition on a reactive surface. These parameters are discussed in the following section on widely used amine-modified (APTES) SiO_2 surfaces, which give intermediate wetting properties with an advancing contact angle of about 45° for glycerol.

12.3.2.2 Tip chemistry

The significant impact of substrate wettability on the liquid spreading is equally expected from the tip outer surface. While hydrophilic NADIS tips yield 1 μm diameter deposits irrespective of the aperture size, the same tips derivatized with dodecanethiol lead to much smaller sizes of deposits. In this later case, the chemical functionalization renders the tip outer surface hydrophobic, thus confining the liquid meniscus in the vicinity of the aperture at the tip apex, by limiting the spreading of the liquid on the tip (Fig. 12.9a). In this case, the deposit size is directly related to the aperture diameter.

To gain further insight into the predominance of the tip surface chemistry on other parameters, we fabricated tips with more complex surface functionalizations by combining FIB machining and surface chemistry. The gold coating was removed by FIB to define a hydrophilic oxidized silicon nitride zone around the tip apex. Figure 12.9b shows an array of deposits performed with such a

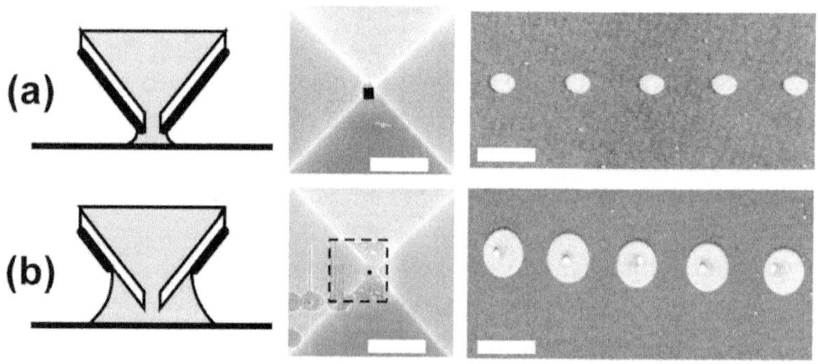

Figure 12.9 Deposition on an amine-treated surface with functionalized tips. The liquid meniscus is schematized on the left, with the thick black part of the tip corresponding to the hydrophobic gold layer obtained by dodecanethiol functionalization. SEM images of the tip outer wall and AFM images of deposited droplets are shown in the center and on the right, respectively. All images are reported at the same scale (scale bar: 1 μm). (a) Hydrophobic tip with an aperture of 200 nm at the tip apex; (b) mixed tip with an aperture of 80 nm and the gold coating removed within a area of 800 × 800 nm around the tip apex to make it inert to dodecanethiol treatment. (Reprinted with permission from Ref. 43, copyright 2007, American Chemical Society.)

tip having an 80 nm aperture surrounded by an 800 × 800 nm hydrophilic zone. In spite of a much smaller aperture size than in the previous case (Fig. 12.9b), the deposited droplets are much larger with a diameter of about 780 ± 20 nm. The deposit dimension is commensurate with the size of the hydrophilic area, indicating that the liquid contact line is pinned on the boundary between hydrophilic and hydrophobic zones, as represented schematically in Fig. 12.9b. A comparison of the results in Fig. 12.9 illustrates how a careful design of the tip surface wetting properties contributes to the control of the deposition dimensions.

12.3.2.3 Tip geometry

The ultimate resolution of NADIS was probed by optimizing the FIB milling process for sub-100 nm apertures and the hydrophobic surface treatment of the tip outer surface. Figure 12.10a shows an array of droplets deposited on an amine surface through am 80 × 120 nm aperture. The hydrophobic treatment of the tip ensures a uniform droplets size of 260 ± 20 nm, that is, about twice the aperture diameter. When conditions close to the current resolution limit of the FIB milling process are performed, a further reduction of the aperture size to 35 nm is achieved, which yields ultrasmall droplets with an average diameter of 70 ± 20 nm on the same surface under identical conditions (Fig. 12.10b).

In the case of hydrophobic tips, the deposit size is therefore closely related to the aperture size. The diameter of the droplets is about twice of the aperture diameter (Fig. 12.10c). This may be explained by the presence of hydrophilic defects or slits on the tip outer surface around the aperture due to residual ion beam exposure inherent to our protocol, which may allow the liquid to wet the tip surface just around the aperture.

12.3.2.4 Contact time

The time of contact between the tip and the surface is an interesting control parameter since it allows changing the size of the spots for a given system. It can also give insights into the mechanism of liquid transfer, a better understanding of which is crucial to optimize the technique.

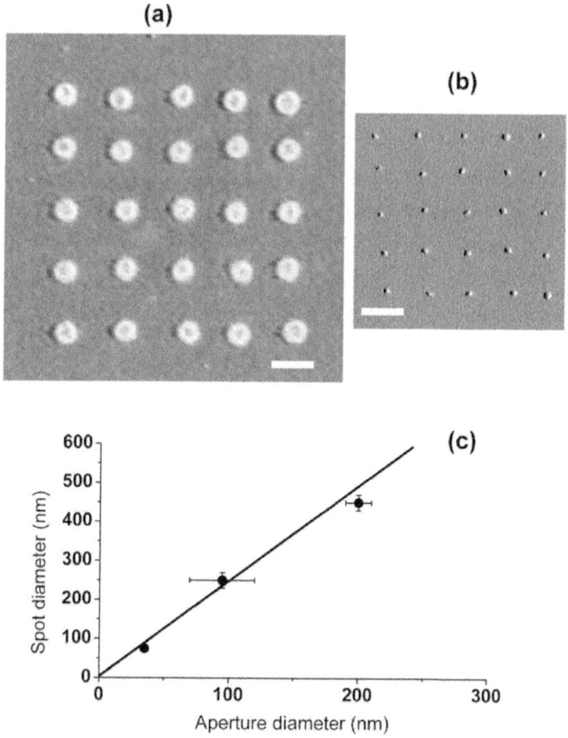

Figure 12.10 Arraying ultrasmall droplets on amine surface. AFM images of 5 × 5 arrays of nanodroplets deposited with hydrophobic tips. Scale bar: 500 nm. (a) 250 nm spots obtained using a tip with an oblong aperture 70 nm × 120 nm; (b) 70 nm spots deposited with a 35 nm aperture tip (see Fig. 12.7e); (c) Plot of the spot diameter as a function of aperture size. Errors bars in *y* correspond to statistics on 25 spots. Errors bars in *x* reflect the hole anisotropy. The solid line is a guide to the eyes. (Reprinted with permission from Ref. 43, copyright 2007, American Chemical Society.)

As shown in Ref. 43 the deposited volume can be deduced from the force curves measured during the deposition process. In the case of femtoliter droplets (diameter in the 1 μm range), a sharp increase of droplet size is observed for short times ($t < 2$ s typically), followed by a saturation regime where prolonged contact time has no obvious impact on the liquid flow. This dynamics indicates a nondiffusive process for the liquid transfer in striking a contrast to molecule transfer in DPN[69] and confirms the fundamental difference between these two methods. The saturation regime corresponds to the situation where equilibrium is reached. In this case, the liquid meniscus between the tip and the sample fulfills the relationship between the equilibrium contact angles on the tip and the substrate with a hydrostatic pressure imposed by the Laplace pressure of the reservoir. Note that all reported results concerning the influence of surface properties and aperture size are obtained in the saturation regime. The short time evolution is determined by both the flow of liquid in the nanochannels and the spreading dynamics on the surface. This is point is detailed in section 12.3.3.

In the case of hydrophobic tips with smaller aperture (<200 nm), which lead to deposited volumes in the attoliter range, no dynamics is observed. The equilibrium situation is reached with the first contact: the small volume needed to reach equilibrium is supplied from the reservoir within a shorter time than that of a single contact.

Consequently, in the femtoliter regime, the contact time can be used to adjust the size of the depositions, whereas in the attoliter regime, the size is entirely defined by the tip and substrate characteristics.

12.3.2.5 Contact force

The contact force was also varied in a wide range from positive values where the tip is pressed on the surface to negative values corresponding to the case where only the liquid at the end of the tip is touching the surface with no solid contact between the tip and the substrate. We observed in both cases that the size of the deposited droplets does not depend on the applied force. This is an important result toward a parallelization of the deposition when the simultaneous control of the same force for all applied tips is virtually impossible.

12.3.3 *Mechanism of Liquid Transfer*

The mechanism of liquid transfer from the reservoir to the surface through the aperture is rather complex. It involves both a flow of the liquid in the nanochannel and the spreading of the liquid on the

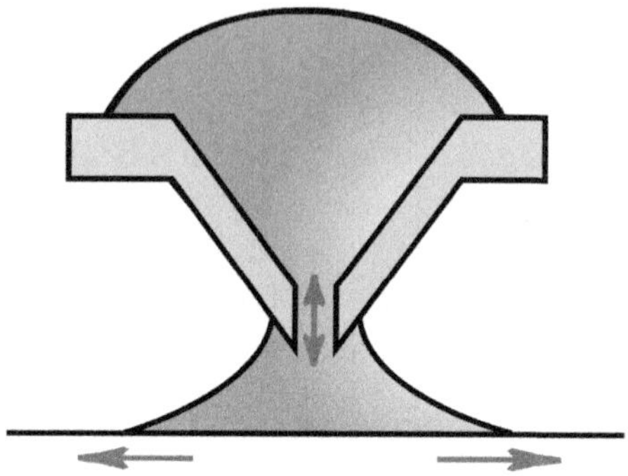

Figure 12.11 Sketch of the deposition mechanism. The droplet size is controlled by the flow of liquid through the nanochannels and by the spreading of the liquid on the substrate.

surface. In order to get more insight into the process, we used two different methods: (i) The capillary force measured during the tip surface separation was analyzed with the help of simulations, and (ii) the dynamics of spreading was studied by varying the contact time for spot deposition or by writing lines at various speeds.

12.3.3.1 Analysis of force curves

The force curves measured during the retraction of the NADIS tip is the only information available during the deposition. It corresponds to the capillary force exerted by the liquid meniscus bridging the tip and the surface. These curves strongly differ from the ones obtained with a standard dry atomic force microscope tip on a surface. Examples of curves obtained using various NADIS tips are reported in Fig. 12.12.

As shown in Fig. 12.12a, the shape of the curves strongly depends on the conditions used during the deposition. They contain information on the liquid transfer and can be used as an *in situ* control during deposition. To make a quantitative interpretation of the force curves, we performed simulations using the Surface Evolver software.[70] Using this method, we were able to simulate quantitatively the different types of curves shown in Fig. 12.12a.[71] We distinguish the case of large apertures (diameter larger than 200 nm), for which the liquid can flow in the nanochannel and allows

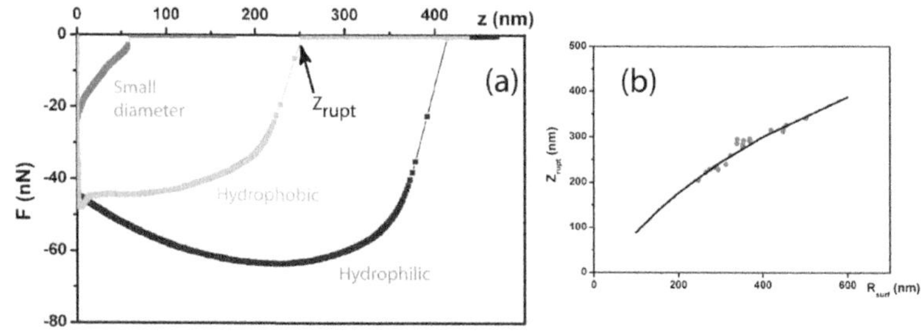

Figure 12.12 (a) Force curves obtained during the retraction of the tip in different conditions. The "hydrophilic" and "hydrophobic" curves correspond to the same tip with an aperture of 400 nm before and after surface functionalization with dodecanethiol. The "small aperture" curve corresponds to a hydrophobic tip with a 35 nm aperture. (b) Plot of Z_{rupt} as a function of the spot radius R_{surf} (dots) for a series of deposits obtained by changing contact time and fit of the curve obtained using Surface Evolver.

a permanent equilibrium of pressure between the meniscus and the reservoir, from the case of small channels (diameter smaller than 200 nm), for which the flow is not possible, leading to a curve with constant volume. For more details on these calculations and more generally on capillary forces, refer to chapter 4.

To understand the value of the transition diameter (around 200 nm from the experiments), we estimate the flow in the nanochannel generated by a pressure difference between the reservoir and the meniscus. Using a simple Poiseuille flow, one gets $Q = \Delta P \frac{\pi R^4}{8\eta L}$, where Q is the flow rate, ΔP the pressure difference, R the radius of the nanochannel, L its length, and η the liquid viscosity. Using a channel length of 200 nm, we obtained a flow rate of 3 fL/s for the large aperture, a value that is one order of magnitude larger than the flow rate needed to maintain a constant pressure in the meniscus. This validates the use of a constant pressure constraint in the calculations. As for the 35 nm nanochannels, a flow rate $Q = 5$ aL/s is required to maintain a constant pressure, while the one provide by the pressure gradient is limited to 0.7 aL/s. The assumptions used in the modelization are therefore justified by this simple estimate.

The adequate description provided by this model for the capillary force exerted on the tip by the liquid meniscus can also be used to establish a direct correlation between the shape of the force curve and the size of the droplet. On the force curve, a characteristic point is the altitude Z_{rupt} at which the liquid meniscus breaks (Fig. 12.12a). For a series of depositions with different contact times, we measured, for each dot, the size of the deposit R_{surf} on the AFM image and the rupture point Z_{rupt} on the force curve. The data, plotted in Fig. 12.12b, shows a correlation between these two values. We performed calculations in order to simulate this relationship. We observe a very good agreement between experiments and simulation, reinforcing the validity of this approach. Therefore, from the curve obtained in real time during the deposition process, it is possible to know the size of the deposited droplet by a direct reading of Fig. 12.12b.

12.3.3.2 Spreading dynamics

As shown in section 12.3.2.4, the use of short contact times limits the extension of the meniscus. In the case of channel apertures larger than 200 nm, the pressure balance inside the meniscus is fast, so that the pressure may be considered constant. In these conditions, the increase in meniscus size is due to the spreading of the liquid on the surface rather than limited by the flow in the nanochannel. A precise determination of the spreading dynamics by measuring the influence of contact time on the size of deposited droplets is nevertheless difficult since contact times lower than 500 ms are difficult to obtain by AFM. To increase the available range of times, we used a different method consisting in writing lines (Fig. 12.13).[72]

The lines are created by moving the tip on the surface at controlled velocity (ranging from 0.1 to 580 μm/s) with a nanopositioning table (NPS3330, Queensgate Instruments) incorporated to an atomic force microscope (Dimension 3000, Veeco).[64] A highly concentrated ruthenium complexes solution in glycerol is used to visualize by AFM the tracks left on the surface (Fig. 12.13b). The deposition of lines is performed with controlled velocity, while maintaining a constant force.

The results obtained for three tips with different channel diameters ($2R_0 = 220$ nm, 500 nm, and 760 nm) are reported in Fig. 12.13c, where the width w is plotted as a function of the velocity v.

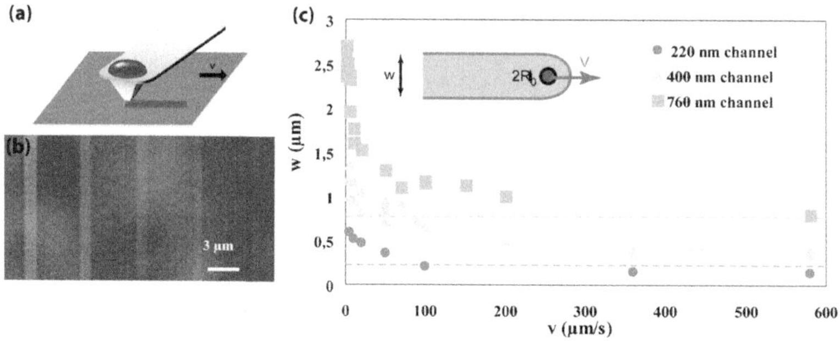

Figure 12.13 (a) Tapping-mode AFM image of lines deposited with increasing velocity (5 µm/s, 10 µm/s, 20 µm/s, and 50 µm/s from left to right); (b) line width as a function of deposition velocity for three different NADIS tips with channel sizes of 220 nm, 500 nm, and 760 nm. Dotted lines represent the channel dimensions of these tips (inset: a sketch presenting the different parameters ruling line deposition); (c) logarithmic curve of width vs. time corresponding to the same experiments as Fig. 12.4.

The width decreases when velocity increases until reaching a constant value, which corresponds to the channel diameter. In the latter situation, the liquid has no time to spread from the aperture and stays confined at the tip apex. When decreasing the velocity, liquid spreads, leading to a larger width. To get the temporal spreading dynamics, we define a time t as the ratio of the width w to the velocity v, that is, $t = w/v$. This transformation is checked by the comparison with the deposition of droplets with different contact time; the evolution of the spot diameter $2r(t)$ scales well with the dynamics deduced from lines $w(t = w/v)$, as shown in Fig. 12.14, where both values are reported in two situations corresponding to hydrophilic and hydrophobic tips. This conversion is also confirmed in different situations where both $r(t)$ and $w(v)$ are obtained analytically.[30,73] Due to the large range of available velocities, the deposition of lines gives access to the spreading dynamics over more than three orders of magnitude in time. The logarithmic plot of $R-R_0(t)$ reveals, at short times, a well-defined evolution, characterized by a $t^{0.26\pm0.04}$ law. This regime is followed by saturation for times larger than about one second. These two regimes have been observed systematically when changing tip or substrate properties.

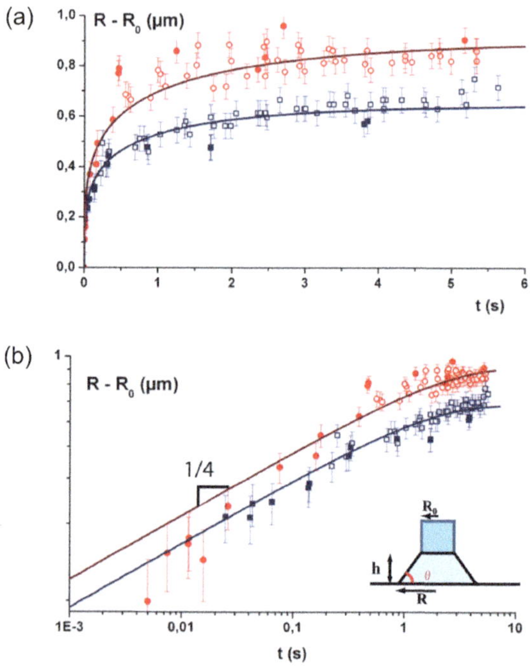

Figure 12.14 (a) Plot of R–$R_0(t)$ deduced from spots (empty symbols) and lines (filled symbols) deposition for a 760 nm aperture tip before (red circles) and after (blue squares) functionalization by dodecanethiol intended to make it hydrophobic; (b) same data in logarithmic scales (inset: a sketch of the situation used in the model).

To interpret the experimental power law, we consider that the spreading dynamics is governed by the Cox–Voinov equation[74,75] resulting from the equilibrium between capillary and viscous forces:

$$\theta^3 - \theta_m^3 = 9\frac{\eta}{\gamma}\ln(L/l) \qquad (12.1)$$

where θ is the macroscopic dynamic macroscopic contact angle observed at a large cutoff length L, θ_m is the microscopic equilibrium contact angle at a short cutoff length l, η is the liquid viscosity, and γ is its surface tension. Contrary to the spreading of a sessile droplet that corresponds to a constant volume situation leading to the well-known Tanner's law ($R \approx t^{1/10}$), the deposition of liquid by NADIS is characterized by a meniscus fed by the reservoir. We deduce from capillary force measurements (section 12.3.3.1) that the pressure inside the meniscus is constant and defined by the Laplace pressure of the reservoir.[71]

A new model has been developed to analyze our experiments, combining the Cox–Voinov Eq. 12.2 with a constant pressure constraint in the meniscus. For the sake of simplicity, we consider the 2D case of a meniscus with straight edges, meaning with no pressure (see inset of Fig 12.14). In these conditions, the relation linking the contact angle θ to the meniscus height h and radius R is $\theta = \frac{h}{R-R_0} = \frac{h}{u}$, where $u = R - R_0$. Inserting this in Eq. 12.1 leads to the general nonlinear equation:

$$\left(\frac{h}{u}\right)^3 - \theta_{\mathrm{m}}^3 = \frac{9\eta \ln L/l}{\gamma}\frac{\mathrm{d}u}{\mathrm{d}t} \tag{12.2}$$

For $\theta_{\mathrm{m}} = 0$, Eq. 12.2 can be integrated analytically, leading to $u \approx t^{1/4}$ in very good agreement with the experimental observations at short time scales. The case $\theta_{\mathrm{m}} \neq 0$, studied numerically with the Matlab software, reproduces also well the saturation observed at longer times. In Fig. 12.14, the results of the model adjusted to fit the experimental measurements obtained with hydrophilic and hydrophobic tips are plotted in solid lines.

The two simulated curves have been set out with the same parameters for the logarithmic term $\ln(L/l) = 10$ and the microscopic contact angle ($\theta_{\mathrm{m}} = 0.062$ rad $= 3.5°$). We only changed the meniscus height h ($h = 56$ nm for the untreated tip and $h = 40$ nm for the hydrophobic tip) consistent with the fact that liquid wets less the tip after hydrophobic treatment. We see that the model fits well the two observed regimes (spreading and saturation). For a short time $R-R_0$ evolves with a power law $t^{1/4}$ as in experiments, and for a long time it saturates at $(R-R_0)_{\mathrm{final}} = h/\theta_{\mathrm{m}}$. The values of the different parameters set in the model are coherent with the experimental conditions in all situations studied. Nevertheless, the microscopic contact angle θ_{m} has a lower value than the macroscopic one ($\theta \approx 30°$), which may be attributed to the effect of solute molecules that could modify the local wetting. A study of this issue is underway.

Additional experiments have been performed to study the influence of tip and substrate surface properties on line width. It has been concluded that by varying the tip contact angle θ_{tip}, the $w(t)$ curve is only shifted vertically in log scale without any slope modification (Fig. 12.14). Smaller sizes are found for hydrophobic tip as expected from the fact that changing the tip-wetting properties may modify the wetted part of the tip and hence the R_0 value. When the liquid contact angle on surface θ_{surf} is modified, the width at a short time stays unchanged but the saturation happens earlier.

This study shows that NADIS is a unique tool to study liquid behavior at small scales. A nonconventional spreading regime has been evidenced in the submicron range and down to a milliseconds time scale, conditions that are difficult to investigate with other methods.

The combination of force curve analysis and deposition experiments allowed us to propose a description of the liquid transfer from the tip to the surface. For apertures larger than 200 nm, the liquid can flow in the channel and maintain a constant pressure in the meniscus. This leads to an original spreading law with a radius scaling as $t^{1/4}$, which was not observed previously. In this situation, a control of the spots sizes or line width is possible by decreasing contact time or increasing tip velocity, respectively. On the contrary, for channel diameters smaller than 200 nm, the liquid has no time to flow through the channel. The droplet diameter is given by the amount of liquid present at the tip aperture and does not increase with contact time. The spot diameter is then fixed and cannot be adjusted.

12.3.4 *Applications of NADIS as a Nanopatterning Tool*

In previous sections, we have demonstrated how a careful design of NADIS tips and deposition conditions allows controlling the size of droplets from femto- to attoliters, corresponding to lateral dimensions ranging from several microns down to 70 nm. These studies are performed using highly concentrated solutions of small molecules that leave spots on the surface reproducing the droplet sizes. However, NADIS was designed as a general tool enabling to locate any solute molecule or nano-object on a surface with high lateral resolution and circumvent the limitation of the DPN technique for large objects such as proteins or NPs. Indeed, the size of deposited droplets should be rather independent of the nature of solute molecules. In order to demonstrate the advantage of liquid manipulation with this respect, we used the NADIS technique to pattern proteins and NP arrays.

The realization of organized arrays of biomolecules has attracted a lot of attention during the past years due to their potential applications in biology and life sciences (genomics, proteomics, drug screening, and biosensing). DPN provides a useful method to achieve biomolecules immobilization,[76] mainly by indirect patterning, as shown, for example, in Fig. 12.4. Direct deposition is less developed since it is a very slow technique, which also requires specific surface functionalization. The drying of solutions on the tip may also lead to proteins denaturation and loss of biological activity. In order to demonstrate that these drawbacks can be circumvented

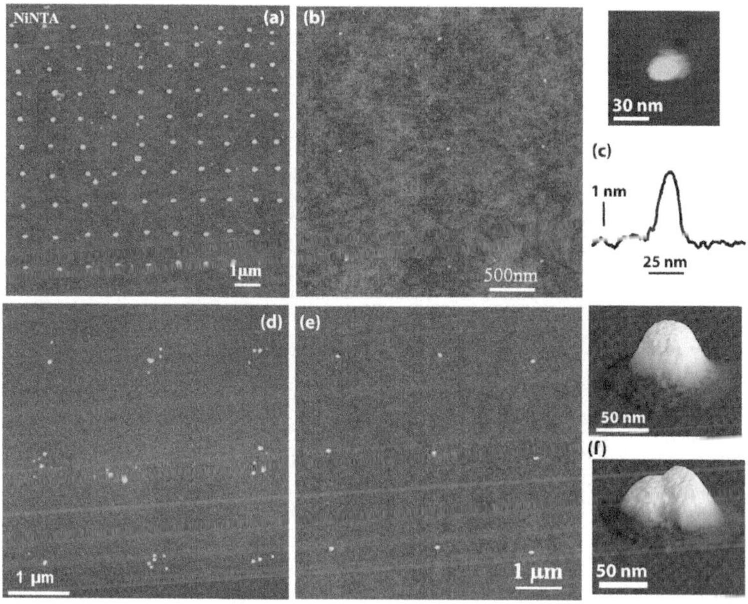

Figure 12.15 AFM images of (a) a GFP protein array deposited with a 220 nm aperture NADIS tip (spot diameters: 150 nm); (b) the same with a 110 nm aperture NADIS tip (spot diameters: 30 nm); (c) zoom on a 25 nm spot and height profile; (d) NP array obtained with a 300 nm aperture NADIS tip; (e) the same with a 130 nm aperture tip; and (f) zoom on spots with one and two NPs.

using NADIS, the deposition of various protein solutions in glycerol have been studied with the same procedure as described above for small molecules. The sizes of spots can be controlled by tip and substrate properties in the same range of sizes as for small molecules, that is, from a few microns down to 25 nm, independently of the precise nature of the proteins. GFP, S65T, and mStrawberry proteins (provided by Blum and Subramaniam, the BioPhysics group, University of Twente, the Netherlands) are patterned with identical lateral dimensions. Examples of GFP arrays with spot sizes of 150 nm and 30 nm are shown in Fig. 12.15a,b, together with a zoom and height profile of a 25 nm spot (Fig. 12.15c). Besides the generality of the method, liquid manipulation may also help to prevent the protein denaturation by keeping the molecules in solution. A fluorescence study of deposited proteins is underway to confirm this fact.

The recent achievements in the synthesis of a wide range of NPs with controlled composition, size, and shape open the way to the realization of devices using these objects as functional unit for electronics, optics, or magnetic applications. The positioning of these objects on the active part of devices remains a technological challenge for the development of nanosciences. The performance of NADIS was tested by patterning polystyrene (PS) NPs 25 nm in diameter. With these rather big objects, the number of NPs per spot can be unambiguously determined by AFM or SEM imaging, allowing assessing the ability of NADIS to perform single-particle deposition. An example of an array deposited using a NADIS tip with a diameter of 300 nm is reported in Fig. 12.15d. The number of NPs per spot varies from two to six, coherent with the number expected from the solution concentration (about one NP per 50 aL). Interestingly, using a 100 nm tip, we realized an 8×8 array constituted of single or double NPs per spot, with no missing spot. A portion of this array is reported in Fig. 12.15e, together with zooms on spots with a single NP and a pair of NPs. This result, which is not expected if one considers the concentration of NPs in the solution, may be explained by the evaporation of glycerol, which is fast at this scale and leads to an accumulation of NPs at the tip apex.

These two examples illustrate the advantage of NADIS with respect to other techniques such as DPN. The size of deposited

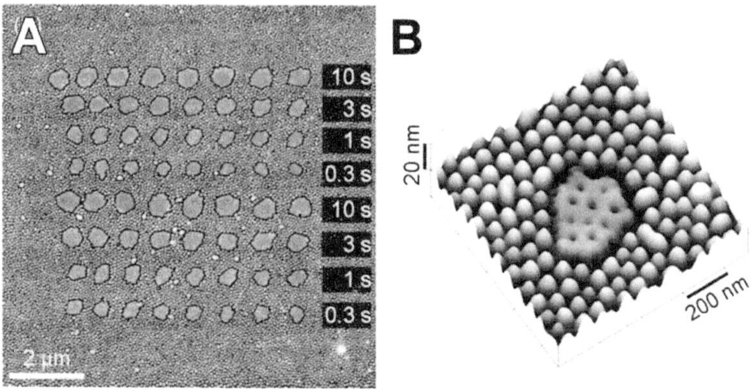

Figure 12.16 Local modification of self-assembled PS-P2VP block copolymer micellar layers using NADIS. (A) AFM image of an array of spots made of reconfigured micelles after exposure to glycerol droplets. Numbers on the right side of the array indicate the contact times of the NADIS probe for each line. (B) Detail view of one of the smaller spots. (Adapted from Ref. 77, copyright 2006, Elsevier B.V.)

droplets only depends on tip and surface properties, which allow to pattern any solute (from small molecules to large biological objects or NPs) with the same resolution. The liquid manipulated by NADIS can also directly induce a local modification of surface properties. A striking example was shown by Meister *et al.*,[77] who used the hydrophilic character of glycerol to restructure, at submicron scale, a micellar layer of block copolymers. When spin-coated on a silicon dioxide substrate, a solution of polystyrene-moly-2-vinylpyridine (PS-P2VP) block copolymers in toluene leads to the formation of a layer made of closely packed micelles (70 nm in diameter) made of a P2VP core surrounded by a PS corona (Fig. 12.16). The exposition of such a layer to a submicron droplet of glycerol deposited by NADIS leads to a local reconfiguration of the micelles, which exhibit an array of holes in place of an array of bumps. This effect is attributed to the hydrophilic properties of glycerol, which rearrange the micelle structure in order to expose preferentially the P2VP moiety of the block copolymer. Even if the precise mechanism (opening or inversion of the micelles) remains to be clarified, this experiment demonstrates an original use of NADIS as a patterning tool.

12.3.5 *Evaporation of Femto Droplets*

Besides numerous applications in surface nanopatterning, the tools presented in this paper can also be used for more fundamental studies of wetting processes at submicron scale. To illustrate this point, the evaporation of droplets produced by NADIS with volumes smaller than the femtoliter was performed. When manipulating liquids at submicron scale, evaporation becomes an important issue. It is therefore essential to study experimentally the evaporation of submicron droplets since most of the previous comprehensive studies of evaporation have been limited to microliter droplets (i.e., diameters in the millimeter range).[66]

Monitoring the evaporation process of submicron droplets cannot be achieved by standard optical microscopy techniques, which prompted us to measure the mass of the droplet during evaporation using ultrasensitive mass sensors.[78] Such nanomechanical resonators, called quad-beam resonators (QBRs), consist in four beams supporting a 5×5 μm suspended central plate. The resonator is excited electrostatically, and a capacitive detection is performed with an on-chip CMOS-integrated device. The resonance frequency is highly sensitive to a small addition of mass with a sensitivity to point mass accretion in the order of 100 attogram/Hz (10^{-16} g/Hz), more than 1,000 times better than the one of quartz microbalances. The combination of QBR performances with the NADIS method presented above therefore gives a unique way to perform this study.[79]

Droplets with diameters in the micron range are obtained using a NADIS tip with a 300 nm aperture and a highly hydrophilic outer wall (Fig. 12.17a). The NADIS tip is aligned with the QBR using a nanopositioning table incorporated in AFM.[64]

During evaporation, the resonance frequency of the QBR is monitored (Fig. 12.17b inset). A calibration curve obtained with calibrated silica beads is used to convert the resonance frequency shifts into mass changes so that the temporal evolution of the droplets mass is determined down to around 10 fg (10 aL).

Time evolution of the mass of evaporating glycerol droplets with initial volumes ranging from 0.2 fL to 20 fL are shown in Fig. 12.17b. The droplet mass decreases nonlinearly, with a slowing-down of the

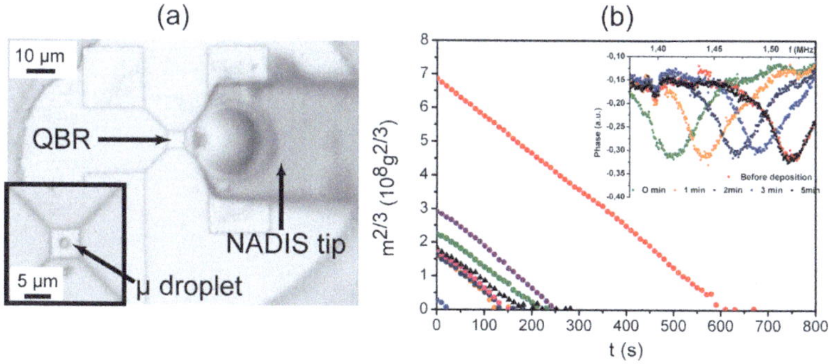

Figure 12.17 (a) Optical microscope image of a loaded NADIS tip close to a QBR resonator before deposition (inset: zoom on the QBR after deposition of a 2 μm droplet). (b) Evaporation curves of droplets with volumes ranging between 0.15 and 15 fl in an $m^{2/3}$ vs. t representation (inset: phase resonance curves before deposition [red] and during the evaporation process after deposition).

evaporation rate along the process. This behavior can be accounted for by macroscopic models describing evaporation on hydrophobic surfaces with a constant contact angle between the droplet and the surface. In this regime at macroscopic scale, the droplet mass m is expected to decrease along the evaporation time t with a power law 2/3:

$$m^{2/3} = m_0^{2/3} - \frac{2}{3}\alpha t \qquad (12.3)$$

where α is a coefficient depending only on physical parameters but is independent of the initial mass m_0.[78] Figure 12.17b shows that the evolution of $m^{2/3}$ as a function of t is linear in agreement with Eq. 12.3 for a series of droplets with different initial volumes. Quantitatively, all the experimental data can be fitted by this simple law with one adjustable parameter α for all the curves, leading to a determination of the diffusion coefficient of glycerol in air $D = 5 \pm 1 \times 10^{-2}$ cm^2/s. This is lower than the $D = 9\times 10^{-2}$ cm^2/s value estimated by the formula given in Ref. 81, but no experimental value is reported in the literature for comparison.

These results indicate that the macroscopic laws are still valid at these scales. The main assumption of the model is that the evaporation process is driven by diffusion, which is correct as long as the mean free path λ of the vapor molecules is smaller than the droplet dimensions (of radius r). This hypothesis should not be valid anymore for dimensions in the submicron range corresponding to masses below 1 picogram,[66] which is at the limit of our mass-sensing device.

12.4 Conclusion

To transcend its own properties, one can bring a single molecule into higher-order architectures of designed molecular materials with tailored morphology, structure, and properties using supramolecular principles. Conversely, one can embed a single molecule into a device designed for the addressing and exploitation of its intrinsic properties. Yet, this still requires the development of tools and principles able to achieve this second goal. The idea of using liquids—and especially water—to carry a few molecules and place them at the core of a surface-borne device has generated a number of fundamental questions and technical proposals which we have tried to survey in this chapter. Challenges are numerous, the most important being spatial resolution down to single-molecular scale, versatility in terms of liquids and solutes, large-area scalability, speed of realization, and multiplexing. To a large extent, each of these challenges has been addressed by at least one of the techniques exposed here (nanopipettes, slit microcantilevers, AFM lithography, NADIS), but one can also clearly see the future challenges resulting from the need to address all these objectives in a unique approach. It appears to us that the direct manipulation of liquid nanodrops holds much of the potential for such a generic method, and therefore we have exposed the technical details of NADIS and the physics of the liquids at the nanoscale that governs the dispensing, evaporation, and spreading of nanodrops using NADIS. The surprising yet stimulating common feature to the presented results is that the nineteenth-century concepts on the macroscopic behavior of liquids remain valid down to a few tens of zeptoliters corresponding to drops of only a million solvent molecules. Will these concepts fail before the assembly of solvent molecules cannot be named a liquid drop anymore?

The handling of nanodrops is now technically feasible, and one can therefore envision its applications in all fields where small volumes of liquids are involved. The emergence of such applied work is the second type of outcome of the recent efforts on the manipulation of liquid nanodrops one should be enthusiastic about!

Acknowledgments

The authors would like to thank the following colleagues for experimental contributions in the nanowetting projects illustrating this chapter, fruitful collaboration, and/or stimulating discussions on liquid nanopatterning of surfaces and manipulation of nanodrops: J. Arcamone, C. Blum, A. Fang, H. Heinzelmann, A. Meister, F. Perez-Murano, V. Subramaniam, and A. Thete. Thierry Ondarçuhu and Erik Dujardin thank the European Community and the European Research Council for the funding of the NADIS activity though the FP6-IP project NaPa (Contract No. NMP4-CT-2003-500120) and as part of an ERC-Starting Grant (Contract FP7-ERC-2007-StG 203872-COMOSYEL).

References

1. E. Descamps, T. Leichle, B. Corso, S. Laurent, P. Mailley, L. Nicu, T. Livache, and C. Bergaud, Fabrication of oligonucleotide chips by using parallel cantilever-based electrochemical deposition in picoliter volumes, *Adv. Mater.*, **19**, 1816–1821 (2007)
2. K. T. Rodolfa, A. Bruckbauer, D. J. Zhou, Y. E. Korchev, and D. Klenerman, Two component graded deposition of biomolecules with a double-barreled nanopipette, *Angew. Chem., Int. Ed.*, **44**, 6854–6859 (2005).
3. B. Rosner, T. Duenas, D. Banerjee, R. Shile, N. Amro, and J. Rendlen, Functional extensions of dip pen nanolithography (TM) : active probes and microfluidic ink delivery, *Smart Mater. Struct.*, **15**, S124–S130 (2006).
4. P. Calvert, Inkjet printing for materials and devices, *Chem. Mater.*, **13**, 3299–3305 (2001).

5. B. J. de Gans, P. C. Duineveld, and U. S. Schubert, Inkjet printing of polymers: state of the art and future developments, *Adv. Mater.*, **16**, 203–213 (2004).
6. E. Tekin, P. J. Smith, and U. S. Schubert, Inkjet printing as a deposition and patterning tool for polymers and inorganic particles, *Soft Matter*, **4**, 703–713 (2008).
7. T. Goldmann and J. S. Gonzalez, DNA-printing: utilization of a standard inkjet printer for the transfer of nucleic acids to solid supports, *J. Biochem. Biophys. Methods*, **42**, 105–110 (2000).
8. T. R. Hughes, M. Mao, A. R. Jones, J. Burchard, M. J. Marton, K. W. Shannon, S. M. Lefkowitz, M. Ziman, J. M. Schelter, M. R. Meyer, S. Kobayashi, C. Davis, H. Y. Dai, Y. D. D. He, S. B. Stephaniants, G. Cavet, W. L. Walker, A. West, E. Coffey, D. D. Shoemaker, R. Stoughton, A. P. Blanchard, S. H. Friend, and P. S. Linsley, Expression profiling using microarrays fabricated by an ink-jet oligonucleotide synthesizer, *Nat. Biotechnol.*, **19**, 342–347 (2001).
9. T. Xu, J. Jin, C. Gregory, J. J. Hickman, and T. Boland, Inkjet printing of viable mammalian cells, *Biomaterials*, **26**, 93–99 (2005).
10. A. Bietsch, J. Y. Zhang, M. Hegner, H. P. Lang, and C. Gerber, Rapid functionalization of cantilever array sensors by inkjet printing, *Nanotechnology*, **15**, 873–880 (2004).
11. J. U. Park, M. Hardy, S. J. Kang, K. Barton, K. Adair, D. K. Mukhopadhyay, C. Y. Lee, M. S. Strano, A. G. Alleyne, J. G. Georgiadis, P. M. Ferreira, and J. A. Rogers, High-resolution electrohydrodynamic jet printing, *Nat. Mater.*, **6**, 782–789 (2007).
12. G. Taylor, Disintegration of water drops in electric field, *Proc. R. Soc. Lond., Ser. A*, **280**, 383–397 (1964).
13. J. U. Park, J. H. Lee, U. Paik, Y. Lu, and J. A. Rogers, Nanoscale patterns of oligonucleotides formed by electrohydrodynamic jet printing with applications in biosensing and nanomaterials assembly, *Nano Lett.*, **8**, 4210–4216 (2008).
14. S. Arscott and D. Troadec, A nanofluidic emitter tip obtained by focused ion beam nanofabrication, *Nanotechnology*, **16**, 2295–2302 (2005).
15. A. Bruckbauer, L. M. Ying, A. M. Rothery, D. J. Zhou, A. I. Shevchuk, C. Abell, Y. E. Korchev, and D. Klenerman, Writing with DNA and protein using a nanopipet for controlled delivery, *J. Am. Chem. Soc.*, **124**, 8810–8811 (2002).
16. http://www.ionscope.com/

17. A. Bruckbauer, D. J. Zhou, L. M. Ying, Y. E. Korchev, C. Abell, and D. Klenerman, Multicomponent submicron features of biomolecules created by voltage controlled deposition from a nanopipet, *J. Am. Chem. Soc.*, **125**, 9834–9839 (2003).
18. L. M. Ying, A. Bruckbauer, D. J. Zhou, J. Gorelik, A. Shevehuk, M. Lab, Y. Korchev, and D. Klenerman, The scanned nanopipette: a new tool for high resolution bioimaging and controlled deposition of biomolecules, *Phys. Chem. Chem. Phys.*, **7**, 2859–2866 (2005).
19. O. T. Guenat, J. F. Dufour, P. D. van der Wal, W. E. Morf, N. F. de Rooij, and M. Koudelka-Hep, Microfabrication and characterization of an ion-selective microelectrode array platform, *Sens. Actuator, B*, **105**, 65–73 (2005).
20. P. Belaubre, M. Guirardel, G. Garcia, J. B. Pourciel, V. Leberre, A. Dagkessamanskaia, E. Trevisiol, J. M. Francois, and C. Bergaud, Fabrication of biological microarrays using microcantilevers, *Appl. Phys. Lett.*, **82**, 3122–3124 (2003).
21. J. T. Xu, M. Lynch, J. L. Huff, C. Mosher, S. Vengasandra, G. F. Ding, and E. Henderson, Microfabricated quill-type surface patterning tools for the creation of biological micro/nano arrays, *Biomed. Microdevices*, **6**, 117–123 (2004).
22. T. Leichle, L. Nicu, E. Descamps, B. Corso, P. Mailley, T. Livache, and C. Bergaud, Copper electrodeposition localized in picoliter droplets using microcantilever arrays, *Appl. Phys. Lett.*, **88**, 254108 (2006).
23. D. Saya, T. Leichle, J. B. Pourciel, C. Bergaud, and L. Nicu, Collective fabrication of an in-plane silicon nanotip for parallel femtoliter droplet deposition, *J. Micromech. Microeng.*, **17**, N1–N5 (2007).
24. J. T. Xu, M. Lynch, S. Nettikadan, C. Mosher, S. Vegasandra, and E. Henderson, Microfabricated "biomolecular ink cartridges"—surface patterning tools (SPTs) for the printing of multiplexed biomolecular arrays, *Sens. Actuator, B*, **113**, 1034–1041 (2006).
25. Y. Roupioz, N. Berthet-Duroure, T. Leichle, J. B. Pourciel, P. Mailley, S. Cortes, M. B. Villiers, P. N. Marche, T. Livache, and L. Nicu, Individual blood-cell capture and 2D organization on microarrays, *Small*, **5**, 1493–1497 (2009).
26. M. Lynch, C. Mosher, J. Huff, S. Nettikadan, J. Johnson, and E. Henderson, Functional protein nanoarrays for biomarker profiling, *Proteomics*, **4**, 1695–1702 (2004).
27. M. Jaschke and H. J. Butt, Deposition of organic material by the tip of a scanning force microscope, *Langmuir*, **11**, 1061–1064 (1995).

28. R. D. Piner, J. Zhu, F. Xu, S. H. Hong, and C. A. Mirkin, "Dip-pen" nanolithography, *Science*, **283**, 661–663 (1999).
29. S. Rozhok, R. Piner, and C. A. Mirkin, Dip-pen nanolithography: what controls ink transport? *J. Phys. Chem. B*, **107**, 751–757 (2003).
30. B. L. Weeks, A. Noy, A. E. Miller, and J. J. De Yoreo, Effect of dissolution kinetics on feature size in dip-pen nanolithography, *Phys. Rev. Lett.*, **88**, 255505 (2002).
31. S. H. Hong and C. A. Mirkin, A nanoplotter with both parallel and serial writing capabilities, *Science*, **288**, 1808–1811 (2000).
32. S. Kramer, R. R. Fuierer, and C. B. Gorman, Scanning probe lithography using self-assembled monolayers, *Chem. Rev.*, **103**, 4367–4418 (2003).
33. K. Salaita, Y. H. Wang, and C. A. Mirkin, Applications of dip-pen nanolithography, *Nat. Nanotechnol.*, **2**, 145–155 (2007).
34. A. A. Tseng, A. Notargiacomo, and T. P. Chen, Nanofabrication by scanning probe microscope lithography: a review, *J. Vac. Sci. Technol., B*, **23**, 877–894 (2005).
35. D. S. Ginger, H. Zhang, and C. A. Mirkin, The evolution of dip-pen nanolithography, *Angew. Chem., Int. Ed.*, **43**, 30–45 (2004).
36. B. Basnar and I. Willner, Dip-pen-nanolithographic patterning of metallic, semiconductor, and metal oxide nanostructures on surfaces, *Small*, **5**, 28–44 (2009).
37. R. A. Vega, D. Maspoch, K. Salaita, and C. A. Mirkin, Nanoarrays of single virus particles, *Angew. Chem., Int. Ed.*, **44**, 6013–6015 (2005).
38. C.-C. Wu, D. N. Reinhoudt, C. Otto, V. Subramaniam, and A. H. Velders, Strategies for patterning biomolecules with dip-pen nanolithography, *Small*, **7**, 982 (2011).
39. C.-C. Wu, H. Xu, C. Otto, D. N. Reinhoudt, R. G. H. Lammertink, J. Huskens, V. Subramaniam, and A. H. Velders, Porous multilayer-coated AFM tips for dip-pen nanolithography of proteins, *J. Am. Chem. Soc.*, **131**, 7526–7527 (2009).
40. A. J. Senesi, D. I. Rozkiewicz, D. N. Reinhoudt, and C. A. Mirkin, Agarose-assisted dip-pen nanolithography of oligonucleotides and proteins, *ACS Nano*, **3**, 2394–2402 (2009).
41. A. Meister, M. Liley, J. Brugger, R. Pugin, and H. Heinzelmann, Nanodispenser for attoliter volume deposition using atomic force microscopy probes modified by focused-ion-beam milling, *Appl. Phys. Lett.*, **85**, 6260–6262 (2004).

42. A. Meister, S. Jeney, M. Liley, T. Akiyama, U. Staufer, N. F. de Rooij, and H. Heinzelmann, Nanoscale dispensing of liquids through cantilevered probes, *Microelectron. Eng.*, **67–8**, 644–650 (2003).

43. A. P. Fang, E. Dujardin, and T. Ondarcuhu, Control of droplet size in liquid nanodispensing, *Nano Lett.*, **6**, 2368–2374 (2006).

44. C. Martin, A. Llobera, T. Leichle, G. Villanueva, A. Voigt, V. Fakhfouri, J. Y. Kim, N. Berthet, J. Bausells, G. Gruetzner, L. Nicu, J. Brugger, and F. Perez-Murano, Novel methods to pattern polymers for microfluidics, *Microelectron. Eng.*, **85**, 972–975 (2008).

45. K. Salaita, Y. H. Wang, J. Fragala, R. A. Vega, C. Liu, and C. A. Mirkin, Massively parallel dip-pen nanolithography with 55000-pen two-dimensional arrays, *Angew. Chem., Int. Ed.*, **45**, 7220–7223 (2006).

46. K. H. Kim, N. Moldovan, and H. D. Espinosa, A nanofountain probe with sub-100 nm molecular writing resolution, *Small*, **1**, 632–635 (2005).

47. N. Berthet-Duroure, T. Leichle, J. B. Pourciel, C. Martin, J. Bausells, E. Lora-Tamayo, F. Perez-Murano, J. M. Francois, E. Trevisiol, and L. Nicu, Interaction of biomolecules sequentially deposited at the same location using a microcantilever-based spotter, *Biomed. Microdevices*, **10**, 479–487 (2008).

48. S. Deladi, N. R. Tas, J. W. Berenschot, G. J. M. Krijnen, M. J. de Boer, J. H. de Boer, M. Peter, and M. C. Elwenspoek, Micromachined fountain pen for atomic force microscope-based nanopatterning, *Appl. Phys. Lett.*, **85**, 5361–5363 (2004).

49. S. Deladi, J. W. Berenschot, N. R. Tas, G. J. M. Krijnen, J. H. de Boer, M. J. de Boer, and M. C. Elwenspoek, Fabrication of micromachined fountain pen with in situ characterization possibility of nanoscale surface modification, *J. Micromech. Microeng.*, **15**, 528–534 (2005).

50. N. Moldovan, K. H. Kim, and H. D. Espinosa, A multi-ink linear array of nanofountain probes, *J. Micromech. Microeng.*, **16**, 1935–1942 (2006).

51. F. Heuck, T. Hug, T. Akiyama, P. Frederix, A. Engel, A. Meister, H Heinzelmann, N. F. de Rooij, and U. Staufer, Evaporation based micro pump integrated into a scanning force microscope probe, *Microelectron. Eng.*, **85**, 1302–1305 (2008).

52. B. Wu, A. Ho, N. Moldovan, and H. D. Espinosa, Direct deposition and assembly of gold colloidal particles using a nanofountain probe, *Langmuir*, **23**, 9120–9123 (2007).

53. O. Loh, R. Lam, M. Chen, N. Moldovan, H. Huang, D. Ho, and H. D. Espinosa, Nanofountain-probe-based high-resolution patterning and single-cell injection of functionalized nanodiamonds, *Small*, **5**, 1667–1674 (2009).

54. K. H. Kim, R. G. Sanedrin, A. M. Ho, S. W. Lee, N. Moldovan, C. A. Mirkin, and H. D. Espinosa, Direct delivery and submicrometer patterning of DNA by a nanofountain probe, *Adv. Mater.*, **20**, 330–334 (2008).
55. O. Y. Loh, A. M. Ho, J. E. Rim, P. Kohli, N. A. Patankar, and H. D. Espinosa, Electric field-induced direct delivery of proteins by a nanofountain probe, *Proc. Natl. Acad. Sci. USA*, 105, 16438–16843 (2008).
56. A. Meister, M. Gabi, P. Behr, P. Studer, J. Voros, P. Niedermann, J. Bitterli, J. Polesel-Maris, M. Liley, H. Heinzelmann, and T. Zambelli, FluidFM: combining atomic force microscopy and nanofluidics in a universal liquid delivery system for single cell applications and beyond, *Nano Lett.*, **9**, 2501–2507 (2009).
57. P. Vettiger, M. Despont, U. Drechsler, U. Durig, W. Haberle, M. I. Lutwyche, H. E. Rothuizen, R. Stutz, R. Widmer, and G. K. Binnig, The "millipede"—more than one thousand tips for future AFM data storage, *IBM J. Res. Dev.*, **44**, 323–340 (2000).
58. K. Salaita, S. W. Lee, X. F. Wang, L. Huang, T. M. Dellinger, C. Liu, and C. A. Mirkin, Sub-100 nm, centimeter-scale, parallel dip-pen nanolithography, *Small*, **1**, 940–945 (2005).
59. K.-H. Kim, C. Ke, N. Moldovan, and H. D. Espinosa, Massively parralel multi-tip nanoscale writer with fluidic capabilities—fountain pen nanolithography (FPN), *Proc. 4th Intern. Symp. MEMS Nanotechnol.*, 235–238 (2003).
60. D. Bullen, S. W. Chung, X. F. Wang, J. Zou, C. A. Mirkin, and C. Liu, Parallel dip-pen nanolithography with arrays of individually addressable cantilevers, *Appl. Phys. Lett.*, **84**, 789–791 (2004).
61. X. F. Wang and C. Liu, Multifunctional probe array for nano patterning and imaging, *Nano Lett.*, **5**, 1867–1872 (2005).
62. C. A. Mirkin, The power of the pen: development of massively parallel dip-pen nanolithography, *ACS Nano*, **1**, 79–83 (2007).
63. S. W. Chung, D. S. Ginger, M. W. Morales, Z. F. Zhang, V. Chandrasekhar, M. A. Ratner, and C. A. Mirkin, Top-down meets bottom-up: dip-pen nanolithography and DNA-directed assembly of nanoscale electrical circuits, *Small*, **1**, 64–69 (2005).
64. T. Ondarcuhu, L. Nicu, S. Cholet, C. Bergaud, S. Gerdes, and C. Joachim, A metallic microcantilever electric contact probe array incorporated in an atomic force microscope, *Rev. Sci. Instrum.*, **71**, 2087–2093 (2000).
65. M. Ben Ali, T. Ondarcuhu, M. Brust, and C. Joachim, Atomic force microscope tip nanoprinting of gold nanoclusters, *Langmuir*, **18**, 872–876 (2002).

66. R. G. Picknett and R. Bexon, Evaporation of sessile or pendant drops in still air, *J. Colloid Interface Sci.*, **61**, 336–350 (1977).
67. S. Uemura, M. Stjernstrom, J. Sjodahl, and J. Roeraade, Picoliter droplet formation on thin optical fiber tips, *Langmuir*, **22**, 10272–10276 (2006).
68. C. Gupta, G. A. Mensing, M. A. Shannon, and P. J. A. Kenis, Double transfer printing of small volumes of liquids, *Langmuir*, **23**, 2906–2914 (2007).
69. P. E. Sheehan and L. J. Whitman, Thiol diffusion and the role of humidity in "dip pen nanolithography," *Phys. Rev. Lett.*, **88**, 156104 (2002).
70. K. A. Brakke, The surface evolver, *Exp. Math.*, **1**, 141–165 (1992).
71. L. Fabie, H. Durou, and T. Ondarcuhu, Capillary forces during liquid nanodispensing, *Langmuir*, **26**, 1870–1878 (2010).
72. L. Fabie and T. Ondarcuhu, Ink writing at sub-micron scale: spreading dynamics at constant pressure, *Soft Matter*, **8**, 4995–5001 (2012).
73. J. Sebilleau, L. Lebon, L. Limat, L. Quartier, and M. Receveur, The dynamics and shapes of a viscous sheet spreading on a moving liquid bath, *EPL*, **92**, 14003 (2010).
74. O. G. Voinov, Hydrodynamics of wetting, *Fluid Dyn.*, **11**, 714–721 (1976).
75. R. G. Cox, The dynamics of the spreading of liquids on a solid-surface. 1. Viscous-flow, *J. Fluid Mech.*, **168**, 169–194 (1986).
76. C. C. Wu, D. N. Reinhoudt, C. Otto, V. Subramaniam, and A. H. Velders, Strategies for patterning biomolecules with dip-pen nanolithography, *Small*, **7**, 989–1002 (2011).
77. A. Meister, S. Krishnamoorthy, C. Hinderling, R. Pugin, and H. Heinzelmann, Local modification of micellar layers using nanoscale dispensing, *Microelectron. Eng.*, **83**, 1509–1512 (2006).
78. J. Arcamone, G. Rius, G. Abadal, J. Teva, N. Barniol, and F. Perez-Murano, Micro/nanomechanical resonators for distributed mass sensing with capacitive detection, *Microelectron. Eng.*, **83**, 1216–1220 (2006).
79. J. Arcamone, E. Dujardin, G. Rius, F. Perez-Murano, and T. Ondarcuhu, Evaporation of femtoliter sessile droplets monitored with nanomechanical mass sensors, *J. Phys. Chem B*, **111**, 13020–13027 (2007).
80. H. Y. Erbil, G. McHale, and M. I. Newton, Drop evaporation on solid surfaces: constant contact angle mode, *Langmuir*, **18**, 2636–2641 (2002).
81. E. N. S. Fuller, P. D. Schettler, and J. C. Giddings, A new method for prediction of binary gas-phase diffusion coefficients, *Ind. Eng. Chem.*, **58**, 18–27 (1966).

Chapter 13

Field-Induced Nanoscale Water Bridges and Tip-Based Oxidation Nanolithography

Ricardo Garcia
Instituto de Ciencia de Materialesde Madrid, CSIC, 28049 Madrid, Spain
r.garcia@csic.es

What once was considered a nuisance in surface science, the spontaneous adsorption of water on surfaces, is now a key element to develop some nanolithographies based on the spatial confinement of chemical and/or electrochemical reactions within the limits of a nanoscale meniscus. In this chapter, I introduce some of the fundamental aspects of the field-induced formation of water and other liquid bridges between two solid surfaces. The role of the different energy contributions such as the condensation, surface, van der Waals, and electrostatic energies are discussed. Special attention is devoted to describe the formation of field-induced nanoscale liquid bridges in a force microscope interface. The last section of the chapter is devoted to introducing tip-based oxidation nanolithography. There, the control and manipulation of water bridges has enabled the fabrication of sophisticated electronic and

Nanoscale Liquid Interfaces: Wetting, Patterning, and Force Microscopy at the Molecular Scale
Edited by Thierry Ondarçuhu and Jean-Pierre Aimé

ISBN 978-981-4316-45-3 (Hardcover), 978-981-4364-48-5 (eBook)
www.panstanford.com

mechanical nanodevices, as well as the construction of molecular architectures.

13.1 Introduction

Water adsorbs spontaneously on surfaces. The thickness of the wetting layer is controlled by the relative humidity, the temperature, and the surface energies.[1-6] When two surfaces are brought together, the above phenomenon could lead to the formation of a water bridge between the surfaces. The bridge, in equilibrium with the vapor, forms a concave meniscus of a negative mean curvature defined by the Kelvin equation.[1-4,7-8] This equation is valid for systems down to menisci of radius of 3 nm.[7-9] The formation of wetting layers and small capillaries are issues of great scientific and technological interest because they are present in a wide range of phenomena such as friction, lubrication, dissipation, dispensing, nanofabrication, and high-resolution imaging.[10-23] In particular, atomic force microscopy (AFM) imaging in air could be largely affected by the formation of a water meniscus.[22-29] The capillary force could increase the net force exerted on the substrate. This, in turn, could imply its modification and/or a deterioration of the spatial resolution. Evidence of the meniscus formation can be inferred by taking force versus distance curves. The presence of a large hysteresis curve is indicative of the meniscus formation.[22-23,28] In fact, the minimization of the lateral force between the tip and the sample to avoid the formation of a stable water meniscus was one of the motivations to develop amplitude modulation AFM.[29-31]

However, the presence of a water bridge turns out to be very useful to perform local electrochemical processes or to control the deposition of molecules, and thus novel lithography and patterning methods have been proposed[19-21,32-40] (see also chapter 12). In particular, AFM (tip-based) oxidation nanolithography has emerged as a highly reliable patterning method.[19,41-58] Tip-based oxidation experiments are performed by applying an external voltage. As a consequence, the external voltage has a dual role. On the one hand, it induces the formation of the liquid meniscus, and on the other hand, it activates the electrochemical process.

The chapter is divided into two sections. In the first section, the main factors that participate in the formation of a nanoscale water bridge in the presence of an electric field are described. In the second section, I introduce some examples of the use of water bridges for patterning and fabricating nanolectronic devices.

13.2 Field-Induced Formation of a Nanoscale Water Bridge

13.2.1 *Spontaneous Formation of a Water Meniscus: Kelvin Equation*

In ambient conditions water spontaneously condenses from vapor into the cavities of hydrophilic surfaces. Water condensation is a thermodynamic process that links the vapor pressure of a liquid with its curvature in the condensed form. It is described by the Kelvin equation,[1-4,7-9]

$$R_g T \ln \frac{p}{p_0} = \frac{\gamma_L v_m}{r_K} \tag{13.1}$$

where the Kelvin radius r_K gives information on the size of the meniscus,

$$\frac{1}{r_K} = \frac{1}{r_1} + \frac{1}{r_2} \tag{13.2}$$

where R_g is the gas constant, γ_L is the surface tension of liquid, p is the actual vapor pressure, p_0 is the vapor pressure at saturation, v_m is the molar volume of the liquid, and r_1 ($r_1 < 0$) and r_2 ($r_2 > 0$) are the principal radii of curvature of the meniscus (Fig. 13.1). In the case of water condensation, p/p_0 is the relative humidity. The Kelvin equation states that in equilibrium the size of the capillary where condensation occurs is determined by the vapor pressure. As a consequence, condensation occurs at a relative humidity below 100% for curved surfaces. In an AFM interface a water meniscus could be formed whenever the tip and the sample gap distance is equal or smaller than the Kelvin radius. For example, at ambient conditions ($p/p_0 = 0.5$ or 50% relative humidity and $T = 293$ K), Eq. 13.1 gives a Kelvin radius of -0.78 nm, which indicates that pores or cavities of sizes below 2 nm should be filled with water.

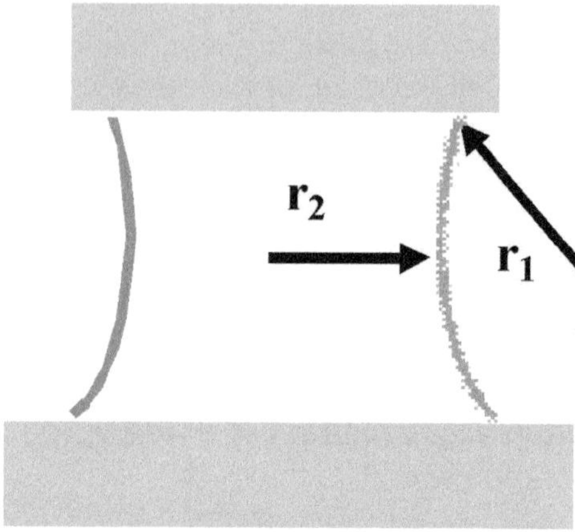

Figure 13.1 Model of the geometry of a water meniscus bridging two flat surfaces. The main curvature radii are indicated.

13.2.2 *Water Bridge Formation Induced by an Electric Field*

At ambient conditions the formation of a water bridge between two surfaces requires very small separations. For gap distances above 3 nm, the formation of a water bridge requires the participation of an additional force such as the one provided by an electric field. The process of forming a water bridge is a dynamic one, where water molecules condense and evaporate at the same time. In the absence of an electric field, the times needed to reach the equilibrium configuration for a capillary condensate vary with its size.[59–61] For example, formation times of 5 ms have been reported for a water meniscus with a curvature radius in the sub-10 nm range.[13,61]

Some aspects of the formation of capillaries under the presence of an electric field are often studied in electrowetting.[62–66] There are a significant number of simulations at the continuum level to describe the field-induced formation of liquid bridges.[67–73] The understanding of the elemental aspects of a water bridge formation is simplified by examining the different contributions to the total energy of the system. The energy involved in the formation of a nanoscale liquid bridge in the presence of an electric field has contributions from surface U_s, condensation U_c, van der Waals U_{vdW}, and electrostatic U_e energies. The different contributions to the energy are calculated by the following expressions:[69,73–74]

$$U_{\mathrm{s}} = \gamma \int_S \mathrm{d}\vec{s} \tag{13.3}$$

where γ is the surface energy of the relevant interface (liquid–vapor, liquid–solid, or solid–vapor), in a first approximation[4] $\gamma \approx \gamma_{\mathrm{LV}} \approx \gamma_{\mathrm{SV}} - \gamma_{\mathrm{LS}}$, where γ_{LV}, γ_{SV}, and γ_{LS} are, respectively, the energies per unit surface of the liquid-vapor, solid-vapor, and liquid-solid surfaces.

$$U_{\mathrm{c}} = \frac{R_{\mathrm{g}}T}{v_{\mathrm{m}}} \ln\left(\frac{1}{RH}\right) \int_V \mathrm{d}\vec{r} \tag{13.4}$$

where v_{m} is the molar volume and $R_{\mathrm{g}} = 8.31\ \mathrm{Jmol^{-1}K^{-1}}$ the gas constant. The condensation energy is proportional to the volume of the liquid and the temperature T. It also depends logarithmically on the ratio p/p_0, which for water is identified with the relative humidity RH expressed in percentage values. The electrostatic energy associated with the polarization of the water induced by the electric field is calculated by

$$U_{\mathrm{e}} = -\frac{\varepsilon_0}{2} \int_V \varepsilon E^2 \mathrm{d}\vec{r} \tag{13.5}$$

where E is the electric field just before water condensation, ε_0 the dielectric constant of the air, and ε the relativity dielectric constant of the liquid. The energy associated with the van der Waals forces between water molecules is given by

$$U_{\mathrm{vdW}} = \frac{H}{12\pi} \int_S \frac{1}{\xi^2} \mathrm{d}\vec{s} \tag{13.6}$$

where H is the Hamaker constant and ξ is a function that depends on the shape of the liquid and can be approximated to the distance between the bottom and top layers of the liquid. Given the relatively short-range character of van der Waals forces, this contribution only becomes important for very small separations (1–4 nm).

The formation of a stable water bridge requires a decrease in the energy of the condensed state with respect to the noncondensed state U_0, that is, the sign of Eq. 13.7 for a given electric field, relative humidity, and temperature has to be negative.

$$\Delta U = U(\text{bridge}) - U_0(\text{vapor}) = \Delta U_{\mathrm{s}} + \Delta U_{\mathrm{c}} + \Delta U_{\mathrm{vdW}} + \Delta U_{\mathrm{e}} \tag{13.7}$$

In Eq. 13.7, U_e and U_c are the larger contributions; however, U_s and U_{vdW} have a pivotal role to determine the surface profile of the water in the condensed configuration.

13.2.3 *Field-Induced Water Bridge Formation between Two Parallel Metallic Plates*

To provide a realistic description of the steps involved in the formation of a nanoscale water bridge requires determining the shape of the water protrusion induced by the field. This is a nonlocal problem that requires, among other things, determining the electric field for the relevant interface geometry, in particular the tip–surface interface. However, insight into the physical parameters that influence the formation of a water bridge for a curved surface can be gained by studying the formation of a water bridge between two parallel and conducting surfaces under the application of a voltage V.[69] For a parallel geometry the polarity of the voltage does not have any influence on the process.

The energy required to condense a cylindrical liquid column of diameter W, bridging the gap D between the metallic surfaces (Fig. 13.2), is the sum of the surface, condensation, van der Waals, and electrostatic contributions. From Eq. 13.3 it can be deduced that the surface energy change involve in the condensation of a

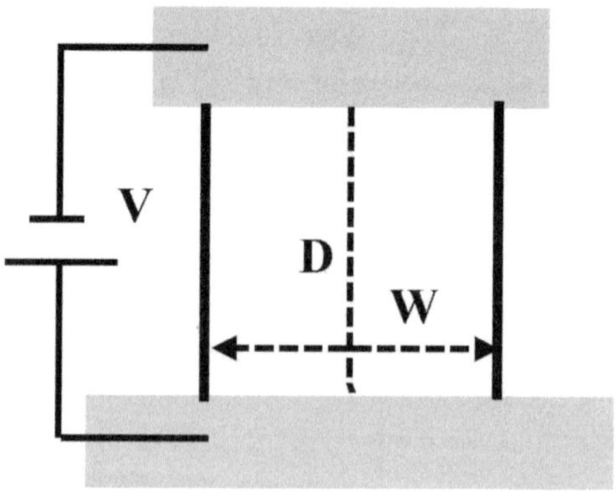

Figure 13.2 Scheme of the interface for a cylindrical liquid column formed between two flat surfaces in the presence of an applied voltage.

cylindrical water bridge is given by

$$\Delta U_{\mathrm{s}} = \pi W D \gamma_{\mathrm{LV}} - \pi \frac{W^2}{2}(\gamma_{\mathrm{SV}} - \gamma_{\mathrm{LS}}). \qquad (13.8)$$

On the other hand, at temperature T, the condensation energy from the vapor phase is proportional to the bridge volume:

$$\Delta U_{\mathrm{c}} = \frac{R_{\mathrm{g}} T}{v_{\mathrm{m}}} \ln\left(\frac{1}{RH}\right) \frac{\pi W^2 D}{4} \qquad (13.9)$$

In the presence of a voltage, the electrostatic energy ΔU_{e} associated with the polarization of the condensed liquid is given by

$$\Delta U_{\mathrm{s}} = \pi W D \gamma_{\mathrm{LV}} - \frac{\pi W^2}{2}(\gamma_{\mathrm{SV}} - \gamma_{\mathrm{LS}}), \qquad (13.10)$$

while the energy associated to the van der Waals forces is given by

$$\Delta U_{\mathrm{vdW}} = \frac{H}{12\pi} \int_S \frac{r}{r^2} dr = \frac{h_0}{2} \left(\frac{h_0^2}{D^2} - 1\right) \frac{R_{\mathrm{g}} T}{v_{\mathrm{m}}} \ln\left(\frac{1}{RH}\right) \qquad (13.11)$$

$$h_0^3 = \frac{|A|}{6\pi} \left[\frac{RT}{v_{\mathrm{m}}} \ln\left(\frac{1}{RH}\right)\right]^{-1} \qquad (13.12)$$

where h_0 is the thickness of the wetting layer on the surface in the absence of an applied voltage. Here, the change of the van der Waals energy is calculated between two configurations, the water bridge and a thin film of water adsorbed on the bottom surface. Then, the total energy change due to the condensation of water is given by

$$\Delta U \approx \frac{\pi W^2 D}{4} \left[\frac{R_{\mathrm{g}} T}{v_{\mathrm{m}}} \ln\left(\frac{1}{RH}\right) - \frac{2\gamma}{D} - \frac{1}{2D^2}(\varepsilon - 1)\varepsilon_0 V^2\right] + \pi \gamma W D - \frac{h_0}{2} \frac{R_{\mathrm{g}} T}{v_{\mathrm{m}}} \ln\left(\frac{1}{RH}\right). \qquad (13.13)$$

In general, the formation of a water bridge is favored by the electrostatic and van der Waals energies and opposed by the surface and condensation energies. Remarkably, the above expression shows that for any separation between the plates, D, there is a threshold voltage that corresponds to the liquid filling the gap between the plates. An expression of the threshold voltage is obtained by neglecting the thickness of the water layer at zero field ($h_0 \approx 0$), which could be 1–3 water molecules in thickness, and then[69]

$$V_{\mathrm{th}} \approx \frac{2D}{\varepsilon_0(\varepsilon - 1)} \left[D \frac{R_{\mathrm{g}} T}{v_{\mathrm{m}}} \ln\left(\frac{1}{RH}\right) - 2\gamma\right]. \qquad (13.14)$$

13.2.4 *Field-Induced Water Bridge Formation in a Nonplanar Interface*

Numerical simulations that include the geometry for a nonplanar interface as well as the shape of the water protrusion give a more detailed picture of the different regimes involved in the formation of a field-induced nanoscale water bridge.[69,72–74]

Figure 13.3 shows a model to describe an interface characterized by a parabolic tip and a flat surface. In this model, the droplet is described by a hyperbolic function that has radial symmetry ρ, and it is parameterized by the height at its maximum h and the width at half-maximum w. The surface profile of the droplet is described by[73]

$$z(\rho) = h_0 + \frac{h}{\cosh(\rho/w)}. \qquad (13.15)$$

The electric field generated by applying a voltage V is approximated by the one produced by a sphere of radius R. Furthermore, the field has a radial symmetry, as shown in Fig. 13.3, and is calculated by using a spherical capacitor with the tip placed at its center.[69] The above geometry has special relevance because it contains the main elements that characterize an AFM interface. The simulations show the existence of three main regimes.[73] Two of them are

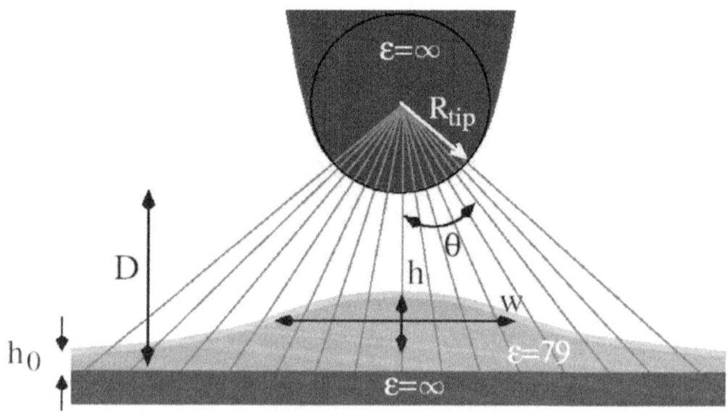

Figure 13.3 Model for simulating the formation of a nanoscale water bridge under an electric field. The field is calculated by assuming a radial distribution; h_0 is the thickness of the adsorbed water layer in the absence of an electric field; h is the maximum height of the droplet induced by the field (total droplet height $= h_0 + h$); and w is width of the liquid protrusion induced by the field at half height.

characterized by the existence of a single and stable configuration, while the other shows the coexistence of two stable configurations or geometries for the water condensate.

Figure 13.4 shows the minimum of the energy ΔU as a function of the height of the protrusion formed in the vicinity of the tip apex for several applied voltages. For low voltages, the interface is characterized by a single stable water shape. The total energy

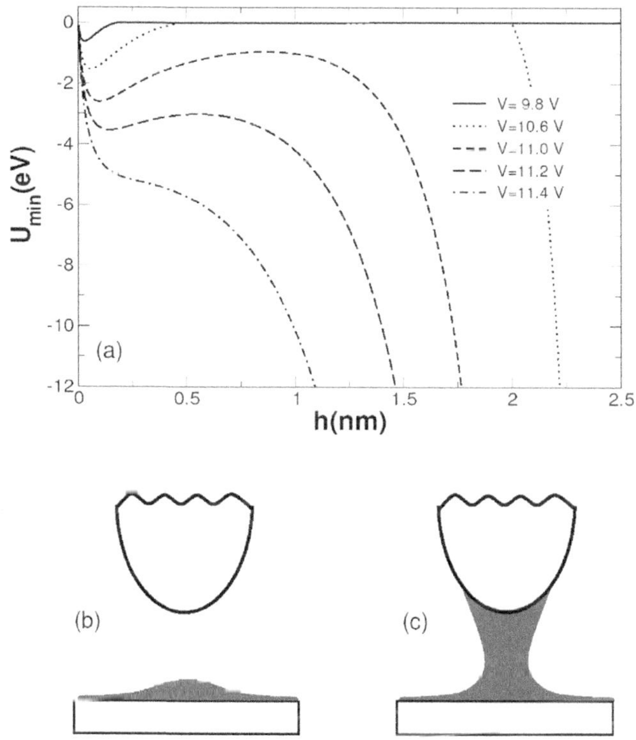

Figure 13.4 Potential energy curves for several voltages as a function of the height of the water droplet. For low voltages, the energy curve shows a minimum very close to the sample surface. Above a certain critical voltage, the curve shows two minima, a local minimum close to the surface and an absolute minimum where the liquid fills the gap. The calculations are performed for a gap distance of 6 nm and $RH = 37\%$. Schematic representation of the two stable configurations associated with, respectively, (b) the first local minimum (droplet) and (c) the liquid bridge. (Data adapted from Ref. 73.)

has a minimum very close to the sample surface (Fig. 13.4a). This minimum corresponds to the formation of a very small protrusion of the water film (droplet) underneath the strongest field lines. The total thickness of the protrusion is $z = h_0 + h$. For voltages above 11.3 V ($D = 6$ nm, and $R_t = 30$ nm), it is not possible to find a droplet geometry that minimizes the energy. The droplet becomes unstable, and a liquid meniscus bridges the gap between tip and surface. However, there is a range of intermediate voltages ($\approx$11–11.2 V) where the energy has two minima with respect to h. The existence of two minima implies the coexistence of two stable water shapes. The local minimum close to the surface is associated with the formation of a droplet under the tip apex (Fig. 13.4b), while the absolute minimum is associated with a configuration where the liquid fills the tip–surface gap (Fig. 13.4c). By increasing the voltage, the local minimum decreases and eventually disappears. The droplet becomes unstable, and the only stable condensate is the water bridge.

The existence of single and bimodal regimes implies the existence of two characteristic voltages and a hysteresis behavior. The lower voltage V_{m} is defined as the minimum voltage that sustains a liquid bridge. It marks the transition from a single configuration (droplet) to a bimodal configuration (droplet and water bridge). This voltage depends on the relative humidity, temperature, and tip–surface geometry. At higher voltages, another characteristic voltage is found. This voltage is called the threshold voltage V_{th}. It defines the transition from the coexistence of the droplet and the water bridge condensates to a single configuration characterized by the water bridge.

The coexistence of two water shapes for some voltages and distances and the activation barrier that separates them introduces a history-dependent behavior or hysteresis. For example, if a voltage sequence is applied from low to high voltages, the most favorable configuration is the formation of a water protrusion under the tip. When the voltage is above the threshold value, the droplet collapses and a water bridge is formed. However, once the liquid bridge has been formed, the voltage could be lowered to an intermediate range of values between V_{m} and V_{th} without evaporating the bridge.

The above discussion has been focused on describing the field-induced formation of water bridges; however, the same equations and methodology can be used to describe the formation of other liquid bridges. The only requirement is to use the values of the dielectric constant, the surface energy, the molar volume, and the Hamaker constant that characterize the liquid.

Figure 13.5 shows the comparison between theory and experiment for the formation of water and ethanol nanoscale liquid bridges. The dependence of the threshold voltage as a function of the average tip–surface separation shows a remarkable agreement between the measured and calculated V_{th} values for the formation of water and ethanol water bridges. The threshold voltage increases monotonically with the average tip–surface separation in both cases. For the same tip–surface separation, the formation of nanoscale water bridges requires slightly higher voltages because the condensation energy for water is higher—v_m(water) $<$ v_m(ethanol). Furthermore, the comparison of surface energies ($\gamma_{eth} < \gamma_{water}$) also favors, for the same conditions, the formation of ethanol protrusions. Those effects offset the influence of the dielectric constant in the electrostatic energy ($\varepsilon_{water} > \varepsilon_{eth}$).

The electric fields associated with the formation of a water bridge in a force microscope interface are extremely high (≈ 1 V nm^{-1} $=$ 10^9 V/m). The electric field at the solid–liquid interface for a droplet just before the formation of a water bridge is about one to two orders of magnitude smaller than the one inside the bridge. The dependence of the maximum electric field at the silicon interface is depicted in the insets of Fig. 13.5. Two fields are calculated, the electric field at the solid–liquid interface underneath the tip's axis for a bias just below V_{th} (solid line) and the field for a voltage just above V_{th} (discontinuous line).

13.2.5 *Field-Induced Water Bridge Formation: Molecular Dynamics Simulations*

Information at the molecular level of the changes experienced by the water molecules in the presence of a field is provided by molecular dynamics simulations.[74] For a water drop deposited on a solid surface (droplet), the dipole moments of the molecules situated

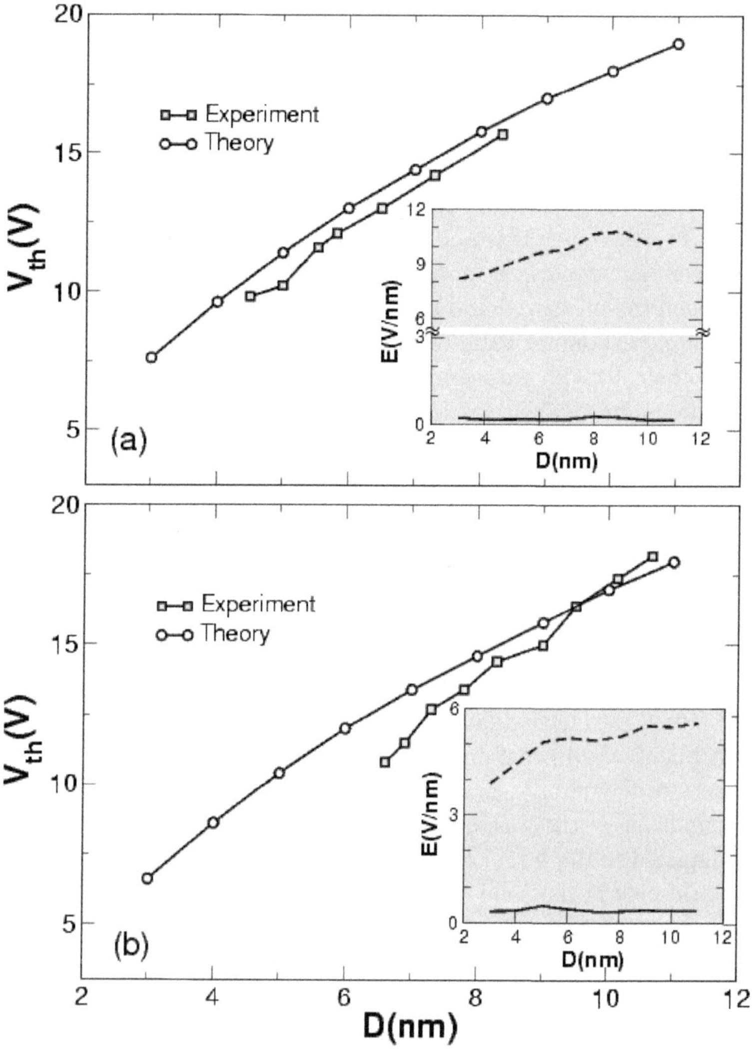

Figure 13.5 Threshold voltage dependence on the tip–surface separation. (a) Water, $RH = 30\%$. (b) Ethanol. The inset shows the dependence of the maximum electric field on the tip–surface separation just before (solid line) and after (dotted line) the liquid bridge formation. The model uses a tip radius of 30 nm. (Data adapted from Ref. 73.)

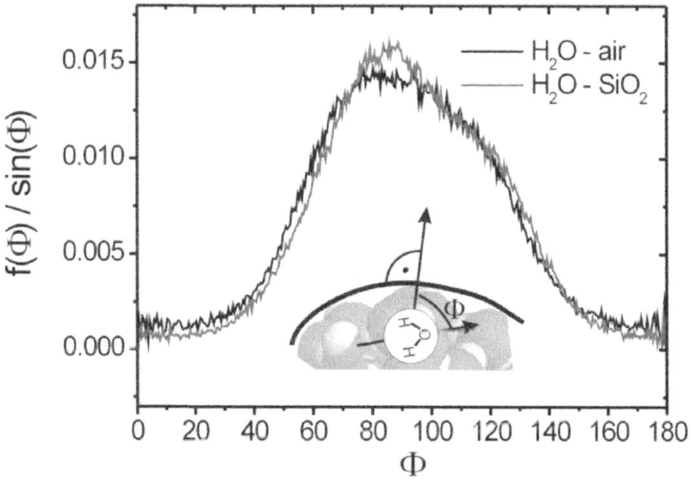

Figure 13.6 Molecular dynamics simulations of the orientation of water molecules. Alignment of the water molecules at the air–drop and silicon dioxide–drop interfaces at 0 V: gray for the silicon dioxide surface and black for the air interface. The angle Φ is between the dipole moment of a water molecule and the normal to the surface. The distribution is normalized by sin(Φ) to give the statistical probability. (Data adapted from Ref. 74.)

at the water–air and water–solid interfaces are not randomly oriented. Figure 13.6 shows that in the absence of an external field the distribution of dipole moments for the interfacial molecules has a maximum for the dipoles parallel to the interface. This preferred alignment of the water dipoles has also been found at the surface of water films on hydrophobic surfaces. It is due to the formation of a hydrogen bond network at the interface.[75,76] In this alignment, three of the four hydrogen bond acceptors/donors are oriented toward the bulk and participate in the hydrogen bond network. The resulting orientational restriction leads to an anisotropic polarizability of the droplet: if the field is perpendicular to the droplet surface, orientational polarization is accompanied by a breaking of the hydrogen bond network. The calculations show that the formation of a water bridge requires the application of a threshold field of 1.2 V/nm, which agrees with the range of values measured in AFM, $E_{th} = 0.7–1.9\ \text{V nm}^{-1}$.[19,53,69,72] The experimental uncertainty is attributed to the tip's curvature, which gives rise to an inhomogeneous field decay.

In the droplet most of the water molecules at the interface are oriented perpendicular to the field. As a consequence, the orientation of the dipoles is restricted and they cannot align with the field. The polarizability of the droplet is low. At the threshold field, the outermost dipoles start aligning with the field. The alignment weakens the hydrogen bonds with the molecules underneath. However, it also triggers a process that ends in the formation of the water bridge. Once the threshold field is exceeded, most of the dipole moments are parallel to the field, the polarizability reaches its maximum and the water bridge is formed. The interfacial molecules that are forced to rotate must also minimize the interactions with the other molecules inside the drop, which would like them oriented parallel to the interface. In the bridge configuration, most of the interfacial water molecules fulfill the double requirement of forming an interfacial hydrogen bond network and being oriented in the direction of the external field.

Figure 13.7 shows four snapshots from the formation of a pillar of water molecules induced by a field of 2 V nm^{-1}. The simulation starts from a droplet of 1,014 molecules equilibrated in the absence of the field (a), where the average contact angle on a silicon dioxide surface is 40°.[53] The field perpendicular to the surface

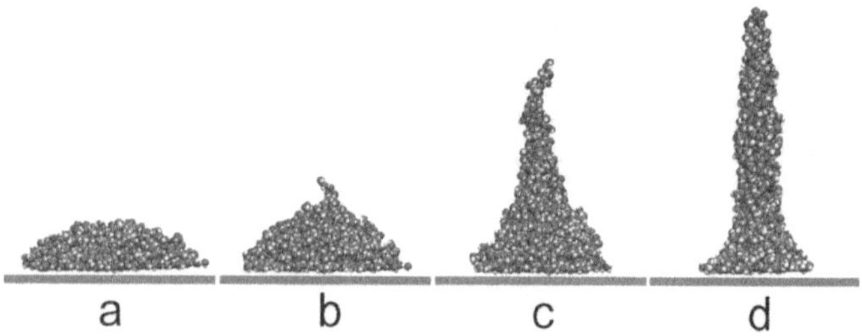

Figure 13.7 Top: molecular dynamics evolution of 1,014 water molecules in the presence of an external electric field on a hydrophilic surface ($E_z = 2.0$ V nm^{-1}). Panel *a* represents the equilibrated state of the droplet without an external field. Panels *b* and *c* represent, respectively, the droplet shape after 25 and 50 ps. Panel *d* represents the final shape after switching on the field. The final shape is reached at $t = 75$ ps. (Data adapted from Ref. 74.)

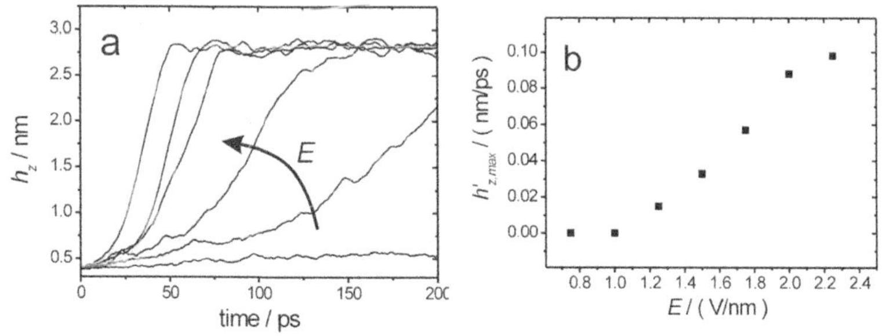

Figure 13.8 Influence of the electric field strength on the evolution of the water droplet. (a) Height increase as a function of time for several values of the electric field (2.25, 2.0, 1.75, 1.5, 1.25, 1.0 V nm^{-1}) (black lines) and fit to sigmoidal function (thin gray lines). (b) Maximum growth rate as a function of the field. The simulations were performed by using the SPC/E potential for water.[78] (Data adapted from Ref. 74.)

polarizes and distorts the outermost water layer. The drop changes shape gradually with time (b and c), and finally a water bridge is formed (d). The snapshots have been taken at 0, 25, 50, and 75 ps, respectively. The sequence of snapshots indicates that the process that leads to the formation of a water bridge, once the field strength is above the threshold value, is in the subnanosecond range.

The change of the drop height h_z perpendicular to the surface for different field strengths is shown in Fig. 13.8. At 2.25, 2.00, 1.75, 1.50, and 1.25 V/nm, the shape of the droplet starts to change almost immediately. In contrast no tendency to form a column was observed for fields below 1.0 V/nm even at longer time scales. Remarkably, below the threshold field the distortion observed in the droplet is of a few tenths of angstrom. The kinetics of the displacement, h_z, is sigmoidal in time. This behavior follows from the presence of the limiting factors during the growth process[77] such as the tensile strength of water, which exerts a force beyond which the pillar cannot extend.

13.3 Observation of Field-Induced Water Bridges with a Force Microscope

A force microscope operated in the amplitude modulation mode[31] offers a method to determine some of the key aspects leading to the formation of nanoscale water bridges under the influence of an electric field. The presence of a water bridge, in fact, its associated capillary force, can be measured by detecting its effects on the cantilever tip's dynamics.[19,53]

A common experimental setup to observe the formation of field-induced water bridges involves enclosing the atomic force microscope head and base into a closed box with inlets for dry and H_2O-saturated nitrogen.[53] The use of doped n^+-type silicon cantilevers are recommended to avoid resistance losses in the tip. Typical values of the force constant and the resonant frequency are, respectively, about 30 N/m and 300 kHz. The cantilever is excited at its resonant frequency to benefit from the maximum amplification conditions. Silicon surfaces such as *p*-type Si(100) with a resistivity of 14 Ωcm are good candidates to observe water bridges. Due to exposure to air, those surfaces have a very thin native oxide layer of about 1 nm. That surface is slightly hydrophilic due to the presence of OH groups. On the other hand, the native oxide layer does not interfere with the observations. To monitor the cantilever movement in real time, the signal from the photodiode is analyzed in an oscilloscope. To avoid an initial tip–sample surface contact, and thus the unintended formation of a water bridge, the atomic force microscope is operated in the noncontact regime of amplitude modulation AFM.[31,79] The protocol to form field-induced water bridges with an atomic force microscope requires several steps. First, the relative humidity is stabilized at a fixed value, usually in the 40%–50% range. Then, the force microscope tip is oscillated at a fixed distance from the surface. The closest tip–surface distance at the bottom part of the oscillation is 1–2 nm. Finally, a voltage pulse is applied between the tip (grounded) and the sample surface.

Figure 13.9 shows the tip's oscillation before, during, and after the application of a voltage pulse. Two situations are shown. The top panel shows the tip's oscillation for a voltage pulse just below V_{th}, while the bottom panel shows the tip's oscillation for a voltage pulse just above V_{th}. In the plot the instantaneous cantilever deflection is represented by a dot and the black line represents the voltage pulse. The top panel shows that when the pulse is on (II), the electric field deflects the mean position of the tip and reduces its amplitude. Once the pulse is off, the tip oscillation recovers its initial amplitude (I).

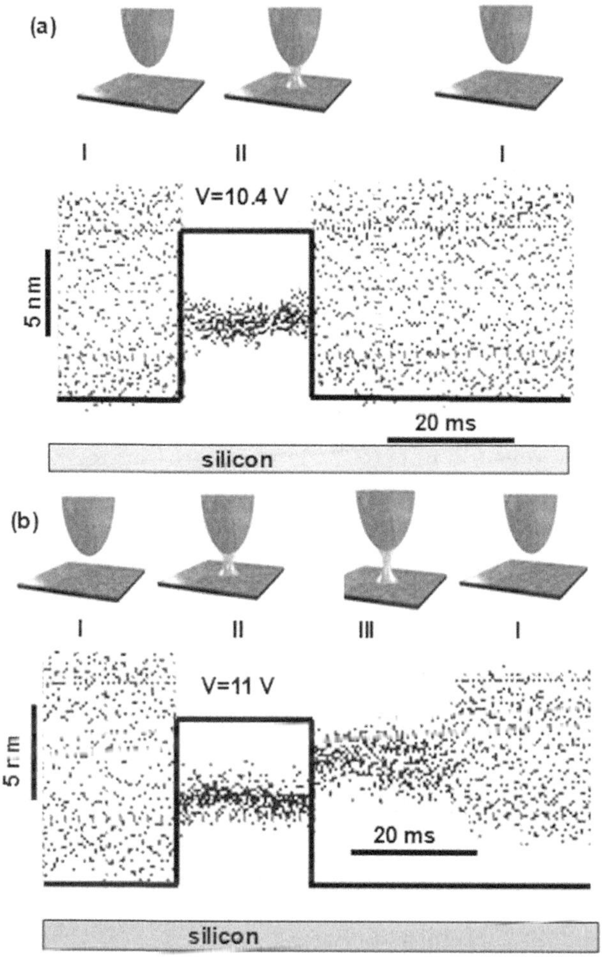

Figure 13.9 Observation of a field-induced water bridge with amplitude modulation AFM. (a) Depicts the tip's oscillation before, during, and after the application of a voltage pulse for 20 ms that did not involve the formation of a water bridge ($V = 10.4$ V). (b) Depicts the tip's oscillation before, during, and after the application of a voltage pulse for 20 ms that involved the formation of a water bridge ($V = 11$ V). Step III is indicative of the formation of a water bridge because the oscillation does not recover its initial value once the pulse is off. Each dot in the oscilloscope traces represents an instantaneous position of the atomic force microscope tip. (Data adapted from Refs. 19, 53, 69, and 81.)

However, when the voltage pulse is slightly increased from 10.4 V to 11 V, a different behavior is observed. During the application of pulse (bottom panel) the electric field deflects the mean position of the tip and reduces its amplitude (II) similar to the previous case. However, here after turning the pulse off, the amplitude remains reduced and the tip mean position is slightly deflected toward the surface (III). This indicates the presence of an additional attractive force between the tip and the silicon surface. Because the effect disappears by lowering the relative humidity, it is deduced that the capillary force is the origin of the behavior observed in step III. Consequently, step III is the signature of the formation of a liquid bridge. Finally when the meniscus is broken, for example, by retracting the tip, the amplitude, and the average deflection recover their initial values in time $(QT/\pi) \approx 1$ ms (I). This method enables to determine the threshold voltage for a given interface, relative humidity, and temperature by finding the minimum voltage that gives rise to the presence of step III. The method also allows to observe the hysteresis of the water bridge as a function of the voltage. The same method has been applied to observe the formation of ethanol, propanol, and octane bridges by filling the atomic force microscope chamber with their respective vapors.[80,81]

The above method also enables to determine the size of the water bridge. This is achieved by measuring the separation at which the water bridge becomes unstable (snap-off separation). First, the water bridge is formed as explained previously. Then the tip and the surface are separated while the oscillation is monitored with an oscilloscope. At the snap-off separation there is a sudden increase of the amplitude. This indicates that the water bridge has been broken. Then the microscope operation is reestablished, and the tip recovers its initial amplitude.[53] Figure 13.10 shows the protocol to measure the snap-off separation. From the snap-off separation, the size of the water bridge can be determined by using the expression,[53,81,82]

$$W = 2\,(R + r_1 + h_0)\sin\phi \qquad (13.16)$$

with

$$\sin\varphi = \frac{r_1 + r}{R + r_1 + h_0} \qquad (13.17)$$

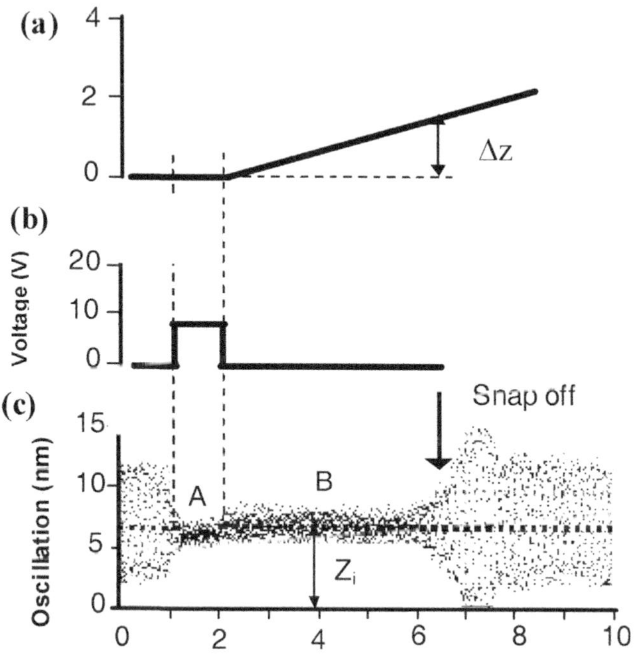

Figure 13.10 Measurement of the snap-off separation with an amplitude modulation AFM. (a) Piezo displacement Δz vs. time. (b) Voltage pulse. (c) Tip oscillation. Z_i represents the average tip–sample distance. The snap-off separation is given by $D_{snap-off} = Z_i + \Delta z_{snap-off}$. The zero position in the y-axis represents the position of the surface. Relative humidity, $RH = 55\%$. Voltage pulse of 9 V and 1 ms. (Data adapted from Ref. 53.)

where ϕ is the filling angle and r is the bridge radius in its narrowest section. Figure 13.11 provides an explanation of the different parameters that appear in Eqs. 13.16 and 13.17.

Figure 13.12 shows the dependence of the snap-off separation and the bridge diameter as a function of the pulse duration (V = 24 V). The snap-off separation and the bridge diameter grow logarithmically with the time. This indicates that the pulse times used to form the bridge are smaller than those required to achieve its equilibrium size. This result might be at variance with some kinetics studies of capillary condensation in nanopores in the absence of an external field.[13,59–61] Some of those studies show that the

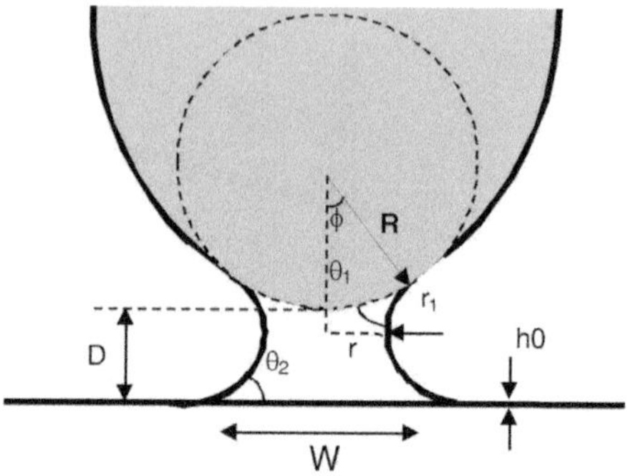

Figure 13.11 Schematics of a concave water meniscus formed between a sphere and a flat surface; θ_1 and θ_2 are the wetting angles, ϕ is the filling angle, D is the tip–sample separation, R is the tip radius, r_1 is one of the main curvature radius, r is the water neck radius at its narrowest section, and h_0 is the water thickness adsorbed on the sample surface.

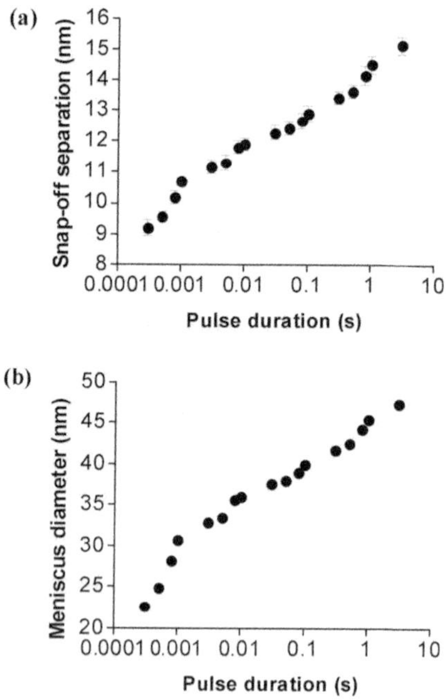

Figure 13.12 (a) Dependence of the snap-off separation on the voltage pulse duration. (b) Dependence of the water bridge diameter on the pulse duration. Data obtained from the snap-off separation measurements presented in (a). The sample is biased negatively with respect to the tip. $RH = 40\%$, $V = 24$ V, and $R = 30$ nm. (Data adapted from Ref. 53.)

time scale over which spontaneously formed capillaries reach their equilibrium sizes may be below one second.[13,61]

13.4 Tip-Based Oxidation Nanolithography

Tip-based oxidation nanolithography, also known as AFM or local oxidation nanolithography, provides a remarkable example of the use of nanoscale water bridges for patterning and device fabrication.[51] AFM oxidation nanolithography is based on the spatial confinement of an anodic oxidation between the tip and the sample surface. The oxidation process is mediated by the formation of a nanoscale water bridge (Fig. 13.13). In fact, due to the sequential character of AFM, the generation of a nanopattern might involve the formation of multiple water bridges. This bridge acts as a nanoscale electrochemical cell that usually contains more than 10^5 water molecules (Fig. 13.14). The polarity of the voltage is in such a way that the tip acts as the cathode (negative) and the sample surface

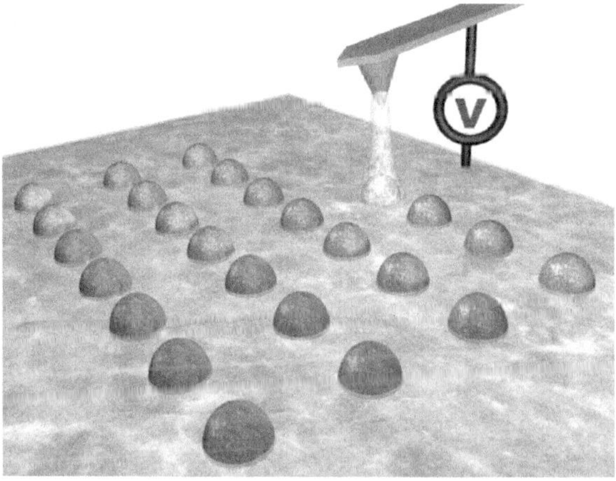

Figure 13.13 Scheme of patterning in AFM oxidation nanolithography. The atomic force microscope tip is placed 1–5 nm above the surface. The application of a voltage pulse induces the formation of a water bridge and activates the anodic oxidation of the surface. During the nanolithography, the atomic force microscope switches back and forward from the imaging mode to the nanolithography mode.

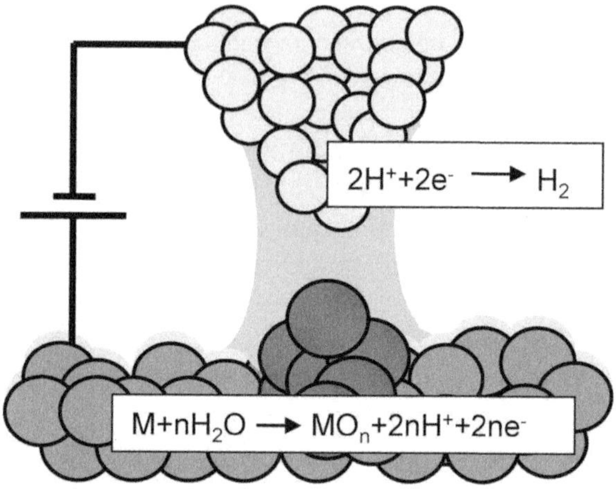

Figure 13.14 Some of the electrochemical reactions involved in a local oxidation experiment. The water bridge provides the oxyanions and confines the spatial extent of the reaction.

is the anode (positive). Tip-based oxidation nanolithography can be either performed with the tip in contact with the sample surface or in a noncontact mode.[84] In the later case, the formation of a water bridge requires the application of a voltage (see previous sections). Noncontact AFM oxidation enhances the tip's lifetime for nanofabrication purposes with respect to experiments performed with AFM in contact mode.

The electric field has three roles in tip-based oxidation. First, it induces the formation of the water bridge. Second, it generates the oxyanions needed for the oxidation by decomposing water molecules. Third, it drives the oxyanions to the sample interface and facilitates the oxidation process. The technique generates ultrasmall silicon oxide nanostructures with a lateral size of about 5 nm and a height in the 1 to 5 nm range.

Tip-based oxidation experiments were first performed on Si(111) and polycrystaline tantalum faces.[85,86] Since then a large number of materials have been locally oxidized, such as compound III–V semiconductors; silicon carbide; several metals such as titanium, tantalum, aluminum, molybdenum, nickel, and niobium; perovskite manganite thin films; dielectrics such as silicon nitride films as well as organosilane self-assembled monolayers; dendritic objects; and carbonaceous films.[51,87]

In AFM oxidation, the height and width of the oxide depend linearly on the voltage. The height also shows a power law dependence with the pulse time t of the type

$$\text{height} \propto (t/t_0)^{\alpha} \tag{13.18}$$

where α is in the 0.1–0.3 range for Si(100) surfaces. Voltage pulses usually change from experiment to experiment, but they are, respectively, in the 10–30 V and 0.005–1 second ranges. The local oxidation process is accompanied by an extremely small current.[45,88] The value depends on the final dot size. Common values are in the subpicoampere regime. Experimental observations have shown that the current measured during the oxidation process matches the current calculated from the volume of the nanostructure. To achieve this result, it was considered that four elementary charges are needed to oxidize one Si atom.

13.4.1 *Nanopatterning and Nanodevices*

Masks, thin insulating barriers, or templates are the elements directly fabricated by AFM oxidation nanolithography. However, a combination of these elements enables the fabrication of a wide variety of nanoscale patterns and nanomechanical or nanolectronic devices.

Figures 13.15 and 13.16 show two examples that underline the flexibility and robustness of this technique for nanofabrication. Figure 13.15a shows the steps to build molecular architectures by combining top-down nanolithography and self-assembled methods. In this case, single molecules of ferritin have been deposited on silicon surfaces with an accuracy similar to the size of the molecules ($\approx$10 nm).[89] First, the silicon surface is covered with a self-assembled monolayer of aminopropyltriethoxysilane (APTES), and then a region is locally oxidized with the atomic force microscope tip. The oxidation process also removes the monolayer under the tip. The nanostripe before the deposition of ferritin molecules is shown in Fig. 13.15b, while Fig. 13.15c shows a densely packed distribution of proteins on the nanostripe. The

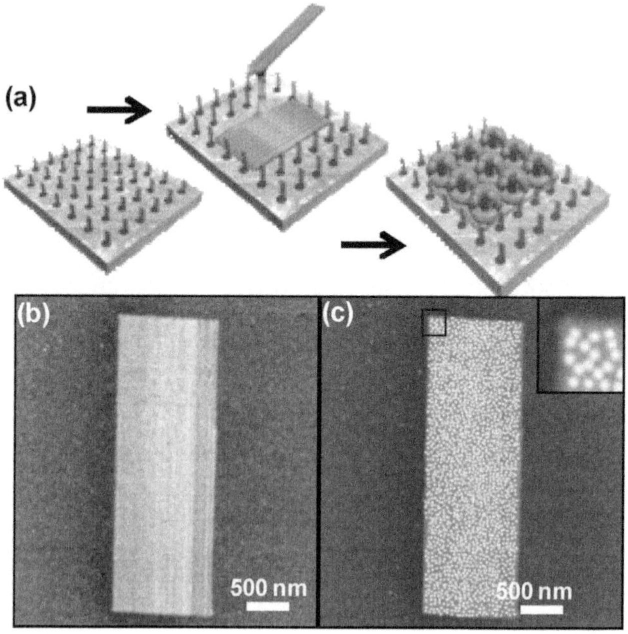

Figure 13.15 Patterning of ferritin molecules by combining AFM oxidation nanolithography and silicon functionalization at low pH values. (a) Steps of the nanopatterning process. (b) AFM image of a local oxide pattern. (c) High-density packing of ferritin molecules on the nanopattern. The inset shows the distribution of the proteins in the marked region. (Data adapted from Ref. 89.)

ferritin molecules form a nonperiodic structure with an average molecule–molecule distance of 20 nm. The total absence of ferritin molecules outside the patterned stripe is remarkable. This result underlines the strong selectivity of the patterning process. In other experiments one-dimensional arrays of individual ferritin molecules have been accurately patterned over a submicrometer region of the macroscopic silicon surface.

Sophisticated nanolectronic devices such as SiNW transistors and circuits have been fabricated by AFM oxidation nanolithography.[54,90] In this application, AFM oxidation generates a narrow oxide mask on top of the active layer of a silicon-on-insulator substrate. The unmasked silicon layer is then removed by using

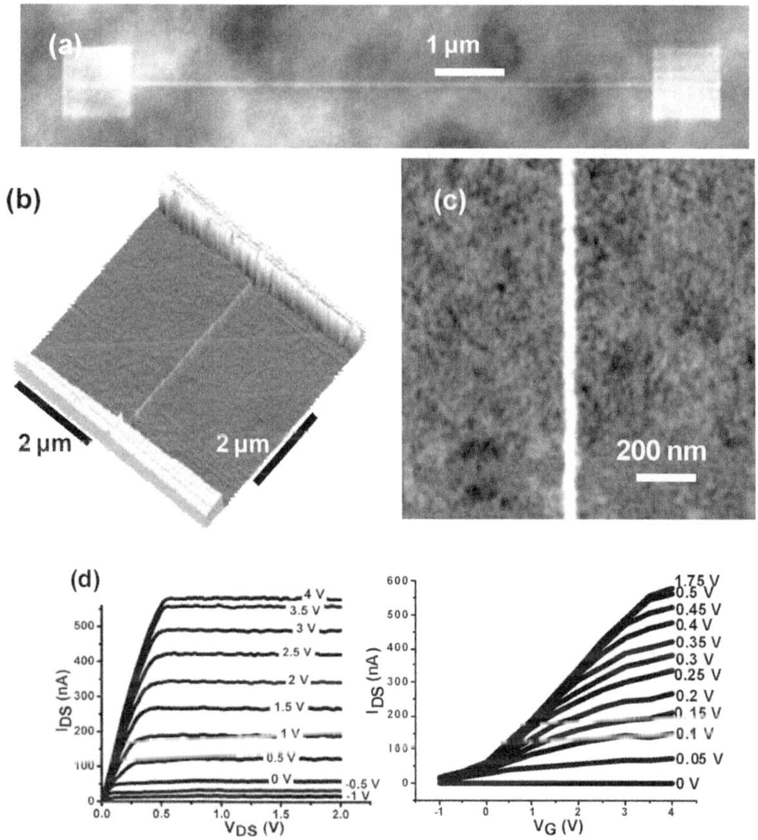

Figure 13.16 SiNW transistor fabricated by AFM oxidation nanolithography. (a) AFM image of a local oxide pattern to be used as mask. (b) AFM image of the SiNW connected to two platinum electrodes. (c) High-resolution image of the SiNW shown in (b). (d) Output (left) and transfer (right) and characteristics of the SiNW transistor (Data adapted from Ref. 90.) *Abbreviation*: SiNW, silicon nanowire.

wet or dry chemical etching procedures. The local oxide protects the underneath silicon from the etching. This leaves a single-crystalline SiNW with a top width that matches the width of the oxide mask. The bottom width of the nanowire will depend on the etching method used. Sharper features are obtained by using plasma etching methods. Finally, the SiNW is contacted to a micrometer-size platinum source and drain contacts by either photolithography or electron beam lithography. After metallization, the sample might be annealed for 30 minutes at 500°C to improve the electrical properties of the contacts. In this way very small SiNWs have been fabricated, in particular SiNWs with a channel width of 4 nm.[54]

Figure 13.16 shows a SiNW 4 μm in length and 9.5 nm in width. Figure 13.16a shows the silicon oxide mask. The thickness of the fabricated mask is about 3 nm. After etching, the SiNW is contacted to two platinum electrodes (Fig. 13.16b). The fabricated SiNW is the main element of a field-effect transistor formed by introducing a gate electrode. Here, the gate electrode is situated at the back of a silicon-on-insulator wafer. The output and transfer characteristics of the transistor formed with the SiNW described above are shown in Fig. 13.16d. The output curve (left panel) shows a clear dependence on the gate voltage. The off-state drain current leakage is about 10^{-11} A. The device shown above has an on/off current ratio of 10^5, and it can be used to develop very sensitive biomolecular sensors.

13.4.2 *Parallel Patterning by Using Liquid Bridges*

The sequential character of AFM somehow limits the technological applications of tip-based nanolithographies, in particular AFM oxidation. In addition, the patterning area is limited by the AFM scanner, which usually has a range of 10–50 microns. To overcome these limitations, several methods have been proposed to upscale the local oxidation process by using a conductive stamp with multiple protrusions as the cathode electrode[91–95] (Fig. 13.17a). Then, in a single step a large area could be patterned with millions of nanostructures.

Figure 13.17 shows AFM images of the stamp (b) and its replica (c). In particular, the replica shows a region of 20.8 × 20.8 μm^2 that contains 29 stripes with a periodicity of 740 nm. The lines were generated by pressing the stamp on the silicon substrate in an atmosphere saturated with octane vapor and by applying a voltage pulse of 36 V for one minute. In this case, the pattern periodicity was established by the stamp periodicity, a gold-coated DVD sample with a periodicity of 740 nm. Because the liquid bridges were made of octane, the chemical composition of the fabricated nanostructures

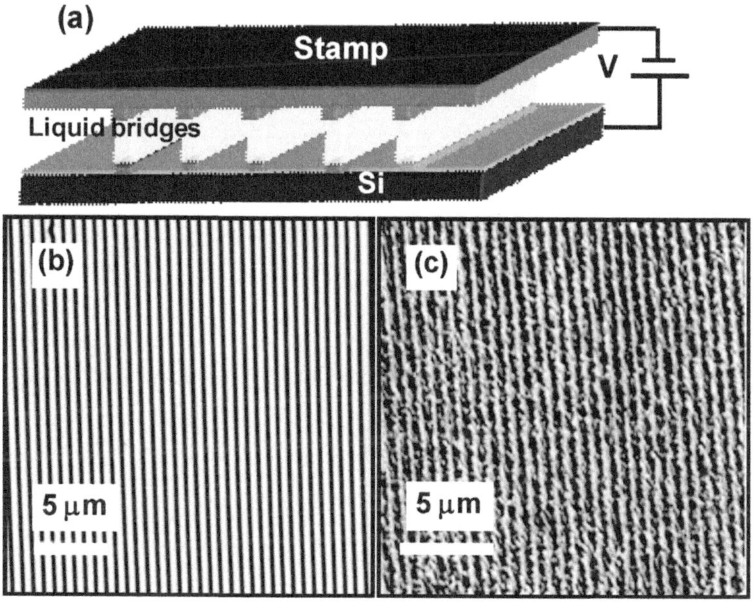

Figure 13.17 Scheme of parallel nanopatterning by using field-induced liquid bridges. (a) AFM image of a region of the gold-coated DVD stamp used for parallel patterning. (b) AFM image of an array of polymer lines fabricated in parallel by using the stamp shown in (a). The pattern periodicity of the replica is 740 nm, which coincides with the stamp periodicity. In this case, the bridges were formed by octane. (Data adapted from Ref. 81.)

are no longer oxides but made of a carbon-rich material. The use of liquid bridges of different chemical compositions illustrates the general character of the method exposed in this chapter to form and manipulate liquid bridges by using electric fields.

13.5 Summary

The control and manipulation of nanoscale liquid bridges or capillaries, in particular water capillaries, plays a crucial role in several high-resolution lithography methods. There the liquid bridge provides the spatial confinement of a chemical reaction, which causes a local modification of the surface. The bridge also supplies the chemicals for the reaction to happen. Electric fields provide useful methods to generate, control, and manipulate the geometry of nanoscale liquid bridges.

Acknowledgments

I am very grateful to Francesco Zerbetto, Tobias Cramer, Juan José Saenz, Monserrat Calleja and Antonio Garcia-Martin for the fruitful and stimulating discussions aimed to elucidate the behavior of water in electric fields. I acknowledge the financial supports from Ministerio de Ciencia, Investigación e Innovación (MAT2009-08650).

References

1. A. W. Adamson, *Physical Chemistry of Surfaces*. New York: John Wiley and Sons (1990).
2. P. G. de Gennes, F. Brochard-Wyart, and D. Queré, *Capillarity and Wetting Phenomena: Drops, Bubbles, Pearls, Waves*. New York: Springer Science (2004).
3. P. W. Atkins, *Physical Chemistry*. 4th ed. New York: W.H. Freeman (1990).
4. J. N. Israelachvili, *Intermolecular and Surface Forces*. London: Academic Press (1994).
5. P. J. Feibelman, The first wetting layer on a solid, *Phys. Today*, **63**, 34–39 (2010).
6. A. Verdaguer, G. M. Sacha, H. Bluhm, and M. Salmeron, Molecular structure of water at interfaces: wetting at the nanometer scale, *Chem. Rev.*, **106**, 1478 (2006).
7. L. R. Fisher and J. N. Israelachvili, Experimental studies on the applicability of the Kelvin equation to highly curved concave menisci, *J. Colloid Interf. Sci.*, **80**, 528 (1981).
8. H. K. Christenson, Capillary condensation due to van-der-Waals attraction in wet slits, *Phys. Rev. Lett.*, **73**, 1821–1824 (1994).
9. R. Digilov, Kelvin equation for meniscuses of nanosize dimensions, *Langmuir*, **16**, 1424–1427 (2000).
10. C. M. Mate, Application of disjoining and capillary pressure to liquid lubricant films in magnetic recording, *J. Appl. Phys.*, **72**, 3084 (1992).

11. R. W. Carpick, D. F. Ogletree, and M. Salmeron, Lateral stiffness: a new nanomechanical measurement for the determination of shear strengths with friction force microscopy, *Appl. Phys. Lett.*, **70**, 1548–1550 (1997).
12. M. Binggeli and C. M. Mate, Influence of capillary condensation of water on nanotribology studied by force microscopy, *Appl. Phys. Lett.*, **65**, 415–417 (1994).
13. L. Sirghi, R. Szoszkiewicz, and E. Riedo, Volume of a nanoscale water bridge, *Langmuir*, **22**, 1093 (2006).
14. R. C. Major, J. E. Houston, M. J. McGrath, J. I. Siepmann, and X. Y. Zhu, Viscous water meniscus under nanoconfinement, *Phys. Rev. Lett.*, **96**, 177803 (2006).
15. L. Nony, T. Cohen-Bouhacina, and J. P. Aime, Dissipation induced by attractive interaction in dynamic force microscopy: contribution of adsorbed water layers, *Surf. Sci.*, **499**, 152–160 (2002).
16. H. Gau, S. Herminghaus, P. Lenz, and R. Lipowsky, Liquid morphologies on structured surfaces: from microchannels to microchips, *Science*, **283**, 46–49 (1999).
17. R. Lipowsky, Morphological wetting transitions at chemically structured surfaces, *Curr. Opin. Colloid Interface Sci.*, **6**, 40–48 (2001).
18. L. Fabie, H. Durou, and T. Ondarcuhu, Capillary forces during liquid nanodispensing, *Langmuir*, **26**, 1870–1878 (2010).
19. R. Garcia, M. Calleja, and H. Rohrer, Patterning of silicon surfaces with noncontact atomic force microscopy: field-induced formation of nanometer-size water bridges, *J. Appl. Phys.*, **86**, 1898–1903 (1999).
20. H. Heinzelmann, A. Meister, P. Niedermann, J. Bitterli, J. Polesel-Maris, M. Liley, M. Gabi, P. Behr, P. Studer, J. Voros, and T. Zambelli, NADIS: a novel AFM-based tool for dispensing fluids into single cells, *Microsc. Anal.*, **26**, 11–13 (2009).
21. M. Cavallini, C. Albonetti, and F. Biscarini, Nanopatterning soluble multifunctional materials by unconventional wet lithography, *Adv. Mater.*, **21**, 1043–1053 (2009).
22. A. L. Weisenhorn, P. K. Hansma, T. R. Albrecht, and C. F. Quate, Forces in AFM in air and water, *Appl. Phys. Lett.*, **54**, 2651–2653 (1989).
23. T. Thundat, X. Y. Zheng, G. Y. Chen, and R. J. Warmack, Role of the relative humidity in atomic force microscopy imaging, *Surf. Sci. Lett.*, **294**, L939–L943 (1993).
24. T. Stifter, O. Martl, and B. Bhushan, Theoretical investigation of the distance dependence of capillary and van der Waals forces in scanning force microscopy, *Phys. Rev. B*, **62**, 13667–13673 (2000).

25. J. Colchero, A. Storch, M. Luna, J. Gómez-Herrero, and A. M. Baró, Observation of liquid neck formation with scanning force microscopy techniques, *Langmuir*, **14**, 2230–2234 (1998).
26. L. Zitzler, S. Herminghaus, and F. Mugele, Capillary forces in tapping mode atomic force microscopy, *Phys. Rev. B*, **66**, 155436 (2002).
27. H. Choe, M. H. Hong, Y. Seo, K. Lee, G. Kim, Y. Cho, J. Ihm, and W. Jhe, Formation, manipulation, and elasticity measurement of a nanometric column of water molecules, *Phys. Rev. Lett.*, **95**, 187801 (2005).
28. Y. M. Men, X. R. Zhang, and W. C. Wang, Capillary liquid bridges in atomic force microscopy: formation, rupture and hysteresis, *J. Chem. Phys.*, **131**, 184702 (2009).
29. Q. Zhong, D. Inniss, K. Kjoller, and V. B. Elings, Fractured polymer/silica fiber surface studied by tapping mode atomic force microscopy, *Surf. Sci. Lett.*, **290**, L688–L692 (1993).
30. J. Tamayo and R. Garcia, Deformation, contact time and phase contrast in tapping mode scanning force microscopy, *Langmuir*, **12**, 4430–4435 (1999).
31. R. Garcia and R. Perez, Dynamic atomic force microscopy methods, *Surf. Sci. Rep.*, **47**, 197–301 (2002).
32. R. D. Piner, J. Zhu, F. Xu, S. H. Hong, and C. A. Mirkin, Dip-pen nanolithography, *Science*, **283**, 661–663 (1999).
33. Y. Li, B. W. Maynor, and J. Liu, Electrochemical AFM "dip-pen" nanolithography, *J. Am. Chem. Soc.*, **123**, 2105–2106 (2001).
34. M. Cavallini and F. Biscarini, Nanostructuring conjugated materials by lithographically controlled wetting, *Nano Lett.*, **3**, 1269–1271 (2003).
35. M. Cavallini, I. Bergenti, S. Milita, G. Ruani, I. Salitros, Z. R. Qu, R. Chandrasekar, and M. Ruben, Micro- and nanopatterning of spin-transition compounds into logical structures, *Angew. Chem., Int. Ed.*, **47**, 8596–8600 (2008).
36. W. L. Cheng, N. Y. Park, M. T. Walter, M. R. Hartman, and D. Luo, Nanopatterning self-assembled nanoparticle superlattices by moulding microdroplets, *Nat. Nanotech.*, **3**, 682–690 (2008).
37. R. V. Martinez, J. Martinez, M. Chiesa, R. Garcia, E. Coronado, E. Pinilla-Cienfuegos, and S. Tatay, Large-scale nanopatterning of single proteins used as carriers of magnetic nanoparticles, *Adv. Mater.*, **22**, 588 (2010).
38. A. Meister, M. Liley, J. Brugger, R. Pugin, and H. Heinzelmann, Nanodispenser for attoliter volume deposition using atomic force microscopy probes modified by focused-ion-beam milling, *Appl. Phys. Lett.*, **85**, 6260–6262 (2004).

39. A. Meister, M. Gabi, P. Behr, P. Studer, J. Voros, P. Niedermann, J. Bitterli, J. Polesel-Maris, M. Liley, H. Heinzelmann, and T. Zambelli, FluidFM: Combining atomic force microscopy and nanofluidics in a universal liquid delivery system for single cell applications and beyond, *Nano Lett.*, **9**, 2501–2507 (2009).

40. J. A. Dagata, Device fabrication by scanned probe oxidation, *Science*, **270**, 1625–1626 (1995).

41. J. A. Dagata, F. Perez-Murano, C. Martin, H. Kuramochi, and H. Yokoyama, Current, charge, and capacitance during scanning probe oxidation of silicon. I. Maximum charge density and lateral diffusion, *J. Appl. Phys.*, **96**, 2386–2392 (2004).

42. E. S. Snow and P. M. Campbell, AFM fabrication of sub-10-nanometer metal-oxide devices with in situ control of electrical properties, *Science*, **270**, 1639–1641 (1995).

43. O. P. Khatri, J. Han, T. Ichii, K. Murase, and H. Sugimura, Self-assembly guided one-dimensional arrangement of gold nanoparticles: a facile approach, *J. Phys. Chem. C*, **112**, 16182–16185 (2008).

44. H. Kuramochi, K. Ando, T. Tokizaki, and H. Yokoyama, In situ detection of faradaic current in probe oxidation using a dynamic force microscope, *Appl.Phys. Lett.*, **84**, 4005–4007 (2004).

45. V. Bouchiat, M. Faucher, C. Thirion, W. Wernsdorfer, T. Fournier, and B. Pannetier, Josephson junctions and superconducting quantum interference devices made by local oxidation of niobium ultrathin films, *Appl. Phys. Lett.*, **79**, 123–125 (2001).

46. C. F. Chen, S. D. Tzeng, H. Y. Chen, and S. Gwo, Silicon microlens structures fabricated by scanning-probe gray-scale oxidation, *Opt. Lett.*, **30**, 652–654 (2005).

47. L. Pellegrino, M. Biasotti, E. Bellingeri, C. Bernini, A. S. Siri, and D. Marré, All-oxide crystalline microelectro mechanical systems: Bending the functionalities of transition-metal oxide thin films, *Adv. Mater.*, **21**, 2377–2381 (2009).

48. L. Pellegrino, Y. Yanagisawa, M. Ishikawa, T. Matsumoto, H. Tanaka, and T. Kawai, $(Fe, Mn)_3O_4$ nanochannels fabricated by AFM local-oxidation nanolithography using Mo/poly(methyl methacrylate) nanomasks, *Adv. Mater.*, **18**, 3099 (2006).

49. N. Rochdi, D. Tonneau, F. Jandard, H. Dallaporta, V. Safarov, and J. Gautier, Electrical conductivity of ultra-thin silicon nanowires, *J. Vac. Sci. Technol. B*, **26**, 159–163 (2008).

50. R. Garcia, R. V. Martinez, and J Martinez, Nano-chemistry and scanning probe nanolithographies, *Chem. Soc. Rev.*, **35**, 29–38 (2006) and references therein.
51. M. Tello and R. Garcia, Nano-oxidation of silicon surfaces: comparison of noncontact and contact atomic force microscopy methods, *Appl. Phys. Lett.*, **79**, 424–426 (2001).
52. M. Calleja, M. Tello, and R. Garcia, Size determination of field-induced water menisci in noncontact atomic force microscopy, *J. Appl. Phys.*, **92**, 5539–5542 (2002).
53. J. Martinez, R. V. Martinez, and R. Garcia, Silicon nanowire transistors with a channel width of 4 nm fabricated by atomic force microscope nanolithography, *Nano Lett.*, **8**, 3636–3639 (2008).
54. C. R. Kinser, M. J. Schmitz, and M. C. Hersam, Kinetics and mechanism of atomic force microscope local oxidation on hydrogen-passivated silicon in inert organic solvents, *Adv. Mater.*, **18**, 1377–1380 (2005).
55. R. Maoz, E. Frydman, S. R. Cohen, and J. Sagiv, "Constructive nanolithography": inert monolayers as patternable templates for in-situ nanofabrication of metal-semiconductor-organic surface structures—a generic approach, *Adv. Mater.*, **12**, 725 (2000).
56. S. Hoeppener, R. Maoz, S. R. Cohen, L. F. Chi, H. Fuchs, and J. Sagiv, Metal nanoparticles, nanowires, and contact electrodes self-assembled on patterned monolayer templates—a bottom-up chemical approach, *Adv. Mater.*, **14**, 1036 (2002).
57. S. T. Liu, R. Maoz, G. Schmid, and J. Sagiv, Template guided self-assembly of [Au_{55}] clusters on nanolithographically defined monolayer patterns, *Nano Lett.*, **2**, 1055–1060 (2002).
58. M. M. Kohonen, N. Maeda, and H. K. Christenson, Kinetics of capillary condensation in a nanoscale pore, *Phys. Rev. Lett.*, **82**, 4667–4670 (1999).
59. M. M. Kohonen and H. K. Christenson, Capillary condensation of water between rinsed mica surfaces, *Langmuir*, **16**, 7285–7288 (2000).
60. R. Szoszkiewicz and E. Riedo, Nucleation time of nanoscale water bridges, *Phys. Rev. Lett.*, **95**, 135502 (2005).
61. A. Klingner, S. Herminghaus, and F. Mugele, Self-excited oscillatory dynamics of capillary bridges in electric fields, *Appl. Phys. Lett.*, **82**, 4187–4189 (2003).
62. F. Mugele and J. C. Baret, Electrowetting: from basics to applications, *J. Phys.: Condens. Matter*, **17**, R705–R774 (2005).

63. C. Quilliet and B. Berge, Electrowetting: a recent outbreak, *Curr. Opin. Colloid Interface Sci.*, **6**, 34–39 (2001).

64. R. Gupta, G. K. Olivier, and J. Frechette, Invariance of the solid–liquid interfacial energy in electrowetting probed via capillary condensation, *Langmuir*, **26**, 11946–11950 (2010).

65. N. Kurra, A. Scott, and G. U. Kulkarni, Electrodeposition and evaporation of attoliter water droplets: direct visualization using atomic force microscopy, *Nano Res.*, **3**, 307–316 (2010).

66. I. C. Yeh and M. L. Berkowitz, Dielectric constant of water at high electric fields: molecular dynamics study, *J. Chem. Phys.*, **110**, 7935–7942 (1999).

67. J. Buehrle, S. Herminghaus, and F. Mugele, Interface profiles near three-phase contact lines in electric fields, *Phys. Rev. Lett.*, **91**, 086101 (2003).

68. S. Gomez-Monivas, J. J. Saenz, M. Calleja, and R. Garcia, Field-induced formation of nanometer-sized water bridges, *Phys. Rev. Lett.*, **91**, 056101 (2003).

69. S. Djurkovic, C. B. Clemons, D. Golovaty, and G. W. Young, Effects of the electric field shape on nano-scale oxidation, *Surf. Sci.*, **601**, 5340–5358 (2007).

70. P. B. Paramonov and S. F. Lyuksyutov, Density-functional description of water condensation in proximity of nanoscale aperity, *J. Chem. Phys.*, **123**, 084705 (2005).

71. G. M. Sacha, A. Verdaguer, and M. Salmeron, Induced water condensation and bridge formation by electric fields in atomic force microscopy, *J. Phys. Chem. B*, **110**, 14870–14873 (2006).

72. A. Garcia-Martin and R. Garcia, Formation of nanoscale liquid menisci in electric fields, *Appl. Phys. Lett.*, **88**, 123115 (2006).

73. T. Cramer, F. Zerbetto, and R. Garcia, Molecular mechanism of water bridge buildup: field-induced formation of nanoscale menisci, *Langmuir*, **24**, 6116–6120 (2008).

74. X. Xia and M. L. Berkowitz, Electric-field-induced restructuring of water at a platinum water interface—a molecular-dynamics computer-simulation, *Phys. Rev. Lett.*, **74**, 3193–3196 (1995).

75. S. J. Suresh, A. V. Satish, and A. Choudhary, Influence of electric field on the hydrogen bond network of water, *J. Chem. Phys.*, **124**, 074506 (2006).

76. F. Lugli, S. Höfinger, and F. Zerbetto, The collapse of nanobubbles in water, *J. Am. Chem. Soc.*, **127**, 8020–8021 (2005).

77. H. J. C. Berendsen, J. R. Grigera, and T. P. Straatsma, The missing term in effective pair potentials, *J. Phys. Chem.*, **91**, 6269–6271 (1987).

78. R. Garcia and A. San Paulo, Attractive and repulsive tip-sample interaction regimes in tapping-mode atomic force microscopy, *Phys. Rev. B*, **60**, 4961–4967 (1999).

79. M. Tello and R. Garcia, Giant growth rate in nano-oxidation of p-silicon surfaces by using ethyl alcohol liquid bridges, *Appl. Phys. Lett.*, **83**, 2339–2341 (2003).

80. R. V. Martinez and R. Garcia, Nanolithography based on the formation and manipulation of nanometer-size organic liquid menisci, *Nano Lett.*, **5**, 1161–1164 (2005).

81. F. M. Orr, L. E. Scriven, and A. P. Rivas, Pendular rings between solids. Meniscus properties and capillary force, *J. Fluid. Mech.*, **67**, 723–742 (1975).

82. E. Bayramli, A. Abou-Obeid, and T. G. M. Van De Ven, Liquid bridges between spheres in a gravitational-field, *J. Colloid. Interface Sci.*, **116**, 490–502 (1987).

83. M. Tello and R. Garcia, Nano-oxidation of silicon surfaces: comparison of non-contact and contact atomic force microscopy methods, *Appl. Phys. Lett.*, **79**, 424–426 (2001).

84. J. A. Dagata, J. Schneir, H. H. Harary, C. J. Evans, M. T. Postek, and J. Bennett, Modification of hydrogen-passivated silicon by a scanning tunneling microscope operated in air, *Appl. Phys. Lett.*, **56**, 2001–2003 (1990).

85. T. Thundat, L. A. Nagahara, P. I. Oden, S. M. Lindsay, M. A. George, and W. S. Glaunsinger, Modification of tantalum surfaces by STM in an electrochemical cell, *J. Vac. Sci. Technol., A*, **8**, 3537–3541 (1990).

86. A. A. Tseng, A. Notargiocomo, and T. P. Chen, Nanofabrication by scanning probe microscope lithography: a review, *J. Vac. Sci. Technol., B*, **23**, 877–894 (2005).

87. P. Avouris, T. Hertel, and R. Martel, Atomic force microscope tip-induced local oxidation of silicon: kinetics, mechanism and nanofabrication, *Appl. Phys. Lett.*, **71**, 285–287 (1997).

88. R. V. Martinez, J. Martinez, and R. Garcia, Silicon nanowire circuits fabricated by AFM oxidation nanolithography, *Nanotechnology*, **21**, 245301 (2010).

89. M. Cavallini, P. Mei, F. Biscarini, and R. Garcia, Parallel writing by local oxidation nanolithography with submicrometer resolution, *Appl. Phys. Lett.*, **83**, 5286–5288 (2003).

90. N. S. Losilla, J. Martinez, and R. Garcia, Large area nanoscale patterning of silicon surfaces by parallel local oxidation, *Nanotechnology*, **20**, 475304 (2009).

91. C. Albonetti, J. Martinez, N. S. Losilla, P. Greco, M. Cavallini, F. Borgatti, M. Montecchi, L. Pasquali, R. Garcia, and F. Biscarini, Parallel-local anodic oxidation of silicon surfaces by soft stamps, *Nanotechnology*, **19**, 435303 (2008).

92. S. Hoeppener, R. Maoz, and J. Sagiv, Constructive microlithography: electrochemical printing of monolayer template patterns extends constructive nanolithography to the micrometer-millimeter dimension range, *Nano Lett.*, **3**, 761–767 (2003).

93. A. Zeira, D. Chowdhury, R. Maoz, and J. Sagiv, Contact electrochemical replication of hydrophilic-hydrophobic monolayer patterns, *ACS Nano*, **2**, 2554–2568 (2008).

Chapter 14

Technological Applications of Dewetting

Fabio Biscarini,* Denis Gentili, Emanuela Margapoti, and Massimiliano Cavallini**

CNR—Istituto per lo Studio dei Materiali Nanostrutturati (ISMN), Via P. Gobetti 101, 40129 Bologna, Italy

*f.biscarini@bo.ismn.cnr.it and **m.cavallini@bo.ismn.cnr.it

14.1 Introduction

Dewetting is a spontaneous phenomenon where a fluid spread on a surface forms an ensemble of separated objects instead of a continuous film. An example of dewetting encountered frequently is the rupture of a water layer into droplets on a hydrophobic surface like a dirty window or a Teflon-coated frying pan. In several instances dewetting phenomena are undesirable, as in the rupture of a protective coating film or paint. Dewetting can also be exploited in several applications and industrial processes when fast drying of the surface is required. For instance, in aviation the aircrafts' fuselage and wings are made nonwettable to water by spraying them with anti-ice fluid. In metallurgy, drying a film by evaporation leaves behind residues and stains, so dewetting is exploited to remove the liquid along with all the impurities. Dishwasher detergents

Nanoscale Liquid Interfaces: Wetting, Patterning, and Force Microscopy at the Molecular Scale
Edited by Thierry Ondarçuhu and Jean-Pierre Aimé

ISBN 978-981-4316-45-3 (Hardcover), 978-981-4364-48-5 (eBook)
www.panstanford.com

are formulated to spontaneously promote water dewetting from the surfaces. Likewise, a fluorinated shampoo facilitates drying by making hair more hydrophobic.[1]

Dewetting has an impact on our everyday life, but only recently its mechanism has started to be fully understood. From this knowledge stems the technological control that is needed for applications in large-area micro- and nanofabrication, organic electronics, heterogeneous catalysis, and flow in mesoscopic channels.

An advantage of exploiting dewetting in technology is the possibility to obtain objects of predefined size and spatial distribution, starting from a homogeneous continuous system, such as a thin film, and avoiding complicated and expensive lithographical subtractive processes at the same time. This is particularly attractive owing to the universal occurrence of dewetting phenomena in all types of materials and the inherent difficulty to pattern important classes of materials by consolidated lithographical methods.

In a broad definition dewetting is the spontaneous rupture of a continuous thin film on a substrate, leading to the formation of a discontinuous deposit of material. It can be equivalently seen as the reverse process of spreading a film on a substrate.[2–5] Dewetting stems from the instability of the thin films where competing intermolecular and surface interactions coexist (see chapter 4 for a detailed description of film stability). The patterns of separated structures form through various mechanisms schematically depicted in Fig. 14.1: nucleation, growth and coalescence of holes,[6] spinodal dewetting of a thin film,[7] and Ostwald ripening of droplets.[8] The former involves the nucleation of holes at defect sites of the film and their subsequent growth by displacement of the material from the three-phase line toward the rims of a spatially correlated Voronoi mosaic of holes.[9] Further rupture yields droplets arranged along the rims of the Voronoi cells. Spinodal dewetting occurs by fluctuations of the film interface whose increasing amplitude leads to the rupture of the film into an array of droplets, or ripples in the case of an anisotropic thin film. This phenomenon is associated with the emergence in time of a characteristic wavelength that, for a given material/substrate, is solely controlled by the film thickness h. In Ostwald ripening, smaller droplets, which are formed either by growth or dewetting on a substrate, coalesce by diffusing across the surface until they

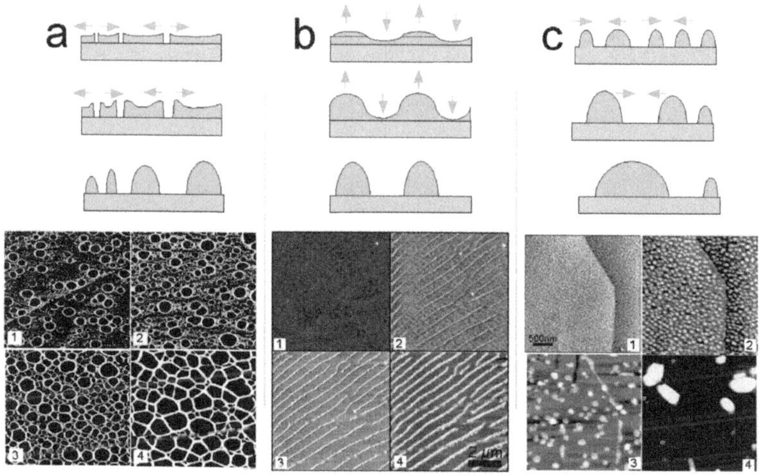

Figure 14.1 Schematic representation of different dewetting mechanisms and corresponding experimental examples. (a) Holes are formed within the film (upper panel), leading to separation of the structures as holes grow. Withdrawing mass accumulates at the rims of the holes (lower panels) (image adapted from Ref. 10). (b) Spinodal dewetting (image adapted from Ref. 11). (c) Partial wetting followed by Ostwald ripening. The mass transfer from smaller droplets to larger ones can occur either collectively by drop migration or by diffusion of the individual molecules or atoms forming the smaller droplets (image adapted from Ref. 12).

are incorporated into larger ones. The process is driven by a size-dependent chemical potential gradient and stops when a distribution of droplets with comparable size and distance is attained.

In Fig. 14.1 a sketch of the different mechanisms of dewetting is shown. They also coexist or occur at different times in different combinations.[7]

Dewetting phenomena have been observed on a variety of materials. The idea of exploiting dewetting into a straightforward large-area technique for patterning nanoparticles, small molecules, and polymers lies on the spontaneous spatial correlations in size and distance of the droplets[13] and stripes formed upon dewetting.[14–16] The characteristic length scales of the features generated by dewetting are not always evident by inspection, and analysis of the pair correlation function of the image might be necessary.

In spontaneous patterns the position of the individual droplet is not precisely defined; thus it is difficult to address each droplet a priori. These limitations can be overcome by controlling the local surface tension of the substrate, for instance, by creating a template with a chemical pattern via soft lithography,[17,18] local electrochemistry,[19–22] or overlaying of a topological pattern.[23,24] Another approach consists of modifying the surface tension of the sample by ultraviolet (UV) irradiation through phase contact masks.[25] A further approach is based on inducing the dewetting by a controlled local perturbation of a metastable thin film either with an atomic force microscopy (AFM) probe,[11,26] a stamp, or confining of the fluid inside micro- and nanochannels with tailored surface tension at the channel walls.

In this chapter, we describe and comment on some of the emerging technological applications based on dewetting at the micro- and nanometer scales. The following sections are organized according to the different types of materials that can be patterned by means of dewetting: inorganic and hybrid materials (section 14.2) and organic and polymeric materials (section 14.3). In section 14.4 a general scheme for patterning soluble materials based on a combination of stamps and dewetting will be described.

14.2 Patterning Inorganic and Hybrid Materials

In this section we consider some relevant applications of inorganic and hybrid materials, including nonvolatile memories based on semiconductor nanocrystals, arrays of nanoparticles as potential sensing elements in biochips, nanorings of single-molecule magnets (SMMs) for studying nanoscale magnetism, and quantum magnetic effects.

14.2.1 *Nonvolatile Memories Based on Nanocrystals*

It is known that if material A is grown on top of material B and is thermodynamically unstable or metastable, layer A is rough and subsequent agglomeration or nanocrystallization occurs. This frequently happens when postgrowth thermal annealing is performed.

Nanocrystallization via dewetting of materials has been observed in processing techniques, including ion implantation followed by thermal annealing, laser ablation, high-vacuum deposition, and molecular beam epitaxy (MBE).[27–29] In this section we focus on agglomeration-induced dewetting.

Semiconductor nanocrystals of group IV elements (Si, Ge) grown on or embedded in thin dielectric layers have been investigated intensely, due to their potential for use in electronics and optoelectronics as single-electron memories,[30–32] quantum dots, transistors,[33] and nanocrystal flash memories.[27,34] An advantage of nonvolatile memory devices would be to substitute the metal or Si-polycrystalline floating gate with a nanoparticles array. For this reason the ability to form a well-ordered array of nanocrystals is desirable. Self-assembly, agglomerated nanocrystals formed by dewetting on a substrate are usually randomly distributed.[29] For their technological implications, several studies aimed at the control of self-assembly of ordered structures.[35,36]

One of the approaches that are emerging in semiconductor technology is the use of a crystalline layer as the thermodynamically unstable film. Silicon-on-insulator (SOI) exhibits a rather complex self-organization behavior. As the thin-Si-layer dewets from the underlying amorphous oxide substrate, the Si nanocrystals self-organize into a characteristic pattern,[37–40] as shown in Fig. 14.2. Nanocrystals are distributed in square regions, with their edges oriented along (110) directions. A careful study within the squares by scanning electron microscopy (SEM) reveals an ordered array of self-organized nanocrystals, distributed oriented along the <130> direction (Fig. 14.2b).

Yang *et al.*[40] followed the dynamics of dewetting and self-organized agglomeration by low energy electron microscopy (LEEM). Their results point to a selective surface-faceting process as the origin of self-organization. It was recognized that in order to achieve a well-defined pattern of nanocrystals, the Si surface must be clean and the dewetting must occur only in an ultrahigh vacuum environment. Also, they demonstrated that dewetting originates predominantly at defect sites, indicating a heterogeneous process. Since defects are most likely at the sidewalls of a pinhole, during thermal annealing the vertical sidewall tilts by the faceting angle θ

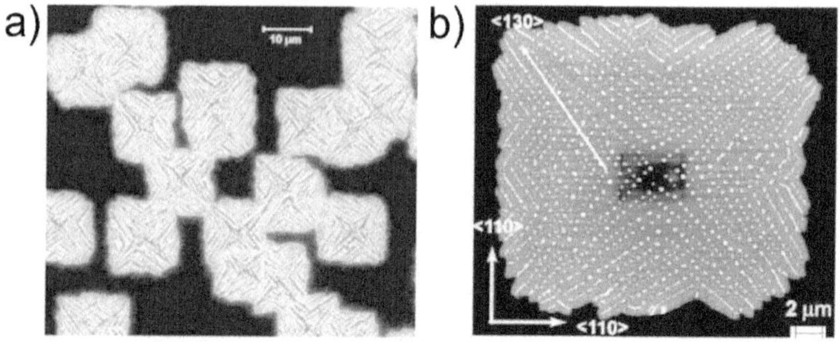

Figure 14.2 Self-organization from dewetting on SOI. Self-organized Si nanocrystals formed when a thin Si-template layer dewets from the oxide in SOI after thermal annealing at 900°C for 5 min. The dewetting has its origin at the defects in the Si layer. (a) Optical image showing several square regions where dewetting on SOI occurs. (b) SEM image showing the ordered array of Si nanocrystals within the square region. The bright spots are the nanocrystals. (Images taken from Ref. 40.)

in order for the Si atoms to reach the lowest-energy configuration. First-principle surface energy calculations confirmed that faceting is controlled by surface energy anisotropy. The *formation energy* for a given low-energy configuration is calculated as:

$$E = \gamma_{\mathrm{f}} \cos\theta + (1/2)(\gamma_{\mathrm{i}} - \gamma_{\mathrm{Si}} - \gamma_{\mathrm{SiO_2}})$$

where γ_{f}, γ_{i}, γ_{Si}, and $\gamma_{\mathrm{SiO_2}}$ are the surface energies of the facet, Si–SiO_2 interface, Si (001), and SiO_2 surfaces, respectively. For the same facet energy, γ_{f}, the smaller the angle the facet makes with the surface, the larger its formation energy. The term in parentheses is called the *spreading coefficient S* and relates to the thermodynamic stability of the film.[41]

More recently, the thermodynamic stability of the Ge/SOI layer was investigated by studying the Ge growth on an Si-template-layer at different coverages.[39,42] In these studies Zhang *et al.*[42] demonstrated that contrary to the case of the silicon grown on SOI, the long-range ordering of three-dimensional (3D) nanocrystals along the <130> direction is lost when more than six monolayers (MLs) are deposited. Authors explain this observation as a decrease of the surface energy anisotropy due to a Ge-induced weaker bond. Figure 14.3a shows the formation of ordered structures of Si (001) in SOI,

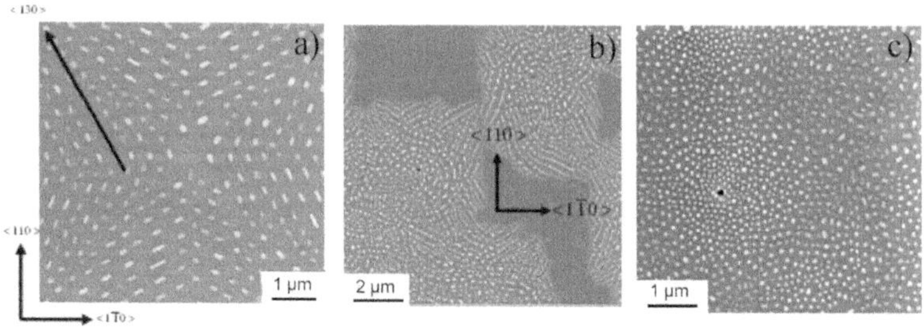

Figure 14.3 Thermal dewetting and agglomeration of a thin layer of Si (001) in SOI, with different amounts of Ge deposited by MBE. (a) SEM image of 1 ML Ge annealed at 900°C for 5 min. Islands are self-aligned along the <130> direction. (b) Higher-resolution SEM image for 2 MLs Ge coverage annealed at 850°C for 20 s. Islands are still faceted but have lost their long-range ordering. (c) Higher-resolution SEM image for 6 MLs Ge coverage annealed at 850°C for 20 s. Islands are now not faceted, and the <130> directionality and long-range order of islands are completely lost. (Images taken from Ref. 42.)

by dewetting, whereas Fig. 14.3b,c shows the evolution of the morphology upon thermal annealing depositing 2 MLs and 6 MLs of Ge, respectively. The order is completely lost for 6 MLs of Ge coverage.

Beyond the formation and self-assembling of nanocrystals, the control of density, size, and distribution of nanostructures fabricated by dewetting remains an open challenge. In order to address this problem several methods for the positioning of Ge dots have been proposed by Stoica *et al.*[43] and Olzierski *et al.*[44] In their studies they considered the deposition of two-dimensional (2D) nanoarrays of Ge dots, on thermal SiO_2, demonstrating that upon thermal annealing, after the formation of a 2D wetting layer, thickness-dependent stress drives the formation of 3D islands.[39] Figure 14.4 shows the evolution of the dewetting process: after the initial rupture (Fig. 14.4a), isolated Ge nanocrystals are formed on SiO_2 (Fig. 14.4b) and finally freestanding nanocrystals appear (Fig. 14.4c).[39]

14.2.2 *Patterning Nanoparticles*

To integrate nanoparticles in functional devices,[45] other approaches based on dewetting, exploiting the spontaneous agglomeration in stripes, have been proposed[46,47] (see also chapter 15).

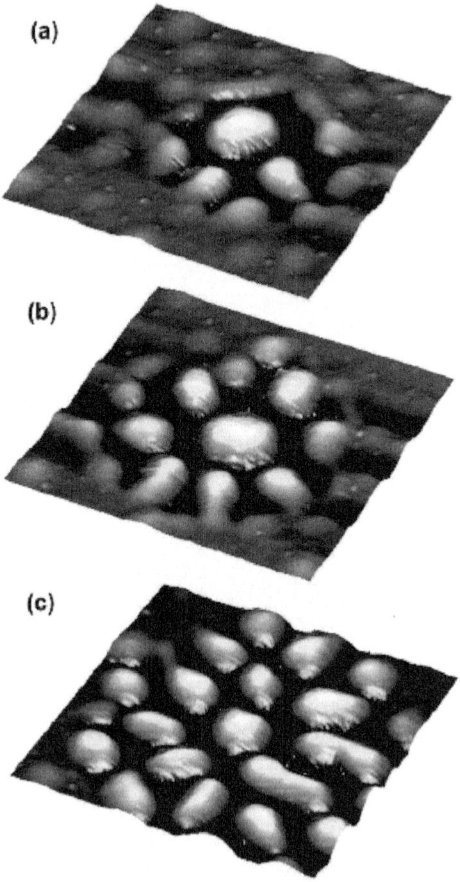

Figure 14.4 AFM images obtained in the early stages of the dewetting of ultrathin SOI (a) during the initial breakup and (b) the resulting isolated Ge nanocrystals on SiO_2. (c) Initial stages of the dewetting process producing clusters of freestanding Ge nanocrystals. (Images taken from Ref. 39.)

Recent works take advantage of the application of external electric fields,[48–51] deposition on prepatterned substrates,[52,53] and deposition on diblock copolymer microphases.[54] Lee *et al.*[47] observed that a thin film of aqueous charged polymer solution containing gold nanoparticles dewets when deposited on a hydrophilic solid substrate to form complex morphologies.[55] Starting from spinodal dewetting, polygonal networks, discontinuous structures, and droplets are spontaneously formed. By increasing the polymer charge density, novel elongated structures (stripes) with embedded nanoparticles were observed during the rupture of the film.

A more promising method to obtain extended arrays of stripes formed by nanoparticles was demonstrated by Huang *et al.*,[46] who reported the spontaneous formation of gold and silver nanoparticles in stripes upon dewetting of a film of polymer blend with nanoparticles floating on a water surface. This new technique is a simple lithography-free method for patterning of nanoparticles arrays, which is achieved by the use of a Langmuir–Blodgett method.[56,57] Here nanoparticles precipitate as agglomerates at the water–substrate contact line. Nanoparticles self-organize by dewetting during the drying process (Fig. 14.5a–d), forming some aggregates that act as nucleation sites. These aggregates collect more nanoparticles during the pullout, forming continuous stripes. By changing the velocity of pulling out the substrate, the periodicity of the stripes

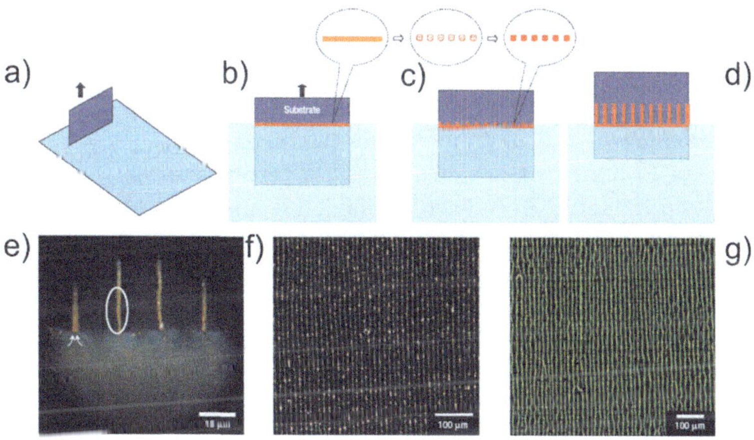

Figure 14.5 Stripe pattern formation by dip coating. (a–d) Schematic illustration of the formation of an aligned nanoparticles-stripe pattern by vertical deposition. During raising the substrate slowly (a, b) the water evaporates and the wet contact line breaks up into aggregates of nanoparticles (b, c). (e) Optical microscopy image of the waterfront, leading to 1D stripes across the entire substrate, formed both for silver (f) and gold (g) nanoparticles. (Images taken from Ref. 46.) *Abbreviation*: 1D, one-dimensional.

can be modified. Therefore, different particles, substrates, velocities, and concentrations could modify the condition of aggregation and therefore the periodicity and size of the wires. A direct optical microscopy observation of the waterfront during the stripes formation is shown in Fig. 14.5e. Furthermore, two optical microscopy images of the stripes covering the entire substrate are shown for gold and silver nanoparticles in Fig. 14.5f and Fig. 14.5g, respectively.

14.2.3 *Patterning Magnetic Nanorings in Polymer Films*

Nanorings of conductive and magnetic materials are important structures for the study of quantum effects, like the Aharanov-Bohm effect[58] and superconductivity.[59] Their fabrication requires the use of sophisticated methods like electron-beam lithography (EBL) and chemical etching. Conversely, dewetting provides a simple and reproducible method to obtain nanorings. In particular, dewetting processes have been employed in the fabrication of nanorings of self-ordered nanoparticles and combined with demixing. In the first case, inorganic semiconductor material in the shape of a nanoring, smaller than 100 nm in diameter, was spontaneously formed during the capping deposition[58] of a thin deposit of quantum dots. The capping layer must be thinner than the quantum dot height, so it dewets, forming ring-like structures with a diameter smaller than 100 nm. Lorke *et al.*[58] have verified the Aharonov–Bohm effect by observing the stepwise increase of the ground-state angular momentum with an increasing magnetic field as more and more flux quanta penetrate the ring.

A new patterning process that takes place as a result of the demixing and dewetting[60] process of a binary polymer/solute mixture was recently proposed by Cavallini *et al.* for the fabrication of ordered patterning of nanometric rings of Mn_{12}-based single-molecular magnets.[61]

Demixing leads to surface-oriented phase separation, giving rise to the formation of a double layer consisting of the polymer and the solute. This process can be triggered, for instance, by solvent annealing. The thickness of the layers is determined by the volume ratio of the two phases in the initial blend. When a component of the blend is considerably lesser in percentage, it

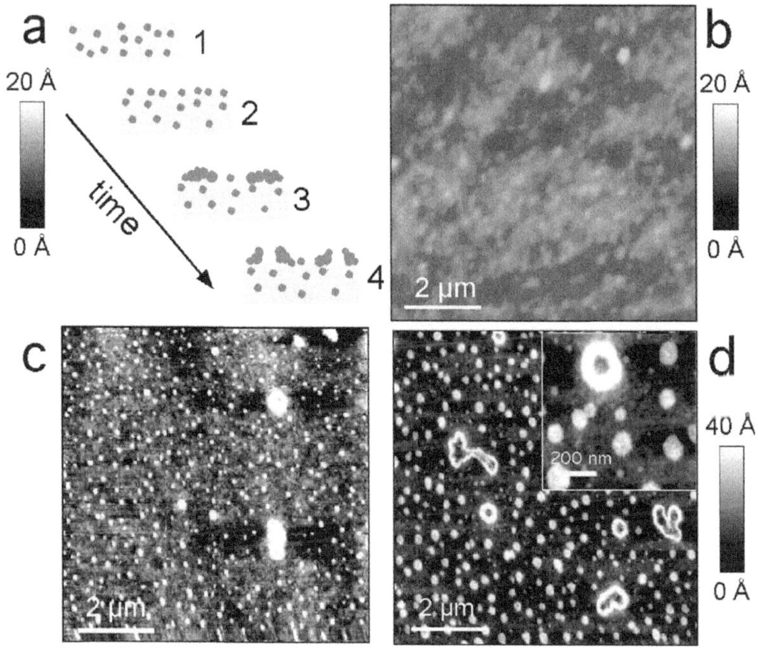

Figure 14.6 Evolution of the surface morphology of an Mn_{12}/PC mixture film prepared by drop-casting. (a) Scheme of the process, (b) the as-cast film, (c) morphology after 2 min of solvent exposure, with the solute emerging as droplets at the surface, and (d) the film after 3 min of solvent exposure. Droplets undergo ripening and, as they reach a critical diameter, form rings. These can then coalesce. The inset shows a ring of 180 nm in diameter. (Images taken from Ref. 61.) *Abbreviation*: PC, polycarbonate.

forms a thin film on the surface of the polymer matrix. If the spreading coefficient of the thin film generated by demixing is negative, the thin film is thermodynamically unstable, leading to the fragmentation into smaller aggregates (droplets) in order to achieve the equilibrium conditions. The thin-film rupture results in the dewetting of segregated material from the polymer surface that can be controlled by the temperature, the partial pressure, and the nature of the solvent. Figure 14.6a depicts the schematics of this process. The corresponding morphological evolution of the thin film is presented in the sequence (Fig. 14.6b–d) upon solvent treatment. Starting from the continuous film (Fig. 14.6b) the solute molecules emerge to the surface and form small aggregates similar to droplets (Fig. 14.6c). These grow and coalesce during the solvent treatment, forming larger droplets. Upon longer solvent treatment, the droplets ripen and, as they reach a critical diameter of about 100 nm, transform into circular rings (Fig. 14.6d, inset). Eventually the rings coalesce with each other, forming larger rings with irregular shape (Fig. 14.6d). By interrupting the exposure to the solvent the transformation is stopped and the morphology remains frozen in the metastable form.

The transformation from droplets to rings during dewetting has been observed in a variety of systems and at different length scales.[62] Figure 14.7 shows a possible mechanism of this transformation. When the film of Mn_{12} clusters emerged to the surface is exposed to a solvent-saturated atmosphere, a thin layer of the solvent, which is in equilibrium with its vapor, condenses on the surface. In correspondence of the droplets' edges the solvent tends to evaporate faster.[1] This is due to the higher Laplace pressure, which scales as the curvature of the droplet. This different evaporation rate induces a solvent flux (σ_e) that carries the solute to the droplet (Fig. 14.7a), where it concentrates and eventually precipitates. The small droplets, which have a higher curvature, act as sites of nucleation, which grow by adsorbing the material that flows from the outer parts of the surface. As the droplet grows, it reduces its curvature at the center (and hence the Laplace pressure) until it reaches a critical diameter where the evaporation rate in the center is lower (Fig. 14.7b) with respect to the boundaries. On the other hand, the droplet boundaries are forced to maintain the contact angle (Θ) constant[63] and hence the Laplace pressure and the associated higher evaporation rate. When the differential evaporation rate between the center and the boundaries of the droplet becomes relevant, an outward capillary flow (σ_i) compensates the amount of evaporated liquid. This flux transports soluble material and accumulates it at the droplet boundaries (Fig. 14.7c), where it reaches supersaturation and reprecipitates. At the end of the process the droplet transforms itself into the ring (Fig. 14.7d). A schematic view of this process is shown in Fig. 14.7a–d.

Based on this process the authors used a patterned film of molecular magnets on polycarbonate (PC) in order to confine the ring

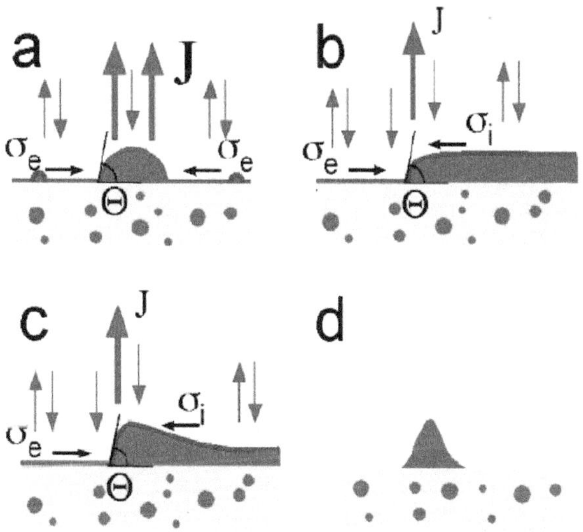

Figure 14.7 Schematic representation of ring growth. (a) At the droplet edges the solvent tends to evaporate faster. This different evaporation rate induces a solvent flux (σ_e) that carries the solute to the droplet. (b) As the droplet grows, it reduces its curvature in the center and reaches a critical diameter, where the evaporation rate in the center reduces. (c) When the differential evaporation rate between the center and the boundaries of the droplet becomes relevant, an outward capillary flow (σ_i) compensates the amount of evaporated liquid. (d) This flux transports soluble material, which accumulates at the droplet boundaries. (Images taken from Ref. 61.)

formation inside the patterned regions. The molecular magnets/PC film was patterned in parallel lines by replica molding. In this case, since the emerging material is anisotropically distributed, the rings are forced to grow and to coalesce following the same anisotropy of the printed stripes, and they self-organize along the printed stripes. Figure 14.8 shows a typical example of nanoring distribution along submicrometric stripes.

14.3 Patterning Polymeric and Organic Materials

Patterning of polymeric materials by dewetting can be obtained either by phase separation or by local modification of a surface by

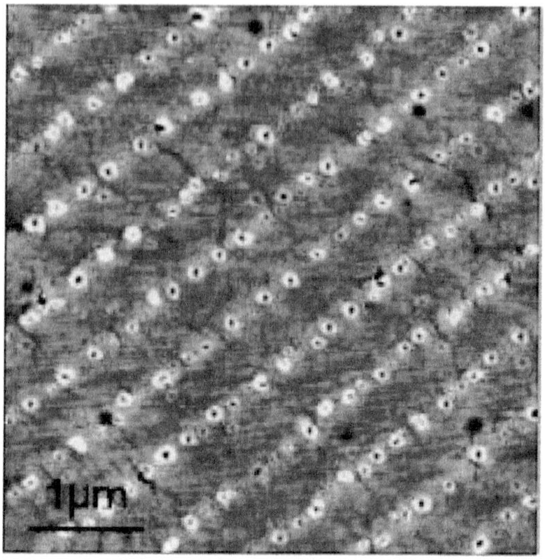

Figure 14.8 Patterning of nanorings by dewetting. AFM image of the early stage of ring formation (image taken from Ref. 61).

chemical lithography.[64] Furthermore it has been proved that the application of an electrical field can influence dewetting processes,[65] and in some case, it allows the control of residual thin-film thickness at the level of a single monolayer.[66] Several examples are presented in this section.

14.3.1 *Lithographically Induced Self-Assembly*

Self-formation of a periodic-pillar array can be induced by single-homopolymer film melt. In 1999 Chou and Zhuang[67] and later in 2002 Russell and Steiner[65,68] proposed self-patterning of micropillars by either a thermal process or by applying an external bias voltage. Placing a mask plate above a polymer film and raising the melting temperature trigger self-formation of well-defined pillars. In their method a mask with protruding patterns is placed above a thin film of a single homopolymer at a distance predefined by means of a preevaporated spacer (see Fig. 14.9) that allows the finer control of the polymer–film distance.

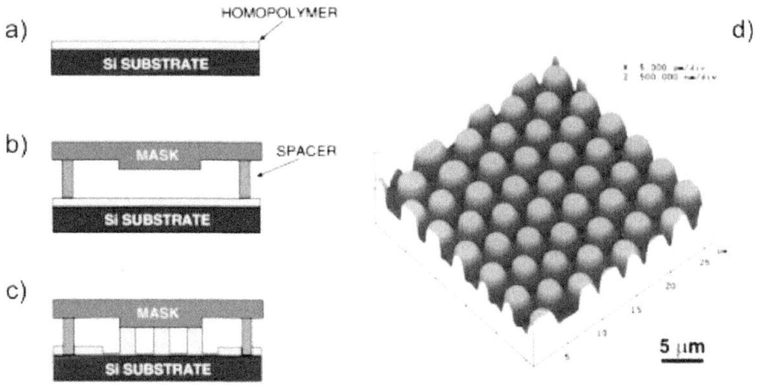

Figure 14.9 Schematic view of the LISA process. (a) Thin layer of polymer spin-coated on a flat silicon wafer. (b) Mask of protruding patterns separated by the polymer film using a spacer. (c) During the heat cooling cycle the polymer film self-assembles into a periodic supramolecular pillar array. (Images taken from Ref. 67.) *Abbreviation*: LISA, lithographically induced self-assembly.

When the system is uniformly heated above the polymer glass transition temperature, then cooled down to room temperature, the polymer film self-assemble into periodic array of pillars (Figure 14.9d). This technique was named LISA. Chou *et al.* explain their observations by building a simple model called image charge–induced electrohydrodynamic (ICE) instability. This model assumes that electrostatic force is the driving force of the process. The patterns are formed as a result of interplay and instability of charges in a polymer melt, image charges in the mask, and hydrodynamic force in the melt polymer. They show that due to the ultrathin polymer thickness, LISA is not induced by instabilities from the thermal convection (i.e., Rayleigh–Benard instability[69]) or by surface tension–driven Benard convention.[70] Upon heating the polymer above its glass transition temperature, it melts, leading to a significant increase of both charge density and charge mobility. If no mask is placed on top of the polymer melt, the charges are uniformly distributed due to a flat surface and symmetry. However, if a mask with finite conductivity is placed near the thin polymer layer, it allows the induction of the charge image in a periodic structure.

14.3.2 *Pattern Formation by Electric-Field-Induced Dewetting*

It was demonstrated that by applying an electrical field to a metastable thin film,[71] electrostatic pressure could amplify surface fluctuations. This property has been used to tune the size of the self-assembled structures from micron to submicron scale in electrically induced dewetting.[68] Furthermore the possibility to apply a sufficiently strong electric field in defined regions allows the fabrication of a variety of patterns controlling size and positioning.[20]

A method to form structures in polymer bilayers on a large area by applying an electric field was independently proposed by Russell and Steiner in 2002.[65,68] The sequential growth of two layers and their electrohydrodynamic instabilities combined with a topographically patterned electrode is a strategy that was used to replicate patterns of about 100 nm periodicity. In this way two materials can be patterned in one step, thus avoiding a multistep process as in standard lithographic techniques. Lin *et al.*, in 2002, demonstrated that when two liquid-polymeric layers are placed in contact, an electric field induces the dewetting of the upper film, giving rise to ordered structures.

In their work the authors proved that by increasing the electrical field up to a threshold value, the surface waves are amplified until the formation of polymer pillars. Evidence for both nucleated and

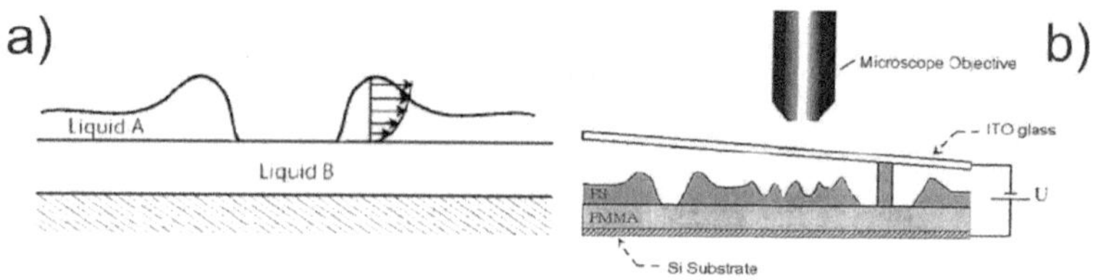

Figure 14.10 Dewetting-induced electric field. (a) Dewetting at the liquid–liquid interface. (b) Typical sample configuration: the air/PS/PMMA system is sandwiched between a silicon wafer (lower electrode) and ITO (upper electrode). A spacer fabricated at the edges of the slides on top of the ITO by evaporation of SiO_x allows the distance between the two electrodes to be controlled. Typically, the electrode spacing varied by a few micrometers over a lateral distance of approximately 1 cm. (Images taken from Ref. 68.) *Abbreviation*: ITO, indium tin oxide.

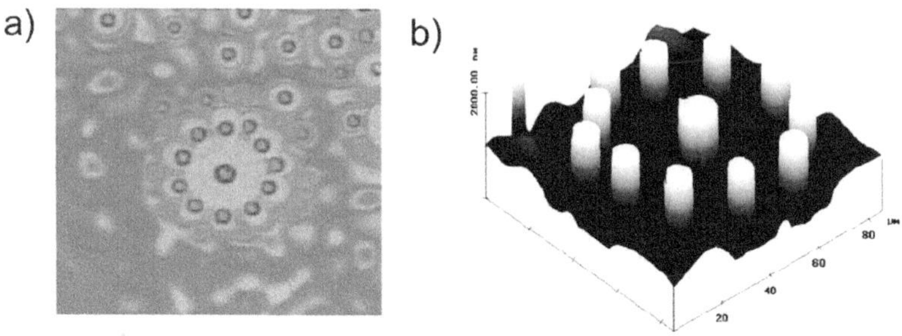

Figure 14.11 Process of hierarchical formation. (a) Optical microscope image recorded at 30 V in the configuration of Fig. 14.10b. The corresponding 3D AFM image shows the columns of PS-96K on top of a PMMA-95K layer. (Images taken from Ref. 68.)

spinodal instabilities were found. Figures 14.10 and 14.11 show examples using the polystyrene (PS)/poly(methyl metacrylate) (PMMA) system.

As PS pillars grow, circular rims around the pillars where an unstable PMMA thin film is exposed to the air are formed. At this point a second instability upon the electrical field occurs at the PMMA–air interface. PMMA dewets from the silicon surface, reorganizing around the PS pillars (Fig. 14.12b). The final morphology consists of PS columns coated by a thin layer of PMMA (Fig. 14.12c).

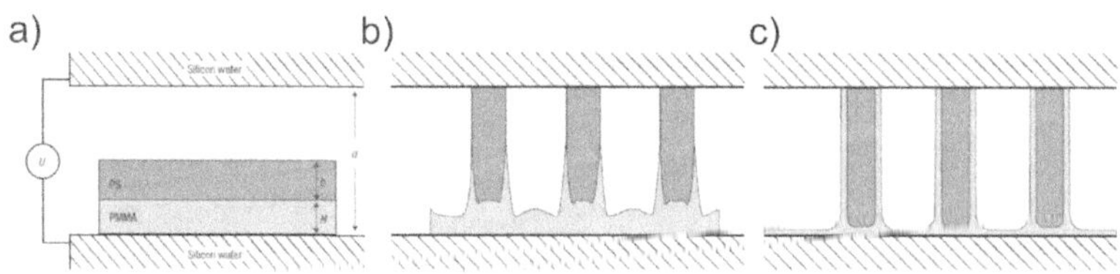

Figure 14.12 Process to form hierarchical bilayer structure. (a) Polymer bilayer of PMMA and PS. The polymer–air surface is first destabilized by an electric field generated between the two Si electrodes. The initial instability results in the formation of columns spanning from the surface of the lower layer to the top electrode. (b) During the column formation process, the bottom polymer layer is deformed at the polymer–polymer–air contact line. (c) In the second instability, the deformation of the lower layer is enhanced by the electric field. The polymer moves upward, surrounding the columns and finally leading to form a mantle-like structure. (Images taken from Ref. 65.)

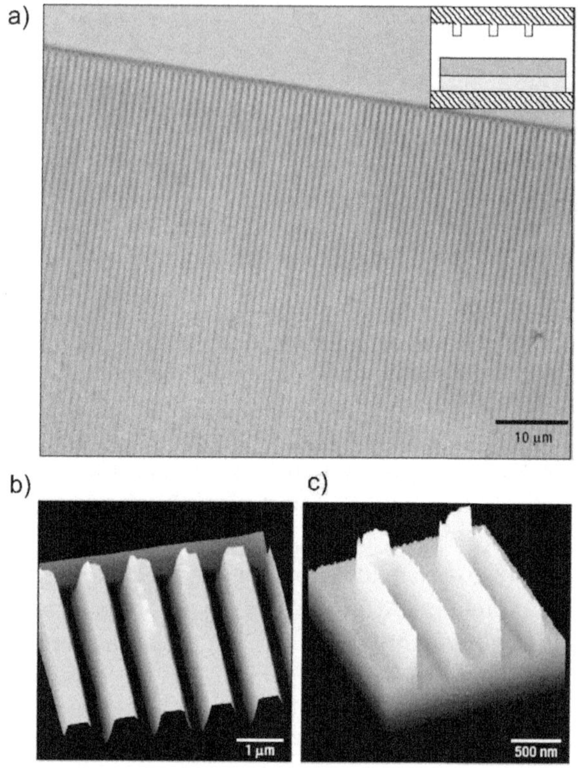

Figure 14.13 Pattern replication by hierarchical instability. (a) Optical micrograph of a line pattern replicated by electrohydrodynamic instability. The use of a structured electrode (inset) leads to a controlled lateral replica of the structure. (b) AFM image of five replicated lines after removing the PS. (Images taken from Ref. 65.)

At the end of the process, a thin layer of PMMA remains, covering the entire substrate. As it is shown in Fig. 14.13b the method was extended from pillars to lines.

14.3.3 *Dewetting and Electromigration of Discotic Liquid Crystals*

In a recent work,[66] Calò *et al.* exploited the electromigration upon dewetting to fabricate ultrathin and highly oriented films of a discotic liquid crystal (DLC) semiconductor (Fig. 14.14). They show that ultrathin films of phthalocyanines (Pc) can be organized to form a conductive channel tens of microns long between metal electrodes with thickness control over a single monolayer. Their approach exploited the electromigration of the isotropic phase formed, starting from the pretransitional region of the columnar-isotropic phase transition. Droplets of dewetted material were displaced by applying a longitudinal electric field of about 1 V/μm and accumulated to the negative electrode. Dewetting of the bulk and electromigration expose an ultrathin film steadily anchored to the substrate, a few monolayers thick and exhibiting columnar liquid crystal order. The layers of this ultrathin film melt progressively above T_C and can be individually exfoliated by electromigration, starting from the ninth down to the first monolayer. Figure 14.14 shows the evolution of the process.

14.3.4 *Self-Aligned Gate Polymer Field-Effect Transistors*

Dewetting finds large consensus in liquid manufacturing due to the continuous demand for controlling the flow and spread of the functional liquid on a surface. The aim is to achieve high-performance electronic devices by downscaling the rule. One important result in which dewetting was used to fabricate polymer field-effect transistors (FETs) with submicrometer channel lengths by inkjet printing was proposed by Wang *et al.*[72] This was achieved with the help of surface energy patterns, which cause the aqueous solution containing the electrode precursor to be repelled and dewetted from hydrophobic prepatterned regions of the substrate. Hydrophobic surface energy barriers of different geometries were used to study the influence of solution viscosity, ink volume, and contact angle of the printed droplets of a water-based conducting polymer that constitutes the electrodes of the device. Wang *et al.*[72] have investigated the dewetting of the water-based conducting polymer poly(3,4-ethylenedioxythiophene)/poly(4-styrenesulphonate) (PEDOT/PSS) inks on a patterned SiO_2 substrate modified with the fluorinated self-assembled monolayer (SAM) perfluorodecyltrichlorosilane (FDTS, $C_{10}F_{17}H_4SiCl_3$). Submicrometer hydrophobic lines whose width ranged from 250 nm to 2 μm by EBL and from 2 to 20 μm by optical lithography were fabricated.

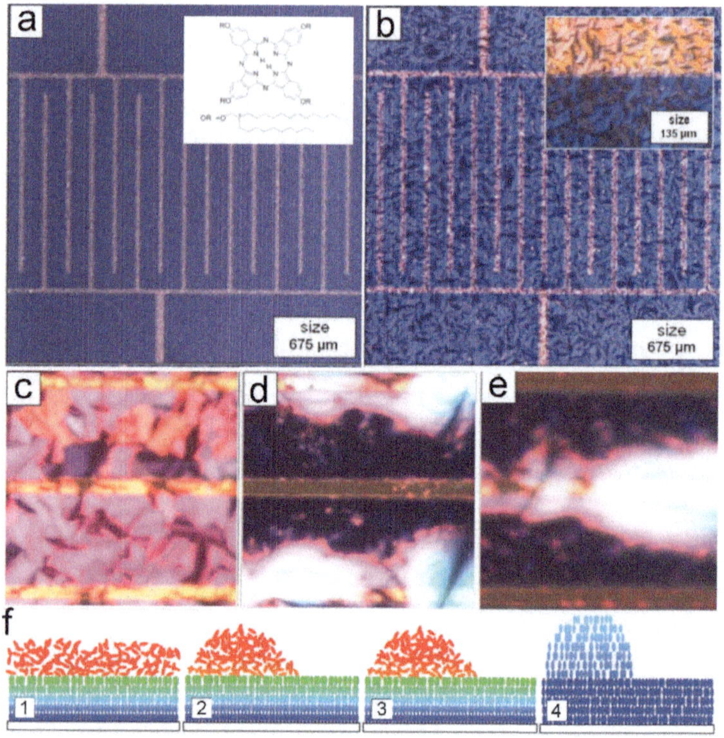

Figure 14.14 Dynamics of DLC film dewetting under an electric field. (a) Morphology of DLC films of H2Pc-O-(14,10)4—chemical structure in the inset of (a)—on a test pattern device with interdigitated electrodes in dark field. (a) Spin-coated film (h = 150 nm) before thermal treatment at T = 460 K. (b) Film after thermal treatment and quenching. Inset shows the difference between the channel and electrode regions. (c–e) Polarized light optical microscopy images acquired with cross polars show the change in contrast upon heating the spin-cast film in the proximity of the melting point T = 460 K. (c) As-cast film in between three interdigitated electrodes. (d) The material is displaced from the channel as a voltage difference is applied from the positive electrode (central) to the adjacent negative finger electrodes (top and bottom); quenching to 433 K. Most of the material accumulates in the form of anisotropic DLC drops at the negatively charged electrodes, whereas an ultrathin film of highly oriented columns remains in the channel. (e) By reverting the applied bias, the drops are displaced and accumulated at the central electrode. The typical birefringent features of the hexagonal columnar Col_h phase appear in the drops and channel in (d and e). (f) Schematic drawing of the migration process of Pc molecules at T = 460 K. The color scale represents the fluidity of molecules, viz., blue: low, red: high (image adapted from Ref. 66).

During the drying of the droplets, deposited on top of the narrow FDTS-mesa structure, it splits into two, so defining the source and drain electrodes of the FET (Fig. 14.15). Top-gate polymer FETs were fabricated using dewetted PEDOT/PSS source and drain electrodes by spin-coating a 50 nm polymer semiconductor layer of poly(9,9'-dioctyl-fluorene-co-bithiophene) (F8T2) and inkjet-printing a PEDOT/PSS top-gate electrode (Fig. 14.15d).

Interestingly, they demonstrated that dewetting also depends on the relative position of the center of the droplet with respect to the hydrophobic barrier. They demonstrated that channel lengths on the order of 500 nm could be obtained by inkjet printing. This is a breakthrough as it shows the possibility of downscaling the lateral resolution of a large-area additive manufacturing technique, thus leading to an increase in the operational bandwidth of the devices.

14.3.5 *Stamp-Assisted Polymer Dewetting*

As already mentioned, dewetting leads to spontaneous patterning with spatial correlations, but a precise positioning of the individual structures is unlikely to occur across a very large area. A possible approach to overcome this limit relies on the spatial confinement of dewetting by means of an external agency, like a stamp.

Yoon *et al.* have demonstrated a kinetically driven method for fabricating ordered microstructures of polymers by combining dewetting with microimprinting lithography.[73,74] This method is based on utilizing soft imprinting to induce a local thickness variation of a polymeric bilayer, which allows the local control of dewetting. In particular, a spin-coated poly-4-vinylpyridine (P4VP)/PS bilayer on a silicon substrate was topographically patterned by microimprinting with a polydimethylsiloxane (PDMS) stamp (with various pattern symmetries) at $T_A = 150°C$. The subsequent annealing above the glass transition temperatures induces the initiation of dewetting of the top layer preferentially at thinner regions, leading to the localized dewetting of the top layer and the partial layer inversion (i.e., PS moves on top of P4VP) of two layers[75] (Fig. 14.16).

The dewetting was guided by the presence of the topographic structure and initiated strictly at the boundary of the thicker and

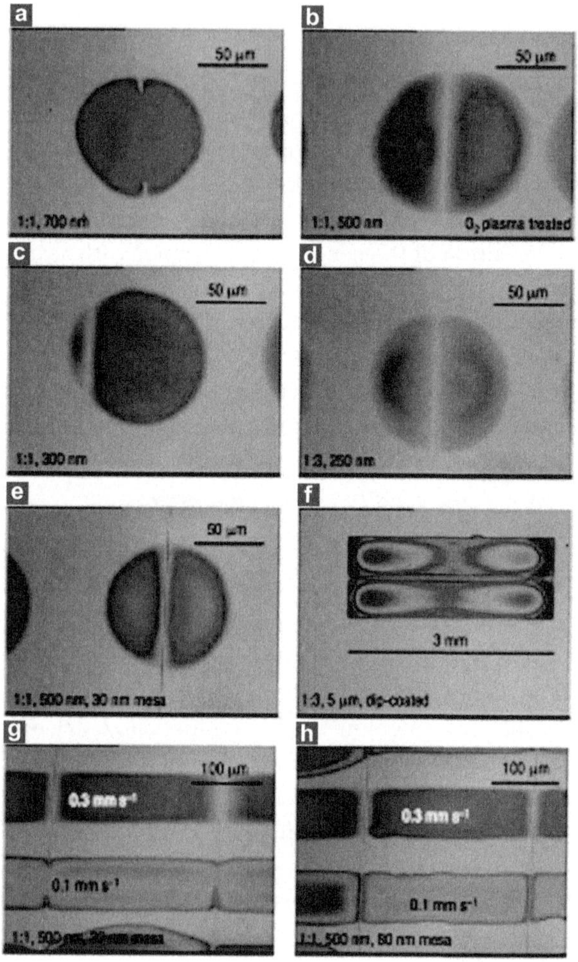

Figure 14.15 Photographs of various dewetted PEDOT/PSS illustrating factors affecting dewetting. (a) PEDOT 1:1, 700 nm FDTS SAM. (b) PEDOT 1:1, 500 nm FDTS SAM, oxygen-plasma-treated surface. (c) PEDOT 1:1, 300 nm FDTS SAM. (d) PEDOT 1:3, 250 nm FDTS SAM. (e) PEDOT 1:1, 500 nm FDTS SAM, 30 nm SiO_2 mesa. (f) 5 μm FDTS SAM gap, dip-coated by PEDOT 1:3. (g) PEDOT 1:1 lines printed with a jet frequency of 4 Hz, printing speed of 0.1 and 0.3 mm s^{-1}, 500 nm FDTS SAM, 30 nm mesa. (h) PEDOT 1:1 lines printed with a jet frequency of 4 Hz, printing speed of 0.1 and 0.3 mm s^{-1}, 500 nm FDTS SAM, 80 nm mesa. (Images taken from Ref. 72.)

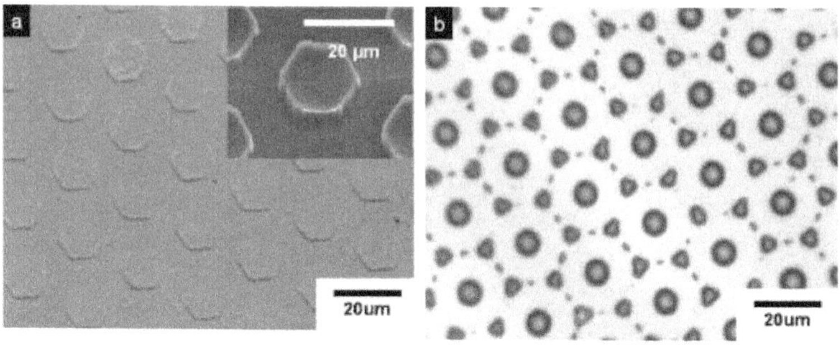

Figure 14.16 Microimprinted controlled dewetting. The controlled dewetting of a microimprinted P4VP/PS bilayer film on an Si substrate. The PDMS mold has hexagonal holes. OM (a) and SEM (inset) images of the imprinted P4VP/PS bilayer pattern. (b) OM image of the imprinted P4VP/PS bilayer after controlled dewetting. The dewetting in the thinner inside and outside regions produced a new pattern with hemispherical and triangular microscopic features. (Images taken from Ref. 74.) *Abbreviation*: PDMS, polydimethylsiloxane.

thinner regions. The dewetting front velocity of the microdomains in the confined regions was linearly proportional to the measured time, enabling to control the size of the dewetted domains with the annealing time. In this way, contrarily to Voronoi polygons typically observed in the dewetting of polymer thin films,[76,77] ordered structures with both chemical and topological heterogeneity in a large area are achieved. Park *et al.* show that the dewetting of a Teflon precursor, confined by saw-toothed stamps, can be used to generate regularly spaced Teflon dot arrays.[78] Teflon precursor solutions were spin-coated onto several substrates with different surface energies (silicon, polyethilene, Teflon) and thermally annealed at $T_A = 120°C$ in contact with saw-toothed PDMS stamps. The contact of PDMS and Teflon, both with very low surface energies, induces a dewetting of the film and the formation of Teflon spherical droplets. Indeed, when Teflon films were thermally annealed at the same condition without PDMS stamps, smooth Teflon films were obtained without formation of holes or polymer droplets. At a very short annealing time (t_A), small Teflon droplets with irregular sizes were formed, but as t_A increases, highly ordered patterns of Teflon nanospheres with a precise size are obtained (Fig. 14.17). The sizes

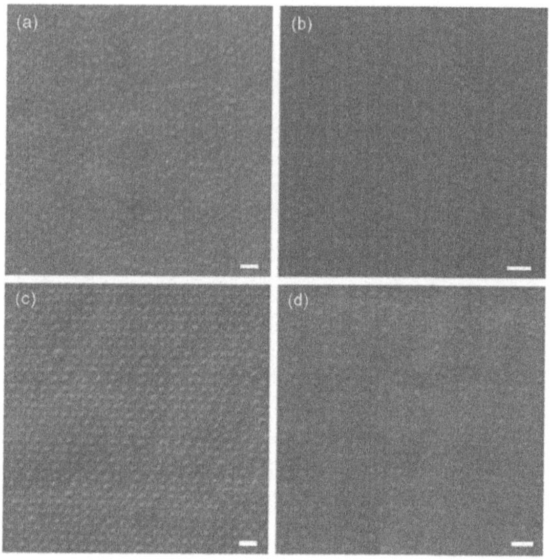

Figure 14.17 SEM images of Teflon nanospheres formed onto silicon surfaces at 120°C and four different annealing times of (a) 5 s, (b) 10 s, (c) 1 min, and (d) 5 min. Teflon nanospheres were formed between saw-toothed PDMS patterns with a pitch of 550 nm and a Teflon sheet in a short time of 5 s. Scale bars are 1 mm. (Images taken from Ref. 78.)

and separation distances of Teflon lines are dictated by the pitch and amplitude of the saw-toothed mold, while sizes of the Teflon droplets can be tuned by modulating the thickness of the Teflon film.

Likewise, Kim *et al.*[79] have shown the ability to achieve polymeric patterns, whose feature size ranges from tens to hundreds of nanometers by using the same stamp while changing the time of the process (see Fig. 14.18). A PDMS stamp is placed on a polystyrene film, and they are heated to a temperature higher than the glass transition temperature of the polymer for a specified period of time. During the annealing the stamp sinks into the polymer melt. When the sharp edges of the mold approach the substrate surface, close enough that dispersion forces take place, dewetting starts and breaks up the film, opening holes or grooves, depending on whether the pattern on the mold has sharp points or lines. After the specified time has elapsed, the polymer is cooled with the mold still in place, and then the mold is simply removed. The instability, which causes

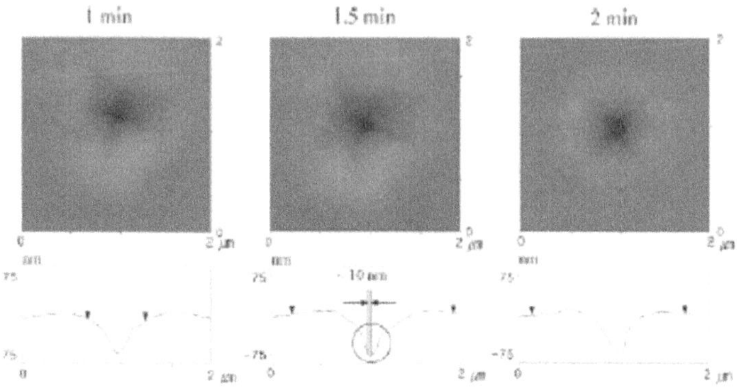

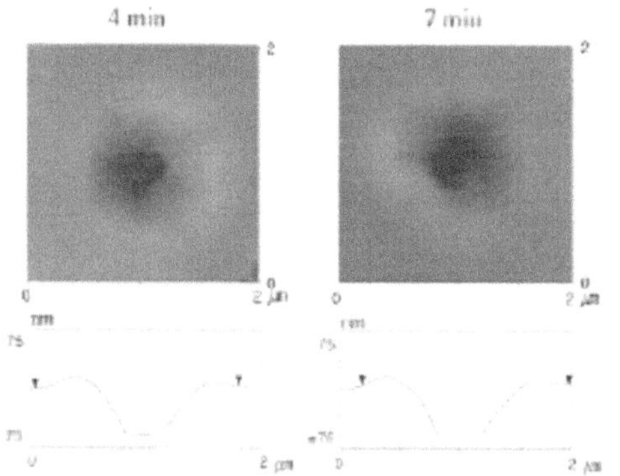

Time	Flat Size
1 min	-
1.5 min	10 nm
2 min	40 nm
4 min	220 nm
7 min	360 nm

Figure 14.18 Temporal evolution of hole opening in a 75 nm thick PS (MW 10 800) film on Si wafer (image size 2 × 2 μm^2) (images taken from Ref. 79).

the film breakup, occurs at the location where the film thickness is the thinnest.[80,81] Once dry spots (holes) form, the hole radius grows (nonlinearly) with time, which in turn allows the hole size to be controlled by simply controlling t_A.

14.3.6 *Time–Temperature Integrating Devices*

Recently, our group reported a new application of dewetting of polymer thin films as time-temperature integrator.[82] The idea stems from the knowledge that dewetting of a solid film is an irreversible process exhibiting an activation energy and whose characteristic length scales depend on time. The device is based on a thin film of polyisobutylene (PIB) cast on a silicon surface that is patterned by irradiation of spatially defined zones using an electron beam. The irradiation fragments the PIB chains, forming a zone in which the molecular weight, and thus the temperature of induced dewetting, is reduced in a controlled manner. It turns out that the temperature of induced dewetting is modulated by the electrons dose received. In particular, the zones containing PIB with a lower molecular weight dewet faster than the zones with a higher molecular weight. The time–temperature integration is demonstrated by observing the evolution of the percentage film area exposed by dewetting versus time and by modeling dewetting according to Brochard-

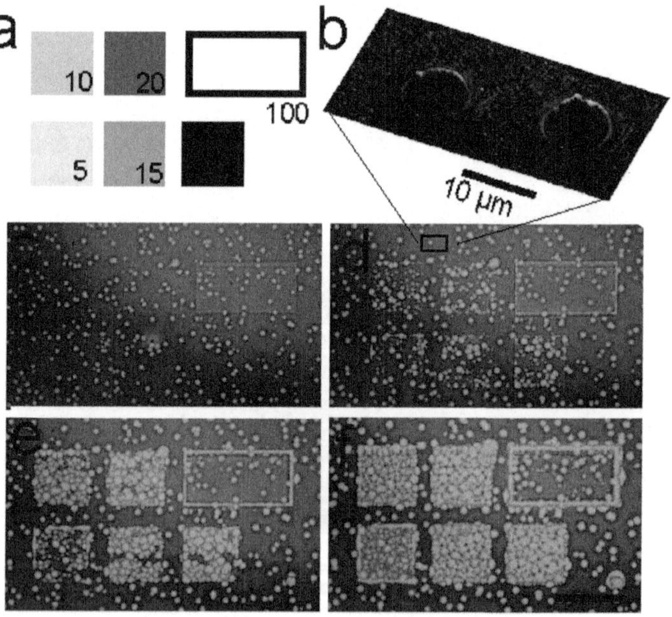

Figure 14.19 Time–temperature integration functionality (a) Scheme of the irradiated zones of a PIB thin film. The gray level is proportional to the electron dose; the number indicates the exact doses expressed in $\mu C \cdot cm^{-2}$. (b) AFM topography of holes formed at the beginning of dewetting. (c–f) Optical images of the evolution of an irradiated PIB film at $T = 60°C$ after (c) 253 s, (d) 387 s, (e) 507 s, and (f) 927 s. The bar is 50 μm. (Images taken from Ref. 82.)

Wyart theory.[83] Figure 14.19 shows the scheme of irradiated zones of a PIB thin film and the time evolution of an irradiated PIB film at a temperature $T = 60°C$.

The approach is general and can be extended to a variety of polymers other than PIB, provided their molecular weight can be modified upon irradiation with electrons by fragmentation or cross-linking. The possibility of fine-tuning the average molecular weight by irradiation allows the method to be applied in wide temperature ranges and, for this reason, exploited in applications based on time–temperature-integrating tags targeted to drug and food distribution or to time lapse monitoring (e.g. security).[84]

14.4 Stamp-Assisted Dewetting: A Versatile Method for Patterning Solutions

In stamp-assisted dewetting a precise amount of material from a solution is deposited into a spatially defined region between a stamp and a substrate. The process of dewetting takes place in a confined system, where characteristic time scales for self-organization depend on the volume and dimensionality. Since the physical dimension of the system, where the materials self-organize, is reduced in size, a quasi-equilibrium condition can be attained in shorter time scales. This may imply enhancement of surface nucleation effects with respect to bulk aggregation or depletion of diffusion-limited processes. Therefore, the use of stamps to assist the dewetting process of soluble materials deposited on substrates is a useful strategy to overcome some of the limitations of thin-film technology and improve the properties of multifunctional materials. Both soft (PDMS, polycarbonate, etc.) and rigid (aluminum, gold, copper, etc.) stamps can be used for controlling the dewetting processes, their most stringent limitation being due to their stability upon exposure to the solvent used. The soft elastomeric polymer stamps, such as PDMS, have demonstrated to be one of the most versatile and technologically viable tools for unconventional micro- and nanomanufacturing. In stamp assisted dewetting, the spatial correlation of the features generated is enhanced: the smaller length scales (size and distance of the molecular aggregates) are intrinsic to the dewetting, and the larger ones are imposed by the motif of the stamp protrusions. As it will be shown later by different examples, a remarkable outcome of the stamp-assisted dewetting approach is the possibility to downsize the printed feature size with respect to the lateral size of the stamp protrusions. Several patterning techniques have been developed in combination with the dewetting process. Among all, lithographic techniques, such as micromolding in capillaries (MIMIC), lithographically controlled wetting (LCW), and grid-assisted LCW (GA-LCW),[85] have been used to fabricate multifunctional materials exploiting dewetting phenomena.

14.4.1 *Dewetting in MIMIC*

MIMIC is a simple and versatile soft lithographic method introduced by Whitesides *et al.*[86] It is suitable for fabricating complex microstructures on both planar and curved surfaces (Fig. 14.20a). The elastomeric stamp or mold is usually made of PDMS and is prepared by replica molding[87] of a structured master fabricated by traditional lithographic methods such as photolithography or EBL.

In particular, in the masters for MIMIC the motif consists of protruding lines that become grooves in the replica. When the stamp is placed in contact with a surface, the grooves form capillary channels. The size of the channels together with the concentration of the solution and the self-organization properties of the solute determine the length scale that can be achieved with this process. When a solution is poured at the open end of the stamp, the liquid spontaneously fills the channels under the effect of capillary pressure.[2,88,89] In order to apply MIMIC at a large scale, control over the experimental conditions is very important. Bystrenova *et al.*[90] reported multiple-length-scale patterning of DNA on a mica surface using MIMIC and dewetting. This approach yielded an array of DNA nanodots on a 1 mm^2 area, where each dot comprised a few molecules of DNA (see Fig. 14.21). The method exploits the self-organization of DNA molecules when the solution was confined between the stamp and the surface. The stamp features (the periodicity and volume of the capillary) control the large-scale arrangement, while the self-organization of the confined molecules introduces the smaller, characteristic length scale. The

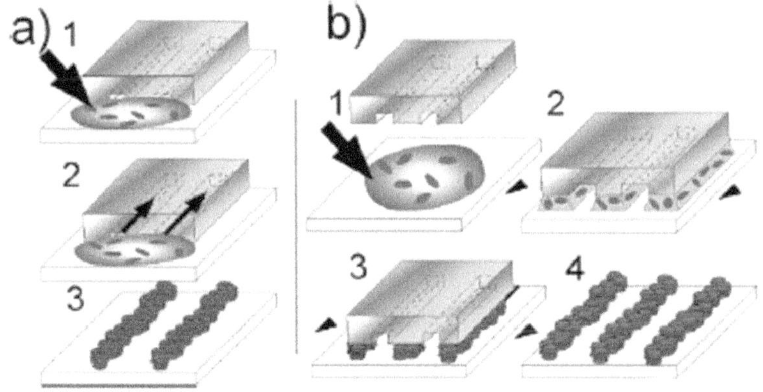

Figure 14.20 Schematic representation of MIMIC and LCW techniques. (a) Schematic representation of micromolding in capillaries: (a1) A PDMS stamp is placed onto a surface to effectively form microchannels. (a2) When a solution is poured at the open end of the stamp, the liquid filled in the microchannels by capillary force. (a3) After complete evaporation of the solvent the stamp is gently removed and the pattern remains on the surface. (b) Schematic illustration of printing nanostructures by lithographically controlled wetting: (b1) A drop of solution is applied to the surface of the substrate, and the stamp is placed on top of it. (b2) The solution is pinned under the stamp protrusions. (b3–b4) When the solvent is evaporated the stamp is gently removed. (Images taken from Ref. 85.)

latter, together with the dot size and interdot distances, depends on the initial DNA concentration of the solution. For a diluted solution of DNA (<1.2 μg/mL), the resulting pattern consists of homogeneous lines. Using a more concentrated solution, dots emerge on top of layers a few angstroms thick, stretching along the direction of the channels. In the range of DNA concentrations between 2 and 4 μg/mL, the dots are roughly aligned but with no precise pitch in their spacing.

When the concentration exceeds 5 μg/mL, spatial correlations emerge among the dots and the dot size increases. In this case, the dots are perfectly aligned along the edges of the channel due to the pinning of the solution and the convective flow from the center toward the edges. The DNA molecules, with a negatively charged shell, have low affinity for the negatively charged mica surface, and the surface tension is the major driving force for dot formation.

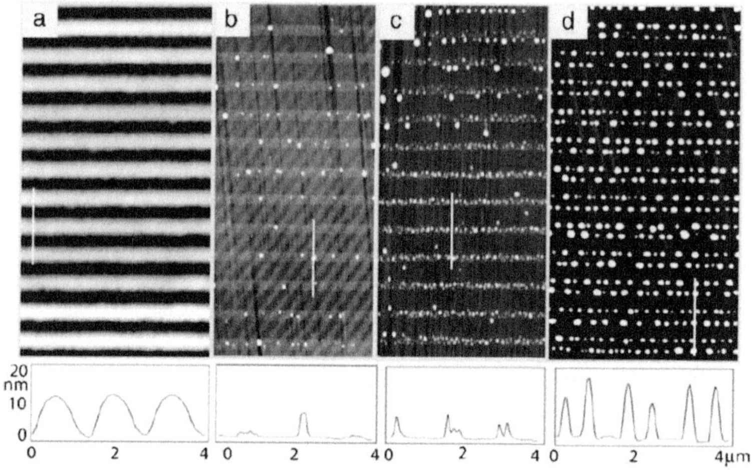

Figure 14.21 AFM images of λ_{DNA} (100–8,000 base pairs) patterned on a mica surface for different concentrations of the infilling solution: (a) 1.25 μg mL^{-1}, (b) 2.0 μg mL^{-1}, (c) 2.8 μg mL^{-1}, and (d) 5.0 μg mL^{-1}. (Image taken from Ref. 90.)

In diluted DNA solutions the repulsive electrostatic interaction, between DNA molecules and the surface, is shielded by the buffer, therefore a homogeneous continuous layer with salt and DNA can be deposited. The transition from continuous lines to split strings of dots at higher DNA concentration arises either from the dewetting of initially continuous stripes[91] or nucleation and growth in a partial wetting regime.[92]

14.4.2 *LCW*

LCW[93] is a particularly attractive for soluble materials because it allows one to exploit self-organization of soluble functional materials with the spatial control provided by the stamp features (Fig. 14.20b). LCW is suitable for large-area nanopatterning and is sustainable because of its simplicity, high transfer rate, and low cost of materials and equipment. The stamp can be composed of either rigid materials, such as metals, ceramics, or rigid polymers, or a soft material, such as polycarbonate or PDMS.[94] When a stamp is gently placed in contact with a liquid thin film spread on a substrate, the fluid layer develops instability where the capillary forces pin the solution to the stamp protrusions, giving rise to an array of menisci. As the critical concentration is reached, the solute precipitates from the solution onto the substrate within the menisci, giving rise to a structured thin film that replicates the protrusion of the stamp. In a recent work,[95] a convergent approach based on the stamp-assisted deposition of molecules from a solution and dewetting phenomena, arising from competing interactions between the molecules and the substrate, has been used to achieve the patterning of single-molecule magnet (SMM) aggregates. Dilute solutions of $Mn_{12}O_{12}(O_2CC_{12}H_9)_{16}(H_2O)_4$, an SMM cluster characterized by a hydrophobic outer shell,[96] were deposited onto a hydrophilic surface (SiO_x/Si), and a stamp was placed on top of the solution layer. Arrays of nanometer-sized aggregates, each made of a few hundred molecules derived from complexes, were achieved (Fig. 14.22). The motif of the stamp protrusions (consisting of parallel lines) imposes the larger length scale, while dewetting phenomena control the spatial correlations on the smaller length scales (size and distance of the molecular aggregates). As a consequence, the pattern exhibits characteristic length scales much smaller than those of the features present in the stamp. Starting from an extremely dilute solution, ultranarrow structures as small as tens of nanometers wide can be obtained by using stamp protrusions that are a few hundred nanometers. These nanoscale materials are attractive for several technological applications,[97] such as high-density magnetic storage, memories, sensors, quantum computing, and spintronics. Moreover, SMMs exhibit quantum physical properties such as quantum tunneling of magnetization[98] and quantum phase interference.[99]

Cheng *et al.* have developed a new simple and efficient method to achieve large-area nanoparticle superlattice patterns controlling the local dewetting dynamics by molding pressures and mold geometry.[100] Molded microdroplets with precisely defined shapes, locations, and contact line boundaries are obtained by sandwiching a nanoparticle solution between a micropatterned PDMS stamp and a silicon substrate. Under different pressures, these artificial microdroplets display distinct and highly controllable dewetting dynamics. Under high pressure, a single microdroplet is formed in

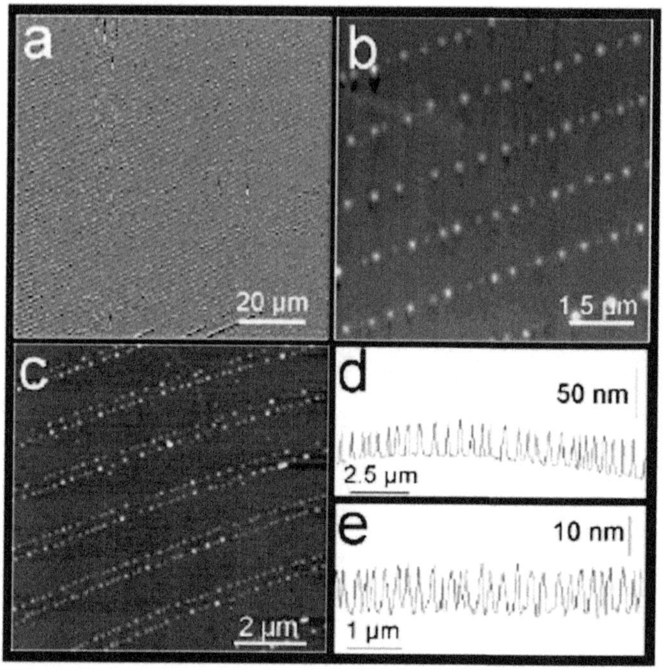

Figure 14.22 (a) AFM image (error signal) of patterns of $Mn_{12}O_{12}(O_2CC_{12}H_9)_{16}(H2O)_4$ fabricated on a large area (100 × 100 mm^2). (b) Image of printed features using a dilute solution. (c) Image of printed features using a very dilute solution. Typical line profile along a line obtained using (d) a dilute solution, corresponding to (b), and (e) a very dilute solution, corresponding to (c). (Image taken from Ref. 95.)

the middle of a micromold (center-dewetting process). Conversely, under lower pressures, the microdroplet collapses in the middle of the micromold, forming a microdroplet corral that remains attached to the edge of the micromold (edge-dewetting process). In center dewetting, water was slowly drawn out of the bottom of the micromold, inducing a slow deformation of the meniscus. Under these conditions, internal currents within the solution are presumably sufficient to keep the solution, near the contact point, dilute and allow it to slip down the side of the micromold. Conversely, during edge dewetting the meniscus deforms rapidly, concentrating solutes at the edges and enhancing evaporation near the edges, which further concentrates the solutes and helps pin the contact point to the micromold edge. Patterning of water-soluble single-stranded DNA-capped gold nanoparticles (12 nm) onto a silicon substrate, tuning the interparticle potential by adjusting the salt concentration (0.1–10 mM NaCl),[101] results in highly ordered nanoparticle superlattices. Under low molding pressure (edge-dewetting process), in which nucleation and growth of superlattices were confined to the micromold edges, edge-defined corrals whose height is one nanoparticle monolayer are observed.

Under intermediate molding pressure, as a result of competing center and edge dewetting, well-formed, edge-defined corrals and de-pinned corrals on the same substrate are observed. Under high molding pressure (center-dewetting process), in which nucleation and growth of superlattices were confined to the center of the micromold, discs one monolayer high or shrunken microcorral or a 3D supracrystal are formed (see Fig. 14.23).

Noteworthy, by controlling the parameters of dewetting dynamics, this method allows to pattern on smaller length scales than those of the features present in the stamp; indeed 12 nm wide microcorrals were obtained from 5 μm diameter micromolds. This method can be generalized to regulate the drying-mediated deposition of a variety of materials, such as CdSe/ZnS quantum dots, conducting polymers, and single crystals, and to micropattern electrode arrays with proper alignments.

14.4.2.1 GA-LCW

LCW can also be applied using a floating grid as a stamp (GA-LCW), in which case the process has been also named "grid-assisted deposition" (Fig. 14.24).[102] The process consists of a metal grid placed on top of a solution cast on a substrate. The stamp grid floats on the liquid film, and the stamp–substrate distance adjusts as the solvent evaporates, until supersaturation is reached and the solute precipitates. The critical distance corresponds to the final stamp–substrate distance. The use of metal grids (including noble metal grids, such as Au, Ag) and the possibility to cover the grids with a protective layer, as a SAM, makes it possible to use GA-LCW with the most aggressive solvents, which cannot be normally used with

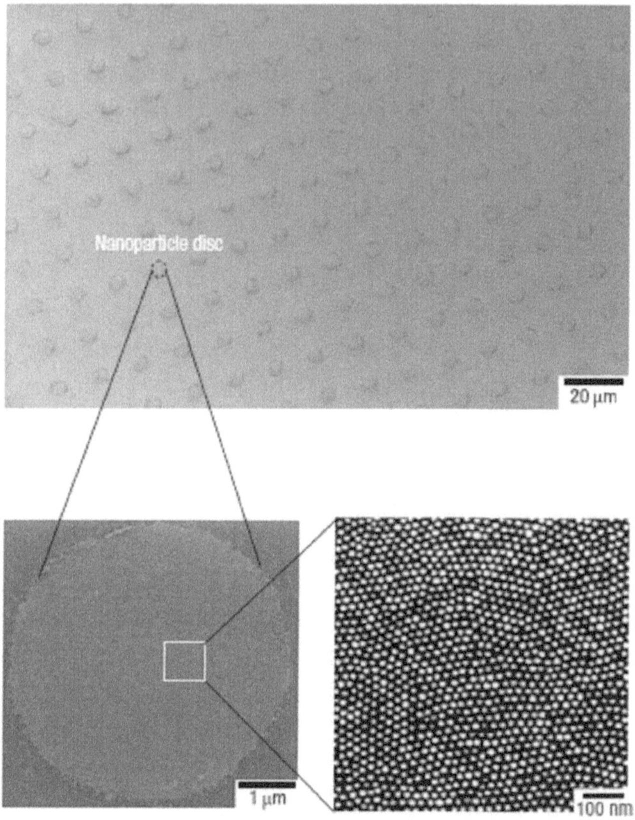

Figure 14.23 SEM micrographs of patterned nanoparticles discs at different magnifications. Concentration of nanoparticles = 1.0 mM; molding pressure = 3.0 N cm^{-2} (images taken from Ref. 100).

soft lithography.[103] Cavallini *et al.* reported on the fabrication of microscopic arrays of rotaxanes on silicon (SiO_x/Si) surfaces based on the dewetting of a thin film confined inside microscopic cavities of copper grids.[102]

Rotaxanes, which are mechanically interlocked molecular architectures, are particularly interesting for their rich phenomenology of organization at a surface controlled by their intrinsic multistability.[104] A series of rotaxanes and their respective threads were investigated. A spontaneous pattern formation occurs and

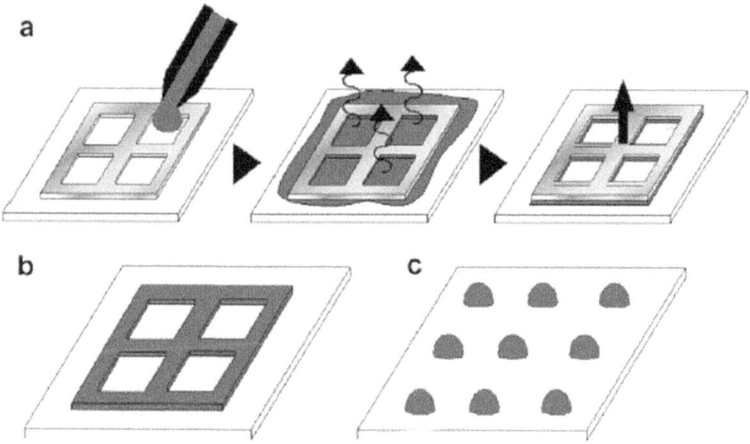

Figure 14.24 Scheme of grid-assisted deposition. (a) General scheme. (b) Patterning in a wetting regime. (c) Patterning in a dewetting regime. (Images taken from Ref. 105.)

exhibits both micro- and submicrometer length scales, in a single step and a few seconds, starting from a template mesh. Dewetting of rotaxane solutions on a copper grid gives rise to patterns with spatial correlations, with both the droplet size and the interdroplet distance exhibiting a radial dependence from the center outward. In addition, on the basis of their evidence, authors show that the dewetting mechanism of rotaxanes is different from that of their respective threads. In particular, thread solutions form a convex meniscus within each grid hole, and during the dewetting process, since the highest evaporation losses occur in the proximity of the contact line and the contact angle is kept constant, an outward radial capillary flow of the solution accumulates material at the border.[63] Conversely, rotaxane solutions form an inverted meniscus, the fastest loss of solvent occurs at the center of the holes, and the constraint of constant contact angle at the boundary gives rise to an inward flow of material (see Fig. 14.25). The origin of the different menisci is due to the different interactions between rotaxane and thread solutions with the stamp, indeed changing the nature of the grid material (polymeric grid); the behavior of two solutions is similar. These results highlight the critical role of interfacial

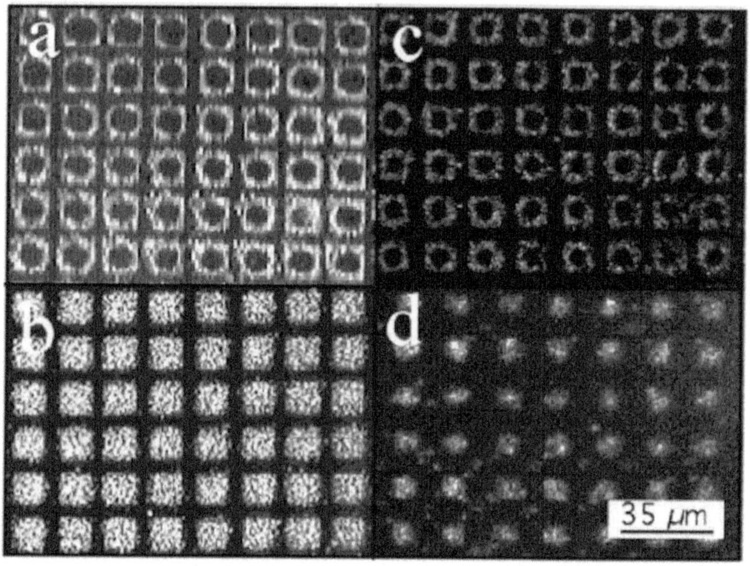

Figure 14.25 Patterns from rotaxane solutions. Optical micrograph (200 times magnification) of an array obtained by deposition of rotaxane, (a) 20 mL and (b) 60 mL droplet of a solution 0.15 mg/mL, and the thread of rotaxane (c) 20 mL and (d) 60 mL droplet of a solution 0.17 mg/mL onto a Cu grid. The grid has been removed. (Image taken from Ref. 102.)

energies between the solutions and the stamp as key factors in the dewetting process. More recently, Massi *et al.*[105] have shown the GA-LCW of tris-(8-hydroxyquinolinato) aluminum(III) (Alq3) onto silicon surfaces controlling wetting/dewetting phenomena changing the substrate hydrophilicity. Alq3 is a polar molecule[106] extensively used in organic light-emitting devices.[107] Growth of Alq3 droplets on hydrophobic surfaces has been studied, its mechanism of growth being explained in terms of a Voronoi mosaic generated by Ostwald ripening.[108,109] Using copper grids for transmission electron microscopy, as a template mesh for the deposition, and dichloromethane solutions of Alq3 (0.1–0.001 g/L), a continuous film, replicating the bars, is obtained on a hydrophilic silicon oxide (water contact angle $12 \pm 2°$). Also, by decreasing the amount of material deposited, the final morphology is a thinner continuous film (wire-like architectures) with no evidence of

Figure 14.26 Semispherical droplets of Alq3. AFM images of printed samples replicating a squared grid. Deposition of an Alq3 solution on a hydrophobic silicon surface (image taken from Ref. 105).

dewetting. Conversely, on hydrophobic silicon oxide (water contact angle 83 ± 3°), the film architecture dramatically changes due to dewetting phenomena, and if the thickness of the deposited film was kept below 50 nm, formation of separated periodic droplets was observed.

In particular, Alq3 is deposited below the crossbars, forming semispherical droplets <10 μm in diameter and 91 ± 5 nm thick, as shown in Fig. 14.26. Since the wetting/dewetting properties of the substrate become relevant during the last step of the solvent evaporation, when the residual solution remains only under the mesh of the grid, the position and periodicity of the droplets are still imposed by its shape.

14.5 Conclusions

We have shown how dewetting, when properly controlled, can become a versatile and useful technology process for fabrication of microscopic, mesoscopic, and even nanoscopic structures across large areas. This approach is not limited only to the fabrication of patterns but is also used in the demonstration of optical, (opto-) electronic, and magnetic devices. Here we focused our attention mostly on multifunctional materials, but it is clear that biomedical applications would benefit enormously by a technology based on control of dewetting phenomena. More applications, such as bioarrays, scaffolds for tissue growth, and micro- and nanodroplet dispensers in microfluidics can be envisioned. The fact that dewetting is ubiquitous makes the approach suitable for any class of materials, from inorganic ones to soft matter, and for a variety of aggregation phases beyond the liquid state. The attractive aspects rely on the simplicity of the approach and the mild conditions relative to the ones required in the processing of a bulk material. Moreover, the manufacturing approach is inherently additive as no material is subtracted or lost during the process.

The multiscale phenomena inherent to the dewetting phenomena can be transferred into a fabrication technique once the complexity of the system is mastered. It has to be said that most of the techniques presented here are interesting and technologically promising but still at the level of proof-of-concept. A closer interaction between experiment and multiscale modeling is needed for a complete understanding and control of the experimental parameters governing the processes. This will lead to standardization of the fabrication steps that serve as a prelude to the technological success of dewetting-based approaches in real applications.

References

1. P.-G. de Gennes, F. Brochard-Wyart, and D. Quéré, *Capillarity and Wetting Phenomena Drops, Bubbles, Pearls, Waves.* Springer (2004).
2. P.-G. de Gennes, Wetting: statics and dynamics, *Rev. Mod. Phys.*, **57**, 827–863 (1985).
3. A. Sharma, Relationship of thin film stability and morphology to macroscopic parameters of wetting in the apolar and polar systems, *Langmuir*, **9**, 861–869 (1993).
4. G. Reiter, Unstable thin polymer films: rupture and dewetting processes, *Langmuir*, **9**, 1344–1351 (1993).

5. O. Karthaus, L. Gråsjö, N. Maruyama, and M. Shimomura, Formation of ordered mesoscopic polymer arrays by dewetting, *Chaos*, **9**, 308–314 (1999).
6. K. Kargupta, R. Konnur, and A. Sharma, Spontaneous dewetting and ordered patterns in evaporating thin liquid films on homogeneous and heterogeneous substrates, *Langmuir*, **17**, 1294 (2001).
7. R. Xie, A. Karim, J. F. Douglas, C. C. Han, and R. A. Weiss, Spinodal dewetting of thin polymer films, *Phys. Rev. Lett.*, **81**, 1251–1255 (1998).
8. A. Stabel, R. Heinz, F. C. De Schryver, and J. P. Rabe, Ostwald ripening of 2-dimensional crystals at the solid–liquid interface, *J. Phys. Chem.*, **99**, 505–507 (1995).
9. M. Schaefer, C. Bartels, F. Leclerc, and M. Karplus, Effective atom volumes for implicit solvent models: comparison between Voronoi volumes and minimum fluctuation volumes, *J. Comp. Chem.*, **22**, 1857–1879 (2001).
10. U. Thiele, M. Mertig, and W. Pompe, Dewetting of evaporating thin liquid film: heterogeneous nucleation and surface instability, *Phys. Rev. Lett.*, **80,** 2869–2873 (1998).
11. F. Biscarini, M. Cavallini, R. Kshirsagar, D. A. Leigh, S. León, and F. Zerbetto, Self-organization of nano-lines and dots triggered by a local mechanical stimulus, *Proc. Natl. Acad. Sci. U S A*, **103**, 17650–17654 (2006).
12. J. F. Moulin, J. C. Kengne, R. Kshirsagar, M. Cavallini, F. Biscarini, S. Léon, F. Zerbetto, G. Bottari, and D. A. Leigh, Self-organization of rotaxane thin films into spatially correlated nanostructures: morphological and structural aspects, *J. Am. Chem. Soc.*, **128**, 526–532 (2006).
13. N. J. Suematsu, Y. Ogawa, Y. Yamamoto, and T. Yamaguchi, Dewetting self-assembly of nanoparticles into hexagonal array of nanorings, *J. Colloid Interface Sci.*, **310**, 648–652 (2007).
14. V. Palermo and P. Samori, Molecular self-assembly across multiple length scales, *Angew. Chem., Int. Ed. Engl.*, **46**, 4428–4432 (2007).
15. R. van Hameren, P. Schon, A. M. van Buul, J. Hoogboom, S. V. Lazarenko, J. W. Gerritsen, H. Engelkamp, P. C. M. Christianen, H. A. Heus, J. C. Maan, T. Rasing, S. Speller, A. E. Rowan, J. Elemans, and R. J. M. Nolte, Macroscopic hierarchical surface patterning of porphyrin trimers via self-assembly and dewetting, *Science*, **314**, 1433–1436 (2006).

16. J. X. Huang, F. Kim, A. R. Tao, S. Connor, and P. D. Yang, Spontaneous formation of nanoparticle stripe patterns through dewetting, *Nat. Mater.*, **4**, 896–900 (2005).
17. Z. X. Zhang, Z. Wang, R. B. Xing, and Y. C. Han, Patterning thin polymer films by surface-directed dewetting and pattern transfer, *Polymer*, **44**, 3737–3743 (2003).
18. A. Sehgal, V. Ferreiro, J. F. Douglas, E. J. Amis, and A. Karim, Pattern-directed dewetting of ultrathin polymer films, *Langmuir*, **18**, 7041–7048 (2002).
19. M. Cavallini, P. Mei, F. Biscarini, and R. Garcia, Parallel writing by local oxidation nanolithography with submicrometer resolution, *App. Phys. Lett.*, **83**, 5286–5289 (2003).
20. C. Albonetti, J. Martinez, N. S. Losilla, P. Greco, M. Cavallini, F. Borgatti, M. Montecchi, L. Pasquali, R. Garcia, and F. Biscarini, Parallel-local anodic oxidation of silicon surfaces by soft stamps, *Nanotechnology*, **19**, 435303–435303 (2008).
21. D. Wouters and U. S. Schubert, Nanolithography and nanochemistry: probe-related patterning techniques and chemical modification for nanometer-sized devices, *Angew. Chem., Int. Ed. Engl.*, **43**, 2480–2495 (2004).
22. R. Maoz, E. Frydman, S. R. Cohen, and J. Sagiv, Constructive nanolithography: site-defined silver self-assembly on nanoelectro-chemically patterned monolayer templates, *Adv. Mater.*, **12**, 424–429 (2000).
23. R. Mukherjee, A. Sharma, G. Patil, D. Faruqui, and P. S. G. Pattader, Soft lithography meets self-organization: some new developments in meso-patterning, *Bull. Mater. Sci.*, **31**, 249–261 (2008).
24. R. Mukherjee, D. Bandyopadhyay, and A. Sharma, Control of morphology in pattern directed dewetting of thin polymer films, *Soft. Matter*, **4**, 2086–2097 (2008).
25. K. Nakata, H. Kimura, M. Sakai, T. Ochiai, H. Sakai, T. Murakami, M. Abe, and A. Fujishima, UV/thermally driven rewritable wettability patterns on TiO2-PDMS composite films, *Appl. Mater. Interfaces*, **2**, 2485–2488 (2010).
26. M. Cavallini, F. Biscarini, S. Leon, F. Zerbetto, G. Bottari, and D. A. Leigh, Information storage using supramolecular surface patterns, *Science*, **299**, 531 (2003).
27. C. J. Park, K. H. Cho, W.-C. Yang, H. Y. Cho, Suk-Ho Choi, R. G. Elliman, J. H. Han, and K. Chungwoo, Large capacitance-voltage hysteresis loops

in SiO_2 films containing Ge nanocrystals produced by ion implantation and annealing, *Appl. Phys. Lett.*, **88**, 071916–071919 (2006).

28. D. Riabinina, C. Durand, M. Chaker, and F. Rosei, Photoluminescent silicon nanocrystals synthesized by reactive laser ablation, *Appl. Phys. Lett.*, **88**, 073105–073108 (2006).
29. D. J. Srolovitz and M. G. Goldiner, The thermodynamics and kinetics of film agglomeration, *JOM*, **47**, 31–36 (1995).
30. K Yano, T. Ishii, T. Hashimoto, T. Kobayashi, F. Murai, and K. Seki, Room-temperature single electron memory, *IEEE Stans. Electron Devices*, **41**, 1628–1638 (1994).
31. M Kanoun, A. Souifi, T. Baron, and F. Mazen, Electrical study of Ge-nanocrystal-based metal-oxide-semiconductor structures for *p*-type nonvolatile memory applications, *Appl. Phys. Lett.*, **84**, 5079–5082 (2004).
32. M. Yang, T. P. Chen, Z. Liu, J. I. Wong, W. L. Zhang, S. Zhang, and Y. Liu, Effect of annealing on charge transfer in Ge nanocrystal based nonvolatile memory structure, *J. Appl. Phys.*, **106**, 103701(1–7) (2009).
33. K. Gacem, A. El Hdiy, M. Troyon, I. Berbezier, and A. Ronda, Conductive AFM microscopy study of the carrier transport and storage in Ge nanocrystals grown by dewetting, *Nanotechnology*, **21**, 065706(1–6) (2010).
34. A. Campera, and G. Iannaccone, Modelling and simulation of charging and discharging processes in nanocrystal flash memories during program and erase operations, *Solid-State Electron.*, **49**, 1745–1753 (2005).
35. C. S. Lent and P. D. Tougaw, Device architecture for computing with quantum dots, *Proc. IEEE*, **85**, 541–557 (1997).
36. C. T. Black, C. B. Murray, R. L. Sandstrom, and S. Sun, Spin-dependent tunneling in self-assembled cobalt-nanocrystal superlattices, *Science*, **290**, 1131–1134 (2000).
37. R Nuryadi, Y Ishikawa, and M. Tabe, Formation and ordering of self-assembled Si islands by ultrahigh vacuum annealing of ultrathin bonded silicon-on-insulator structure, *Appl. Surf Sci.*, **159**, 121–126 (2000).
38. R Nuryadi, Y. Ishikawa, Y. Ono, and M. Tabe, Thermal agglomeration of single crystalline Si layer on buried SiO_2 in ultrahigh vacuum, *J. Vac. Sci. Technol., B*, **20**, 167–172 (2002).

39. E. Sutter and P. Sutter, Assembly of Ge nanocrystals on SiO_2 via a stress-induced dewetting process, *Nanotechnology*, **17**, 3724–3727 (2006).

40. B. Yang, P. Zhang, D. E. Savage, G.-H. Lu, M. Huang, F. Liu, and M. G. Lagally, Self-organization of semiconductor nanocrystals by selective surface faceting, *Phys. Rev. B*, **72**, 235413(1–4) (2005).

41. F. Brochardwyart, J. M. Dimeglio, D. Quere, and P. G. Degennes, Spreading of nonvolatile liquids in a continuum picture, *Langmuir*, **7**, 335–338 (1991).

42. P. P. Zhang, B. Yang, P. P. Rugheimer, M. M. Roberts, D. E. Savage, F. Liu and M. G. Lagally, Influence of germanium on thermal dewetting and agglomeration of the silicon template layer in thin silicon-on-insulator, *J. Phys. D: Appl. Phys.*, **42**, 175309 (2009).

43. T. Stoica, L. Vescan, and E. Sutter, Absence of island–island interaction during formation of isolated Ge islands in small windows, *J. Appl. Phys.*, **95**, 7707–7712 (2004).

44. A. Olzierski, A. G. Nassiopoulou, I. Raptis, and T. Stoica, Two-dimensional arrays of nanometre scale holes and nano-V-grooves in oxidized Si wafers for the selective growth of Ge dots or Ge/Si hetero-nanocrystals, *Nanotechnology*, **15**, 1695–1700 (2004).

45. A. P. Alivisatos, Semiconductor clusters, nanocrystals, and quantum dots, *Science*, **271,** 933–937 (1996).

46. J. Huang, F. Kim, A. R. Tao, S. Connor, and P. Yang, Spontaneous formation of nanoparticle stripe patterns through dewetting, Nature Mat., **4** 896–900 (2005).

47. L.-T. Lee, C. A. P. Leite, and F. Galembeck, Controlled Nanoparticle Assembly by Dewetting of Charged Polymer Solutions Lamgmuir, **20**, 4430–4435 (2004).

48. M. Trau, D. A. Saville, and I. A Aksay, Field-induced layering of colloidal crystals, *Science*, **272**, 706–709 (1996).

49. S. O. Lumsdon, E. W. Kaler, J. P. Williams, and O. D. Velev, Dielectrophoretic assembly of oriented and switchable two-dimensional photonic crystals, *Appl. Phys. Lett.*, **82**, 949–952 (2003).

50. K. D. Hermanson, S. O. Lumsdon, J. P. Williams, E. W. Kaler, and O. D. Velev, Dielectrophoretic assembly of electrically functional microwires from nanoparticle suspensions, *Science*, **294**, 1082–1086 (2001).

51. S. R. Yeh, M. Seul, and B. I. Shraiman, Assembly of ordered colloidal aggregates by electric-field-induced fluid flow, *Nature*, **386**, 57–59 (1997).

52. A. Kumar and G. M. Whitesides, Magnetization process observation of a CoNiCr film in the field of a weakly excited objective lens in transmission electron microscope, *Appl. Phys. Lett.*, **63**, 1993–1996 (2002).
53. J. Aizenberg, P. V. Braun, and P. Wiltzius, Patterned colloidal deposition controlled by electrostatic and capillary forces, *Phys. Rev. Lett.*, **84**, 2997–3000 (2000).
54. R. W. Zehner, W. A. Lopes, T. L. Morkved, H. Jaeger, and L. R. Sita, Selective decoration of a phase-separated diblock copolymer with thiol-passivated gold nanocrystals, *Langmuir*, **14**, 241–244 (1998).
55. L. T. Lee, M. D. C. V. da Silva, and F. Galembeck, Dewetting patterns of thin films of charged polymer solutions, *Langmuir*, **19**, 6717–6722 (2003).
56. I. Langmuir and K. B. Blodgett, Über einige neue Methoden zur Untersuchung von monomolekularen Filmen, *Kolloid-Zeitschrift*, **73**, 257–263 (1935).
57. K. B. Blodgett, Monomolecular films of fatty acids on glass, *J. Am. Chem. Soc.*, **56**, 495 (1934).
58. A. Lorke, R. J. Luyken, J. M. Garcia, and P. M. Petroff, Growth and electronic properties of self-organized quantum rings, *Jpn. J. Appl. Phys.*, **40**, 1857–1859 (2001).
59. T. T. Hongisto and K. Yu Arutyunov, Persistent currents in superconducting nanorings, *J. Phys.: Conf. Ser.*, **97**, 012114 (2008).
60. S. Walheim, M. Boltau, J. Mlynek, G. Krausch, and U. Steiner, Structure formation via polymer demixing in spin-cast films, structure formation via polymer demixing in spin-cast films, *Macromolecules*, **30**, 4995–5003 (1997).
61. M. Cavallini, J. Gomez-Segura, C. Albonetti, D. Ruiz-Molina, J. Veciana, and F. Biscarini, Ordered patterning of nanometric rings of single molecule magnets on polymers by lithographic control of demixing, *J. Phys. Chem. B*, **110**, 11607–11610 (2006).
62. R. Blossey and A. Lorke, Wetting problem instability and quantum ring formation, *Phys. Rev. E*, **65**, 0216031(1–3) (2002).
63. R. D. Deegan, O. Bakajin, T. F Dupont, G. Huber, S. R. Nagel, and Th. A. Witten, Capillary flow as the cause of ring stains from dried liquid drops, *Nature*, **389**, 827–829 (1997).
64. J. C. Meiners, H. Elbs, A. Ritzi, J. Mlynek, and G. Krausch, Absence of island–island interaction during formation of isolated Ge islands in small windows, *J. Appl. Phys.*, **80**, 2224–2228 (1996).

65. D. Morariu, N. E. Voicu, E. Schaeffer, Z. Lin, T. P. Russell, and U. Steiner, Hierarchical structure formation and pattern replication induced by an electric field, *Nat. Matter*, **2**, 48–52 (2003).

66. A. Calò, P. Stoliar, M. Cavallini, S. Sergeyev, Y. H. Geerts, and F. Biscarini, Monolayer control of discotic liquid crystal by electromigration of dewetted layers in thin film devices, *J. Am. Chem. Soc.*, **130**, 11953–11958 (2008).

67. S. Y. Chou and L. Zhuang, Lithographically induced self-assembly of periodic polymer micropillar arrays, *J. Vac. Sci. Technol., B*, **17**, 3197–3202 (1999).

68. Z. Lin, T. Kerle, E. Schaeffer, U. Steiner, and P. Russell, Electric field induced dewetting at polymer/polymer interfaces, *Macromolecules*, **35**, 6255–6262 (2002).

69. R. Rayleigh, On convection currents in a horizontal layer of fluid, when the higher temperature is on the under side, *Philos. Mag.*, **32**, 529–546 (1916).

70. H. Benard, Les tourbillons cellularies dans une nappe liquide, *Rev. Gen. Sci. Pure Appl.*, **11**, 1261–1309 (1900).

71. E. Schaeffer, T. Thurn-Albrecht, T. P. Russell, and U. Steiner, Electrically induced structure formation and pattern transfer, *Nature*, **403**, 874–877 (2000).

72. J. Z. Wang, Z. H. Zheng, H. W. Li, W. T. S. Huck, and H. Sirringhaus, Dewetting of conducting polymer inkjet droplets on patterned surfaces, *Nat. Matter*, **3**, 171–176 (2004).

73. B. K. Yoon, J. Huh, H. C. Kim, J. M. Hong, and C. Park, Ordered patterns of microimprinted bilayer polymer films with controlled dewetting and layer inversion, *Macromolecules*, **39**, 901–903 (2006).

74. G. Lee, B. Yoon, H. Acharya, C. Park, and J. Huh, Ordered micropatterns by confined dewetting of an imprinted polymer thin film and their microlens application, *Macromol. Res.*, **17**, 181–186 (2009).

75. H. Kang, S. H. Lee, S. Kim, and K. Char, Dewetting and layer inversion of inverted PVP/PS bilayer films, *Macromolecules*, **36**, 8579–8583 (2003).

76. P. F. Green, Wetting and dynamics of structured liquid films, *J. Polym. Sci., Part B: Polym. Phys.*, **41**, 2219–2235 (2003).

77. F. Brochard-Wyart, G. Debregeas, R. Fondecave, and P. Martin, Dewetting of supported viscoelastic polymer films: birth of rims, *Macromolecules*, **30**, 1211–1213 (1997).

78. H. Park, T. P. Russell, and S. Park, Spatial control of dewetting: highly ordered Teflon nanospheres, *J. Colloid Interface Sci.*, **348**, 416–423 (2010).

79. Y. S. Kim and H. H. Lee, Selective dewetting for general purpose patterning, *Adv. Mater.*, **15**, 332–334 (2003).

80. A. Ghatak, R. Khanna, and A. Sharma, Dynamics and morphology of holes in dewetting of thin films, *J. Colloid Interface Sci.*, **212**, 483–494 (1999).

81. J.-L. Masson and P. Green, Hole formation in thin polymer film: a two-stage process, *Phys. Rev. Lett.*, **88**, 205504(1–4) (2002).

82. A. Calò, P. Stoliar, F. C. Matacotta, M. Cavallini, and F. Biscarini, Time-temperature integrator based on the dewetting of polyisobutylene thin films, *Langmuir*, **26**, 5312–5315, (2010).

83. C. Redon, F. Brochard-Wyart, and F. Rondelez, Dymanics of dewetting, *Phys. Rev. Lett.*, **66,** 715–718 (1991)

84. M. Cavallini, A. Calò, P. A. Stoliar, F. Biscarini, and F. C. Matacotta, Italian patent: MI2009A000208 of 17/02/2009.

85. M. Cavallini, C. Albonetti, and F. Biscarini, Nanopatterning soluble multifunctional materials by unconventional wet lithography, *Adv. Mater.*, **21**, 1043–1053 (2009).

86. E. Kim, Y. Xia, and G. M. Whitesides, Polymer microstructures formed by moulding in capillaries, *Nature*, **376**, 581–584 (1995).

87. Y. Xia and G. M. Whitesides, Soft lithography, *Angew. Chem., Int. Ed.*, **37**, 551–575 (1998).

88. D. Myers, *Surfaces, Interfaces, and Colloids*. New York: VCH. 87 (1991).

89. M. Dong, F. A. Dullien, and I. Chatzis, Imbibition of oil in film form over water present in edges of capillaries with an angular cross section, *J. Colloid Interface Sci.*, **172**, 21–36 (1995).

90. E. Bystrenova, M. Facchini, M. Cavallini, M. G. Cacace, and F. Biscarini, Multiple length-scale patterning of DNA by stamp-assisted deposition, *Angew. Chem., Int. Ed.*, **45**, 4779–4782 (2006).

91. G. Reiter, Unstable thin polymer films: rupture and dewetting processes, *Langmuir*, **9** 1344–1351 (1993).

92. D. Beysens and C. M. Knobler, Growth of breath figures, *Phys. Rev. Lett.*, **57**, 1433–1436 (1986).

93. M. Cavallini and F. Biscarini, Nanostructuring conjugated materials by lithographically controlled wetting, *Nano Lett.*, **3**, 1269–1271 (2003).

94. M. Cavallini, M. Murgia, and F. Biscarini, Patterning a conjugated molecular thin film at submicron scale by modified microtransfer molding, *Nano Lett.*, **1**, 193–195 (2001).

95. M. Cavallini, F. Biscarini, J. Gomez-Segura, D. Ruiz, and J. Veciana, Multiple length scale patterning of single-molecule magnets, *Nano Lett.*, **3**, 1527–1530 (2003).

96. D. Ruiz-Molina, Ph. Gerbier, E. Rumberger, D. B. Amabilino, I. A. Guzei, K. Folting, J. C. Huffman, A. L. Rheingold, G. Christou, J. Veciana, and D. N. Hendrickson, Characterisation of nanoscopic [Mn12O12(O2CR)(16)(H2O)(4)] single-molecule magnets: physicochemical properties and LDI- and MALDI-TOF mass spectrometry, *J. Mater. Chem.*, **12**, 1152–1161 (2002).

97. M. Cavallini, M. Facchini, C. Albonetti, and F. Biscarini, Single molecule magnets: from thin films to nano-patterns, *Phys. Chem. Chem. Phys.*, **10**, 784–793 (2008).

98. J. R. Friedman, M. P. Sarachik, J. Tejada, and R. Ziolo, Macroscopic measurements of resonant magnetization tunnelling in high spin molecules, *Phys. Rev. Lett.*, **76**, 3830–3836 (1996).

99. W. Wernsdorfer and R. Sessoli, Quantum phase interference and parity effects in magnetic molecular clusters, *Science*, **284**, 133–135 (1999).

100. W. L. Cheng, N. Y. Park, M. T. Walter, M. R. Hartman, and D. Luo, Nanopatterning self-assembled nanoparticle superlattices by moulding microdroplets, *Nat. Nanotechnol.*, **3**, 682–690 (2008).

101. O. D. Velev, Self-assembly of unusual nanoparticle crystals, *Science*, **312**, 376–377 (2006).

102. M. Cavallini, F. Biscarini, A. Farran-Morales, M. Massi, D. A. Leigh, and F. Zerbetto, Spontaneous fabrication of microscopic arrays of molecular structures with submicron length scales, *Nano Lett.*, **2**, 635–639 (2002).

103. J. Ng Lee, C. Park, and G. M. Whitesides, Solvent compatibility of poly(dimethylsiloxane)-based microfluidic devices, *Anal. Chem.*, **75**, 6544–6554 (2003).

104. W. Clegg, C. Gimenez-Saiz, D. A. Leigh, A. Murphy, A. M. Z. Slawin, and S. J. Teat, "Smart" rotaxanes: shape memory and control in tertiary amide peptido[2]rotaxanes, *J. Am. Chem. Soc.*, **121**, 4124–4129 (1999).

105. M. Massi, M. Cavallini, and F. Biscarini, Influence of the substrate hydrophilicity on the grid assisted deposition of tris-(8-hydroxyquinolinato) aluminum(III) thin films, Surf. Sci., 603, 503–506 (2009).

106. A. Curioni, M. Boero, and W. Andreoni, Alq(3): ab initio calculations of its structural and electronic properties in neutral and charged states, *Chem. Phys. Lett.*, **294**, 263–271 (1998).

107. R. H. Friend, R. W. Gymer, A. B. Holmes, J. H. Burroughes, R. N. Marks, C. Taliani, D. D. C. Bradley, D. A. Dos Santos, J. L. Brédas, M. Lögdlund, and W. R. Salaneck, Electroluminescence in conjugated polymers, *Nature*, **397**, 121–128 (1999).

108. M. Brinkmann, F. Biscarini, C. Taliani, I. Aiello, and M. Ghedini, Growth of mesoscopic correlated droplet patterns by high-vacuum sublimation, *Phys. Rev. B, Rapid Commun.*, **61**, R16339–R16342 (2000).

109. M. Brinkmann, S. Graaf, and F. Biscarini, Mechanism of nonrandom pattern formation of polar-conjugated molecules in a partial wetting regime, *Phys. Rev. B*, **66**, 165430(1–8) (2002).

Chapter 15

Directed Assembly of Particles through Convective Flows and Capillary Forces

François-Damien Delapierre,* Jean-Louis Viovy, and Laurent Malaquin*****

Institut Curie/CNRS/UPMC UMR168, 26 rue d'Ulm, 75 005 Paris, France

*françois-damien.delapierre@curie.fr, **jean-louis.viovy@curie.fr, and ***laurent.malaquin@curie.fr

15.1 Introduction

Significant progress has been made in the world of colloid and cluster science in the last decades. The ability to synthesize nanometer- to micrometer-sized building blocks with controllable shape, material, functionalities, and size plays a key role in the development of nanosciences and guarantees a continuation in the miniaturization of functional devices. Although many of these particles have been known for a long time, the development of new engineering methods of synthesis and characterization has opened the door to new disciplines.[1–7] Indeed nanoparticles exhibit unique functions and properties (electronic, optical, or catalytically properties) originating from their small dimensions

Nanoscale Liquid Interfaces: Wetting, Patterning, and Force Microscopy at the Molecular Scale
Edited by Thierry Ondarçuhu and Jean-Pierre Aimé

ISBN 978-981-4316-45-3 (Hardcover), 978-981-4364-48-5 (eBook)
www.panstanford.com

and that differ strongly from those of bulk materials. Nevertheless, one of the main remaining challenges in the development of devices that integrate such micro- and nanoparticles is still their one-dimensional (1D), two-dimensional (2D), or three-dimensional (3D) spatial arrangement on surfaces. Some of the most intriguing properties do not appear at the single particle level but only when particles are arranged into well-ordered structures. For instance, periodic particle arrays can provide interesting optical, magnetic, or transport properties that are essential to develop photonic structures,[8-10] chemical sensors,[11] or data storage devices.[12] In such applications, the symmetry and packing quality of particles will impact directly performances of the devices. Other applications, including electronics,[13] plasmonics,[14-16] and biology,[17-20] also rely either on the accurate placement or particle on a surface or on the specific properties that arise due to coupling effects in regularly spaced arrays of individual objects.

Self-assembly is a fundamental mechanism that could be defined as the autonomous organization of building blocks into ordered structures to form a material. Self-assembly results from the interaction forces that appear spontaneously between particles themselves or between particles and surfaces. In order to control the placement of particles at a desired location or to drive the formation of regular arrays with specific packing symmetry, self-assembly is often combined with templates with chemical, electrostatic, magnetic or topographic patterns. The introduction of one or several directing forces allows the user to benefit from spontaneous and parallel construction mechanisms, while ensuring structure fabrication according to coding instructions.

This so-called directed assembly method has received an increasing interest during the last decade.[21-24] A large variety of directing forces has been investigated, such as magnetic forces, electrostatic forces, hydrodynamic forces, gravitational forces, and chemical interactions. Here, we will focus our study on a set of methods involving convective or capillary forces, which are well suited for the assembly of particles, starting from colloidal suspension. Both are using the confinement effect induced at the three-phase contact line of a droplet that is dragged over a substrate. These assembly mechanisms called convective assembly[25-29] and capillary assembly[23,30,31] are based on the drying of a colloidal suspension and thus must be performed under controlled evaporation conditions. They both differ from the wetting properties of the assembly substrate: considering aqueous suspensions, convective assembly takes place on wetting hydrophilic surfaces, while capillary assembly is encountered on hydrophobic ones. All along this study, we will illustrate the applicability of these assembly mechanisms for the fabrication of 1D, 2D, and 3D structures and their compatibility with a large range of micro and nanoparticles.

15.2 Assembly Mechanisms

15.2.1 *Using Convective Flows for the Growth of 2D Layers: Convective Assembly*

Convective assembly is usually obtained on substrates with good wetting properties, that is, for contact angle values below 20° (Fig. 15.1a). The assembly mechanism relies on the convective flow of a solvent induced by the evaporation of a droplet close to the meniscus.[28,29,32] From this evaporation a resulting hydrodynamic flow drags the suspended particles toward the contact line. The profile of the liquid meniscus at the three-phase contact line leads to the formation of a very thin film of solvent. This causes a vertical confinement of the particles. The assembly process starts when the thickness of the solvent layer becomes approximately equal to the particle size.[29] The combined effects of i) convective flow and ii) attractive capillary forces that arise when the top of the particles protrude from the solvent layer during drying result in the compaction of beads into closely packed arrangements. The formation of extended layers can be initiated by displacing the relative position of the meniscus, that is, by dragging the liquid droplet over the sample.

An example of an experimental setup for such an experiment will be described in section 15.2.3. On a flat surface, convective assembly leads spontaneously to extended ordered layers or multilayers (Fig. 15.2). In many applications, the control of the position and lattice parameters would be of great interest. Intuitively, combining

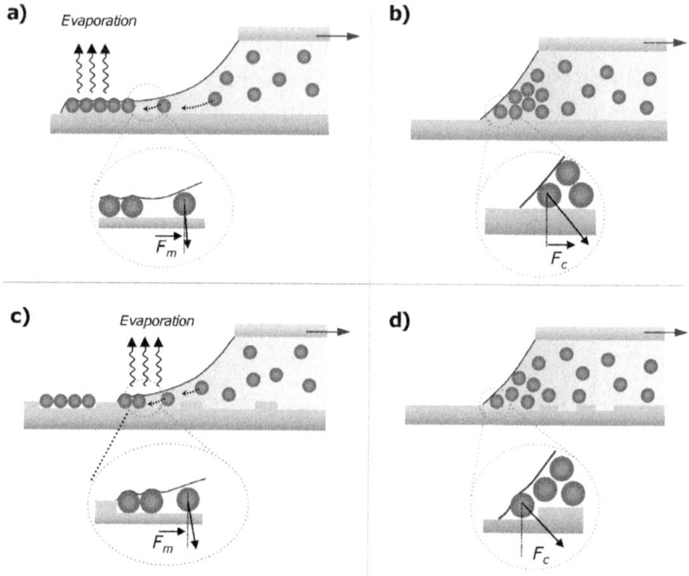

Figure 15.1 Illustration of the various assembly mechanisms based on particle confinement at the contact line of a colloidal suspension droplet. (a) Convective assembly is obtained for low contact angle values (typically <20°) and leads to the assembly of continuous 2D structures on flat surfaces or 2D discontinuous arrangements on patterned surfaces (c). Capillary assembly can be initiated for high values (typically above 60°) of the receding contact angle. Whereas no deposition occurs on flat surfaces (b), the capillary forces and the geometrical confinement resulting from meniscus pinning on the structures results in single- or multiple-particle deposition (d).[33]

convective assembly with patterned substrates appears as a simple and straightforward solution (Fig. 15.1c). When considering the assembly mechanism, however, this idea seems incompatible with convective assembly as the introduction of structures close to the meniscus would perturb the hydrodynamic flow and thus compromise the assembly mechanism. It turns out that using shallow patterns, this approach extends the range of structures achievable and allows for the creation of 2D layers with controlled lattice parameters or discontinuous 2D structures. This approach will be discussed in section 15.3.3.

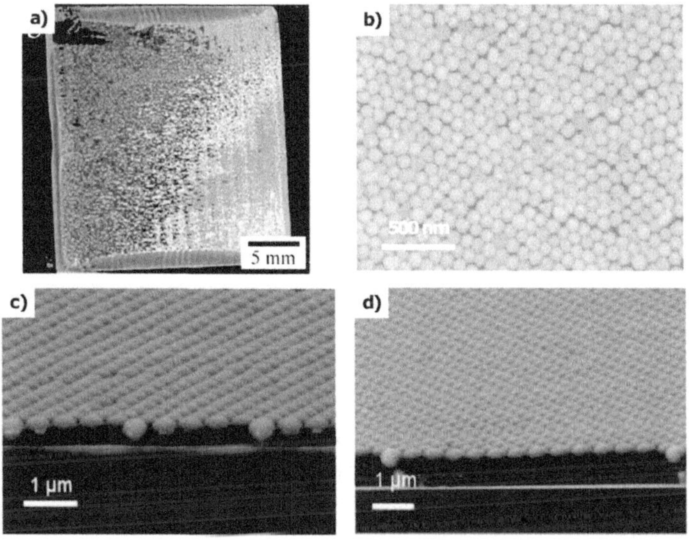

Figure 15.2 Examples of 2D layers fabricated by convective assembly. (a) Examples of a crystal monolayer obtained from 1.1 μm PS latex particles. The crystalline order leading to bright diffraction is revealed when illuminated with a white light from behind.[26] (b) SEM images of a 100 nm gold nanoparticles monolayer assembled on a silicon surface (unpublished results), a monolayer (c), and a bilayer (d) obtained with 500 nm PS particles. The structures were first assembled on a flat PDMS substrate and then transferred onto a silicon wafer by printing.[33] *Abbreviations*: SEM, scanning electron microscopy; PDMS, polydimethylsiloxane; PS, polystyrene.

15.2.2 *Using Capillary Forces for Directing Single-Particle Placement: Capillary Assembly*

As we have seen, on hydrophilic surfaces, the meniscus profile induces a vertical confinement and results in the formation of ordered 2D continuous structures. When the contact angle of the liquid suspension increases above a critical contact angle value, the horizontal force exerted by the liquid meniscus becomes high enough to prevent particles from depositing onto a flat substrate (Fig. 15.1b). On patterned surfaces, however, this force component can be countered by a vertical structure: while the contact line is being dragged over the patterned substrate, the liquid meniscus is pinned and deforms around the structure.

The combined effects of capillary forces (resulting from the local distortion of the meniscus when the contact line is dragged over the structures) and geometrical confinement (induced by the structures) causes one or more particle being trapped close to the obstacle, whereas no particles are deposited in the surrounding areas (Fig. 15.1d). This capillary assembly mechanism is well suited for accurate placement of individual particles to create discontinuous 1D structures (Fig. 15.3) and can be extended to 3D patterns.

The exact mechanism by which particles are immobilized on a patterned surface is quite complex. It is generally assumed that at the instance when the meniscus breaks up, the capillary forces created can overcome the thermal fluctuations as well as the other repulsive forces exerted on the particles.[30,23] Latest experiments suggest that these capillary forces are important for but not sufficient to explain particle trapping.

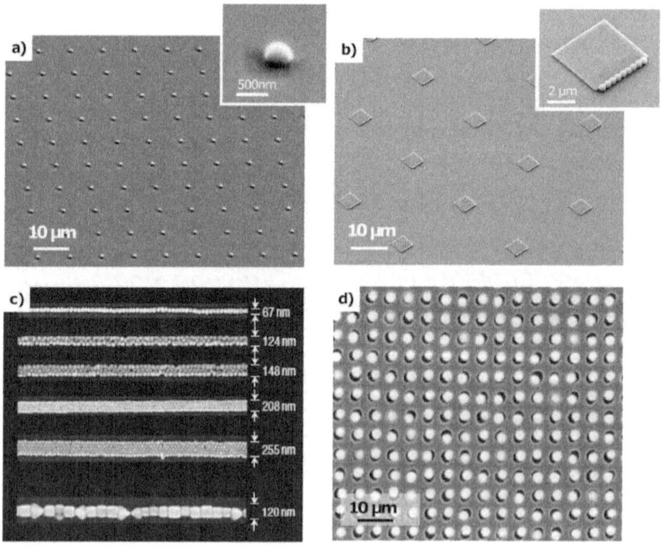

Figure 15.3 Examples of 500 nm PS particles (a and b) and 60–80 nm gold particles immobilized on a PDMS template using capillary assembly.[33–35]

15.2.3 *Experimental Approaches*

The choice of the experimental setup is determinant and has to be done according to the complexity of the structure to be achieved. Many simple methods can provide straightforward routes for the manufacturing of extended 2D layers through convective assembly. However, the control of the precise positioning of objects in the capillary assembly mode is more restrictive and usually requires a more accurate control of the evaporation rate and/or control of the meniscus velocity. Systems that are capable of decoupling these two parameters are more flexible and offer versatile platforms assembly methods. The objective of this paragraph is to present an overview of the systems reported in literature and to help the reader to select the approach the more appropriate to its application with a balanced investment/performance ratio.

First demonstrations of the convective assembly mechanism were performed on static evaporating droplets.[27,36–38] In most experiments using this simple but still efficient approach (Fig. 15.4), the control of the shape of the meniscus is usually used to define the number of layers of the assembled 2D array.[27,28,39–42]

Spin coating offers also a simple way of preparing 2D crystalline layers of particles on surfaces.[43] The spin velocity can be adjusted to control the number of layers, but the lack of homogeneity of the assembled structure remains a limiting factor in this method.

In order to improve the control on the evaporation rate, strategies relying on enclosed cells were proposed. Such cells allow to tune the evaporation rate of the liquid either in a passive way by reducing vapor exchange with the environment and/or by controlling the temperature of the environment itself.[23,45–50] In both cases, the goal is to stabilize and to get a control over the velocity of the meniscus. Such a method is thus well adapted to capillary assembly where the control of the dewetting speed plays a key role in the assembly yield. However, it does not provide a direct tuning of the velocity, which is a main limitation in regard with the yield of the process (see section 15.1.3). Similar approaches were also integrated in microfluidic device channels.[55–59] The assembly of particles in such confined geometries provides efficient ways of creating 3D ribbons that can be used in optical (e.g., wave

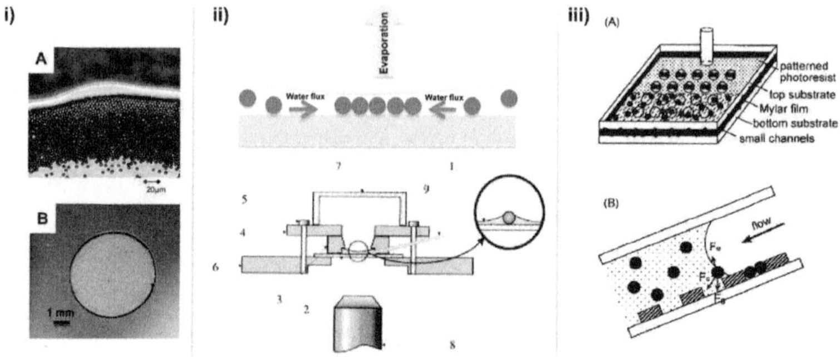

Figure 15.4 (i) Example of ring stain demonstration. (A) The edge of a suspension drop of 100 nm fluorescent and 5 μm white particles. During the drying process, particles concentrate at the edge of the drop.[36,88] (B) When the drying is completed, this specific concentration on the edge leads to the formation of a ring. Here is another example—a typical coffee stain. (ii) Schematic illustration of the process and experimental setup used by Denkov *et al.* for the assembly of particles entrapped in wetting films.[28,86] (iii) Schematic outline of the procedure used by Xia *et al.* to assemble colloidal particles into well-defined aggregates in a fluidic cell. Balance between gravitational, electrostatic, and capillary forces exerted by the meniscus leads to a selective deposition of particles in the template patterns.[23]

guides)[10,51,57,58] or biological applications (e.g., separation in a sieving matrix).[49,59]

Substrate dipping offers a better control of dewetting and seems to be well adapted to capillary assembly (Fig. 15.5a). It is usually derived from Langmuir-Blodgett methods. The template is usually placed vertically and dipped directly into the colloidal suspension.[32,51–53,55,56,60] By means of a translation stage, the sample is then vertically withdrawn from the solution. By adjusting the speed of the downstroke and upstroke of the substrate through the air–water interface, the transfer of a single layer of particles at each cycle can yield 3D colloidal crystals. An additional control of the evaporation rate can be done by setting the whole suspension temperature or by performing the assembly in low vacuum.[30,61] Nevertheless, the tuning of the evaporation is restrictive, as it requires a control of the whole environment.

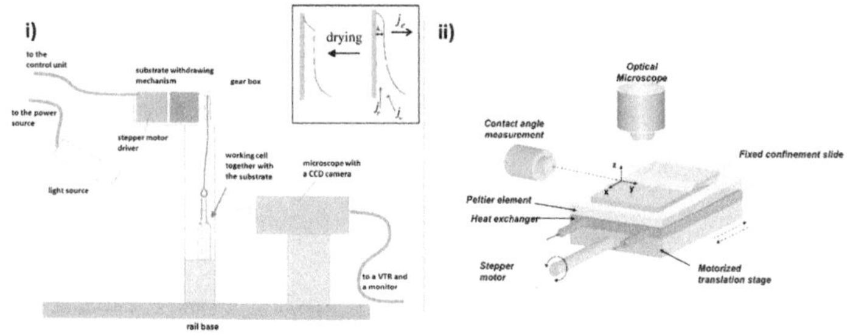

Figure 15.5 Schematic illustrations of the experimental setups used for convective and capillary assembly (i) by withdrawing a template substrate immerged vertically in a colloidal suspension[32] and (ii) by dragging a droplet of a colloidal suspension on a thermalized template using a translation stage.[33]

Finally, methods that are capable of decoupling control of the evaporation rate from the velocity of the liquid suspension are probably the most versatile solutions for convective and capillary assembly even though they are the most demanding ones. Most approaches use horizontal translation stage systems to withdraw a droplet of a liquid suspension at a well-controlled velocity (Fig 15.5.ii). This configuration is advantageous: i) it can be easily mounted on a microscope stage for process monitoring, ii) it allows a simple integration of temperature control systems (e.g., peltier elements), and iii) time and material consumption can be drastically decreased in comparison to conventional dip coating as it involves small volume of concentrated colloidal suspensions (down to 20 µL).[30]

However, due to the small volume involved in such a configuration, the assembly process is very sensitive to the evolution of concentrations as evaporation goes on. One possible implementation of the technique may be to supply continuously the suspension via an injection line to the meniscus, keeping the volume fraction exactly constant.

Several solutions were proposed in literature to improve the stability of the process. One solution relies on the control of the humidity rate around the sample.[26] Diguet *et al.* reported an alternative approach where the evaporation rate of the solution was driven by aspiration system pumping air around the meniscus at a controlled flow rate.[62] This approach was coupled to a temperature regulation system and showed to improve the evaporation stability over time and its homogeneity over large surfaces.

15.3 Convective Assembly

Controlling the mass transfer of particles toward the meniscus is fundamental to create extended continuous layers. Indeed, as the layer of particles grows, the substrate is withdrawn together with the already formed layer. The condition to build a continuous layer is to make the withdrawal speed v_s equal to the speed of layer formation v_c. For steady-state assembly, a simple equation describing the speed of layer formation depending on solvent flux and accumulation of particles in the drying region was proposed by Dimitrov and Nagayama.[32]

$$v_c = \frac{\beta j_e l \phi}{h(1-\varepsilon)(1-\phi)} \qquad (15.1)$$

This equation establishes a relation between the growth velocity of the layer v_c, the porosity ε, the height of the assembled layer h, the volume fraction ratio of the particles in the suspension φ, a coefficient that relates the solvent velocity to the particle velocity $\beta(0 < \beta < 1)$, and the evaporation rate of pure water j_e. The variable $l(x)$ is defined as $l = J_{evap}/j_e(x)$, where J_{evap} is the integral of water evaporation over the x-axis and $j_e(x)$ is the evaporation rate at any point of the drying region (Fig. 15.6a). The x-axis is parallel to the substrate and perpendicular to the contact line.

Equation 15.1 shows that three key process parameters can be used to control the coating thickness and thus its structure: (i) the particle volume fraction Φ, (ii) the solvent evaporation rate J_{evap}, and (iii) the substrate velocity v_s This latter parameter does not appear explicitly in Eq. 15.1 which refers only to a steady state situation. v_s is an experimental parameter. Its value should be set close to the v_c value to ensure a stable assembly process. For a given evaporation rate and volume fraction, a uniform crystalline monolayer with a defined thickness is obtained if $v_s = v_c$ so that the particle influx toward the meniscus exactly compensates

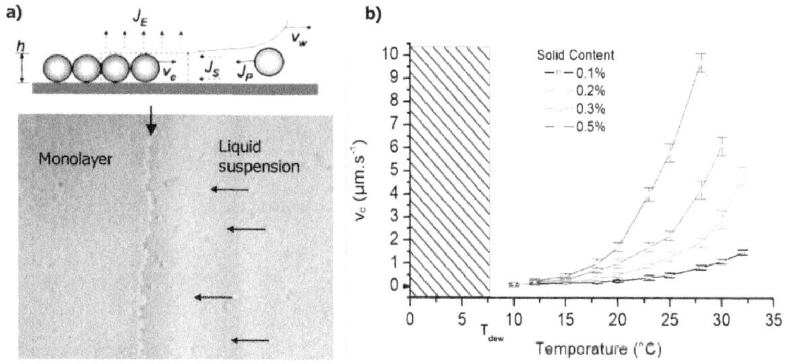

Figure 15.6 (a) Typical optical image acquired during the assembly of 500 nm PS particles (0.2% solid content, $T_s \approx 18°C$) on a flat PDMS substrate treated with oxygen plasma. The left-hand side of the image shows the assembled layer, whereas the right-hand side shows the liquid suspension of particles. A schematic illustration of the evaporation process mechanism in convective assembly is given.[26] (b) Temperature and solid content dependence of the convective assembly speed (obtained for 500 nm PS particles).[33]

for the particle consumption due to layer growth. Increasing the substrate velocity above the natural assembly speed would result in an incomplete layer as the assembly speed becomes lower than the meniscus velocity. In contrast decreasing the substrate velocity would result to an accumulation of nonassembled particles close to the meniscus and thus to the formation of multiple layers. At a given substrate velocity $v_{S,}$ the same effect would be observed by decreasing or increasing the volume fraction of particles, respectively. Besides substrate velocity and volume fraction, the evaporation rate J_{evap} of the liquid is the driving mechanism in the assembly process and influences directly the growth velocity. The exact relationship between these parameters will be discussed in the next paragraphs.

Figure 15.6a shows an example of 2D layers obtained with PS particles on an oxidized PDMS surface. Because of their narrow size distribution (1.3% coefficient of variation), the assembly of such calibrated particles gives rise to a dense, hexagonally packed monolayer (Fig. 15.6b). Uniform layers with areas of a few square centimeters can be prepared using this method.[26,32,33,63] It is important to notice that the whole coating is made of hexagonal crystalline domains of different orientations. The domain size of a typical crystal in the monolayer is typically in the range of 100–250 μm. Indeed, during the crystallization process, the assembled layers located close to the meniscus remains wetted by the solvent. The drying length, l, defines the distance over which the liquid film penetrates into the porous colloidal crystal before it is completely evaporated. Far from l, as the layer is drying, the volume loss due to solvent evaporation induces defects such as dislocations. While the quality of the layers is slightly better at lower v_c values, it seems that there is no noticeable influence of the humidity rate on the formation of dislocations or on the number of vacancies (around $2.5 \times 10^2/\text{mm}^2$, i.e., 100 ppm).[33]

15.3.1 *Controlling the Growth Speed: Evaporation Rate and Temperature Dependence*

The evaporation rate around the drying region can be influenced both by the temperature of the colloidal suspension close to the meniscus and by the relative humidity of the environment. These two experimental approaches were explored in literature either by controlling the substrate temperature or by controlling both the environment humidity and the temperature. Nevertheless, control of the assembly process through humidity has shown to be experimentally difficult, and studies reported in literature lead to differing conclusions.[26]

Acting on temperature appears as a more simple, efficient and reactive way to control J_{evap}. Figure 15.7 shows an illustration of the temperature dependence of the convective flows. In this experiment, the substrate temperature was used to control the evaporation rate J_{evap}. Experiments were carried on using 500 nm diameter PS particles suspended in water. The assembly was performed on oxidized PDMS surfaces treated with oxygen plasma.

Figure 15.6b shows the evolution of the assembly rate as a function of temperature in the range $T_{dew} < T < 30°C$, where T_{dew} is the dew point temperature. The assembly rate was measured by adjusting the substrate velocity in order to keep the edge of the assembled layer fixed relative to the observation field.

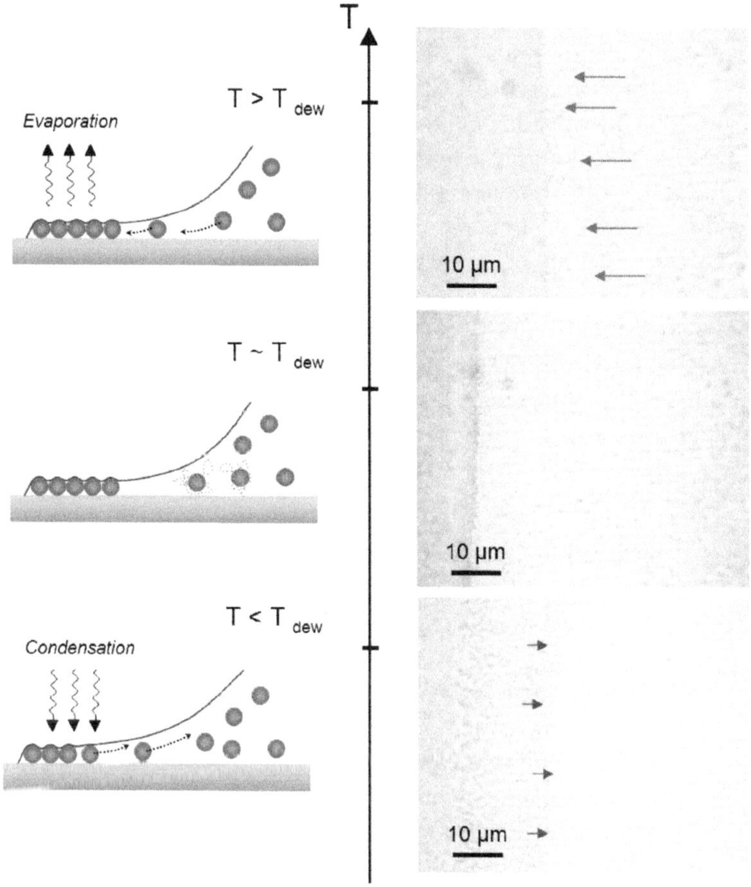

Figure 15.7 Influence of the evaporation rate on the convective flow involved in convective assembly. Experiment was performed on an oxygen plasma-treated PDMS surface (500 nm PS particles, 0.2% solid content).[33]

This method allows for the creation of a uniform monolayer and provides an estimation of the growth rate of the layer since $v_c \approx v_S$. The experiment was performed for different solid content values Φ ranging from 0.1% to 0.5%. The critical influence of Φ on the assembly rate is clearly observable on the curves, as demonstrated in previous studies.[26,32] According to Eq. 15.1, v_c increases nonlinearly together with Φ.

The assembly process has shown to be very sensitive to the substrate temperature. It is possible to speed up the assembly process by a factor 10 by switching temperature from 10°C to 30°C. The investigation of high temperatures (typically above 30°C) remains difficult because of the high evaporation rate that modifies quickly the solid content of the suspension. The use of larger droplet volume or the development of a continuous liquid supplying system would be a possible way for extending the application range of the method.

Investigation of the influence of temperature was also carried out for temperature below the dew point. Figure 15.7 shows a typical example of optical images obtained during the assembly of 500 nm PS particles corresponding to three different regimes $T > T_{\text{dew}}$, $T \approx T_{\text{dew}}$, and $T < T_{\text{dew}}$.

- $T > T_{\text{dew}}$: Above the dew point, convective flows, caused by evaporation, lead to particle influx toward the contact line.
- $T \approx T_{\text{dew}}$: Decreasing temperature induces an assembly slowdown. Near the dew point, solvent evaporation in the regions located close to the substrate is almost null. As no particle influx is created, the assembly process is stopped. The mean position of particles is fixed and only Brownian motion is observable.
- $T < T_{\text{dew}}$: When temperature decreases further, condensation takes over from evaporation. As a consequence water now flows in from the former drying zone back toward the bulk liquid. This causes a progressive disassembly of the particles. The assembly process has thus been shown to be fully reversible.

The existence of three different regimes, namely, assembly, stagnation, and disassembly, and the reversibility of the whole mechanism are clearly a convenient feature in convective assembly.

The three regimes can be directly accessed by temperature control, which is experimentally easy to implement. This provides the possibility to not only switch on and off the assembly process but also allow to modify or repair an already existing layer (e.g., removing defects such as voids or switching back multilayers to monolayers).

15.3.2 *Controlling Multiple-Layer Growth*

The control of the film thickness and in particular the adjustment of the number of assembled layers may have critical influence for many applications since it influences directly the structural optical or electrical properties of the film.[26,63–65] According to Eq. 15.1, the coating thickness h, which directly reflects the number of stacked layers, is a function of the particle solid content, the substrate velocity and the evaporation rate as well. Whereas many studies report on the control of the solid content as a driven parameter, it seems that, once again, the adjustment of the evaporation rate may provide more flexibility.

Considering a given solid content and a given temperature (i.e., a given evaporation rate J_e), the withdrawal speed of the substrate can be used to switch from a single layer to multiple layers.

At the equilibrium, the particle influx at the contact line compensates exactly for layer formation. Increasing the deposition speed above the natural assembly speed would result in incompletely filled layers (also referred as submonolayers). However, decreasing the withdrawal velocity below this value would lead to the formation of multilayers: the withdrawal speed does not compensate for monolayer formation, and stacking takes place.[26,32,33] It is obvious that the transition between single- and multiple-layer growth results in a significant decrease of the assembly speed. Here also, it is possible to take advantage of temperature control to compensate for this assembly slowdown by increasing the evaporation rate. As an illustration, Fig. 15.2c,d shows side-view images of a single layer and a bilayer prepared from a 500 nm PS particle suspension (0.2% solid content). In this experiment, the substrate velocity was fixed to 0.6 μm.s^{-1}, while the substrate temperature was switched from 20°C to 26.5°C to switch from a stable single-layer formation regime to a double-layer one.

The number of layers, N, can be associated to color changes that can be observed through optical microscopy (Fig. 15.8b). Indeed, a modification in the layer thickness induces an abrupt color change due to interference phenomena occurring between the light reflected from the top and bottom interfaces.

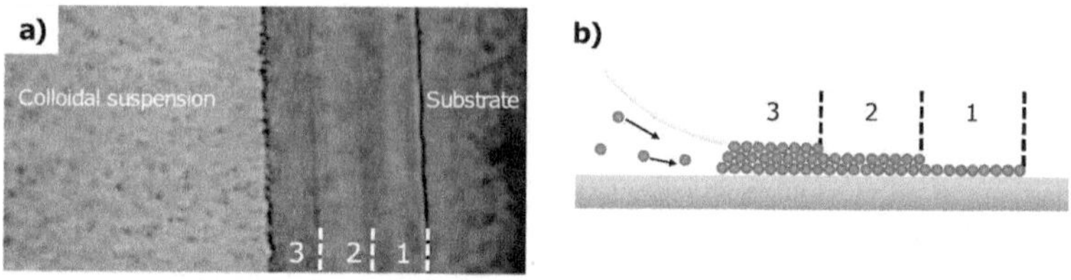

Figure 15.8 Illustration and optical images showing the assembly of 500 nm PS particle multilayers. The structures were first assembled on a flat PDMS substrate and then transferred onto a silicon wafer by printing (unpublished results).

15.3.3 *Controlling the Layer Structure*

15.3.3.1 Influencing the crystal growth mechanism

The assembly of 2D layers by convective phenomena leads, for an assembly speed equal to the natural growth speed of the crystal, to polycrystalline coating with crystal domains of different orientations. Indeed, particles tend to maximize the packing density and thus to form naturally hexagonal packing. Nevertheless, it has been shown that the deposition speed and the particle volume fraction can be used to influence the packing structure. Quite surprisingly, the evaporation rate was not found to affect the process to a large extent.

The influence of the balance between evaporation rate and deposition speed on the lattice properties was carefully studied by Prevo *et al.*[26] They could establish phase diagrams relating the crystal layer thickness and packing symmetry to the process parameters. It was demonstrated that during multilayer growth, the symmetry of the multilayer was not always hexagonal but was alternating between hexagonal close-packed (hcp) and square as the coating thickness was changing. Experiments have also shown the difficulty to stabilize square-packed layers over long distances. First, particles tend to maximize packing density and thus switch back to hexagonal packing. Second, the operational phase diagram that was established showed that the operating parameters for assembling square lattices were very close to those leading to hexagonal ones. The window of opportunity for stable growth was much more restricted than for hexagonal layers.

Note: The influence of the assembly speed on the layer quality is not clearly established. It seems that while the homogeneity and overall quality is slightly better for low v_c values, the number of voids (typically values around 100 ppm) or dislocation cannot be directly coupled to this parameter.

15.3.3.2 Directing the assembly using templates

Convective assembly can be combined with templated substrates either to define areas of assembly or to modify the structural properties of the assembled layers.

15.3.3.2.1 Localized assembly

The first way is to perform the assembly into confined geometries such as microchannels.[55–59] The structure and packing of the crystal can be dictated by the ratio between particle and channel dimensions.[66] Approaches relying on wetting contrast or chemical interactions have also been proposed to create continuous patterns such as stripes.[53,60] It is important to notice that the use of localized interactions limits the mobility of particles in the vicinity of the surface and thus their capacity to self-organize into compact and crystalline layers.

The use of patterned templates with small-dimension structures is usually not associated to convective assembly: indeed, the introduction of topology may disturb the assembly process as each pattern can potentially behave as an obstacle that impedes particles motion toward the contact line. Nevertheless some experiments have demonstrated that the use of shallow patterns preserves the convective assembly mechanism as long as i) the liquid film above the structures is high enough and ii) the pattern density is sufficiently low to allow particles to move around the patterns. Figure 15.9 shows an example of structures assembled on a PDMS layer.

15.3.3.3 Directing the crystal orientation

The use of shallow patterns with lateral size comparable to the dimension of particles is a way to direct the position of particles individually and thus to dictate the parameters for crystal growth.[67]

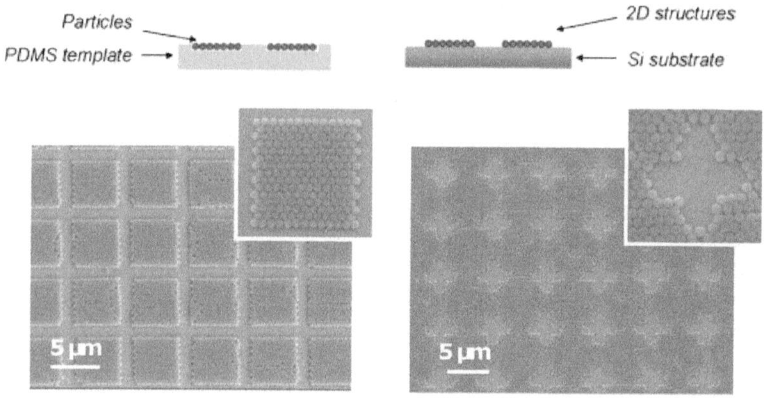

Figure 15.9 Illustration of the templated convective assembly mechanism: 500 nm PS particles were assembled on a PDMS substrate (T_s = 20°C; v_s = 1.5 µm.s^{-1} with 200 nm deep patterns) (unpublished results). The assembled structures were transferred on a silicon wafer by printing.[23,26]

Xia *et al.* have developed a simple and convenient procedure for the growth of large colloidal crystals with predetermined crystal orientation using a 2D patterned surface.[45,68] In their work, they have shown the construction of (100)-oriented crystal assemblies on regular arrays of square pyramidal pits or V-shaped grooves etched in the surfaces of Si(100) substrates. They have demonstrated the crystallization of spherical colloids (>250 nm in diameter) over areas as large as several square centimeters. Figure 15.10 shows an example of a structure obtained with this approach. The confinement of particles is dictated by the geometric shape of the pyramidal pits, leading to cubic close-packed (ccp) lattices oriented with their (100) planes parallel to the surface of the substrate.

Such templates have been successfully applied to spherical colloids with diameter ranging from 300 nm to several microns. For particles smaller than 300 nm, the relative weak templating influence of the (100) planes and the strong Brownian motion usually resulted in the growth of (111) planes having a higher packing density. Such structures could only generate a small domain size (typically ≈100 µm across) due to the lack of tight registration between the crystals nucleated simultaneously at different locations

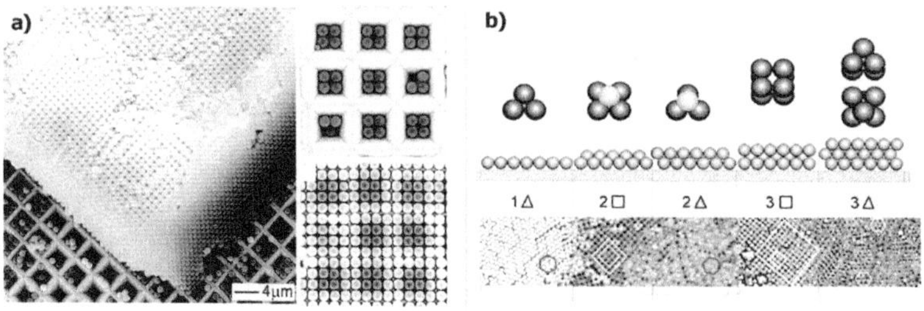

Figure 15.10 (a) SEM images of 1 μm PS beads on the surface of a template containing an array of 4 μm wide pyramidal pits. Assembly was performed by evaporating the solvent slowly at room temperature.[45,68] (b) Evolution of the packing symmetry in 2D multiple layers of 1.1 μm PS particles as a function of the number of layers.[26]

on the substrate. Similarly, another limitation can also be found when using structure dimensions too small compared to the dimension of the particles. Indeed, as the confinement effect of the structures is weak, capillary forces appearing during dewetting are dominating and the effect of the template is lost over long distances.

15.3.4 *Influence of Particle Size and Mobility*

To be effective, convective assembly requires free suspended particles that can be easily dragged by hydrodynamic forces. In the previous paragraphs, we have already shown that the convective assembly mechanism is perfectly compatible with PS particles in the micrometer range.

Figure 15.2b shows another example assembly carried out with 100 nm Au particles assembled on the surface of an oxidized PDMS sample. Uniform monolayers could be created over large surfaces ($>5 \times 5$ mm^2). Even if the particle size and shape distribution is broad and not adapted for the creation of crystalline arrays of particles, the assembly process remains useable to produce layers with controlled thickness.

However, contrary to capillary assembly (see the next section), the assembly of particles sedimenting quickly compared to the duration of the experiments (<10 min) or particles interacting strongly with surfaces is nearly impossible (results not shown).

15.4 Capillary Assembly

As we have seen, the deposition of a colloidal suspension droplet causes, on wetting surfaces (typical angle $<20°$), the convective assembly of particles into ordered 2D continuous structures. On less wetting substrates, however, the capillary force exerted by the meniscus exhibits a horizontal component that is usually sufficient to prevent particles from deposition on flat surfaces. It is possible to take advantage of the high contact value in combination with appropriately patterned surfaces. Indeed, structures that are prone to create meniscus pinning or deformation can be used to force, locally, the deposition of particles. The process by which particles are immobilized on the patterned substrate is usually a combination of geometrical confinement and capillary forces appearing while the meniscus is distorted and breaks on the features. This combination of forces overcomes thermal fluctuations or repulsive forces and causes particle deposition. Adapting the feature dimension to the particle diameter allow to deposit selectively a well-defined number of objects at a precise location.

Whereas the mechanism leading to particle immobilization was known for a long time, the parameters controlling particle deposition are numerous and their origin is not yet clear.[23,30] In the following section, we propose an overview of the main parameters experimentally accessible to control the deposition process, as well as a qualitative study of their influence on the deposition mechanism.

15.4.1 *Controlling Evaporation*

Whereas the behavior of the meniscus is different in both assembly mechanisms, evaporation plays a critical role in capillary assembly as well. Hydrodynamic flows created by evaporation tend to bring particles from the bulk suspension toward the contact line. In convective assembly, this accumulation of particles close to the contact line results directly in layer formation. In capillary assembly, even if capillary forces prevent deposition, it was recently shown that this process was essential to initiate particle deposition.[33,34]

Figure 15.11 illustrates experiments performed on a patterned PDMS surface to investigate the influence of evaporation on the yield

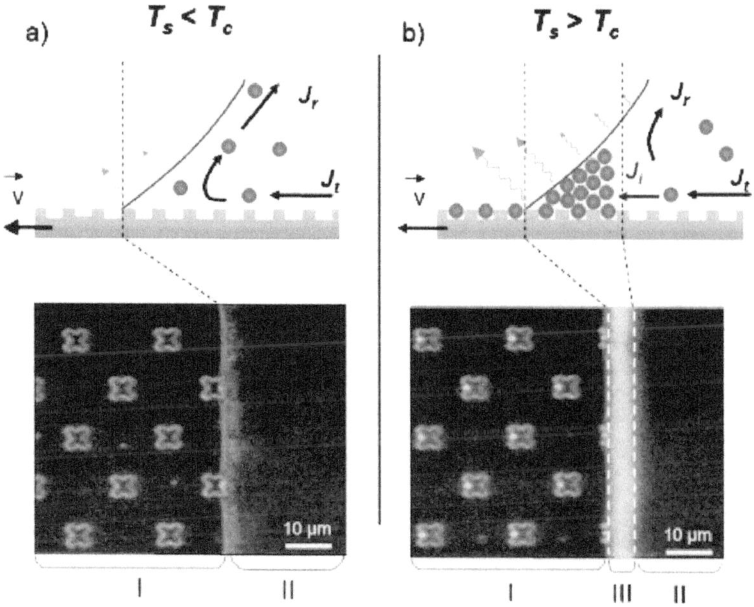

Figure 15.11 Images taken during the assembly of 500 nm PS particles on a patterned PDMS substrate (200 nm deep features) for two different temperatures, T_S, 27°C and 32°C, respectively. This experiment was carried out with a colloidal suspension of 0.1% solid content (estimated dew point: 7°C; measured contact angle ≈ 53°). (Region I) Patterned substrate; (Region II) liquid suspension. Experiments were carried out with a constant substrate velocity of 1 μm.s^{-1}. The assembly yield was measured by subsequent analysis of image sequences taken during the experiment.[33]

of the assembly process. These experiments were performed at two different substrate temperatures. At 27°C, no deposition occurs on the substrate. When increasing temperature up to 32°C, however, a selective trapping of a single particle in each cross-shaped pattern is observed (yield 99%). In addition to the immobilization process, careful investigations through optical microscopy also revealed the influence of temperature on the accumulation of particles close to contact line. This phenomenon is observable through the formation of bright opalescent region on micrographs (Fig. 15.11, region III). This observation suggests the presence of polycrystalline arrangements of particles in this region.

A close correlation between the formation of this accumulation region and the deposition of particles could be demonstrated. One hypothesis suggests that the increase of the local particle concentration is required to initiate the assembly process: i) as the local density of particles is very high, the probability of trapping particles increases, and ii) the accumulation of particles contributes to the reduction of particle fluctuation and mobility during the immobilization process and thus increases the trapping efficiency. Indeed, the self-diffusion coefficient of the particle is strongly sensitive to the volume fraction of the suspension which seems to reach a maximum close to the contact line.[69-71]

The creation of this accumulation region seems to be strongly dependent on the evaporation of the solvent and thus can be directly controlled by an appropriate tuning of the temperature close to the contact line. Indeed, as it was observed for convective assembly, evaporation induces a particle flow J_{i} from the suspension to the drying region. Accumulation of particles occurs when particle consumption through deposition does not compensate for particle income. This mechanism is illustrated in Fig 15.11 in the case of 500 nm PS particles.

Once again, it is possible to establish a temperature threshold value for the assembly mechanism. However, whereas this threshold value was matching dew point temperature in the case of convective assembly, experimental observations of capillary assembly show that this threshold value is well above the dew point value. This means that considering both the influence of particle influx due to evaporation and particle deposition is not sufficient to illustrate mass transfer balance.

Careful investigations of the particle transport phenomena close to the meniscus showed that an additional mechanism is competing with particle accumulation. Indeed, due to a no-slip condition at the surface of the substrate, a recirculation flow J_{r} of liquid is induced while the droplet is dragged over the substrate. As a consequence, because of stokes drag, particles are dragged from the meniscus region back to the suspension (See Fig. 15.11). This particle flow competes with the accumulation process created by evaporation. To start accumulation, J_{i} has to compensate for the depletion of the accumulation region created both by deposition particles on the surface and by recirculation (and eventually diffusion) of particles from the assembly region to the bulk suspension. Experimentally, this condition can be fulfilled by increasing the substrate temperature above a threshold value T_c.

According to this description, the control of the substrate temperature directly affects the balance between J_i and J_r: it is an effective way of controlling the accumulation of particles and, thus, switching the assembly process on and off. Temperature is not the only experimental parameter accessible. Indeed, the particle flow can be influenced by the initial solid content of the suspension. In addition, the recirculation flow can be tuned, to a certain extent, by adjusting the substrate withdrawal velocity.[33]

15.4.2 *Controlling Wetting Properties*

As we already said, the immobilization of particles on the surface is driven by the combined effects of capillary forces, both that drive meniscus deformation (and consequently particle confinement in the trapping structure) and that are exerted on the particles during meniscus break-up. Consequently, the assembly process must be sensitive to both the wetting properties of the colloidal suspension and the shape and dimension of the trapping structures.

We will first consider the influence of the wetting properties of the colloidal suspension relative to a given substrate. The role of surface tension on the yield of the assembly process can be easily investigated by adding a well-defined concentration of surfactants in the initial suspension. As an example, it was shown that the assembly of 500 nm PS particles on a PDMS substrate was optimum for a contact angle ranging from 30° and 60°.[a]

This result was demonstrated by monitoring the influence of adding a sodium dodecyl sulfate (SDS)/TritonX45 surfactant mixture, while keeping all other experimental parameters constant.[33] Increasing the surfactant concentration resulted in a decrease of the contact angle. For contact angle values below 15°–20°, the process switches from capillary to convective assembly below 20°. Indeed,

[a] In these experiments, contact angles were measured at a constant substrate velocity of 1 μm.s^{-1}. No significant influence of the velocity on the contact angle could be observed in the 0.5–10 μm.s^{-1} range.

in such situations, the horizontal component of the capillary forces becomes too weak to prevent particle deposition on flat areas, and the assembly of 2D structures is initiated. For values above 60°, no deposition is observed, even if accumulation of particles close to the contact line can be observed as well. This threshold value can be directly connected to the horizontal component of capillary forces that, even for deep structures, are sufficient to drag particles and keep them suspended.

The question of pattern shape and its influence on the wetting properties and trapping efficiency is more complex. In the case of a high-aspect-ratio structure, wetting properties are affected by capillary pressure, which can lead, in some extreme cases, to the conversion of capillary assembly into convective assembly. This case will be discussed in section 15.4.4.

In addition to pattern dimension and shape, the influence of pattern density has to be taken into account when considering capillary assembly. Indeed, as each pattern may induce a local pinning of the meniscus, the use of high-density structures may affect the dewetting process and thus modify the contact angle. In the case of a patterned surface, the trapping structures can be considered obstacles that will induce a local anchoring of the contact line and work against dewetting. The impact of such surface roughness on dewetting is now described in literature.[34,72] Two models can be used to predict the influence of surface roughness on the contact angle. The first model was proposed by Wenzel. It considers a surface chemically homogenous where the surface roughness corresponds to the ratio between the real surface area and its projection in the horizontal plane. This model is well adapted to patterned templates with lower-density protruding structures. A second model, named the Cassie Baxter model, considers a flat surface made of two different materials. This model can be used in the case of enclosed patterns prone to induce locally complete wetting due to capillary pressure. In this situation, the ratio between wetted and nonwetted areas is used to predict the contact angle evolution.

The applicability of these models remains limited. Indeed, they are only devoted to the calculation of static contact angles. They do not consider friction forces that will affect the dynamic dewetting process. In addition, the impact of the accumulation of particles on the meniscus motion is hard to predict. Even if these models can be used to give general tendencies, quantitative calculation remains hazardous and experimental investigations are usually essential.

15.4.3 *Tuning Meniscus Velocity*

15.4.3.1 Controlling depletion at the contact line

We have shown previously that the existence of a temperature threshold T_c results from the balance between particle influx induced by evaporation and particle depletion due to diffusion, re-circulation, and deposition of particles. The velocity v_s of the contact line is directly influencing the deposition rate and recirculation and thus the T_c value. Figure 15.12 shows the evolution of T_c in the case of 500 nm PS particles assembled on PDMS templates. The influence of the particle solid content was also added in the figure in order to illustrate the impact of mass transfer on the accumulation process.

As expected, T_c increases with the substrate velocity almost linearly. When increasing the recirculation flow, the evaporation rate has to be increased in order to compensate for particle loss in the accumulation region. This figure also shows that increasing the solid content seems to help lower T_c. Indeed, the particle influx is nearly proportional to the solid content, whereas diffusion and recirculation are not directly connected to this parameter. Increasing solid content thus results in pushing the balance toward particle accumulation.

15.4.3.2 Velocity influences contact angle

In addition to hydrodynamics considerations, the influence of the meniscus velocity on wetting in the dynamic regime has to be considered. No observable influence of v_s on the contact angle could be observed for values ranging from 0 to 10 μm.s^{-1} (e.g., for aqueous suspensions on PDMS substrates). Indeed, for low speeds, the angle values remain close to the value of the receding angle (quasistatic regime). This assumption is not true for high velocity values. As the speed increases, the friction forces and interactions between the liquid and the substrate work against meniscus motion and tend to lower the contact angle value and thus to modify capillary forces while dewetting occurs.

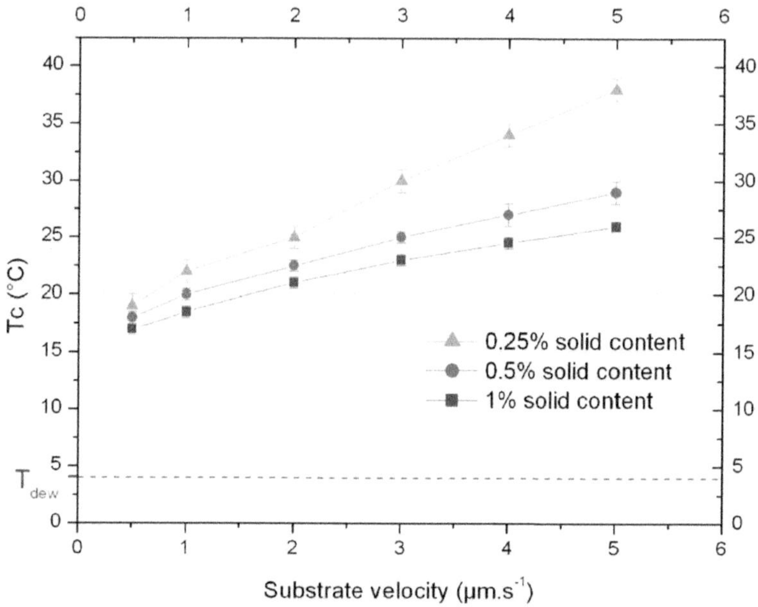

Figure 15.12 Evolution of the threshold temperature T_c as a function of velocity v_s and solid content.[33] In this experiment, the solid content of the dispersion was adjusted between 0.25% and 1%, whereas the concentration of the solutes in the continuous phase was set to 25% of the original concentration (by appropriate dilution and centrifugation steps). The substrate velocity was varied from 0.5 to 5 $\mu m.s^{-1}$.

Figure 15.13 shows a study of the assembly yield for 4.5 μm beads on a hexagonal patterned array (periodicity 30 μm) consisting of 2.5–3 μm depth cavities.[a] The evolution of the yield with both velocity and surfactant concentration has been monitored.

For high surfactant concentrations (>5%) the yield is almost stable and remains well above 95% for velocities ranging from 0 to 80 $\mu m.s^{-1}$. Same observations can be done down to 5% of surfactant concentration. When reaching this limit, the yield is not stable anymore in the 0–80 $\mu m.s^{-1}$ velocity range. At low speed (typically below 30 $\mu m.s^{-1}$), no deposition is observed, meaning that the assembly mechanism is ineffective, when increasing the value above 30 $\mu m.s^{-1}$, the yield jumps suddenly above 95% and remains

[a]The assembly mechanism of such particles will be described in section 15.4.5.

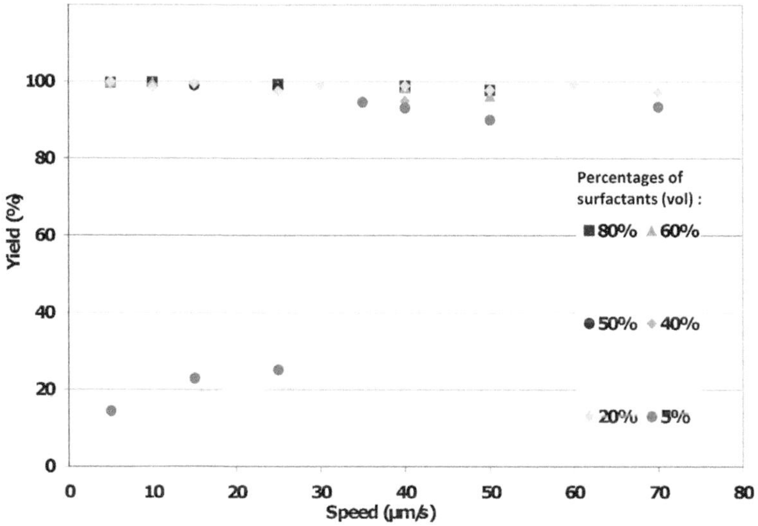

Figure 15.13 Evolution of the assembly yield as a function of substrate velocity and surfactant concentration (proportions are given in volume) for 4.5 μm beads trapped on an hexagonal patterned array (periodicity 30 μm; diameter of cavities 6 μm). The surfactant mixture consists of 50% of a 10 mM aqueous solution of SDS and 50% of a 0,1% in weight aqueous solution of Triton X45.

stable for higher velocities. As described before, the emergence of a leap of performance can be explained by the dynamic nature of wetting. Indeed, experimental observations have shown that this transition was occurring for a contact angle around 60° (results not shown). Once again, these results strengthen the premise of a capillary forces-directed mechanism. Below this threshold value, the forces of vertical confinement induced by the meniscus become sufficient to cause immobilization of beads on the surface, whereas confinement is not possible above this value.

15.4.4 *Influence of Pattern Size and Shape*

15.4.4.1 Influence of pattern geometry and orientation

As the immobilization of particles on surfaces is due to the combination of geometrical confinement and capillary forces, it

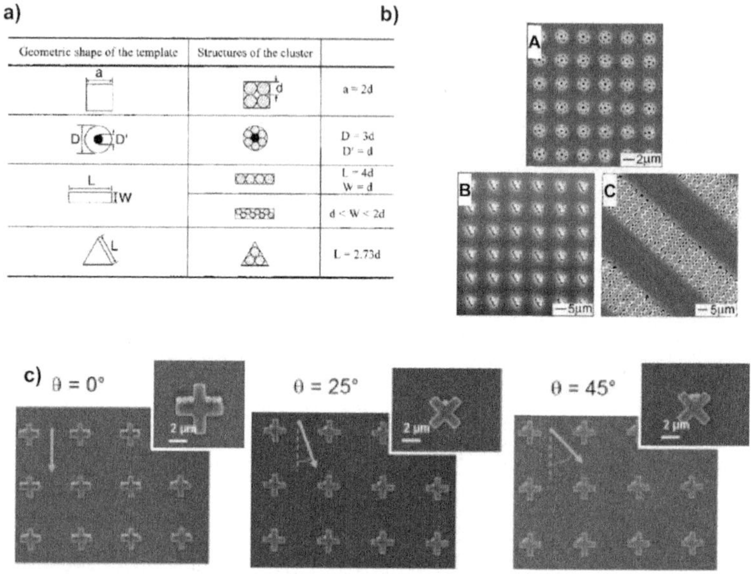

Figure 15.14 (a) Illustration of the influence of pattern geometry on the position and packing structures of particles assembled through capillary assembly.[23] (b) SEM images of colloidal assembly in holes (A), triangles (B) and lines (C).[87] (c) Illustration of the effect of the pattern orientation on capillary assembly. The experiment was performed on a PDMS substrate with 500 nm PS nanoparticles (0.5% solid content) at a constant substrate velocity of 1 $\mu m.s^{-1}$ and at a constant temperature of 25°C. Patterns are 200 nm deep.[33,34]

is obvious that the dimension and structure of the patterns used as traps play a crucial role on the efficiency of the assembly as well as on particle arrangements. In particular, the magnitude and direction of the capillary forces induced by the contact line pinning depend strongly on the geometry of patterns. This effect can be observed in Fig. 15.14, which shows the assembly of 500 nm PS particles in 200 nm deep cross-shaped patterns. Three different orientations of the patterns relative to the contact line were investigated.

Changing the orientation of the patterns with respect to the contact line direction modifies the pinning mechanism as well as the geometrical confinement effect of particles before meniscus breakup. This effect can be used to predict the number of deposited particles as well as their position on the pattern.

15.4.4.2 High-aspect-ratio structures

The influence of the pattern aspect ratio on particle assembly will be illustrated in two different situations.

The first example concerns deep, enclosed structures such as those presented in Fig. 15.15a,b. As illustrated in the previous example, the dewetting mechanism in such structures is modified

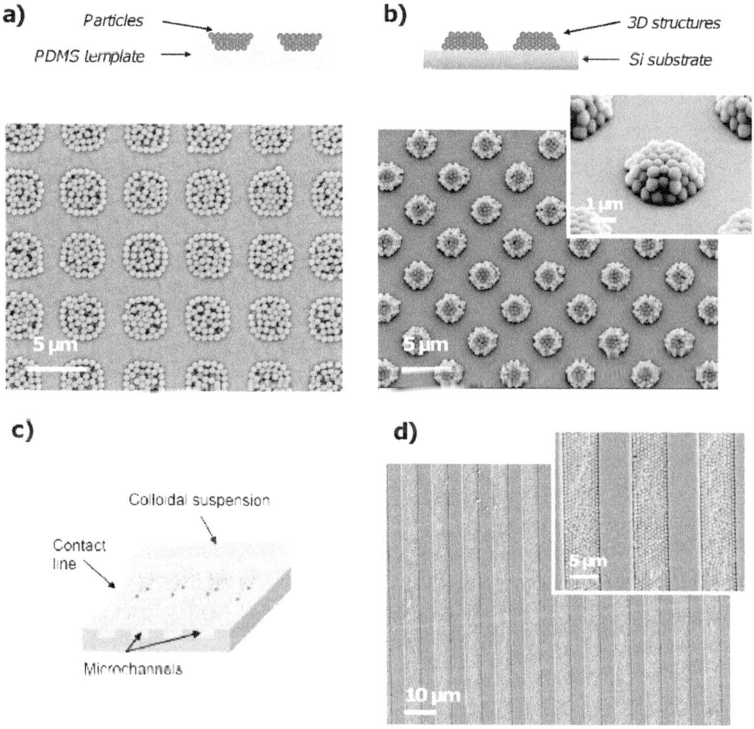

Figure 15.15 (a) 500 nm PS nanoparticles were assembled into 2.5 µm deep V-shaped grooves on a PDMS template. (b) The resulting 3D structures were further transferred by printing on a silicon wafer. (c and d) The same particles were assembled into 500 nm deep and around 5 µm wide trenches patterned on a PDMS substrate. In these experiments, the withdrawal speed was varied between 1 and 10 µm/s.[35]

by capillary pressure, which tends to keep the structures wetted. While the meniscus is dragged over the structures, a well-defined volume of liquid remains trapped. Because of the accumulation of particles close to the contact line, this results in a complete filling of the structures with particles and in the creation of well-defined 3D stacks of particles.

The second example deals with deep 2D trenches such as those presented in Fig. 15.15c,d. Because of capillary pressure, such structures tend to behave as microfluidic channels and are filled spontaneously by colloidal dispersion.[23] The liquid is pinned at the top edges of the structures, and like an enclosed channel, the volume of liquid is defined by the lateral and vertical dimensions of the trench. As solvent evaporation takes place, a particle influx is created into the channel. The solvent is progressively replaced by particles, and the formation of a closely packed 3D arrangement is initiated. Observations revealed that the assembly mechanism in the channel was similar to convective assembly, in particular considering temperature influence. On the unpatterned areas, however, the wetting is not modified and capillary forces prevent particle deposition. This method is thus extremely interesting to create crystalline stripes or ribbons that can be used, for instance as wave guides in optical applications or as artificial gels in separation or capture applications (see section 15.4.6).

15.4.5 *Influence of Particle Properties*

Both convective and capillary assemblies have been shown to be compatible with macromolecule-sized nanoparticles with diameters <10 nm.[23] Unfortunately such small particles do not form well-ordered structures as compared to larger ones, typically above 50 nm. These results bring up important questions for further studies, in particular how solvent fluctuations, substrate interactions, and thermodynamic states affect drying patterns in a confined geometry.

The compatibility of the assembly mechanism with particles over a few micrometers in size has to be investigated carefully. Indeed, as it has been described all along in this chapter, particle mobility is a key factor in both assembly mechanisms. Capillary and convective

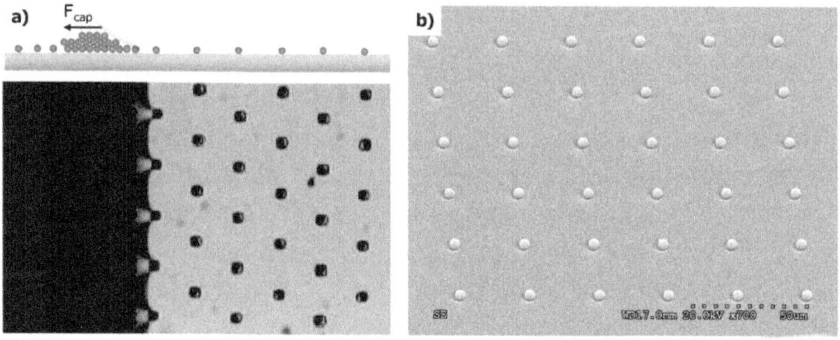

Figure 15.16 (a) Optical image taken during the assembly of 4.5 μm magnetic particles (Dynal) on a PDMS template. The accumulation region is visible at the contact line (b) SEM images of the immobilized particles (unpublished results).

assemblies are usually restricted to a colloidal suspension, where sedimentation remains negligible during the time frame of the experiment.

This criterion is critical for convective assembly. It seems that it is less restrictive for capillary assembly. Recent studies have demonstrated that even in the case of sedimenting particles, it is still possible to perform capillary assembly, as long as particles are not interacting strongly with the template surface and can be dragged by the mean of capillary forces. Figure 15.16 shows experimental results obtained with 4.5 μm magnetic particles (Dynal).

These particles tend to sediment within a few minutes and are no longer sensitive to solvent fluxes created by evaporation. Nevertheless, the capillary forces exerted by the meniscus are sufficient to drag particles deposited on flat areas. This mechanism induces an accumulation of particles close to the contact line and finally leads to a selective immobilization of particles in the recessed areas of the template. Here also, the accumulation process is required to induce particle deposition.

The possibility of assembling anisotropic particles is an interesting feature and has been mainly investigated in capillary assembly Complex particles such as tetrahedrally shaped CdTe nanotetrapods were successfully assembled.[30] The assembly of nanorods and nanowires was also reported.[73–75]

15.4.6 *Examples of Applications*

The flexibility and compatibility of convective and capillary assembly with a wide range of particle materials and sizes provide an efficient route for complex structure integration in devices. Beyond direct patterning applications (e.g., using PS particles as a resist for pattern transfer), these methods offer an opportunity of exploiting both the intrinsic particle properties and those emerging from specific arrangement of particles. The objective of this section is to demonstrate, through a short list of selected examples, the flexibility of these techniques to be used in routine to integrate 1D, 2D, and 3D structures. The scientific fields covered include optics, electronics, microfluidics, and biology. Of course this list of examples is not exhaustive, but to our point, these representative achievements demonstrate the versatility of the method.

15.4.6.1 Photonic band gap structures

Among all possible applications of colloidal crystals, the manufacturing of photonic devices has probably still attracted a strong interest. Photonic band gap (PBG) structures are probably the best example of the application of the colloidal assembly technique in this field. These crystalline structures are able to block the propagation of light through the crystal in one or more directions. The periodicity of the structures is in the same order as the wavelength of light. Bragg diffraction by the PBG crystal results in a forbidden range of frequencies (stop bands). The colloidal particles of interest for photonic crystal engineering usually have a spherical shape and are often made of silica or polymers such as poly(methyl methactylate) (PMMA) or PS.

The Ozin group pioneered the fabrication of planarized microphotonic structures and the growth of ordered crystals using evaporation techniques derived from convective assembly. Figure 15.17 illustrates the methods used in these experiments and some examples of PBG structures.[51,57,76] Opened microchannel templates were immersed vertically in a suspension of silica microspheres. The assembly was then driven by the evaporation of the colloidal suspension. Whereas the assembly of 100–500 nm

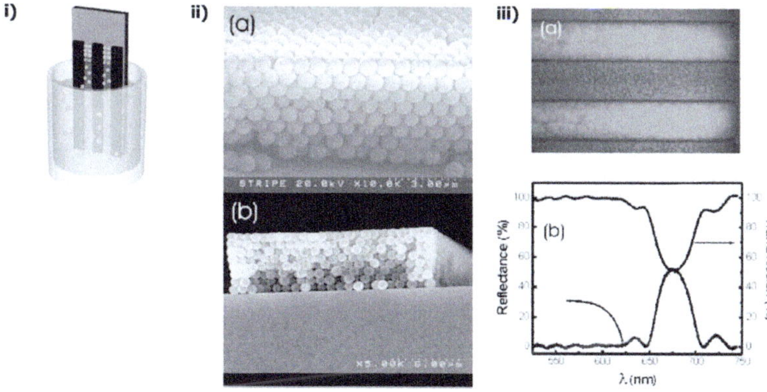

Figure 15.17 Illustration of the assembly mechanism leading to the growth of PBG structures (i) Opened microchannel templates are immersed vertically in a suspension of silica microspheres. Upon evaporation, channels are selectively filled with particles that self-organize into crystalline 3D structures (ii). Illustration of the wave-guiding properties of such opal channels is given in (iii) through an optical micrograph and reflectance/transmittance spectra.[51]

particles was straightforward, an additional stirring system was added in the suspension of microspheres over 1 μm to compensate for sedimentation.

These pioneering works were later extended by different research groups in order to optimize the control of the packing symmetry, which is of fundamental importance in the growth of PBGs. For this purpose, combination of convective or capillary-based assembly methods with topologically patterned substrates has been successfully implemented.[33]

15.4.7 *Integration of Gold Nanoparticles: Application in Optics and Electronics*

Metal nanocrystals also interact strongly with electromagnetic waves and offer remarkable properties due to the localized surface plasmon resonance (SPR) that induces, through optical excitation, very intense local electrical fields. This property can be exploited for surface-enhanced Raman spectroscopy (SERS) and SPR-based agglomeration assays.[16,77,78]

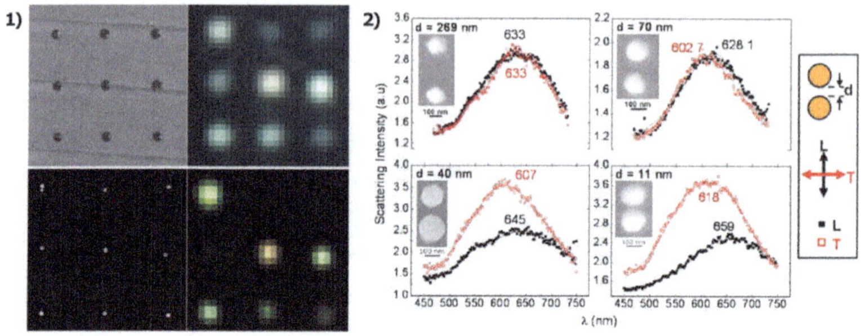

Figure 15.18 (1) Demonstration of the optical activity of printed gold particles (60 nm) particles.[34] AFM topography and the corresponding dark-field image of particles before transfer (top row) and SEM image and corresponding dark-field micrograph after printing on a silicon substrate. (2) Spectra of longitudinal (L) and transversal (T) scattered light polarizations of gold colloidal dimers with various spacings.[79] *Abbreviation*: AFM, atomic force microscopy.

Capillary assembly has shown its potentiality to create Au nanoparticle assemblies with single-particle resolution. Fabrication of 2D crystalline arrays, regularly spaced arrays of sparsed particles, 1D chains, and dimers could be successfully demonstrated (Fig. 15.18) and their electrical and optical properties analyzed.[64,79,80] In such applications capillary assembly offers many advantages compared to more conventional fabrication techniques, for instance e-beam lithography and transfer by etching. Even if a transfer step by printing is usually required in addition to capillary assembly,[34,72,80–82] the use of commercially available nanoparticle suspensions, with narrow polydispersity, offers a straightforward and reproducible approach for generating homogeneous structures at low cost.

The use of gold nanoparticles as a catalyst also offers interesting opportunities, for instance for nanowire growth. In such applications it is crucial that printed nanoparticles retain their properties during integration. Using such an approach, a silicon nanowire array has been obtained through the vapor-liquid-solid (VLS) process. Gold nanoparticles were first assembled on a PDMS substrate through capillary assembly and then transferred by printing on a silicon surface.[34] Nanowires nucleated and grew from the gold particles, and their arrangement was preserved.

Such nanowires are regarded as novel building blocks for future transistors, memory cells and sensors.

15.4.8 *Extending Capillary Assembly to Biological Systems: Selective Docking of Cells*

To study the response of a large population of cells at the individual cell level, many efforts have been dedicated to the development of parallel and simple methods to localize precisely individual cells on a surface or create assemblies. It has been recently demonstrate that capillary assembly can be adapted to biological objects to create cells arrays.

The first example of the application of capillary assembly to biology was reported by Ressier *et al. Escherichia coli* (*E. coli*) bacteria in aqueous suspensions were successfully assembly on chemical patterns.[83] It seems that the bacterial cell immobilization method induces very weak physical tension and no chemical contamination on the bacterial cell membrane.

Park *et al.*[84,85] have also used capillary force created by a receding meniscus to capture nonadherent yeast cells onto microwells inside a microfluidic channel. The resulting cells arrays are shown in Fig. 15.19.

As illustrated before, both control of meniscus velocity and cell density were critical in regard with the efficiency and uniformity of the process. Here the mobility of the cells in the suspension is also critical to preserve the assembly mechanism. The complexity of the buffers used in biology and the interaction forces between the cells themselves and the surfaces are clearly the limitation of this method.

In their work, the authors demonstrated the viability of cells after deposition. As the array was integrated in a microfluidic channel, docked cells could be exposed to chemicals and their responses studied. In these two examples, the authors could analyze the response of α-factor-dependent green fluorescent protein (GFP) expression cells or monitor the response of the mating mitogen-activated protein kinase (MAPK) pathway. These studies could be performed in real time and at the single cell resolution.

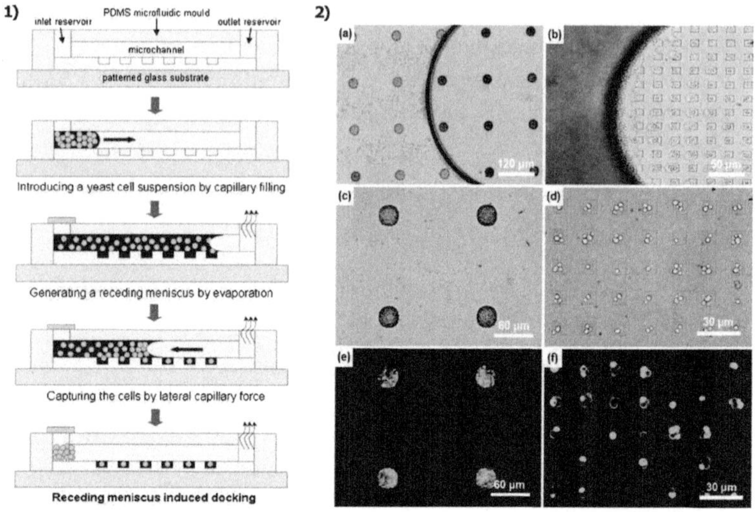

Figure 15.19 Schematic illustration of the cell-docking method within a patterned microfluidic channel. (1) Modified yeast cells were introduced into the microfluidic channel. After closing one inlet, the meniscus was swept over the microwells. Remaining cells were washed via capillary action. After the washing process, the microfluidic channel was further filled with culture or analysis media. (2) Optical observation during the assembly revealed the importance of cell density and pattern dimension on the deposition process.[85]

15.5 Conclusion

The combination of dewetting of colloidal dispersions, capillary forces, and geometric templating can provide an effective approach for assembling colloidal particles into complex structures. In this chapter we have shown that capillary forces and convective flows can be exploited on flat or patterned surfaces to reach a large range of structures achievable from single-particle placement to almost any 1D, 2D, and even some 3D structures. These processes are compatible with a large range of particle materials, shapes, and sizes. The ultimate limit to the dimensions of colloidal particles might come from the interplay of particle mobility and capillary forces.

It has been shown that particles as small as 2 nm can be handled and immobilized selectively on a surface. The applicability of these assembly processes to larger or heavier particles is limited by particle mobility. In the case of quickly sedimenting particles, convective flow is inefficient to ensure mass transfer toward the assembly region. The creation of 2D layers is thus difficult. Capillary forces offer more flexibility. First results with micron-sized particles have showed that they compensate for particle sedimentation, but it is still not clear what the limit is for capillary assembly above a few micrometers.

Of course, maintaining control over the structure and number of particle layers is a major step toward large-scale applicability. The choice of the experimental setup is essential and has to be done according to the complexity of the targeted structures.

It seems that two parameters are essential to benefit from these assembly mechanisms. Obviously, the control of the *substrate velocity* is needed to control the meniscus motion and to perform the assembly over large areas. In the two processes investigated, it has been shown also that controlling the hydrodynamic flows created by *evaporation* is critical and can easily be achieved by adjusting the substrate temperature. In convective assembly, the temperature allows fast and accurate control of the deposition speed, including reversal of the assembly process, whereas in capillary assembly, a distinct temperature threshold has been found that triggered the assembly.

Considering applications, convective and capillary assembly methods offer an opportunity to build complex devices exploiting the intrinsic material and surface properties of particles or the more specific properties that appear in well-ordered particle arrangements. In combination with printing techniques such as those proposed recently,[34,35] these nonlithographic approaches based on directed organization processes are especially appealing. Indeed, they allow for large-scale production with very small infrastructure investments. Applications in optics or electronics could benefit from the creation of closely packed metallic particle arrangements. For instance, it could be applied to the fabrication of PBG structures, transistors or even plasmon-guiding structures that would rely on the coupling effects between regularly spaced particles. Another promising field of application is biology. Combining the optical, magnetic or even catalytic properties of particles with specific recognition through surface functionalization renders particles very appealing for detection or analysis purposes. Moreover, it has been recently shown that capillary assembly was also applicable to cells, in particular yeasts and bacteria.[84,85] The compatibility of these techniques with real, living samples offers unique opportunities for biological studies down to the single-cell level.

References

1. A. N. Shipway, E. Katz, and I. Willner, Nanoparticle arrays on surfaces for electronic, optical, and sensor applications, *ChemPhysChem*, **1**, 18–52 (2000).
2. E. Katz and I. Willner, Integrated nanoparticle-biomolecule hybrid systems: synthesis, properties, and applications, *Angew. Chem., Int. Ed.*, **43**, 6042–6108 (2004).
3. W. J. Parak, D. Gerion, T. Pellegrino, D. Zanchet, C. Micheel, S. C. Williams, R. Boudreau, M. A. L. Gros, C. A. Larabell, and A. P. Alivisatos, Biological applications of colloidal nanocrystals, *Nanotechnology*, **14**, R15 (2003).
4. Y. Xia, B. Gates, Y. Yin, and Y. Lu, Monodispersed colloidal spheres: old materials with new applications, *Adv. Mater.*, **12**, 693–713 (2000).
5. C. M. Niemeyer, Nanoparticles, proteins, and nucleic acids: biotechnology meets, *Mater. Sci., Angew. Chem., Int. Ed.*, **40**, 4128–4158 (2001).
6. I. Willner and B. Willner, Functional nanoparticle architectures for sensoric, optoelectronic, and bioelectronic applications, *Pure Appl. Chem.*, **74**, 1773–1783 (2002).
7. P. Alivisatos, The use of nanocrystals in biological detection, *Nat. Biotechnol.*, **22**, 47–52 (2003).
8. R. E. Schaak, R. E. Cable, B. M. Leonard, and B. C. Norris, Colloidal crystal microarrays and two-dimensional superstructures: a versatile approach for patterned surface assembly, *Langmuir*, **20**, 7293–7297 (2004).
9. H. Fan, K. Yang, D. M. Boye, T. Sigmon, K. J. Malloy, H. Xu, G. P. López, and C. J. Brinker, Self-assembly of ordered, robust, three-dimensional gold nanocrystal/silica arrays, *Science*, **304**, 567–571 (2004).

10. G. Ozin, The photonic opal–the jewel in the crown of optical information processing, *Chemical Commun. (Cambridge, Engl.)*, **21**, 2639–2643 (2003).
11. J. H. Holtz and S. A. Asher, Polymerized colloidal crystal hydrogel films as intelligent chemical sensing materials, *Nature*, **389**, 829–832 (1997).
12. I. Gourevich, H. Pham, J. E. N. Jonkman, and E. Kumacheva, Multidye nanostructured material for optical data storage and security labeling, *Chem. Mater.*, **16**, 1472–1479 (2004).
13. S. I. Khondaker and Z. Yao, Fabrication of nanometer-spaced electrodes using gold nanoparticles, *Appl. Phys. Lett.*, **81**, 4613 (2002).
14. Q.-H. Wei, K. H. Su, S. Durant, and X. Zhang, Plasmon resonance of finite one-dimensional Au nanoparticle chains, *Nano Lett.*, **4**, 1067–1071 (2004).
15. S. A. Maier, M. L. Brongersma, P. G. Kik, S. Meltzer, A. A. G. Requicha, and H. A. Atwater, Plasmonics—a route to nanoscale optical devices, *Adv. Mater.*, **13**, 1501–1505 (2001).
16. E. Hutter and J. H. Fendler, Exploitation of localized surface plasmon resonance, *Adv. Mater.*, **16**, 1685–1706 (2004).
17. N. A. Kotov, Y. Liu, S. Wang, C. Cumming, M. Eghtedari, G. Vargas, M. Motamedi, J. Nichols, and J. Cortiella, Inverted colloidal crystals as three-dimensional cell scaffolds, *Langmuir*, **20**, 7887–7892 (2004).
18. C. Wang, F. L. Yap, and Y. Zhang, Micropatterning of polystyrene nanoparticles and its bioapplications, *Colloids Surf. B, Biointerfaces*, **46**, 255–260 (2005).
19. S. Pautot, C. Wyart, and E. Y. Isacoff, Colloid-guided assembly of oriented 3D neuronal networks, *Nat. Methods*, **5**, 735–740 (2008).
20. K. L. Gunderson, S. Kruglyak, M. S. Graige, F. Garcia, B. G. Kermani, C. Zhao, D. Che, T. Dickinson, E. Wickham, J. Bierle, D. Doucet, M. Milewski, R. Yang, C. Siegmund, J. Haas, L. Zhou, A. Oliphant, J.-B. Fan, S. Barnard, and M. S. Chee, Decoding randomly ordered DNA arrays, *Genome Res.*, **14**, 870–877 (2004).
21. N. V. Dziomkina and G. J. Vancso, Colloidal crystal assembly on topologically patterned templates, *Soft Matter*, **1**, 265 (2005).
22. S. C. Glotzer, M. J. Solomon, and N. A. Kotov, Self-assembly: from nanoscale to microscale colloids, *AIChE J.*, **50**, 2978–2985 (2004).
23. Y. Yin, Y. Lu, B. Gates, and Y. Xia, Template-assisted self-assembly: a practical route to complex aggregates of monodispersed colloids with well-defined sizes, shapes, and structures, *J. Am. Chem. Soc.*, **123**, 8718–8729 (2001).

24. J. Dutta and H. Hofmann, Self-organization of colloidal nanoparticles, *Encyclopedia Nanosci. Nanotechnol.*, **9**, 617–640 (2003).
25. B. G. Prevo, J. C. Fuller, and O. D. Velev, Rapid deposition of gold nanoparticle films with controlled thickness and structure by convective assembly, *Chem. Mater.*, **17**, 28–35 (2005).
26. B. Prevo and O. D. Velev, Controlled, rapid deposition of structured coatings from micro-and nanoparticle suspensions, *Langmuir*, **20**, 2099–2107 (2004).
27. N. Denkov, O. Velev, P. Kralchevski, I. Ivanov, H. Yoshimura, and K. Nagayama, Mechanism of formation of two-dimensional crystals from latex particles on substrates, *Langmuir*, **8**, 3183–3190 (1992).
28. N. D. Denkov, O. D. Velev, P. A. Kralchevsky, I. B. Ivanov, H. Yoshimura, and K. Nagayama, Two-dimensional crystallization, *Nature*, **361**, 26–26 (1993).
29. E. Adachi, A. S. Dimitrov, and K. Nagayama, Stripe patterns formed on a glass surface during droplet evaporation, *Langmuir*, **11**, 1057–1060 (1995).
30. Y. Cui, M. Björk, J. Liddle, C. Sönnichsen, B. Boussert, and A. P. Alivisatos, Integration of colloidal nanocrystals into lithographically patterned devices, *Nano Lett.*, **4**, 1093–1098 (2004).
31. F. Juillerat, H. H. Solak, P. Bowen, and H. Hofmann, Fabrication of large-area ordered arrays of nanoparticles on patterned substrates, *Nanotechnology*, **16**, 1311–1316 (2005).
32. A. S. Dimitrov and K. Nagayama, Continuous convective assembling of fine particles into two-dimensional arrays on solid surfaces, *Langmuir*, **12**, 1303–1311 (1996).
33. L. Malaquin, T. Kraus, H. Schmid, E. Delamarche, and H. Wolf, Controlled particle placement through convective and capillary assembly, *Langmuir*, **23**, 11513–11521 (2007).
34. T. Kraus, L. Malaquin, H. Schmid, W. Riess, N. D. Spencer, and H. Wolf, Nanoparticle printing with single-particle resolution, *Nat. Nanotechnol.*, **2**, 570–576 (2007).
35. T. Kraus, L. Malaquin, E. Delamarche, H. Schmid, N. D. Spencer, and H. Wolf, Closing the gap between self-assembly and microsystems using self-assembly, transfer, and integration of particles, *Adv. Mater.*, **17**, 2438–2442 (2005).
36. R. D. Deegan, O. Bakajin, and T. F. Dupont, Capillary flow as the cause of ring stains from dried liquid drops, *Nature*, **389**, 827–829 (1997).

37. R. D. Deegan, Pattern formation in drying drops, *Phys. Rev. E*, **61**, 475–485 (2000).

38. S. Hayashi, Y. Kumamoto, T. Suzuki, and T. Hirai, Imaging by polystyrene latex particles, *J. Colloid Interface Sci.*, **144**, 538–547 (1991).

39. J. Zhang, L. Xue, and Y. Han, Fabrication of gradient colloidal topography, *Langmuir*, **21**, 5667–5671 (2005).

40. Q. Yan, Z. Zhou, and X. S. Zhao, Inward-growing self-assembly of colloidal crystal films on horizontal substrates, *Langmuir: ACS J. Surf. Colloids*, **21**, 3158–64 (2005).

41. I. Karakurt, P. Leiderer, and J. Boneberg, Size-dependent self-organization of colloidal particles on chemically patterned surfaces, *Langmuir*, **22**, 2415–2417 (2006).

42. Y. Ikezoe, Y. Kumashiro, K. Tamada, T. Matsui, I. Yamashita, K. Shiba, and M. Hara, Growth of giant two-dimensional crystal of protein molecules from a three-phase contact line, *Langmuir*, **24**, 12836–12841 (2008).

43. P. Jiang and M. J. McFarland, Large-scale fabrication of wafer-size colloidal crystals, macroporous polymers and nanocomposites by spin-coating, *J. Am. Chem. Soc.*, **126**, 13778–13786 (2004).

44. A. Hadjiiski, R. Dimova, N. D. Denkov, I. B. Ivanov, and R. Borwankar, Film trapping technique: precise method for three-phase contact angle determination of solid and fluid particles of micrometer size, *Langmuir*, **12**(26), 6665–6675 (1996).

45. Y. Yin and Y. Xia, Growth of large colloidal crystals with their (100) planes orientated parallel to the surfaces of supporting substrates, *Adv. Mater.*, **14**, 605 (2002).

46. Z. Chen, P. Zhan, Z. L. Wang, J. H. Zhang, W. Y. Zhang, N. B. Ming, C. T. Chan, and P. Sheng, Two- and three-dimensional ordered structures of hollow silver spheres prepared by colloidal crystal templating, *Adv. Mater.*, **16**, 417–422 (2004).

47. M. Ishii, H. Nakamura, H. Nakano, A. Tsukigase, and M. Harada, Large-domain colloidal crystal films fabricated using a fluidic cell, *Langmuir*, **21**, 5367–5371 (2005).

48. F. Ramiro-Manzano, F. Meseguer, E. Bonet, and I. Rodriguez, Faceting and commensurability in crystal structures of colloidal thin films, *Phys. Rev. Lett.*, **97**, 28304 (2006).

49. Y. Zeng and D. J. Harrison, Confinement effects on electromigration of long DNA molecules in an ordered cavity array, *Electrophoresis*, **27**, 3747–3752 (2006).

50. F. Ramiro-Manzano, E. Bonet, I. Rodriguez, and F. Meseguer, Colloidal crystal thin films grown into corrugated surface templates, *Langmuir*, **26**, 4559–4562 (2010).
51. S. M. Yang, H. Miguez, and G. A. Ozin, Opal circuits of light-planarized microphotonic crystal chips, *Adv. Funct. Mater.*, **12**, 425–431 (2002).
52. M. Marquez and B. P. Grady, The use of surface tension to predict the formation of 2D arrays of latex spheres formed via the Langmuir-Blodgett-like technique, *Langmuir*, **20**, 10998–11004 (2004).
53. P. Maury, M. Escalante, D. N. Reinhoudt, and J. Huskens, Directed assembly of nanoparticles onto polymer-imprinted or chemically patterned templates fabricated by nanoimprint lithography, *Adv. Mater.*, **17**, 2718–2723 (2005).
54. J. Huang, F. Kim, A. R. Tao, S. Connor, and P. Yang, Spontaneous formation of nanoparticle stripe patterns through dewetting, *Nat. Mater.*, **4**, 896–900 (2005).
55. J. Huang, A. R. Tao, S. Connor, R. He, and P. Yang, A general method for assembling single colloidal particle lines, *Nano Lett.*, **6**(3), 524–529 (2006).
56. Y. Fu, Z. Jin, Z. Liu, Y. Liu, and W. Li, Self-assembly of colloidal crystals from polystyrene emulsion at elevated temperature by dip-drawing method, *Mater. Lett.*, **62**, 4286–4289 (2008).
57. S. Yang and G. Ozin, Opal chips: vectorial growth of colloidal crystal patterns inside silicon wafers, *Chem. Commun.*, 2507–2508 (2000).
58. J. Y. Shiu, C. W. Kuo, and P. Chen, Actively controlled self-assembly of colloidal crystals in microfluidic networks by electrocapillary forces, *J. Am. Chem. Soc.*, **126**, 8096–8097 (2004).
59. A. B. Jemere, R. D. Oleschuk, and D. J. Harrison, Microchip-based capillary electrochromatography using packed beds, *Electrophoresis*, **24**, 3018–3025 (2003).
60. C.-A. Fustin, G. Glasser, H. W. Spiess, and U. Jonas, Parameters influencing the templated growth of colloidal crystals on chemically patterned surfaces, *Langmuir*, **20**, 9114–9123 (2004).
61. S. Watanabe, K. Inukai, S. Mizuta, and M. T. Miyahara, Mechanism for stripe pattern formation on hydrophilic surfaces by using convective self-assembly, *Langmuir: ACS J. Surf. Colloids*, **25**, 7287–7295 (2009).
62. A. Diguet, M. Le Berre, Y. Chen, and D. Baigl, Preparation of phospholipid multilayer patterns of controlled size and thickness by capillary assembly on a microstructured substrate, *Small (Weinheim, Ger.)*, **5**, 1661–1666 (2009).

63. B. G. Prevo, J. C. Fuller, and O. D. Velev, Rapid deposition of gold nanoparticle films with controlled thickness and structure by convective assembly, *Chem. Mater.*, **17**, 28–35 (2005).

64. J. Grisolia, B. Viallet, C. Amiens, S. Baster, A. S. Cordan, Y. Leroy, C. Soldano, J. Brugger, and L. Ressier, 99% random telegraph signal-like noise in gold nanoparticle micro-stripes, *Nanotechnology*, **20**, 355303 (2009).

65. P. Jiang, J. F. Bertone, K. S. Hwang, and V. L. Colvin, Single-crystal colloidal multilayers of controlled thickness, *Chem. Mater.*, **11**, 2132–2140 (1999).

66. E. Kumacheva, P. Garstecki, H. Wu, and G. M. Whitesides, Two-dimensional colloid crystals obtained by coupling of flow and confinement, *Phys. Rev. Lett.*, **91**, 128301 (2003).

67. J. Zhang, A. Alsayed, K. H. Lin, S. Sanyal, F. Zhang, W. J. Pao, V. S. K. Balagurusamy, P. A. Heiney, and A. G. Yodh, Template-directed convective assembly of three-dimensional face-centered-cubic colloidal crystals, *Appl. Phys. Lett.*, **81**, 3176 (2002).

68. Y. Yin, Z. Y. Li, and Y. Xia, Template-directed growth of (100)-oriented colloidal crystals, *Langmuir*, **19**, 622–631 (2003).

69. W. Russel, D. Saville, and W. R. Schowalter, *Colloidal Dispersions*. Cambridge: Cambridge University Press (1992).

70. T. O. Shin Dosho, N. Ise, K. Ito, S. Iwai, H. Kitano, H. Matsuoka, H. Nakamura, and H. Okumura, Recent study of polymer latex dispersions, *Langmuir*, 394–411 (1993).

71. N. Ise, H. Matsuoka, K. Ito, H. Yoshida, and J. Yamanaka, Ordering of latex particles and ionic polymers in solutions, *Langmuir*, **6**, 296–302 (1990).

72. K. Lhoste, L. Malaquin, L. Billot, A. M. Haghiri-Gosnet, and Y. Chen, Fabrication of high density gold nanoparticle arrays on glass for high sensitivity bio-detection, *Microelectron. Eng.* (2010).

73. K. J. Stebe, E. Lewandowski, and M. Ghosh, Oriented assembly of metamaterials, *Science*, **325**, 159–160 .

74. Y. Huang, X. Duan, Q. Wei, and C. M. Lieber, Directed assembly of one-dimensional nanostructures into functional networks, *Science*, **291**, 630 (2001).

75. B. Messer, J. H. Song, and P. Yang, Microchannel networks for nanowire patterning, *J. Am. Chem. Soc.*, **122**, 10232–10233 (2000).

76. H. Míguez, S. M. Yang, N. Tétreault, and G. a Ozin, Oriented free-standing three-dimensional silicon inverted colloidal photonic crystal microfibers, *Adv. Mater.*, **14**, 1805–1808 (2002).

77. R. G. Freeman, K. C. Grabar, K. J. Allison, R. M. Bright, J. A. Davis, A. P. Guthrie, M. B. Hommer, M. A. Jackson, P. C. Smith, D. G. Walter, and M. J. Natan, Self-assembled metal colloid monolayers: an approach to SERS substrates, *Science*, **267**, 1629–1632 (1995).

78. J. J. Storhoff, A. A. Lazarides, R. C. Mucic, C. A. Mirkin, R. L. Letsinger, and G. C. Schatz, What controls the optical properties of DNA-linked gold nanoparticle assemblies? *J. Am. Chem. Soc.*, **122**, 4640–4650 (2000).

79. T. P. Rivera, O. Lecarme, J. Hartmann, R. L. Inglebert, and D. Peyrade, Spectroscopic studies of plasmonic interactions in colloidal dimers fabricated by convective-capillary force assembly, *Microelectron. Eng.*, **86**, 1089–1092 (2009).

80. A. Cerf, G. Molnár, and C. Vieu, Novel approach for the assembly of highly efficient SERS substrates, *ACS Appl. Mater. Interfaces*, **1**, 2544–2550 (2009).

81. A. Cerf and C. Vieu, Transfer printing of sub-100 nm nanoparticles by soft lithography with solvent mediation, *Colloids Surf. A: Physicochem. Eng. Aspects*, **342**, 136–140 (2009).

82. J. Aizenberg, P. V. Braun, and P. Wiltzius, Patterned colloidal deposition controlled by electrostatic and capillary forces, *Phys. Rev. Lett.*, **84**, 2997–3000 (2000).

83. L. Ressier, B. Viallet, A. Beduer, D. Fabre, L. Fabie, E. Palleau, and E. Dague, Combining convective/capillary deposition and AFM oxidation lithography for close-packed directed assembly of colloids, *Langmuir*, **24**, 13254–13257 (2008).

84. M. C. Park, J. Y. Hur, H. S. Cho, S.-H. Park, and K. Y. Suh, High-throughput single-cell quantification using simple microwell-based cell docking and programmable time-course live-cell imaging, *Lab Chip*, **11**, 79–86 (2011).

85. M. C. Park, J. Y. Hur, K. W. Kwon, S. H. Park, and K. Y. Suh, Pumpless, selective docking of yeast cells inside a microfluidic channel induced by receding meniscus, *Lab Chip*, **6**, 988–994 (2006).

86. A. Hadjiiski, R. Dimova, N. D. Denkov, I. B. Ivanov, R. Borwankar, Film trapping technique: precise method for three-phase contact angle determination of solid and fluid particles of micrometer size, *Langmuir*, **12**, 6665–6675 (1996).

87. Y. Xia, Y. Yin, Y, and Y. Lu, Template-assisted self-assembly of spherical colloids into complex and controllable structures, *Adv. Funct. Mater.*, **13**, 907–918 (2003).

88. C. Monteux and F. Lequeux, Packing and sorting colloids at the contact line of a drying drop, *Langmuir*, **27**, 2917–2922 (2011).

Chapter 16

Molecular Combing of DNA and Carbon Nanotubes by a Moving Meniscus

C. Khripin,[a,b,*] A. Jagota,[b] and M. Zheng[a]

[a]*Polymers Division, National Institute of Standards and Technology, Gaithersburg, Maryland 20899, USA*

[b]*Department of Chemical Engineering and Bioengineering Program, Lehigh University, Bethlehem, Pennsylvania 18015, USA*

*constantine.khripin@gmail.com

Molecular combing, also called meniscus alignment, is a technique for stretching long molecules in a uniform direction on a substrate. Molecules several micrometers long can be stretched out, with directional control, on a scale limited only by the size of the substrate. The secret is that the end of a molecule deposited on the surface binds strongly, and the remainder is pulled tight by the passing liquid meniscus, lining up in a direction perpendicular to the meniscus itself. Here, we outline the major work that has been carried out to understand and utilize this phenomenon and discuss the mechanism of the process as it is presently understood, with the goal of advancing the accessibility of this technique.

Although molecular combing has been demonstrated with a variety of high-aspect ratio molecules, here we focus on the combing of two such molecules, DNA and carbon nanotubes (CNTs). The stretching of DNA molecules is attractive for sequencing applications

Nanoscale Liquid Interfaces: Wetting, Patterning, and Force Microscopy at the Molecular Scale
Edited by Thierry Ondarçuhu and Jean-Pierre Aimé

ISBN 978-981-4316-45-3 (Hardcover), 978-981-4364-48-5 (eBook)
www.panstanford.com

as well as for templating molecular wires; CNTs, on the other hand, are combed to facilitate their integration into various electronic devices.

The mechanism of molecular combing, as presently understood, is discussed, from the factors affecting the initial deposition of the object on the surface to the interplay of forces that result in its alignment. Factors that govern both the density of deposition and the quality of alignment are discussed. The most important conclusion is that alignment occurs in a direction perpendicular to the liquid meniscus, as opposed to the direction of fluid flow; the implication is that the direction of alignment can be precisely controlled by controlling the angle of the meniscus.

Molecular combing is the extension of a high-aspect-ratio object, for example, double-stranded DNA, on a flat surface by the surface tension of a passing liquid meniscus. The discovery of molecular combing was largely made possible by recently developed single-molecule imaging methods, specifically scanning probe microscopy (SPM). This group of molecular imaging techniques, discussed at length in other chapters of this book, employs an atomically sharp probe that is raster-scanned over the substrate surface. Measurements of probe interaction with the surface, whether mechanical, as in atomic force microscopy (AFM),[1] or electrical, as in scanning tunneling microscopy (STM),[2] are recorded to create an image. In this way, it is possible, for example, to record the outline of a single DNA molecule deposited on a substrate. Therefore, for the first time it became possible to study directly the shape of an adsorbed macromolecule.[3,4]

SPM enabled the discovery of molecular combing and also drove it. It was expected that AFM or STM could eventually be used to directly determine the sequence of the DNA molecule through straightforward imaging (see, for example, Hansma and Hoh, 1994).[5] In what appeared to be a large step toward this goal, it was discovered that under certain conditions the DNA molecule was stretched into a straight line.[6] To explain this phenomenon, it was suggested that the passing of the liquid meniscus aligns the molecule (which is initially in a random coil state) in a direction perpendicular to the meniscus. Thus the long DNA molecule is "combed" in a way that is superficially similar to combing hair—

hence the term "molecular combing." Although this has become a mature technique, attempts to sequence combed DNA have not been successful, although a recent work reports the identification of at least the "G" nucleotide in single-stranded DNA (ssDNA).[7]

Molecular combing has evolved into a popular technique for taking a molecule from a solution and placing it on a substrate, for example, for imaging or for integration into an electronic device. It has been applied to a variety of objects such as viruses,[8] nanofibers,[9] and nanowires;[10,11] here we focus on two extensively investigated molecules, DNA and CNTs. In this way we hope to address the mechanistic aspects of the phenomenon and to provide guidelines by means of which the method might be adapted to other applications.

16.1 Overview of Molecular Combing

Here we outline the work that has been carried on molecular combing of DNA and single-walled CNTs (SWCNTs). These studies describe many useful methods and present applications and move toward understanding the mechanism (see also Table 16.1 for a summary of molecular combing studies).

16.1.1 *Molecular Combing of DNA*

Molecular combing of DNA was developed in order to extend the molecule for imaging by SPM.[6] As shown schematically in Fig. 16.1a, the DNA deposits onto a flat surface as a random coil.[12] When the liquid is removed, the passing of the liquid–air meniscus exerts a force on the hydrophilic DNA molecule, stretching it into a straight line (Fig. 16.1b,c). At the extreme, when a force sufficient to break the DNA strand is applied, the double helix is stretched to nearly double its relaxed length.[13] The imaging resolution was not sufficient to distinguish features of the double helix to identify the nucleotide sequence. These investigations did, however, probe the conditions, such as pH and ionic strength, which determine whether the DNA molecule deposits on the surface or is combed.[14] In addition, the fluorescent marking of select domains on the stretched

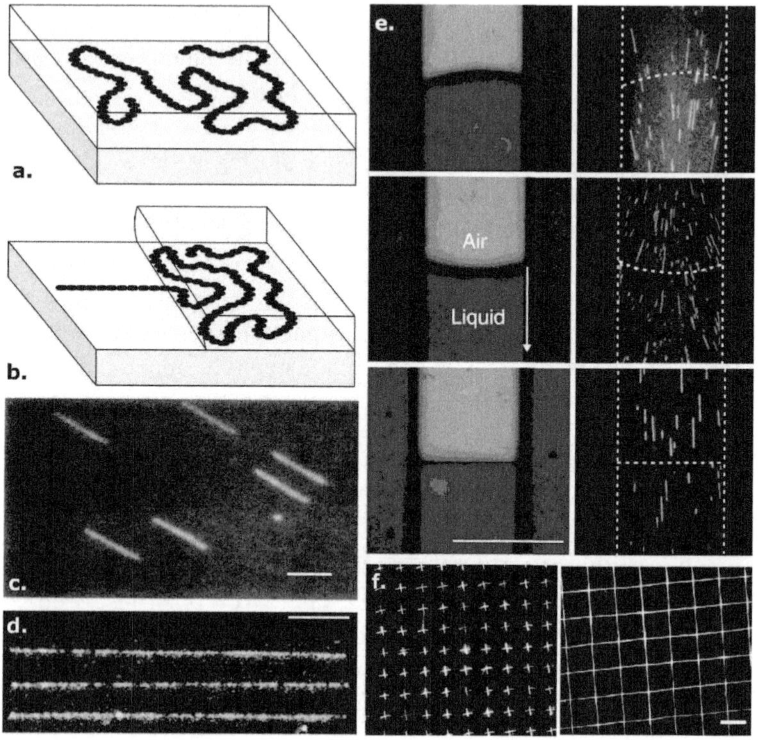

Figure 16.1 Molecular combing of DNA begins with the deposition of the molecule on the surface (a). When the liquid is removed, the passing of the meniscus pulls the molecule straight (b). Fluorescent labeling (c) reveals a large number of DNA molecules aligned on the surface (image 80 μm wide, from Allemand *et al.*).[14] A variety of useful phenomena has been demonstrated using this technique. When reimmersed in water, fluorescently tagged labels can be employed to locate specific DNA sequences (d) (DNA green, labels yellow, scale bar: 5 μm; from Geron-Landre *et al.*).[17] Furthermore, DNA can be aligned inside the controlled flow environment of microchannels (e). This case demonstrates that DNA aligns perpendicular to the meniscus, not with the direction of the flow (from Petit *et al.*,[19] scale bar: 20 μm). When combined with soft lithography, molecular combing can produce a variety of well-controlled patterns (f) (from Guan *et al.*,[20] scale bar: 10 μm).

DNA (Fig. 16.1d) was shown to be a feasible means of learning the general location of genes.[15-17] Many insights into the relative locations of genes have been produced; for a detailed review see Herrick and Bensimon.[18]

Later investigations aimed to perfect the technique of molecular combing through improvements such as combing on unmodified hydrophilic surfaces,[21] inside microchannels (Fig. 16.1e),[19] and, together with soft lithography, in precisely defined arrays (Fig. 16.1f).[20] Such precise placement of DNA has also become important for microelectronics applications, since chemical methods have been devised to transform combed DNA to nanowires of various materials, such as copper and platinum; for a review, see Stoltenberg and Woolley.[22]

16.1.2 *Molecular Combing of Carbon Nanotubes*

SWCNTs.[23,24] are high-aspect-ratio macromolecules with a range of potentially unique properties.[25] and proposed applications,[26] for example, as field-effect transistor sensors.[27] However, the synthetic mixture contains nanotubes of varying length and electronic properties, in addition to nonnanotube impurities. One route to isolating the nanotubes of desired properties is solution-based purification;[28-31] in this way it is possible to select for nanotubes of desired electronic properties[29,30] and lengths.[32] After obtaining a nanotube of desired properties, it must be placed in contact with electrodes to be integrated into a device, and molecular combing methods have been investigated for this purpose.

Initially, molecular combing was used to comb a nanotube over a pair of gold electrodes.[33,34] Self-assembled monolayers (SAMs) on the electrodes and the silicon substrate were used to help induce the SWCNT (dispersed *via* sodium dodecyl sulphate [SDS]) to deposit at the correct location; the liquid–air meniscus was then used to orient the nanotube across the two electrodes (Fig. 16.2b). This method has the advantage of being easily scalable.[34] However, contact with the nanotube is superior when the electrode is deposited with e-beam lithography over the nanotube. Because molecular combing is a very convenient way to extend nanotubes on the substrate, it is often used in this case as well.[34-37] These studies have demonstrated

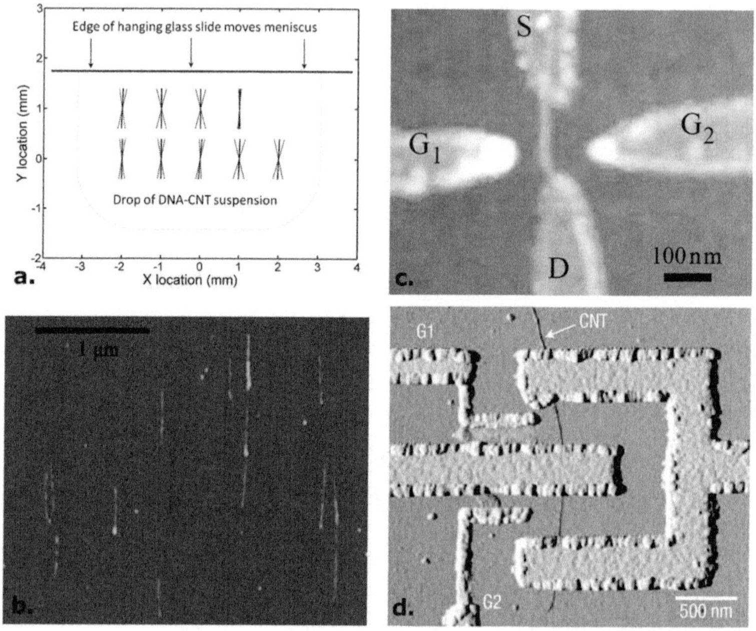

Figure 16.2 By controlling the shape of the receding meniscus, DNA-dispersed SWCNTs can be aligned in the same direction over a large area of a substrate (a and b). The solid line represents the observed nanotube angle, and the dashed line spans an interval of 6σ.[41] The uniform size distribution of SWCNTs in (b) was achieved by sorting the raw material with SEC.[42] To make a device, an SWCNT (in this case, SDS dispersed) can be deposited directly onto an electrode and aligned across the gap (c); many devices can be made in parallel this way.[34] It is also possible to deposit electrodes on top of a nanotube aligned on a substrate as in (b), by locating the nanotube using AFM and then depositing the electrodes with e-beam lithography (d).[37] *Abbreviation*: SEC, size-exclusion chromatography.

that nanotubes deposited from solution can make functional field-effect transistors[34-36,38,39] and SQUID devices[37] (Fig. 16.2d), as well as help to study nanotube behavior in high magnetic fields.[35]

The phenomenon of molecular combing of SWCNTs dispersed with SDS was studied further by Ko *et al.*[40] They investigated how surface preparation and solution conditions can be used to control the density of nanotubes and demonstrated that, just like DNA, nanotubes were aligned perpendicular to the meniscus, not parallel to the flow direction.

A special case arises when DNA is used to disperse SWCNTs. The hybrid nanoparticle is similar to double-stranded DNA (dsDNA) in several ways. AFM observations have suggested that DNA wraps around nanotubes in a helical fashion, forming a tight structure with a diameter of 2 nm[29,43] similar to dsDNA; this is supported by molecular modeling.[44] Capillary electrophoresis measurements have shown that the charge density of DNA-SWCNT is similar to dsDNA.[45] Therefore detailed studies of molecular combing of DNA-SWCNT have shown that the hybrid behaves in many ways similar to dsDNA, for example, having similar dependence of deposition density on the pH of the medium.[14,41]

Further mechanistic investigation of DNA-SWCNT molecular combing was carried out by Khripin *et al.*[41] It was shown that the density of nanotubes on the surface, as imaged by AFM (Fig. 16.2a,b), could be controlled by means of incubation time, solution pH, and ionic strength. The data suggests that the deposition is an activated process, and a kinetics model was proposed. In all cases, DNA-SWCNTs were aligned perpendicular to the meniscus, and it was possible to control alignment on a global scale by controlling the shape of the meniscus. Layer-by-layer deposition was demonstrated to achieve crisscross patterns of nanotubes (see section 16.3.2). The maximum density of nanotubes accessible through this method was also investigated,[41,46] and it was found that there are some intrinsic limitations, with the maximum density on the order of $\pi/2l$, where l is the nanotube length in units of length/length^2. This phenomenon may limit applications of molecular combing in cases where very high densities are needed.

16.1.3 *Molecular Combing as a General Technique*

In this chapter, we try to present a mechanistic view of molecular combing so that this process can be adapted to new applications. The theoretical formulation for deposition and alignment,[41] which we develop further here, originally described the deposition and alignment of DNA-SWCNTs. Due to the similarities between dsDNA and DNA-SWCNTs, the theory should be applicable to both cases.

Molecular combing occurs in two steps, the deposition of the object on the surface and its alignment by a passing meniscus (Fig. 16.1a,b). We discuss first the deposition and then the alignment, addressing the techniques that have been developed to control both processes. The theoretical framework developed for the system will be presented.

The deposition of a charged entity, such as a DNA molecule, is governed by the strength of electrostatic and Van der Waals interactions as per Derjaguin and Landau, Verwey and Overbeek (DLVO) theory.[47] In certain cases, such as with silica particles, where these forces can be weak, other forces such as interaction with the solvent through hydrogen bonding can play a role, but with polyelectrolytes such as DNA this is not expected to be the case because of strong electrostatic forces.

When it comes to controlling the deposition process, electrostatic interactions are of particular interest because unlike Van der Waals forces they can be easily controlled by modifying the charge on the target surface and by changing the properties of the solvent. Thus the deposition of varying SAMs has been used to control where SWCNTs do and do not deposit.[33,40,41] Nanolithography techniques enable the charge of nanoscale areas to be precisely defined, controlling the location of the deposited object on the nanoscale.[40] Alternatively, the ionic strength of the solvent may be modified to modulate the strength of electrostatic interactions in general and thus control the rate of object deposition.[41]

After deposition on the surface, the liquid–air meniscus must pass over deposited objects when the liquid is removed. During this process the molecule aligns perpendicular to the meniscus—rather than parallel to the direction of motion, as intuition might suggest, due to hydrodynamic forces. This is particularly evident during molecular combing of DNA in microchannels (Fig. 16.1e).[19] In order for this to occur, certain conditions must be satisfied. First, the adsorption of the molecule must be strong enough to not be dragged excessively by the meniscus, as this can remove molecules or produce curious roping effects (which could be useful in its own right; see below and Fig. 16.6c). Second, the adsorption must be weak enough to allow the meniscus to pivot the molecule. At first glance this might seem paradoxical: if the molecule is bound

Table 16.1 Summary of molecular combing investigations for DNA and DNA-SWCNTs

Study	Molecules combed	Substrate	Main findings
		DNA combing studies	
Bensimon *et al.*, 1994.[6]	Bacteriophage λ DNA (10^6bp)	SAM of vinyl-terminated alkenes on glass	Over 70% of DNA molecules extended by combing; density of molecules increases with DNA concentration and at low pH (where surface charge is reduced).
Bensimon *et al.*, 1995[13]	Long dsDNA	SAM of vinyl-terminated alkenes, antidigoxigenine-treated glass	DNA attached at both ends results in loops; combed DNA can be stretched >2x relaxed length, depending on hydrophobicity of the substrate; estimations of meniscus force on the DNA molecule.
Allemand *et al.*, 1997[14]	Long dsDNA	Vinyl SAM, polystyrene, poly-L-lysine on glass	Variation of combing properties with surface treatment and solution pH investigated.
Michalet *et al.*, 1997[15]	Genomic DNA from eukaryotes	Vinyl-terminated SAM on glass	Demonstrated the combing of entire eukaryotic genomes; demonstrated mapping of labeled DNA segments by fluorescence microscopy.
Li *et al.*, 1998[21]	λDNA	Clean mica surface	Demonstration of molecular combing on a clean mica surface, without end attachment to vinyl groups.
Petit *et al.* 2003[19]	ΛDNA	Polystyrene-coated glass	Demonstrated combing of DNA in microchannels, with alignment strictly perpendicular to the meniscus; demonstrated various alignment patterns by controlling the shape of the meniscus.
Guan *et al.*, 2005[20]	λDNA	PDMS	Combined molecular combing with soft lithography, generating a large variety of aligned, well-ordered DNA arrays.
		CNT combing studies	
Gerdes *et al.*, 1999[33]	SDS-dispersed SWCNT	CH_3-, CF_3−, and NH_2-modified silicon and gold electrodes	Combed an SWCNT across two gold electrodes and measured its resistance.

(*Contd.*)

Table 16.1 *(Contd.)*

Study	Molecules combed	Substrate	Main findings
Ondarcuhu *et al.*, 2000[34]	SDS-dispersed SWCNT	CH_3-modified gold electrodes and silicon	Demonstrated the parallel fabrication of a large number of transistors through molecular combing.
Sagnes *et al.*, 2003[35] Sagnes *et al.*, 2004[36] Cleuziou *et al.*, 2006[37]	SDS-dispersed SWCNT	CH_3-modified gold electrodes and silicon	Demonstrated that electronic devices made from molecularly combed SWCNTs can be used to study complicated electronic properties of the SWCNTs.
Ko *et al.*, 2004[40]	SDS-dispersed SWCNT	NH_2 SAM-modified silicon	Demonstrated a variety of SWCNT densities and showed that SWCNTs are aligned perpendicular to the meniscus; proposed COOH groups at nanotube ends are responsible for anchoring.
Khripin, 2008[46]	ssDNA-dispersed SWCNT	CH_3-modified silicon	Proposed a modified random sequential adsorption theory to predict density of combed molecules.
Khripin *et al.*, 2009[41]	ssDNA-dispersed SWCNT	CH_3-modified silicon	Demonstrated that the deposition of SWCNTs for molecular combing follows the standard DLVO-activated barrier mechanism.

Abbreviation: PDMS, polydimethylsiloxane.

weakly enough to be pivoted by meniscus forces, would it not also be susceptible to being washed away? However, in reality, this process is rather robust, and by carefully controlling surface preparation one can reliably obtain molecular combing.

The molecular combing of SWCNTs is an example of such adaptation. When dispersed in solution, for example, by SDS, the SWCNT becomes a negatively charged object. This object does not deposit on silica, which carries a strongly negative charge at neutral pH. However, when the surface is covered with a SAM, such as of chlorodimethyloctyl silane, the charge is partly negated, resulting in deposition and making alignment possible. A fluorinated monolayer is insufficiently adhesive, and most nanotubes are washed away. On the other hand, a cationic monolayer produces copious deposition, but the binding of the SWCNT is too strong and no alignment occurs.[33]

16.2 The Process of Deposition

The process of molecular combing consists of object deposition and subsequent alignment by the moving meniscus. Here we present experimental results on deposition kinetics and an accompanying theoretical framework.[41,46] The experiments were carried out with DNA-SWCNTs (to our knowledge the kinetics of DNA absorption for molecular combing have not been well studied), but because of similarities between the two molecules,[45] we expect the results and theory to be applicable to DNA.

In molecular combing, deposition is generally kept relatively slow because too high a surface concentration of molecules inhibits the alignment process. Under such conditions, an energy barrier exists, which depositing molecules must "hop" over in order to deposit. This energy barrier can be described by DLVO theory;[47] it can arise when the attractive van der Waals force between the object and the surface is approximately balanced by a repulsive electrostatic force. Next we discuss how such surface conditions can be achieved and how this affects the deposition on the surface.

The following discussion is based on experiments conducted on alkene SAM-modified silicon (with a "native oxide" layer of silica). However, if the electrostatic and van der Waals interactions are known, this discussion can be readily adapted for other surfaces, such as gold and mica, which are in some cases used for polyelectrolyte deposition (e.g., Li *et al.*[21])

16.2.1 *The Interaction Potential of a Long Charged Molecule with a Charged Surface*

DNA is often modeled as a hydrophilic cylinder with a constant charge density of 5.8 e^-/nm.[48] Recent experiments have shown that DNA-dispersed SWCNTs have similar electrostatic properties, with a charge density on the order of 5.8 to 6.2 e^-/nm.[45] Thus a similar theoretical framework can be applied to both. In the case of surfactant-dispersed nanotubes less information is available, but presumably the micelle enclosing the SWCNT can possess an even higher charge due to the large number of surfactant molecules (of which SDS is commonly used) surrounding the nanotube. It should be noted that surfactant systems are complicated by the presence of empty surfactant micelles and by interaction of the surfactant with certain surfaces.

The kinetics of deposition can be modeled as an activated process, and the rate of deposition will depend on the energy barrier height, ΔE. To examine the functional dependence of ΔE on various experimental parameters, we consider that the interaction potential between the charged DNA-SWCNTs and the substrate consists of two components, electrostatic repulsion and van der Waals attraction, as in DLVO theory. The interaction potential depends on separation between, and relative orientation of, the DNA-SWCNT and the surface. The potential barrier is lowest when the axis of the DNA-SWCNT is perpendicular to the surface. This has been confirmed numerically: we have previously examined deposition by thermal hopping over an electrostatic energy barrier of DNA-SWCNT rods aligned parallel to the substrate[49] in order to explain a different set of experiments,[50] and the potential barrier can be far higher.

The interaction energy e (in units of k_bT) per unit length of DNA-SWCNTs with the surface is, in the Debye-Hückel approximation,[47]

$$e = \frac{\phi_o \lambda}{k_b T} \exp(-x/l_d) - \frac{A r_c t_c}{3x^3 k_b T}, \tag{16.1}$$

where the first term represents electrostatic repulsion and the second term represents van der Waals attraction, ϕ_o is the surface potential, x is the distance from the surface, λ is the effective linear charge density of the DNA-SWCNT (which includes the counterion condensation effect, which can be described by several theoretical formulations, including the Manning counterion condensation theory[51] and the standard nonlinear Poisson-Boltzmann theory,[52]) A is the Hamaker constant, r_c is the tube radius, t_c is the tube wall thickness, k_b is the Boltzmann constant, and T is the temperature. Finally, l_d is the Debye length, given by

$$l_d = \sqrt{\frac{\varepsilon_w \varepsilon_o k_b T}{2 N_a e^2 c}}, \tag{16.2}$$

where ε_w is the relative dielectric constant of water, ε_o is the permittivity of free space, N_a is the Avogadro number, e is the charge of the electron, and c is the concentration of a 1-1 salt. We integrate Eq. 16.1 along the length of a DNA-SWCNT rod aligned perpendicular to the substrate to find the interaction energy of the entire nanotube. Because the DNA-SWCNT is long (hundreds of nanometers) compared to the decay lengths of the forces involved, we can assume that it extends from the surface to infinity. The tip of the DNA-SWCNT is a distance x away from the surface and the integration variable $x' = 0$ at that point. The integral becomes:

$$\begin{aligned} E(x) &= \int_0^\infty \left(\frac{\phi_o \lambda}{k_b T} \exp\left(-\left(x + x'\right)/l_d\right) - \frac{A r_c t_c}{3\left(x + x'\right)^3 k_b T} \right) dx' \\ &= \frac{\phi_o \lambda l_d}{k_b T} \exp\left(-x/l_d\right) - \frac{A r_c t_c}{6 x^2 k_b T} \end{aligned} \tag{16.3}$$

In Fig. 16.3 we plot the interaction potential using values for parameters relevant to experimental conditions.[a] For example, surface potential of 80 mV at $l_d = 3.1$ nm in (a) corresponds to a

[a]The following parameters were used to calculate interaction potentials:
Dielectric constants for water: $\varepsilon_w = 80$ and $\varepsilon_o = 8.81 \cdot 10^{-12}$ C^2/Nm^2
Thermal energy of at room temperature: $k_b T = 4.1 \cdot 10^{-21}$ J
Length of the nanotube, confirmed with AFM, controlled by SEC: 560 ± 50 nm
Approximate nanotube radius and thickness: $r_c = 0.5$ nm and $t_c = 0.34$ nm
Hamaker constant: $A = 2 \cdot 10^{-20}$ J
Charge density, SWCNT saturated with DNA, including the effect of counterion condensation: $\lambda = 2.2 \cdot 10^{-10}$ C/m

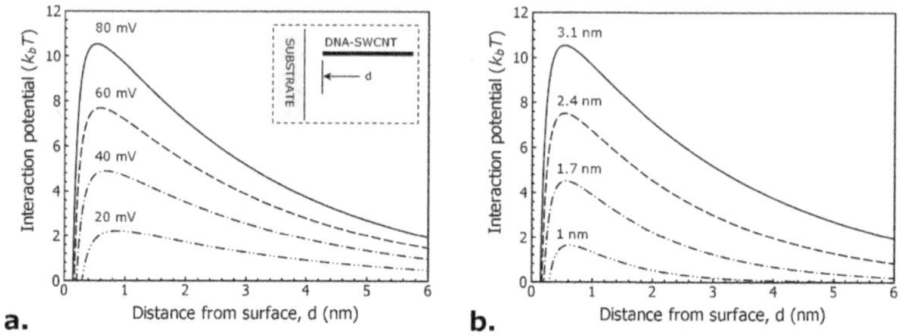

Figure 16.3 The interaction potential between a high-aspect-ratio object and a surface of like electrostatic charge contains a barrier. This barrier height varies with surface potential (a) and Debye length (b). Thus these two factors can be employed to modulate deposition rate. Modeling parameters reflect experimental conditions; for a complete set see the text.[41]

silica surface at pH 8 and 10 mMol/L salt and 20 mV to a surface at pH 3.5. In Fig. 16.3b a Debye length of 3.1 nm corresponds to 10 mMol/L NaCl solution and 1 nm to 100 mMol/L NaCl solution. As expected, there is an electrostatic potential barrier, ΔE, which the DNA-SWCNT must overcome in order to reach the deep potential well near the surface where the DNA-SWCNT will presumably zip into contact and stick. Note that the height of this barrier varies approximately linearly both with (a) surface potential and (b) Debye length. We now use these results to derive the expression for deposition kinetics.

16.2.2 *Deposition Kinetics*

As discussed before, the deposition during molecular combing is usually controlled by means of an energy barrier. Thus, deposition kinetics are governed by activated, or Arrhenius, kinetics and the flux, J, is given by

$$J = J_{\mathrm{d}} \exp\left(-\Delta E\right), \tag{16.4}$$

where J_{d} is the flux in the absence of a barrier, which is proportional to the bulk density of DNA-SWCNT.[53] This can also summarized as follows: The stability ratio, which determines the reduction in deposition rate compared to the reference flux, increases exponentially with the potential barrier height.[54]

We consider two models for deposition kinetics. The first (case I) assumes that depositing nanotubes increase surface charge homogeneously. The corresponding surface potential and potential barrier, ΔE, also increase. As ΔE increases, the deposition rate is reduced. This model is useful for cases where deposition is extensive or the salt concentration is very low, although as we will see below, it was not applicable to molecular combing of DNA-SWCNTs.

The second kinetics formulation is applicable to cases where deposition density is low. In this case (case II), we assume that the surface has a maximum number of sites that the DNA-SWCNT can occupy and that deposition gradually slows as sites are filled. The number of such sites is limited because deposited molecules, by virtue of not aggregating in solution, must necessarily repel other molecules and thus block part of the surface by depositing. Such processes are governed by random sequential absorption (RSA) theory.[55,56] Kinetics formulations have been developed for some special geometries,[57,58] but these are not applicable to molecular combing of long molecules.

16.2.2.1 Case I (charge accumulation–limited deposition)

As noted above, we find that this kinetics formulation is applicable to cases where the salt concentration is low, and deposition extensive, so that the "smeared charge" approximation holds. Assuming that depositing DNA-SWCNTs form a homogenous layer on the surface implies that the surface charge will increase with deposited density ρ_s (in μm of nanotubes per μm^2 of surface). As surface charge increases, surface potential relative to bulk solution will also increase. Within the linearized Debye-Hückel approximation, the surface potential is related to surface charge as[47]

$$\phi_o = \frac{l_d \sigma}{\varepsilon_w \varepsilon_o}, \tag{16.5}$$

where σ is the surface charge density. When the nanotubes deposit, this model assumes that the surface charge density increases according to $\sigma = \sigma_i + \rho_s \lambda$, where σ_i is the initial charge density on the surface. Thus the surface potential will increase with deposition according to

$$\phi_o = \phi_o^i + \frac{l_d \rho_s \lambda}{\varepsilon_w \varepsilon_o}, \qquad (16.6)$$

where ϕ_o^i is the initial surface potential. Surface potential thus varies linearly with ρ_s. We now demonstrate that ΔE varies approximately linearly with ϕ_o, as shown qualitatively in Fig. 16.3.

We start by finding x_{peak}, the location of the barrier, by setting $dE/dx = 0$. E in Eq. 16.3 has two extrema, one at the barrier ΔE and a secondary minimum further away from the surface. For small x_{peak}/l_d we find the location of the barrier to be:

$$x_{peak} \approx \left(\frac{Ar_c t_c}{3\phi_o \lambda}\right)^{1/3} \qquad (16.7)$$

For our parameters, x_{peak}/l_d is significantly smaller than unity, which justifies the approximation. Substituting Eq. 16.7 in Eq. 16.3 and again employing the approximation that x_{peak}/l_d is small yields:

$$\Delta E \approx \frac{\phi_o \lambda l_d}{k_b T} \qquad (16.8)$$

Substituting Eq. 16.6 in Eq. 16.8 shows that ΔE varies linearly with deposition density, ρ_s. Combining these two equations with Eq. 16.4 gives

$$J = \frac{d\rho_s}{dt} = \underbrace{J_d \exp\left(-\Delta E_o\right)}_{J_o} \exp\left(-\frac{\rho_s}{\rho *}\right), \qquad (16.9)$$

where ΔE_o is the initial barrier height, J_o is the initial flux, and ρ^* is a reference surface density. Equation 16.9 can be integrated to yield:

$$\rho_s = \rho^* \ln\left(\frac{J_o t}{\rho^*} + 1\right) \qquad (16.10)$$

Such logarithmic kinetics are expected when the deposition process results in an increase in the barrier and has been observed in oxidation processes.[59] Equation 16.10 can be used to describe deposition in the case of varying surface charge (such as for a silica surface at different pH.[60]) or in the case of varying Debye length (e.g., when the salt concentration is varied). We will revisit this point when we consider experimental data.

16.2.2.2 Case II (site-limited deposition)

Now we consider the second case, when deposition is limited by the number of sites available on the surface. The flux of nanotubes onto the surface will change as the number of available sites decreases, and the differential equation becomes.[61]

$$J = \frac{d\rho_s}{dt} = \left(1 - \frac{\rho_s}{\rho_t}\right) J_d \exp\left(-\frac{\Delta E}{k_b T}\right), \tag{16.11}$$

where ρ_t is the number of available surface sites. This yields:

$$\rho_S = \rho_t\left(1 - \exp\left(\frac{-J_o t}{\rho_t}\right)\right) \tag{16.12}$$

Note that J_o, the initial deposition rate, has the same form as in Eq. 16.9. At this point, we do not address what exactly determines the number of sites ρ_t; practically speaking, it is the largest density of deposited objects that can be attained for a particular set of experimental conditions. Below we will attempt to explain the nature of these sites for the case of DNA-SWCNTs, which can be approximated as rigid rods, at least in cases where the molecule is not very long compared to the persistence length.

16.2.3 *Experimental Studies of Deposition Kinetics*

More work exploring the deposition kinetics of DNA is needed; here we discuss studies on the deposition kinetics of DNA-SWCNTs. The focus of these studies is exploring the three experimental parameters that can be adjusted to control the number of objects deposited on the surface: time, pH, and ionic strength.

16.2.3.1 Measuring deposition rate

With advances in AFM, *in situ* imaging has become easier and more accessible, with many techniques available for imaging the deposition of macromolecules in real time.[62] Probably it would not be difficult to conduct a thorough *in situ* study of deposition kinetics on dsDNA or DNA-SWCNTs. Unfortunately, to our knowledge such a study has not been undertaken. One reason is that perhaps the final density of objects observed on the surface after molecular combing is reduced by the combing process,[46] and it is more relevant to consider the number of molecules observed on the surface in the dry state after combing. This is the approach we present here.

DNA-SWCNTs at a low concentration of 1 μg/mL were incubated on a surface treated with chlorodimethyloctyl silane. This treatment creates a SAM of alkene chains on the surface. It favors the deposition of negatively charged DNA and DNA-SWCNTs on the surface by reducing the number of ionisable siloxane groups.[63] This process also creates the highly hydrophobic, low-hysteresis surface necessary to avoid evaporation at the meniscus and to allow molecular combing to take place. Note that molecular combing has been demonstrated on clean, hydrophilic mica, but the set of conditions was very narrow.[21]

After allowing the DNA-SWCNT solution to incubate on the surface for some time, the meniscus was set in motion using the motorized stage at a rate of $\approx$1 mm/s; based on the mechanism proposed here, we do not anticipate a significant effect of velocity on alignment, but this question was not investigated further. The sample was subsequently imaged by AFM, and the number and alignment of SWCNTs were measured (see Fig. 16.2a,b). Note that the DNA-SWCNTs used here were length-sorted via SEC, and the length of the SWCNTs was 500 nm $\pm$ 10% (standard deviation).[42]

16.2.3.2 Increase in the number of objects with incubation time

Figure 16.4a shows the increase of measured length of combed DNA-SWCNTs with time. Sample AFM images are shown in Fig. 16.4b. This experiment was conducted under standard conditions, at pH 8, 0.1 mMol/L phosphate buffer, and 10 mMol/L NaCl.

Localized deposition near the meniscus during meniscus motion has been reported for general colloidal particles[64] and for SWCNTs.[40] In those studies particles deposit chiefly at the meniscus, where drying effects increase their local concentration (see chapter 15 for a detailed description of this process). Here, kinetics data shows that deposition increases with time and that there is negligible deposition for very small incubation times. This latter fact indicates that in our case no significant deposition occurs at the meniscus. Furthermore, on examining areas on the substrate where the drop had a total residence time of less than five seconds, we found that there was only an occasional deposited nanotube, at a concentration of less than 0.05 μm/μm^2 (length of SWCNTs deposited/area).

We find that the observed deposition kinetics agree well with our activated kinetics formulation, with the site-limited model (Eq. 16.12) plotted in Fig. 16.4a. A reference deposition flux (number depositing per unit time per unit surface area) can be estimated for conditions when there is a negligible potential barrier based on the idea that it is limited only by diffusion.[53] To obtain a conservative estimate for this flux, we can use the relation $J_d = D\rho_b/L$, where J_d is the flux, D is the diffusivity, ρ_b is the bulk density of nanotubes (1 nM), and L is some length scale. We use a depletion argument to find L; assuming a linear concentration profile, we define $L \equiv \rho_s/\rho_b$, where we take $\rho_s = 2$ μm/μm^2 as a high observed surface density of nanotubes. Finally as an estimate of diffusivity we take the diffusivity of the tomato bushy stunt virus, which is an object roughly 10 times heavier with an aspect ratio of 5:1, $D = 1\cdot10^{-7}\text{cm}^2/\text{s}$.[65] Thus we arrive at flux $J_d = 1$ μm/μm^{2}s. The fact that this quantity is much larger than measured deposition rates strongly suggests that deposition is barrier limited. Experiments on the pH and salt concentration dependence of deposition, discussed next, establish that the barrier is electrostatic in nature.

16.2.3.3 Effect of pH

We have seen that the energy barrier to deposition (Fig. 16.3a) depends linearly on the surface potential. Thus, we may control the rate of deposition by controlling the surface potential, which can in turn be controlled by solution pH, since the concentration of ionized groups on a silica surface depends on pH. For a silica surface with octyl dimethyl chlorosilane (ODC), Huang *et al.* found that the surface charge increases approximately linearly from 0 at pH 2 to full ionization at pH 8.[63] This implies that surface hydroxide groups must have a range of pK_a values. This is a known phenomenon for polysilica; while silicic acid has a pK_a of 9, fully conjugated silica is thought to have a pK_a of about 2.[60]

The density of observed DNA-SWCNTs as a function of pH, at a fixed incubation time and ionic strength, is shown in Fig. 16.4c.

As expected, deposition is faster at low pH values and slows at increased pH. We now fit our kinetics models to the experimental data.

For charge accumulation limited deposition (case I), using the fact that the initial surface potential, ϕ_o^i, varies approximately linearly with pH from nearly 0 mV at pH 2 to about 80 mV at pH 8 in Eq. 16.6 and applying it to Eq. 16.8, we conclude that the potential barrier will vary linearly in both the pH and surface density of deposited nanotubes, which we may write as

$$\Delta E = \overbrace{C_1 pH + C_2}^{\Delta E_o} + \frac{\rho_s}{\rho^*}, \tag{16.13}$$

where C_1, C_2, and ρ^* are constants, making the differential equation

$$J = \frac{d\rho_s}{dt} = \underbrace{J_d \exp\left(-\left(C_1 pH + C_2\right)\right)}_{J_o} \exp\left(-\frac{\rho_s}{\rho *}\right), \tag{16.14}$$

producing the solution in Eq. 16.10 with J_o now a function of pH. The constant C_2 can be absorbed into J_d to give

$$J_o = J_d' \exp\left(-C_1 pH\right). \tag{16.15}$$

We substitute Eq. 16.15 in Eq. 16.10. If we now expand Eq. 16.10 about $t = 0$, we expect an exponential decay in deposition density with increasing pH for small times ($J_o t/\rho^* <<1$), which is qualitatively consistent with the data in Fig. 16.4c. Shown is a fit using Eq. 16.10 with the value for $\rho^* = 0.8$ μm/μm^2, obtained by fitting the kinetics experiment. It can be seen that the fit is able to capture well the observed reduction in deposited density.

We now consider site-limited (case II) kinetics. In order to fit the equation to the pH dependence data, we use a relation similar to Eq. 16.13 (but without the surface accumulation term), $\Delta E = C_1 \cdot \text{pH} + C_2$. The fit is made using the value $\rho_t = 0.8$μm/μm^2 from the kinetics experiment (Fig. 16.4a). Note that just as in case I, this model predicts exponential decay of ρ_s with pH for small $J_o t/\rho_t$. Both models fit the data adequately here, although only case II is shown.

16.2.3.4 Effect of ionic strength

We have shown that the energy barrier opposing deposition is linear with Debye length (Fig. 16.3b), which depends in turn on the ionic

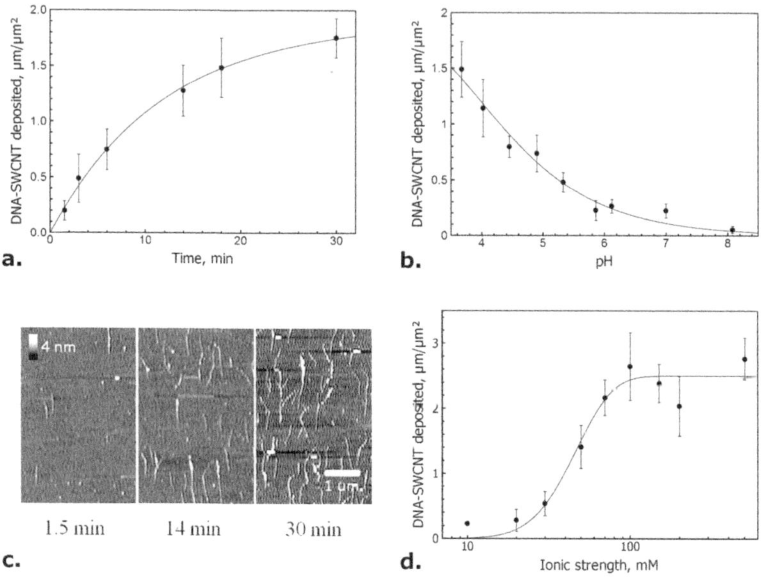

Figure 16.4 The density of combed objects depends on the number of objects allowed to deposit on the surface prior to combing. Thus an increase in incubation time leads to an increase in density (a), based on measurements of AFM images (b). The deposition rate can be increased by lowering the energy barrier, which in this case is done by lowering pH (c) or raising ionic strength (d). The kinetics model (see text) is fitted in (a), (b), and (c); error bars represent the 95% confidence interval in the mean.[41]

strength of the medium. Thus the salt concentration is another parameter that can be adjusted to control the number of molecules observed on the surface, and the data for DNA-SWCNTs is shown in Fig. 16.4d. This data was collected at pH 8 in 0.1 mMol/L phosphate buffer, with the ionic strength controlled by adding NaCl up to a concentration of 500 mMol/L.

This experiment is significant because it shows that there is a maximum deposition density that can be achieved. This maximum could not be reached in the kinetics experiment because of the very long time period that would be required and in the pH experiment at pH below 3 the DNA-SWCNTs coagulate. Thus, increasing the salt concentration provides a simple way of maximizing deposition density, while still producing aligned DNA-SWCNTs.

We now apply our kinetics models to this experiment. Under conditions where the energy barrier is linear in surface potential, that is, for small x/l_d, it follows from Eq. 16.3 that $\Delta E \propto l_\mathrm{d}$. If we represent this linear dependence by

$$\Delta E = \overbrace{D_1 l_\mathrm{d} + D_2}^{\Delta E_\mathrm{o}} + \frac{\rho_\mathrm{s}}{\rho *} \tag{16.16}$$

then $J_\mathrm{o} = J_\mathrm{d}'' \exp(-D_1 l_\mathrm{d})$ in Eq. 16.10. For low salt concentration, the model captures well the observed increase in deposition density with increasing salt concentration. For higher concentrations, experiments show saturation, whereas the model predicts that deposition would keep increasing since in this model there is no natural limit to deposition. Thus we conclude that charge-limited deposition kinetics probably are not the right choice for describing the present experimental situation.

In case II kinetics, on the other hand, the density of deposited objects is limited by the number of available sites. The fit of Eq. 16.12 in the Fig. 16.4d fit was made with $\rho_\mathrm{t} = 2.5\ \mu\mathrm{m}/\mu\mathrm{m}^2$, since the deposition appears to reach this clearly defined limit. This is slightly greater than the site density obtained at a lower salt concentration from the fit of the kinetics data in Fig. 16.4a, $\rho_\mathrm{t} = 1.9 \mu\mathrm{m}/\mu\mathrm{m}^2$. In this case, $\Delta E = D_1 l_\mathrm{d} + D_2$.

Since case II kinetics, but not case I kinetics, successfully predict a limit to the deposition density, we have plotted only the case II fits in Fig. 16.4. We will now consider the possible nature of these sites.

16.2.3.5 The nature of limits to deposition density

The quantity ρ_t reflects this maximum density of surface sites. Its value was determined by the above experiments to be in the range of $\rho_\mathrm{t} = 1.9$ to $2.5\ \mu\mathrm{m}/\mu\mathrm{m}^2$. Given the small area occupied by a DNA-SWCNT ($2\ \mathrm{nm} \times 2\ \mu\mathrm{m} = 0.004\ \mu\mathrm{m}^2$, or just 0.4% of the surface occupied), what could be the nature of the sites to give such a small surface density? Here we consider both the traditional RSA theory and a new formulation to answer this question.

Let us first consider the possibility that these surface sites are a result of electrostatic exclusion by the deposited nanotubes. As a conservative estimate, suppose that each nanotube excludes an area $2l_d$ wide around itself. Then, knowing $l_d = 3$ nm (at low salt concentration, less at higher concentrations) and the nanotube length is 500 nm, the nanotube can be considered a rectangle with a 40:1 aspect ratio. Each of these rectangles deposits on the surface and occupies some amount of space, preventing other rectangles from depositing in that area. According to a study of RSA of rectangles,[56] this aspect ratio gives a jamming density of $\rho_t = 27\ \mu m/\mu m^2$. At higher salt concentrations, the aspect ratio would be even higher and the jamming density would be even greater. Thus our measured limiting site density is an order of magnitude smaller than would be expected by ordinary site exclusion.

We propose another mechanism, "thwarted RSA," or "TRSA." This model[46] assumes the following: DNA-SWCNTs deposit randomly on the surface and are allowed to intersect. When molecular combing occurs, DNA-SWCNTs experience a strong force from the passing meniscus (see below) and are washed away unless they are flush with the surface. In other words, all DNA-SWCNTs that landed on top of another object are removed. Using a numerical model, we obtain the limiting deposition density of $\rho_t = 3.1414 \pm 0.0001\ \mu m/\mu m^2$ ($\pm$ standard error). This is an interesting result considering that it is very close to π, which has not emerged in the density predicted by standard RSA theory, although it appears, under somewhat different circumstances, in Buffon's needle problem.[66] In any case, this result is much closer to experimental results for maximum density than the conventional RSA of rectangles predicts.

16.3 Molecular Combing by the Liquid–Air Interface

DNA and DNA-SWCNTs deposit on a surface with random orientation. The subsequent passage of the liquid–air meniscus orients the objects perpendicular to the meniscus. Some workers have suggested that this occurs because the DNA binds strongly at the end;[13] however, this does not explain, for example, how DNA can be combed on mica[21] and inert hydrophobic surfaces.[19] Here we consider the forces that are involved in alignment and present a preliminary explanation.

16.3.1 *The Forces Responsible for Meniscus Alignment*

The DNA-SWCNTs or DNA deposited on the surface are exposed to meniscus and fluid forces. As suggested by the fact that in experiments alignment is perpendicular to the meniscus, not parallel to fluid flow,[19,40,41] the alignment is due to forces applied by the meniscus on DNA-SWCNT rods, not due to fluid friction. One can estimate the fluid and friction forces on the object to see if indeed the meniscus force is much greater.

16.3.1.1 The hydrodynamic and surface tension forces exerted during combing

We begin by estimating the force exerted by the meniscus on the nanotube, F_m, given by.[67]

$$F_m \approx \frac{\pi \gamma \alpha_o^2}{\log L/D} x, \tag{16.17}$$

where γ is the surface tension, α_o is the contact angle with the surface far away from the nanotube, and x is how far the contact at the defect lags behind the contact line where distortion is zero. L is the length scale at which distortion is zero, which we take here to be a typical distance between nanotubes on the surface, and $D = 1$ nm is the radius of the DNA-SWCNT complex. We estimate x in terms of models for the critical force F_c at which the contact line will release the defect. Assuming that we have a Gaussian defect, that is, a small circular defect where adhesion energy is strongest at the center, we may use the relation

$$F_c = \sqrt{2\pi} h_o D \tag{16.18}$$

to estimate the critical force F_c at which the contact line will let go. Here h_o is the work of adhesion of the defect. In the limit of the defect contact angle $\alpha_d \rightarrow 0$, $h_o = 1/2\gamma\alpha_o^2$. Using $\gamma = 0.072$ N/m, $\alpha_o = \pi/2$ rad, and $D = 1$ nm, we find $F_c = 220$ pN. Others have estimated this force as well; Bensimon *et al.* put the stretching force on the DNA at 400 pN.[6] and 500 pN;[13] Gerdes *et al.* put it at 350 pN.[33]

We can use several models to estimate the hydrodynamic frictional force. A simple way is to start with the fluid force on the cylinder immersed in perpendicular laminar flow; this would be an upper limit to the force since our cylinder is on the surface. We use a formula for the kinetic force F_k exerted on length L of a cylinder by perpendicular laminar flow, valid at Re <1.[68]

$$F_k = \frac{4\pi \mu v_{DNA-SWCNT} L}{\ln(7.4/Re)} \quad (16.19)$$

where μ is the fluid viscosity and $Re = D_{DNA-SWCNT} v_{DNA-SWCNT} \rho_w/\eta$ with $D_{DNA-SWCNT}$ the diameter of the DNA-SWCNT complex and ρ_w the density of water. For a velocity of 1 mm/s, $Re = 2.5 \cdot 10^{-6}$, this gives a friction force on the order of 0.5 pN.

Thus the surface tension force is on the order of 0.1 nN, while the hydrodynamic force is at most on the order of 1 pN. This implies that the alignment is probably due to surface tension forces, which explains why alignment is perpendicular to the meniscus rather than parallel to fluid flow.

16.3.1.2 The process of alignment

Let us consider how the meniscus force aligns the DNA or DNA-SWCNTs (for purposes of discussion we will use our group's original subject, DNA-SWCNTs, but the same argument applies to DNA). When the meniscus encounters the molecule on the surface (Fig. 16.5), it will be initially pinned by it because the DNA-SWCNT and DNA are hydrophilic, while the surface is hydrophobic (Fig. 16.5, meniscus position b). At some point, the meniscus will reach the position of greatest distortion (Fig. 16.5, meniscus position c). The meniscus will then exert a force, F_m, on the object. If this force is greater than some static frictional force F_f, the DNA-SWCNT will be moved. If, on the other hand, F_m is less than F_f, the nanotube will remain stationary. As argued by Gerdes *et al.*,[33] alignment requires the friction between the DNA-SWCNT and the surface to be neither too high nor too low. We point out that this is a paradox, because this simplified argument implies that the DNA-SWCNT is either washed away or unmoved.

The paradox can be overcome by a simple modification of the argument. Suppose that the meniscus reaches position b in Fig. 16.5. Thermal vibration (of the DNA-SWCNT or the meniscus), evaporation at the meniscus, or some other factor causes the tip of the DNA-SWCNT to emerge into the dry region (Fig. 16.5d) before

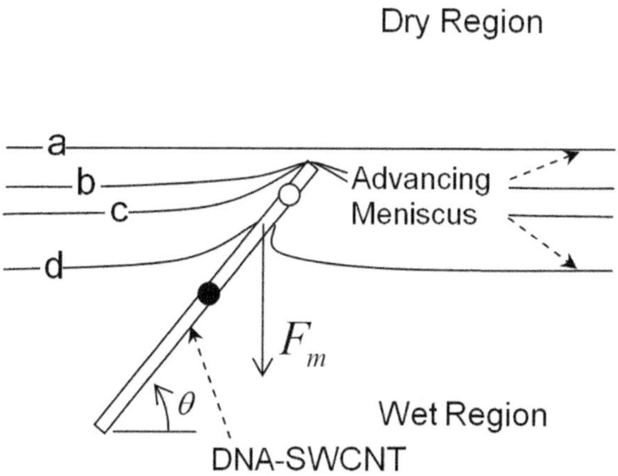

Figure 16.5 Here we consider the forces that act on the DNA-SWCNT or DNA molecule that produce the combing effect. At point (a) the meniscus approaches the object; at points (b) and (c) it begins to exert a force F_m, which would tend to misalign the object by rotating it around the dark circle. However, if the meniscus advances to point (d) on the molecule, then the force becomes aligning, rotating the rod about the empty circle.[41]

the meniscus force can move the nanotube or perhaps while the nanotube is moving. If the dry frictional force is much stronger than the wet frictional force, which is a reasonable expectation, the nanotube will not be swept away, although it might move slightly. At this point the force exerted by the meniscus will rotate the DNA-SWCNT about the hollow circle in Fig. 16.5. Bensimon *et al.* explain meniscus alignment of DNA in a similar manner by arguing that the ends of DNA have higher adhesion than the rest of the molecule.[6,13,14] In those studies, it was also asserted that the DNA was chemically bound at the end, whether through a vinyl group or by means of surface absorbed proteins. However, molecular combing of DNA is also possible on inert hydrophobic[19] or in some cases even on hydrophilic mica,[21] so chemical attachment of the end does not appear to be necessary.

We now demonstrate this more quantitatively. We assume that the friction between DNA-SWCNTs and the substrate is characterized by a coefficient, η, defined such that the friction force per unit length, f, is

$$\mathbf{f} = -\eta\,\mathbf{v}, \qquad (16.20)$$

where v is the local velocity of the DNA-SWCNT with respect to the substrate. Consider two limiting cases: (a) η is uniform everywhere, and (b) η is high in the "dry" region of the rod above the meniscus and is much smaller in the "wet" region (Fig. 16.5d). As expected from the discussion in the previous paragraph, we will find that in case (a) the motion of the DNA-SWCNT rod would reduce alignment, whereas in case (b) the rod would move to align itself normal to the meniscus.

From force balance of the rod (neglecting inertial terms) we conclude that

$$\begin{aligned} v_{cx} &= 0 \\ v_{cy} &= F_m/(\eta l) \end{aligned} \qquad (16.21)$$

where l is the length of the rod and v_{cx} and v_{cy} are the "x" and "y" components of the velocity of the center of mass of the region that has uniform friction.

For case (a), the center of mass is the filled circle in Fig. 16.5; for case (b) it is the unfilled circle. Moment balance of the rod (again neglecting inertia) requires

$$\eta\omega I = \mp F_m(s/2)\cos\theta, \qquad (16.22)$$

where I is the moment of inertia of the region of the rod that has nonzero friction and s is the length of this region. The "$-$" sign applies in case (a); the "+" sign applies in case (b), that is, in case (a) the moment about the center of mass is clockwise and Eq. 16.22 tells us that the rod will rotate with rate $\omega = d\theta/dt$, which is negative so that θ will reduce. This would be a misaligning tendency.

For case (b), where the friction in the dry region is considerably greater than in the wet region, the rod will rotate to align with the force. One can integrate the moment equation to determine how the angle will change in time:

$$\ln\left(\frac{\sec\theta + \tan\theta}{\sec\theta_o + \tan\theta_o}\right) = \frac{F_m s}{2\eta I}t, \qquad (16.23)$$

where θ_o is the initial orientation of the rod.

Our model for alignment of rods relies essentially on forces due to distortion of the meniscus. In this regard it is similar to examples of colloidal ordering at a liquid–vapor interface due to distortion of the interface[69,70] or ordering of membrane-bound DNA due to distortion of the membrane.[71] It would be interesting to carry out a study of molecular combing while varying the surface tension of the meniscus; to our knowledge such a study has not been done.

16.3.2 *Effects of Meniscus other than Simple Alignment*

Molecular combing can be used to demonstrate a number of interesting, and potentially useful, geometries of the combed molecule. These are summarized in Fig. 16.6.

For some applications it is necessary to produce a transparent conductor. An efficient way to create a percolating network is by using a layer of orthogonally aligned rods. Because DNA can

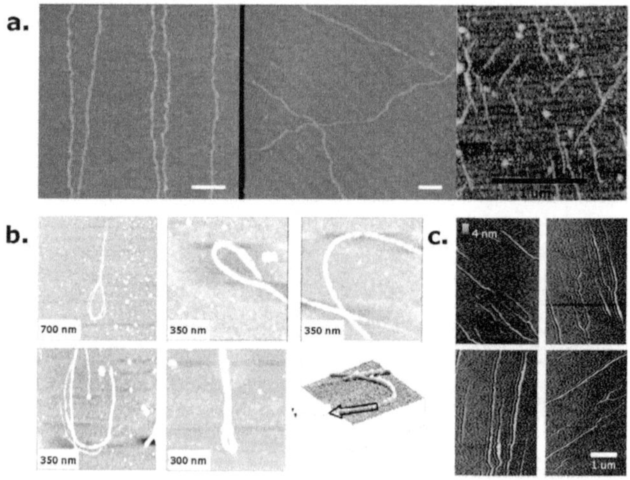

Figure 16.6 (a) Repeated deposition with a change in combing direction can create cross patterns of DNA on mica (left and centre panels) and of DNA-SWCNT (right panel; scale bar: 200 nm).[21,41] (b) Combing on a slightly hydrophilic surface can produce loops in the SWCNT due to small drops left behind by the meniscus.[40] (c) Finally, at large surface concentrations, braiding of DNA-SWCNTs into ropes has been observed, although it is not clear precisely what conditions produce this effect.[41]

be made conductive through metallisation,[22] and SWCNTs are naturally conducting or semiconducting, a percolating network of orthogonal molecules may be useful in a situation where maximum transparency is desired. Figure 16.6a shows crisscross patterns of DNA and DNA-SWCNTs. These patterns were created via layer-by-layer deposition in orthogonal directions; by adding more layers, greater surface density may be achieved. As discussed above, deposition does not favor objects depositing on other objects and intersections are not allowed. However, drying could alter the structure of both DNA and DNA-SWCNTs, allowing for intersections and thus a percolating network. One implication of the crossed patterns is that drying also "affixes" the objects to the surface, meaning that the same object cannot be combed twice if the surface is allowed to dry.

Although molecular combing generally produces straight, individual molecules, other shapes have been demonstrated. By using a slightly hydrophilic surface that leaves drops behind during combing, Ko *et al.* showed that "hooked" SWCNTs can be produced (Fig. 16.6b). Khripin *et al.* reported that under some conditions, the DNA-SWCNTs were drawn into long, continuous ropes (Fig. 16.6c). These effects could be of interest for problems involving the creation of SWCNT electronic devices.

16.4 Conclusion

Molecular combing has emerged as a mature technique for stretching individual molecules on a surface. Originally intended for stretching DNA so that it could be sequenced by AFM, this technique has been applied to many different high-aspect ratio objects (DNA-SWCNTs, nanowires, viruses, etc.) for a variety of applications. The main advantage of molecular combing is excellent control of alignment angle though control of the meniscus and the potentially large scale of alignment. One disadvantage is lack of control in spatial distribution of objects, although by combining this technique with lithography this can be overcome.[20,40]

Molecular combing is a flexible technique, in that many different substrates have been used, including gold,[33] silicon,[34,40,41] silica,[6,19] and mica.[21] In many of these cases, surface modification by means of SAMs was used to enhance deposition and combing. Although it is clear how these layers improve deposition by controlling surface charge, many factors remain unknown, for example, whether there is adhesion between the end of the combed molecule and the surface and what frictional forces exist during alignment.

We expect that as technology calls for ever greater control over molecules at the nanoscale, molecular combing will find extensive application for the creation of devices. Perhaps the most prominent of these are DNA-SWCNT transistors, which have shown potential as molecular sensors; however, other molecular wires, such as poly-G DNA[72] and bacterial wires,[73] could be combed and made into devices as well for specialized applications.

In a general sense, molecular combing is the link between interesting molecules in suspension and their utilization. Nanowires, nanorods, and CNTs that are available in solution form must be placed on a surface in order to be useful; DNA must be placed on a surface in order to be visualized. It is as a link between 3D and 2D that molecular combing has made the greatest contribution.

References

1. G. Binnig, C. F. Quate, and C. Gerber, Atomic force microscope, *Phys. Rev. Lett.*, **56**, 930–933 (1986).
2. G. Binnig, H. Rohrer, C. Gerber, and E. Weibel, Surface studies by scanning tunneling microscopy, *Phys. Rev. Lett.*, **49**, 57–61 (1982).
3. G. Travaglini, H. Rohrer, M. Amrein, and H. Gross, Scanning tunneling microscopy on biological matter, *Surf. Sci.*, **181**, 380–390 (1987).
4. A. L. Weisenhorn, M. Egger, F. Ohnesorge, S. A. C. Gould, S. P. Heyn, H. G. Hansma, R. L. Sinsheimer, H. E. Gaub, and P. K. Hansma, Molecular-resolution images of Langmuir-Blodgett films and DNA by atomic force microscopy, *Langmuir*, **7**, 8–12 (1991).
5. H. G. Hansma and J. H. Hoh, Biomolecular imaging with the atomic force microscope, *Annu. Rev. Bioph. Biom.*, **23**, 115–140 (1994).
6. A. Bensimon, A. Simon, A. Chiffaudel, V. Croquette, F. Heslot, and D. Bensimon, Alignment and sensitive detection of DNA by a moving interface, *Science*, **265**, 2096 (1994).

7. H. Tanaka and T. Kawai, Partial sequencing of a single DNA molecule with a scanning tunnelling microscope, *Nat. Nanotechnol.*, **4**, 518–522 (2009).
8. J. Hu, Z. H. Zhang, Z. Q. Ouyang, S. F. Chen, M. Q. Li, and F. J. Yang, Stretch and align virus in nanometer scale on an atomically flat surface, *J. Vac. Sci. Technol., B* **16**, 2841–2843 (1998).
9. T. Ondarcuhu and C. Joachim, Combing a nanofibre in a nanojunction, *Nanotechnology*, **10**, 39 (1999).
10. J. Guan, B. Yu, and L. J. Lee, Forming highly ordered arrays of functionalized polymer nanowires by dewetting on micropillars, *Adv. Mater.*, **19**, 1212–1217 (2007).
11. W. Salalha and E. Zussman, Investigation of fluidic assembly of nanowires using a droplet inside microchannels, *Phys. Fluids*, **17**, 063301 (2005).
12. H. G. Hansma, J. Vesenka, C. Siegerist, G. Kelderman, H. Morrett, R. L. Sinsheimer, V. Elings, C. Bustamante, and P. K. Hansma, Reproducible imaging and dissection of plasmid DNA under liquid with the atomic force microscope, *Science*, **256**, 1180–1184 (1992).
13. D. Bensimon, A. J. Simon, V. Croquette, and A. Bensimon, Stretching DNA with a receding meniscus: experiments and models, *Phys. Rev. Lett.*, **74**, 4754–4757 (1995).
14. J. F. Allemand, D. Bensimon, L. Jullien, A. Bensimon, and V. Croquette, pH-dependent specific binding and combing of DNA, *Biophys. J.*, **73**, 2064–2070 (1997).
15. X. Michalet, R. Ekong, F. Fougerousse, S. Rousseaux, C. Schurra, N. Hornigold, M. Slegtenhorst, J. Wolfe, S. Povey, and J. Beckmann, Dynamic molecular combing: stretching the whole human genome for high-resolution studies, *Science*, **277**, 1518 (1997).
16. Z. Gueroui, C. Place, E. Freyssingeas, and B. Berge, Observation by fluorescence microscopy of transcription on single combed DNA, *Proc. Natl. Acad. Sci. USA*, **99**, 6005–6010 (2002).
17. B. Geron-Landre, T. Roulon, P. Desbiolles, and C. Escude, Sequence-specific fluorescent labeling of double-stranded DNA observed at the single molecule level, *Nucleic Acids Res.*, **31**, e125 (2003).
18. J. Herrick and A. Bensimon, Introduction to molecular combing: genomics, DNA replication, and cancer. In eds., S. Vengrova and J. Z. Dalgaard, *DNA Replication*. Vol. 521. Humana Press. 1–32 (2009).
19. C. A. P. Petit and J. D. Carbeck, Combing of molecules in microchannels (COMMIC): a method for micropatterning and orienting stretched molecules of DNA on a surface, *Nano Lett.*, **3**, 1141–1146 (2003).

20. J. Guan and L. J. Lee, Generating highly ordered DNA nanostrand arrays, *Proc. Natl. Acad. Sci. USA* **102**, 18321 (2005).
21. J. Li, C. Bai, C. Wang, C. Zhu, Z. Lin, Q. Li, and E. Cao, A convenient method of aligning large DNA molecules on bare mica surfaces for atomic force microscopy, *Nucleic Acids Res.*, **26**, 4785–4786 (1998).
22. R. M. Stoltenberg and A. T. Woolley, DNA-templated nanowire fabrication, *Biomed. Microdevices*, **6**, 105–111 (2004).
23. S. Iijima and T. Ichihashi, Single shell carbon nanotubes of one nanometer diameter, *Nature*, **363**, 503–605 (1993).
24. D. S. Bethune, C. H. Klang, M. S. De Vries, G. Gorman, R. Savoy, J. Vazquez, and R. Beyers, Cobalt-catalysed growth of carbon nanotubes with single-atomic-layer walls, *Nature*, **363**, 605–607 (1993).
25. R. Saito, G. Dresselhaus, and M. S. Dresselhaus, *Physical Properties of Carbon Nanotubes*. Imperial College Press (1998).
26. R. H. Baughman, A. A. Zakhidov, and W. A. De Heer, Carbon nanotubes—the route toward applications, *Science*, **297**, 787 (2002).
27. J. Kong, M. G. Chapline, and H. Dai, Functionalized carbon nanotubes for molecular hydrogen sensors, *Adv. Mater.*, **13**, 1384–1386 (2001).
28. M. Hersam, Progress towards monodisperse single-walled carbon nanotubes, *Nat. Nanotechnol.*, **3**, 387–394 (2008).
29. M. Zheng, A. Jagota, M. Strano, A. Santos, P. Barone, S. Chou, B. Diner, M. Dresselhaus, R. Mclean, and G. Onoa, Structure-based carbon nanotube sorting by sequence-dependent DNA assembly, *Science*, **302**, 1545 (2003).
30. M. Zheng, A. Jagota, E. Semke, B. Diner, R. McLean, S. Lustig, R. Richardson, and N. Tassi, DNA-assisted dispersion and separation of carbon nanotubes, *Nat. Mater.*, **2**, 338–342 (2003).
31. X. Tu, S. Manohar, A. Jagota, and M. Zheng, DNA sequence motifs for structure-specific recognition and separation of carbon nanotubes, *Nature*, **460**, 250–253 (2009).
32. L. Huang, Z. Jia, and S. O'Brien, Orientated assembly of single-walled carbon nanotubes and applications, *J. Mater. Chem*, **17**, 3863–3874 (2007).
33. S. Gerdes, T. Ondarçuhu, S. Cholet, and C. Joachim, Combing a carbon nanotube on a flat metal-insulator-metal nanojunction, *Europhys. Lett.*, **48**, 292 (1999).
34. T. Ondarçuhu, C. Joachim, and S. Gerdes, Co-planar carbon nanotube hybrid molecular transistors fabricated in parallel, *Europhys. Lett.*, **52**, 178 (2000).

35. M. Sagnes, B. Raquet, B. Lassagne, J. M. Broto, E. Flahaut, C. Laurent, T. Ondarçuhu, F. Carcenac, and C. Vieu, Probing the electronic properties of individual carbon nanotube in 35 T pulsed magnetic field, *Chem. Phys. Lett.*, **372**, 733–738 (2003).
36. M. Sagnes, J. M. Broto, B. Raquet, C. Vieu, V. Conedera, P. Dubreuil, T. Ondarçuhu, C. Laurent, and E. Flahaut, Nanodevices for correlated electrical transport and structural investigation of individual carbon nanotubes, *Microelectron. Eng.*, **73**, 689–694 (2004).
37. J. P. Cleuziou, W. Wernsdorfer, V. Bouchiat, T. Ondarçuhu, and M. Monthioux, Carbon nanotube superconducting quantum interference device, *Nat. Nanotech.*, **1**, 53–59 (2006).
38. L. Zhang, S. Zaric, X. Tu, X. Wang, W. Zhao, and H. Dai, Assessment of chemically separated carbon nanotubes for nanoelectronics, *J. Am. Chem. Soc.*, **130**, 2686–2691 (2008).
39. L. Zhang, X. Tu, K. Welsher, X. Wang, M. Zheng, and H. Dai, Optical characterizations and electronic devices of nearly pure (10, 5) single-walled carbon nanotubes, *J. Am. Chem. Soc.*, **131**, 2454–2455 (2009).
40. H. Ko, S. Peleshanko, and V. V. Tsukruk, Combing and bending of carbon nanotube arrays with confined microfluidic flow on patterned surfaces, *J. Phys. Chem. B*, **108**, 4385–4393 (2004).
41. C. Khripin, M. Zheng, and A. Jagota, Deposition and meniscus alignment of DNA-CNT on a substrate, *J. Colloid Interface Sci.*, **330**, 255–265 (2009).
42. X. Huang, R. S. Mclean, and M. Zheng, High-resolution length sorting and purification of DNA-wrapped carbon nanotubes by size-exclusion chromatography, *Anal. Chem.*, **77**, 6225–6228 (2005).
43. A. Jagota, B. A. Diner, S. Boussaad, and M. Zheng, Carbon nanotube—biomolecule interactions: applications in carbon nanotube separation and biosensing. In eds., P. Avouris, B. Bhushan, D. Bimberg, K. Klitzing, H. Sakaki, and R. Wiesendanger, *Applied Physics of Carbon Nanotubes*. Heidelberg, Berlin: Springer. 253–271 (2005).
44. S. Manohar, T. Tang, and A. Jagota, Structure of homopolymer DNA-CNT hybrids, *J. Phys. Chem. C*, **111**, 17835–17845 (2007).
45. C. Y. Khripin, S. Manohar, M. Zheng, and A. Jagota, Measurement of electrostatic properties of DNA-carbon nanotube hybrids by capillary electrophoresis, *J. Phys. Chem. C*, **113**, 13616–13621 (2009).
46. C. Khripin, Random sequential adsorption of line segments with a finite jamming limit, *J. Colloid Interface Sci.*, **323**, 213–215 (2008).
47. W. B. Rüssel, D. A. Saville, and W. R. Schowalter, *Colloidal Dispersions*. New York: Cambridge (1989).

48. S. A. Allison and S. Mazur, Modeling the free solution electrophoretic mobility of short DNA fragments, *Biopolymers*, **46**, 359–373 (1998).
49. C. Khripin, T. Tang, and A. Jagota, Deformation of a sheet with coupling between elasticity and concentration, *J. Phys. Chem. B*, **112**, 722–726 (2008).
50. R. S. McLean, X. Huang, C. Khripin, A. Jagota, and M. Zheng, Controlled two-dimensional pattern of spontaneously aligned carbon nanotubes, *Nano Lett.*, **6**, 55–60 (2006).
51. G. S. Manning, The molecular theory of polyelectrolyte solutions with applications to the electrostatic properties of polynucleotides, *Quart. Rev. Biophys.*, **11**, 179–246 (1978).
52. S. A. Allison and D. Stigter, A commentary on the screened-Oseen, counterion-condensation formalism of polyion electrophoresis, *Biophys. J.*, **78**, 121–124 (2000).
53. T. G. M. van de Ven, The capture of colloidal particles on surfaces and in porous material: basic principles, *Colloid Surf. A*, **138**, 207–216 (1998).
54. J. W. Goodwin, *Colloids and Interfaces with Surfactants and Polymers: An Introduction*. Wiley (2004).
55. J. Feder, Random sequential adsorption, *J. Theor. Biol.*, **87**, 237–254 (1980).
56. R. Vigil and R. Ziff, Random sequential adsorption of unoriented rectangles onto a plane, *J. Chem. Phys.*, **91**, 2599–2602 (1989).
57. P. Schaaf and J. Talbot, Kinetics of random sequential adsorption, *Phys. Rev. Lett.*, **62**, 175–178 (1989).
58. R. D. Vigil and R. M. Ziff, Kinetics of random sequential adsorption of rectangles and line segments, *J. Chem. Phys.*, **93**, 8270–8272 (1990).
59. D. M. Lipkin, H. Schaffer, F. Adar, and D. R. Clarke, Lateral growth kinetics of -alumina accompanying the formation of a protective scale on (111) NiAl during oxidation at 1100° C, *Appl. Phys. Lett.*, **70**, 2550 (1997).
60. R. K. Iler, *The Chemistry of Silica*. New York: Wiley (1979).
61. K. A. Dill and S. Bromberg, *Molecular Driving Forces*. New York: Garland Science (2003).
62. Z. Liu, Z. Li, H. Zhou, G. Wei, Y. Song, and L. Wang, Imaging DNA molecules on mica surface by atomic force microscopy in air and in liquid, *Microsc. Res. Tech.*, **66**, 179–185 (2005).
63. X. Huang, J. Kovaleski, and M. Wirth, Spectroscopic probing of mixed-mode adsorption of Ru (bpy) 32+ to silica, *Anal. Chem.*, **68**, 4119–4123 (1996).

64. R. D. Deegan, O. Bakajin, T. F. Dupont, G. Huber, S. R. Nagel, and T. A. Witten, Contact line deposits in an evaporating drop, *Phys. Rev. E*, **62**, 756–765 (2000).

65. H. Neurath and G. R. Cooper, The diffusion constant of tomato bushy stunt virus, *J. Biol. Chem.*, **135**, 455 (1940).

66. H. Dorrie, *100 Great Problems of Elementary Mathematics: Their History and Solution*. New York: Dover (1989).

67. J. F. Joanny and P. G. de Gennes, A model for contact angle hysteresis, *J. Chem. Phys.*, **81**, 552 (1984).

68. R. B. Bird, W. E. Stewart, and E. N. Lightfoot, *Transport Phenomena*. 2nd ed. New York: John Wiley & Sons (2002).

69. S. H. Im, Y. T. Lim, D. J. Suh, and O. O. Park, Three-dimensional self-assembly of colloids at a water–air interface: a novel technique for the fabrication of photonic bandgap crystals, *Adv. Mater.*, **14**, 1367–1369 (2002).

70. N. D. Denkov, O. D. Velev, P. A. Kralchevsky, I. B. Ivanov, H. Yoshimura, and K. Nagayama, Two-dimensional crystallization, *Nature*, **361**, 26–26 (1993).

71. N. Dan, Formation of ordered domains in membrane-bound DNA, *Biophys. J.*, **71**, 1267–1272 (1996).

72. R. G. Endres, D. L. Cox, and R. R. P. Singh, Colloquium: the quest for high-conductance DNA, *Rev. Mod. Phys.*, **76**, 195–214 (2004).

73. G. Reguera, K. D. McCarthy, T. Mehta, J. S. Nicoll, M. T. Tuominen, and D. R. Lovley, Extracellular electron transfer via microbial nanowires, *Nature*, **435**, 1098–1101 (2005).

Chapter 17

High-Resolution AFM Imaging of Native Biological Membranes

Lu-Ning Liu and Simon Scheuring*

U1006 INSERM / Aix-Marseille Université, Campus Scientifique de Luminy, Marseille 13288, France

*simon.scheuring@inserm.fr

17.1 Introduction

Studies on biological membranes and proteins have been advanced along with the development of research methodologies. Atomic force microscopy (AFM) has several trumps that render the technique particularly attractive in membrane biology research (Fig. 17.1).[1,2]

- The biological samples studied by AFM only need little prior biochemical treatment. In contrast, most other methods available nowadays to study membrane proteins at high resolution, that is, three-dimensional (3D) X-ray crystallography and electron microscopy (EM), require extensive pretreatments of the samples, such as solubilization, purification, or fixation, drying, freezing and/or metal coating; these may cause to different extents artifacts to the samples.

Nanoscale Liquid Interfaces: Wetting, Patterning, and Force Microscopy at the Molecular Scale
Edited by Thierry Ondarçuhu and Jean-Pierre Aimé

ISBN 978-981-4316-45-3 (Hardcover), 978-981-4364-48-5 (eBook)
www.panstanford.com

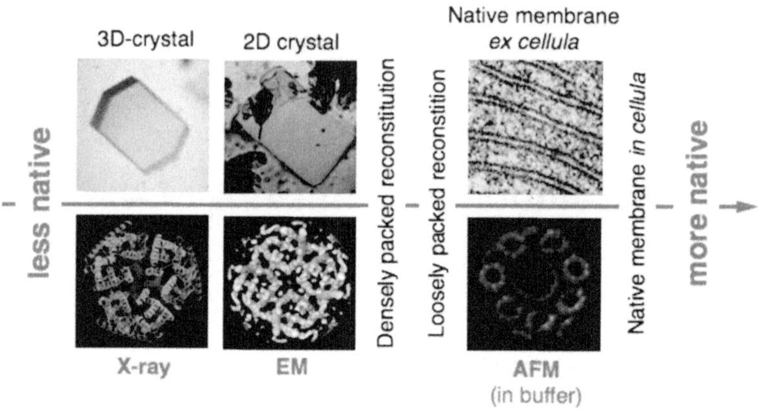

Figure 17.1 The nativeness of studied samples from left (less native) to right (more native): membrane protein 3D crystal, membrane protein 2D crystal, densely packed reconstitution, loosely packed reconstitution, native membrane *ex cellula*, and native membrane *in cellula*. *Abbreviation*: 2D, two dimensional.

- Of significant importance are the native conditions that AFM can confer during imaging. The AFM measurement is accomplished under near-physiological conditions, in physiological buffer, at room temperature, and under atmospheric pressure. This is a technical breakthrough compared to other high-resolution techniques that often demand vacuum and/or low-temperature conditions.
- The high signal-to-noise ratio of AFM allows single-membrane protein structures or substructures to be directly discerned. This ensures its capacity to study single molecules, without following molecule averaging, and allows us to study variability of individual membrane proteins what no other technique can survey (Fig. 17.1).
- The high signal-to-noise ratio allows, furthermore, a molecule to be observed in the context of its molecular environment. Hence, while the information is surface restricted and somewhat lower compared to X-ray and electron crystallography, AFM visualizes molecular partners and nonordered assemblies.

To date, we are able to image "native membranes *ex cellula*" using AFM, with only minimal biochemical pretreatments employed (cell breakage and density gradient centrifugation) to isolate membranes from other cellular constituents. Ultimately, the long-term goal for AFM study is the exploration of membrane proteins in native membranes directly on a live cell. This is, however, to date impossible due to some physical difficulties, for example, the softness of the cell membrane and the diffusion dynamics, which seem to occlude imaging of individual membrane molecules on a cell.

This chapter overviews some experimental examples of high-resolution AFM imaging of native biological membranes, with an emphasis on prokaryotic membranes and the assembly of multiple functionally coupled proteins. Then, the focus is set on the achievements acquired from AFM imaging on the specialized bacterial photosynthetic membranes, a field in which AFM has contributed major advances in the understanding of the highly efficient machinery involved in the conversion of solar energy. Finally, we highlight issues that merit further investigations by means of high-resolution AFM imaging.

17.2 High-Resolution AFM Imaging of Biological Membranes

Despite the fact that AFM is currently the only tool that allows native biological membranes and large-scale organization of membrane proteins to be studied at molecular resolution, the acquisition of high-resolution topographs of biological membranes is a daunting endeavor because of the fragility of biological membranes. Several factors should be considered during high-resolution AFM imaging.

- A crucial limitation for high-resolution AFM topograph acquisition is related to the tip radius and geometry. In principle, the sharper the tip is, the higher the resolution one may be able to acquire.
- Though sharp tips are always preferably selected, the AFM tip is still vast compared to target protein molecules.

Therefore, the flat adsorption of biological membranes is advantageous to high-resolution imaging.

- In addition, optimized buffer conditions are required for sample absorption and high-resolution imaging: for sample absorption, attractive Van der Waals forces must be favored; for imaging, electrostatic repulsions must balance Van der Waals attraction. AFM topographs with spatial resolutions of $\approx$1 nm allow single-membrane proteins to be observed and reveal details of their surface structures. However, the force applied during the imaging process must be carefully balanced to contour native membrane surface to avoid reversible and irreversible structural deformation. Tremendous progress has been achieved by optimizing sample preparation and imaging conditions of soft biological materials (forces applied to the cantilever below 100 pN, optimization of the scanning speed and feedback parameters).
- Other crucial parameters are related to the physics of the surface contouring mechanism employed, that is, the feedback loop that should properly react on cantilever deflection using minimal forces applied.

It is worthy to emphasize that many of the available high-resolution AFM imaging techniques were acquired in contact mode, though operating AFM in contact mode induces friction forces; therefore, samples need to adhere firmly to freshly cleaved mica substrates. In contrast, lateral friction forces acting on the samples can be minimized by using oscillating mode that may allow weakly adsorbed samples to be observed, although the resolution achieved by oscillating mode is often slightly lower (see chapters 18 and 19).

Topographs of membrane proteins at subnanometer resolution were first acquired on highly ordered 2D reconstituted systems, that is, OmpF,[3] bacteriorhodopsin (BR),[4] water channels,[5] potassium channels,[6] halorhodopsin,[7] outer membrane (OM) porins,[8] adenosine triphosphate synthase (ATPase)[9–14] and light-harvesting (LH) complexes.[15–19] Target proteins were initially isolated from biological membranes and then reconstituted into lipid bilayers to form regular arrays. However, as described before, there is a certain risk that isolation and reconstitution processes may disturb the native state of membrane proteins. More importantly, the function of membrane proteins depends on their structure and organization in the native molecular environment in the membrane. Remarkable progress has been derived from high-resolution AFM imaging of membrane proteins that assemble in regular lattices in native membranes, that is, BR in purple membrane and surface layers (S-layers). This has paved the way for AFM analysis of membrane proteins directly in native membranes.

In addition to high-resolution imaging, another key advantage of AFM is its excellent capacity to nanomanipulate the sample, mediated by applying additional loading forces to the imaging tip, along with adapting scan rates and feedback parameters to deliberately act on the surface of the biological object. Under higher additional loading forces, stacked membrane layers can be dissected to give access to underlying membranes.[3,20–23] Moreover, individual protein subunits of multiprotein complexes can be dissected at slightly increased forces, allowing the analysis of underlying protein structures.[22,24,25] At low additional loading forces, individual protein domains can be manipulated. This last process is nondestructive and provides access for the analysis of flexible protein surface domains.[23,26,27] High-resolution imaging and nanomanipulation often benefit each other during the study of membrane protein structure.

17.2.1 *Early Achievements—Purple Membrane Topographs*

The plasma membrane of the archaebacteria *Halobacterium salinarium* (*H. salinarium*) consists of 75% BR and 25% lipids (w/w). As a light-driven proton pump containing seven transmembrane α-helices,[28] BR absorbs light ($\lambda_{max} = 560$ nm) and builds up an electrochemical potential across the inner cell membrane of *H. salinarium* BR forms a 2D trigonal crystal in purple membranes (Fig. 17.2A).

The topography of native purple membranes has been extensively studied at nanometer resolution using AFM in contact mode[4,27,29] and oscillating mode[30,31] in aqueous conditions (see

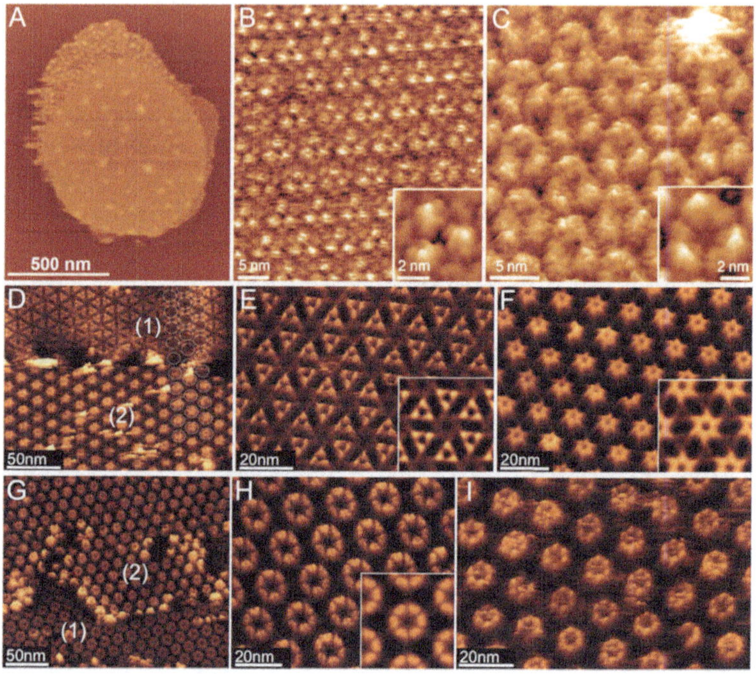

Figure 17.2 (A) Purple membrane directly adsorbed onto a mica surface.[34] (B) Extracellular and (C) cytoplasmic surfaces showing the substructures and trimeric assembly of BR.[27] Contact-mode AFM topographs were recorded in buffer solution at forces of $\approx$100 pN. (D) Nanodissection of the top outer surface of the S-layer of *C glutamicum* using AFM (1), providing access to the bottom S-layer and revealing the topography of the inner surface (2).[21] (E) High-resolution topograph of the outer S-layer surface (inset: average). (F) High-resolution topograph of the inner S-layer surface (inset: average). (G) Nanodissection and high-resolution imaging of S-layers of *D. radiodurans* using AFM, showing the bottom outer surface of the S-layer (1) and the top S-layer which exposes the inner surface (2).[35] (H) High-resolution topograph of the outer S-layer surface. The high imaging contrast allows detection of the substructure on each individual subunit, revealing V-shaped units with a slight left-handed twist (inset: average). (I) High-resolution topograph of the inner S-layer surface. *Abbreviations*: *C. glutamicum, Corynebacterium glutamicum*; *D. radiodurans, Deinococcus radiodurans*.

also chapters 18 and 19). On its extracellular surface, the purple membrane exhibited a major and a minor tripartite protrusion per BR trimer on a trigonal lattice ($a = b = 6.2 \pm 0.2$ nm) (Fig. 17.2B). The average height above the lipid bilayer was about 0.2 ± 0.02 nm. Distinct features on the extracellular surface can be assigned to the loops of BR.[32]

Specifically, the distinct conformations of the cytoplasmic surface can be recorded in a single topography when the force is changed during imaging.[27] On the cytoplasmic surface of the purple membrane, two different conformations have been detected depending on the scanning force applied during imaging.[33] When imaged at 250–500 pN, the BR trimers revealed a distinct doughnut-like shape.[27] The tripartite morphology was pronounced, and the trimers appeared to be connected by fibrous arms. Another conformation was visualized when a 100–150 pN scanning force was applied. The BR trimers revealed a more tripartite topography with three major protrusions at the periphery of the doughnuts (Fig. 17.2C).[27] The most prominent loop connecting the membrane-spanning α-helices E and F of BR in its extended conformation was assigned to this major protrusion with a height of 0.6 ± 0.02 nm above the lipid bilayer.

The capability of AFM to measure forces in the range of tens of picoNewtons has stimulated its application to study intermolecular and intramolecular forces stabilizing the fold of membrane proteins. The protein of interest is pulled out of the biological membrane by the AFM probe, and the force–distance spectrum depicts the force required to unfold the molecule in distinct steps as well as the contour length of the unfolded polypeptide. BR is one of the bestcharacterized membrane proteins by single-molecule force spectroscopy (SMFS).[36,37] When force spectroscopy measurements were applied to polytopic membrane proteins,[36] the technique was shown to be sufficiently sensitive to detect forces between the transmembrane helices.[38,39]

17.2.2 *Surface Layers*

S-layers are regular, 2D protein networks, functioning as the outermost cell wall layer of many bacteria and archaea.[40,41] These layers withstand nonphysiological pH, radiation, temperature, proteolysis, pressure and detergent treatment, thus protecting the cell from such hostile factors.[35,42]

PS2 is the protein that forms the S-layer of *C glutamicum*.[43,44] Native *C. glutamicum* S-layers stick together via their inner surfaces in aqueous condition, with the outer surfaces exposed on opposite sides (Fig. 17.2D). When imaging at a minimal loading force of 100 pN, the outer S-layer surface of the top layer was imaged at high resolution, showing the triangle-shaped protrusions with unit cell dimensions of $a = b = 16.0 \pm 0.2$ nm and $\gamma = 60 \pm 1^\circ$ (Fig. 17.2E).[21] The hexameric core between the triangles can be identified as the other side of the flower-shaped hexameric core visible on the inner surface. During image acquisition, the loading force to the tip was increased to 500 pN. This mechanical treatment punctured the top S-layer and gave the AFM tip access to the inner surface of the layer below without severe damage, exposing a flower-shaped surface (Fig. 17.2F). It was deduced that the flower-shaped hexameric S-layer cores of the lower layer faced exactly the flower-shaped cores of the top layer. On the basis of these observations, a 3D reconstruction of the S-layer architecture has been calculated from the topographies of both surfaces.[21]

The hexagonally packed intermediate (HPI) layer in *D radiodurans* is another typical bacterial S-layer.[45,46] There are six identical protomers that form the HPI layer core. High-resolution AFM images together with manipulation exhibit both outer and inner layer surfaces (Fig. 172G). The HPI layer is a hexagonal lattice with unit cell dimensions of 18 nm (Fig. 172H).[47] This arrangement presents relatively large openings around the sixfold axis. The inner surface of the HPI layer shows conformational dynamics of pores: a "closed" with and an "open" without a central "plug" (Fig. 172I).[46]

Furthermore, the possibility to image the S-layer *in vivo* without invasive sample preparation has been demonstrated.[48] More recently, the S-layer on living *C. glutamicum* bacterial cells has been imaged.[49]

17.2.3 *Outer Membrane*

The OM of gram-negative bacteria protects the cell against bactericidal substances. Passage of nutrients and waste is assured by porins, the β-barrel transmembrane channels in OMs.

The first high-resolution AFM view of a bacterial OM revealed that porins cover 70% of the membrane surface and form locally regular lattices in *Roseobacter denitrificans* (*R denitrificans*).[23] After nanomanipulation to remove peptidoglycan remnants or stacked layers, both extracellular and periplasmic surfaces were visualized. The extracellular surface of porins existed in two distinct conformations one with separated protrusions and another with protrusions contracted at the threefold axis (Fig. 17.3A). Their occurrence was correlated with their position within the membrane: trimers which were positioned in the center of a membrane displayed the "relaxed" conformation (Fig. 17.3B) whereas the trimers at the membrane edges displayed the "contracted" conformation (Fig. 17.3C), a phenomenon attributed to trapped ions under the central membrane region, in agreement with observations on OmpF 2D crystals.[8] The periplasmic surface of the porins exhibit oval-shaped cavities separated by walls crowned by three major protrusions (Fig. 17.3D). Though not all particles were well aligned in the native membrane, an "average unit cell" of $a = b = 81 \pm 2.5$ Å, $\gamma = 60^\circ$ could be determined. The high-resolution topography average could be compared in considerable detail to molecular surface representations of the X-ray structure[50] (Fig. 17.3EF). Whether the observed dense porin packing is a common morphological feature of OM in different bacterial species remains to be determined.

17.2.4 *Bacterial Photosynthetic Membrane*

Due to the high signal-to-noise ratio, AFM has emerged as an indispensable tool that allows the structural identification of individual membrane proteins, not only with regular protein arrangements, but also in "nonordered" assemblies. A breakthrough is high-resolution AFM imaging of bacterial photosynthetic membranes in contact mode, which have substantially provided the first surface views of the organization of multicomponent biological membranes at molecular resolution.[51,52]

In purple bacterial photosynthesis, light capture and energy transfer are efficiently accomplished by the strong cooperativity among different photosynthetic components, that is, LH complexes (LH1 and LH2), reaction center (RC), cytochrome (cyt) bc_1

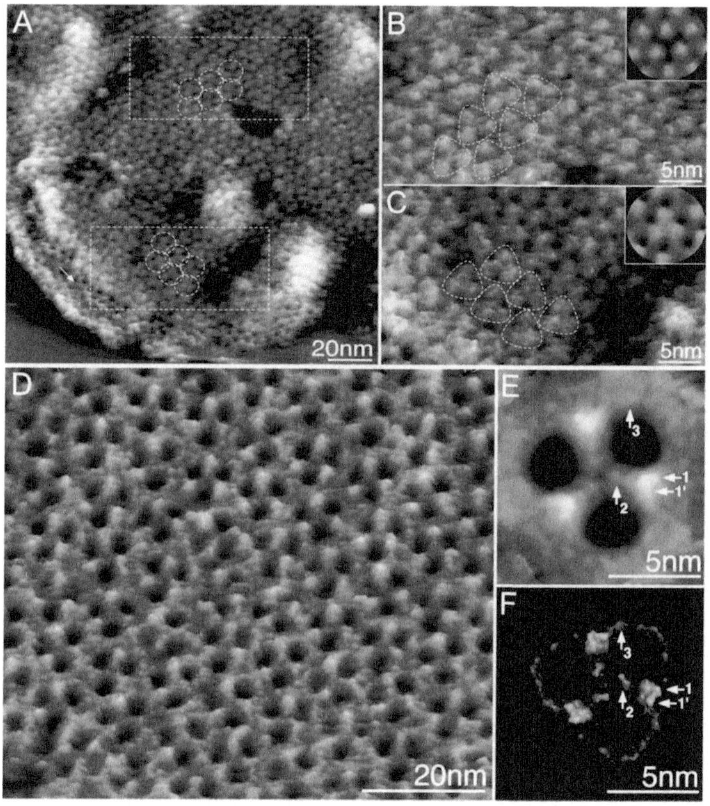

Figure 17.3 AFM images of the native extracellular and the periplasmic OM surfaces of *R. denitrificans*.[23] (A) Topograph of the top layer exposing the extracellular surface to the AFM tip. The arrow marks a border strip, which exposes the underlying layer showing the periplasmic surface. Encircled are trimers representing two different conformations of the extracellular domains. (B and C) The two conformations of the extracellular protrusions correlate with their location within the membrane. The "relaxed" conformation (top) was found inside the membrane patch (dashed outlines). The threefold symmetrized average ($n = 30$) discloses a well-preserved, central indentation. The "contracted" conformation (bottom) displayed by porins located close to the membrane border (dashed outlines). (D) High-resolution analysis of the periplasmic surface of the OM. (E) At highresolution the substructure is visible on individual trimers. (F) Average ($n = 18$) reveals major peripheral protrusions, fine protrusions close to the trimer axis, and peripheral protrusion corresponding to beta turns of just a few amino acids in size.

complex and ATPase which are embedded in the photosynthetic membranes. The shape of bacterial photosynthetic membranes is species dependent. They may be arranged in regular parallel layers (*Blastochloris viridis* [*B. viridis*], *Rhodopseudomonas palustris* [*R. palustris*], *Rhodospirillum photometricum* [*Rsp photometricum*], *Phaeospirillum molischianum* [*Ph. molischianum*]) or form vesicles (*Rhodobacter sphaeroides* [*R. sphaeroides*], *Rhodobacter blasticus* [*R. blasticus*]) or small regular stacks of lamellae (*Rhodospirillum fulvum* [*Rsp. fulvum*]*Thiocapsa spp.*) resembling the grana of chloroplasts. The information concerning protein domain formation, complex assembly, and protein heterogeneity, derived from high-resolution AFM imaging, provided important insights into the structure–function relationship of the photosynthetic machinery.

17.2.4.1 RC-LH1 core complex

AFM imaging of native photosynthetic membranes has provided detailed information of the architecture of RC-LH1 core complexes. It is evident that the core complex architecture varies considerably between species (Fig. 17.4). In many species like *B. viridis*, *Rsp. photometricum*, and *Ph. molischianum*, a monomeric core complex is surrounded by a complete ring of 16 LH1 α/β heterodimers ($LH1_{16}$ $RC_{L,M,H}$) The LH1 assembly around the RC is elliptical, with axes having a length difference of $\approx$10%. The orientation of the long axis of the LH1 ellipsis coincides with the long axis of the RC.[25] AFM images showed that there are nonmembranous tetraheme cytochrome (4Hcyt) subunits on top of *B. viridis* RCs, protruding strongly into the periplasm with a 4.8 nm height.[25] After scraping away the 4Hcyt subunits by applying increased scanning force and speed, high-resolution images of an unobstructed periplasmic surface view of the underlying RC-LH1 core complexes could be acquired It was found that the ellipticity of the LH1 was induced by the presence of the membranous RC subunits.[25]

In other *Rhodobactor* species, dimeric core complexes were seen in high-resolution AFM images. In native membranes of *R. blasticus* the core complexes possess $(PufX_2\text{-}LH1_{13}\text{-}RC_{L,M,H})_2$ conformation, comprising two RCs surrounded by an S-shaped LH1 assembly with two times 13 LH1 α/β heterodimers along with 2 PufX polypeptides

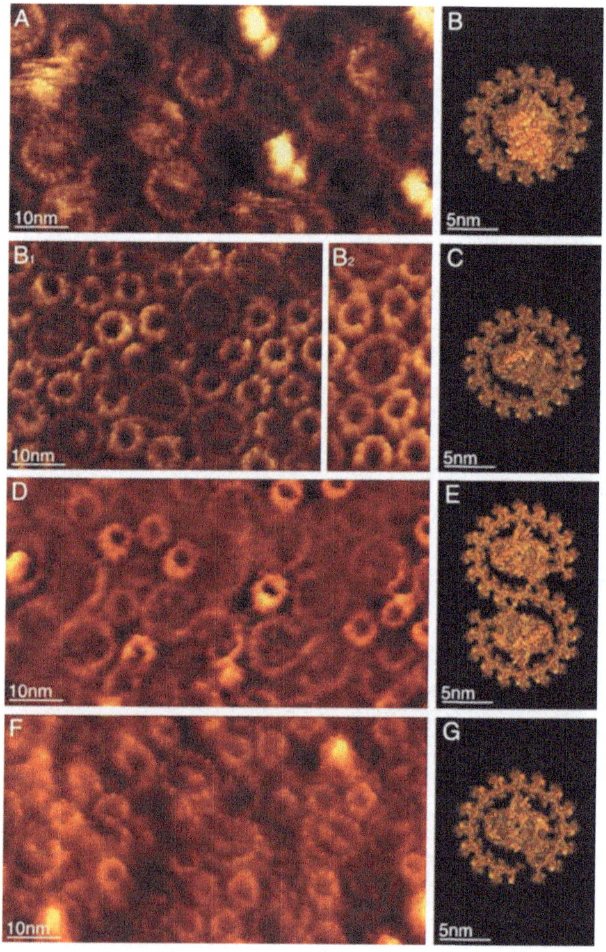

Figure 17.4 Diversity of core-complex architectures among purple photosynthetic bacteria species. (A) Core complexes ($LH1_{16}$-$RC_{L,M,H}$-4Hcyt) in a native membrane of *B. viridis*. (B) Core complexes ($LH1_{16}$-$RC_{L,M,H}$) in native membranes of *Rsp. photometricum* (B1) and *Ph. molischianum* (B2). (C) Core complexes ($PufX_2$-$LH1_{13}$-$RC_{L,M,H}$)$_2$ in a native membrane of *R. blasticus*. (D) Core complexes (W-$LH1_{15}$-$RC_{L,M,H}$) in a native membrane of *R. palustris*.

located at the dimer center.[53,54] The LH1 assembly of each core is also elliptical (100 Å vs. 90 Å).[55] This is probably in line with the *R. sphaeroides* core,[53] but the precise *Rhodobacter* core complex architecture is still under debate.[53,54,56,57]

Core complexes in *R. palustris* membranes had W-$LH1_{15}$-$RC_{L,M,H}$ architecture: the LH1 ring with 15 subunits is broken by the inclusion of a W subunit, resulting in a partially open ellipsis (99 Å vs. 83 Å).[58] Core complexes, from which the RC was removed by the AFM tip, revealed "relaxed" circular W-LH1 assemblies. The core complex of *Rhodobacter veldkampii* shares this architecture, being monomeric with a PufX-$LH1_{15}$ assembly.[59,60] The RC-LH1-PufX cores possess a unique C-shaped structure comprising an enlarged gap within the 15 LH1 $\alpha\beta$ pairs.[61] AFM study on the native photosynthetic membrane showed that such a gap was found to be larger than those in core complexes of *R. palustris*,[58] though its physiological significance remains to be further addressed.

17.2.4.2 LH2 complex

High-resolution AFM has the unique capability of imaging individual membrane proteins in their native membranes, thus enabling the study of LH complex architecture and molecular heterogeneity *in situ*. The structure of LH complexes in native membranes was found to comprise complexes of variable size (Fig. 17.5). In the survey of LH2 in *Rsp. photometricum* membranes, in addition to the majority of normal nonameric LH2 complexes about 80% several different types of complexes were observed:[62] about 10% of the complexes were octameric and another 10% were decameric. This size heterogeneity was related to the known spectral heterogeneity of LH2. Moreover, some individual complexes were small rings containing only 6 or 7 α/β heterodimers, open C-shaped complexes, or large complexes containing 14 α/β heterodimers.[62] Such heterogeneous stoichiometry appears likely an inherent feature of LH2, as it has also been observed in *Ph. molischianum*[63] and *R. palustris*.[58]

One of the characteristics of *R. palustris* is the multiplicity of LH2 complexes. High-light-adapted membranes comprise $\approx$50% of standard B800–850 LH2 and $\approx$50% of specific B800 LH2, while in low-light-adapted membranes only $\approx$10% of B800–850

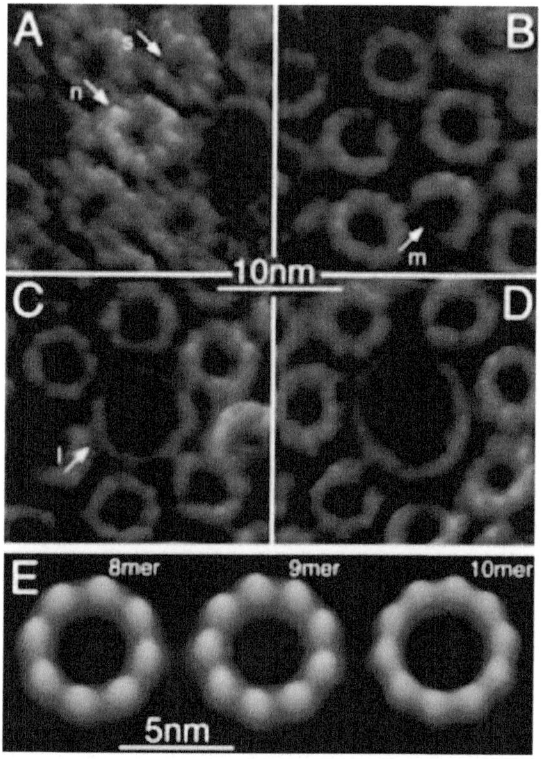

Figure 17.5 LH2 complexes of *Rsp. photometricum.* (A) A small LH2 complex (s) neighbors a normal sized (*n*) LH2. (B) A moon-shaped (m) LH2 complex consisting of 6 subunits adjacent to normal LH2 nonamers. (C) A large (l) antenna complex of 80 Å in diameter. (D) A core complex for comparison. (E) Averages calculated by the use of reference-free single-particle analysis. The 8mer, the 9mer and 10mer averages comprise about 10%, 80% and 10% of the total of 1251 particles analyzed.

LH2 and ≈90% B800 LH2 occur. The low-light LH2 complexes form an octameric ring, on average smaller in diameter than all standard nonameric B800–850 LH2 complexes.[58] The fact that LH2 complexes with both eight- and nine-fold symmetries form hexagonal packed assemblies strongly indicates that their assembly is driven by high packing densities and solvation effects within these domains, rather than specific protein–protein interaction as dictated by their inherent symmetry differences.[63]

17.2.5 Multiprotein assembly

The capacity of AFM to study individual protein complexes and gather molecular structural information gives the exciting possibility of studying protein assembly at an individual molecule level. Analysis of the distribution of distinct photosynthetic complexes showed strong clustering of LH2 and RC-LH1 core complexes in native membranes (Fig. 17.6). Clustering of complexes is a functional necessity, as each LH component must pass its harvested energy to a neighboring complex and eventually to an RC. Each type of bacterial photosynthetic membranes investigated by AFM to date revealed particularities concerning the organizations of photosynthetic apparatus, though most of them house a common underlying assembly principle.[52] With the exception of *R. sphaeroides*,[55] no fixed structural assembly of LH2 and core complexes were observed, and photosynthetic complexes were rather randomly oriented with respect to each other in the native membranes. Core complexes completely surrounded by several LH2 and core complexes making multiple core–core contacts were visualized[22,64,65] (Fig. 17.6A). About 30% of core complexes are completely surrounded by LH2, and about 70% are connected to other cores.[63] However, their organization is far from random. Pair correlation function analysis has revealed that there is a most favorable assembly within the native membranes, which is core-LH2-core-LH2- and so on. The photosynthetic apparatus of *Ph. molischianum* resembles *Rsp. photometricum*.[63] LH complexes are segregated into two structurally different types of domains, consisting of a mixture of core and octameric LH2 complexes, as well as paracrystalline, hexagonally packed octameric LH2 rings. One specific example is the photosynthetic membrane of *B. viridis* which has no LH2 complexes, and core complexes form a 2D hexagonal lattice in the photosynthetic membranes.[25]

Topographs of the subsolubilizing concentration detergent-treated photosynthetic membranes in *R. sphaeroides* derived from oscillatingmode AFM in aqueous condition revealed a relatively ordered network of both LH2 and dimeric RC-LH1 core complexes[55] (Fig. 17.6B). The dimeric RC-LH1 core complexes formed linear arrays of up to six dimers. The specific core organization may allow the LH1 excitation to migrate along a series of dimers until

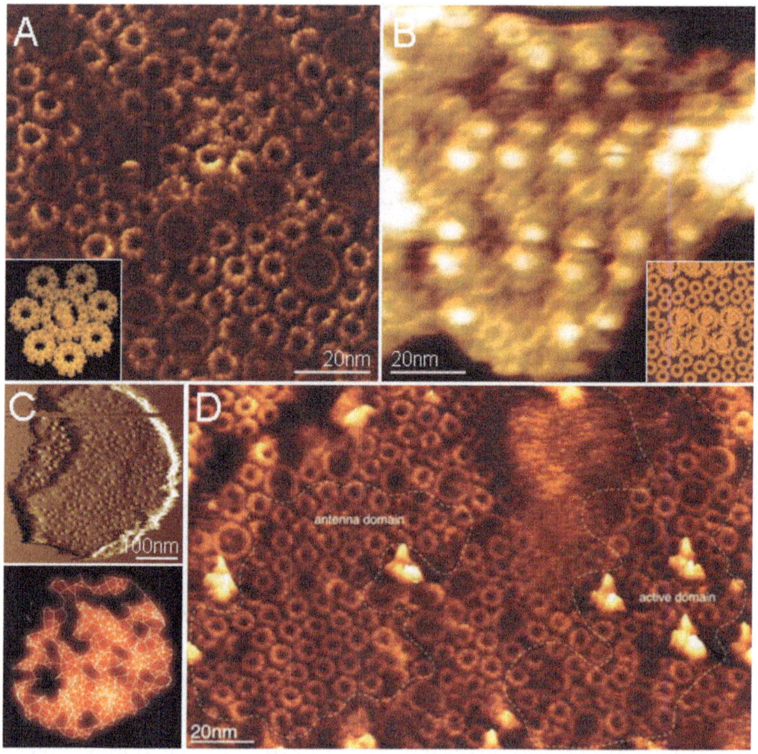

Figure 17.6 Assembly of LH2 and core complexes in native photosynthetic membranes. (A) AFM topograph of the high-light-adapted photosynthetic membrane of *Rsp. photometricum*. Inset: Structural model of core complexes surrounded by seven LH2. (B) AFM topograph of the photosynthetic membrane of *R. sphaeroides* Inset: Structural model of rows of core dimers arrangement with rows of LH2 dimers. (C) AFM deflection image of entire native chromatophores of *Rsp. photometricum* and calculated quinone pathways. (D) AFM topograph of the low-light-adapted photosynthetic membrane of *Rsp. photometricum*. The LH2–RC ratio is about 7. An antenna domain (LH2-only) and the photosynthetically active domain (LH2/RC = 3.5) are outlined.

an "open" RC is found, thus favoring the energy transfer. These arrays were interconnected by light-capturing domains of LH2 rings, also arranged in rows, fulfilling the basic requirement for efficient trapping and transmission of light energy.[55] In addition, specialized domains are made up exclusively of LH2 complexes. It is interesting that the core complexes in *R. blasticus* do not form rows as observed in *R. sphaeroides* membranes, although they are also open S-shape dimmers.[54]

Photosynthetic membranes are densely packed with photosynthetic proteins favorable for excitation energy transfer However such a dense packing represents an obstacle for efficient quinone/quinol diffusion between cores and cyt bc_1 complex in membranes. Pair correlation function analysis of the entire photosynthetic membranes showed that core complexes influence their molecular environment within a critical radius of 250 Å, with high probability that two-core complexes are separated by one LH2.[22] Analysis of the molecular environment further indicated that the local environment of core complexes contains more lipid space due to the geometric mismatch of cores and LH2. Short-range core interaction and enlarged lipid spaces finally establish long-range quinone pathways through the entire photosynthetic membranes (Fig. 17.6C).[22]

The organization of the photosynthetic protein assembly was found to vary depending on the light intensity during cell growth, derived from high-resolution AFM imaging of the photosynthetic membrane in *Rsp. photometricum*. In low-light-adapted membranes, additional LH2 complexes form paracrystalline hexagonally packed antenna domains, whereas core complexes remained in domains in which they were locally concentrated as under high-light cell growth (Fig. 17.6D), while the medium-light-adapted membrane was observed to have an intermediate organization of the photosynthetic apparatus.[22] The various organizations of photosynthetic apparatus of *R. palustris* were also studied under different light conditions.[58] In membranes from high-light adapted cells, paracrystalline domains of hexagonally packed core complexes were detected. These findings provided direct evidence and a rationale of the chromatic adaptation of the photosynthetic apparatus.[52]

17.3 Summary and Perspectives

Due to the significant ability in imaging the native architecture of biological membranes comprising supramolecular assemblies in the organization of membrane proteins at molecular resolution, AFM has matured into an important technique in membrane protein research, proving to be greatly complementary to X-ray crystallography and electron microscopy.

On the basis of high-resolution AFM imaging, attractive progress has been achieved passing over developments and improved sample preparation for visualizing the molecular machineries of cellular membranes,[66] such as measurements of local membrane elasticity, adhesive forces, and conformational flexibility of membrane proteins. Single-molecule force measurements have been executed to unveil quantitatively the molecular folding and interactions that drive membrane protein assembly,[36] LH2 intermolecular interaction, and LH2 intramolecular interaction within native photosynthetic membrane.[67] It will be a challenge to analyze by AFM the protein communication, to understand the assembly of membrane proteins into rafts and to investigate some processes involved in trafficking and signal transduction. Measurements of forces within supramolecular aggregates or forces dictating the fold of proteins can be executed, delivering quantitative information previously not accessible. Furthermore, high-resolution imaging combined with high-speed scanning will shed light on the protein dynamics and lateral diffusion in membranes. Accordingly, one can envision that the existing and developing AFM-based approaches will contribute multifunctional tools that reveal novel insights into questions in membrane biology and dramatically advance our understanding of numerous fundamental biological processes.

Despite the success of many AFM studies of membrane proteins, several obstacles still remain prior to the stark implementation of high-resolution AFM imaging. At present, high-resolution AFM imaging on biological materials is still a time-consuming work (including sample preparation and imaging) and valuable experimental experience is required. However, evaluations on a large number of data are necessary for extensive comprehension of one biological event. Therefore, high-throughout AFM screening is important and awaits development. In addition, not all the membrane proteins are visible in AFM images, as many membrane types or some membrane area are not well adapted to AFM imaging, particularly for those membrane proteins that are highly mobile or reside in strongly curved membranes. For instance, in the study of bacterial photosynthetic membranes, a number of questions have not been addressed by AFM, like the location of the cyt bc_1 and ATPase complexes.[51] The development of recognition imaging techniques, such as immuno-AFM, and the combination of AFM and confocal microscope are anticipated to overcome the difficulties.

Acknowledgments

Work in our team was supported by Institut Curie, Institut National de la Santé et Recherche Médicale (INSERM), Centre National de la Recherche Scientifique (CNRS), Agence Nationale de la Recherche (ANR), and the "City of Paris."

References

1. A. Engel, and H. E. Gaub, Structure and mechanics of membrane proteins, *Annu. Rev. Biochem.*, **77**, 127–148 (2008).
2. P. L. Frederix, P. D. Bosshart, and A. Engel, Atomic force microscopy of biological membranes, *Biophys. J.*, **96**, 329–338 (2009).
3. F. A. Schabert, C. Henn, and A. Engel, Native *Escherichia coli* OmpF porin surfaces probed by atomic force microscopy, *Science*, **268**, 92–94 (1995).
4. D. J. Müller, F. A. Schabert, G. Büldt, and A. Engel, Imaging purple membranes in aqueous solutions at sub-nanometer resolution by atomic force microscopy, *Biophys. J.*, **68**, 1681–1686 (1995).
5. S. Scheuring, P. Ringler, M. Borgnia, H. Stahlberg, D. J. Müller, P. Agre, and A. Engel, High resolution AFM topographs of the *Escherichia coli* water channel aquaporin Z, *EMBO J.*, **18**, 4981–4987 (1999).
6. S. Jaroslawski, B. Zadek, F. Ashcroft, C. Venien-Bryan, and S. Scheuring, Direct visualization of KirBac3.1 potassium channel gating by atomic force microscopy, *J. Mol. Biol.*, **374**, 500–505 (2007).
7. N. Persike, M. Pfeiffer, R. Guckenberger, M. Radmacher, and M. Fritz, Direct observation of different surface structures on high-resolution images of native halorhodopsin, *J. Mol. Biol.*, **310**, 773–780 (2001).
8. D. J. Müller, and A. Engel, Voltage and pH-induced channel closure of porin OmpF visualized by atomic force microscopy, *J. Mol. Biol.*, **285**, 1347–1351 (1999).

9. D. J. Müller, A. Engel, U. Matthey, T. Meier, P. Dimroth, and K. Suda, Observing membrane protein diffusion at subnanometer resolution, *J. Mol. Biol.*, **327**, 925–930 (2003).
10. I. Arechaga, and D. Fotiadis, Reconstitution of mitochondrial ATP synthase into lipid bilayers for structural analysis, *J. Struct. Biol.*, **160**, 287–194 (2007).
11. D. Matthies, L. Preiss, A. L. Klyszejko, D. J. Müller, G. M. Cook, J. Vonck, and T. Meier, The c13 ring from a thermoalkaliphilic ATP synthase reveals an extended diameter due to a special structural region, *J. Mol. Biol.*, **388**, 611–618 (2009).
12. D. Pogoryelov, J. Yu, T. Meier, J. Vonck, P. Dimroth, and D. J. Müller, The c15 ring of the *Spirulina platensis* F-ATP synthase: F1/F0 symmetry mismatch is not obligatory, *EMBO Rep.*, **6**, 1040–1044 (2005).
13. H. Stahlberg, D. J. Müller, K. Suda, D. Fotiadis, A. Engel, T. Meier, U. Matthey, and P. Dimroth, Bacterial Na(+)-ATP synthase has an undecameric rotor, *EMBO Rep.*, **2**, 229–233 (2001).
14. H. Seelert, A. Poetsch, N. A. Dencher, A. Engel, H. Stahlberg, and D. J. Müller, Proton-powered turbine of a plant motor, *Nature*, **405**, 418–419 (2000).
15. S. Bahatyrova, R. N. Frese, K. O. van der Werf, C. Otto, C. N. Hunter, and J. D. Olsen, Flexibility and size heterogeneity of the LH1 light harvesting complex revealed by atomic force microscopy: functional significance for bacterial photosynthesis, *J. Biol. Chem.*, **279**, 21327–21333 (2004).
16. S. Scheuring, J. Seguin, S. Marco, D. Lévy, C. Breyton, B. Robert, and J. L. Rigaud, AFM characterization of tilt and intrinsic flexibility of *Rhodobacter sphaeroides* light harvesting complex 2 (LH2), *J. Mol. Biol.*, **325**, 569–580 (2003).
17. S. Scheuring, F. Reiss-Husson, A. Engel, J. L. Rigaud, and J. L. Ranck, High-resolution AFM topographs of *Rubrivivax gelatinosus* light-harvesting complex LH2, *EMBO J.*, **20**, 3029–3035 (2001).
18. R. P. Gonçalves, J. Busselez, D. Lévy, J. Seguin, and S. Scheuring, Membrane insertion of *Rhodopseudomonas acidophila* light harvesting complex 2 investigated by high resolution AFM, *J. Struct. Biol.*, **149**, 79–86 (2005).
19. L. N. Liu, T. J. Aartsma, and R. N. Frese, Dimers of light-harvesting complex 2 from *Rhodobacter sphaeroides* characterized in reconstituted 2D crystals with atomic force microscopy, *FEBS J.*, **275**, 3157–3166 (2008).

20. J. H. Hoh, R. Lal, S. A. John, J. P. Revel, and M. F. Arnsdorf, Atomic force microscopy and dissection of gap junctions, *Science*, **253**, 1405–1408 (1991).

21. S. Scheuring, H. Stahlberg, M. Chami, C. Houssin, J. L. Rigaud, and A. Engel, Charting and unzipping the surface layer of *Corynebacterium glutamicum* with the atomic force microscope, *Mol. Microbiol.*, **44**, 675–684 (2002).

22. L. N. Liu, K. Duquesne, J. N. Sturgis, and S. Scheuring, Quinone pathways in entire photosynthetic chromatophores of *Rhodospirillum photometricum*, *J. Mol. Biol.*, **393**, 27–35 (2009).

23. S. Jarosawski, K. Duquesne, J. N. Sturgis, and S. Scheuring, High-resolution architecture of the outer membrane of the Gram-negative bacteria *Roseobacter denitrificans*, *Mol. Microbiol.*, **74**, 1211–1222 (2009).

24. D. Fotiadis, D. J. Muller, G. Tsiotis, L. Hasler, P. Tittmann, T. Mini, P. Jeno, H. Gross, and A. Engel, Surface analysis of the photosystem I complex by electron and atomic force microscopy, *J. Mol. Biol.*, **283**, 83–94 (1998).

25. S. Scheuring, J. Seguin, S. Marco, D. Lévy, B. Robert, and J. L. Rigaud, Nanodissection and high-resolution imaging of the *Rhodopseudomonas viridis* photosynthetic core complex in native membranes by AFM, *Proc. Natl. Acad. Sci. USA*, **100**, 1690–1693 (2003).

26. S. Scheuring, P. Ringler, M. Borgnia, H. Stahlberg, D. J. Müller, P. Agre, and A. Engel, High resolution AFM topographs of the *Escherichia coli* water channel aquaporin Z, *EMBO J.*, **18**, 4981–4987 (1999).

27. D. J. Müller, G. Büldt, and A. Engel, Force-induced conformational change of bacteriorhodopsin, *J. Mol. Biol.*, **249**, 239–243 (1995).

28. R. Henderson, J. M. Baldwin, T. A. Ceska, F. Zemlin, E. Beckmann, and K. H. Downing, Model for the structure of bacteriorhodopsin based on high-resolution electron cryo-microscopy, *J. Mol. Biol.*, **213**, 899–929 (1990).

29. D. J. Muller, C. A. Schoenenberger, G. Büldt, and A. Engel, Immuno-atomic force microscopy of purple membrane, *Biophys. J.*, **70**, 1796–1802 (1996).

30. C. Möller, M. Allen, V. Elings, A. Engel, and D. J. Müller, Tapping-mode atomic force microscopy produces faithful high resolution images of protein surfaces, *Biophys. J.*, **77**, 1150–1158 (1999).

31. M. Stark, C. Möller, D. J. Müller, and R. Guckenberger, From images to interactions: high-resolution phase imaging in tapping-mode atomic force microscopy, *Biophys. J.*, **80**, 3009–3018 (2001).

32. D. J. Müller, J. B. Heymann, F. Oesterhelt, C. Müller, H. Gaub, G. Büldt, and A. Engel, Atomic force microscopy of native purple membrane, *Biochim. Biophys. Acta*, **1460**, 27–38 (2000).
33. D. J. Müller, C.-A. Schoenenberger, F. Schabert, and A. Engel, Structural changes in native membrane proteins monitored at subnanometer resolution with the atomic force microscope: a review, *J. Struct. Biol.*, **119**, 149–157 (1997).
34. D. J. Müller, and A. Engel, Atomic force microscopy and spectroscopy of native membrane proteins, *Nat. Protoc.*, **2**, 2191–2197 (2007).
35. R. P. Gonçalves, and S. Scheuring, Manipulating and imaging individual membrane proteins by AFM, *Surf. Interface Anal.*, **38**, 1413–1418 (2006).
36. F. Oesterhelt, D. Oesterhelt, M. Pfeiffer, A. Engel, H. E. Gaub, and D. J. Müller, Unfolding pathways of individual bacteriorhodopsins, *Science*, **288**, 143–146 (2000).
37. D. J. Müller, M. Kessler, F. Oesterhelt, C. Moller, D. Oesterhelt, and H. Gaub, Stability of bacteriorhodopsin alpha-helices and loops analyzed by single-molecule force spectroscopy, *Biophys. J.*, **83**, 3578–3588 (2002).
38. H. Janovjak, M. Kessler, D. Oesterhelt, H. Gaub, and D. J. Müller, Unfolding pathways of native bacteriorhodopsin depend on temperature, *EMBO J.*, **22**, 5220–5229 (2003).
39. F. Oesterhelt, and S. Scheuring, High-resolution imaging and force measurement of individual membrane proteins by AFM, *Curr. Nanosci.*, **2**, 329–335 (2006).
40. U. B. Sleytr, and M. Sara, Bacterial and archaeal S-layer proteins: structure-function relationships and their biotechnological applications, *Trends Biotechnol.*, **15**, 20–26 (1997).
41. W. Baumeister, I. Wildhaber, and B. M. Phipps, Principles of organization in eubacterial and archaebacterial surface proteins, *Can. J. Microbiol.*, **35**, 215–227 (1989).
42. H. Engelhardt, and J. Peters, Structural research on surface layers: a focus on stability, surface layer homology domains, and surface layer-cell wall interactions, *J. Struct. Biol.*, **124**, 276–302 (1998).
43. J. L. Peyret, N. Bayan, G. Joliff, T. Gulik-Krzywicki, L. Mathieu, E. Schechter, and G. Leblon, Characterization of the cspB gene encoding PS2, an ordered surface-layer protein in *Corynebacterium glutamicum*, *Mol. Microbiol.*, **9**, 97–109 (1993).
44. M. Chami, N. Bayan, J. Dedieu, G. Leblon, E. Shechter, and T. Gulik-Krzywicki, Organization of the outer layers of the cell envelope

of *Corynebacterium glutamicum*: a combined freeze-etch electron microscopy and biochemical study, *Biol. Cell*, **83**, 219–229 (1995).

45. S. Karrasch, R. Hegerl, J. H. Hoh, W. Baumeister, and A. Engel, Atomic force microscopy produces faithful high-resolution images of protein surfaces in an aqueous environment, *Proc. Natl. Acad. Sci. USA*, **91**, 836–838 (1994).
46. D. J. Müller, W. Baumeister, and A. Engel, Conformational change of the hexagonally packed intermediate layer of *Deinococcus radiodurans* monitored by atomic force microscopy, *J. Bacteriol.*, **178**, 3025–3030 (1996).
47. D. J. Müller, W. Baumeister, and A. Engel, Controlled unzipping of a bacterial surface layer with atomic force microscopy, *Proc. Natl. Acad. Sci. USA*, **96**, 13170–13174 (1999).
48. T. E. Lister, and P. J. Pinhero, *In vivo* atomic force microscopy of surface proteins on *Deinococcus radiodurans*, *Langmuir*, **17**, 2624–2628 (2001).
49. V. Dupres, D. Alsteens, K. Pauwels, and Y. F. Dufrene, *In vivo* imaging of S-layer nanoarrays on *Corynebacterium glutamicum*, *Langmuir*, **25**, 9653–9655 (2009).
50. A. Kreusch, and G. E. Schulz, Refined structure of the porin from *Rhodopseudomonas blastica*. Comparison with the porin from *Rhodobacter capsulatus*, *J. Mol. Biol.*, **243**, 891–905 (1994).
51. J. N. Sturgis, J. D. Tucker, J. D. Olsen, C. N. Hunter, and R. A. Niederman, Atomic force microscopy studies of native photosynthetic membranes, *Biochemistry*, **48**, 3679–3698 (2009).
52. S. Scheuring, and J. N. Sturgis, Atomic force microscopy of the bacterial photosynthetic apparatus: plain pictures of an elaborate machinery, *Photosynth. Res.*, **102**, 197–211 (2009).
53. S. Scheuring, F. Francia, J. Busselez, B. A. Melandri, J. L. Rigaud, and D. Lévy, Structural role of PufX in the dimerization of the photosynthetic core complex of *Rhodobacter sphaeroides*, *J. Biol. Chem.*, **279**, 3620–3626 (2004).
54. S. Scheuring, J. Busselez, and D. Lévy, Structure of the dimeric PufX-containing core complex of *Rhodobacter blasticus* by in situ atomic force microscopy, *J. Biol. Chem.*, **280**, 1426–1431 (2005).
55. S. Bahatyrova, R. N. Frese, C. A. Siebert, J. D. Olsen, K. O. van der Werf, R. van Grondelle, R. A. Niederman, P. A. Bullough, C. Otto, and C. N. Hunter, The native architecture of a photosynthetic membrane, *Nature*, **430**, 1058–1062 (2004).

56. P. Qian, P. A. Bullough, and C. N. Hunter, Three-dimensional reconstruction of a membrane-bending complex: the RC-LH1-PufX core dimer of *Rhodobacter sphaeroides*, *J. Biol. Chem.*, **283**, 14002–14011 (2008).
57. P. Qian, C. N. Hunter, and P. A. Bullough, The 8.5 Å projection structure of the core RC-LH1-PufX dimer of *Rhodobacter sphaeroides*, *J. Mol. Biol.*, **349**, 948–960 (2005).
58. S. Scheuring, R. P. Gonçalves, V. Prima, and J. N. Sturgis, The photosynthetic apparatus of *Rhodopseudomonas palustris*: structures and organization, *J. Mol. Biol.*, **358**, 83–96 (2006).
59. J. Busselez, M. Cottevieille, P. Cuniasse, F. Gubellini, N. Boisset, and D. Lévy, Structural basis for the PufX-mediated dimerization of bacterial photosynthetic core complexes, *Structure*, **15**, 1674–1683 (2007).
60. L. N. Liu, J. N. Sturgis, and S. Scheuring, Native architecture of the photosynthetic membrane from *Rhodobacter veldkampii*, *J. Struct. Biol.*, **173**, 138–145 (2011).
61. E. Shimoni, O. Rav-Hon, I. Ohad, V. Brumfeld, and Z. Reich, Three-dimensional organization of higher-plant chloroplast thylakoid membranes revealed by electron tomography, *Plant Cell*, **17**, 2580–2586 (2005).
62. S. Scheuring, J. L. Rigaud, and J. N. Sturgis, Variable LH2 stoichiometry and core clustering in native membranes of *Rhodospirillum photometricum*, *EMBO J.*, **23**, 4127–4133 (2004).
63. R. P. Gonçalves, A. Bernadac, J. N. Sturgis, and S. Scheuring, Architecture of the native photosynthetic apparatus of *Phaeospirillum molischianum*, *J. Struct. Biol.*, **152**, 221–228 (2005).
64. S. Scheuring, J. N. Sturgis, V. Prima, A. Bernadac, D. Lévy, and J. L. Rigaud, Watching the photosynthetic apparatus in native membranes, *Proc. Natl. Acad. Sci. USA*, **101**, 11293–11297 (2004).
65. S. Scheuring, and J. N. Sturgis, Chromatic adaptation of photosynthetic membranes, *Science*, **309**, 484–487 (2005).
66. D. J. Müller, AFM: a nanotool in membrane biology, *Biochemistry*, **47**, 7986–7998 (2008).
67. L. N. Liu, F. Oesterhelt, K. Duquesne, J. N. Sturgis, and S. Scheuring, Intramolecular and intermolecular forces of LH2 complexes in native membranes, *Proc. Natl. Acad. Sci. USA*, 108, 9455–9459 (2011).

Chapter 18

Molecular-Resolution FM-AFM Imaging of Biological Systems

Takeshi Fukuma
Frontier Science Organization, Kanazawa University, Kakuma-machi, Kanazawa 920-1192, Japan
fukuma@staff.kanazawa-u.ac.jp

Frequency modulation atomic force microscopy (FM-AFM) has been traditionally used in ultrahigh vacuum for subnanometer-scale studies on various materials. One of the most remarkable features of FM-AFM is the capability of imaging atomic-scale structures, even on insulating surfaces with a small loading force. However, this capability was available only in vacuum, and hence its applications in air and liquid were very limited. Recently, there were tremendous advancements in the liquid-environment FM-AFM technique, which made it possible to operate FM-AFM in liquid with true atomic resolution. This advancement triggered subsequent molecular-scale studies on biological molecules by FM-AFM in physiological solution. In this chapter, the basic principle of FM-AFM and the special requirements for its operation in liquid are described. Furthermore, high-resolution FM-AFM studies on biological systems reported so far are also summarized.

Nanoscale Liquid Interfaces: Wetting, Patterning, and Force Microscopy at the Molecular Scale
Edited by Thierry Ondarçuhu and Jean-Pierre Aimé

ISBN 978-981-4316-45-3 (Hardcover), 978-981-4364-48-5 (eBook)
www.panstanford.com

18.1 Introduction

In 1991, Albrecht *et al.* invented frequency modulation atomic force microscopy (FM-AFM) for operating dynamic-mode AFM in vacuum environments.[1] Before this invention, it was common to operate dynamic-mode AFM with the amplitude detection method, which is referred to as amplitude modulation AFM (AM-AFM).[2] In AM-AFM, the tip–sample distance is regulated such that the oscillation amplitude of the cantilever (A) is kept constant.

AM-AFM was mainly used in air and liquid, and its applications in vacuum were very limited. This is because of the high Q factor in vacuum. For a typical cantilever, $Q = 1{,}000–100{,}000$ in vacuum, $Q = 100–1{,}000$ in air and $Q = 1–10$ in liquid. In dynamic-mode AFM, a high Q factor provides higher force sensitivity. In addition, the vacuum environment provides a clean surface and a well-defined environment, which are essential for the surface science studies. Thus, there were strong demands for operating dynamic-mode AFM in vacuum. However, the time response of A is inversely proportional to the Q factor, so the use of AM-AFM in vacuum was difficult in most of the applications because of the slow time response.

To resolve this issue, Albrecht *et al.* proposed to detect the frequency shift Δf of the cantilever resonance instead of A. In FM-AFM, the cantilever is always oscillated at its resonance frequency using a self-oscillation circuit. Thus, the Δf induced by the tip–sample interaction results in the frequency shift of the cantilever vibration. The Δf is detected from the cantilever deflection signal and used for the tip–sample distance regulation feedback. As the time response of Δf is free from the influence of Q, the time response of FM-AFM in vacuum is much faster than that of AM-AFM. Owing to the high force sensitivity obtained by the high Q factor and the atomically well-defined environment provided by vacuum, true atomic-resolution imaging by dynamic-mode AFM has become possible in 1995.[3,4]

Since the first demonstration of true atomic-resolution imaging by FM-AFM in vacuum, FM-AFM has been used for investigating various materials, including inorganic and organic surfaces. The distinctive advantages of FM-AFM include atomic-scale resolution even on an insulating surface and a small loading force during the imaging. These two features are closely related. The Δf signal steeply changes in the atomic-scale contact/noncontact regime, while the A signal gradually changes or shows a discontinuous jump in the same z range. Therefore, FM-AFM is suitable for the precise control of the vertical tip position in the atomic-scale contact/noncontact regime. This capability is essential for obtaining true atomic-resolution images as well as for preventing a hard contact with the sample.

Owing to these features, FM-AFM was considered suitable for investigations on organic samples. Most of the organic systems are electrically insulating and mechanically weak. In addition, there were increasing interests in organic electronics where the structures and properties of individual molecules play important roles in the function of electronic devices. Therefore, there were strong demands for the high-resolution imaging tools for organic systems. Because of this background, FM-AFM was applied to the investigations on various functional organic systems such as self-assembled monolayers,[5] ferroelectric thin films,[6] semiconducting molecules,[7] and conductive polymers.[8] These results revealed the wide applicability of FM-AFM to the molecular-scale investigations on organic systems.

In spite of these results, the number of FM-AFM users in molecular science was not increasing rapidly. This was mainly because all the high-resolution FM-AFM images were obtained in vacuum. This limitation in the operating environment had prevented its applications in biology and chemistry. In particular, investigations in liquid are essential in biological applications. The main difficulty in operating FM-AFM in air or liquid was the low Q factor of the cantilever resonance in these environments. The low Q factor results in low force sensitivity and makes it difficult to excite the cantilever vibration with a self-oscillation circuit. In addition, as FM-AFM was originally developed for operation in vacuum to take advantage of the high Q factor, not many researchers found any benefits of using FM-AFM in other environments. Because of these reasons, the development of liquid-environment FM-AFM had been impeded.

However, many of these ideas were not a conclusion derived from the detailed experimental or theoretical studies. Firstly, although there was no doubt that the low Q factor in liquid deteriorates FM-AFM performance, there was no evidence showing that it is the only reason for poor FM-AFM performance in liquid. Secondly, although FM-AFM has been mainly used in vacuum, there was no evidence showing that FM-AFM has no advantages over other AFM operation modes in air or liquid.

To clarify these points, the author and coworkers carried out detailed experiments to compare the noise measured with the actual instruments and the noise expected from the operation principle. The results revealed that FM-AFM performance in air and liquid was limited not only by the Q factor but also by the noise from the cantilever deflection sensor. By developing a low-noise cantilever deflection sensor,[9] true atomic-resolution imaging by FM-AFM became possible even in liquid in 2005.[10] This dramatic improvement of FM-AFM performance in liquid triggered subsequent applications of FM-AFM to the studies in biology and chemistry.

In this chapter, the instrumentation and biological applications of FM-AFM in liquid are described. In section 18.2, the basic principle of FM-AFM and the special requirements for the operation in liquid are described. In section 18.3, applications of FM-AFM to the molecular-resolution imaging of biological systems are reviewed. In section 18.4, the present state of liquid-environment FM-AFM and the future prospects are summarized.

18.2 Instrumentation

18.2.1 *Basic Principle*

In FM-AFM, a cantilever is mechanically oscillated at its resonance frequency using a piezo actuator placed adjacent to the cantilever. The induced cantilever vibration is detected typically with the optical beam deflection (OBD) method, as shown in Fig. 18.1.

The detected cantilever deflection signal is routed back to the piezo actuator through a phase shifter and an automatic gain control (AGC) circuit to form a self-oscillation loop. In this self-oscillation circuit, the cantilever works as a mechanical resonator and determines the oscillation frequency of this circuit.

Figure 18.1b shows the amplitude and phase versus frequency curves plotted around the cantilever resonance frequency (f_0).

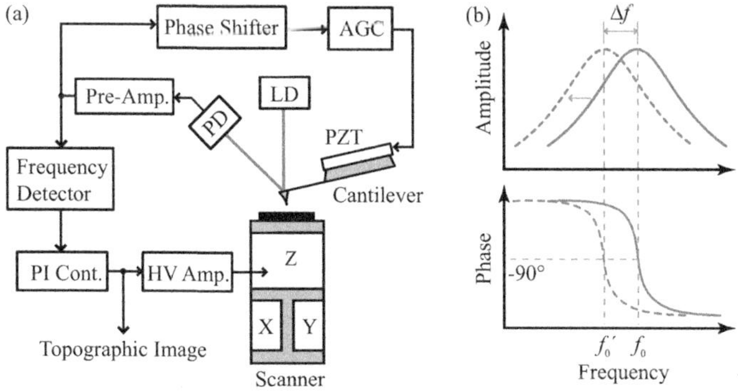

Figure 18.1 (a) The experimental setup for FM-AFM and (b) amplitude and phase vs. frequency characteristics of a cantilever plotted around the resonance frequency. f_0 and f_0' are the resonance frequency of the cantilever without and with the tip–sample interaction, respectively.

At the cantilever resonance, the phase difference between the cantilever deflection and excitation signals (ϕ) becomes $-90°$. At the steady state, the total phase delay in the self-oscillation loop becomes a multiple of $-360°$. Thus, ϕ is always kept at $-90°$ by adjusting the phase delay at the phase shifter to an appropriate value. Consequently, the oscillation frequency f is always kept at the cantilever resonance.

As the tip approaches the sample surface, the tip–sample interaction force induces the frequency shift (Δf) of the cantilever resonance. Accordingly, the cantilever oscillation frequency is also shifted by the same amount. The deflection signal is fed into the frequency detector, which produces a voltage signal proportional to Δf. A phase-locked loop (PLL) circuit is typically used as the frequency detector. The Δf signal is fed into the proportional-integral (PI) controller, which outputs Z signal to control the tip–sample distance.

18.2.2 *Spuriousness-Free Cantilever Excitation*

The cantilever excitation in liquid is more complicated than in air or vacuum. As described before, the cantilever vibration is typically excited by the piezo actuator integrated in the cantilever holder.

This piezo excitation method works fine in air and vacuum, while it causes problems in liquid. In the method, an alternating current (AC) voltage signal is applied across the piezo actuator to induce its mechanical vibration. This vibration, in turn, gives rise to an acoustic wave, which propagates to the cantilever and excites its vibration.

18.2.2.1 Piezo excitation

In a typical design, the acoustic wave propagates to any components mechanically coupled with the piezo actuator. This uncontrolled propagation excites vibrations of spurious resonances in the propagation paths. In air and vacuum, the influence from the spurious resonances is often negligible since their Q factors are lower than that of the cantilever resonance. In liquid, the Q factor of the cantilever drops to less than 10, which is often lower than that for a spurious resonance. Hence, the amplitude and phase versus frequency curves obtained in liquid are heavily distorted because of the influence of the spurious resonances.

Clean amplitude and phase curves are desirable in dynamic-mode AFM, especially with the FM detection method. In FM-AFM, a cantilever is always oscillated at its resonance frequency using the self-oscillation circuit. This self-oscillation loop works as a feedback circuit to keep ϕ at -90°. This feedback control operates on the linear slope of the phase versus frequency curve around the resonance frequency, as shown in Fig. 18.1b. Thus, large distortions in the phase curve result in instability of the cantilever excitation and deteriorate the accuracy of the force measurements by FM-AFM.

18.2.2.2 Magnetic excitation

To avoid this problem, other methods for cantilever excitation have been proposed. Among the most popular methods is magnetic excitation (see also chapter 1).[11] In the method, magnetic sensitivity is given to a cantilever by coating the cantilever backside with a magnetic thin film or attaching a magnetic particle on it. A coil is placed near the cantilever for inducing an AC magnetic field to drive the cantilever. In this case, the magnetic force is applied directly to the cantilever. Hence, the amplitude and phase curves are free from the influence of the spurious resonances.

There are several practical issues in magnetic excitation. The magnetic film or magnetic particle is made of a material that is not necessarily passive to an aqueous solution. Thus, the use of such a cantilever may lead to an increase in contamination. Attaching a magnetic particle on a cantilever backside is difficult for a cantilever with small dimension. Besides, the increase of the cantilever resonance frequency makes it difficult to produce a magnetic field strong enough to drive a cantilever. This is because the impedance of a coil increases with increasing the driving frequency. Hence, driving a high-frequency cantilever requires a high AC voltage signal. It is often difficult to deposit a magnetic thin film only on the backside of the cantilever, especially when the sputtering method is used. Consequently, the tip apex is often covered with a magnetic film. This may lead to an increase of the tip radius and low spatial resolution.

18.2.2.3 Photothermal excitation

Another method is photothermal excitation.[12] In the method, a cantilever is coated with an Au thin film. A laser beam with a power modulated at f_0 is irradiated to the cantilever. The modulated laser beam causes thermal expansion of the Si (or Si_3N_4) cantilever and the Au film. Owing to the difference in the thermal expansion coefficient between the two materials, the laser beam generates a stress at the cantilever–film interface, which induces the cantilever vibration synchronized with the laser power modulation. In this case, the cantilever is directly driven by the laser beam so that the amplitude and phase curves are free from the influence of the spurious resonances.

Figure 18.2 shows the amplitude and phase versus frequency curves obtained by the photothermal excitation in liquid.[13]

The frequency dependence of the experimentally measured amplitude curve shows a good agreement with that of the thermal PSD spectrum. The experimentally observed phase curve also shows good agreement with the ideal curve expected from the thermal PSD spectrum. These results show that ideal amplitude and phase responses can be obtained by photothermal excitation.

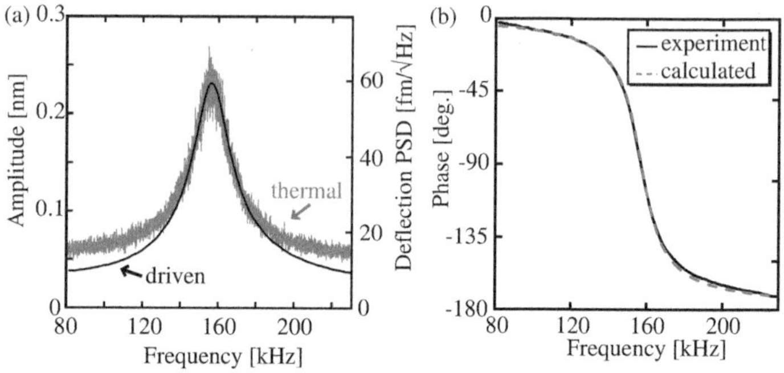

Figure 18.2 (a) Amplitude and (b) phase vs. frequency curves measured with the photothermal excitation in water. The PSD spectrum of the deflection signal measured without cantilever excitation is superposed on the amplitude curve obtained with the driven cantilever in (a). The values for Q and f_0 obtained from the thermal PSD spectrum are 8.7 and 157.079 kHz, respectively. The dotted line in (b) is calculated with these parameters, while the solid line shows the experimentally measured values.[13]

There are several practical issues in photothermal excitation. Although the Au thin film is not dissolved into an aqueous solution, the Cr thin film, which is often used as an adhesion layer, is not necessarily passive to all aqueous solutions. Instead of using Cr for an adhesion layer, one can use a Ti thin film, which is more resistive to an aqueous solution. The coating method is also important. When Au is deposited on the backside of a cantilever by the sputtering method, Au is often deposited on the tip apex as well. The use of thermal evaporation in a high-vacuum condition and a careful alignment of the cantilever backside to the evaporation source are required to solve this problem. An ultraviolet laser beam gives a higher efficiency in driving the cantilever than an infrared laser beam. However, it is highly likely to damage biological samples when it is irradiated to them. Thus, the use of the infrared laser beam is more desirable. In this case, most of the laser beam is reflected by the Au thin film. Hence, a relatively high laser power is required. Therefore, it is important to use a lower laser power while the beam spot is aligned to the cantilever backside. Once the alignment process is finished, the power can be safely increased for obtaining a sufficient driving force.

18.2.2.4 Improved piezo excitation

In spite of the clean amplitude and phase curves obtained by the magnetic and photothermal excitation methods, the piezo excitation method has been the most widely used, even in liquid. This is mainly because of its simple setup and wide applicability. The method does not require any modifications on the cantilever, which is very attractive for commercial instruments. Therefore, considerable efforts have been made for improving the performance of the piezo excitation method.

A typical strategy for suppressing the influence from the spurious resonances in piezo excitation is to control the propagation of the acoustic wave. For example, Degertekin *et al.* proposed to use an ultrasonic wave.[14,15] Owing to the short wavelength of the ultrasonic wave, its propagation path can be controlled with a Fresnel lens. Carrasco *et al.* proposed to use an accoustic absorber.[16] In their design, a piezo actuator is placed close to the cantilever and an acoustic absorber is placed in the propagation path from the piezo actuator to the other mechanical components.

Recently, another approach has been proposed by Asakawa *et al.* Instead of using an acoustic wave, they used the deformation of a flexure hinge prepared in the cantilever holder to drive a cantilever. To illustrate the effect of the flexure drive mechanism, they made two cantilever holders with the same structure but made of different materials. One of them is made of SS316, and the other is made of polyetheretherketone (PEEK). The former is too hard to be deformed by the impulsive force generated by the vibration of the piezo actuator, while the latter is compliant enough to be deformed by that. This difference is illustrated in the simulation results obtained by the finite element method (Fig. 18.3a,b).

The amplitude and phase curves obtained with the SS316 holder are distorted by the random peaks caused by the spurious resonances (Fig. 18.3c,e). On the contrary, such peaks are suppressed in the curves obtained with the PEEK holder (Fig. 18.3d,f). While the amplitude curve obtained by the PEEK holder shows almost an ideal response, the phase curve deviates from the ideal response. This is because of the linear phase delay caused by the finite delay time required for the deformation of the flexure hinge. However,

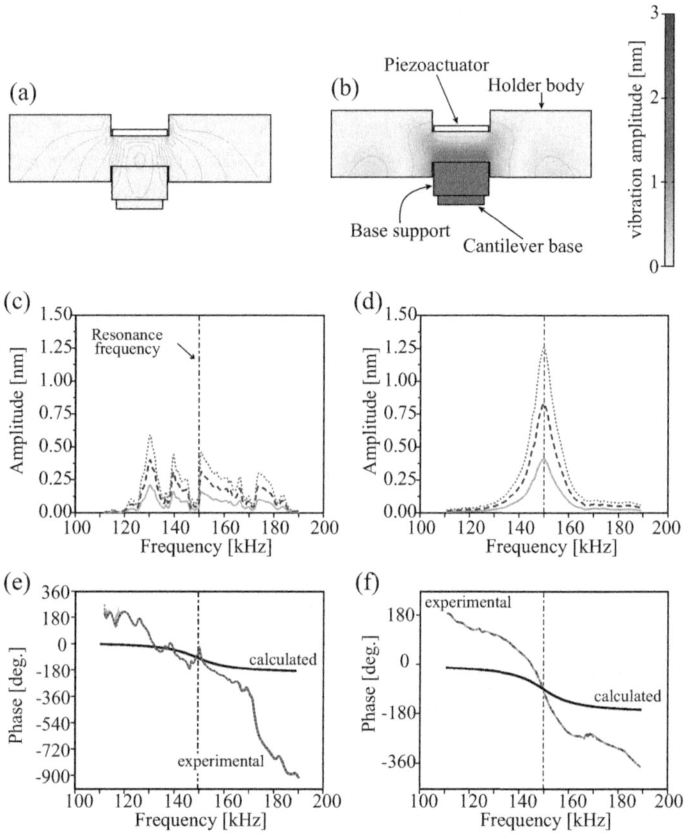

Figure 18.3 The 2D distribution of vibration amplitude in the simplified model of the flexure hinge made of (a) SS316 and (b) PEEK simulated by the finite element method. Amplitude and phase curves obtained in water by the cantilever holder with the holder body made of (c, e) SS316 and (d, f) PEEK. The cantilever was driven with an excitation signal having an amplitude of 50, 100, or 150 mV. Note that the measured phase curves in (e) and (f) appear as a single curve due to their small dependence on the excitation amplitude.[17] *Abbreviation*: 2D, two dimensional.

such a constant propagation delay does not prevent the stable self-oscillation of the cantilever. In addition, the error caused by the linear phase delay in the force measurement is easily compensated by the following method.

The relationship between the Q factor and the phase curve is given by

$$\frac{\partial\phi(f_0)}{\partial f} = -\frac{2Q}{f_0}. \tag{18.1}$$

Owing to the linear phase delay, the apparent Q factor (Q_a) measured from the experimentally observed phase curve is larger than the true Q factor of the cantilever resonance. As a result, the apparent frequency shift Δf_a observed in the experiment is smaller than the true Δf expected from the ideal phase curve. The true Δf can be recovered with the following equation.

$$\Delta f = \frac{Q_a}{Q}\Delta f_a. \tag{18.2}$$

Note that this equation is valid for the conservative force but invalid for the dissipative force. The recovery of the dissipative force is more complicated, so it is highly recommended that either magnetic or photothermal excitation be used for that purpose.

18.2.3 *Low-Noise Cantilever Deflection Sensor*

The cantilever deflection measurement system is one of the most important elements in AFM. In FM-AFM, the cantilever deflection signal is used for producing the cantilever excitation signal as well as for detecting Δf. Thus, having a clean deflection signal is important for stable cantilever oscillation as well as for sensitive force detection.

To date, various methods have been proposed for cantilever deflection measurement. Among the most widely used methods is the OBD method, owing to its simple setup and high sensitivity. In the method, a focused laser beam is irradiated to the cantilever backside and the reflected beam is detected with a position-sensitive photodetector (PSPD). The PSPD is typically made of a two- or four-divided-photodiode array.

The cantilever deflection induces the displacement of the laser spot on the PSPD. Thus, the difference between the photo-induced currents from the upper and lower elements varies in proportion to the cantilever deflection. The current signals are converted to the voltage signals using current-to-voltage (I–V) converters. These

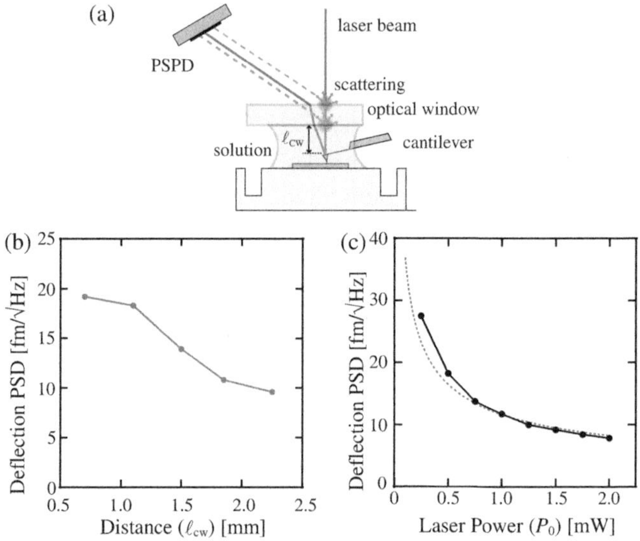

Figure 18.4 (a) Schematic illustration of the OBD sensor showing laser beam interference at the PSPD surface, caused by the scattering at the air–glass and glass–liquid interfaces. (b) Dependence of n_{zs} on the distance between the optical window and the cantilever backside (ℓ_{cw}). (c) Laser power (P_0) dependence of n_{zs}.[13]

voltage signals are fed into a differential amplifier, which produces a cantilever deflection signal.

There are several noise sources in the OBD method, which includes the Johnson noise from the transimpedance resistor in the I–V converter, the shot noise from the photodiodes, and the noise arising from the laser beam. In most of the designs, the ultimate detection limit is determined by the photodiode shot noise. However, the laser beam noise is often predominant in actual AFM. This is particularly evident in the case of liquid environment AFM.

In liquid-environment AFM, the laser beam is reflected or scattered at the air–glass and glass–liquid interfaces, as shown in Fig. 18.4a.

Some of the reflected beam goes back to the laser diode to induce the optical feedback noise, while some of the scattered laser beam goes to the PSPD to interfere with the beam reflected at the backside of the cantilever. These optical feedback and interference noises are often predominant in the case of liquid-environment AFM.

Such laser beam noises can be suppressed by modulating the laser power at a high frequency (typically 300–500 MHz). The laser modulation changes the laser oscillation mode from the single mode to the multimode. As the optical feedback noise arises from the competition between the different laser oscillation modes (i.e., mode hopping), multimode oscillation is effective to suppress the optical feedback noise. In addition, multimode oscillation reduces the coherence length (ℓ_{ch}) of the laser beam, which is effective to suppress the interference noise at the PSPD surface.

To suppress the interference noise, ℓ_{ch} must be reduced to a value less than the difference between the optical paths ($\Delta\ell$) of the laser beam reflected at the cantilever backside and that scattered at the glass–air or glass–liquid interface. $\Delta\ell$ depends on the distance between the optical window and the cantilever backside (ℓ_{cw}). For example, in the design shown in Fig. 18.4a, $\Delta\ell \simeq 2\ell_{cw}$. Thus, ℓ_{cw} should be longer than the half of ℓ_{ch}. This condition is not necessarily met even with laser power modulation. Figure 18.4b shows ℓ_{cw} dependence of the PSD of the cantilever deflection noise arising from the deflection sensor (n_{zs}).[13] By increasing ℓ_{cw} from 0.5 to 2.5 mm, n_{zs} was reduced by a factor of two, owing to the reduced influence from the laser beam interference. From the result, ℓ_{ch} was estimated to be approximately 4–6 mm in this particular case.

The ultimate detection limit in the OBD method is determined by the photodiode shot noise. The PSD of the deflection noise arising from the photodiode shot noise (n_{zp}) is given by[18]

$$n_{zp} = \frac{a_0 \ell_c}{6\chi n_m \ell_f} \sqrt{\frac{2e}{\eta\alpha P_0}}, \tag{18.3}$$

where a_0, χ, ℓ_f, e, η, α, ℓ_c, and n_m are the diameter of the collimated laser beam, the correction factor for a Gaussian laser beam profile, the focal length of the focusing lens, the elementary electric charge, the efficiency of the light-to-current conversion at the PSPD, the efficiency of the optical transmission from the focusing lens to the PSPD, the cantilever length, and the refractive index of the environment, respectively.

When the laser beam noise is reduced enough to reach the shot noise limit, Eq. 18.3 provides a direct guideline for reducing the deflection noise. For example, the use of smaller a_0 and longer ℓ_f gives a better performance in the OBD sensor. Figure 18.4c shows the power dependence of n_{zp}.[13] The good agreement between the experimentally measured noise values and the theoretically calculated values with Eq. 18.3 reveals that the noise performance of this OBD sensor is limited by the photodiode shot noise.

The n_{zs} values of an OBD sensor typically ranges from 100 to 1,000 fm/$\sqrt{\text{Hz}}$. However, the use of the high-frequency laser modulation technique and the careful design of the optical system made it possible to reduce n_{zs} to a value less than 10 fm/$\sqrt{\text{Hz}}$.[13,18] Such an excellent noise performance in the deflection measurement has been found to be essential for obtaining subnanometer-scale spatial resolution by FM-AFM in liquid.

18.2.4 *Operating Conditions*

The true (i.e., nonaverage) subnanometer-resolution imaging by AFM typically requires vertical and lateral spatial resolutions of 10–50 pm and 100–500 pm, respectively. To achieve vertical resolution, the vertical position of the tip front atom should be controlled with a precision better than 50 pm. A cantilever is always vibrating because of the thermal energy determined by the temperature (T) of the environment. The root mean square (rms) amplitude of the cantilever thermal vibration ($\langle z_{\text{th}} \rangle$) is given by

$$\langle z_{\text{th}} \rangle = \sqrt{\frac{k_{\text{B}} T}{k}}, \tag{18.4}$$

where k_{B} and k are Boltzmann's constant and spring constant of the cantilever, respectively. The equation shows that the thermal vibration increases with decreasing k. For cantilevers with k = 0.4 and 40 N/m, $\langle z_{\text{th}} \rangle$ values are 100 and 10 pm, respectively. To suppress the thermal vibration amplitude less than 50 pm, k should be higher than 1.6 N/m. In addition, the uncertainty will be increased by the feedback regulation noise. Therefore, it is recommended that a cantilever having a spring constant higher than 10 N/m be used.

Although this value is higher than the typical values (0.01–0.1 N/m) used for biological imaging in liquid, the use of a stiff cantilever does not necessarily lead to a high loading force as long as the tip–sample distance regulation is working with enough precision. This has been confirmed by the subnanometer-resolution imaging of biological samples by FM-AFM with a stiff cantilever.

To achieve a lateral resolution better than 500 pm, Δf should be sensitive enough to the short-range interaction forces having a decay length less than 500 pm. For this purpose, it is effective to reduce the cantilever oscillation amplitude to a value less than 500 pm.[19] With a large oscillation amplitude, the short-range interaction force is applied to the tip only for a fraction of the oscillation period. With a small oscillation amplitude, the short-range interaction force is always applied to the tip. Thus, the small amplitude operation is desirable to achieve a high spatial resolution.

The minimum detectable force (F_{min}) required for detecting the short-range force is typically 10–100 pN. The rough estimate of F_{min} in FM-AFM using the small amplitude approximation is given by[1]

$$F_{\text{min}} = \sqrt{\frac{4kk_{\text{B}}TB}{\pi f_0 Q}}, \tag{18.5}$$

where B is the bandwidth of the measurement. For a typical stiff cantilever (NCH: Nanoworld) in liquid, $k = 30$ N/m, $f_0 = 130$ kHz, $Q = 8$, and $F_{\text{min}} = 12$ pN at $B = 1$ kHz. This F_{min} value is just as much as required for true subnanometer-resolution imaging.

Equation 18.5 is valid only when the force sensitivity is limited by the thermal vibration of the cantilever. To satisfy this condition, the deflection noise density arising from the deflection sensor (i.e., n_{zs}) must be sufficiently smaller than that arising from the thermal Brownian motion ($n_{z\text{B}}$) around f_0. Owing to the low Q factor in liquid, $n_{z\text{B}}$ around f_0 is nearly constant. Thus, the above condition is described by

$$n_{zs} \ll n_{z\text{B}}(f_0) = \sqrt{\frac{2k_{\text{B}}TQ}{\pi f_0 k}}. \tag{18.6}$$

For a typical stiff cantilever mentioned before, $n_{z\text{B}}(f_0) = 74$ fm/$\sqrt{\text{Hz}}$. If we set the criterion at half of this value, n_{zs} should be smaller than 37 fm/$\sqrt{\text{Hz}}$. Experimentally, the author confirmed that an n_{zs} value smaller than 40 fm/$\sqrt{\text{Hz}}$ is good enough to obtain true atomic and molecular-resolution images by FM-AFM in liquid.[9,10]

In summary, a stiff cantilever is necessary for high vertical resolution and high stability, while a small amplitude operation is desirable for achieving high lateral and vertical resolution. To achieve the force sensitivity required for detecting the short-range force with a stiff cantilever, a low-noise cantilever deflection sensor is necessary. With these operating conditions, it has become possible to obtain true atomic and molecular resolution even in liquid by FM-AFM.

18.3 Biological Applications

18.3.1 *Bacteriorhodopsin*

In 2005, using a stiff cantilever, a small oscillation amplitude, and a low-noise cantilever deflection sensor, true molecular-resolution images of a polymer single crystal[20] and true atomic-resolution images of a mica surface[10] were obtained in water. These results showed that subnanometer-resolution imaging by FM-AFM is possible even in liquid. However, the spring constant of the cantilevers used in these experiments was much higher ($k \simeq 30$ N/m) than that of a typical cantilever used in AFM imaging of biological systems. Thus, there was a concern about the possibility of damaging soft biological samples.

To clarify this point, the author and coworkers imaged purple membranes by FM-AFM in phosphate buffer solution.[21] A purple membrane consists of closely packed trimers of bacteriorhodopsins (bRs) embedded in a lipid bilayer. This sample has been often used as a model biological system to be imaged by contact-mode AFM and AM-AFM. Since molecular-resolution images obtained by these existing methods were already reported before that time, it was a good model sample for testing the applicability of FM-AFM to the investigations on biological systems (see also chapters 17 and 19).

Figure 18.5 shows FM-AFM images of purple membranes.[21] In these images, individual bR molecules constituting the membrane are clearly resolved (Fig. 18.5b).

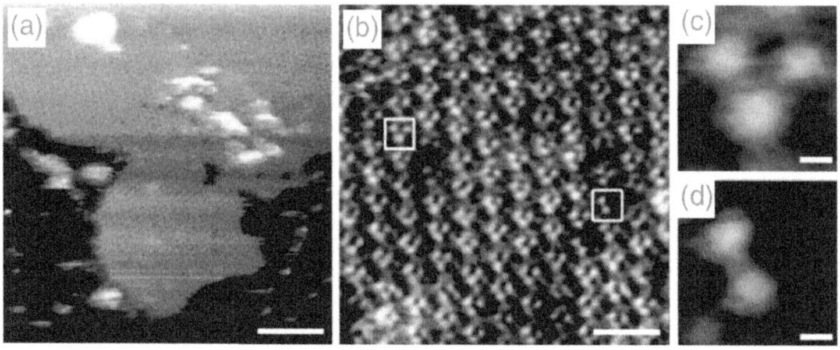

Figure 18.5 FM-AFM images of a purple membrane including bR protein molecules adsorbed on muscovite mica in phosphate buffer solution. (a) Scale bar: 60 nm. (b) Scale bar: 16 nm. (c, d) Scale bars: 1.4 nm. (c) and (d) are the magnified images of the areas indicated by the white squares in (b).[21]

In addition, a single molecular defect (Fig. 18.5d) is clearly resolved in the image, which demonstrates that FM-AFM is capable of imaging soft biological samples with true molecular resolution, even with a stiff cantilever.

This result was soon confirmed by another research group. Hoogenboom *et al.* developed a low-noise cantilever deflection sensor using a Fabry–Perot interferometer.[22] With this deflection sensor, they obtained molecular-scale images of bR trimers and atomic-scale images of mica by FM-AFM in liquid.[23]

18.3.2 *GroELs*

The applicability of FM-AFM to the imaging of a soft biological system was shown by imaging bR molecules.[21] However, since a purple membrane can be imaged by other existing methods, such as contact mode AFM and AM-AFM, these results did not show any advantages over these methods. In particular, contact-mode AFM had been widely used for high-resolution imaging of membrane proteins before that time. Thus, there was a strong interest in the advantages of FM-AFM over contact-mode AFM.

One of the obvious advantages of dynamic-mode AFM over contact-mode AFM is the capability of imaging isolated molecules weakly bound to the substrate. In contact-mode AFM, a tip is laterally scanned in contact with a sample so that a lateral friction force often damages the sample unless it is closely packed to form a rigid crystal structure, such as bRs in a purple membrane. In dynamic-mode AFM, the vertical motion of the cantilever greatly reduces the lateral force during the scan so that an isolated molecule can be imaged without destruction.

To confirm this capability, the author and coworkers imaged isolated GroEL molecules on mica in buffer solution.[21] In this experiment, no chemical modifications were made either to the molecules or to the substrate. Hence, the molecules were physically adsorbed on a mica surface with a weak interaction force. Even with this condition, the obtained images clearly showed individual GroEL molecules constituting a ring-shaped structure. The result revealed that FM-AFM is capable of imaging isolated biomolecules weakly bound to a substrate.

18.3.3 *Lipid Bilayer*

Although FM-AFM imaging of bR and GroEL molecules demonstrated the applicability of FM-AFM to biological systems, these images have not been obtained with subnanometer-scale resolution. This was not because of the limited resolution of FM-AFM but because of the fluctuations of biological molecules themselves. However, as one of the major advantages of FM-AFM is its high spatial resolution, there was a strong interest in whether such high-resolution imaging is possible even on a biological system.

The author and coworkers obtained the first subnanometer-resolution FM-AFM image of biological molecules by imaging a lipid bilayer consisting of dipalmitoylphosphatidylcholine (DPPC) molecules (Fig. 18.6a).[24]

A DPPC bilayer consists of amphiphilic DPPC molecules having a hydrophilic head group and a hydrophobic tail group. They form a bilayer in an aqueous solution with the head groups next to water and the tail groups inside the membrane (Fig. 18.6b). The DPPC bilayer takes a stable gel-phase structure at room temperature. Thus, it was possible to resolve individual molecules periodically arranged with a spacing of about 0.5 nm, as shown in Fig. 18.6c. The result demonstrated that subnanometer-resolution

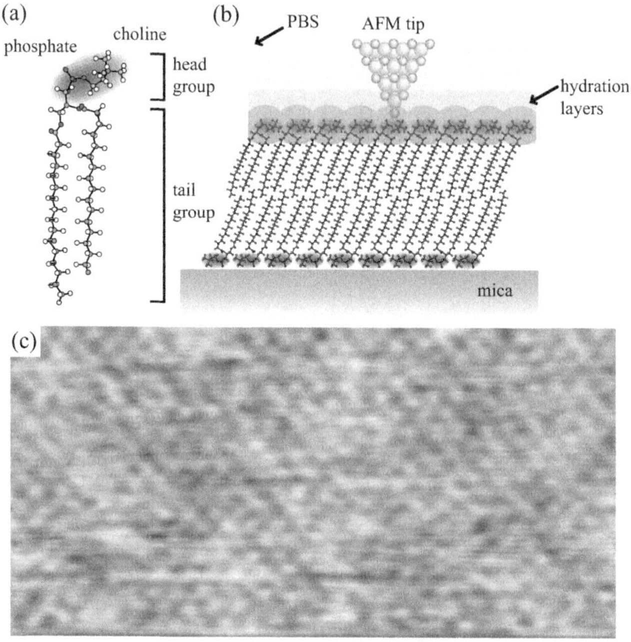

Figure 18.6 (a) Model of a DPPC molecule, (b) model of a DPPC bilayer formed on mica in water, and (c) FM-AFM image of a DPPC bilayer in water (15 nm × 7.5 nm).[24]

FM-AFM imaging of biological molecules is possible even with a stiff cantilever.

18.3.4 *Amyloid Fibrils*

The capability of imaging isolated biomolecules weakly bound to a substrate was shown by imaging GroEL molecules on mica with molecular-scale resolution. However, the resolution achieved in the imaging was on the order of nanometers. Thus, there was an interest in whether it is possible to obtain subnanometer resolution even on isolated biological molecules weakly bound to a substrate.

To confirm this possibility, the author and coworkers imaged amyloid fibrils in buffer solution.[25] Amyloid fibrils are formed from a variety of proteins, which are normally soluble in aqueous solution. Under certain conditions, the precursor proteins misfold to form

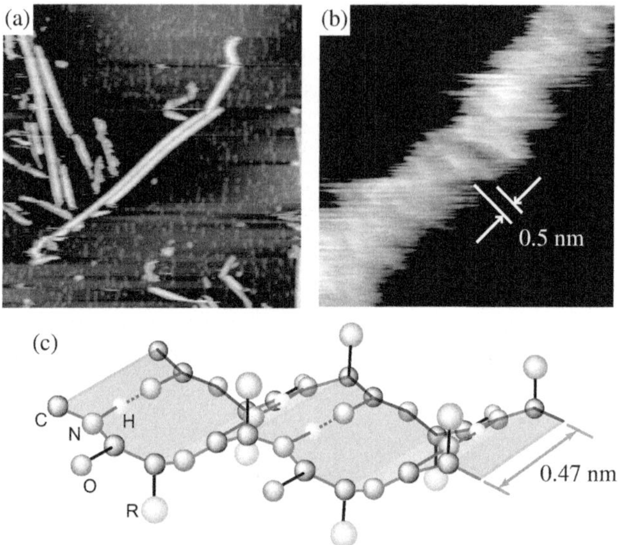

Figure 18.7 FM-AFM images of IAPP fibrils on mica in PBS solution. (a) 800 nm × 800 nm, $\Delta f = -55$ Hz, tip velocity = 1 μm/s. (b) 10 nm × 10 nm, $\Delta f = +50$ Hz, tip velocity = 195 nm/s. (c) Schematic model of the β-strands.[25] *Abbreviations*: IAPP, islet amyloid polypeptide; PBS, phosphate buffer solution.

amyloid fibrils, which are typically 10–30 nm in diameter and up to several micrometers in length, as shown in Fig. 18.7a.

The fibrils are insoluble β-sheet structures that can form aggregates in tissues and organs, known as amyloid deposits. These deposits are found in a range of neurodegenerative diseases such as Alzheimer's, Parkinson's, and Huntington's. Understanding the structure of amyloid fibrils is essential for clarifying the nucleation and kinetics of fibrillation in relation to the pathogenic pathway of amyloidoses.

Figure 18.7a,b shows FM-AFM images of amyloid fibrils formed from the IAPP deposited on mica in PBS solution.[25] IAPP fibrils are associated with type 2 diabetes when deposited in pancreatic islets. The image shows finely striped features perpendicular to the fibril axis. The distance between adjacent stripes is approximately 0.5 nm. A previous X-ray diffraction study suggested the existence of periodic structure along IAPP fibril axes with a spacing of 0.47 nm.

Since this periodicity agrees with hydrogen bond spacing between the β-strands (Fig. 18.7c), the periodic structure was attributed to the alignment of the β-strands perpendicular to the fibril axis. The fine stripes observed in the FM-AFM image have comparable spacing aligned perpendicular to the fibril axis. Thus, the striped features are attributed to the β-strands of IAPP fibrils. The result demonstrates the possibility of FM-AFM to image subnanometer-scale structures of isolated biomolecules in liquid.

18.3.5 *Lipid-Water Complexes*

The imaging of bRs, GroELs, DPPCs, and amyloid firbrils in liquid revealed that FM-AFM is applicable to high-resolution imaging of biological systems with subnanometer resolution. After these experiments for understanding the capability of FM-AFM, the next step was to apply this method to the investigation of biological issues. Many of the important biological issues have been intensively studied by the existing methods. Therefore, it is important to make use of the unique capability of FM-AFM to provide additional information that has not been available by other methods.

AFM is a method for imaging surface structures at a solid–liquid interface with high spatial resolution. Thus, it is almost impossible to image internal structure of biological systems. However, many of the important biological processes and phenomena take place at the interface between the biological systems and the surrounding physiological solution. Hence, this capability can be a strong advantage over other existing methods. In fact, tremendous efforts have been made for enhancing the sensitivity to the interfacial phenomena in traditional methods such as optical microscopy, X-rays, and neutron spectroscopy.

The author and coworkers investigated the interfacial phenomena happening at the interface between a biological membrane and physiological solution. A biological membrane consists of various molecules such as lipids, proteins, and cholesterols. The physiological solution mainly consists of water molecules but contains various ions and small organic molecules. At the interface between these two domains, they interact with each other to form various complexes. Such interfacial phenomena have a strong influence on the structure and function of biological membranes.

Biological membranes are so complicated that a lipid membrane is often used as a model sample for their studies. The author and coworkers investigated the interface between a DPPC bilayer and PBS buffer solution. At the lipid–water interface, water molecules interact with lipid head groups to show nonuniform distribution. Since an AFM tip interacts with water molecules as well as the lipid head groups, the force distribution detected by AFM should be influenced by the water distribution as well as by the surface corrugation.

18.3.5.1 Force curve measurement

Figure 18.8a shows a Δf versus distance curve measured on a DPPC bilayer formed on mica in PBS solution.[26]

The Δf curve shows an oscillatory profile with two peaks separated by 0.25 nm. This distance corresponds to the diameter of a water molecule, which suggests that the oscillatory force is caused by the hydration layers formed on a DPPC bilayer.

The hydration force on a lipid membrane has been investigated by various methods due to its importance in the biological processes. However, the force measurements by surface force apparatus (SFA) and the osmotic pressure method showed no oscillatory profiles.

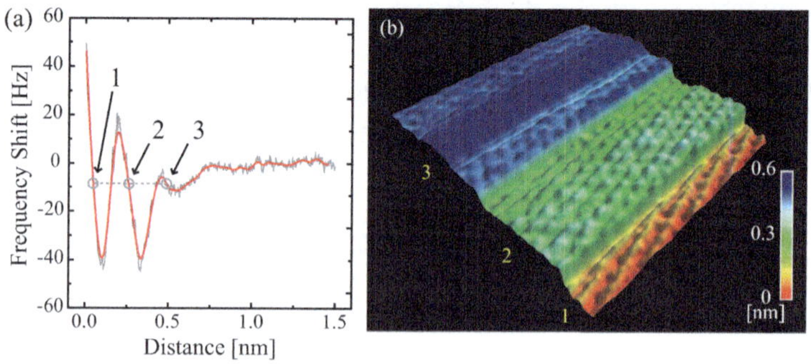

Figure 18.8 (a) Δf vs. distance curve measure on a DPPC bilayer formed on mica in PBS. (b) FM-AFM image of a DPPC bilayer in PBS solution (8 nm × 8 nm).[26]

Thus, it was questioned if there were any intrinsic hydration layers on a lipid membrane. The oscillatory profile observed in the Δf versus distance curve obtained by FM-AFM revealed the existence of the hydration layer on a DPPC bilayer (Fig. 18.8a). In FM-AFM, the force is measured with a tip having a nanometer-scale cross section, while the force measured by SFA and the osmotic pressure method is averaged over a micrometer-scale area. Such global averaging may smear out the local distance dependence, showing the oscillatory profile. The result clearly showed the importance of having local spatial resolution in the investigations on interfacial phenomena.

The operating conditions required for the true atomic-resolution imaging by FM-AFM were found to be also suitable for high-sensitivity measurement of hydration forces. The hydration force has an interaction length on the order of the diameter of a water molecule (i.e., 0.25 nm). For detecting such short-range forces, it is desirable to operate FM-AFM with a oscillation amplitude less than 0.25 nm, which corresponds to the requirement for high-resolution imaging by FM-AFM.

Another important condition is the use of a stiff cantilever. In addition to the small thermal vibration provided by a stiff cantilever, the adhesion events are prevented by using a stiff cantilever. The oscillatory profile caused by the hydration layer often shows an attractive force branch with a large force gradient. When the magnitude of the force gradient exceeds the spring constant of the cantilever, the tip jumps to contact with the surface, which is referred to as jump-to-contact. Thus, a stiff cantilever (typically $k > 1$ N/m) is required for the measurement of hydration forces. This condition is also met in a typical operating condition in high-resolution FM-AFM.

18.3.5.2 FM-AFM imaging

Owing to the oscillatory force profile caused by the hydration layers, there is more than one stable tip position for the tip–sample distance regulation, as indicated by the dotted line in Fig. 18.8a. Consequently, the tip jumps between these feedback positions during the imaging, as shown in Fig. 18.8b. In the image, the tip was scanned from the lowest terrace (Terrace 1) corresponding to the DPPC bilayer surface. During the imaging, the tip jumped to the Terrace 2 corresponding to the primary hydration layer. The height of the jump was 0.26 nm, which corresponds to the diameter of a water molecule. Even on the primary hydration layer, the image shows molecular-scale corrugations, which suggests that the spatial distribution of the water molecules reflects the surface corrugations corresponding to the lipid head groups. Further scan results in the tip jump from Terrace 2 to the secondary hydration layer (Terrace 3). The height of the jump was 0.3 nm, which also agreed with the size of a water molecule.

The results obtained in this experiment revealed the existence of stable hydration layers with a nanometer-scale lateral extent. Although the vertical distribution of the hydration layer at a single position can be probed by the force curve measurement, the investigations on the lateral distribution of a hydration layer was very difficult. In this experiment, the high force sensitivity and the precise control of the vertical tip position made it possible to visualize lateral distribution of hydration layers. The result also showed that the loading force during the FM-AFM imaging is small enough to image individual hydration layers without destruction. This result strongly supports that the use of a stiff cantilever does not necessarily result in a high loading force. Furthermore, the result showed that FM-AFM can visualize not only the structures of solid surfaces but also the distribution of water molecules interacting with the surface.

18.3.6 *Lipid-Ion Complexes*

In addition to water molecules, metal cations in the physiological solution play important roles in the biological processes taking place at the lipid–water interface. The addition of salts into water triggers aggregation and fusion processes of lipid membranes. The metal cations enhance the mechanical strength of the lipid bilayer and reduce the fluidity of lipid molecules. To explain such influence of metal cations on the membrane properties, it has been postulated that the metal cations interact with negatively charged moieties of the lipid molecule to form complicated lipid-ion networks (Fig. 18.9a). However, it has been a great challenge to

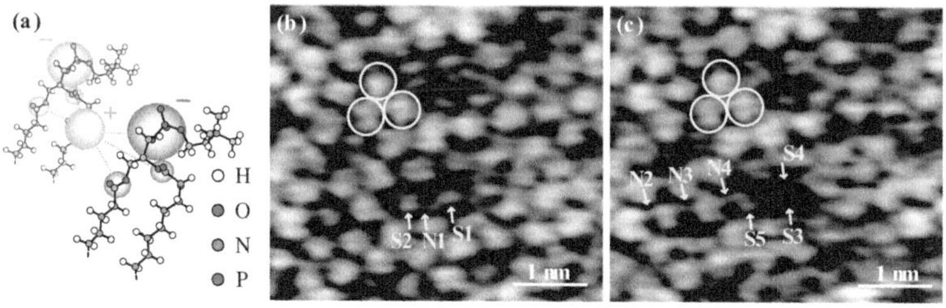

Figure 18.9 (a) Schematic model of the lipid-ion complex formed at the lipid–water interface. (b) and (c) Sequential FM-AFM images of the same area of the DPPC bilayer in PBS solution. Height range: 0.1 nm. Tip velocity: 120 nm/s. Imaging speed: 85 s/image.[27]

experimentally confirm the existence of lipid-ion networks due to the lack of a method to directly visualize the distribution of ions with atomic-scale resolution.

Figure 18.9b,c shows sequential FM-AFM images taken on a DPPC bilayer in PBS.[27]

The image shows hexagonally packed surface groups separated by about 0.5 nm. The individual surface group consists of two oval-shaped subunits separated by 0.1–0.2 nm. The detailed analyses of the FM-AFM images, together with the previously reported spectroscopic studies, revealed that the pair of subunits found in the image represents the spatial distribution of the Na^+ ions interacting with the negatively charged two oxygen atoms in the lipid head group. The result reveals that FM-AFM is capable of imaging distribution of ions at a solid–liquid interface.

While some of the surface groups did not show any changes during the imaging, as indicated by the circles, some of the lipid-ion networks disappeared or appeared, as indicated by the arrows. In this experiment, one image was taken in 85 seconds. Thus, it is likely that the ions were moving much faster than the scan speed of the tip. The observed changes should correspond to the dynamic changes of the averaged spatial distribution of ions, namely, lipid-ion networks.

If we define the spatial resolution as the closest distance resolved in one image, these images show that the spatial resolution of FM-AFM is as small as 90 pm. This is by far the highest spatial resolution obtained in liquid so far. Such high-resolution imaging was already reported for the experiments in vacuum.[28] However, the result revealed that it is also possible even in liquid.

18.3.7 *Lipid-Cholesterol Complexes*

Cholesterols are one of the major components constituting a biological membrane. The insertion of cholesterols into a lipid membrane dramatically changes the molecular fluidity and mechanical strength of the membrane. However, the molecular-scale mechanisms for such influence of cholesterols have not been well understood. This is partly because of the lack of a method to visualize the molecular-scale structure of the lipid-cholesterol complexes formed in the membrane.

The author and coworkers applied FM-AFM to the investigation of the DPPC-cholesterol (50-50 mol%) mixed bilayer. Figure 18.10 shows an FM-AFM image of a DPPC-cholesterol mixed bilayer formed on mica in PBS.

The image shows bright rows from top left to the bottom right. Each row consists of bright spots separated by 0.46 nm. Since this spacing agrees with the molecular spacing observed in the DPPC bilayer, the bright spots are attributed to the DPPC molecules.

The distance between the adjacent molecular rows is 0.71 nm. This is much longer than the molecular spacing (0.46 nm). The

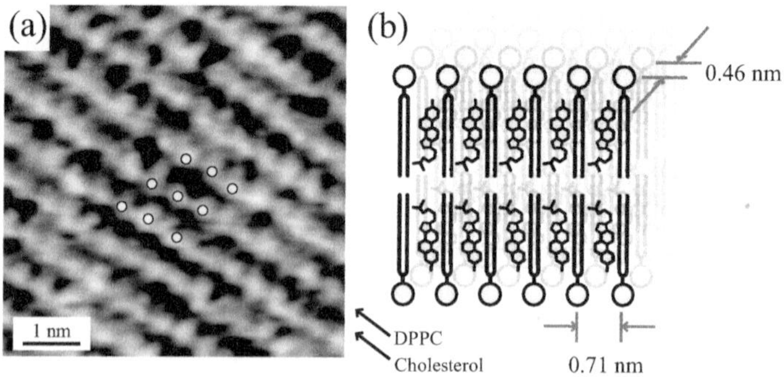

Figure 18.10 (a) FM-AFM image of a DPPC-cholesterol (50–50 mol%) mixed bilayer formed on mica in PBS solution. (b) The molecular-scale model of the lipid-cholesterol complex proposed based on the FM-AFM image shown in (a).[29]

previous studies suggested that cholesterols are inserted between the lipid molecules to interact with their hydrophobic tail groups. Therefore, the lower area observed between the DPPC rows should correspond to the cholesterol rows. In fact, as cholesterol molecules are smaller than DPPC molecules, if they interact with the tail groups, the cholesterol rows should be lower than the DPPC rows.

On the basis of these discussions, the molecular-scale model for the lipid-cholesterol complex is proposed, as shown in Fig. 18.10b. Before that time, several models for the lipid-cholesterol complexes were proposed on the basis of the results obtained by spectroscopic methods and molecular dynamics simulations. However, it had been difficult to unambiguously determin the molecular-scale structure. The result obtained in this experiment showed that FM-AFM can be a powerful means to determine the molecular-scale structure of biomolecular complexes.

18.4 Summary and Future Prospects

There have been tremendous advancements in the liquid-environment FM-AFM technique in the last 10 years. Using a stiff cantilever, a small oscillation amplitude, and a low-noise cantilever deflection sensor,[9] it has become possible to obtain true atomic-resolution images even in liquid.[10] Its applicability to the biological systems is demonstrated by imaging bRs and GroELs.[21,23] The subnanometer-resolution imaging of biological systems is also demonstrated by imaging lipid bilayers[24] and amyloid fibrils [25]. In addition to the structure of the molecular surfaces, the distribution of water[26] and ions[27] interacting with the surface is also visualized by FM-AFM. Finally, the molecular-scale arrangement of the unknown biomolecular complex (i.e., the lipid-cholesterol complex) was determined by FM-AFM imaging.[29]

Until recently, the use of FM-AFM in liquid was very limited. However, several AFM research groups have already started to use it for their own applications. Kawakatsu *et al.* developed a low-noise cantilever deflection sensor using a Doppler interferometer[30] and obtained atomic-resolution images of mica in liquid.[31] Umeda *et al.* reduced the deflection noise by modifying the cantilever deflection sensor of a commercial AFM and applied the instrument to the studies in electrochemistry.[32] Nagashima *et al.* obtained molecular-resolution images of tetragonal lysozyme (110) surface in liquid using FM-AFM with a low-noise OBD sensor.[33] These recent results demonstrated the wide applicability of liquid-environment FM-AFM. Besides, these applications also highlighted the need for a low-noise cantilever deflection sensor. At present, FM-AFM users are limited to the researchers who are able to modify the AFM instruments to reduce the deflection noise. However, the author expects that the use of FM-AFM will rapidly increase in the near future as soon as scanning probe microscope manufacturers implement the FM detection mode in their commercial products.

Because of the existence of the hydration layers, a solid–liquid interface has a vertical extent up to a few nanometers, which is much broader than that of a solid–vacuum interface. It is possible to investigate vertical and lateral distribution of water by taking a one-dimensional (1D) force profile and a 2D height image, as shown in Fig. 18.8.

However, understanding the three-dimensional (3D) interfacial phenomena by taking 1D profiles and 2D images is often too inefficient or too inaccurate to be used in a practical study. The author and coworkers recently developed a method referred to as 3D scanning force microscopy (3D-SFM), where the tip is scanned in the vertical direction as well as in the lateral direction.[34] The method made it possible to image the 3D distribution of water molecules at the mica–water interface with atomic-scale resolution in 53 seconds, as shown in Fig 18.11.

The image revealed the 3D distribution of a hydration layer and adsorbed water molecules. Such direct information on the water distribution has not been available with other methods and hence invaluable in the studies on hydration phenomena. Therefore, the author expects that such a 3D force imaging technique should become a powerful tool in the various research fields related to the solid–liquid interface.

There are several remaining issues in the biological applications of FM-AFM. One of the most serious problems is the insufficient operation speed of FM-AFM (for high-speed imaging refer to chapter 19). The biological systems have much larger corrugations,

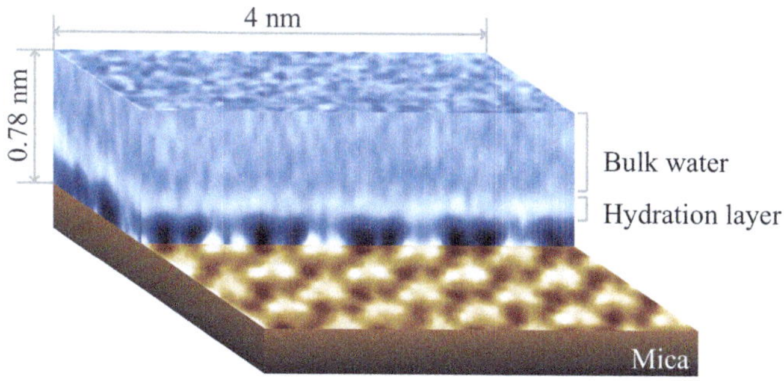

Figure 18.11 A 3D-SFM image obtained at an interface between a cleaved mica surface and PBS solution ($A = 0.62$ nm). The 3D-SFM image ($4 \times 4 \times 0.78$ nm^3) was constructed from the approaching Z profiles and has $64 \times 64 \times 155$ pixels in XYZ. The frequency and amplitude of the Z modulation during 3D-SFM imaging were 200 Hz and 0.78 nm, respectively. The lateral scan speed was 12.2 nm/sec. Each XZ cross-sectional image was obtained in 0.32 sec, while the entire 3D image was obtained in 53 sec.[34]

fluctuations, and complexity than the samples that have been investigated by FM-AFM in vacuum. Thus, the operation speed of FM-AFM is often insufficient. Another problem is the theoretical understanding of the image contrasts. For experiments in vacuum, it has become common to compare the experimental results with theoretically simulated images to understand the origins for the atomic-scale contrasts. However, the theoretical methodology for simulating the FM-AFM images in liquid has not been established yet. In particular, the interpretation of 3D images obtained by 3D-SFM is very complicated. Although the 3D force distribution obtained by 3D-SFM suggested strong correlation between the image contrasts and the distribution of water molecules, an accurate understanding should require a detailed comparison with the results obtained by theoretical simulations. Although the simulation of an atomic-scale AFM image obtained in liquid is very difficult, some of the research groups have already started to deal with this challenging subject. Therefore, the author expects that the collaborative work between experimentalists and theoreticians should rapidly increase in the next a few years in this field and hopes that they will improve the understanding of FM-AFM images obtained in liquid.

References

1. T. R. Albrecht, P. Grütter, D. Horne, and D. Ruger, Frequency modulation detection using high-Q cantilevers for enhanced force microscopy sensitivity, *J. Appl. Phys.* **69**, 668 (1991).
2. Y. Martin, C. C. Williams, and H. K. Wickramasinghe, Atomic force microscope-force mapping and profiling on a sub 100-Å scale, *J. Appl. Phys.* **61**, 4723 (1987).
3. F. J. Giessibl, Atomic resolution of the silicon (111)-(7×7) surface by atomic force microscopy, *Science.* **267**, 68 (1995).
4. S. Kitamura and M. Iwatsuki, Observation of 7 7 reconstructed structure on the silicon (111) surface using ultrahigh vacuum noncontact atomic force microscopy, *Jpn. J. Appl. Phys. Part II.* **34**, L145 (1995).
5. T. Fukuma, K. Kobayashi, T. Horiuchi, H. Yamada, and K. Matsushige, Alkanethiol self-assembled monolayers on Au(111) surfaces investigated by non-contact AFM, *Appl. Phys. A.* **72**, S109 (2001).
6. T. Fukuma, K. Kobayashi, K. Noda, K. Ishida, T. Horiuchi, H. Yamada, and K. Matsushige, Molecular-scale non-contact AFM studies of ferroelectric organic thin films epitaxially grown on alkali halides, *Surf. Sci.* **516**, 103 (2002).
7. T. Fukuma, K. Kobayashi, H. Yamada, and K. Matsushige, Non-contact atomic force microscopy study of copper-phthalocyanines: Submolecular-scale contrasts in topography and energy dissipation, *J. Appl. Phys.* **95**, 4742 (2004).
8. T. Ichii, H. Kawabata, T. Fukuma, K. Kobayashi, H. Yamada, and K. Matsushige, Molecular-scale investigations of semi-insulating polymer single crystals by noncontact atomic force microscopy, *Nanotechnology.* **16**, S22 (2005).
9. T. Fukuma, M. Kimura, K. Kobayashi, K. Matsushige, and H. Yamada, Development of low noise cantilever deflection sensor for multienvironment frequency-modulation atomic force microscopy, *Rev. Sci. Instrum.* **76**, 053704 (2005).
10. T. Fukuma, K. Kobayashi, K. Matsushige, and H. Yamada, True atomic resolution in liquid by frequency-modulation atomic force microscopy, *Appl. Phys. Lett.* **87**, 034101 (2005).
11. S. P. Jarvis, A. Oral, T. P. Weihs, and J. B. Pethica, A novel force microscope and point contact probe, *Rev. Sci. Instrum.* **64**, 3515 (1993).
12. N. Umeda, S. Ishizaki, and H. Uwai, Scanning attractive force microscope using photothermal vibration, *J. Vac. Sci. Technol. B.* **9**, 1318 (1991).

13. T. Fukuma, Wideband low-noise optical beam deflection sensor with photothermal excitation for liquid-environment atomic force microscopy, *Rev. Sci. Instrum.* **80**, 023707 (2009).

14. F. L. Degertekin, B. Hadimioglu, T. Sulchek, and C. F. Quate, Actuation and characterization of atomic force microscope cantilevers in fluids by acoustic radiation pressure, *Appl. Phys. Lett.* **78**, 1628 (2001).

15. A. G. Onaran and F. L. Degertekin, A fluid cell with integrated acoustic radiation pressure actuator for atomic force microscopy, *Rev. Sci. Instrum.* **76**, 103703 (2005).

16. C. Carrasco, P. Ares, P. J. de Pablo, and J. Gómez-Herrero, Cutting down the forest of peaks in acoustic dynamic atomic force microscopy in liquid, *Rev. Sci. Instrum.* **79**, 126106 (2008).

17. H. Asakawa and T. Fukuma, Spurious-free cantilever excitation in liquid by piezoactuator with flexure drive mechanism, *Rev. Sci. Instrum.* **80**, 103703 (2009).

18. T. Fukuma and S. P. Jarvis, Development of liquid-environment frequency modulation atomic force microscope with low noise deflection sensor for cantilevers of various dimensions, *Rev. Sci. Instrum.* **77**, 043701 (2006).

19. F. J. Giessibl, H. Bielefeldt, S. Hembacher, and J. Mannhart, Calculation of the optimal imaging parameters for frequency modulation atomic force microscopy, *Appl. Surf. Sci.* **140**, 352 (1999).

20. T. Fukuma, K. Kobayashi, K. Matsushige, and H. Yamada, True molecular resolution in liquid by frequency-modulation atomic force microscopy, *Appl. Phys. Lett* **86**, 193108 (2005).

21. H. Yamada, K. Kobayashi, T. Fukuma, Y. Hirata, T. Kajita, and K. Matsushige, Molecular resolution imaging of protein molecules in liquid using frequency modulation atomic force microscopy, *Appl. Phys. Express.* **2**, 095007 (2009).

22. B. W. Hoogenboom, P. L. T. M. Frederix, J. L. Yang, S. Martin, Y. Pellmont, M. Steinacher, S. Zäch, E. Langenbach, and H.-J. Heimbeck, A fabry perot interferometer for micrometer-sized cantilevers, *Appl. Phys. Lett.* **86**, 074101 (2005).

23. B. W. Hoogenboom, H. J. Hug, Y. Pellmont, S. Martin, P. L. T. M. Frederix, D. Fotiadis, and A. Engel, Quantitative dynamic-mode scanning force microscopy in liquid, *Appl. Phys. Lett.* **88**, 193109 (2006).

24. M. Higgins, M. Polcik, T. Fukuma, J. Sader, Y. Nakayama, and S. P. Jarvis, Structured water layers adjacent to biological membranes, *Biophys. J.* **91**, 2532 (2006).

25. T. Fukuma, A. S. Mostaert, L. C. Serpell, and S. P. Jarvis, Revealing molecular-level surface structure of amyloid fibrils in liquid by means of frequency modulation atomic force microscopy, *Nanotechnology*. **19**, 384010 (2008).
26. T. Fukuma, M. J. Higgins, and S. P. Jarvis, Direct imaging of individual intrinsic hydration layers on lipid bilayers at Ångstrom resolution, *Biophys. J.* **92**, 3603 (2007).
27. T. Fukuma, M. J. Higgins, and S. P. Jarvis, Direct imaging of lipid-ion network formation under physiological conditions by frequency modulation atomic force microscopy, *Phys. Rev. Lett.* **98**, 106101 (2007).
28. F. J. Giessibl, S. Hembacher, H. Bielefeldt, and J. Mannhart, Subatomic features on the silicon (111)-(7×7) surface observed by atomic force microscopy, *Science*. **289**, 422 (2000).
29. H. Asakawa and T. Fukuma, The molecular-scale arrangement and mechanical strength of phospholipid/cholesterol mixed bilayers investigated by frequency modulation atomic force microscopy in liquid, *Nanotechnology*. **20**, 264008 (2009).
30. S. Kawai, D. Kobayashi, S. Kitamura, S. Meguro, and H. Kawakatsu, An ultrahigh vacuum dynamic force microscope for high resonance frequency cantilevers, *Rev. Sci. Instrum.* **76**, 083703 (2005).
31. S. Nishida, D. Kobayashi, T. Sakurada, T. Nakazawa, Y. Hoshi, and H. Kawakatsu, Photothermal excitation and laser doppler velocimetry of higher cantilever vibration modes for dynamic atomic force microscopy in liquid, *Rev. Sci. Instrum.* **79**, 123703 (2008).
32. K. Umeda and K. Fukui, Potential dependent change in local structure of ferrocenyl-terminated molecular islands by electrochemical frequency modulation atomic force microscopy, *J. Vac. Sci. Technol. B*. **28**, C4D40 (2010).
33. K. Nagashima, M. Abe, S. Morita, N. Oyabu, K. Kobayashi, H. Yamada, M. Ohta, R. Kokawa, R. Murai, H. Matsumura, H. Adachi, K. Takano, and S. Murakami, Molecular resolution investigation of tetragonal lysozyme (110) face in liquid by frequency-modulation atomic force microscopy, *J. Vac. Sci. Technol. B*. **28**, C4C11 (2010).
34. T. Fukuma, Y. Ueda, S. Yoshioka, and H. Asakawa, Atomic-scale distribution of water molecules at the mica-water interface visualized by three-dimensional scanning force microscopy, *Phys. Rev. Lett.* **104**, 016101 (2010).

Chapter 19

High-Speed AFM and Imaging of Biomolecular Processes

Toshio Ando* and Takayuki Uchihashi
Department of Physics, Kanazawa University, Kakuma-machi Kanazawa, Ishikawa 920-1192, Japan
*tando@staff.kanazawa-u.ac.jp

This article presents recent progress in high-speed atomic force microscopy for dynamic imaging of biomolecular processes, including the theoretical basis for the highest possible imaging rate, instrumentation, hydrodynamic effects on cantilever oscillation, and various visualization studies on the dynamic molecular process and structure of proteins.

19.1 Introduction

The structure of proteins has thus far been studied by X-ray crystallography, electron microscopy (EM), nuclear magnetic resonance (NMR), or atomic force microscopy (AFM), but the obtained structures are substantially static. The dynamic behavior of protein molecules has been studied at the single-molecule level using fluorescence microscopy, but the protein molecules themselves are

Nanoscale Liquid Interfaces: Wetting, Patterning, and Force Microscopy at the Molecular Scale
Edited by Thierry Ondarçuhu and Jean-Pierre Aimé

ISBN 978-981-4316-45-3 (Hardcover), 978-981-4364-48-5 (eBook)
www.panstanford.com

invisible in the observations, even using super-resolution fluorescence microscopy. This means that the concomitant assessment of structure and dynamics is infeasible, a situation prevailing throughout biological research. This lack of techniques has been preventing us from comprehensive understanding of the functional mechanism of proteins.

AFM only lacks temporal resolution among various capabilities required for the simultaneous observation of the structure and dynamics of proteins. In fact, it can directly visualize individual molecules under aqueous solutions at high spatial resolution (see chapters 17 and 18). To afford the additional capability (i.e., high temporal resolution) to AFM, we have developed high-speed tapping-mode AFM that can capture an image within 40–70 ms, without damaging fragile biological samples.[1,2] This new microcopy has a potential to transform our style of studying the functional mechanism of proteins.

High-speed AFM was materialized by increasing the response speed of all devices contained in the feedback loop and by reducing the tip-sample interaction force without deteriorating the imaging rate. Moreover, several issues associated with "in-liquid" high-speed imaging were taken into consideration. In this chapter, we first present quantitative relationships between the imaging rate and the feedback bandwidth and then briefly describe the instrumentation and several means contrived to reduce hydrodynamic effects. In the last part, we show imaging results of various dynamic processes of proteins.

19.2 Imaging Speed and Feedback Bandwidth

It is mandatory in bio-AFM imaging that the tip–sample distance (hence, the tip–sample interaction force) be maintained constant by feedback control. When a sample is scanned over an area of $W \times W$ with scan velocity V_s in the x-direction and N scan lines in the y-direction, the image acquisition time T is given by $T = 2WN/V_s$. Supposing that the sample surface is characterized with a single spatial frequency $1/\lambda$, then the feedback scan is executed in the z-direction with frequency $f = V_s/\lambda$ to trace the sample surface.

The feedback bandwidth f_B of the microscope should be equal to or higher than f, and thus, we obtain the relationship $T \geq 2WN/(\lambda f_B)$.

The feedback bandwidth is usually defined by a feedback frequency at which a closed-loop phase delay of $\pi/4$ occurs in tracing the sample surface. The phase delay in the closed loop (ϕ_{closed}) is approximately twice that in the open loop (ϕ_{open}), provided the feedback gain is maintained at ≈1.[2] The phase delay is caused by the limited response speed of devices contained in the feedback loop and by the time delay due to a below-mentioned "parachuting" effect. By denoting the total time delay by $\Delta\tau_{total}$, ϕ_{open} is given by $\phi_{open} = 2\pi f \Delta\tau_{total}$. The phase delay can be compensated by a factor of α (2–3) using the (P + D) operation of the proportional-integral-derivative (PID) feedback controller. f_B is thus given by $f_B = \alpha/(16\Delta\tau_{total})$. Hence, the highest possible imaging rate (V_{image}) is given by $V_{image} = \alpha\lambda/(32WN\Delta\tau_{total})$, when the sample fragility is not taken into account. To materialize, for example, an imaging rate of 25 fps for $\alpha = 2.5$, $\lambda = 10$ nm, $W = 200$ nm, and $N = 100$, we have to achieve $f_B > 100$ kHz and thus $\Delta\tau_{total} < 1.6$ μs. *Abbreviation*: fps, frame per second.

19.3 Instrument and Techniques for High-Speed AFM

The main time delays occur by the limited response speed of a cantilever and the *z*-scanner, the detection speed of cantilever oscillation amplitude, and the parachuting time. Here, we briefly describe various techniques to shorten these time delays.

19.3.1 *Cantilevers*

The response time of a cantilever (τ_c) is expressed by $\tau_c = Q_c/(\pi f_c)$, where Q_c and f_c are the quality factor and resonant frequency of a cantilever, respectively. To shorten the response time, we need a high f_c and a small Q_c. A high f_c together with a small spring constant k_c can be achieved by reducing the cantilever size.[3] We developed small cantilevers collaborating with Olympus.[1] The most recent small cantilevers made of Si_3N_4 are ≈6 μm long, 2 μm wide, and 90 nm thick, resulting in f_c in air 3.5 MHz, f_c in water 1.2 MHz, $k_c \approx 0.2$ N/m, and Q_c in water 2–3.[4] Therefore, its response time in water is about 0.66 μs. Somewhat larger cantilevers (BL-AC10DS-A2, Olympus: f_c in air 1.5 MHz, f_c in water 600 kHz, $k_c \approx 0.1$ N/m) are already commercially available (Atomic Force F&E GmbH). We designed an optical beam deflection detector for small cantilevers; a laser beam reflected back from the rear side of a cantilever is collected and collimated using the same objective lens (CFI Plan Fluor ELWD 20×C, NA, 0.45, Nikon) as those used for focusing the incident laser beam onto the cantilever.[1]

19.3.2 *Scanner*

The response time of the *z*-scanner (τ_z) is expressed by $\tau_z = Q_z/(\pi f_z)$, where Q_z and f_z are the quality factor and resonant frequency of the *z*-scanner, respectively. f_z is almost solely determined by the resonant frequency of a piezoactuator used and by the way to hold the piezoactuator, as far as the structural resonance arising from the designed framework is completely suppressed. In the most recent *z*-scanner, a piezoactuator with a resonant frequency of ≈500 kHz is held at the four rims parallel to the displacement direction.[5] In this way of holding, the center of mass of the piezoactuator is not displaced (hence, no impulsive force is exerted on the supporting mechanism), and importantly, the resonant frequency of the *z*-piezoactuator is nearly unchanged. The maximum displacement is ≈1 μm.

To reduce Q_z, we developed an active damping technique based on a Q-control method with a "mock *z*-scanner."[6] In a conventional Q-control method, the displacement or velocity (V_z) of the *z*-scanner has to be measured and a signal proportional to $-V_z$ is added to the driving signal. However, it is difficult to measure the displacement or velocity. In our method, the mock *z*-scanner (an LRC circuit) characterized with a transfer function similar to that of the real *z*-scanner is used. A driving signal for the *z*-scanner is also input to the mock *z*-scanner, and the output from the mock *z*-scanner is used for Q-control (i.e., its minus derivative is added to the driving signal). As a result, Q_z is reduced to ≈0.8 and thus τ_z is shortened to ≈0.5 μs.

19.3.3 *Fast Amplitude Detector*

We developed two types of fast detectors of cantilever oscillation amplitude (peak-hold type and Fourier type). In the peak-hold analog circuit, the peak and bottom voltages of the output signal from a two-segment (or four-segment) photodiode sensor are held and their difference is output as an amplitude signal.[1] The amplitude signal is renewed every half cycle of cantilever oscillation (hence, time delay $\approx$0.4 μs). In the Fourier type (hybrid analog/digital system), Fourier cosine and sine coefficients (a_1, b_1) are calculated every one cycle of cantilever oscillation and a signal proportional to $(a_1^2 + b_1^2)^{1/2}$ is output.[2] The time delay is only slightly longer than 0.8 μs. The Fourier method is less susceptible to the cantilever's thermal deflection fluctuations than the peak-hold method.

19.3.4 *Suppression of Parachuting*

"Parachuting" means that a cantilever tip completely detaches from a sample surface at a steep downhill region of the sample and does not reland on the surface for a while. Parachuting more frequently occurs when the amplitude set point (A_s) is set closer to the cantilever-free oscillation amplitude (A_0). The error signal is saturated at (A_s–A_0) during parachuting, and the parachuting period is more prolonged when the saturated error signal is smaller. However, for fragile biological samples, the tapping force exerting from an oscillating cantilever tip on the sample has to be minimized by using A_s very close to A_0. To overcome this difficulty, we developed a dynamic PID controller in which the gain parameters are automatically and dynamically changed depending on the cantilever oscillation amplitude (A).[7]

The sign of (A–A_s) is a good indicator of whether the sample is being scanned at the downhill region or uphill region; the plus and minus signs approximately correspond to downhill and uphill regions, respectively (note that A cannot be maintained exactly at A_s because of the delay of feedback control). In the dynamic PID controller, the gain parameters are increased when A exceeds A_s (or a level A_t slightly higher than A_s). The gain increment can be constant or proportional to (A–A_s) or (A_t–A_s). The dynamic

PID controller is effective, and the feedback bandwidth becomes independent of A_s as long as A_s is smaller than $\approx 0.95A_0$, indicating that no parachuting occurs under the condition of $A_s < 0.95A_0$.

19.4 Hydrodynamic Effects

In high-speed AFM, the sample stage that is in contact with a buffer solution is displaced very quickly and a cantilever that is positioned very close to the sample surface is oscillating at a high frequency. Therefore, hydrodynamic effects have to be taken into consideration to fully bring out the high-speed imaging capability of the instrument.

19.4.1 *Size of Sample Stage*

In our high-speed AFM instrument, the sample stage, rather than a cantilever tip, is scanned. At the highest possible imaging rate, the sample stage is displaced in the z-direction at $\approx$100 kHz, which generates considerable hydrodynamic pressure. A cantilever is positioned very close to the sample surface and is therefore affected by the hydrodynamic pressure, resulting in a slow cantilever deflection response to the tip–sample interaction. We have to contrive to minimize this effect by various means. The first means is to minimize the size of sample stage.[8] Figure 19.1a shows the response of a small cantilever to the quick displacement of a sample stage with a dimension of 5×5 mm^2. The cantilever is vibrated at $\approx$20 kHz due to the hydrodynamic force. When the size of the sample stage is reduced to 1 mm in diameter, almost no vibrations appear (Fig. 19.1b).

19.4.2 *Relative Position of Cantilever and Sample Stage*

The hydrodynamic pressure generated by quick displacement of a sample stage in the z-direction also exerts a force onto a supporting base of a cantilever. The supporting base positions between the cantilever and the main body of cantilever chip. The supporting base is relatively thin and much larger than the cantilever itself and is

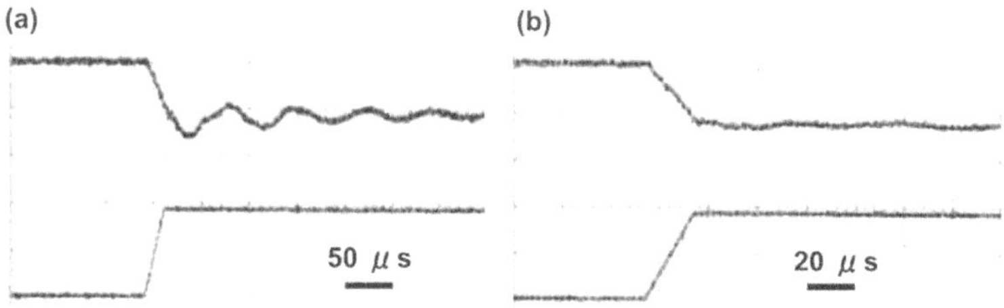

Figure 19.1 Response of a cantilever to a hydrodynamic force and its dependence on the size of the sample stage. The sample stage is quickly displaced toward the cantilever, whose tip is in contact with a mica surface in water. The lower signals represent the voltage applied to the z-piezoactuator. The upper signals represent the cantilever deflection. (a) The size of the sample stage is 5×5 mm^2. (b) The size of the sample stage is 1 mm in diameter.

therefore apt to be moved by the hydrodynamic force. When the cantilever tip is placed around the center of a sample stage of 1 mm diameter, the response of cantilever oscillation amplitude to the quick displacement of the sample stage is delayed by the movement of the supporting base (gray line at the top of Fig. 19.2). In the worst case, this delay extends to a few microseconds, which is significantly longer than the minimized time delays at devices contained in the feedback loop. When the cantilever tip is placed at an edge region of the small sample stage so that the entire cantilever chip is minimally overlapped with the sample stage, no time delay due to hydrodynamic pressure is observed in the response of cantilever oscillation amplitude (black line at the top of Fig. 19.2). The angle between the sample stage and a cantilever (the entire cantilever chip) is also an important factor. When the angle is small ($\approx 5°$), the supporting base is apt to undergo the hydrodynamic pressure. We usually set the angle at $\approx 10°$.

19.4.3 *Other Factors*

The design of small cantilevers is an important factor in minimizing the hydrodynamic effect. Figure 19.3 shows previous (a) and recent (b) designs for the supporting base. The recent design is much less susceptible to the hydrodynamic force and does not show a time

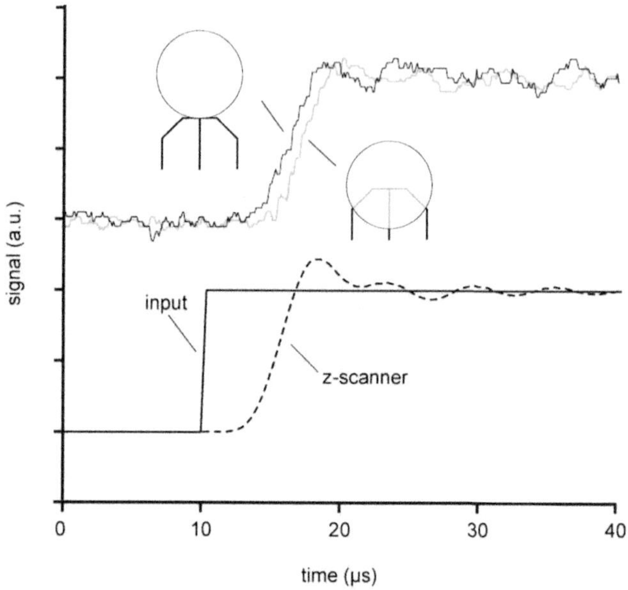

Figure 19.2 Effect of hydrodynamic pressure on cantilever response. The *z*-scanner was moved (displacement is shown by the broken line in the bottom) toward a cantilever using an input signal (solid line in the bottom) through a 100 kHz low-pass filter. The response of cantilever oscillation amplitude is more significantly delayed when the cantilever is placed at the center of a sample stage of 1 mm diameter (gray line at the top) than when placed at the edge of the sample stage (black line at the top).

delay as observed in Fig. 19.2. The shape and length of the cantilever tip are also important factors to consider. When the tip is shorter, the cantilever gets closer to the sample surface at the bottom of the swing and thus a solution confined between them is more squeezed, resulting in oscillation damping. This squeeze effect also occurs at the space confined by the tip and the sample surface when the tip is wider. Not only damping but also deterioration of sensitivity to the tip–sample interaction is caused by the squeeze effect. Figure 19.4 shows force curves obtained using small cantilevers with different tip shapes and lengths. When the 1.5 μm long tip is bird-bill shaped, the cantilever oscillation amplitude is getting smaller as approaching the substrate surface, whereas when the tip has an additional 1 μm long needle-shaped tip, the oscillation amplitude is

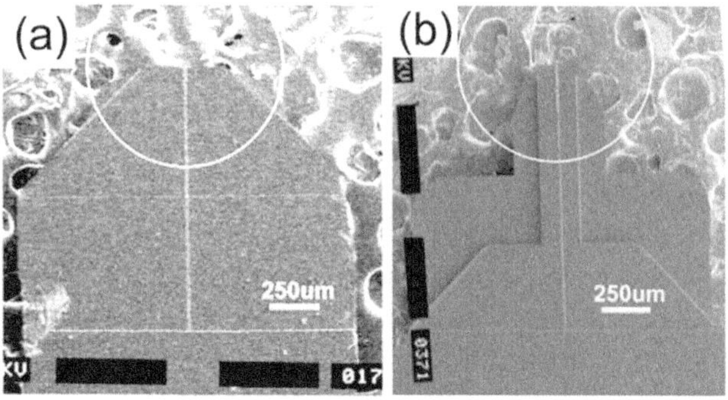

Figure 19.3 SEM images of previous (a) and recent (b) designs of a supporting base for a small cantilever. There is a small cantilever at the center of each circle (not visible on the images). *Abbreviation*: SEM, scanning electron microscopy.

maintained constant until the tip comes in contact with the substrate surface. Moreover, when contacted with the surface, the oscillation amplitude more sharply decreases with the longer tip than with the shorter tip.

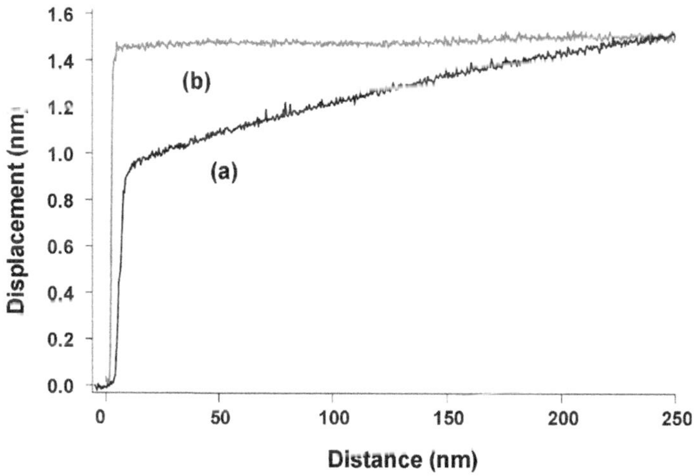

Figure 19.4 Squeeze effect on cantilever oscillation amplitude and sensitivity to tip–sample interaction. The amplitude–distance curves were obtained using a cantilever with a 1.5 μm long, bird-bill-shaped tip (a) or a cantilever having an additional 1.0 μm long needle-shaped tip (b).

19.5 Imaging of Biomolecular Processes

Recent technical advances of high-speed AFM have enabled visualizing various biological phenomena such as conformational change of proteins, molecular interactions, and assembling/disassembling processes of proteins. Further, high-speed AFM can provide structural information of proteins with unstructured domains, which cannot be clearly resolved by slow AFM imaging. Owing to gradual accumulation of successful imaging results on known molecular processes, high-speed AFM has recently been recognized as a reliable tool. Thus, newly filmed data on unexplored biological processes can be widely accepted. In this section, we introduce the availability of high-speed AFM through some successful observations of dynamics in two-dimensional (2D) protein crystals, conformational change of membrane protein triggered by an external stimulus, and intrinsically disordered (ID) regions of proteins.

19.5.1 *Defect Diffusion in a 2D Streptavidin Crystal*

For protein crystal formation, the protein-protein association energy must be in an appropriate range. However, it was impossible to experimentally assess the association energy at each contact point, especially at the single-molecular level. Here, we show that high-speed AFM imaging is useful to estimate the association energy between proteins using streptavidin as a model sample.[9]

Streptavidin is a protein consisting of four identical subunits, each of which specifically binds to one biotin.[10] It is easily crystallized into 2D forms on biotinylated lipid layers, and hence, the 2D crystals are considered to be ideal model systems to investigate 2D crystal grown on a substrate, in general. On the biotinylated lipid layers, two biotin-binding sites are occupied by the biotin moiety of a lipid layer, while the other two are exposed to an aqueous environment and therefore free from biotin, as depicted in Fig. 19.5a. Two-dimensional crystals of streptavidin were formed on the supported lipid bilayer (SLB). As shown in Fig. 19.5b, in the $C222$ crystal, the intermolecular contacts between biotin-bound subunits are contiguously aligned along one crystal axis (a-axis), while the

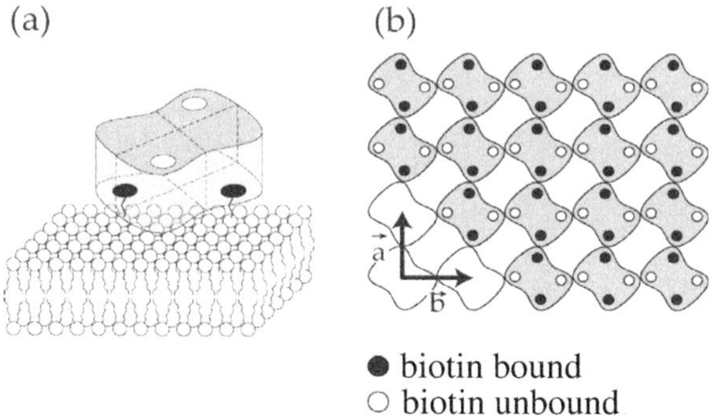

Figure 19.5 (a) Schematic of a streptavidin molecule on a biotinylated lipid bilayer and (b) schematic of streptavidin arrays in a *C* 222 crystal.

contacts between biotin-unbound subunits are contiguously aligned along the other axis (*b*-axis).

The lipid composition used in this study was dioleoylphosphatidylcholine (DOPC), dioleoylphosphatidylserine (DOPS), and 1,2-dioleoyl-*sn*-glycero-3-phosphoethanolamine-*N*-(cap biotinyl) (biotin-cap-DOPE) (7:2:1 weight ratio). Dried lipid films were produced by mixing lipids dissolved in chloroform followed by evaporating the solvent with argon. The lipid films were further dried in a desiccator by aspirating for more than 30 minutes. To obtain multilamellar vesicles (MLVs), the dried lipid films were resuspended in a buffer (10 mM HEPES-NaOH, 150 mM NaCl, 2 mM $CaCl_2$, pH 7.4) by vortexing. Small unilamellar vesicles (SUVs) were produced from the MLV suspension by sonications with a tip-sonicator for a few seconds. SLBs were prepared by depositing 0.1 mg/mL SUVs onto a freshly cleaved mica surface and incubated for 30 minutes in a chamber with saturated humidity at room temperature. After that, the excess lipids were washed out with the buffer. Two-dimensional crystallization of streptavidin on biotin-containing SLBs was performed by injecting streptavidin dissolved in an appropriate buffer at a final concentration of 0.1 mg/mL and by incubating for 2 hours in a chamber with saturated humidity at room temperature. The buffer used for streptavidin crystallization has a same composition as the one used for the SLB formation. After that, excess streptavidin molecules were washed out with the buffer.

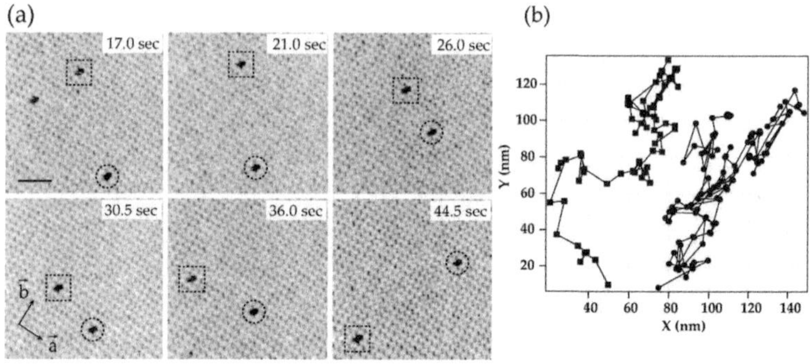

Figure 19.6 Migration of monovacancy defects in a streptavidin 2D crystal. (a) High-speed AFM images of a streptavidin 2D crystal and monovacancy defects. The monovacancy defects are enclosed by dashed squares and circles. The directions of the lattice vectors of the crystal are also indicated. Successive images were obtained at a frame rate of 2 fps with a scan area of 150×150 nm^2. (b) Trajectories of individual monovacancy defects. Closed squares and circles correspond to defects indicated by open squares and circles, respectivel,y shown in (a).

Figure 19.6a shows AFM images of the streptavidin 2D crystal with mobile monovancancy defects, which are clipped from successive high-speed AFM images captured at a frame rate of 2 fps. Here, monovacancy defects in the streptavidin 2D crystals were produced by increasing the tapping force. In Fig. 19.6b, the trajectories of two monovancancy defects observed in Fig. 19.6a are shown. The mobility of the monovacancy defects was obviously anisotropic with respect to the two crystalline axes. The mobility along the *b*-axis is much higher than that along the *a*-axis.

Figure 19.7 plots the mean-square displacements (MSDs) of the monovacancy defects along each axis of the $C222$ crystal as a function of time. The linear increase in MSDs with time indicates that the monovacancy defect moves in a random walk. From the slope of the MSDs, the diffusion rate constants of the migration were determined to be $D_{\mathrm{a}} = 20.5$ nm^2/sec along the *a*-axis, which includes rows of contiguous biotin-bound subunits, and $D_{\mathrm{b}} = 48.8$ nm^2/sec along the *b*-axis, which includes rows of contiguous biotin-unbound subunits. This anisotropy in lateral mobility (i.e., $D_{\mathrm{b}} > D_{\mathrm{a}}$) arises from a difference of association energies between

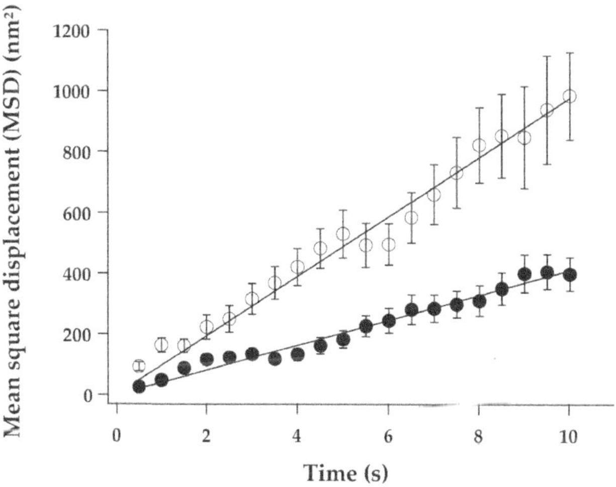

Figure 19.7 Plot of MSDs of monovacancy defects as a function of time. Closed circle, MSDs along the *a*-axis that includes rows of contiguous biotin-bound subunits; open circle, MSDs along the *b*-axis that includes rows of contiguous biotin-unbound subunits. Data fitted to a linear function yield diffusion constants $D_a = 20.5\ \text{nm}^2/\text{sec}$ and $D_b = 48.8\ \text{nm}^2/\text{sec}$ for the directions along the *a*- and *b*-axis, respectively.

the biotin-bound subunit–subunit and biotin-unbound subunit–subunit interactions. When a streptavidin molecule adjacent to a monovacancy defect moves to the defect site along the *a*-axis, two intermolecular bonds between biotin-unbound subunits ("u–u bond") and one intermolecular bond between biotin-bound subunits ("b–b bond") should be broken, as shown in Fig. 19.5b. On the other hand, when it moves to the defect site along the *b*-axis, one u–u bond and two b–b bonds should be broken. Therefore, the difference in the activation energies E_a and E_b for the step movement of a monovacancy defect along the respective *a*- and *b*-axis simply corresponds to the difference between the free energy changes G_{u-u} and G_{b-b} produced by the formation of the respective u–u bond and b–b bond; namely, $E_b - E_a = G_{u-u} - G_{b-b}$. Therefore, the observed relationship $D_b > D_a$ indicates $G_{u-u} < G_{b-b}$; namely, the u–u bond affinity is higher than the b–b bond affinity. The ratio of the two diffusion rate constants (D_b/D_a) can be expressed by

$$D_b/D_a = \exp\left[-(E_b - E_a)/(k_B T)\right] \qquad (19.1)$$

where k_B is the Boltzman constant and T is the absolute temperature. Thus, from the observed value of $D_b/D_a \approx 2.4$, the free energy difference G_{u-u}- G_{b-b} is estimated to be approximately $-0.88\ k_BT$ ($T \approx 300$ K), which corresponds to -0.52 kcal/mol.[9]

19.5.2 *Dynamic Equilibrium at the Bacteriorhodopsin Crystal Edge*

The purple membrane (PM) exists in the plasma membrane of *Halobacterium halobium*, and its constituent protein, bacteriorhodopsin (bR), functions as a light-driven proton pump. In the PM, bR monomers are associated to form a trimeric oligomer and the trimers are arranged in a hexagonal lattice (see also chapters 17 and 18).[11] However, several aspects in the crystal formation remain open; for example, (1) trimer–trimer interaction sites and association energy and (2) existence of preformed trimers in the fluidic noncrystal region. In the 2D crystal of bR and any crystals in general, they are in dynamic equilibrium with the constituents at the interface between the crystal and the liquid phase. Here, we visualized dynamic events at the interface in the PM to provide information of the crystal formation and intermolecular interactions.[12]

The PM adsorbed on a mica surface in a buffer solution (10 mM Tris-HCl, pH 8.0, and 300 mM KCl) exhibits flat, round-shaped patches (Fig. 19.8a). A 2D crystal lattice of bR is formed over the inner region surrounded by a dotted line in Fig. 19.8a, whereas in the peripheral outer region, there are no bR crystals. Figure 19.8b,c shows magnified images of an edge region of the PM, which were successively captured at 1 fps. There is a distinct border between the crystal and noncrystal areas. As indicated by the arrows, the border shape is different between both images, indicating that the border region of the crystal is unstable and appears to be in dynamic equilibrium with bR molecules in the noncrystal area.

To investigate the details of the dynamic structural changes at the crystal edge of PM, higher-magnification images were acquired at the boundary region. Figure 19.9 shows typical AFM images taken at 3 fps. The bR trimers in the crystal are indicated by the thin-lined triangles. At 0.6 seconds, two bR trimers (red triangles) have newly

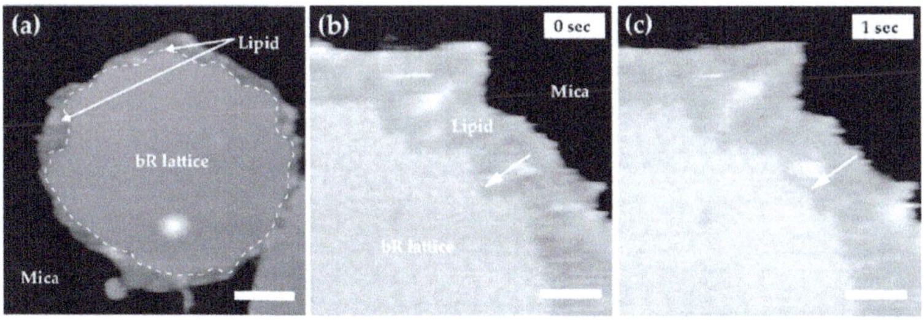

Figure 19.8 (a) Low-magnified AFM images of PM adsorbed onto a mica surface. The PM patch consists of a crystal area (encircled with the dotted line) and a noncrystal area (the periphery of the crystallized area). Scale bar: 80 nm. (b) and (d) are successive AFM images of an edge region of a PM captured at 1 fps. The arrows indicate shape variation of the crystal edge with time. Scale bars: 20 nm.

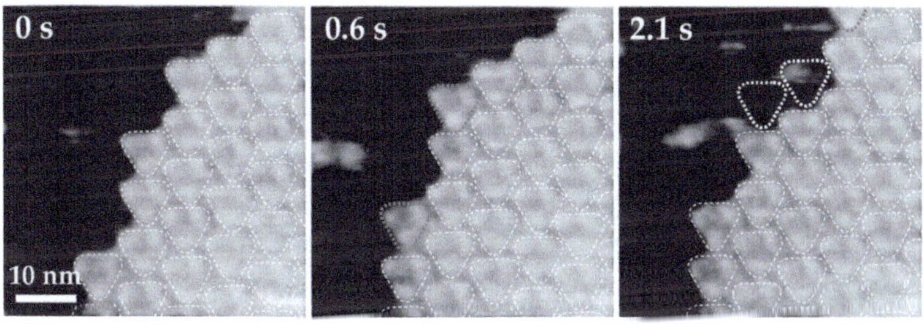

Figure 19.9 Time-lapse high-magnification AFM images of PMs at the interface between crystalline and noncrystalline areas captured at 3 fps. The thin-lined triangles show bR trimers. The bR trimers encircled by the red dotted line at 0.6 sec indicate newly bound bR trimers. The white dotted triangles at 2.1 sec indicate the previously bound trimers.

bound to the crystal edge. Subsequently, two bR trimers (white triangles) have dissociated and another trimer (red triangle) has bound to the crystal edge at 2.1 seconds. One of the two dissociated trimers remains in the crystal area for ≈0.9 seconds.

Although we sometimes observed association of a dimer and a monomer to the edge, most of binding events of bR molecules occurred as being a trimer. According to the analysis for 239 observed binding events, the binding of trimeric bR occurred predominantly (82%), whereas binding of dimeric bR (6.7%) was

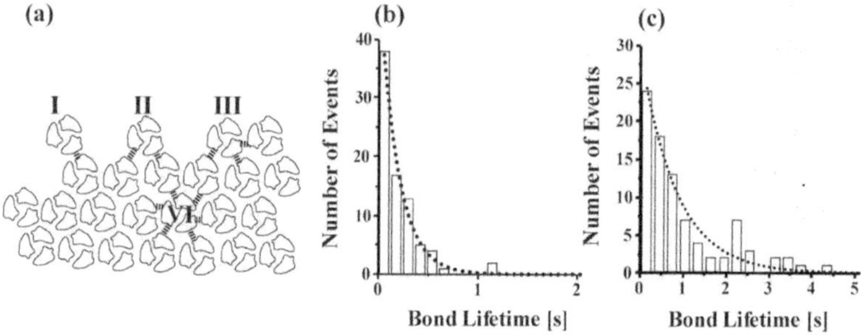

Figure 19.10 (a) Schematic representing the binding manner of a bR trimer at the crystal edge (I; II; III) and in the crystal interior (VI). Roman numerals indicate the number of interaction bonds (dotted lines) containing a W12 residue. (a) Histogram showing the type II binding events vs. lifetime. The histogram was fitted by a single exponential function (dotted line). (c) Histogram showing the type III binding events vs. lifetime. The dotted line indicates the best fit with a single exponential function.

only about half that of monomeric bR (11.3%). This means that the trimer is a stable oligomeric unit formed prior to the crystal formation.[12]

To estimate the intertrimer interaction energy, we analyzed the residence time of newly bound bR trimers at the crystal edge and its dependence on the number of interaction sites. For the analysis, we assumed that within the 2D bR crystal, a trimer can interact with the surrounding trimers through six sites, which are depicted by the dotted lines around the bR trimer marked by "VI" in Fig. 19.10a. Intertrimer interactions around W12 residues participate in lattice formation.[13,14] Following the same model, the number of interaction sites at the crystal edge is reduced, depending on the binding position, as marked by "I," "II," and "III" in Fig. 19.10a. Successive AFM images, as exemplified in Fig. 19.9, showed many binding and dissociation events in which bR trimers bound to different sites at the border between the crystal and noncrystal areas. These events can be classified into types I, II, and III, depending on the number of interaction sites involved. Type II binding events are predominant (≈74%), whereas type I (≈6 %) and type III (≈20%) bindings are minor events. The lifetime of the type I bonds was too short to obtain clear images of the corresponding event, preventing reliable statistics.

Figure 19.10b shows a histogram of the lifetime for a type II bond, which was measured using AFM images taken at 10 fps. The histogram can be well fitted by a single exponential (correlation coefficient, $r = 0.9$), from which the average lifetime τ_2 was estimated to be 0.19 ± 0.01 seconds. Figure 19.10c shows a histogram of the type III bond lifetime, from which the average lifetime τ_3 was estimated to be 0.85 ± 0.08 seconds. The longer lifetime of type III bonds compared with type II obviously arises from a relationship of $E_{\text{III}} < E_{\text{II}} < 0$, where E_{II} and E_{III} are the association energies responsible for type II and type III interactions, respectively. Assuming that the prefactor is the same for both interactions, the average lifetime ratio, τ_2/τ_3, is given by

$$\tau_2/\tau_3 = \exp\left[-\left(E_{\text{III}} - E_{\text{II}}\right) / \left(k_{\text{B}}T\right)\right] \qquad (19.2)$$

Because the type II interaction contains two elementary bonds, whereas the type III interaction contains three, the energy difference $E_{\text{III}}-E_{\text{II}}$ corresponds to the association energy of the single elementary bond. From the ratio $\tau_2/\tau_3 = 0.22$ and Eq. 19.2, this elementary association energy is estimated to be about -1.5 $k_{\text{B}}T$, which corresponds to -0.9 kcal/mol at 300 K.[12] This value is approximately consistent with that estimated by differential scanning calorimetry[15,16]

19.5.3 *Dynamics of a 2D Crystal of Annexin V*

Annexin V is a soluble protein belonging to a protein family that binds to negatively charged phospholipids in the presence of calcium ions, in particular phospholipids that contain serine at the polar head, including DOPS. It forms 2D crystals on lipid monolayers.[17] The property of annexin V self-organization has been proposed to be functionally relevant in its biological function.[18] The structure of its soluble form has been obtained by X-ray crystallography, while that of the membrane-bound form was extensively investigated by electron crystallography and conventional AFM.[19] The fundamental oligomeric state of annexin V is a trimer, and the trimers assemble into two common crystal forms with $p3$ or $p6$ symmetry. Here, we demonstrate some dynamic events such as dynamic equilibrium between the 2D crystal and the liquid phase, crystal growth, and de-crystallization.

As lipids, we used DOPC, DOPS, and DOPE (5:2:3 weight ratio). The lipid bilayer supported on a mica surface was prepared by the same method described in section 19.5.1. Two-dimensional crystallization of annexin V on the lipid bilayer was performed by injecting an annexin V solution into the observation buffer containing 50 mM Tris-HCl pH 8.0, 5 mM KCl, 2.5 mM $MgCl_2$, and 3mM $CaCl_2$.

The annexin V crystal with *p*6 symmetry exhibits a honeycomb structure. The "holes" in the honeycomb lattice tend to be filled with a relatively mobile trimer. The trimer captured in a hole more loosely interacts with the surrounding cage than a molecule forming part of the honeycomb lattice. Due to this mobility, the central trimer shows a less sharply defined density in the EM images of the crystal.[20,21] Figure 19.11 shows successive AFM images captured at an imaging rate of 2 fps. The hole indicated by the arrow (0 second image) is filled by a trimer in the next frame (0.5 second image). This trimer is weakly bound, and therefore dissociates soon at one second. The hole is filled again at 1.5 seconds, but the trimer is bound more stably at this time. The dissociation and association of center trimers occur at several places in the crystal. This observation also indicates that unbound trimers exist on the crystal surface and are rapidly diffusing on it.

Figure 19.12 shows high-speed AFM images showing the crystal growth of annexin V. In this observation, the buffer solution initially

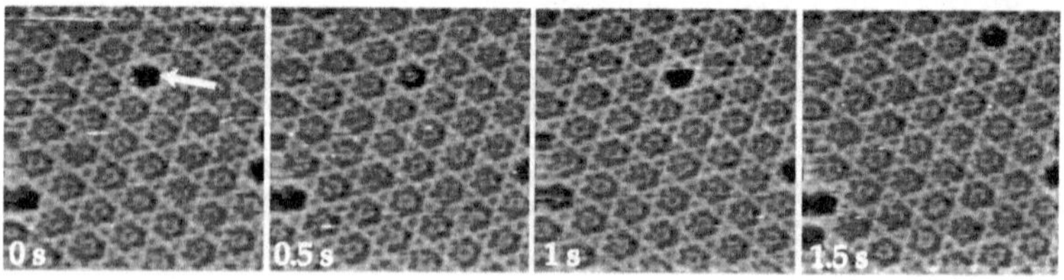

Figure 19.11 Binding and dissociation dynamics of annexin V trimers on the 2D crystal with *p*6 symmetry. For example, the hole in the honeycomb lattice indicated by an arrow in the 0 sec image is filled by a trimer diffusing on the crystal surface at 0.5 sec and then dissociates in the next 0.5 sec. Successive images were obtained at 2 fps for a scan area of 150×150 nm^2.

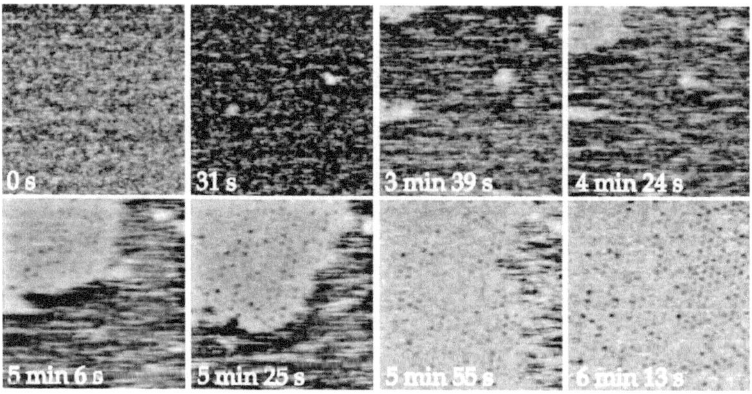

Figure 19.12 Crystal growth of annexin V. At zero seconds, the image shows only the lipid surface and a large noise induced by diffusing molecules on the surface. $CaCl_2$ solution was injected at seven seconds. Successive images were obtained at a frame rate of 1 fps with a scan area of $400 \times 400\ nm^2$.

does not contain $CaCl_2$. During imaging, $CaCl_2$ solution was injected into the sample chamber (a final concentration, 3 mM). At zero seconds, the lipid surface is primarily observed because of the absence of Ca^+ ions. This image shows large noise presumably caused by annexin V molecules rapidly diffusing on the surface. A $CaCl_2$ solution was injected at seven seconds. Soon after the injection, small particles appear (see the image at 31 seconds). This is probably the first stage of assembly in which three molecules cluster together in an almost irreversible manner to form a trimer. Surface-diffusing annexin molecules come into contact with the surface-bound trimer, and the cluster size gradually increases with time. However, in the captured images, crystallization progresses more predominantly from the top left in the images. The precursor protein clusters observed at the first stage are incorporated into the crystal during the progress of crystallization. Eventually, the lipid surface is completely covered by the crystal in a few minutes.

Next, disassembling process of the 2D crystal was imaged by chelating Ca^+ ions. Figure 19.13 shows high-speed AFM images clipped from the successive images. An ethylene glycol tetraacetic acid (EGTA) solution was injected into the observation buffer at zero seconds during imaging. At eight seconds, the 2D crystal form is still

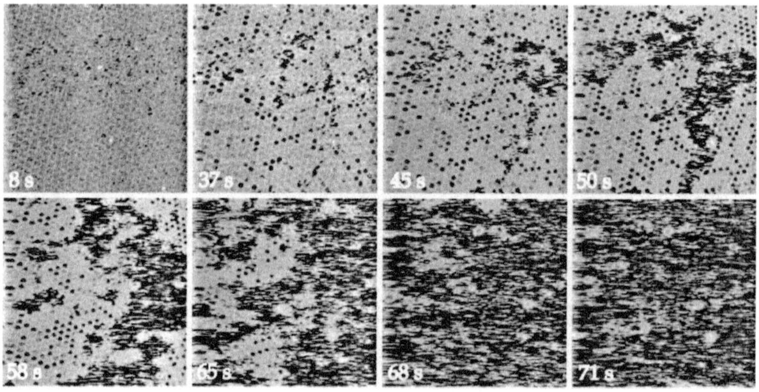

Figure 19.13 De-crystallization of an annexin V 2D crystal. After the crystallization, EGTA solution was injected into the sample chamber at zero seconds. Successive images were obtained at a frame rate of 1 fps with a scan area of 500×500 nm^2.

kept. After a while, some trimers, especially weakly bound trimers in the holes, are desorbed from the lipid membrane. Then the number of vacancy increases with time. The 2D crystal rapidly collapses from the disrupt regions and completely disappears in a few minutes.

19.5.4 *Structural Changes of Bacteriorhodopsin under Light*

bR is one of well-known examples showing the association between stimulus-triggered structural dynamics and biological function,[22,23] and the direct visualization of its structural change has long been a goal. bR contains seven transmembrane α-helices (named A–G) enclosing the chromophore retinal.[24,25] Absorption of light by the retinal triggers a cyclic reaction (photocycle) that comprises a series of spectroscopically identified intermediates, designated as the J, K, L, M, N, and O states.[23] The light-induced conformational changes in bR have been investigated by various methods,[26–30] leading to a consensus that the proton channel at the cytoplasmic surface is opened by the tilting of helix F away from the protein center.[29,30] This movement of helix F leads to a rearrangement of the E-F loop, which results in large-scale conformation changes in the M

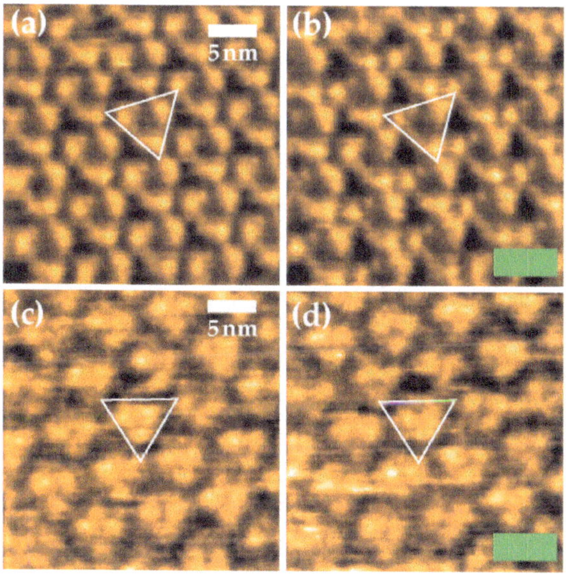

Figure 19.14 High-speed AFM images of the cytoplasmic surface (a, b) and extracellular surface (c, d) of D96N under dark or illuminated conditions. bR trimers are highlighted by white triangles. The green bars indicate illumination of 532 nm green light with 0.5 μW. The successive images were captured at a frame rate of 1 fps.

and N intermediates. The photo-induced conformational change at cytoplasmic surface of bR was recently captured by high-speed AFM.[31]

The photocycle of wild-type bR completes in ≈10 ms at pH 7.0, and therefore the photo-activated intermediate states are too brief to be clearly captured by high-speed AFM. Thus, we used the D96N mutant that has a longer photocycle (≈10 sec) at pH 7.0 but retains an ability of proton pumping. Figure 19.14a,b shows successive images of D96N at the cytoplasmic surface captured at 1 fps. Under the unphotolyzed state (Fig. 19.14a), one can clearly see individual bR molecules forming well-ordered trimers in the 2D hexagonal lattice. A trimer is encircled by the triangle. Upon illumination with green light, bR drastically changes its structure, as shown in Fig. 19.14b. This activated state with a unique structure returns to the prior unexcited state in a few seconds after light-off.

This cyclic structural change is highly reproducible in repeated dark-illumination cycles. On the other hand, significant conformational changes of bR are not observed on the extracellular surface, as shown in Fig. 19.14c,d.

We analyzed the center of mass for each monomer during the dark-illumination cycles. All activated bR monomers exhibited displacement of their centers of mass by 0.69 ± 0.15 nm on average. In Fig. 19.15a, the trajectories of the centers of mass during the dark-illumination cycles are superimposed on the AFM image in the unphotolyzed state (red, blue, and black marks). They show highly centrosymmetric trajectories, oriented outward from the trimer centers. Figure 19.15b,c shows magnified images obtained before

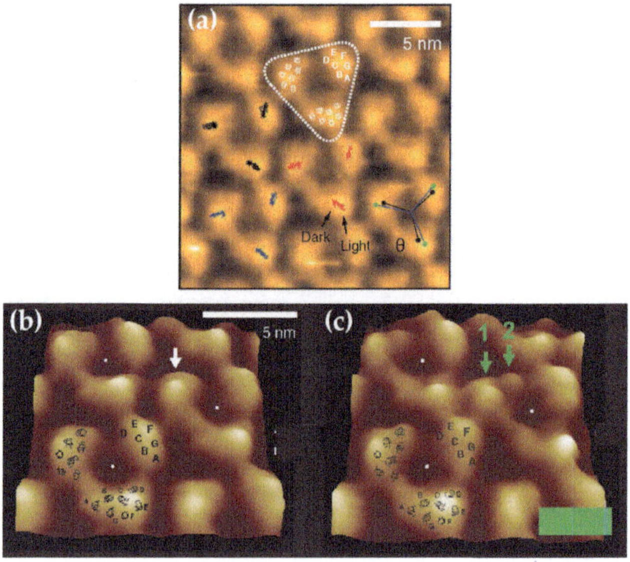

Figure 19.15 (a) Traces (red, blue, and black marks) of the mass-center positions under the dark-illumination cycles are superimposed on the image of D96N in the dark. The photo-induced movement of bR includes counterclockwise rotation ($7.4 \pm 2.2^\circ$) around the trimer center. Surface maps of the magnified images (b) in the dark and (c) under illumination. The position of each trimer center is denoted by the white dots. A monomer in the dark is indicated by the single white arrow. Under illumination, the topography of the monomer splits into major and minor protrusions, as indicated by the green arrows 1 and 2, respectively.

and during illumination, respectively. The prominent protrusion in the AFM topographs reflects interhelical loop structures on the cytoplasmic surface.[32] Compared with the atomic model of the α-helical cytoplasmic ends in the unphotolyzed state[25] (A–G in Fig. 19.15b,c), the protruding areas around helices E and F shift outward from the trimer center under illumination. Therefore, the protrusion movement visualized here is ascribed to the displacement of the E-F loop. As a result of E-F loop displacement, the protrusion is divided into two sections (the green arrows 1 and 2 in Fig. 19.15c), and the minor protrusion (green arrow 2) likely corresponds to the position around helices A and B. To ensure that the conformational changes observed here are not artifacts such as tip-force–induced structural alterations, we measured the decay of the active state after flash illumination at various pH values. Ultraviolet-visible (UV-vis) spectroscopy[33] has shown that the lifetime of the M-intermediate of D96N and WT is prolonged at alkaline pH; in concordance with this fact, the decay detected by high-speed AFM showed strong pH dependence.[31]

19.5.5 *Intrinsically Disordered Region of Proteins*

In addition to biological molecular processes, high-speed AFM imaging is also useful for identifying structurally flexible regions of proteins, such as ID regions. Most unstructured segments comprise flexible linkers that play important roles in the assembly of macromolecular complexes and in the recognition of smaller biomolecules. The functional importance of ID regions has recently been recognized, particularly in transcription, translation, and cellular signal transduction.[34,35] However, there are no useful techniques for visualizing and analyzing unstructured segments of this sort at the single-molecule level. Indeed, X-ray crystallography and EM do not allow us to directly observe the unstructured ID regions. Conventional slow AFM also cannot reveal such a thin and flexible entity. Under ambient conditions, thin and flexible unstructured polypeptides tend to form lumps or become flattened because of their strong attachment to a surface. Such flattened polypeptides are difficult to visualize by slow AFM. In addition, they

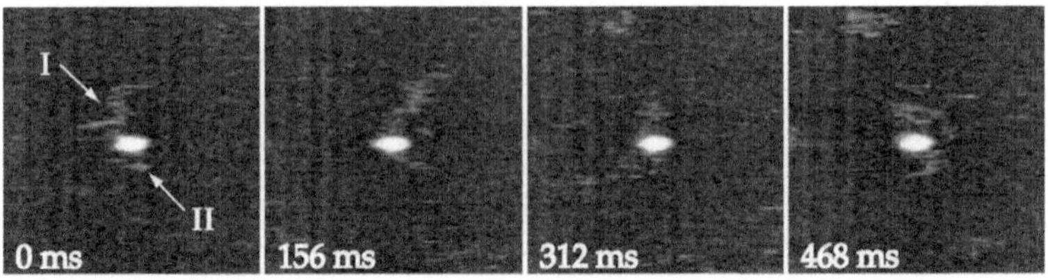

Figure 19.16 Successive AFM images of wild-type FACT protein captured at a frame rate of 6 fps with a scan area of 150 × 150 nm^2. The lump and two tail-like structures indicated by I and II can be seen. The two tail-like structures fluctuate with time.

are not easily adsorbed onto substrate surfaces in solution, even with the assistance of adhesive chemicals.

Recently, we successfully observed the facilitates chromatin transcription (FACT) protein in solution using high-speed AFM.[36] FACT is a heterodimer and involved in displacing histone H2A/H2B dimers from nucleosomes, thereby facilitating RNA polymerase II transcription[37,38] and chromatin remodeling.[39] Figure 19.16 shows successive AFM images of a FACT monomer adsorbed on a mica surface. These images of FACT clearly revealed two undulating tail-like structures protruding from the main body of FACT (arrows I and II in Fig. 19.16). The two tail-like structures have different lengths: the tail-like structure indicated by the arrow I is longer than that indicated by the arrow II. This finding coincides with the ID regions predicted from the amino acid sequences. In fact, the AFM images of deletion mutants lacking either of the two predicted ID regions showed either the shorter or longer tail-like structure. Their macroscopic contour lengths of the shorter and longer ID regions showed wide distributions and were on average 17.8 nm and 26.2 nm, respectively.

We investigated the effect of the AFM imaging rate on the appearance of an ID region of FACT. We used the deletion mutant in which an ID region with shorter length is deleted. At a frame rate of 12.5 fps, the tail-like ID region is clearly seen. Even at 5 fps, the ID region is still observed as a string. However, the free end of the ID region is observed broader, presumably due to its rapid change in position. At 2 fps, the ID region is still observable but is blurry. Finally, at 1 fps, the ID region is no longer discernible as a string but appears hazy.

These experiments suggest that an imaging rate greater than about 5 fps is required to clearly visualize ID regions of proteins.

We analyzed mechanical properties (stiffness) of the observed tail-like structures. The stiffness of a polymer chain is described by the persistence length p of the chain, that is, the length over which the average correlation in the tangent direction decays by 1/e. In two dimensions, the mean square point-to-point distance of the chain r^2 is given by

$$< r^2 (l) >_{2D} = 4pl \left[1 - \frac{2p}{l} \left(1 - e^{-l/2p} \right) \right] \qquad (19.3)$$

where l is the contour length between two points on the chain.[40] Here, note that the persistence length depends on the hierarchy of the concerned polymer structures. In the AFM images obtained for the tail-like structures of FACT, polypeptide chains are not resolved. Therefore, the persistence lengths that can be obtained from the images are those for overall (macroscopic) string structures (see inset in Fig. 19.17a). We analyzed the tail-like structures of the deletion mutants. The mean square point-to-point distance r^2 as a function of l was well fitted by Eq. 19.3 for both cases, giving

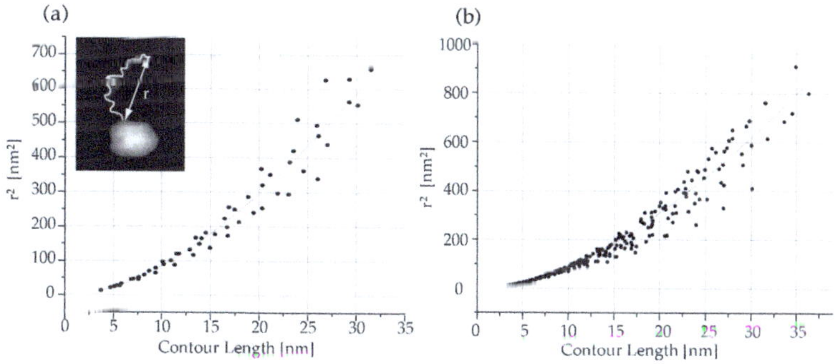

Figure 19.17 Persistence-length determination of the macroscopic structures of the tail segments of mutants with (a) longer and (b) shorter ID regions. The mean square point-to-point distance is plotted as a function of the macroscopic contour length between the corresponding two points. The solid lines indicate the best-fit curves of Eq. 19.3 with (a) $p = 10.5$ nm and (b) $p = 12.1$ nm. The inset image in (a) indicates the relationship between the microscopic contour and the macroscopic structure. The microscopic contour is schematically drawn as a thin white line.

persistence lengths of 10.5 nm and 12.1 nm for the shorter and longer tail-likes structures, respectively (Fig. 19.17). These similar values indicate that the two ID regions have a physically similar property except for the length.

The Young modulus E is a material property that represents its stiffness and is related to the macroscopic persistence length as $E = pk_B T/I$, where I is the momentum of inertia of the cross section of the object.[41] Assuming that the tail-like structures of FACT are thin rods with a radius (ρ) of 0.3 nm, their value of Young modulus was estimated to be $\approx$4.5 MPa from their similar macroscopic persistence lengths. Here, a relationship of $I = \pi\rho^4/4$ was used. This value for E estimated as a macroscopic stiffness of the ID regions is two to three orders of magnitude smaller than those reported for globular proteins such as an actin monomer ($\approx$2 GPa)[42] and lysozyme (0.5–2 GPa).[43,44] On the other hand, the Young modulus estimated as macroscopic stiffness of bovine carbonate anhydrase II is reported to be in a range of 2–6 MPa when the sample is incubated in 2–6 M guanidine HCl.[45] As these values of Young modulus are closed to the value for the ID regions of FACT, the structures of the ID regions are considered to be similar to those of denatured proteins.

The unique high-speed AFM capability to detect unstructured regions of proteins is also useful for successful protein crystallization, since the predicted ID regions can be confirmed and deleted before crystallization to facilitate crystal growth.

References

1. T. Ando, N. Kodera, E. Takai, D. Maruyama, and A. Toda, A high-speed atomic force microscope for studying biological mactomolecules, *Proc. Natl. Acad. Sci. USA*, **98**, 12468 (2001).
2. T. Ando, T. Uchihashi, and T. Fukuma, High-speed atomic force microscopy for nano-visualization of dynamic biomolecular processes, *Prog. Surf. Sci.*, **83**, 337 (2008).
3. D. A. Walters, J. P. Cleveland, N. H. Thomson, P. K. Hansma, M. A. Wendman, G. Gurley, and V. Elings, Short cantilevers for atomic force microscopy, *Rev. Sci. Instrum.*, **67**, 3583 (1996).

4. M. Kitazawa, K. Shiotani, and A. Toda, Batch fabrication of sharpened silicon nitride tips, *Jpn. J. Appl. Phys.*, **42**, 4844 (2003).

5. T. Fukuma, Y. Okazaki, N. Kodera, T. Uchihashi, and T. Ando, High resonance frequency force microscope scanner using inertia balance support, *Appl. Phys. Lett.*, **92**, 243119 (2008).

6. N. Kodera, H. Yamashita, and T. Ando, Active damping of the scanner for high-speed atomic force microscopy, *Rev. Sci. Instrum.*, **76**, 053709 (2005).

7. N. Kodera, M. Sakashita, and T. Ando, Dynamic proportional-integral-differential controller for high-speed atomic force microscopy, *Rev. Sci. Instrum.*, **77**, 083704 (2006).

8. T. Ando, N. Kodera, D. Maruyama, E. Takai, K. Saito, and A. Toda, A high-speed atomic force microscope for studying biological macromolecules in action, *Jpn. J. Appl. Phys.*, **41**, 4851 (2002).

9. D. Yamamoto, T. Uchihashi, N. Kodera, and T. Ando, Anisotropic diffusion of point defects in two-dimensional crystal of streptavidin observed by high-speed atomic force microscopy, *Nanotechnology*, **19**, 384009 (2008).

10. P. C. Weber, D. H. Ohlendorf, J. J. Wendoloski, and F. R. Salemme, Structural origins of high-affinity biotin binding to streptavidin, *Science*, **243**, 85–88 (1989).

11. R. Henderson, J. M. Baldwin, T. A. Ceska, F. Zemlin, E. Beckmann, and K. H. Downing, Model for the structure of bacteriorhodopsin based on high resolution electron cryo-microscopy, *J. Mol. Biol.*, **213**, 899 (1990).

12. H. Yamashita, K. Voïtchovsky, T. Uchihashi, S. A. Contera, J. F. Ryan, and T. Ando, Dynamics of bacteriorhodopsin 2D crystal observed by high-speed atomic force microscopy, *J. Struc. Biol.*, **167**, 153 (2009).

13. M. Weik, H. Patzelt, G. Zaccai, and D. Oesterhelt, Localization of glycolipids in membranes by in vivo labeling and neutron diffraction, *Mol Cell*, **1**, 411 (1998).

14. K. T. Sapra, H. Besir, D. Oesterhelt, and D. J. Müller, Characterizing molecular interactions in different bacteriorhodopsin assemblies by single-molecule force spectroscopy, *J. Mol. Biol.*, **355**, 640 (2006).

15. M. B. Jackson and J. M. Sturtevant, Phase transitions of the purple membranes of Halobacterium halobium, *Biochemistry*, **17**, 911 (1978).

16. I. Koltover, J. O. Raedler, T. Salditt, K. J. Rothschild, and C. R. Safinya, Phase behavior and interactions of the membrane-protein bacteriorhodopsin, *Phys. Rev. Lett.*, **82**, 3184 (1999).

17. R. P. Richter, J. Lai Kee Him, B. Tessier, C. Tessier, A. R. Brisson, On the kinetics of adsorption and two-dimensional self-assembly of annexin A5 on supported lipid bilayers, *Biophys. J.*, **89**, 3372 (2005).
18. F. Oling, J. Sopkova-de Oliveira Santos, N. Govorukhina, C. Mazères-Dubut, W. Bergsma-Schutter, G. Oostergetel, W. Keegstra, O. Lambert, A. Lewit-Bentley, and A. Brisson, Structure of membrane-bound annexin A5 trimers: a hybrid cryo-EM-X-ray crystallography study, *J. Mol. Biol.*, **304**, 561 (2000).
19. I. Reviakine, W. Bergsma-Schutter, A. N. Morozov, and A. Brisson, Two-dimensional crystallization of annexin A5 on phospholipid bilayers and monolayers: a solid-solid phase transition between crystal forms, *Langmuir*, **17**, 1680 (2001).
20. I. Reviakine, W. Bergsma-Schutter, C. Mazeres-Dobut, N. Govorukhina, and A. Brisson, Surface topography of the p3 and p6 annexin V crystal forms determined by atomic force microscopy, *J. Struct. Biol.*, **131**, 234 (2000).
21. G. Mosser, C. Ravanat, J.-M. Freyssinet, and A. Brisson, Sub-domain structure of lipid-bound annexin-V resolved by electron image analysis, *J. Mol. Biol.*, **217**, 241 (1991).
22. U. Haupts, J. Tittor, and D. Oesterhelt, Closing in on bacteriorhodopsin: progress in understanding the molecule, *Annu. Rev. Biophys. Biomol. Struct.*, **28**, 367 (1999).
23. J. K. Lanyi, Bacteriorhodopsin, *Annu. Rev. Physiol.*, **66**, 665 (2004).
24. Y. Kimura, D. G. Vassylyev, A. Miyazawa, A. Kidera, M. Matsushima, K. Mitsuoka, K. Murata, T. Hirai, and Y. Fujiyoshi, Surface of bacteriorhodopsin revealed by high-resolution electron crystallography, *Nature*, **389**, 206 (1997).
25. H. Luecke, B. Schobert, H. T. Richter, J. P. Cartailler, and J. K. Lanyi, Structure of bacteriorhodopsin at 1.55 Å resolution, *J. Mol. Biol.*, **291**, 899 (1999).
26. N. A. Dencher, D. Dresselhaus, G. Zaccai, and G. Büldt, Structural changes in bacteriorhodopsin during proton translocation revealed by neutron diffraction, *Proc. Natl. Acad. Sci. USA*, **86**, 7876 (1989).
27. S. Subramaniam, M. Gerstein, D. Oesterhelt, and R. Henderson, Electron diffraction analysis of structural changes in the photocycle of bacteriorhodopsin, *EMBO J.*, **12**, 1 (1993).
28. L. S. Brown, R. Needleman, and J. K. Lanyi, Conformational change of the E-F interhelical loop in the M photointermediate of bacteriorhodopsin, *J. Mol. Biol.*, **317**, 471 (2002).

29. H. J. Sass, G. Büld, R. Gessenich, D. Hehn, D. Neff, R. Schlesinger, J. Berendzen, and P. Ormos, Structural alterations for proton translocation in the M state of wild-type bacteriorhodopsin, *Nature*, **406**, 649 (2000).
30. S. Subramaniam, and R. Henderson, Molecular mechanism of vectorial proton translocation by bacteriorhodopsin, *Nature*, **406**, 653 (2000).
31. M. Shibata, H. Yamashita, T. Uchihashi, H. Kandori, and T. Ando, High-speed atomic force microscopy shows dynamic molecular processes in photo-activated bacteriorhodopsin, *Nat. Nanotech.*, **5**, 208 (2010).
32. D. J. Müller, J. B. Heymann, F. Oesterhelt, C. Möller, H. Gaub, G. Büldt, and A. Engel, Atomic force microscopy of native purple membrane, *Biochim. Biophys. Acta*, **1460**, 27 (2000).
33. H. Otto, T. Marti, T. Holz, T. Mogi, M. Lindau, H. G. Khorana, and M. P. Heyn, Aspartic acid-96 is the internal proton donor in the reprotonation of the Schiff base of bacteriorhodopsin, *Proc. Natl. Acad. Sci. USA*, **86**, 9228 (1989).
34. S. J. Demarest, M. Martinez-Yamout, J. Chung, H. Chen, W. Xu, H. J. Dyson, R. M. Evans, and P. E. Wright, Mutual synergistic folding in recruitment of CBP/p300 by p160 nuclear receptor coactivators, *Nature*, **415**, 549 (2002).
35. Y. Minezaki, K. Homma, A. R. Kinjo, and K. Nishikawa, Human transcription factors contain a high fraction of intrinsically disordered regions essential for transcriptional regulation, *J. Mol. Biol.*, **359**, 1137 (2006).
36. A. Miyagi, Y. Tsunaka, T. Uchihashi, K. Mayanagi, S. Hirose, K. Morikawa, and T. Ando, Visualization of intrinsically disordered regions of proteins by high-speed atomic force microscopy, *Chem. Phys. Chem.*, **9**, 1859 (2008).
37. R. Belotserkovskaya, S. Oh, V. A. Bondarenko, G. Orphanides, V. M. Studitsky, and D. Reinberg, FACT facilitates transcription-dependent nucleosome alteration, *Science*, **301**, 1090 (2003).
38. D. Reinberg, and R. J. Sims, III de FACT of nucleosome dynamics, *J. Biol. Chem.*, **281**, 23297 (2006).
39. T. Shimojima, M. Okada, T. Nakayama, H. Ueda, K. Okawa, A. Iwamatsu, H. Handa, and S. Hirose, Drosophila FACT contributes to Hox gene expression through physical and functional interactions with GAGA factor, *Genes Dev.*, **17**, 1605 (2003).
40. C. Rivetti, M. Guthold, and C. Bustamante, Scanning force microscopy of DNA deposited onto mica: equilibration versus kinetic trapping studied by statistical polymer chain analysis, *J. Mol. Biol.*, **264**, 919 (1996).

41. G. S. Manning, The persistence length of DNA is reached from the persistence length of its null isomer through an internal electrostatic stretching force, *Biophys. J.*, **91**, 3607 (2006).
42. H. Kojima, A. Ishijima, and T. Yanagida, Direct measurement of stiffness of single actin filaments with and without tropomyosin by in vitro nanomanipulation, *Proc. Natl. Acad. Sci. USA*, **91**, 12962 (1994).
43. M. Radmacher, M. Fritz, J. P. Cleveland, D. A. Walters, and P. K. Hansma, Imaging adhesion forces and elasticity of lysozyme adsorbed on mica with the atomic force microscope, *Langmuir*, **10**, 3809 (1994).
44. M. Tachibana, H. Koizumi, and K. Kojima, Effect of intracrystalline water on longitudinal sound velocity in tetragonal hen-egg-white lysozyme crystals, *Phys. Rev. E*, **69**, 051921.
45. R. Afrin, M. Alam, and A. Ikai, Pretransition and progressive softening of bovine carbonic anhydrase II as probed by single molecule atomic force microscopy, *Protein Sci.*, **14**, 1447 (2005).

Index

For Product Safety Concerns and Information please contact our EU representative GPSR@taylorandfrancis.com Taylor & Francis Verlag GmbH, Kaufingerstraße 24, 80331 München, Germany

Batch number: 10733699

Printed by Printforce, the Netherlands